Hopf Algebras, Tensor Categories and Related Topics

International Workshop on Hopf Algebras and Tensor Categories
Sep 9-13, 2019
Nangjing China

CONTEMPORARY MATHEMATICS

771

Hopf Algebras, Tensor Categories and Related Topics

International Workshop on
Hopf Algebras and Tensor Categories
September 9–13, 2019
Nanjing University
Nanjing, China

Nicolás Andruskiewitsch
Gongxiang Liu
Susan Montgomery
Yinhuo Zhang
Editors

AMERICAN MATHEMATICAL SOCIETY

2020 *Mathematics Subject Classification.* Primary 14M30, 16G20, 16S40, 16T05, 16T25, 17B37, 18D10, 18D25, 19A22, 20G42.

For additional information and updates on this book, visit
www.ams.org/bookpages/conm-771

Library of Congress Cataloging-in-Publication Data

Names: Andruskiewitsch, Nicolás, 1958- editor. | Liu, Gongxiang, 1979- editor. | Montgomery, Susan, editor. | Zhang, Yinhuo, 1963- editor.

Title: Hopf algebras, tensor categories and related topics : International Workshop on Hopf Algebras and Tensor Categories, September 9-13, 2019, Nanjing University, China / Nicolás Andruskiewitsch, Gongxiang Liu, Susan Montgomery, Yinhuo Zhang, editors.

Description: Providence, Rhode Island : American Mathematical Society, [2021] | Series: Contemporary mathematics, 0271-4132 ; volume 771 | Includes bibliographical references.

Identifiers: LCCN 2020043147 | ISBN 9781470456245 (paperback) | 9781470464271 (ebook)

Subjects: LCSH: Hopf algebras. | Tensor algebra. | Geometry, Algebraic. | AMS: Algebraic geometry – Special varieties – Supervarieties [See also 32C11, 58A50]. | Associative rings and algebras – Representation theory of rings and algebras – Representations of quivers and partially ordered sets. | Associative rings and algebras – Rings and algebras arising under various constructions – Smash products of general Hopf actions [See also 16T05]. | Associative rings and algebras – Hopf algebras, quantum groups and related topics – Hopf algebras and their applications [See also 16S40, 57T05]. | Associative rings and algebras – Hopf algebras, quantum groups and related topics – Yang-Baxter equations. | Nonassociative rings and algebras – Lie algebras and Lie superalgebras – Quantum groups (quantized enveloping algebras) and related deformations [See also 16T20, 20G42, 8 | Category theory; homological algebra | Group theory and generalizations – Linear algebraic groups and related topics

Classification: LCC QA613.8 .H695 2021 | DDC 512/.55–dc23

LC record available at https://lccn.loc.gov/2020043147

---

10 9 8 7 6 5 4 3 2 1   26 25 24 23 22 21

Contents

Preface

The International Workshop on Hopf Algebras and Tensor Categories was held at Nanjing University, China, during September 9-13, 2019. About one hundred specialists and students attended the meeting. The participants enjoyed not only the beauty of mathematics but also the wonderful lodging and facilities in the newly built "International Conference Center of Nanjing University". The list of speakers and the titles of the talks is included below.

The main goal of this meeting was to bring the researches together and to exchange their recent achievements and advances in Hopf algebras, Tensor categories and related topics. Twenty seven researchers from nine different countries gave their talks, which covered a wide variety of topics including classification of Hopf algebras, structures and actions of (weak) Hopf algebras, algebraic supergroups, representations of quantum groups, quasi-quantum groups, algebras in tensor categories, construction method of fusion categories. From those topics, we have seen not only advances in the theory of tensor categories motivated by the theory of Hopf algebras, but also beautiful applications of categorical methods in the study of Hopf algebras and Mathematical Physics.

The organizers were Nicolás Andruskiewitsch (Universidad Nacional de Córdoba, Argentina) Gongxiang Liu (Nanjing University, China), Susan Montgomery (University of Southern California) and Yinhuo Zhang (University of Hasselt, Belgium). We would like to thank all participants and speakers at the workshop for making it such an interesting and fruitful event.

The present volume contains thirteen articles contributed by the invited speakers of the session. These articles cover a wide variety of topics in the theory of tensor categories and Hopf algebras and contain a number of original results:

- Böhm and Vercruysse consider a special class of monoidal categories and present BiHom Hopf algebras as Hopf monoids in such categories.
- The paper of Brown and Zhang is a survey on the ongoing program of classifying Hopf algebras of Gelfand-Kirillov dimension 1 or 2.
- Bulacu and Torrecillas show that braided double-biproduct Hopf algebras in the sense of Majid can be identified to braided biproduct Hopf algebras using a categorical interpretation by Bespalov and Drabant.
- Fuchs, Schaumann and Schweigert establish a variant of the categorical Eilenberg-Watts calculus for module categories over a finite tensor category.
- Guo and Li explore braided Hopf algebra structures on free objects from braided operads by the method of coherent unit actions.
- Hu and Zhuang construct a class of admissible quantum affine algebras of type $A_1^{(1)}$, which have the Drinfel'd double structure.

- Huang, Liu, Yang and Ye give a survey on the results obtained recently on finite quasi-quantum groups over abelian groups.
- Kashina-Sommerhäuser describe in which ways the Radford biproducts of certain eight-dimensional Yetter-Drinfel'd Hopf algebras over the elementary abelian group of order four can be written as extensions of Hopf algebras.
- Natale presents an overview of the notions of exact sequences of Hopf algebras and tensor categories and their connections. Examples illustrating their main features are given.
- Shibata gives a review on the framework of characteristic-free study of quasireductive supergroups.
- Shimizu gives a survey on the character and integral theories for Hopf algebras extended to finite tensor categories, as well as recent techniques for finite abelian categories and finite tensor categories.
- Wakui investigates the direct sum construction of weak bialgebras and the indecomposability of them as weak bialgebras.
- Yang and Zhang construct a class of finite-dimensional quasi-Hopf algebras including the natural generalizations of the small quantum groups.

We are grateful to the editorial staff of *Contemporary Mathematics* and especially Christine Thivierge for the help with the preparation of this volume and the American Mathematical Society for supporting this project. Last but not least we thank the contributors for their articles and the anonymous referees for their valuable time refereeing the papers.

List of speakers and titles of talks

(1) Nicolás Andruskiewitsch (Universidad Nacional de Córdoba–CONICET).
An approach to the classification of pointed Hopf algebras with finite GK-dimension.

(2) Iván Angiono (Universidad Nacional de Córdoba–CONICET).
On Nichols algebras over basic Hopf algebras.

(3) Yanhong Bao (Anhui University).
From Weak Bialgebras to Algebraic Operads.

(4) Gabriella Böhm (Wigner RCP Budapest).
Multimonoidal monads.

(5) Stefaan Caenepeel (Vrij Universiteit Brussels).
Galois theory for Hopf categories II: Application to partial actions by groupoids.

(6) Michael Cuntz (Leibniz Universität Hannover).
A bound for finite Weyl groupoids.

(7) Jincheng Dong (Nanjing University of Information Science and Technology).
Examples of crossed product fusion categories.

(8) Jürgen Fuchs (Karlstads Universitet).
Module categories, relative Serre functors and conformal field theory.

(9) João Matheus Jury Giraldi (Universidade Federal do Rio Grande do Sul).
On Nichols algebras over dual Radford algebras.

(10) Vladimir Bavula (University of Sheffield).
The Poisson universal enveloping algebra and the Poisson differential operators of a generalized Weyl Poisson algebra.

(11) Jiwei He (Hangzhou Normal University).
Singularities of noncommutative quadric hypersurfaces.

(12) Naihong Hu (East China Normal University).
Some Hopf algebras from QDO on Manin superspaces and simple $U_q(gl(m|n))$-modules.

(13) Hualin Huang (Huajiao University).
On non-diagonal finite quasi-quantum groups over abelian groups.

(14) Yevgenia Kashina (DePaul University).
Cores in Yetter-Drinfel'd Hopf Algebras.

(15) Simon Lentner (Hamburg University).
Nichols algebra in braided tensor categories and screening operators in conformal field theory.

(16) Susan Montgomery (University of Southern California).
Actions of pointed Hopf algebras on matrices.

(17) Sonia Natale (Universidad Nacional de Córdoba–CONICET).
Extension of tensor categories by finite group fusion categories.

(18) Dmitri Nikshych (University of New Hampshire).
Classifying braidings on fusion categories.
(19) Christoph Schweigert (Hamburg University).
String-net models and invariants for mapping class groups.
(20) Taiki Shibata (Okayama University of Science).
Algebraic supergroups and their representations over a field.
(21) Kenichi Shimizu (Shibaura Institute of Technology).
Class functions and the action functor.
(22) York Sommerhäuser (Memorial University of Newfoundland).
Cores in Yetter-Drinfel'd Hopf Algebras: Examples.
(23) Blas Torrecillas (University of Almería).
On sovereign, balanced and ribbon quasi-Hopf algebras.
(24) Michihisa Wakui (Kansai University).
Indecomposability of weak Hopf algebras.
(25) Zhengfang Wang (Max Planck Institute for Mathematics).
B-infinity algebras and Keller's conjecture.
(26) Milen Yakimov (Louisiana State University).
Irreps and Poisson geometry of large quantum groups.
(27) Guisong Zhou (Ningbo University).
The structure of connected (graded) Hopf algebras.

Contemporary Mathematics
Volume **771**, 2021
https://doi.org/10.1090/conm/771/15506

BiHom Hopf algebras viewed as Hopf monoids

Gabriella Böhm and Joost Vercruysse

ABSTRACT. We introduce monoidal categories whose monoidal products of any positive number of factors are lax coherent and whose nullary products are oplax coherent. We call them $\mathsf{Lax}^+\mathsf{Oplax}^0$-monoidal. Dually, we consider $\mathsf{Lax}_0\mathsf{Oplax}_+$-monoidal categories which are oplax coherent for positive numbers of factors and lax coherent for nullary monoidal products. We define $\mathsf{Lax}_0^+\mathsf{Oplax}_+^0$-duoidal categories with compatible $\mathsf{Lax}^+\mathsf{Oplax}^0$- and $\mathsf{Lax}_0\mathsf{Oplax}_+$-monoidal structures. We introduce comonoids in $\mathsf{Lax}^+\mathsf{Oplax}^0$-monoidal categories, monoids in $\mathsf{Lax}_0\mathsf{Oplax}_+$-monoidal categories and bimonoids in $\mathsf{Lax}_0^+\mathsf{Oplax}_+^0$-duoidal categories.

Motivation for these notions comes from a generalization of a construction due to Caenepeel and Goyvaerts. This assigns a $\mathsf{Lax}_0^+\mathsf{Oplax}_+^0$-duoidal category D to any symmetric monoidal category V. The unital BiHom-monoids, counital BiHom-comonoids, and unital and counital BiHom-bimonoids in V, due to Grazianu et al., are identified with the monoids, comonoids and bimonoids in D, respectively.

Introduction

In recent years, lead by diverse motivations, several different generalizations of Hopf algebra have been proposed. Their similar features naturally raise the question whether they are instances of the same, more general notion. In [**6**] several examples, such as groupoids, Hopf monoids in braided monoidal categories, Hopf algebroids over central base algebras and weak Hopf algebras, were unified as Hopf monoids in the duoidal (called 2-monoidal in [**1**]) endohom category of a naturally Frobenius map monoidale in a suitable monoidal bicategory. In [**4**] also Hopf group algebras [**13**], Hopf categories [**2**] and Hopf polyads [**7**] were shown to fit this framework.

The aim of the current paper is an interpretation of the (unital and counital) BiHom-bimonoids in [**11**] as bimonoids in a category with some generalized duoidal structure. Recall that a BiHom-*monoid* in a monoidal category $(\mathsf{V},\otimes,I)$ (whose coherence natural isomorphisms are omitted) consists of an object a together with morphisms $\alpha,\beta : a \to a$ and $\mu : a\otimes a \to a$ such that α and β commute, both of them preserve the multiplication μ, and, instead of the associativity of μ, the first

The first author is grateful for the financial support by the Hungarian National Research, Development and Innovation Office – NKFIH (grant K124138).

The second author thanks the FNRS (National Research Fund of the French speaking community in Belgium) for support via the MIS project 'Antipode' (grant F.4502.18).

diagram of
(*)

$$
\begin{array}{ccccc}
a\otimes a\otimes a & \xleftarrow{1\otimes 1\otimes \beta} & a\otimes a\otimes a & \xrightarrow{\alpha\otimes 1\otimes 1} & a\otimes a\otimes a \\
{\scriptstyle \mu\otimes 1}\downarrow & & & & \downarrow{\scriptstyle 1\otimes \mu} \\
a\otimes a & \xrightarrow[\mu]{} & a & \xleftarrow[\mu]{} & a\otimes a
\end{array}
\qquad
\begin{array}{ccccc}
a & \xrightarrow{\beta} & a & \xleftarrow{\alpha} & a \\
{\scriptstyle \eta\otimes 1}\downarrow & & \| & & \downarrow{\scriptstyle 1\otimes \eta} \\
a\otimes a & \xrightarrow[\mu]{} & a & \xleftarrow[\mu]{} & a\otimes a
\end{array}
$$

commutes. A BiHom-monoid (a,α,β,μ) is said to possess a *unit* $\eta : I \to a$ if it is preserved by α and β and also the second diagram above commutes. Diagrams with reversed arrows define *(counital)* BiHom-*comonoids*. The *(unital and counital)* BiHom-*bimonoids* in a *symmetric* monoidal category consist of a (unital) BiHom-monoid structure and a (counital) BiHom-comonoid structure on the same object a (with possibly different endomorphism parts α,β and κ,ν) such that the multiplication μ (and the unit η) are morphisms of (counital) BiHom-comonoids; equivalently, the comultiplication (and the counit) are morphisms of (unital) BiHom-monoids.

One can see several attempts in the literature aiming at a description of (unital) BiHom-monoids, (counital) BiHom-comonoids and (unital and counital) BiHom-bimonoids, respectively, as monoids, comonoids and bimonoids in a suitable category. All of these ideas originate from the same construction in [**8**]. In [**8**], to any monoidal category $(\mathsf{V},\otimes,I)$, a category is associated whose objects are pairs consisting of an object a of V and an automorphism α of a. The morphisms $(a,\alpha)\to(a',\alpha')$ are morphisms $\phi : a \to a'$ in V such that $\phi.\alpha = \alpha'.\phi$. It is equipped with a monoidal structure with monoidal unit $(I,1)$, monoidal product $(a,\alpha)\otimes(a',\alpha') := (a\otimes a',\alpha\otimes\alpha')$ and associativity and unitality coherence isomorphisms

$$
\begin{array}{rcl}
\alpha\otimes 1\otimes \alpha''^{-1} : ((a,\alpha)\otimes(a',\alpha'))\otimes(a'',\alpha'') & \to & (a,\alpha)\otimes((a',\alpha')\otimes(a'',\alpha'')), \\
\alpha : (I,1)\otimes(a,\alpha) & \to & (a,\alpha), \\
\alpha : (a,\alpha)\otimes(I,1) & \to & (a,\alpha).
\end{array}
$$

The monoids (respectively, comonoids) in this monoidal category are those unital BiHom-monoids (respectively, those counital BiHom-comonoids) in V whose two constituent endomorphisms are equal automorphisms. They are called in [**8**] *monoidal* Hom-monoids (respectively, *monoidal* Hom-comonoids). Any possible symmetry on $(\mathsf{V},\otimes,i)$ is inherited by this monoidal category of [**8**]. Then the bimonoids therein are those unital and counital BiHom-bimonoids in V whose comonoid part contains two copies of an automorphism and the monoid part contains two copies of its inverse. They are called in [**8**] *monoidal* Hom-bimonoids.

The above construction from [**8**] was widely generalized in [**11**, Section 2]. In the particular case which is relevant here, the objects are triples consisting of an object a of V and *two* commuting automorphisms α and β of a. The resulting category also admits a similar (symmetric) monoidal structure (where half of the occurring morphisms α is replaced by β). The monoids, comonoids and bimonoids in this (symmetric) monoidal category cover also those unital BiHom-monoids, counital BiHom-comonoids and unital and counital BiHom-bimonoids, respectively, in which two possibly unrelated automorphisms occur. Such BiHom-objects are called in [**11**] *monoidal* BiHom-monoids, *monoidal* BiHom-comonoids and *monoidal* BiHom-bimonoids respectively.

So the essential observation, originated in [**8**], is that these *monoidal* Hom- and BiHom-structures are monoids, comonoids and bimonoids in a suitable (symmetric) monoidal category. Consequently, the standard machinery applies to them. This explains why many aspects of this theory have been obtained in recent literature with a mild adaptation of the classical proofs.

However, more general BiHom-structures, with not necessarily invertible endomorphism parts, are not covered by this construction, and it is the main aim of this paper to provide a categorical construction that covers these cases.

Firstly, let us remark that the category whose objects consist of an object a of a symmetric monoidal category V together with *four* commuting automorphisms of a carries a *duoidal* structure (termed *2-monoidal* in [**1**]). One monoidal structure is defined in terms of one half of the automorphisms and a second monoidal structure is defined in terms of the other half. The compatibility morphisms are provided by the symmetry of V. The monoids, comonoids and bimonoids in this duoidal category finally cover also those unital BiHom-monoids, counital BiHom-comonoids and unital and counital BiHom-bimonoids, respectively, in which four possibly different, but still invertible endomorphisms occur. (A construction of this flavor occurred in [**14**]. However, that construction is based on four monoidal natural automorphisms of the identity functor on V. This seems to be quite restrictive since the identity functor on the category of vector spaces, for example, admits no other monoidal natural automorphism but the identity.)

In order to deal with BiHom-structures with not necessarily invertible endomorphism parts, one should give up the invertibility of the coherence natural transformations and use the *unbiased* variants of monoidal category [**12**]. Although there is a meaningful theory of duoidal categories with one lax and one oplax monoidal structure [**10**, Section 4], for the description of BiHom-structures neither the lax nor the oplax variant looks suitable but rather a certain mixture of both. Indeed, for the formulation of the diagrams of $(*)$, one needs coherence morphisms of the kind

$$(a\otimes b)\otimes c \longleftarrow a\otimes b\otimes c \longrightarrow a\otimes (b\otimes c) \qquad I\otimes a \longrightarrow a \longleftarrow a\otimes I.$$

Those on the left are of the oplax type, while those on the right are of the lax type. In this paper we define unbiased monoidal categories of the above mixed type.

As a first step, we consider Lax^+- (respectively, Oplax_+-) monoidal categories with monoidal products of only positive number of factors with lax (respectively, oplax) coherence morphisms. We define cosemigroups (respectively, semigroups) in such monoidal categories. We also define $\mathsf{Lax}^+\mathsf{Oplax}_+$-duoidal categories which possess compatible Lax^+- and Oplax_+-monoidal structures. Semigroups in a $\mathsf{Lax}^+\mathsf{Oplax}_+$-duoidal category are shown to constitute a Lax^+-monoidal category; dually, cosemigroups in a $\mathsf{Lax}^+\mathsf{Oplax}_+$-duoidal category are shown to constitute an Oplax_+-monoidal category. So we can define bisemigroups in a $\mathsf{Lax}^+\mathsf{Oplax}_+$-duoidal category as cosemigroups in the category of semigroups; equivalently, as semigroups in the category of cosemigroups.

Next we introduce so called $\mathsf{Lax}_0\mathsf{Oplax}_+$-monoidal categories, which have n-fold monoidal products for any non-negative integer n such that the monoidal products of positive number of factors are oplax coherent while the 0-fold monoidal product is lax coherent. Forgetting about 0-fold monoidal products, a $\mathsf{Lax}_0\mathsf{Oplax}_+$-monoidal category can be seen Oplax_+-monoidal. We define monoids in $\mathsf{Lax}_0\mathsf{Oplax}_+$-monoidal categories. Dually, we introduce so called $\mathsf{Lax}^+\mathsf{Oplax}^0$-monoidal categories, again with n-fold monoidal products for any non-negative integer n such that the monoidal products of positive number of factors are lax coherent while the 0-fold monoidal product is oplax coherent. We define comonoids in $\mathsf{Lax}^+\mathsf{Oplax}^0$-monoidal categories. Finally we define $\mathsf{Lax}_0^+\mathsf{Oplax}_+^0$-duoidal categories with compatible $\mathsf{Lax}^+\mathsf{Oplax}^0$- and $\mathsf{Lax}_0\mathsf{Oplax}_+$-monoidal structures. Monoids in them are shown to constitute a $\mathsf{Lax}_0\mathsf{Oplax}_+$-monoidal category and, dually, comonoids in them are shown to constitute a $\mathsf{Lax}^+\mathsf{Oplax}^0$-monoidal category. So we can define bimonoids in them as comonoids in the category of monoids; equivalently, as monoids in the category of comonoids.

Generalizing the construction in [**8**], we associate a $\mathsf{Lax}_0^+\mathsf{Oplax}_+^0$-duoidal category D to any symmetric monoidal category V. We identify the semigroups, cosemigroups and bisemigroups in D with the BiHom-monoids, BiHom-comonoids and the BiHom-bimonoids in V, respectively. We also identify the monoids, comonoids and bimonoids in D with the unital BiHom-monoids, counital BiHom-comonoids and the unital and counital BiHom-bimonoids in V, respectively, in the sense of [**11**].

The consequence of our construction is twofold. Firstly, by extending the *microcosm* hosting bimonoids [**3**], it provides a suitable categorical framework where BiHom-objects with not necessarily invertible endomorphisms can be studied. On the other hand, our proposed setting goes beyond the known framework of symmetric monoidal categories, and even of duoidal categories. In this way results for such objects can no longer directly be obtained from the existing theory of monoids, comonoids a bimonoids in duoidal categories, let alone in (symmetric) monoidal categories. Indeed, one would need first to extend this theory to the framework of $\mathsf{Lax}_0^+\mathsf{Oplax}_+^0$-duoidal categories, which is more involved than the classical case. This indicates that the theory of BiHom-objects with not necessarily invertible endomorphisms is itself more involved; no longer a straightforward generalization of the classical theory.

The same axioms defining our various lax and oplax monoidal categories can be used to define monoids of the same lax and oplax type in any Gray monoid. Such monoids can be seen as the 0-cells of a 2-category. If the ambient Gray monoid is a symmetric strict monoidal 2-category, then this 2-category of monoids is again strict monoidal. In this case we can also define duoids of various lax and oplax type, as suitably lax monoids in the strict monoidal 2-category of suitably lax monoids.

However, not to distract attention from the main aim to describe BiHom-structures, we do not work at the level of generality of the previous paragraph. We restrict our study to the symmetric strict monoidal 2-category Cat of categories, functors, and natural transformations.

We do not assume that the monoidal categories in the paper are strict monoidal but — relying on coherence — we omit explicitly denoting their Mac Lane type coherence natural isomorphisms.

1. Some combinatorial background

As a preparation for the constructions of the subsequent sections, we begin with recording some notation and a few technical results that will be heavily applied later.

1.1. In the 2-category Cat of categories, functors and natural transformations we denote by $\cdot$ the composition of functors and the corresponding Godement product of natural transformations. Cat is strict monoidal via the Cartesian product $\times$ and symmetric via the flip maps. We often denote the Cartesian product (of any of categories, functors, and natural transformations) by juxtaposition. The Cartesian product of n copies of the same category A is also denoted by A^n. By A^0 we mean the singleton category $\mathbb{1}$ for any category A. For all non-negative integers n, p, consider the particular components

$$\begin{array}{ccc} \mathsf{Cat}^{np} = (\mathsf{Cat}^p)^n & \xrightarrow{\overset{\times}{n}} & \mathsf{Cat}^p \\ \| & & \\ (\mathsf{Cat}^n)^p & \Downarrow \tau_{np} & \Big\downarrow \overset{\times}{p} \\ \overset{\times}{p}\Big\downarrow & & \\ \mathsf{Cat}^n & \xrightarrow[\overset{\times}{n}]{} & \mathsf{Cat} \end{array}$$

of the symmetry; where $\overset{\times}{n}$ stands for the n-fold Cartesian product. They are 2-natural transformations whose component at any double sequence of categories $\{\mathsf{A}_{ij}\}_{\substack{i=1,\dots,p \\ j=1,\dots,n}}$ is the flip functor

$$\begin{array}{ll} (A_{11},\dots,A_{1n}),\dots,(A_{p1},\dots,A_{pn}) & \overset{\tau_{np}}{\to} (A_{11},\dots,A_{p1}),\dots,(A_{1n},\dots,A_{pn}), \\ (a_{11},\dots,a_{1n}),\dots,(a_{p1},\dots,a_{pn}) & \mapsto (a_{11},\dots,a_{p1}),\dots,(a_{1n},\dots,a_{pn}). \end{array}$$

They satisfy the equalities

$$(1.1)\qquad \tau_{np}\cdot\tau_{pn} = 1 \qquad \text{and} \qquad \tau_{n1} = 1 \qquad \text{and} \qquad \tau_{np}\cdot(\tau_{nk_1}\dots\tau_{nk_p}) = \tau_{n\sum_{i=1}^p k_i}$$

for all non-negative integers n, p and $k_1, \dots, k_p$.

1.2. We adopt the convention that empty sums are equal to zero. The set of non-negative integers is denoted by $\mathbb{N}$ and the set of (strictly) positive integers is denoted by $\mathbb{N}^+$.

The following maps $\mathbb{N} \to \mathbb{N}$ will occur frequently.

$$Z(m) := \begin{cases} 0 & \text{if} \quad m > 0 \\ 1 & \text{if} \quad m = 0 \end{cases} \qquad\qquad \overline{m} := m + Z(m).$$

For $n \in \mathbb{N}$ and a sequence of non-negative integers $\{m_1, \dots, m_n\}$ we denote the sum and the number of zeros, respectively by

$$M := \sum_{i=1}^n m_i \qquad \text{and} \qquad Z := \sum_{i=1}^n Z(m_i).$$

Clearly, $M + Z \geq n$ and thus $M + Z = 0$ iff $n = 0$.

For a double sequence of non-negative integers $\{k_{ij}\}_{\substack{i=1,\dots,n \\ j=1,\dots,m_i}}$ (also in the particular case when $n = 1$), we put

$$K_i := \sum_{j=1}^{m_i} k_{ij} \qquad Z_i := \sum_{j=1}^{m_i} Z(k_{ij}) \qquad K := \sum_{i=1}^n K_i \qquad Z := \sum_{i=1}^n Z_i.$$

We introduce two associated double sequences with $i = 1, \dots, n$ and $j = 1, \dots, \overline{m}_i$:

$$\widetilde{k}_{ij} := \begin{cases} k_{ij} & \text{if} \quad m_i > 0 \\ 0 & \text{if} \quad m_i = 0 \end{cases} \qquad\qquad \widehat{k}_{ij} := \begin{cases} k_{ij} & \text{if} \quad K_i > 0 \\ 1 & \text{if} \quad K_i = 0, m_i > 0, j = 1 \\ 0 & \text{if} \quad K_i = 0, m_i > 0, j > 1 \\ 1 & \text{if} \quad m_i = 0. \end{cases}$$

We put

$$\widetilde{Z} := \sum_{i=1}^{n} \sum_{j=1}^{\overline{m}_i} Z(\widetilde{k}_{ij}) = Z + \sum_{i=1}^{n} Z(m_i) = Z + \sum_{i=1}^{n} Z(K_i + Z_i).$$

Note that

$$\sum_{i=1}^{n} \sum_{j=1}^{\overline{m}_i} (\widehat{k}_{ij} + Z(\widehat{k}_{ij})) = \sum_{i=1}^{n} \sum_{j=1}^{\overline{m}_i} (\widetilde{k}_{ij} + Z(\widetilde{k}_{ij})) = \sum_{i=1}^{n} K_i + Z_i + Z(K_i + Z_i) = K + \widetilde{Z}.$$

For any non-negative integer k, and for any functor F from the singleton category $\mathbb{1}$ to an arbitrary category A, we introduce the functor

$$\mathsf{A}^k \xrightarrow{[k]} \mathsf{A}^{\overline{k}} := \begin{cases} \mathsf{A}^k \xrightarrow{1} \mathsf{A}^k & \text{if} \quad k > 0 \\ \mathbb{1} \xrightarrow{F} \mathsf{A} & \text{if} \quad k = 0 \end{cases} \tag{1.2}$$

The proof of the following easy lemma about such functors in (1.2) is left to the reader.

LEMMA 1.3. *For any sequence of non-negative integers* $\{k_1, \ldots, k_m\}$, *for any functor* F *from the singleton category* $\mathbb{1}$ *to an arbitrary category* A, *and for the associated functors of* (1.2), *the following diagrams commute.*

$$\begin{array}{ccc} \mathsf{A}^K & \xrightarrow{[k_1]\ldots[k_m]} & \mathsf{A}^{K+Z} \\ & \searrow^{[\widetilde{k}_1]\ldots[\widetilde{k}_{\overline{m}}]} & \downarrow {\scriptstyle [K+Z]} \\ & & \mathsf{A}^{K+\widetilde{Z}} \end{array} \qquad \begin{array}{ccc} \mathsf{A} & \xrightarrow{[K]} & \mathsf{A}^{K+Z(K)} \\ & \searrow^{[\widetilde{k}_1]\ldots[\widetilde{k}_{\overline{m}}]} & \downarrow {\scriptstyle [\widehat{k}_1]\ldots[\widehat{k}_{\overline{m}}]} \\ & & \mathsf{A}^{K+\widetilde{Z}} \end{array}$$

2. Cosemigroups in Lax^{+}-monoidal categories

Following the *microcosm* principle [**3**, Section 4.3], in order to define semigroups and cosemigroups internally to some monoidal category, one only needs to have products of positive number of factors, not necessarily nullary ones. In this section we introduce cosemigroups in monoidal categories with monoidal products of arbitrary positive number of factors and lax coherence morphisms between them. The dual situation is discussed in the next section.

Throughout, we use the notation introduced in Section 1.

DEFINITION 2.1. A Lax^{+}-*monoidal category* consists of

- a category L
- for all positive integers n, a functor $\underset{n}{\bullet} : \mathsf{L}^n \to \mathsf{L}$
- for all positive integers $n, k_1, \ldots, k_n$, natural transformations

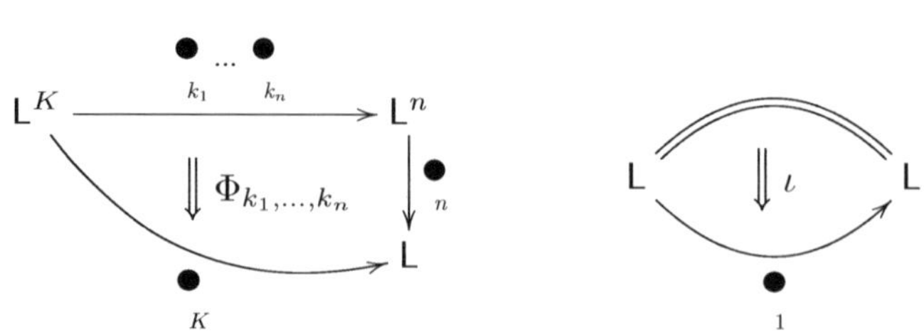

such that the diagrams

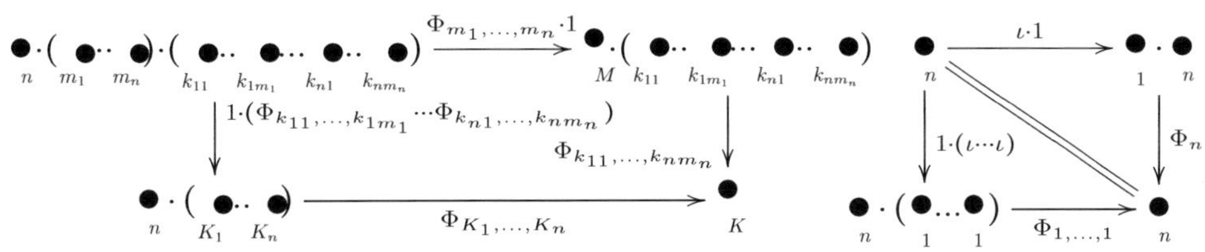

commute, for any positive integers n, $\{m_i\}$ labelled by $i = 1, \ldots, n$, and $\{k_{ij}\}$ labelled by $i = 1, \ldots, n$ and $j = 1, \ldots, m_i$.

A Lax^+-monoidal category is *normal* if ι is invertible and *strict normal* if ι is the identity natural transformation.

More generally, Lax^+-monoids can be defined in any Gray monoid — so in particular in any strict monoidal 2-category — by the same diagrams in Definition 2.1. Lax^+-monoidal categories are then re-obtained by applying it to Cat.

Throughout, in a Lax^+-monoidal category we denote by $\left(a \right)$ the image of any object a under the functor $\underset{1}{\bullet} : \mathsf{L} \to \mathsf{L}$; and we denote by $a_1 \bullet \cdots \bullet a_n$ the image of an object $(a_1, \ldots, a_n)$ under the functor $\underset{n}{\bullet} : \mathsf{L}^n \to \mathsf{L}$.

PROPOSITION 2.2. *The Cartesian product* LL' *of* Lax^+*-monoidal categories* $(\mathsf{L}, \bullet, \Phi, \iota)$ *and* $(\mathsf{L}', \bullet', \Phi', \iota')$ *is again* Lax^+*-monoidal via*

- *the functors* $(\mathsf{LL}')^n \xrightarrow{\tau_{2n}} \mathsf{L}^n\mathsf{L}'^n \xrightarrow{\underset{n}{\bullet}\underset{n}{\bullet}'} \mathsf{LL}'$
- *the natural transformations* $\iota\iota' : 1 \to \underset{1}{\bullet}\underset{1}{\bullet}'$ *and*

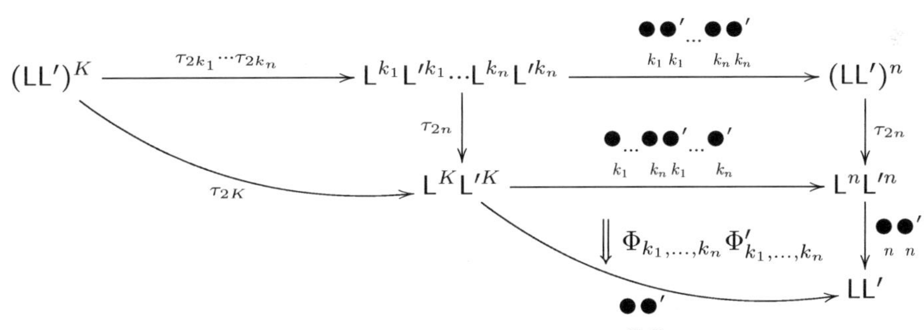

where the unlabelled regions denote identity natural transformations.

PROOF. It is left to the reader to check that the stated datum satisfies the axioms in Definition 2.1 (analogously to Example (6) on page 80 of [9]). $\square$

DEFINITION 2.3. A Lax^+*-monoidal functor* consists of

- a functor $G : \mathsf{L} \to \mathsf{L}'$
- for all positive integers n, a natural transformation $\Gamma_n : G \cdot \underset{n}{\bullet} \to \underset{n}{\bullet}' \cdot (G \cdots G)$

such that for all sequences of positive integers $\{k_1, \ldots, k_n\}$ the following diagrams commute.

$$\begin{array}{ccc} G & \xrightarrow{1\cdot\iota} & G\cdot\underset{1}{\bullet} \\ & {\scriptstyle \iota'\cdot 1}\searrow & \downarrow{\scriptstyle\Gamma_1} \\ & & \underset{1}{\bullet'}\cdot G\end{array}
\qquad
\begin{array}{ccc} G\cdot\underset{n}{\bullet}\cdot(\underset{k_1}{\bullet}\ldots\underset{k_n}{\bullet}) & \xrightarrow{1\cdot\Phi_{k_1,\ldots,k_n}} & G\cdot\underset{K}{\bullet} \\ {\scriptstyle\Gamma_n\cdot 1}\downarrow & & \\ \underset{n}{\bullet'}\cdot(G\cdots G)\cdot(\underset{k_1}{\bullet}\ldots\underset{k_n}{\bullet}) & & \downarrow{\scriptstyle\Gamma_K} \\ {\scriptstyle 1\cdot(\Gamma_{k_1}\cdots\Gamma_{k_n})}\downarrow & & \\ \underset{n}{\bullet'}\cdot(\underset{k_1}{\bullet'}\ldots\underset{k_n}{\bullet'})\cdot(G\cdots G) & \xrightarrow[\Phi'_{k_1,\ldots,k_n}\cdot 1]{} & \underset{K}{\bullet'}\cdot(G\cdots G)\end{array}$$

PROPOSITION 2.4. (1) *The composite of (composable)* Lax⁺*-monoidal functors* (G,Γ) *and* (H,Ξ) *is again* Lax⁺*-monoidal via the natural transformations*

$$H\cdot G\cdot\underset{n}{\bullet} \xrightarrow{1\cdot\Gamma_n} H\cdot\underset{n}{\bullet}\cdot(G\cdots G) \xrightarrow{\Xi_n\cdot 1} \underset{n}{\bullet}\cdot(H\cdots H)\cdot(G\cdots G) = \underset{n}{\bullet}\cdot(H\cdot G)^n.$$

(2) *The Cartesian product of* Lax⁺*-monoidal functors* (G,Γ) *and* (H,Ξ) *is again* Lax⁺*-monoidal via the natural transformations*

$$\begin{array}{ccccc} (\mathsf{LN})^n & \xrightarrow{\tau_{2n}} & \mathsf{L}^n\mathsf{N}^n & \xrightarrow{\underset{n\ n}{\bullet\bullet}} & \mathsf{LN} \\ {\scriptstyle (GH)^n}\downarrow & & {\scriptstyle G^nH^n}\downarrow & \Downarrow\Gamma_n\Xi_n & \downarrow{\scriptstyle GH} \\ (\mathsf{L'N'})^n & \xrightarrow[\tau_{2n}]{} & \mathsf{L'}^n\mathsf{N'}^n & \xrightarrow[\underset{n\ n}{\bullet'\bullet'}]{} & \mathsf{L'N'}\end{array}$$

where the unlabelled region denotes the identity natural transformation.

PROOF. It is left to the reader to check that the stated data satisfy the axioms in Definition 2.3. □

DEFINITION 2.5. A Lax⁺*-monoidal natural transformation* $(G,\Gamma)\to(G',\Gamma')$ is a natural transformation $\omega : G \to G'$ for which the following diagram commutes for all positive integers n.

$$\begin{array}{ccc} G\cdot\underset{n}{\bullet} & \xrightarrow{\omega\cdot 1} & G'\cdot\underset{n}{\bullet} \\ {\scriptstyle\Gamma_n}\downarrow & & \downarrow{\scriptstyle\Gamma'_n} \\ \underset{n}{\bullet'}\cdot(G\cdots G) & \xrightarrow[1\cdot(\omega\cdots\omega)]{} & \underset{n}{\bullet'}\cdot(G'\cdots G')\end{array}$$

PROPOSITION 2.6. *All of the composites, the Godement products, and the Cartesian products of (composable)* Lax⁺*-monoidal natural transformations are again* Lax⁺*-monoidal. Consequently, there is a strict monoidal 2-category* Lax⁺ *of* Lax⁺*-monoidal categories,* Lax⁺*-monoidal functors and* Lax⁺*-monoidal natural transformations.*

PROOF. It is left to the reader to check that the diagram of Definition 2.5 commutes in all of the stated cases. □

DEFINITION 2.7. A *cosemigroup* in a Lax^+-monoidal category $(\mathsf{L}, \bullet, \Phi, \iota)$ consists of

- an object a of L and
- a morphism $\delta : a \to a \bullet a$

rendering commutative the first (coassociativity) diagram below. A *morphism of cosemigroups* $(a, \delta) \to (a', \delta')$ is a morphism $\phi : a \to a'$ rendering commutative the second diagram.

$$
\begin{array}{ccccc}
a\bullet a & \xleftarrow{\ \delta\ } & a & \xrightarrow{\ \delta\ } & a\bullet a \\
{\scriptstyle \delta\bullet\iota}\downarrow & & & & \downarrow{\scriptstyle \iota\bullet\delta} \\
(a\bullet a)\bullet(a) & \xrightarrow[\Phi_{2,1}]{} & a\bullet a\bullet a & \xleftarrow[\Phi_{1,2}]{} & (a)\bullet(a\bullet a)
\end{array}
\qquad
\begin{array}{ccc}
a & \xrightarrow{\ \delta\ } & a\bullet a \\
{\scriptstyle \phi}\downarrow & & \downarrow{\scriptstyle \phi\bullet\phi} \\
a' & \xrightarrow[\delta']{} & a'\bullet a'
\end{array}
$$

THEOREM 2.8. *For any* Lax^+*-monoidal category* $(\mathsf{L}, \bullet, \Phi, \iota)$*, the following categories are isomorphic.*

(i) *The category of cosemigroups and their morphisms in* $(\mathsf{L}, \bullet, \Phi, \iota)$.
(ii) *The category of* Lax^+*-monoidal functors* $(\mathbb{1}, 1, 1, 1) \to (\mathsf{L}, \bullet, \Phi, \iota)$ *and their* Lax^+*-monoidal natural transformations.*

Consequently, Lax^+*-monoidal functors preserve cosemigroups.*

PROOF. A Lax^+-monoidal functor $(\mathbb{1}, 1, 1, 1) \to (\mathsf{L}, \bullet, \Phi, \iota)$ as in part (ii) consists of

- an object a of L
- for all positive integers n, a morphism δ_n from a to the n-fold product $a\bullet\cdots\bullet a$

such that $\delta_1 = \iota$ and for all sequences of positive integers $\{k_1, \ldots, k_n\}$ the following diagram commutes.

$$
\begin{array}{ccc}
a & \xrightarrow{\qquad\qquad \delta_n \qquad\qquad} & a\bullet\cdots\bullet a \\
{\scriptstyle \delta_K}\downarrow & & \downarrow{\scriptstyle \delta_{k_1}\bullet\cdots\bullet\delta_{k_n}} \\
a\bullet\cdots\bullet a & \xleftarrow[\Phi_{k_1,\ldots,k_n}]{} & (a\bullet\cdots\bullet a)\bullet\cdots\bullet(a\bullet\cdots\bullet a)
\end{array}
\tag{2.1}
$$

Comparing (2.1) at $n = 2$, $k_1 = 2$, $k_2 = 1$ and $n = 2$, $k_1 = 1$, $k_2 = 2$, we see that (a, δ_2) is a cosemigroup.

Conversely, starting with a cosemigroup (a, δ), we put $\delta_1 := \iota$, $\delta_2 := \delta$ and for $i > 1$ iteratively

$$
\delta_{i+1} := \big(\, a \xrightarrow{\ \delta\ } a\bullet a \xrightarrow{\ \iota\bullet\delta_i\ } (a)\bullet(a\bullet\cdots\bullet a) \xrightarrow{\ \Phi_{1,i}\ } a\bullet\cdots\bullet a \,\big).
\tag{2.2}
$$

Then the coassociativity of the cosemigroup tells exactly that δ_3 equals

$$
a \xrightarrow{\ \delta\ } a\bullet a \xrightarrow{\ \delta\bullet\iota\ } (a\bullet a)\bullet(a) \xrightarrow{\ \Phi_{2,1}\ } a\bullet a\bullet a.
$$

By induction on i, one proves that the morphism of (2.2) is equal to

$$
a \xrightarrow{\ \delta_i\ } a\bullet\cdots\bullet a \xrightarrow{\ \delta\bullet\iota\bullet\cdots\bullet\iota\ } (a\bullet a)\bullet(a)\bullet\cdots\bullet(a) \xrightarrow{\ \Phi_{2,1,\ldots,1}\ } a\bullet\cdots\bullet a
\tag{2.3}
$$

for all $i > 2$.

Now we check, by induction on K, that the so defined family of morphisms $\{\delta_i\}_{i\in\mathbb{N}^+}$ renders commutative (2.1) for all sequences of positive integers $\{k_1,\dots,k_n\}$. For $K=1$ both n and k_1 must be equal to 1 and thus (2.1) commutes by one of the unitality axioms in Definition 2.1. There are two kinds of induction step:

- replacing $\{k_1,\dots,k_n\}$ with $\{1,k_1,\dots,k_n\}$ and
- replacing $\{k_1,\dots,k_n\}$ with $\{1+k_1,\dots,k_n\}$.

Assuming that (2.1) commutes for the original sequence $\{k_1,\dots,k_n\}$, it is seen to commute for the modified sequence using the Lax^+-monoidal category axioms in Definition 2.1 together with (2.2) in the first case and (2.3) in the second case.

From the constructions, it is obvious that the cosemigroup constructed from the Lax^+-monoidal functor induced by a cosemigroup is exactly the initial cosemigroup. In order to see triviality of the composite of the above constructions in the opposite order, note that (2.1) for $n=2$, $k_1=1$, and $k_2=i$ yields (2.2).

A Lax^+-monoidal natural transformation $(a,\{\delta_i\}_{i\in\mathbb{N}^+})\to(a',\{\delta'_i\}_{i\in\mathbb{N}^+})$ is a morphism $f:a\to a'$ such that for all positive integers i,

$$
\begin{array}{ccc}
a & \xrightarrow{\ \delta_i\ } & a\bullet\cdots\bullet a \\
{\scriptstyle f}\big\downarrow & & \big\downarrow{\scriptstyle f\bullet\cdots\bullet f} \\
a' & \xrightarrow[\ \delta'_i\ ]{} & a'\bullet\cdots\bullet a'
\end{array}
\tag{2.4}
$$

commutes. Such a morphism f is clearly a cosemigroup morphism $(a,\delta_2)\to(a',\delta'_2)$.

Conversely, for a cosemigroup morphism $f:(a,\delta)\to(a',\delta')$, the diagram of (2.4) commutes for $i=1$ by the naturality of $\delta_1=\iota=\delta'_1$. For the morphisms (2.2) for $i>1$, commutativity of (2.4) is easily checked by induction on i. □

We close this section by introducing a natural notion of comodule over a cosemigroup in the sense of Definition 2.7, without entering its deeper analysis.

Definition 2.9. A *comodule* of a cosemigroup (a,δ) in a Lax^+-monoidal category $(\mathsf{L},\bullet,\Phi,\iota)$ consists of

- an object x of L
- a morphism $\varrho:x\to x\bullet a$ (the *coaction*)

such that the (coassociativity) diagram on the left of

$$
\begin{array}{ccccc}
x\bullet a & \xleftarrow{\ \varrho\ } & x & \xrightarrow{\ \varrho\ } & x\bullet a \\
{\scriptstyle \iota\bullet\delta}\big\downarrow & & & & \big\downarrow{\scriptstyle \varrho\bullet\iota} \\
(x)\bullet(a\bullet a) & \xrightarrow[\ \Phi_{1,2}\ ]{} & x\bullet a\bullet a & \xleftarrow[\ \Phi_{2,1}\ ]{} & (x\bullet a)\bullet(a)
\end{array}
\qquad
\begin{array}{ccc}
x & \xrightarrow{\ \varrho\ } & x\bullet a \\
{\scriptstyle \theta}\big\downarrow & & \big\downarrow{\scriptstyle \theta\bullet 1} \\
x' & \xrightarrow[\ \varrho'\ ]{} & x'\bullet a
\end{array}
$$

commutes. A *morphism of comodules* $(x,\varrho)\to(x',\varrho')$ is a morphism $\theta:x\to x'$ in L such that the diagram on the right above commutes.

Example 2.10. Comparing the coassociativity conditions of Definition 5.7 and Definition 2.9, it becomes obvious that (a,δ) is a comodule over an arbitrary cosemigroup (a,δ) in any Lax^+-monoidal category.

Notice, however, that it is not clear if the forgetful functor, from the category of comodules over a cosemigroup (a,δ) in some Lax^+-monoidal category $(\mathsf{L},\bullet)$ to L, is comonadic. In particular, there is no obvious candidate for its right adjoint.

Indeed, the usual construction of 'free' comodules does not seem to work: for an arbitrary object x of L, the comultiplication δ does not seem to induce a coaction on $x \bullet a$.

3. Semigroups in Oplax_+-monoidal categories

Dually to the previous section, here we introduce semigroups in monoidal categories in which monoidal products of positive number of factors are available together with oplax coherence morphisms.

DEFINITION 3.1. An Oplax_+*-monoidal category* consists of

- a category R
- for all positive integers n, a functor $\underset{n}{\circ} : \mathsf{R}^n \to \mathsf{R}$
- for all positive integers $n, k_1, \ldots, k_n$, natural transformations

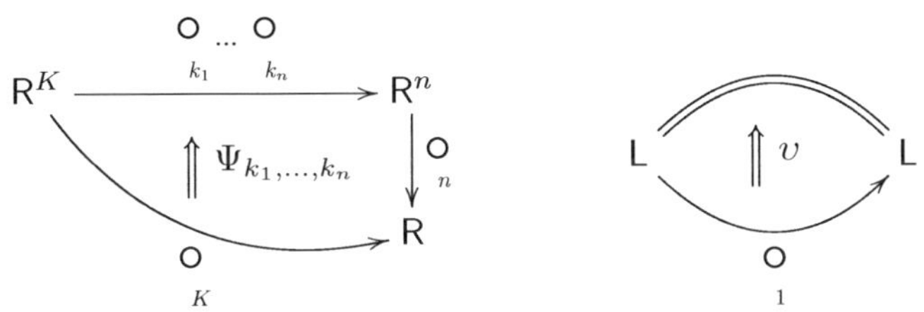

satisfying the so-called coassociativity and counitality axioms encoded in the diagrams which are obtained from the diagrams of Definition 2.1 by reversing the arrows.

Again, Definition 3.1 can be extended in a straightforward way to define Oplax_+-monoids in any Gray monoid — so in particular in any strict monoidal 2-category.

$(\mathsf{R}, \circ, \Psi, \upsilon)$ is an Oplax_+-monoidal category if and only if $(\mathsf{R}^{\mathsf{op}}, \circ^{\mathsf{op}}, \Psi^{\mathsf{op}}, \upsilon^{\mathsf{op}})$ is a Lax^+-monoidal category.

Throughout, in an Oplax_+-monoidal category we denote by $(\!|a|\!)$ the image of any object a under the functor $\underset{1}{\circ} : \mathsf{R} \to \mathsf{R}$; and we denote by $a_1 \circ \cdots \circ a_n$ the image of an object $(a_1, \ldots, a_n)$ under the functor $\underset{n}{\circ} : \mathsf{R}^n \to \mathsf{R}$.

Dually to Proposition 2.2, the Cartesian product of Oplax_+-monoidal categories carries a canonical Oplax_+-monoidal structure.

DEFINITION 3.2. An Oplax_+*-monoidal functor* consists of

- a functor $G : \mathsf{R} \to \mathsf{R}'$
- for all positive integers n, a natural transformation $\underset{n}{\circ}' \cdot (G \cdots G) \to G \cdot \underset{n}{\circ}$

such that the diagrams of Definition 2.3 with reversed arrows commute.

Symmetrically to Proposition 2.4, the composites and the Cartesian products of (composable) Oplax_+-monoidal functors are again Oplax_+-monoidal via the evident natural transformations.

DEFINITION 3.3. An Oplax_+*-monoidal natural transformation* is a natural transformation for which the diagrams of Definition 2.5 with reversed arrows commute.

Symmetrically to Proposition 2.6, all of the composites, the Godement products, and the Cartesian products of (composable) Oplax_+-monoidal natural transformations are again Oplax_+-monoidal. Consequently, there is a strict monoidal

2-category Oplax_+ of Oplax_+-monoidal categories, Oplax_+-monoidal functors and Oplax_+-monoidal natural transformations.

DEFINITION 3.4. A *semigroup* in an Oplax_+-monoidal category $(\mathsf{R}, \circ, \Psi, \upsilon)$ is a cosemigroup in the Lax^+-monoidal category $(\mathsf{R}^{\mathsf{op}}, \circ^{\mathsf{op}}, \Psi^{\mathsf{op}}, \upsilon^{\mathsf{op}})$. That is, it consists of

- an object a of R and
- a morphism $\mu : a \circ a \to a$

rendering commutative the first (associativity) diagram below. A *morphism of semigroups* is a morphism ϕ rendering commutative the second diagram.

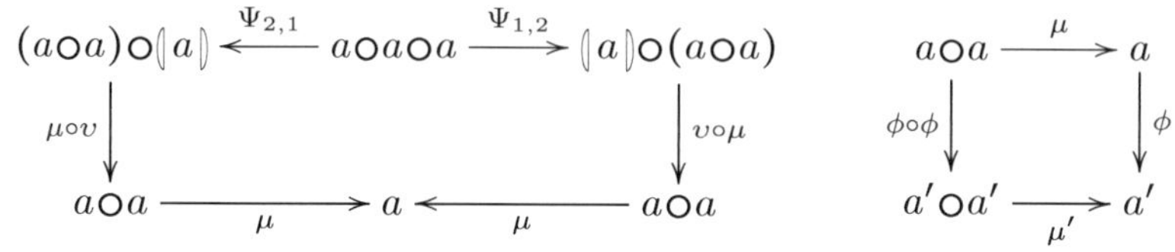

THEOREM 3.5. *For any* Oplax_+*-monoidal category* $(\mathsf{R}, \circ, \Psi, \upsilon)$*, the following categories are isomorphic.*

(i) *The category of semigroups and their morphisms in* $(\mathsf{R}, \circ, \Psi, \upsilon)$.
(ii) *The category of* Oplax_+*-monoidal functors* $(\mathbb{1}, 1, 1, 1) \to (\mathsf{R}, \circ, \Psi, \upsilon)$ *and their* Oplax_+*-monoidal natural transformations.*

Consequently, Oplax_+*-monoidal functors preserve semigroups.*

PROOF. It follows by the application of Theorem 2.8 to the Lax^+-monoidal category $(\mathsf{R}^{\mathsf{op}}, \circ^{\mathsf{op}}, \Psi^{\mathsf{op}}, \upsilon^{\mathsf{op}})$. □

Dually to comodules in Definition 2.9, one may consider modules over semigroups in the following sense.

DEFINITION 3.6. A *module* of a semigroup (a, μ) in an Oplax_+-monoidal category $(\mathsf{R}, \circ, \Psi, \upsilon)$ is a comodule of the cosemigroup (a, μ) in the Lax^+-monoidal category $(\mathsf{R}^{\mathsf{op}}, \circ^{\mathsf{op}}, \Psi^{\mathsf{op}}, \upsilon^{\mathsf{op}})$. Thus it consists of

- an object x of R
- a morphism $\varrho : x \circ a \to x$ (the *action*)

such that the (associativity) diagram on the left of

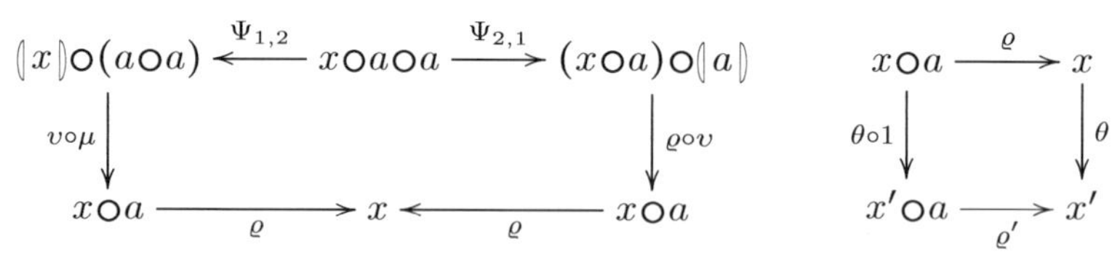

commutes. A *morphism of modules* $(x, \varrho) \to (x', \varrho')$ is a morphism $\theta : x \to x'$ in R such that the diagram on the right above commutes.

4. Bisemigroups in $\mathsf{Lax}^+\mathsf{Oplax}_+$-duoidal categories

In order to define bisemigroups in some category, one needs both a Lax^+-monoidal structure — for the cosemigroup part to live in — and an Oplax_+-monoidal structure — to host the semigroup part. Also a suitable compatibility of these monoidal structures is needed so that the category of cosemigroups inherits the

Oplax_+-monoidal structure and, dually, the category of semigroups inherits the Lax^+-monoidal structure. Then we can define bisemigroups by the usual principle: as cosemigroups in the category of semigroups; equivalently, as semigroups in the category of cosemigroups. In this section we formulate a suitable compatibility between Lax^+- and Oplax_+-monoidal structures which allows us to carry out this program.

DEFINITION 4.1. A $\mathsf{Lax}^+\mathsf{Oplax}_+$-*duoidal category* consists of

- a category D
- a Lax^+-monoidal structure $(\mathsf{D}, \bullet, \Phi, \iota)$
- an Oplax_+-monoidal structure $(\mathsf{D}, \circ, \Psi, \upsilon)$
- for all positive integers n and p, natural transformations

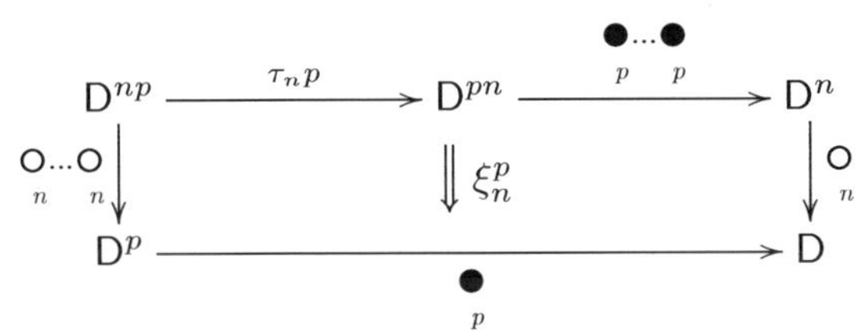

satisfying the following equivalent compatibility conditions, for all positive integers $n, p, k_1, \ldots, k_p$.

(i)
- $(\underset{n}{\circ}, \xi_n)$ is a Lax^+-monoidal functor $(\mathsf{D}, \bullet, \Phi, \iota)^n \to (\mathsf{D}, \bullet, \Phi, \iota)$;
- $\Psi_{k_1,\ldots,k_p}$ and υ are Lax^+-monoidal natural transformations.

Succinctly, $((\mathsf{D}, \bullet, \Phi, \iota), (\circ, \xi), \Psi, \upsilon)$ is an Oplax_+-monoid in the strict monoidal 2-category Lax^+ of Proposition 2.6.

(ii)
- $(\underset{n}{\bullet}, \xi^n)$ is an Oplax_+-monoidal functor $(\mathsf{D}, \circ, \Psi, \upsilon)^n \to (\mathsf{D}, \circ, \Psi, \upsilon)$;
- $\Phi_{k_1,\ldots,k_p}$ and ι are Oplax_+-monoidal natural transformations.

Succinctly, $((\mathsf{D}, \circ, \Psi, \upsilon), (\bullet, \xi), \Phi, \iota)$ is a Lax^+-monoid in the strict monoidal 2-category Oplax_+ of Section 3.

(iii) The following diagrams commute.

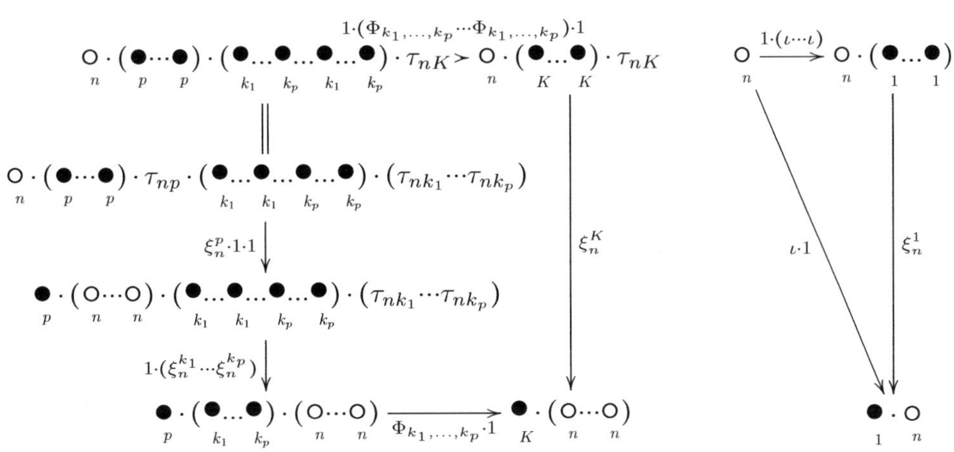

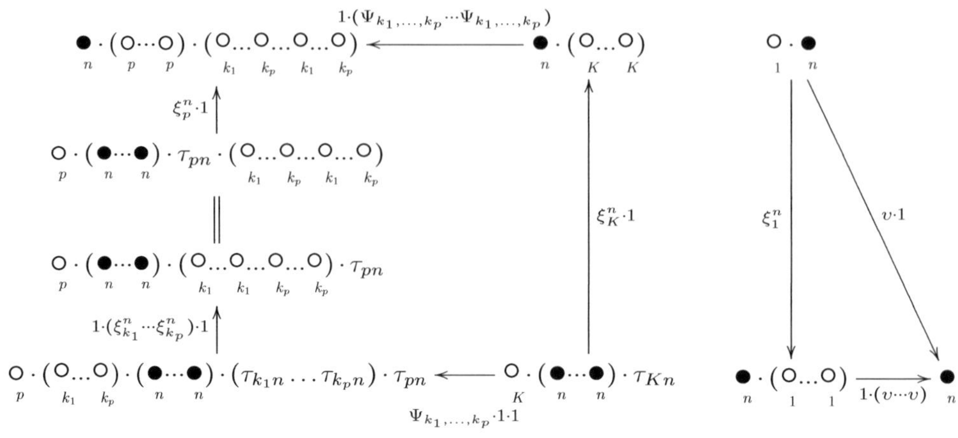

DEFINITION 4.2. A $\mathsf{Lax}^+\mathsf{Oplax}_+$-*duoidal functor* $(\mathsf{D},\bullet,\circ)\to(\mathsf{D}',\bullet',\circ')$ is defined as

(i) an Oplax_+-monoidal 1-cell $((G,\Gamma^\bullet),\Gamma^\circ):((\mathsf{D},\bullet),\circ)\to((\mathsf{D}',\bullet'),\circ')$ in the strict monoidal 2-category Lax^+ of Proposition 2.6; equivalently,

(ii) a Lax^+-monoidal 1-cell $((G,\Gamma^\circ),\Gamma^\bullet):((\mathsf{D},\circ),\bullet)\to((\mathsf{D}',\circ'),\bullet')$ in the strict monoidal 2-category Oplax_+ of Section 3; equivalently,

(iii)
- a functor $G:\mathsf{D}\to\mathsf{D}'$
- a Lax^+-monoidal structure $\{\Gamma^\bullet_n : G\cdot\underset{n}{\bullet}\to\underset{n}{\bullet'}\cdot(G\cdots G)\}_{n\in\mathbb{N}}$
- a Oplax_+-monoidal structure $\{\Gamma^\circ_n : \underset{p}{\circ'}\cdot(G\cdots G)\to G\cdot\underset{p}{\circ}\}_{p\in\mathbb{N}}$

rendering commutative the following diagram for all positive integers n,p.

(4.1)

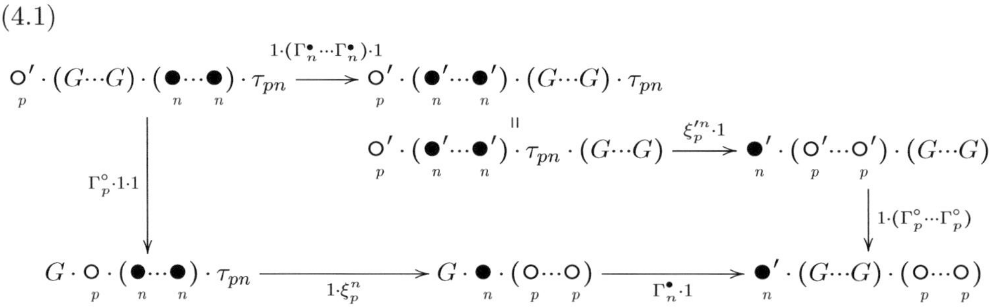

It follows immediately from Definition 4.2 and Proposition 2.4 (1) that the composite of $\mathsf{Lax}^+\mathsf{Oplax}_+$-duoidal functors is again $\mathsf{Lax}^+\mathsf{Oplax}_+$-duoidal via the composite Lax^+-monoidal structure in Proposition 2.4 (1) and the symmetrically constructed composite Oplax_+-monoidal structure.

DEFINITION 4.3. A $\mathsf{Lax}^+\mathsf{Oplax}_+$-*duoidal natural transformation* is defined as

(i) an Oplax_+-monoidal 2-cell in the strict monoidal 2-category Lax^+ of Proposition 2.6; equivalently,

(ii) a Lax^+-monoidal 2-cell in the strict monoidal 2-category Oplax_+ of Section 3; equivalently,

(iii) a natural transformation which is both Oplax_+-monoidal and Lax^+-monoidal. That is, it renders commutative both the diagram of Definition 2.5 and its symmetric counterpart with reversed arrows and inverted colors.

THEOREM 4.4. *For any* $\mathsf{Lax}^+\mathsf{Oplax}_+$*-duoidal category* $(\mathsf{D},(\bullet,\Phi,\iota),(\circ,\Psi,\upsilon),\xi)$, *the following assertions hold.*

(1) *The category of cosemigroups in the* Lax^+*-monoidal category* $(\mathsf{D},\bullet,\Phi,\iota)$ *is* Oplax_+*-monoidal via the structure* $(\circ,\Psi,\upsilon)$.
(2) *The category of semigroups in the* Oplax_+*-monoidal category* $(\mathsf{D},\circ,\Psi,\upsilon)$ *is* Lax^+*-monoidal via the structure* $(\bullet,\Phi,\iota)$.

PROOF. We only prove (1), part (2) follows symmetrically.

By Theorem 2.8 and Definition 4.1 (i), any sequence $\{a_1,\cdots,a_n\}$ of cosemigroups in $(\mathsf{D},\bullet,\Phi,\iota)$ determines a Lax^+-monoidal functor; and any sequence $\{f_1:a_1\to a'_1,\cdots,f_n:a_n\to a'_n\}$ of cosemigroup morphisms determines a Lax^+-monoidal natural transformation

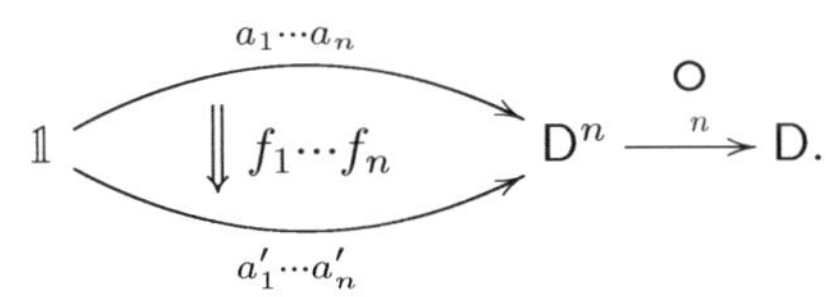

That is, $\overset{n}{\circ}$ determines a functor to the category of cosemigroups in $(\mathsf{D},\bullet,\Phi,\iota)$ from the nth Cartesian power of this category.

Again by Theorem 2.8 and Definition 4.1 (i), the natural transformations Ψ and υ — if evaluated at cosemigroups — yield morphisms of cosemigroups. Thus they can be seen as natural transformations between functors connecting Cartesian powers of categories of cosemigroups. They satisfy the axioms in Definition 3.1 by construction. □

DEFINITION 4.5. A *bisemigroup* in a $\mathsf{Lax}^+\mathsf{Oplax}_+$-duoidal category $(\mathsf{D},\bullet,\circ,\xi)$ is

(i) a semigroup in the Oplax_+-monoidal category of cosemigroups in $(\mathsf{D},\bullet)$ — see Theorem 4.4 (1); equivalently,
(ii) a cosemigroup in the Lax^+-monoidal category of semigroups in $(\mathsf{D},\circ)$ — see Theorem 4.4 (2); equivalently,
(iii)
- an object a
- a semigroup (a,μ) in $(\mathsf{D},\circ)$
- a cosemigroup (a,δ) in $(\mathsf{D},\bullet)$

subject to the compatibility condition encoded in the commutative diagram

$$\begin{array}{ccccc} a\circ a & \xrightarrow{\ \mu\ } & a & \xrightarrow{\ \delta\ } & a\bullet a. \\ {\scriptstyle\delta\circ\delta}\downarrow & & & & \uparrow{\scriptstyle\mu\bullet\mu} \\ (a\bullet a)\circ(a\bullet a) & & \xrightarrow[\ \xi_2^2\ ]{} & & (a\circ a)\bullet(a\circ a). \end{array}$$

A *morphism of bisemigroups* is a semigroup morphism in the category of cosemigroups in $(\mathsf{D},\bullet)$; equivalently, a cosemigroup morphism in the category of semigroups in $(\mathsf{D},\circ)$; equivalently, a morphism in D which is both a cosemigroup morphism in $(\mathsf{D},\bullet)$ and a semigroup morphism in $(\mathsf{D},\circ)$.

THEOREM 4.6. *For any* $\mathsf{Lax}^+\mathsf{Oplax}_+$*-duoidal category* $(\mathsf{D},\bullet,\circ,\xi)$ *the following categories are isomorphic.*

(i) *The category of bisemigroups and their morphisms in* $(\mathsf{D},\bullet,\circ,\xi)$.
(ii) *The category of* $\mathsf{Lax}^+\mathsf{Oplax}_+$*-duoidal functors* $(\mathbb{1},1,1,1)\to(\mathsf{D},\bullet,\circ,\xi)$ *and their* $\mathsf{Lax}^+\mathsf{Oplax}_+$*-duoidal natural transformations.*

Consequently, $\mathsf{Lax}^+\mathsf{Oplax}_+$*-duoidal functors preserve bisemigroups.*

PROOF. By Definition 4.5, the category of part (i) is isomorphic to the category of semigroups in the category of cosemigroups in $(\mathsf{D},\bullet)$. So by Theorem 3.5, it is isomorphic to the category of Oplax_+-monoidal functors and Oplax_+-monoidal natural transformations from $\mathbb{1}$ to the category of cosemigroups in $(\mathsf{D},\bullet)$. Then by Theorem 2.8, it is further equivalent to the category of Oplax_+-monoidal 1-cells and Oplax_+-monoidal 2-cells from $(\mathbb{1},1)$ to $(\mathsf{D},\bullet)$ in Lax^+. By Definition 4.2 and Definition 4.3 this is isomorphic to the category of part (ii). □

THEOREM 4.7. *For any bisemigroup* (a,δ,μ) *in an arbitrary* $\mathsf{Lax}^+\mathsf{Oplax}_+$*-duoidal category* $(\mathsf{D},(\bullet,\Phi,\iota),(\circ,\Psi,\upsilon),\xi)$ *the following assertions hold.*

(1) *The category of modules — in the sense of Definition* 3.6 *— of the semigroup* (a,μ) *in the* Oplax_+*-monoidal category* $(\mathsf{D},\circ,\Psi,\upsilon)$ *admits the* Lax^+*-monoidal structure* $(\bullet,\Phi,\iota)$.
(2) *The category of comodules — in the sense of Definition* 2.9 *— of the cosemigroup* (a,δ) *in the* Lax^+*-monoidal category* $(\mathsf{D},\bullet,\Phi,\iota)$ *admits the* Oplax_+*-monoidal structure* $(\circ,\Psi,\upsilon)$.

PROOF. We only prove part (1), part (2) is verified symmetrically.

For any sequence of (a,μ)-modules $\{(x_i,\varrho_i)\}_{i=1,\ldots,n}$, there is an (a,μ)-module

$$(x_1\bullet\cdots\bullet x_n)\circ a \xrightarrow{1\circ\delta_n} (x_1\bullet\cdots\bullet x_n)\circ a^{\bullet n} \xrightarrow{\xi_2^n} (x_1\circ a)\bullet\cdots\bullet(x_n\circ a) \xrightarrow{\varrho_1\bullet\cdots\bullet\varrho_n} x_1\bullet\cdots\bullet x_n.$$

Indeed, one checks by induction on l that the l-fold comultiplication of (2.2) for $a\circ a$ comes out as $a\circ a \xrightarrow{\delta_l\circ\delta_l} a^{\bullet l}\circ a^{\bullet l} \xrightarrow{\xi_2^l} (a\circ a)^{\bullet l}$. The induction step uses the top right diagram in part (iii) of Definition 4.1 for $n=2$, and the top left diagram of Definition 4.1 (iii) for $n=2$, $p=2$, $k_1=1$ and $k_2=l$. Since by Definition 4.5 the multiplication μ is a morphism of cosemigroups, we obtain the commutative diagram

$$\begin{array}{ccccc} a\circ a & \xrightarrow{\delta_n\circ\delta_n} & a^{\bullet n}\circ a^{\bullet n} & \xrightarrow{\xi_2^n} & (a\circ a)^{\bullet n} \\ {\scriptstyle\mu}\downarrow & & & & \downarrow{\scriptstyle\mu^{\bullet n}} \\ a & & \xrightarrow[\delta_n]{\qquad\qquad} & & a^{\bullet n} \end{array}$$

for all non-negative integers n. Together with the bottom right diagram of Definition 4.1 (iii), this proves the commutativity of the leftmost region of Figure 1. The region labelled by (BR) in Figure 2 also commutes by the bottom right diagram of Definition 4.1 (iii). The regions marked by (BL) in Figure 1 and Figure 2 both commute by the bottom left diagram of Definition 4.1 (iii) at the respective values $(n, p = 2, k_1 = 1, k_2 = 2)$ and $(n, p = 2, k_1 = 2, k_2 = 1)$. The remaining regions commute by evident naturality. Whenever each action ϱ_i is associative, the right verticals of the diagrams in Figure 1 and Figure 2 are equal. Hence also the paths on their left hand sides are equal, proving the associativity of the stated action on $x_1 \bullet \cdots \bullet x_n$.

The $\bullet$-monoidal products of module morphisms are easily seen to be morphisms of modules. The natural transformation ι — if evaluated at a module — is a morphism of modules by the top right diagram of Definition 4.1 (iii). In order to see that for all non-negative integers $p, k_1, \ldots k_p$, the component of $\Phi_{k_1,\ldots,k_p}$ at any modules is a morphism of modules, use (2.1) and the top left diagram of Definition 4.1 (iii) for $n = 2$. □

PROPOSITION 4.8. *For any bisemigroup (a, δ, μ) in an arbitrary* $\mathsf{Lax}^+\mathsf{Oplax}_+$-*duoidal category the following assertions hold.*

(1) *$((a, \delta), \mu)$ is a semigroup in the* Oplax_+-*monoidal category of comodules over the cosemigroup (a, δ) (cf. Theorem 4.7 (2)).*
(2) *$((a, \mu), \delta)$ is a cosemigroup in the* Lax^+-*monoidal category of modules over the semigroup (a, μ) (cf. Theorem 4.7 (1)).*

PROOF. We only prove part (1), part (2) follows symmetrically.

By Example 2.10, (a, δ) is an (a, δ)-comodule. The compatibility diagram of Definition 4.5 (iii) can be interpreted as μ being a morphism of (a, δ)-comodules. Associativity of μ in the category of (a, δ)-comodules follows from the associativity of the monoid (a, μ). □

DEFINITION 4.9. A *Hopf module* over a bisemigroup (a, δ, μ) in a $\mathsf{Lax}^+\mathsf{Oplax}_+$-duoidal category $(\mathsf{D}, \bullet, \circ, \xi)$ is defined by the following equivalent data.

(i) A module over the semigroup $((a, \delta), \mu)$ in the category of comodules over the cosemigroup (a, δ) in $(\mathsf{D}, \bullet)$.
(ii) A comodule over the cosemigroup $((a, \mu), \delta)$ in the category of modules over the semigroup (a, μ) in $(\mathsf{D}, \circ)$.
(iii)
- An object x of D,
- a module (x, ν) over the semigroup (a, μ) in $(\mathsf{D}, \circ)$,
- a comodule (x, ϱ) over the cosemigroup (a, δ) in $(\mathsf{D}, \bullet)$,

rendering commutative the following diagram.

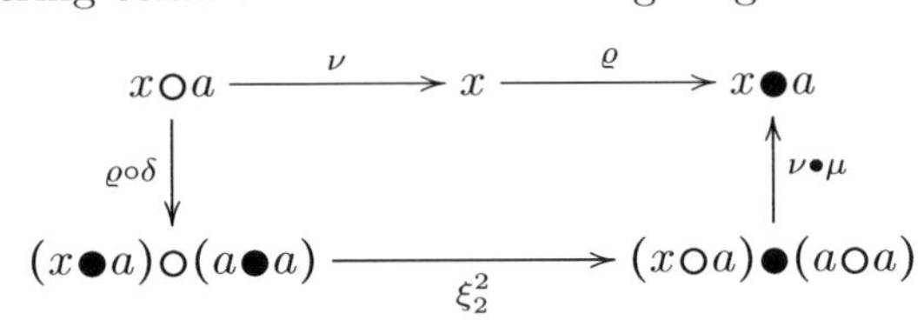

EXAMPLE 4.10. By Example 2.10 and its dual counterpart, and by the compatibility diagram of Definition 4.5 (iii), (a, δ, μ) is a Hopf module over an arbitrary bisemigroup (a, δ, μ) in a $\mathsf{Lax}^+\mathsf{Oplax}_+$-duoidal category.

5. Comonoids in $\mathsf{Lax}^+\mathsf{Oplax}^0$-monoidal categories

For the definition of monoids and comonoids — that is, to define units for multiplications and counits for comultiplications — we also need nullary monoidal

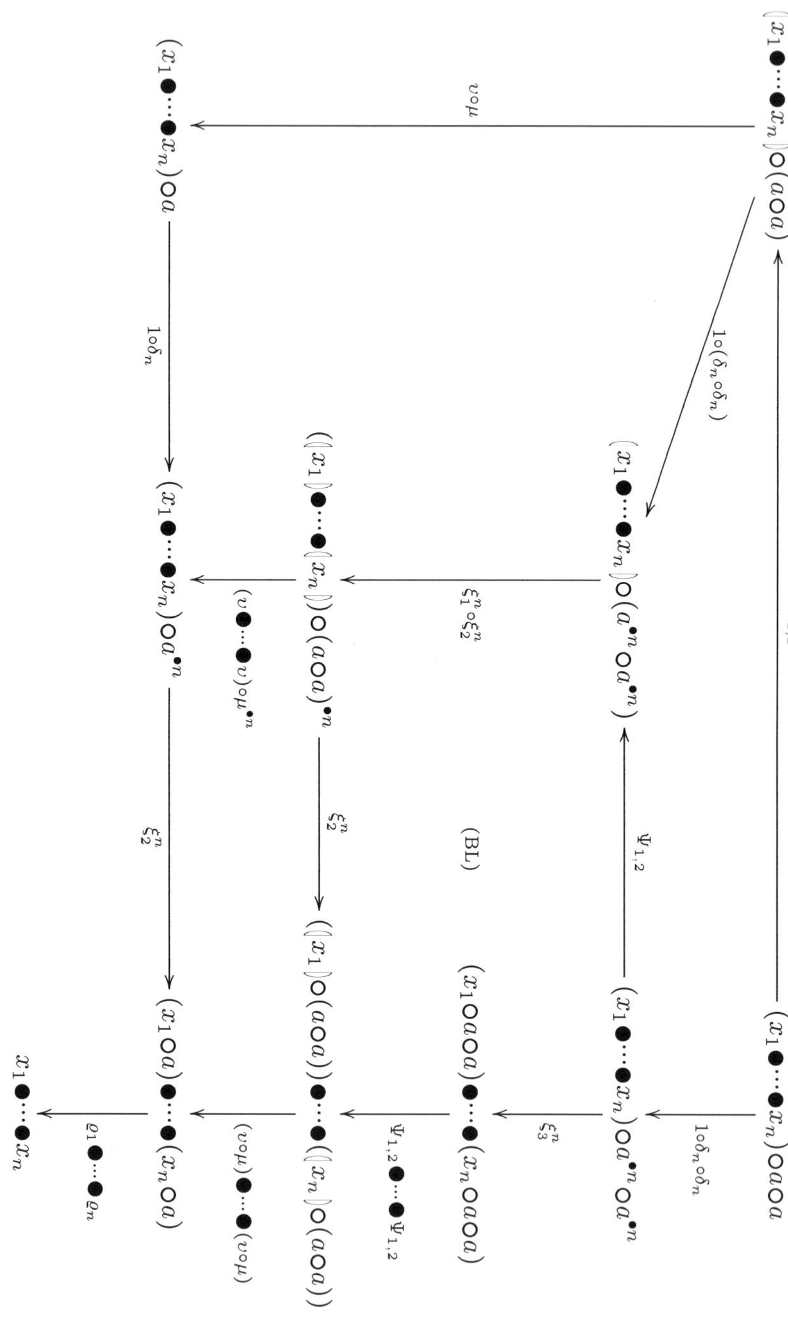

FIGURE 1. Associativity of the action on ●-products of modules — first part

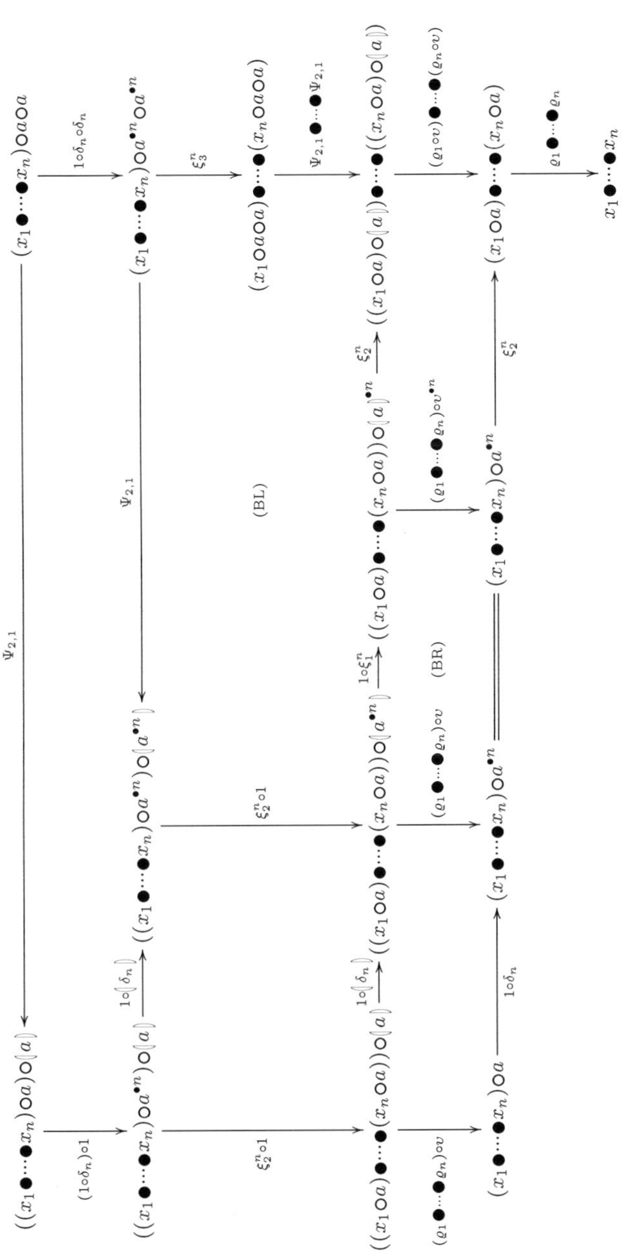

FIGURE 2. Associativity of the action on ●-products of modules — second part

products. They may be related to the higher monoidal products in possibly different ways. There are well-known and well-studied structures called lax and oplax monoidal categories, see e.g. [**9**]. However, the gadgets motivating this paper, namely, unital BiHom-monoids and counital BiHom-comonoids, do not seem to fit these well-studied situations. In this section we show that counital BiHom-comonoids can be described as comonoids in monoidal categories in which the monoidal products of positive numbers of factors come with lax compatibiliy morphisms, but the nullary part is oplax coherent. The dual situation, suitable to describe unital BiHom-monoids as monoids, will be discussed in the next section.

DEFINITION 5.1. A $\mathsf{Lax}^+\mathsf{Oplax}^0$-*monoidal category* consists of

- a category L
- for any non-negative integer n, a functor $\underset{n}{\bullet} : \mathsf{L}^n \to \mathsf{L}$
- for all non-negative integers $n, k_1, \dots, k_n$, and for the functors

$$\mathsf{L}^{k_i} \xrightarrow{[k_i]} \mathsf{L}^{\overline{k_i}} := \begin{cases} \mathsf{L}^{k_i} \xrightarrow{1} \mathsf{L}^{k_i} & \text{if} \quad k_i > 0 \\ \mathbb{1} \xrightarrow{\underset{0}{\bullet}} \mathsf{L} & \text{if} \quad k_i = 0, \end{cases}$$

  and using the notation of Section 1, natural transformations

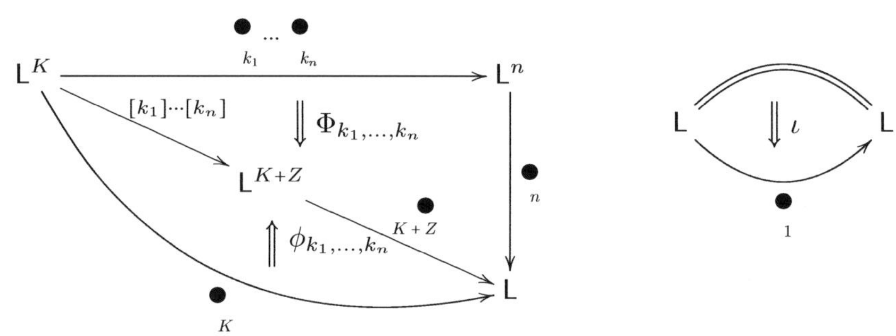

such that

$$\Phi_{k_1,\dots,k_n} = 1 \quad \text{if} \quad K = 0, \qquad \phi_{k_1,\dots,k_n} = 1 \quad \text{if} \quad Z = 0, \qquad \phi_0 = \iota \cdot 1,$$

and the diagram

(5.1)

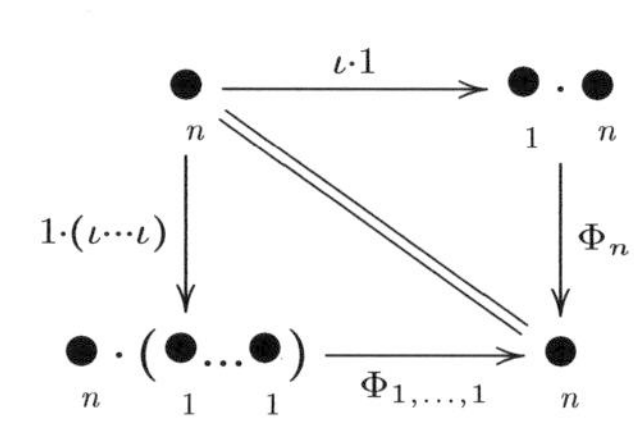

as well as the regions of the diagram of Figure 3 commute, for all non-negative integers n, $\{m_i\}$ labelled by $i = 1, \dots, n$, and $\{k_{ij}\}$ labelled by $i = 1, \dots, n$ and $j = 1, \dots, m_i$.

Remark that the diagram of Figure 3 reduces to that of Definition 2.1 if $Z = 0$. Hence, forgetting about the natural transformations $\phi_{k_1,\dots,k_n}$, and those instances of $\Phi_{k_1,\dots,k_n}$ for which $Z > 0$, a $\mathsf{Lax}^+\mathsf{Oplax}^0$-monoidal category can be regarded as a Lax^+-monoidal category.

A $\mathsf{Lax}^+\mathsf{Oplax}^0$-monoidal category is *(strict) normal* if it is (strict) normal as a Lax^+-monoidal category.

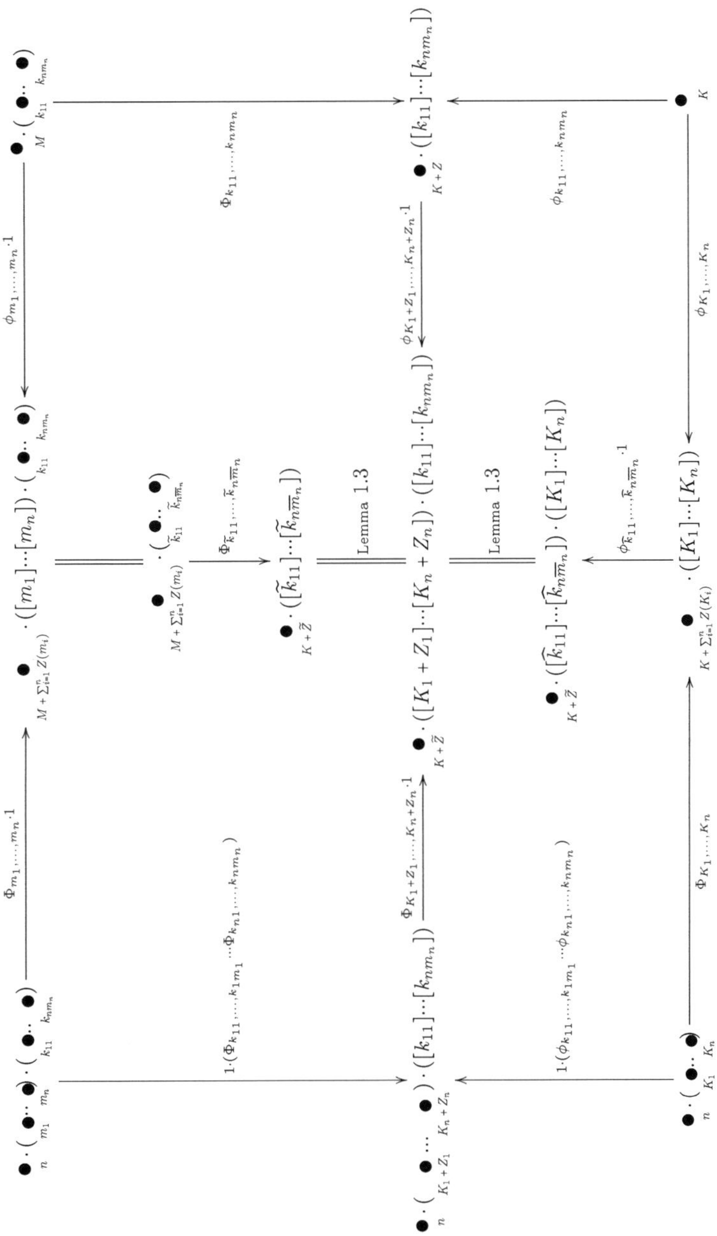

FIGURE 3. Axioms of $\mathsf{Lax}^{+}\mathsf{Oplax}^{0}$-monoidal category

Again, Definition 5.1 can be extended in a straightforward way to define $\mathsf{Lax}^{+}\mathsf{Oplax}^{0}$-monoids in any Gray monoid — so in particular in any strict monoidal 2-category.

Throughout, in a $\mathsf{Lax}^{+}\mathsf{Oplax}^{0}$-monoidal category we denote by j the image of the single object of $\mathbb{1}$ under the functor $\underset{0}{\bullet} : \mathbb{1} \to \mathsf{L}$; and keep the earlier notation $\langle a \rangle$ for the image of any object a under the functor $\underset{1}{\bullet} : \mathsf{L} \to \mathsf{L}$; and $a_1 \bullet \cdots \bullet a_n$ for the image of an object $(a_1, \ldots, a_n)$ under the functor $\underset{n}{\bullet} : \mathsf{L}^n \to \mathsf{L}$.

Analogously to Proposition 2.2, the following holds.

PROPOSITION 5.2. *The Cartesian product* LL' *of* $\mathsf{Lax}^{+}\mathsf{Oplax}^{0}$*-monoidal categories* $(\mathsf{L}, \bullet, \Phi, \phi, \iota)$ *and* $(\mathsf{L}', \bullet', \Phi', \phi', \iota')$ *is again* $\mathsf{Lax}^{+}\mathsf{Oplax}^{0}$*-monoidal via*

- *the functors* $(\mathsf{LL}')^n \xrightarrow{\tau_{2n}} \mathsf{L}^n\mathsf{L}'^n \xrightarrow{\underset{n\ n}{\bullet\bullet'}} \mathsf{LL}'$
- *the natural transformations* $\iota\iota' : 1 \to \underset{1\ 1}{\bullet\bullet'}$,

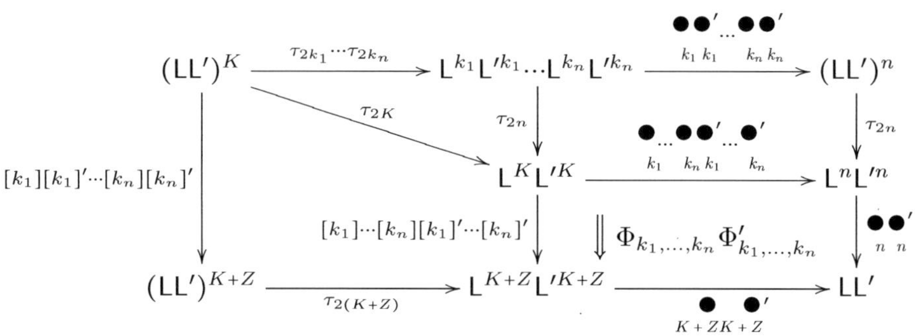

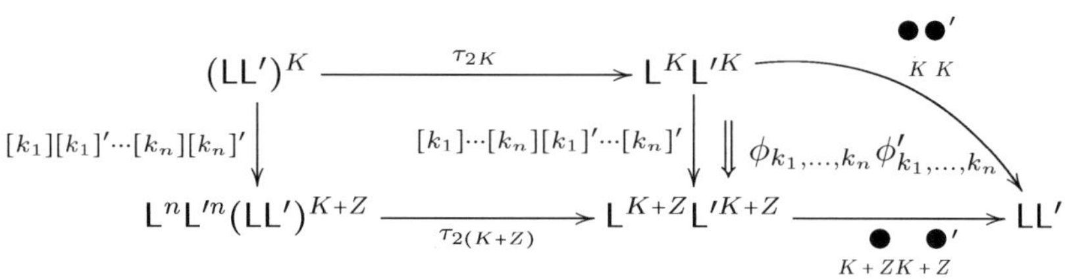

where the unlabelled regions denote identity natural transformations.

PROOF. We leave it to the reader to check that the stated datum satisfies the axioms in Definition 5.1. □

DEFINITION 5.3. A $\mathsf{Lax}^{+}\mathsf{Oplax}^{0}$*-monoidal functor* consists of

- a functor $G : \mathsf{L} \to \mathsf{L}'$
- for all non-negative integers n, a natural transformation $\Gamma_n : G \cdot \underset{n}{\bullet} \to \underset{n}{\bullet}' \cdot (G \cdots G)$

such that for all sequences of non-negative integers $\{k_1, \ldots, k_n\}$, for the sequence of natural transformations

$$G^{\overline{k}_i} \cdot [k_i] \xrightarrow{[\![k_i]\!]} [k_i]' \cdot G^{k_i} := \begin{cases} G^{k_i} \xrightarrow{1} G^{k_i} & \text{if } \ k_i > 0 \\ G \cdot \underset{0}{\bullet} \xrightarrow{\Gamma_0} \underset{0}{\bullet}' & \text{if } \ k_i = 0 \end{cases}$$

for $i = 1, \ldots, n$, and using the notation of Section 1, the following diagrams commute.

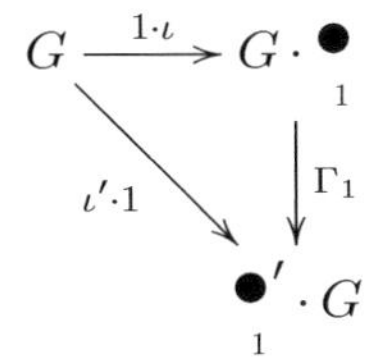

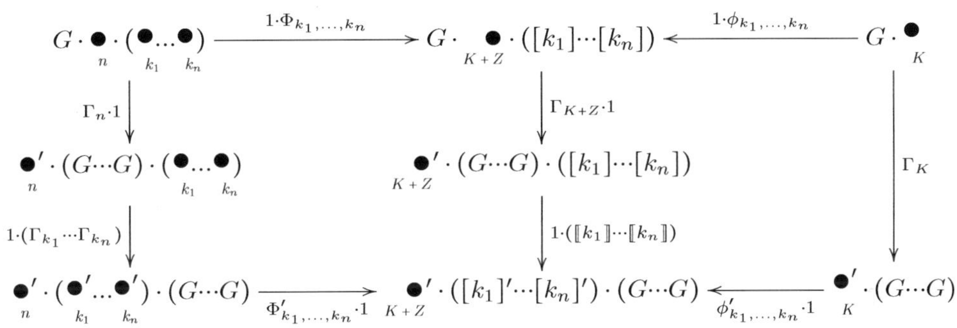

Forgetting about the nullary part of the structure, $\mathsf{Lax}^{+}\mathsf{Oplax}^{0}$-monoidal functors can be seen Lax^{+}-monoidal.

Analogously to Proposition 2.4, the following holds.

PROPOSITION 5.4. (1) *The composite of (composable)* $\mathsf{Lax}^{+}\mathsf{Oplax}^{0}$-*monoidal functors* (G, Γ) *and* (H, Ξ) *is again* $\mathsf{Lax}^{+}\mathsf{Oplax}^{0}$-*monoidal via the natural transformations*

$$H \cdot G \cdot \underset{n}{\bullet} \xrightarrow{1\cdot\Gamma_n} H \cdot \underset{n}{\bullet} \cdot (G \cdots G) \xrightarrow{\Xi_n \cdot 1} \underset{n}{\bullet} \cdot (H \cdots H) \cdot (G \cdots G) = \underset{n}{\bullet} \cdot (H \cdot G)^n.$$

(2) *The Cartesian product of* $\mathsf{Lax}^{+}\mathsf{Oplax}^{0}$-*monoidal functors* (G, Γ) *and* (H, Ξ) *is again* $\mathsf{Lax}^{+}\mathsf{Oplax}^{0}$-*monoidal via the natural transformations*

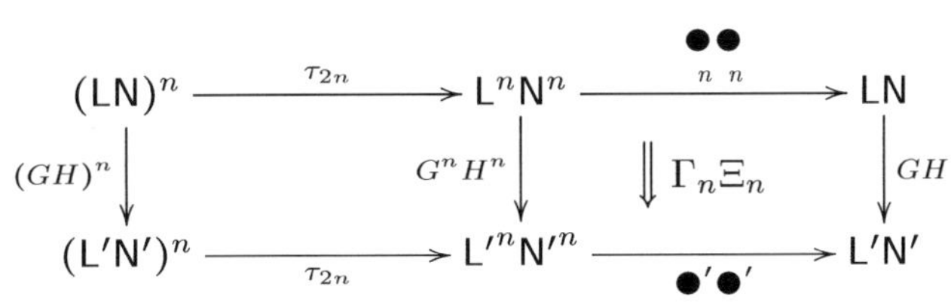

where the unlabelled region denotes the identity natural transformation.

PROOF. It is left to the reader to check that the stated datum in both parts satisfies the axioms in Definition 5.3. □

DEFINITION 5.5. A $\mathsf{Lax}^{+}\mathsf{Oplax}^{0}$-*monoidal natural transformation* $(G, \Gamma) \to (G', \Gamma')$ is a natural transformation $\omega : G \to G'$ for which the following diagram commutes for all non-negative integers n.

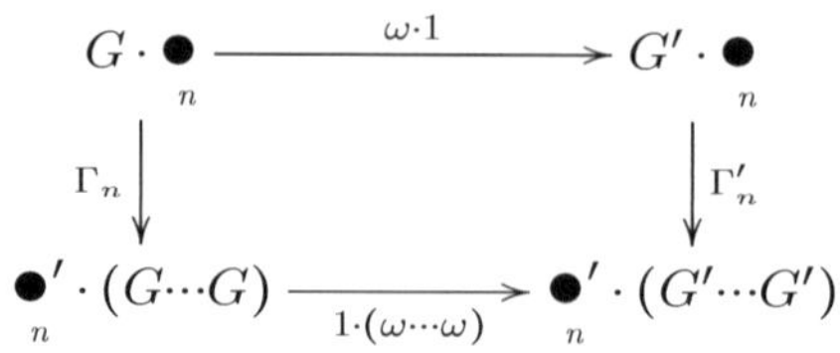

$\mathsf{Lax}^{+}\mathsf{Oplax}^{0}$-monoidal natural transformations are in particular Lax^{+}-monoidal. Analogously to Proposition 2.6, the following holds.

PROPOSITION 5.6. *All of the composites, the Godement products, and the Cartesian products of (composable)* $\mathsf{Lax}^{+}\mathsf{Oplax}^{0}$*-monoidal natural transformations are again* $\mathsf{Lax}^{+}\mathsf{Oplax}^{0}$*-monoidal. Consequently, there is a strict monoidal 2-category* $\mathsf{Lax}^{+}\mathsf{Oplax}^{0}$ *of* $\mathsf{Lax}^{+}\mathsf{Oplax}^{0}$*-monoidal categories,* $\mathsf{Lax}^{+}\mathsf{Oplax}^{0}$*-monoidal functors and* $\mathsf{Lax}^{+}\mathsf{Oplax}^{0}$*-monoidal natural transformations, admitting a strict monoidal forgetful 2-functor to the 2-category* Lax^{+} *of Proposition* 2.6.

PROOF. The straightforward check of the axiom in Definition 5.5 in each of the stated cases is left to the reader. □

DEFINITION 5.7. A *comonoid* in a $\mathsf{Lax}^{+}\mathsf{Oplax}^{0}$-monoidal category $(\mathsf{L},\bullet,\Phi,\phi,\iota)$ consists of

- a cosemigroup (a,δ) in the Lax^{+}-monoidal category $(\mathsf{L},\bullet,\Phi,\iota)$ and
- a morphism $\varepsilon : a \to j$ in L

rendering commutative the first (counitality) diagram below. A *comonoid morphism* $(a,\delta,\varepsilon) \to (a',\delta',\varepsilon')$ is a morphism of cosemigroups $\phi : (a,\delta) \to (a',\delta')$ rendering commutative the second diagram as well.

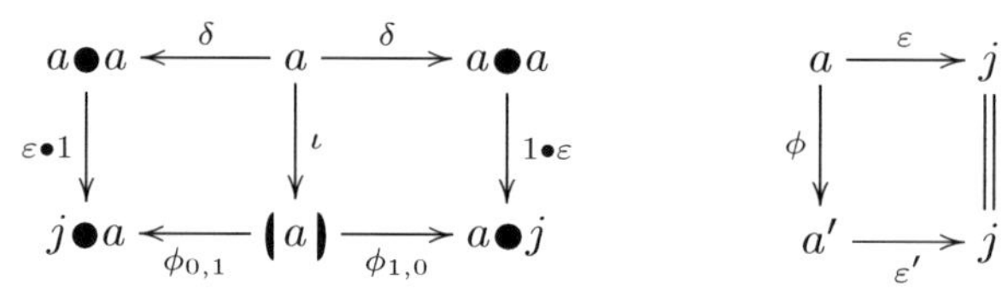

THEOREM 5.8. *For any* $\mathsf{Lax}^{+}\mathsf{Oplax}^{0}$*-monoidal category* $(\mathsf{L},\bullet,\Phi,\phi,\iota)$*, the following categories are isomorphic.*

(i) *The category of comonoids and their morphisms in* $(\mathsf{L},\bullet,\Phi,\phi,\iota)$.
(ii) *The category of* $\mathsf{Lax}^{+}\mathsf{Oplax}^{0}$*-monoidal functors* $(\mathbb{1},1,1,1,1) \to (\mathsf{L},\bullet,\Phi,\phi,\iota)$ *and their* $\mathsf{Lax}^{+}\mathsf{Oplax}^{0}$*-monoidal natural transformations.*

Consequently, $\mathsf{Lax}^{+}\mathsf{Oplax}^{0}$*-monoidal functors preserve comonoids.*

PROOF. A $\mathsf{Lax}^{+}\mathsf{Oplax}^{0}$-monoidal functor $(\mathbb{1},1,1,1,1) \to (\mathsf{L},\bullet,\Phi,\phi,\iota)$ as in part (ii) consists of

- an object a of L
- for all non-negative integers n, a morphism δ_n from a to the n-fold product $a^{\bullet n} := a\bullet\cdots\bullet a$ for $n>1$, to $a^{\bullet 1} := (a)$ for $n=1$, and to $a^{\bullet 0} := j$ for $n=0$,

such that

$$\delta_1 = \iota \tag{5.2}$$

and for all sequences of non-negative integers $\{k_1,\dots, k_n\}$, and for the double sequences $\{\widehat{a}_{pq}\}_{\substack{p=1,\dots,n\\ q=1,\dots,\overline{k}_p}}$ of elements in L, and $\{\langle pq\rangle\}_{\substack{p=1,\dots,n\\ q=1,\dots,\overline{k}_p}}$ of morphisms in L, given by

$$
\widehat{a}_{pq} = \begin{cases} a & \text{if } k_p > 0\\ j & \text{if } k_p = 0, \end{cases} \qquad\qquad a \xrightarrow{\langle pq\rangle} \widehat{a}_{pq} = \begin{cases} a \xrightarrow{1} a & \text{if } k_p > 0\\ a \xrightarrow{\delta_0} j & \text{if } k_p = 0. \end{cases}
$$

the following diagram commutes.
(5.3)

$$
\begin{array}{ccccc}
a^{\bullet n} & \xleftarrow{\quad\delta_n\quad} & a & = & a\\
\Big\downarrow{\scriptstyle \delta_{k_1}\bullet\cdots\bullet\delta_{k_n}} & & \Big\downarrow{\scriptstyle \delta_{K+Z}} & & \Big\downarrow{\scriptstyle \delta_K}\\
 & & a^{\bullet K+Z} & & \\
 & & \Big\downarrow{\scriptstyle \langle 11\rangle\bullet\cdots\bullet\langle 1\overline{k}_1\rangle\bullet\cdots\bullet\langle n1\rangle\bullet\cdots\bullet\langle n\overline{k}_n\rangle} & & \\
a^{\bullet k_1}\bullet\cdots\bullet a^{\bullet k_n} & \xrightarrow{\Phi_{k_1,\dots,k_n}} & \widehat{a}_{11}\bullet\cdots\bullet\widehat{a}_{1\overline{k}_1}\bullet\cdots\bullet\widehat{a}_{n1}\bullet\cdots\bullet\widehat{a}_{n\overline{k}_n} & \xleftarrow{\phi_{k_1,\dots,k_n}} & a^{\bullet K}
\end{array}
$$

If $Z = 0$ then the region on the left of (5.3) reduces to (2.1). So we conclude by Theorem 2.8 that (a, δ_2) is a cosemigroup. Evaluating the region on the right of (5.3) at $n = 2$, $k_1 = 0$, $k_2 = 1$ and $n = 2$, $k_1 = 1$, $k_2 = 0$ we infer that δ_0 is a counit for it.

Conversely, if (a, δ, ε) is a comonoid in $(\mathsf{L}, \bullet, \Phi, \phi, \iota)$ then we put $\delta_0 := \varepsilon$, $\delta_1 := \iota$, $\delta_2 := \delta$ and for $i > 1$ we define δ_{i+1} iteratively as in (2.2) (so that it satisfies (2.3)). By the following similar procedure to that in the proof of Theorem 2.8, one checks by induction on $K + Z$ that (5.3) commutes for the so defined family of morphisms, for all sequences of non-negative integers $\{k_1,\dots, k_n\}$.

If $K + Z = 0$ then $n = 0$ and (5.3) reduces to a trivial identity. If $K + Z = 1$ then $n = 1$ and k_1 is either 0 or 1. If $k_1 = 0$ then (5.3) commutes by the axioms $\Phi_0 = 1$ and $\phi_0 = \iota\cdot 1$ of $\mathsf{Lax}^{+}\mathsf{Oplax}^{0}$-monoidal category. If $k_1 = 1$ then (5.3) commutes by the axiom $\phi_1 = 1$ of $\mathsf{Lax}^{+}\mathsf{Oplax}^{0}$-monoidal category and the upper triangle of (5.1) for $n = 1$.

There are now three kinds of the induction step:

- replacing $\{k_1,\dots,k_n\}$ with $\{0, k_1,\dots,k_n\}$,
- replacing $\{k_1,\dots,k_n\}$ with $\{1, k_1,\dots,k_n\}$, and
- replacing $\{k_1,\dots,k_n\}$ with $\{1 + k_1,\dots,k_n\}$ if $k_1 > 0$.

We leave it to the reader to check that if (5.3) commutes for the original sequence $\{k_1,\dots,k_n\}$ then it also commutes for the modified sequence in each of the listed cases.

These constructions are mutually inverse bijections by (5.2) and by (5.3) at $n = 1$, $k_1 = 1$, and $k_2 = i$ (see the proof of Theorem 2.8).

A $\mathsf{Lax}^{+}\mathsf{Oplax}^{0}$-monoidal natural transformation $(a, \{\delta_i\}_{i\in\mathbb{N}}) \to (a', \{\delta'_i\}_{i\in\mathbb{N}})$ is obviously a comonoid morphism $(a, \delta_2, \delta_0) \to (a', \delta'_2, \delta'_0)$. Conversely, a comonoid morphism $(a, \delta, \varepsilon) \to (a', \delta', \varepsilon')$ is compatible with the unary parts $\delta_1 = \iota = \delta'_1$ of the corresponding $\mathsf{Lax}^{+}\mathsf{Oplax}^{0}$-monoidal functors by the naturality of ι and it is compatible with their higher components by Theorem 2.8. □

DEFINITION 5.9. A *comodule* of a comonoid (a, δ, ε) in a $\mathsf{Lax}^{+}\mathsf{Oplax}^{0}$-monoidal category $(\mathsf{L}, \bullet, \Phi, \phi, \iota)$ is a comodule $(x, \varrho : x \to x\bullet a)$ in the sense of Definition 2.9

over the underlying cosemigroup (a, δ) such that also the following (counitality) diagram commutes.

$$
\begin{array}{ccc}
x & \xrightarrow{\ \varrho\ } & x \bullet a \\
{\scriptstyle \iota}\downarrow & & \downarrow{\scriptstyle 1 \bullet \varepsilon} \\
(\!| x |\!) & \xrightarrow[\ \phi_{1,0}\ ]{} & x \bullet j
\end{array}
$$

A *morphism of comodules* over a comonoid is a morphism of comodules over the underlying cosemigroup in the sense of Definition 2.9.

EXAMPLE 5.10. Comparing the counitality conditions of Definition 5.7 and Definition 5.9, it follows immediately from Example 2.10 that (a, δ) is a comodule over an arbitrary comonoid (a, δ, ε) in any $\mathsf{Lax}^{+}\mathsf{Oplax}^{0}$-monoidal category.

Generalizing a construction in [**8**], we obtain the following example of $\mathsf{Lax}^{+}\mathsf{Oplax}^{0}$-monoidal category (in whose presentation below the notation of Section 1 is used).

THEOREM 5.11. *Associated to any monoidal category* $(\mathsf{V}, \otimes, I)$, *there is a strictly normal* $\mathsf{Lax}^{+}\mathsf{Oplax}^{0}$*-monoidal category* $(\mathsf{L}, \otimes, \Phi, \phi, 1)$ *as follows.*

- *In the category* L, *an* <u>*object*</u> *consists of an object* a *of* V *together with two morphisms* $\alpha, \beta : a \to a$ *such that* $\alpha \cdot \beta = \beta \cdot \alpha$.

  A <u>*morphism*</u> $(a, \alpha, \beta) \to (a', \alpha', \beta')$ *is a morphism* $\vartheta : a \to a'$ *such that* $\vartheta \cdot \alpha = \alpha' \cdot \vartheta$ *and* $\vartheta \cdot \beta = \beta' \cdot \vartheta$.
- *For* $n = 0$ *we put* $(I, 1, 1) : \mathbb{1} \to \mathsf{L}$, *for* $n = 1$ *we take the identity functor* $\mathsf{L} \to \mathsf{L}$ *and for any* $n > 1$ *we take the* n*-fold monoidal product functor* $\mathsf{L}^n \to \mathsf{L}$,

$$\{(a_1, \alpha_1, \beta_1), \ldots, (a_n, \alpha_1, \beta_n)\} \xrightarrow{\{\vartheta_1, \ldots, \vartheta_n\}} \{(a'_1, \alpha'_1, \beta'_1), \ldots, (a'_n, \alpha'_1, \beta'_n)\} \mapsto$$

$$(a_1 \otimes \cdots \otimes a_n, \alpha_1 \otimes \cdots \otimes \alpha_n, \beta_1 \otimes \cdots \otimes \beta_n) \xrightarrow{\vartheta_1 \otimes \cdots \otimes \vartheta_n} (a'_1 \otimes \cdots \otimes a'_n, \alpha'_1 \otimes \cdots \otimes \alpha'_n, \beta'_1 \otimes \cdots \otimes \beta'_n).$$

- *For any sequence of non-negative integers* $\{k_1, \ldots, k_n\}$, *and for any collection of objects* $\{(a_{ij}, \alpha_{ij}, \beta_{ij})\}_{\substack{i=1,\ldots,n \\ j=1,\ldots,k_i}}$, *the natural transformation* $\Phi_{k_1,\ldots,k_n}$ *is taken to be*

  $$\otimes_{i=1}^{n} \otimes_{j=1}^{k_i} \alpha_{ij}^{\sum_{p=i+1}^{n} (\overline{k}_p - 1)} \beta_{ij}^{\sum_{p=1}^{i-1} (\overline{k}_p - 1)} : \otimes_{i=1}^{n} \otimes_{j=1}^{k_i} a_{ij} \to \otimes_{i=1}^{n} \otimes_{j=1}^{k_i} a_{ij},$$

  and the natural transformation $\phi_{k_1,\ldots,k_n}$ *is taken to be*

  $$\otimes_{i=1}^{n} \otimes_{j=1}^{k_i} \alpha_{ij}^{\sum_{p=i+1}^{n} Z(k_p)} \beta_{ij}^{\sum_{p=1}^{i-1} Z(k_p)} : \otimes_{i=1}^{n} \otimes_{j=1}^{k_i} a_{ij} \to \otimes_{i=1}^{n} \otimes_{j=1}^{k_i} a_{ij}.$$

PROOF. Obviously, $\Phi_{k_1,\ldots,k_n} = 1$ if $K = 0$ and $\phi_{k_1,\ldots,k_n} = 1$ if $Z = 0$. Since also Φ_n and $\Phi_{1,\ldots,1}$ are trivial for all positive integers n, we only need to check the commutativity of the regions of the diagram of Figure 3 for the stated natural transformations. For each region of Figure 3 this means a comparison of endomorphisms of $\otimes_{i=1}^{n} \otimes_{j=1}^{m_i} \otimes_{l=1}^{k_{ij}} a_{ijl}$ for non-negative integers n, $\{m_i\}_{i=1,\ldots,n}$, and $\{k_{ij}\}_{\substack{i=1,\ldots,n \\ j=1,\ldots,m_i}}$; and for all collections $\{(a_{ijl}, \alpha_{ijl}, \beta_{ijl})\}_{\substack{i=1,\ldots,n \\ j=1,\ldots,m_i \\ l=1,\ldots,k_{ij}}}$ of objects in L. Each of these endomorphisms is a monoidal product of endomorphisms of the individual factors a_{ijl}, and each of the occurring endomorphisms of a_{ijl} is a composite of copies of the commuting morphisms α_{ijl} and β_{ijl}. So we only need to compare the resulting exponents of α_{ijl} and of β_{ijl} for all values of i, j, l. We only do it for β_{ijl} in each case; the computations for α_{ijl} are symmetric.

Then the four regions of the diagram of Figure 3 amount to the respective conditions

$$\sum_{p=1}^{i-1}(\overline{m}_p - 1) + \sum_{p=1}^{i-1}\sum_{q=1}^{\overline{m}_p}(\overline{\widetilde{k}}_{pq} - 1) + \sum_{q=1}^{j-1}(\overline{\widetilde{k}}_{iq} - 1) = \sum_{q=1}^{j-1}(\overline{k}_{iq} - 1) + \sum_{p=1}^{i-1}(\overline{K_p + Z}_p - 1)$$

$$\sum_{q=1}^{j-1} Z(k_{iq}) + \sum_{p=1}^{i-1}(\overline{K_p + Z}_p - 1) = \sum_{p=1}^{i-1}(\overline{K}_p - 1) + \sum_{p=1}^{i-1}\sum_{q=1}^{\overline{m}_p} Z(\widehat{k}_{pq}) + \sum_{q=1}^{j-1} Z(\widehat{k}_{iq})$$

$$\sum_{p=1}^{i-1} Z(m_p) + \sum_{p=1}^{i-1}\sum_{q=1}^{\overline{m}_p}(\overline{\widetilde{k}}_{pq} - 1) + \sum_{q=1}^{j-1}(\overline{\widetilde{k}}_{iq} - 1) = \sum_{p=1}^{i-1}\sum_{q=1}^{\overline{m}_p}(\overline{k}_{pq} - 1) + \sum_{q=1}^{j-1}(\overline{k}_{iq} - 1) + \sum_{p=1}^{i-1} Z(K_p + Z_p)$$

$$\sum_{p=1}^{i-1} Z_p + \sum_{q=1}^{j-1} Z(k_{iq}) + \sum_{p=1}^{i-1} Z(K_p + Z_p) = \sum_{p=1}^{i-1} Z(K_p) + \sum_{p=1}^{i-1}\sum_{q=1}^{\overline{m}_p} Z(\widehat{k}_{pq}) + \sum_{q=1}^{j-1} Z(\widehat{k}_{iq})$$

for all $i = 1, \ldots, n$ and $j = 1, \ldots, m_i$, which are easily checked to hold. $\square$

A straightforward application of Definition 5.7 yields the following.

THEOREM 5.12. *For any monoidal category* $(\mathsf{V}, \otimes, I)$ *the following categories are pairwise isomorphic.*

(1.i) *The category of cosemigroups in the* Lax^+*-monoidal category* L *of Theorem* 5.11.

(1.ii) *The category of* BiHom*-comonoids in* V.

and

(2.i) *The category of comonoids in the* $\mathsf{Lax}^+\mathsf{Oplax}^0$*-monoidal category* L *of Theorem* 5.11.

(2.ii) *The category of counital* BiHom*-comonoids in* V.

PROOF. In both categories of part (1), an object consists of

- an object (a, α, β) of L
- a morphism $\delta : (a, \alpha, \beta) \to (a \otimes a, \alpha \otimes \alpha, \beta \otimes \beta)$ of L

such that the first diagram of

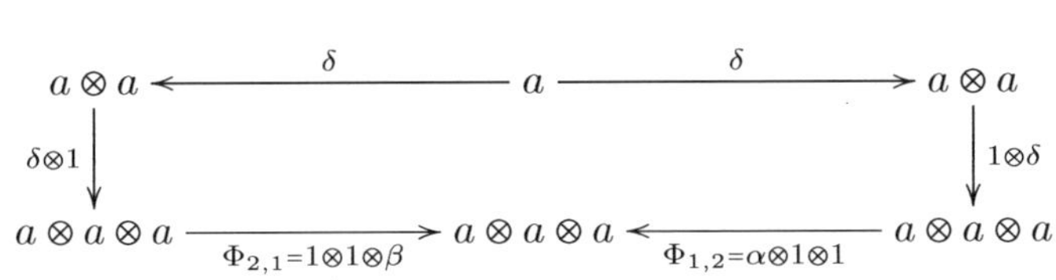

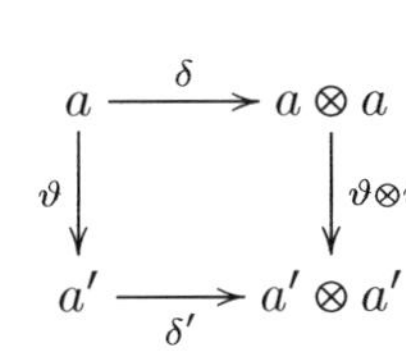

commutes. In both categories of part (1), a morphism $((a, \alpha, \beta), \delta) \to ((a', \alpha', \beta'), \delta')$ is a morphism $\vartheta : (a, \alpha, \beta) \to (a', \alpha', \beta')$ in L such that the second diagram above commutes.

In both categories of part (2), an object consists of

- an object $((a, \alpha, \beta), \delta)$ in the isomorphic categories of part (1)
- a morphism $\varepsilon : (a, \alpha, \beta) \to (I, 1, 1)$ in L

such that the first diagram of

$$
\begin{array}{ccccc}
a\otimes a & \xleftarrow{\delta} & a & \xrightarrow{\delta} & a\otimes a \\
{\scriptstyle \varepsilon\otimes 1}\downarrow & & \Vert & & \downarrow{\scriptstyle 1\otimes\varepsilon} \\
a & \xleftarrow[\phi_{0,1}=\beta]{} & a & \xrightarrow[\phi_{1,0}=\alpha]{} & a
\end{array}
\qquad\qquad
\begin{array}{ccc}
a & \xrightarrow{\varepsilon} & I \\
{\scriptstyle \vartheta}\downarrow & & \Vert \\
a' & \xrightarrow[\varepsilon']{} & I
\end{array}
$$

commutes. In both categories of part (2), a morphism $((a,\alpha,\beta),\delta,\varepsilon) \to ((a',\alpha',\beta'), \delta', \varepsilon')$ is a morphism $\vartheta : ((a,\alpha,\beta),\delta) \to ((a',\alpha',\beta'),\delta')$ in the isomorphic categories of part (1) such that the second diagram above commutes. □

A comodule in the sense of Definition 2.9 over a cosemigroup in the Lax^+-monoidal category L of Theorem 5.11 is precisely the same as a comodule in the sense of [**11**, Definition 5.3] over the corresponding BiHom-comonoid in Theorem 5.12 (1.ii). A comodule in the sense of Definition 5.9 over a comonoid in the $\mathsf{Lax}^+\mathsf{Oplax}^0$-monoidal category L of Theorem 5.11 is precisely the same as a counital comodule in the sense of [**11**, Definition 5.3] over the corresponding counital BiHom-comonoid in Theorem 5.12 (2.ii).

The same category L of Theorem 5.11 also has a 'biased' monoidal structure inherited from $(\mathsf{V},\otimes,I)$. In this monoidal category — let us denote it by M — the monoidal product and the monoidal unit are $\underset{2}{\otimes} = \otimes$, respectively, $\underset{0}{\otimes} = (I,1,1)$ of L. Mac Lane's coherence natural isomorphisms are given by the same (omitted) morphisms as in V. A cosemigroup (respectively, comonoid) in $(\mathsf{M},\underset{2}{\otimes},\underset{0}{\otimes})$ is a triple consisting of a cosemigroup (respectively, comonoid) in $(\mathsf{V},\otimes,I)$ together with two commuting cosemigroup (respectively, comonoid) endomorphisms. Cosemigroup (respectively, comonoid) morphisms in M are those cosemigroup (respectively, comonoid) morphisms in $(\mathsf{V},\otimes,I)$ which are compatible with the endomorphism parts too.

PROPOSITION 5.13. *Consider an arbitrary monoidal category* $(\mathsf{V},\otimes,I)$*, the associated strictly normal* $\mathsf{Lax}^+\mathsf{Oplax}^0$*-monoidal category* L *of Theorem* 5.11*, and the monoidal category* M *of the previous paragraph regarded as a* $\mathsf{Lax}^+\mathsf{Oplax}^0$*-monoidal category. The identity functor* $\mathsf{M}\to\mathsf{L}$ *carries a* $\mathsf{Lax}^+\mathsf{Oplax}^0$*-monoidal structure given by the morphisms*

$$\Gamma_n := \otimes_{i=1}^n \alpha_i^{n-i}\cdot\beta_i^{i-1} : \otimes_{i=1}^n a_i \to \otimes_{i=1}^n a_i,$$

for all non-negative integers n *and any sequence of objects* $\{(a_i,\alpha_i,\beta_i)\}_{i=1,\dots,n}$.

Consequently, it preserves cosemigroups and comonoids.

The action of the $\mathsf{Lax}^+\mathsf{Oplax}^0$-monoidal functor of Proposition 5.13 on cosemigroups and comonoids was termed in [**11**, Proposition 5.9] the *Yau twist*.

PROOF. A similar comparison of exponents as in the proof of Theorem 5.11 shows the commutaivity of the diagrams in Definition 5.3. □

REMARK 5.14. In the category L of Theorem 5.11 take the full subcategory for whose objects (a,α,β) the morphisms α and β are invertible. It is monoidal with the associativity natural isomorphism

$$\alpha^{-1}\otimes 1\otimes\beta'' : ((a\otimes a')\otimes a'',(\alpha\otimes\alpha')\otimes\alpha'',(\beta\otimes\beta')\otimes\beta'') \to (a\otimes(a'\otimes a''),\alpha\otimes(\alpha'\otimes\alpha''),\beta\otimes(\beta'\otimes\beta'')),$$

left unit natural isomorphism

$$\beta^{-1} : (I\otimes a, 1\otimes\alpha, 1\otimes\beta) \to (a,\alpha,\beta)$$

and right unit natural isomorphism

$$\alpha^{-1} : (a \otimes I, \alpha \otimes 1, \beta \otimes 1) \to (a, \alpha, \beta).$$

This is the same monoidal category that was constructed in Section 2 of [**11**] under the name $\mathcal{H}^{(-1,0),(0,-1),1}(\mathbb{Z} \times \mathbb{Z}, \mathsf{V})$. By Theorem 5.12 the cosemigroups (respectively, comonoids) therein are those BiHom-comonoids (respectively, counital BiHom-comonoids) whose endomorphism parts are automorphisms; thus we re-cover [**11**, Remark 3.5].

Let us take also in the monoidal category M of Proposition 5.13 the full (monoidal) subcategory of those objects (a, α, β) whose morphisms α and β are invertible; this is the monoidal category $\mathcal{H}(\mathbb{Z} \times \mathbb{Z}, \mathsf{V})$ of [**11**, Section 2]. As shown in [**11**, Theorem 2.5], the $\mathsf{Lax}^{+}\mathsf{Oplax}^{0}$-monoidal functor of Proposition 5.13 restricts to a strong monoidal isomorphism between these monoidal subcategories; hence it induces isomorphisms between the cosemigroups, as well as the comonoids in these monoidal categories. That is, analogously to [**11**, Claim 3.7], we obtain an isomorphism between

(i) the full subcategory of the isomorphic categories in Theorem 5.12 (1) for whose objects $((a, \alpha, \beta), \delta)$ the morphisms α and β are invertible;
(ii) the category whose objects are triples consisting of a cosemigroup in V together with two commuting cosemigroup automorphisms; and the morphisms are the cosemigroup morphisms which commute with both of these automorphisms.

It induces an analogous isomorphism also between

(i) the full subcategory of the isomorphic categories in Theorem 5.12 (2) for whose objects $((a, \alpha, \beta), \delta, \varepsilon)$ the morphisms α and β are invertible;
(ii) the category whose objects are triples consisting of a comonoid in V together with two commuting comonoid automorphisms; and the morphisms are the comonoid morphisms which commute with both of these automorphisms.

Restricting further to those objects for which $\beta = \alpha^{-1}$, the latter isomorphism reduces to [**8**, Proposition 1.13].

6. Monoids in $\mathsf{Lax}_0\mathsf{Oplax}_+$-monoidal categories

Dually to the previous section, here we introduce monoids in monoidal categories in which all monoidal products with non-negative number of factors are available; and the monoidal products of positive number of factors are oplax coherent and the nullary product is lax coherent. Unital BiHom-monoids are described as monoids in such monoidal categories.

DEFINITION 6.1. A $\mathsf{Lax}_0\mathsf{Oplax}_+$-*monoidal category* consists of

- a category R
- for all non-negative integers n, a functor $\underset{n}{\circ} : \mathsf{R}^n \to \mathsf{R}$

- for all non-negative integers $n, k_1, \ldots, k_n$, natural transformations

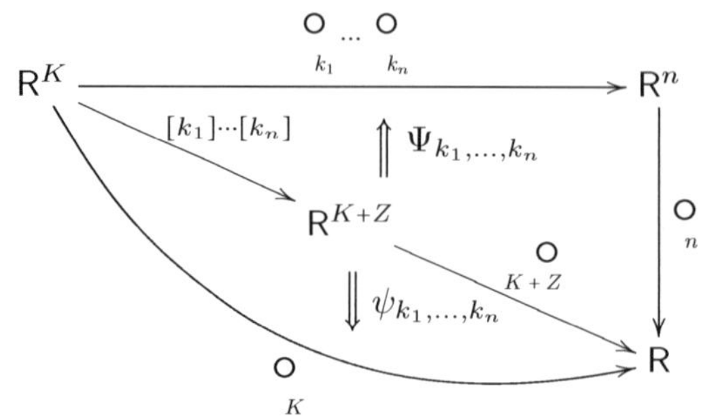

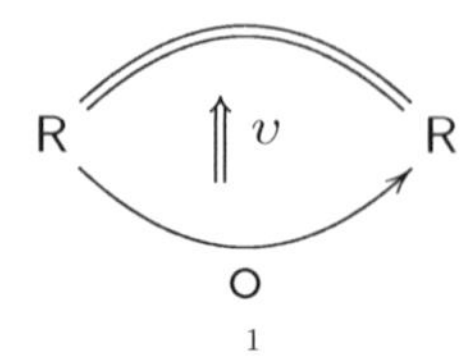

satisfying the so-called coassociativity and counitality axioms encoded in the diagrams which are obtained from the diagrams of Definition 5.1 by reversing the arrows.

Forgetting about the irrelevant structure, any $\mathsf{Lax}_0\mathsf{Oplax}_+$-monoidal category can be regarded as an Oplax_+-monoidal category.

A $\mathsf{Lax}_0\mathsf{Oplax}_+$-monoidal category is *(strictly) normal* if it is (strictly) normal as an Oplax_+-monoidal category.

$(\mathsf{R}, \circ, \Psi, \psi, \upsilon)$ is a $\mathsf{Lax}_0\mathsf{Oplax}_+$-monoidal category if and only if $(\mathsf{R}^{\mathsf{op}}, \circ^{\mathsf{op}}, \Psi^{\mathsf{op}}, \psi^{\mathsf{op}}, \upsilon^{\mathsf{op}})$ is a $\mathsf{Lax}^+\mathsf{Oplax}^0$-monoidal category.

Throughout, in a $\mathsf{Lax}_0\mathsf{Oplax}_+$-monoidal category we denote by i the image of the single object of $\mathbb{1}$ under the functor $\underset{0}{\circ} : \mathbb{1} \to \mathsf{R}$ and we keep our earlier notation $(\!|a|\!)$ for the image of any object a under the functor $\underset{1}{\circ} : \mathsf{R} \to \mathsf{R}$; and $a_1 \circ \cdots \circ a_n$ for the image of an object $(a_1, \ldots, a_n)$ under the functor $\underset{n}{\circ} : \mathsf{R}^n \to \mathsf{R}$.

Symmetrically to Proposition 5.2, the Cartesian product of $\mathsf{Lax}_0\mathsf{Oplax}_+$-monoidal carries a canonical $\mathsf{Lax}_0\mathsf{Oplax}_+$-monoidal structure.

DEFINITION 6.2. A $\mathsf{Lax}_0\mathsf{Oplax}_+$*-monoidal functor* $(\mathsf{R}, \circ, \Psi, \psi, \upsilon) \to (\mathsf{R}', \circ', \Psi', \psi', \upsilon')$ consists of

- a functor $G : \mathsf{R} \to \mathsf{R}'$
- for all non-negative integers n, a natural transformation $\underset{n}{\circ'} \cdot (G \cdots G) \to G \cdot \underset{n}{\circ}$

such that the diagrams of Definition 5.3 with reversed arrows commute.

Analogously to Proposition 5.4, the composite of some (composable) $\mathsf{Lax}_0\mathsf{Oplax}_+$-monoidal functors is again $\mathsf{Lax}^+\mathsf{Oplax}^0$-monoidal via the evident natural transformations; and the Cartesian product of $\mathsf{Lax}_0\mathsf{Oplax}_+$-monoidal functors is $\mathsf{Lax}^+\mathsf{Oplax}^0$-monoidal via the evident natural transformations too.

DEFINITION 6.3. A $\mathsf{Lax}_0\mathsf{Oplax}_+$*-monoidal natural transformation* is a natural transformation for which the diagrams of Definition 5.5 with reversed arrows commute.

Symmetrically to Proposition 5.6, all of the composites, the Godement products, and the Cartesian products of (composable) $\mathsf{Lax}_0\mathsf{Oplax}_+$-monoidal natural transformations are again $\mathsf{Lax}_0\mathsf{Oplax}_+$-monoidal. Consequently, there is a strict monoidal 2-category $\mathsf{Lax}_0\mathsf{Oplax}_+$ of $\mathsf{Lax}_0\mathsf{Oplax}_+$-monoidal categories, $\mathsf{Lax}_0\mathsf{Oplax}_+$-monoidal functors and $\mathsf{Lax}_0\mathsf{Oplax}_+$-monoidal natural transformations, admitting a strict monoidal forgetful 2-functor to the 2-category Oplax_+ of Section 3.

DEFINITION 6.4. A *monoid* in $(\mathsf{R}, \circ, \Psi, \psi, \upsilon)$ is a comonoid in $(\mathsf{R}^{\mathsf{op}}, \circ^{\mathsf{op}}, \Psi^{\mathsf{op}}, \psi^{\mathsf{op}}, \upsilon^{\mathsf{op}})$. That is, it consists of

- a semigroup (a, μ) and
- a morphism $\eta : i \to a$

rendering commutative the first (unitality) diagram below. A *morphism of monoids* $(a, \mu, \eta) \to (a', \mu', \eta')$ is a morphism of semigroups $\phi : (a, \mu) \to (a', \mu')$ rendering commutative the second diagram too.

$$\begin{array}{ccccc} i \circ a & \xrightarrow{\psi_{0,1}} & (\!|a|\!) & \xleftarrow{\psi_{1,0}} & a \circ i \\ {\scriptstyle \eta\circ 1}\downarrow & & \downarrow{\scriptstyle \upsilon} & & \downarrow{\scriptstyle 1\circ\eta} \\ a \circ a & \xrightarrow[\mu]{} & a & \xleftarrow[\mu]{} & a \circ a \end{array} \qquad \begin{array}{ccc} i & \xrightarrow{\eta} & a \\ \| & & \downarrow{\scriptstyle \phi} \\ i & \xrightarrow[\eta']{} & a' \end{array}$$

THEOREM 6.5. *For any* $\mathsf{Lax_0Oplax_+}$*-monoidal category* $(\mathsf{R}, \circ, \Psi, \psi, \upsilon)$*, the following categories are isomorphic.*

(i) *The category of monoids and their morphisms in* $(\mathsf{R}, \circ, \Psi, \psi, \upsilon)$.
(ii) *The category of* $\mathsf{Lax_0Oplax_+}$*-monoidal functors* $(\mathbb{1}, 1, 1, 1, 1) \to (\mathsf{R}, \circ, \Psi, \psi, \upsilon)$ *and their* $\mathsf{Lax_0Oplax_+}$*-monoidal natural transformations.*

Consequently, $\mathsf{Lax_0Oplax_+}$*-monoidal functors preserve monoids.*

PROOF. Apply Theorem 5.8 to the $\mathsf{Lax^+Oplax^0}$-monoidal category $(\mathsf{R}^{\mathsf{op}}, \circ^{\mathsf{op}}, \Psi^{\mathsf{op}}, \psi^{\mathsf{op}}, \upsilon^{\mathsf{op}})$. □

DEFINITION 6.6. A *module* of a monoid (a, μ, η) in a $\mathsf{Lax_0Oplax_+}$-monoidal category $(\mathsf{R}, \circ, \Psi, \psi, \upsilon)$ is a comodule of the comonoid (a, μ, η) in the $\mathsf{Lax^+Oplax^0}$-monoidal category $(\mathsf{R}^{\mathsf{op}}, \circ^{\mathsf{op}}, \Psi^{\mathsf{op}}, \upsilon^{\mathsf{op}})$. Equivalently, it is a module $(x, \varrho : x \circ a \to x)$ in the sense of Definition 3.6 of the semigroup (a, μ) such that also the diagram

$$\begin{array}{ccc} x \circ i & \xrightarrow{\psi_{1,0}} & (\!|x|\!) \\ {\scriptstyle 1\circ\eta}\downarrow & & \downarrow{\scriptstyle \upsilon} \\ x \circ a & \xrightarrow[\varrho]{} & x \end{array}$$

commutes. A *morphism of modules* over a monoid is a morphism of modules over the underlying semigroup in the sense of Definition 3.6.

THEOREM 6.7. *Associated to a monoidal category* $(\mathsf{V}, \otimes, I)$*, there is a strictly normal* $\mathsf{Lax_0Oplax_+}$*-monoidal category* R *as follows.*

- *As a category,* R *is identical to the category* L *of Theorem* 5.11.
- *For all non-negative integers* n*, we take the same functors* $\underset{n}{\otimes} : \mathsf{R}^n \to \mathsf{R}$ *as in Theorem* 5.11.
- *For any sequence of non-negative integers* $\{k_1, \dots, k_n\}$*, and for any collection of objects* $\{(a_{ij}, \kappa_{ij}, \nu_{ij})\}_{\substack{i=1,\dots,n \\ j=1,\dots,k_i}}$*, the natural transformation* $\Psi_{k_1,\dots,k_n}$ *is taken to be*
$$\otimes_{i=1}^n (\otimes_{j=1}^{k_i} \kappa_{ij}^{\Sigma_{p=i+1}^n (\overline{k}_p - 1)} \nu_{ij}^{\Sigma_{p=1}^{i-1} (\overline{k}_p - 1)}) : \otimes_{i=1}^n (\otimes_{j=1}^{k_i} a_{ij}) \to \otimes_{i=1}^n (\otimes_{j=1}^{k_i} a_{ij}),$$
and the natural transformation $\psi_{k_1,\dots,k_n}$ *is taken to be*
$$\otimes_{i=1}^n (\otimes_{j=1}^{k_i} \kappa_{ij}^{\Sigma_{p=i+1}^n Z(k_p)} \nu_{ij}^{\Sigma_{p=1}^{i-1} Z(k_p)}) : \otimes_{i=1}^n (\otimes_{j=1}^{k_i} a_{ij}) \to \otimes_{i=1}^n (\otimes_{j=1}^{k_i} a_{ij}).$$

PROOF. Apply Theorem 5.11 to the monoidal category $(\mathsf{V}^{\mathsf{op}}, \otimes^{\mathsf{op}}, I^{\mathsf{op}})$; and take the opposite of the resulting $\mathsf{Lax}^+\mathsf{Oplax}^0$-monoidal category. $\square$

In complete analogy with Theorem 5.12 we get the following.

THEOREM 6.8. *For any monoidal category* $(\mathsf{V}, \otimes, I)$ *the following categories are pairwise isomorphic.*

(1.i) *The category of semigroups in the* Oplax_+*-monoidal category* R *of Theorem* 6.7.
(1.ii) *The category of* BiHom*-monoids in* V.

and

(2.i) *The category of monoids in the* $\mathsf{Lax}_0\mathsf{Oplax}_+$*-monoidal category* R *of Theorem* 5.11.
(2.ii) *The category of unital* BiHom*-monoids in* V.

A module in the sense of Definition 3.6 over a semigroup in the Oplax_+-monoidal category R of Theorem 6.7 is precisely the same as a module in the sense of [**11**, Definition 4.1] over the corresponding BiHom-monoid in Theorem 6.8 (1.ii). A module in the sense of Definition 6.6 over a monoid in the $\mathsf{Lax}_0\mathsf{Oplax}_+$-monoidal category R of Theorem 6.7 is precisely the same as an unital module in the sense of [**11**, Definition 4.1] over the corresponding unital BiHom-monoid in Theorem 6.8 (2.ii).

Applying Proposition 5.13 to the opposite categories, we obtain the following explanation of the Yau twist of semigroups and monoids in [**11**, Proposition 5.9].

PROPOSITION 6.9. *Consider an arbitrary monoidal category* $(\mathsf{V}, \otimes, I)$, *the associated strictly normal* $\mathsf{Lax}_0\mathsf{Oplax}_+$*-monoidal category* R *of Theorem* 6.7, *and the monoidal category* M *of Proposition* 5.13 *regarded now as a* $\mathsf{Lax}_0\mathsf{Oplax}_+$*-monoidal category. The identity functor* $\mathsf{M} \to \mathsf{R}$ *carries a* $\mathsf{Lax}_0\mathsf{Oplax}_+$*-monoidal structure given by the same morphisms in Proposition* 5.13. *Consequently, it preserves semigroups and monoids.*

REMARK 6.10. In the coinciding categories R of Theorem 6.7 and L of Theorem 5.11 take the full monoidal subcategory described in Remark 5.14. The category of semigroups (resp. monoids) in it is isomorphic to the category whose objects are triples consisting of a semigroup (resp. monoid) in V together with two commuting semigroup (resp. monoid) automorphisms.

7. Bimonoids in $\mathsf{Lax}_0^+\mathsf{Oplax}_+^0$-duoidal categories

In this section we introduce compatibility conditions between $\mathsf{Lax}^+\mathsf{Oplax}^0$-monoidal, and $\mathsf{Lax}_0\mathsf{Oplax}_+$-monoidal structures on the same category. Similarly to Section 4, they imply that the category of comonoids w.r.t. the $\mathsf{Lax}^+\mathsf{Oplax}^0$-monoidal structure inherits the $\mathsf{Lax}_0\mathsf{Oplax}_+$-monoidal structure; symmetrically, the category of monoids w.r.t. the $\mathsf{Lax}_0\mathsf{Oplax}_+$-monoidal structure inherits the $\mathsf{Lax}^+\mathsf{Oplax}^0$-monoidal structure. We define then bimonoids as monoids in the category of comonoids; equivalently, as comonoids in the category of monoids. Unital and counital BiHom-bimonoids are described as bimonoids in this sense.

DEFINITION 7.1. A $\mathsf{Lax}_0^+\mathsf{Oplax}_+^0$-*duoidal category* consists of

- a category D

- a $\mathsf{Lax}^{+}\mathsf{Oplax}^{0}$-monoidal structure $(\mathsf{D}, \bullet, \Phi, \phi, \iota)$
- a $\mathsf{Lax}_0\mathsf{Oplax}_{+}$-monoidal structure $(\mathsf{D}, \circ, \Psi, \psi, \upsilon)$
- for all non-negative integers n and p, natural transformations

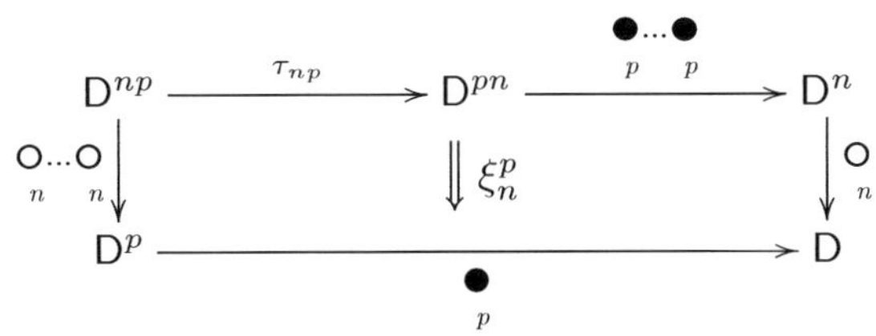

satisfying the following equivalent compatibility conditions, for all non-negative integers $n, p, k_1, \ldots, k_p$.

(i)
- $(\underset{n}{\circ}, \xi_n)$ is a $\mathsf{Lax}^{+}\mathsf{Oplax}^{0}$-monoidal functor $(\mathsf{D}, \bullet, \Phi, \phi, \iota)^n \to (\mathsf{D}, \bullet, \Phi, \phi, \iota)$;
- $\Psi_{k_1,\ldots,k_p}$, $\psi_{k_1,\ldots,k_p}$ and υ are $\mathsf{Lax}^{+}\mathsf{Oplax}^{0}$-monoidal natural transformations.

Succinctly, $((\mathsf{D}, \bullet, \Phi, \phi, \iota), (\circ, \xi), \Psi, \psi, \upsilon)$ is a $\mathsf{Lax}_0\mathsf{Oplax}_{+}$-monoid in the strict monoidal 2-category $\mathsf{Lax}^{+}\mathsf{Oplax}^{0}$ of Proposition 5.6.

(ii)
- $(\underset{n}{\bullet}, \xi^n)$ is a $\mathsf{Lax}_0\mathsf{Oplax}_{+}$-monoidal functor $(\mathsf{D}, \circ, \Psi, \psi, \upsilon)^n \to (\mathsf{D}, \circ, \Psi, \psi, \upsilon)$;
- $\Phi_{k_1,\ldots,k_p}$, $\phi_{k_1,\ldots,k_p}$ and ι are $\mathsf{Lax}_0\mathsf{Oplax}_{+}$-monoidal natural transformations.

Succinctly, $((\mathsf{D}, \circ, \Psi, \psi, \upsilon), (\bullet, \xi), \Phi, \phi, \iota)$ is a $\mathsf{Lax}^{+}\mathsf{Oplax}^{0}$-monoid in the strict monoidal 2-category $\mathsf{Lax}_0\mathsf{Oplax}_{+}$ of Section 6.

(iii) The diagrams

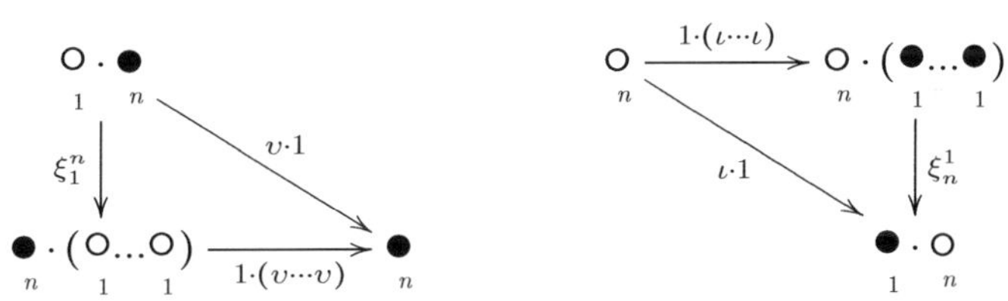

and those of Figure 4 commute, for all non-negative integers $n, p, k_1, \ldots, k_p$, for the families of functors

$$\mathsf{D}^{k_i} \xrightarrow{[k_i]^{\bullet}} \mathsf{D}^{\overline{k_i}} := \begin{cases} \mathsf{D}^{k_i} \xrightarrow{1} \mathsf{D}^{k_i} & \text{if} \quad k_i > 0 \\ \mathbb{1} \xrightarrow{\underset{0}{\bullet}} \mathsf{D} & \text{if} \quad k_i = 0 \end{cases}$$

$$\mathsf{D}^{k_i} \xrightarrow{[k_i]^{\circ}} \mathsf{D}^{\overline{k_i}} := \begin{cases} \mathsf{D}^{k_i} \xrightarrow{1} \mathsf{D}^{k_i} & \text{if} \quad k_i > 0 \\ \mathbb{1} \xrightarrow{\underset{0}{\circ}} \mathsf{D} & \text{if} \quad k_i = 0 \end{cases}$$

and for the families of natural transformations

$$\underset{n}{\circ}^{\overline{k_i}} \cdot ([k_i]^{\bullet})^n \xrightarrow{[\![k_i]\!]^{\bullet}_n} [k_i]^{\bullet} \cdot \underset{n}{\circ}^{k_i} := \begin{cases} \underset{n}{\circ}^{k_i} \xrightarrow{1} \underset{n}{\circ}^{k_i} & \text{if} \quad k_i > 0 \\ \underset{n}{\circ} \cdot \underset{0}{\bullet}^{n} \xrightarrow{\xi^0_n} \underset{0}{\bullet} & \text{if} \quad k_i = 0 \end{cases}$$

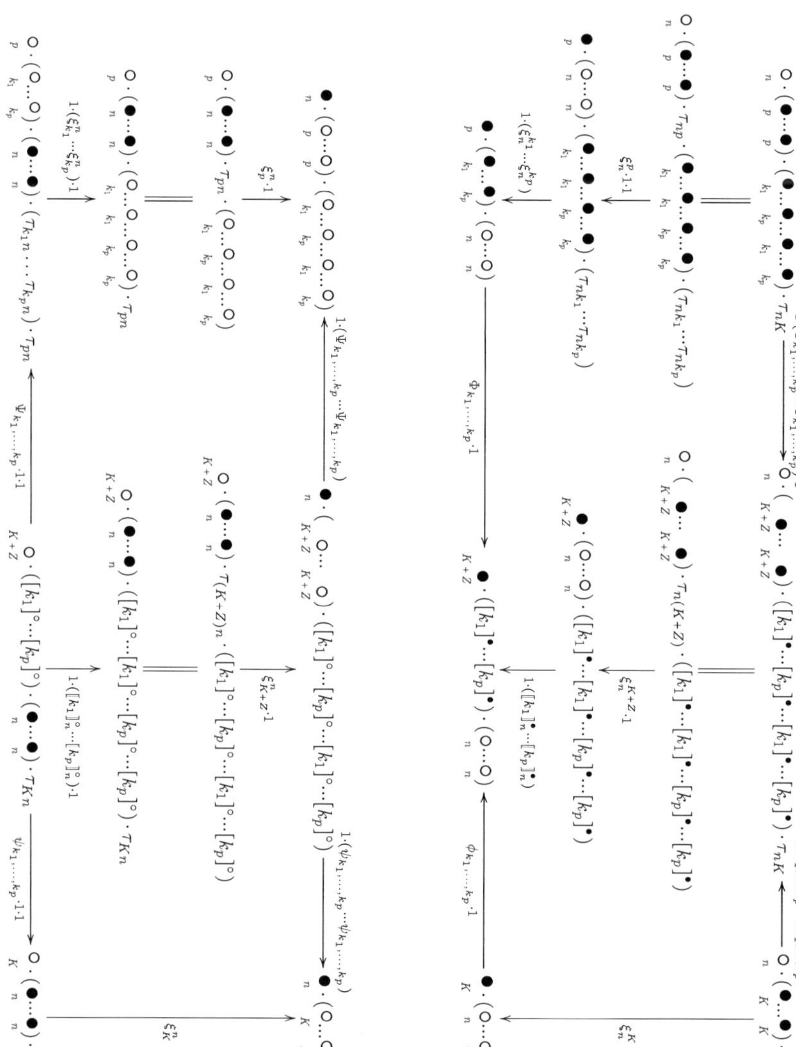

FIGURE 4. Axioms of $\mathsf{Lax}_0^+\mathsf{Oplax}_+^0$-duoidal category

$$[k_i]^\circ \cdot \underset{n}{\bullet}^{k_i} \xrightarrow{[\![k_i]\!]_n^\circ} \underset{n}{\bullet}^{\overline{k}_i} \cdot ([k_i]^\circ)^n := \begin{cases} \underset{n}{\bullet}^{k_i} \xrightarrow{1} \underset{n}{\bullet}^{k_i} & \text{if} \quad k_i > 0 \\ \underset{0}{\circ} \xrightarrow{\xi_0^n} \underset{n}{\bullet} \cdot \underset{0}{\circ}^n & \text{if} \quad k_i = 0. \end{cases}$$

Forgetting about the irrelevant structure, any $\mathsf{Lax}_0^+\mathsf{Oplax}_+^0$-duoidal category can be regarded as a $\mathsf{Lax}^+\mathsf{Oplax}_+$-duoidal category.

It is immediate from Definition 7.1 (i) and Theorem 5.8 that for any $\mathsf{Lax}_0^+\mathsf{Oplax}_+^0$-duoidal category $(\mathsf{D}, \bullet, \circ)$, (i, ξ_0^2, ξ_0^0) is a comonoid in $(\mathsf{D}, \bullet)$. Dually, it follows from Definition 7.1 (ii) and Theorem 6.5 that (j, ξ_2^0, ξ_0^0) is a monoid in $(\mathsf{D}, \circ)$.

DEFINITION 7.2. A $\mathsf{Lax}_0^+\mathsf{Oplax}_+^0$-*duoidal functor* $(\mathsf{D}, \bullet, \circ) \to (\mathsf{D}', \bullet', \circ')$ is defined as

(i) a $\mathsf{Lax}_0\mathsf{Oplax}_+$-monoidal 1-cell $((G, \Gamma^\bullet), \Gamma^\circ) : ((\mathsf{D}, \bullet), \circ) \to ((\mathsf{D}', \bullet'), \circ')$ in the strict monoidal 2-category $\mathsf{Lax}^+\mathsf{Oplax}^0$ of Proposition 5.6; equivalently,

(ii) A $\mathsf{Lax}^+\mathsf{Oplax}^0$-monoidal 1-cell $((G, \Gamma^\circ), \Gamma^\bullet) : ((\mathsf{D}, \circ), \bullet) \to ((\mathsf{D}', \circ'), \bullet')$ in the strict monoidal 2-category $\mathsf{Lax}_0\mathsf{Oplax}_+$ of Section 6; equivalently,

(iii)
- a functor $G : \mathsf{D} \to \mathsf{D}'$
- a $\mathsf{Lax}^+\mathsf{Oplax}^0$-monoidal structure $\{\Gamma_n^\bullet : G \cdot \underset{n}{\bullet} \to \underset{n}{\bullet}' \cdot (G \cdots G)\}_{n \in \mathbb{N}}$
- a $\mathsf{Lax}_0\mathsf{Oplax}_+$-monoidal structure $\{\Gamma_p^\circ : \underset{p}{\circ}' \cdot (G \cdots G) \to G \cdot \underset{p}{\circ}\}_{p \in \mathbb{N}}$

rendering commutative the diagram of (4.1) for all non-negative integers n, p.

Forgetting about the nullary parts of the structures, $\mathsf{Lax}_0^+\mathsf{Oplax}_+^0$-duoidal functors can be seen as $\mathsf{Lax}^+\mathsf{Oplax}_+$-duoidal functors.

It follows immediately from Definition 7.2 (ii) and Proposition 5.4 (1) that the composite of $\mathsf{Lax}_0^+\mathsf{Oplax}_+^0$-duoidal functors is again $\mathsf{Lax}_0^+\mathsf{Oplax}_+^0$-duoidal via the composite $\mathsf{Lax}^+\mathsf{Oplax}^0$-monoidal structure in Proposition 5.4 (1) and the symmetrically constructed composite $\mathsf{Lax}_0\mathsf{Oplax}_+$-monoidal structure.

DEFINITION 7.3. A $\mathsf{Lax}_0^+\mathsf{Oplax}_+^0$-*duoidal natural transformation* is defined as

(i) a $\mathsf{Lax}_0\mathsf{Oplax}_+$-monoidal 2-cell in the strict monoidal 2-category $\mathsf{Lax}^+\mathsf{Oplax}^0$ of Proposition 5.6; equivalently,

(ii) a $\mathsf{Lax}^+\mathsf{Oplax}^0$-monoidal 2-cell in the strict monoidal 2-category $\mathsf{Lax}_0\mathsf{Oplax}_+$ of Section 6; equivalently,

(iii) a natural transformation which is both $\mathsf{Lax}_0\mathsf{Oplax}_+$-monoidal and $\mathsf{Lax}^+\mathsf{Oplax}^0$-monoidal. That is, it renders commutative both the diagram of Definition 5.5 and its symmetric counterpart with reversed arrows and inverted colors.

$\mathsf{Lax}_0^+\mathsf{Oplax}_+^0$-duoidal natural transformations are in particular $\mathsf{Lax}^+\mathsf{Oplax}_+$-duoidal.

THEOREM 7.4. *For any* $\mathsf{Lax}_0^+\mathsf{Oplax}_+^0$-*duoidal category* $(\mathsf{D}, (\bullet, \Phi, \phi, \iota), (\circ, \Psi, \psi, \upsilon), \xi)$, *the following assertions hold.*

(1) *The comonoids in the* $\mathsf{Lax}^+\mathsf{Oplax}^0$-*monoidal category* $(\mathsf{D}, \bullet, \Phi, \phi, \iota)$ *constitute a* $\mathsf{Lax}_0\mathsf{Oplax}_+$-*monoidal category with the structure* $(\circ, \Psi, \psi, \upsilon)$.

(2) *The monoids in the* $\mathsf{Lax}_0\mathsf{Oplax}_+$-*monoidal category* $(\mathsf{D}, \circ, \Psi, \psi, \upsilon)$ *constitute a* $\mathsf{Lax}^+\mathsf{Oplax}^0$-*monoidal category with the structure* $(\bullet, \Phi, \phi, \iota)$.

PROOF. In exactly the same way as we derived Theorem 4.4 from Theorem 2.8 and Definition 4.1 (ii), part (1) follows from Theorem 5.8 and Definition 7.1 (ii). Part (2) follows symmetrically. □

DEFINITION 7.5. A *bimonoid* in a $\mathsf{Lax}_0^+\mathsf{Oplax}_+^0$-duoidal category $(\mathsf{D},\bullet,\circ,\xi)$ is defined by the following equivalent data.

(i) A monoid in the $\mathsf{Lax}_0\mathsf{Oplax}_+$-monoidal category of comonoids in $(\mathsf{D},\bullet)$ — see Theorem 7.4 (1).

(ii) A comonoid in the $\mathsf{Lax}^+\mathsf{Oplax}^0$-monoidal category of monoids in $(\mathsf{D},\circ)$ — see Theorem 7.4 (2).

(iii)
- An object a of D,
- a monoid (a,μ,η) in $(\mathsf{D},\circ)$
- a comonoid (a,δ,ε) in $(\mathsf{D},\bullet)$

such that (a,μ,δ) is a bisemigroup in the $\mathsf{Lax}^+\mathsf{Oplax}_+$-duoidal category $(\mathsf{D},\bullet,\circ,\xi)$ and the following diagrams commute as well.

$$
\begin{array}{ccc}
a\circ a & \xrightarrow{\ \mu\ } & a \\
{\scriptstyle \varepsilon\circ\varepsilon}\downarrow & & \downarrow{\scriptstyle \varepsilon} \\
j\circ j & \xrightarrow[\ \xi_2^0\ ]{} & j
\end{array}
\qquad
\begin{array}{ccc}
i & \xrightarrow{\ \xi_0^2\ } & i\bullet i \\
{\scriptstyle \eta}\downarrow & & \downarrow{\scriptstyle \eta\bullet\eta} \\
a & \xrightarrow[\ \delta\ ]{} & a\bullet a
\end{array}
\qquad
\begin{array}{ccc}
i & \xrightarrow{\ \eta\ } & a \\
& {\scriptstyle \xi_0^0}\searrow & \downarrow{\scriptstyle \varepsilon} \\
& & j
\end{array}
$$

A *morphism of bimonoids* is a monoid morphism in the category of comonoids in $(\mathsf{D},\bullet)$; equivalently, a comonoid morphism in the category of monoids in $(\mathsf{D},\circ)$; equivalently, a morphism in D which is both a comonoid morphism in $(\mathsf{D},\bullet)$ and a monoid morphism in $(\mathsf{D},\circ)$.

THEOREM 7.6. *For any* $\mathsf{Lax}_0^+\mathsf{Oplax}_+^0$*-duoidal category* $(\mathsf{D},\bullet,\circ,\xi)$ *the following categories are isomorphic.*

(i) *The category of bimonoids and their morphisms in* $(\mathsf{D},\bullet,\circ,\xi)$.

(ii) *The category of* $\mathsf{Lax}_0^+\mathsf{Oplax}_+^0$*-duoidal functors* $(\mathbb{1},1,1,1)\to(\mathsf{D},\bullet,\circ,\xi)$ *and their* $\mathsf{Lax}_0^+\mathsf{Oplax}_+^0$*-duoidal natural transformations.*

Consequently, $\mathsf{Lax}_0^+\mathsf{Oplax}_+^0$*-duoidal functors preserve bimonoids.*

PROOF. By Definition 7.5, the category of part (i) is isomorphic to the category of monoids in the category of comonoids in $(\mathsf{D},\bullet)$. So by Theorem 6.5, it is isomorphic to the category of $\mathsf{Lax}_0\mathsf{Oplax}_+$-monoidal functors and $\mathsf{Lax}_0\mathsf{Oplax}_+$-monoidal natural transformations from $\mathbb{1}$ to the category of comonoids in $(\mathsf{D},\bullet)$. Then by Theorem 5.8, it is further equivalent to the category of $\mathsf{Lax}_0\mathsf{Oplax}_+$-monoidal 1-cells and $\mathsf{Lax}_0\mathsf{Oplax}_+$-monoidal 2-cells from $(\mathbb{1},1)$ to $(\mathsf{D},\bullet)$ in $\mathsf{Lax}^+\mathsf{Oplax}^0$. By Definition 7.2 and Definition 7.3 this is isomorphic to the category of part (ii). □

Analogously to Theorem 4.7, we have the following.

THEOREM 7.7. *For any bimonoid* $(a,(\delta,\varepsilon),(\mu,\eta))$ *in a* $\mathsf{Lax}_0^+\mathsf{Oplax}_+^0$*-duoidal category* $(\mathsf{D},(\bullet,\Phi,\phi,\iota),(\circ,\Psi,\psi,\upsilon),\xi)$ *the following assertions hold.*

(1) *The category of modules — in the sense of Definition 6.6 — over the underlying monoid* (a,μ,η) *in the* $\mathsf{Lax}_0\mathsf{Oplax}_+$*-monoidal category* $(\mathsf{D},\circ,\Psi,\psi,\upsilon)$ *admits the* $\mathsf{Lax}^+\mathsf{Oplax}^0$*-monoidal structure* $(\bullet,\Phi,\phi,\iota)$.

(2) *The category of comodules — in the sense of Definition* 5.9 *— over the underlying comonoid* (a, δ, ε) *in the* $\mathsf{Lax}^{+}\mathsf{Oplax}^{0}$*-monoidal category* $(\mathsf{D}, \bullet, \Phi, \phi, \iota)$ *admits the* $\mathsf{Lax}_0\mathsf{Oplax}_+$*-monoidal structure* $(\circ, \Psi, \psi, \upsilon)$.

PROOF. We only prove part (1), part (2) is verified symmetrically.

For any sequence of (a, μ, η)-modules $\{(x_i, \varrho_i)\}_{i=1,\dots,n}$, the object $x_1 \bullet \cdots \bullet x_n$ with the action

$$(x_1 \bullet \cdots \bullet x_n) \circ a \xrightarrow{1\circ\delta_n} (x_1 \bullet \cdots \bullet x_n) \circ a^{\bullet n} \xrightarrow{\xi_2^n} (x_1 \circ a) \bullet \cdots \bullet (x_n \circ a) \xrightarrow{\varrho_1 \bullet \cdots \bullet \varrho_n} x_1 \bullet \cdots \bullet x_n$$

is an (a, μ)-module — in the sense of Definition 3.6 — by Theorem 4.7.

Let us check its unitality whenever all of the actions ϱ_i are unital. The diagram on the right of part (iii) of Definition 7.1 for $n = 0$ says the equality of $\iota : i \to (\!| i |\!)$ and ξ_0^1. By Definition 7.1 (i), $(\underset{0}{\circ}, \xi_0) : \mathbb{1} \to \mathsf{D}$ is a $\mathsf{Lax}^{+}\mathsf{Oplax}^{0}$-monoidal functor. Equivalently, by Theorem 5.8, (i, ξ_0^2, ξ_0^0) is a comonoid in $(\mathsf{D}, \bullet)$. With these facts at hand, from the commutativity of the left half of the upper diagram of Figure 4 at $n = 0$, $p = 2$, $k_1 = 1$ and $k_2 = l$, the l-fold comultiplication $i \to i^{\bullet l}$ as in (2.2) comes out as ξ_0^l. By Definition 7.5, η is a comonoid morphism. Therefore the leftmost region of

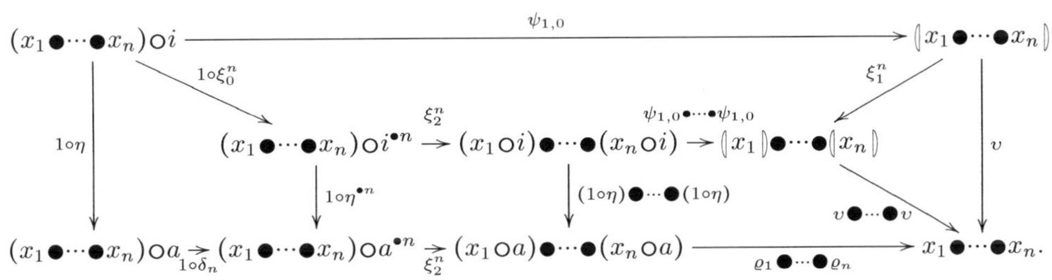

commutes. The triangular region on the right commutes by the diagram on the left in part (iii) of Definition 7.1; and the large pentagonal region at the top commutes by the right half of the lower diagram of Figure 4 for the values n, $p = 2$, $k_1 = 1$ and $k_2 = 0$. The quadratic region at the bottom right commutes by the unitality of each action ϱ_i, and the region on its left commutes by the naturality of ξ_2^n. This proves the unitality of the action in the bottom row.

The $\bullet$-monoidal products of module morphisms, as well as the natural transformations ι and $\Phi_{k_1,\dots,k_p}$ for all $k_i > 0$ — if evaluated at (a, μ, η)-modules — are morphisms of (a, μ)-modules by Theorem 4.7. Then they are morphisms of (a, μ, η)-modules by definition. We leave it to the reader to check — similarly to the proof of Theorem 4.7 — that the remaining natural transformations ϕ and $\Phi_{k_1,\dots,k_p}$ with possibly zero values of k_i — if evaluated at (a, μ, η)-modules — are morphisms of (a, μ, η)-modules too. □

PROPOSITION 7.8. *For any bimonoid* $(a, (\delta, \varepsilon), (\mu, \eta))$ *in an arbitrary* $\mathsf{Lax}_0^{+}\mathsf{Oplax}_+^{0}$*-duoidal category the following assertions hold.*

(1) $((a, \delta), \mu, \eta)$ *is a monoid in the* $\mathsf{Lax}_0\mathsf{Oplax}_+$*-monoidal category of comodules over the comonoid* (a, δ, ε) *(cf. Theorem 7.7 (2)).*
(2) $((a, \mu), \delta, \varepsilon)$ *is a comonoid in the* $\mathsf{Lax}^{+}\mathsf{Oplax}^{0}$*-monoidal category of modules over the monoid* (a, μ, η) *(cf. Theorem 7.7 (1)).*

PROOF. By Example 5.10 (a, δ) is a comodule and by Proposition 4.8 μ is a morphism of comodules. Also η is a morphism of comodules by the second diagram of Definition 7.5 (iii) which completes the proof of part (1). Part (2) follows symmetrically, using now the first diagram of Definition 7.5 (iii). □

DEFINITION 7.9. A *Hopf module* over a bimonoid $(a, (\delta, \varepsilon), (\mu, \eta))$ in an arbitrary $\mathsf{Lax}_0^+\mathsf{Oplax}_+^0$-duoidal category $(\mathsf{D}, \bullet, \circ, \xi)$ is defined by the following equivalent data.

(i) A module over the monoid $((a, \delta), \mu, \eta)$ in the category of comodules over the comonoid (a, δ, ε) in $(\mathsf{D}, \bullet)$ (cf. Proposition 7.8 (1)).

(ii) A comodule over the comonoid $((a, \mu), \delta, \varepsilon)$ in the category of modules over the monoid (a, μ, η) in $(\mathsf{D}, \circ)$ (cf. Proposition 7.8 (2)).

(iii)
- An object x of D,
- a module (x, ν) over the monoid (a, μ, η) in $(\mathsf{D}, \circ)$,
- a comodule (x, ϱ) over the comonoid (a, δ, ε) in $(\mathsf{D}, \bullet)$,

rendering commutative the diagram of Definition 4.9 (iii).

EXAMPLE 7.10. By Example 5.10 and its dual counterpart, and by the compatibility diagram of Definition 4.5 (iii), (a, δ, μ) is a Hopf module over an arbitrary bimonoid $(a, (\delta, \varepsilon), (\mu, \eta))$ in any $\mathsf{Lax}_0^+\mathsf{Oplax}_+^0$-duoidal category.

Our final task is a description of unital and counital BiHom-bimonoids as bimonoids in a suitable $\mathsf{Lax}_0^+\mathsf{Oplax}_+^0$-duoidal category.

THEOREM 7.11. *Associated to any symmetric monoidal category* $(\mathsf{V}, \otimes, I, \sigma)$, *there is a* $\mathsf{Lax}_0^+\mathsf{Oplax}_+^0$*-duoidal category* D *as follows.*

- *The objects of the category* D *consist of an object* a *of* V *and four pairwise commuting endomorphisms* $\alpha, \beta, \kappa, \nu$ *of* a.

  The morphisms in D *are those morphisms in* V *which commute with all of the four endomorphisms of the source and target objects.*
- *The* $\mathsf{Lax}^+\mathsf{Oplax}^0$*-monoidal structure is given by the same data as in* L *of Theorem 5.11; in terms of the first two endomorphisms* α, β *at each object.*
- *The* $\mathsf{Lax}_0\mathsf{Oplax}_+$*-monoidal structure is given by the same data as in* R *of Theorem 6.7; in terms of the last two endomorphisms* κ, ν *at each object.*
- *For any non-negative integers* n, p, *the natural transformation* ξ_n^p *is given by the unique component of the symmetry* $\sigma_{pn} : \underset{n}{\otimes}\cdot(\underset{p}{\otimes}\cdots\underset{p}{\otimes})\cdot\tau_{np} \to \underset{p}{\otimes}\cdot(\underset{n}{\otimes}\cdots\underset{n}{\otimes})$.

PROOF. An easy substitution, using naturality of the symmetry σ, shows that the axioms of Definition 7.1 hold; we leave it to the reader. □

As an immediate consequence of Theorem 5.12 and Theorem 6.8 we obtain the following.

THEOREM 7.12. *For any symmetric monoidal category* $(\mathsf{V}, \otimes, I, \sigma)$ *the following categories are pairwise isomorphic.*

(1.i) *The category of bisemigroups in the* $\mathsf{Lax}^+\mathsf{Oplax}_+$*-duoidal category* D *of Theorem* 7.11.

(1.ii) *The category of* BiHom*-bimonoids in* V.

and

(2.i) *The category of bimonoids in the* $\mathsf{Lax}_0^+\mathsf{Oplax}_+^0$*-duoidal category* D *of Theorem* 7.11.

(2.ii) *The category of unital and counital* BiHom*-bimonoids in* V.

Next we explain the Yau twist of (unital and counital) BiHom-bimonoids in [**11**, Proposition 5.9]. Analogously to the monoidal category M of Proposition 5.13, we associate the following symmetric monoidal category S to any symmetric monoidal category $(\mathsf{V}, \otimes, I, \sigma)$. As a category, S is the same as D of Theorem 7.11. Its monoidal product is $\underset{2}{\otimes} = \otimes$ of D and the monoidal unit is $\underset{0}{\otimes} = (I, 1, 1, 1, 1)$. Mac Lane's coherence natural isomorphisms are given by the same (omitted) morphisms as in V and the symmetry is given by σ. A bisemigroup (respectively, bimonoid) in S is a quintuple consisting of a bisemigroup (respectively, bimonoid) in $(\mathsf{V}, \otimes, I, \sigma)$ together with four commuting bisemigroup (respectively, bimonoid) endomorphisms. Bisemigroup (respectively, bimonoid) morphisms are those bisemigroup (respectively, bimonoid) morphisms in $(\mathsf{V}, \otimes, I, \sigma)$ which are compatible with the endomorphism parts too.

PROPOSITION 7.13. *Consider an arbitrary symmetric monoidal category* $(\mathsf{V}, \otimes, I, \sigma)$*, the corresponding* $\mathsf{Lax}_0^+\mathsf{Oplax}_+^0$*-duoidal category* D *of Theorem* 7.11*, and the symmetric monoidal category* S *in the previous paragraph regarded as a* $\mathsf{Lax}_0^+\mathsf{Oplax}_+^0$*-duoidal category. The identity functor* $\mathsf{S} \to \mathsf{D}$ *is* $\mathsf{Lax}_0^+\mathsf{Oplax}_+^0$*-duoidal via the* $\mathsf{Lax}^+\mathsf{Oplax}^0$*-monoidal structure in Proposition* 5.13 *and the* $\mathsf{Lax}_0\mathsf{Oplax}_+$*-monoidal structure in Proposition* 6.9*. Consequently it preserves bisemigroups and bimonoids.*

PROOF. The easy check of the commutativity of (4.1) for the stated data, using the naturality of σ, is left to the reader. □

REMARK 7.14. In the category D of Theorem 7.11, the full subcategory $\mathsf{D}^\times$ for whose objects $(a, \alpha, \beta, \kappa, \nu)$ all morphisms α, β, κ and ν are invertible, carries a duoidal structure. Its monoidal structures are as in Remark 5.14, and the compatibility morphisms are either trivial or given by the symmetry σ of V.

The $\mathsf{Lax}_0^+\mathsf{Oplax}_+^0$-duoidal functor of Proposition 7.13 restricts to a strong duoidal isomorphism between $\mathsf{D}^\times$ and the full symmetric monoidal subcategory $\mathsf{S}^\times$ of S in Proposition 7.13 with the same objects. This strong duoidal isomorphism $\mathsf{D}^\times \cong \mathsf{S}^\times$ gives rise to an isomorphism of the categories of bisemigroups (respectively, bimonoids) in the isomorphic duoidal categories $\mathsf{D}^\times$ and $\mathsf{S}^\times$. This yields an isomorphism between those BiHom-bialgebras (resp. unital and counital BiHom-bialgebras) whose four endomorphisms are invertible; and the bisemigroups (resp. bimonoids) in V together with four commuting bisemigroup (resp. bimonoid) automorphisms.

This amounts to saying that any bimonoid in $\mathsf{D}^\times$ is a Yau twist of a bimonoid in V. Moreover, for any bimonoid $((a, \alpha, \beta, \kappa, \nu), (\delta, \varepsilon), (\mu, \eta))$ in $\mathsf{D}^\times$; and the corresponding bimonoid $(a, (\widetilde{\mu} = \mu \cdot (\kappa^{-1} \otimes \nu^{-1}), \eta), (\widetilde{\delta} = (\alpha^{-1} \otimes \beta^{-1}) \cdot \delta, \varepsilon))$, the following assertions are equivalent.

(i) $(a, (\widetilde{\delta}, \varepsilon), (\widetilde{\mu}, \eta))$ is a Hopf monoid in V. That is, there is a (unique) morphism $\chi : a \to a$ — the 'antipode' — rendering commutative the following diagram.

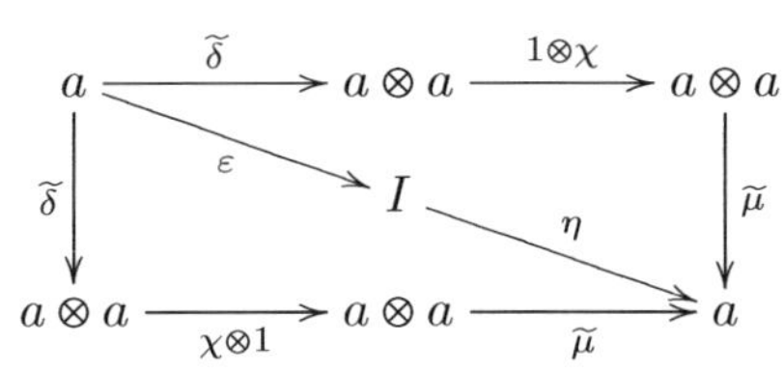

(ii) The following 'canonical morphism' (see [**5**, (1.10)]) is invertible, for all objects y and a-modules (x, ϱ) in $\mathsf{D}^\times$.

$$x \otimes y \otimes a \xrightarrow{1\otimes 1\otimes \delta} x \otimes y \otimes a \otimes a \xrightarrow{1\otimes \sigma\otimes 1} x \otimes a \otimes y \otimes a \xrightarrow{\varrho\otimes 1\otimes 1} x \otimes y \otimes a$$

(iii) $((a, \alpha, \beta, \kappa, \nu), (\mu, \eta), (\delta, \varepsilon))$ satisfies the axioms of [**11**, Definition 6.9]. That is, there is a (unique) morphism $\chi : a \to a$ — the 'BiHom-antipode' — rendering commutative the following diagram.

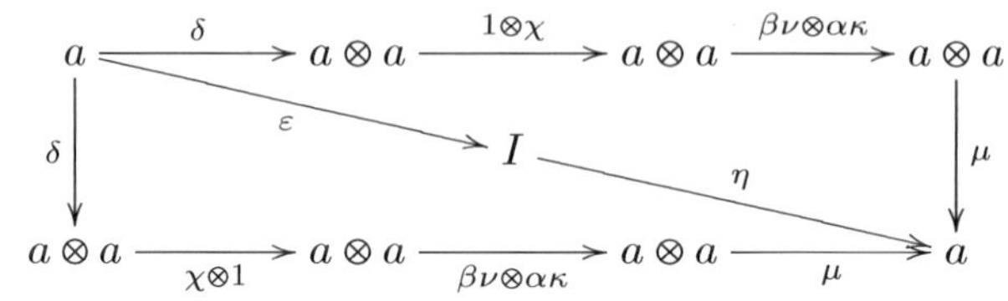

The morphism χ occurring in parts (i) and (iii) is the same. Although the assertions in parts (ii) and (iii) (but not in (i)) are meaningful for any bimonoid in the $\mathsf{Lax}_0^+\mathsf{Oplax}_+^0$-duoidal category D of Theorem 7.11, there seems to be no reason to expect their equivalence any longer.

References

[1] Marcelo Aguiar and Swapneel Mahajan, *Monoidal functors, species and Hopf algebras*, CRM Monograph Series, vol. 29, American Mathematical Society, Providence, RI, 2010. With forewords by Kenneth Brown and Stephen Chase and André Joyal, DOI 10.1090/crmm/029. MR2724388

[2] E. Batista, S. Caenepeel, and J. Vercruysse, *Hopf categories*, Algebr. Represent. Theory **19** (2016), no. 5, 1173–1216, DOI 10.1007/s10468-016-9615-6. MR3551315

[3] John C. Baez and James Dolan, *Higher-dimensional algebra. III. n-categories and the algebra of opetopes*, Adv. Math. **135** (1998), no. 2, 145–206, DOI 10.1006/aima.1997.1695. MR1620826

[4] Gabriella Böhm, *Hopf polyads, Hopf categories and Hopf group monoids viewed as Hopf monads*, Theory Appl. Categ. **32** (2017), Paper No. 37, 1229–1257. MR3709394

[5] Gabriella Böhm, Yuanyuan Chen, and Liangyun Zhang, *On Hopf monoids in duoidal categories*, J. Algebra **394** (2013), 139–172, DOI 10.1016/j.jalgebra.2013.07.016. MR3092715

[6] Gabriella Böhm and Stephen Lack, *Hopf comonads on naturally Frobenius map-monoidales*, J. Pure Appl. Algebra **220** (2016), no. 6, 2177–2213, DOI 10.1016/j.jpaa.2015.11.002. MR3448792

[7] Alain Bruguières, *Hopf polyads*, Algebr. Represent. Theory **20** (2017), no. 5, 1151–1188, DOI 10.1007/s10468-017-9681-4. MR3707910

[8] S. Caenepeel and I. Goyvaerts, *Monoidal Hom-Hopf algebras*, Comm. Algebra **39** (2011), no. 6, 2216–2240, DOI 10.1080/00927872.2010.490800. MR2813174

[9] Brian Day and Ross Street, *Lax monoids, pseudo-operads, and convolution*, Diagrammatic morphisms and applications (San Francisco, CA, 2000), Contemp. Math., vol. 318, Amer. Math. Soc., Providence, RI, 2003, pp. 75–96, DOI 10.1090/conm/318/05545. MR1973511

[10] Brian Day and Ross Street, *Quantum categories, star autonomy, and quantum groupoids*, Galois theory, Hopf algebras, and semiabelian categories, Fields Inst. Commun., vol. 43, Amer. Math. Soc., Providence, RI, 2004, pp. 187–225. MR2075586

[11] Giacomo Graziani, Abdenacer Makhlouf, Claudia Menini, and Florin Panaite, *BiHom-associative algebras, BiHom-Lie algebras and BiHom-bialgebras*, SIGMA Symmetry Integrability Geom. Methods Appl. **11** (2015), Paper 086, 34, DOI 10.3842/SIGMA.2015.086. MR3415909

[12] Tom Leinster, *Higher operads, higher categories*, London Mathematical Society Lecture Note Series, vol. 298, Cambridge University Press, Cambridge, 2004, DOI 10.1017/CBO9780511525896. MR2094071

[13] Vladimir Turaev, *Crossed group-categories* (English, with English and Arabic summaries), Arab. J. Sci. Eng. Sect. C Theme Issues **33** (2008), no. 2, 483–503. MR2500054

[14] Xiaohui Zhang and Dingguo Wang, *Cotwists of Bicomonads and BiHom-bialgebras*, Alg. Represent. Theory (2019) — available online at https://link.springer.com/article/10.1007/s10468-019-09888-2.

WIGNER RESEARCH CENTRE FOR PHYSICS, H-1525 BUDAPEST 114, P.O.B. 49, HUNGARY
Email address: bohm.gabriella@wigner.mta.hu

DÈPARTEMENT DE MATHÈMATIQUES, FACULTÈ DES SCIENCES, UNIVERSITÈ LIBRE DE BRUXELLES, BOULEVARD DU TRIOMPHE, B-1050 BRUXELLES, BELGIUM
Email address: jvercruy@ulb.ac.be

Contemporary Mathematics
Volume **771**, 2021
https://doi.org/10.1090/conm/771/15507

Survey on Hopf algebras of GK-dimension 1 and 2

K. A. Brown and J. J. Zhang

ABSTRACT. This is a brief survey on the current state of play on the programs to classify infinite dimensional Hopf algebras of Gel'fand-Kirillov dimension one or two. We list a number of open questions and suggest directions for future work.

1. Introduction

During the last 20 years, there has been significant progress towards a deeper understanding of infinite dimensional Hopf algebras satisfying some natural finiteness conditions. There has been wide interest in this effort, most notably in the programs concerning (both finite and infinite dimensional) Nichols algebras led by Andruskiewitsch, Angiono, Heckenberger, Schneider and others [**AA06, AS00, AS04, AAH19, Ang13**], and the program concerning affine noetherian Hopf algebras initiated by Brown and Goodearl in 1997 [**BG97, Br98, BZ08, BZ10, LWZ07, WZ03, WLD16**]. The subject of the present short survey is the classification and analysis of infinite dimensional Hopf algebras of low Gel'fand-Kirillov dimension. This topic lies at the intersection of the above two projects. Thus, as well as yielding results which are of interest in their own right, the research which we describe is a rich source both of examples and of methods which will prove useful in further work on both the above long-term endeavors.

The low-dimensional classification project has necessitated the development and refinement over the last 20 years of a number of algebraic concepts and homological invariants, for example AS-regularity, the Nakayama automorphism, the integral order and the integral minor, some of which we shall review here. These have played important roles in the further study of infinite dimensional noetherian Hopf algebras, strengthening connections to noncommutative algebraic geometry and the representation theory of quantum groups, with further applications likely in combinatorics, representation theory, statistical mechanics, topology and mathematical physics. To give one example, the project to classify prime AS-Gorenstein Hopf algebras of GK-dimension 1 in characteristic 0 has culminated in work of Liu [**Li20**] which provides large families of examples ripe for exploitation in the study

2020 *Mathematics Subject Classification.* Primary 16T05; Secondary 17B37, 20G42.
Key words and phrases. Hopf algebra, Gel'fand-Kirillov dimension, homological integral.
The first author was partially supported by Leverhulme Emeritus Fellowship (EM-2017-081-9).
The second author was partially supported by the US National Science Foundation (Nos. DMS-1700825 and DMS-2001015).

of tensor categories, support varieties, Hochschild cohomology, and so on. On the other hand despite (or rather, because of) this recent progress, very many open questions remain, many of which we have listed and discussed in what follows. For these reasons it is therefore a good time to briefly survey these classification programs.

Note that this account does not touch upon the large volume of recent significant research concerning finite dimensional (semisimple or not) Hopf algebras, or Nichols algebras, or tensor categories. For recent surveys on these topics we refer the reader to [**And14**], [**And17**], [**EGNO15**] respectively.

1.1. General setup. Throughout, the base field will be denoted by $\Bbbk$ and will be assumed to be algebraically closed. For many of the results described this hypothesis can undoubtedly be weakened, but we will not pursue this question. On the other hand, the characteristic of the field will often be significant, so we will make clear the hypotheses concerning the characteristic of $\Bbbk$ on a case by case basis; whenever the characteristic is not mentioned, this means that the characteristic is arbitrary. All vector spaces, (co)algebras, tensor products, etc. are taken over $\Bbbk$. An algebra is called *affine* if it is finitely generated over $\Bbbk$. We refer to Montgomery's book [**Mo93**] as a basic reference for Hopf algebras. For any Hopf algebra H, we denote the multiplication, unit map, comultiplication, counit and antipode by m, u, Δ, ϵ and S respectively. We use the same symbol $\Bbbk$ to denote both the trivial $\Bbbk$-algebra, and the trivial module over H, namely, $\Bbbk = H/\ker\epsilon$. Where not otherwise indicated, modules are left modules. Let H^{op} denote the opposite ring of H, so a right H-module can be viewed as a left H^{op}-module and an H-bimodule is identified with a left $H \otimes H^{op}$-module. Many examples and open questions on the topics discussed here can be found in textbooks and in the survey papers [**Br98, Br07, BGi14, Go13**]. In the present paper we will update these sources on their coverage of algebras of small GK-dimension, but we recommend them as references for wider issues concerning noetherian Hopf algebras.

1.2. Gel'fand-Kirillov dimension. Let A be an algebra over $\Bbbk$. Recall that the *Gel'fand-Kirillov dimension* (or *GK-dimension* for short) of A is defined to be

$$\operatorname{GKdim} A := \sup_V \limsup_{n\to\infty} \left(\log_n(\dim_{\Bbbk} V^n)\right)$$

where V runs over all finite dimensional subspaces of A. We refer to [**KL00**] for basic properties of GK-dimension of algebras and modules. Note that finite dimensional algebras have GK-dimension zero. Conversely, every algebra of non-zero GK-dimension is clearly infinite dimensional. Examples of Borho-Kraft and Warfield [**KL00**, Theorem 2.9], together with the Bergman Gap Theorem [**KL00**, Theorem 2.5] show that GK-dimension of an affine algebra can take any value from the set

$$\{0\} \cup \{1\} \cup [2, \infty]. \tag{E0.0.1}$$

The trichotomy (E0.0.1) motivates the study of Hopf $\Bbbk$-algebras of GK-dimension 1 and 2, since these classes constitute a laboratory where classification is feasible and where more general conjectures can be tested. Note, however, that there is currently no known example of a Hopf algebra with finite non-integral GK-dimension. Thus we repeat [**BGi14**, Question F] (which was also asked by Zhuang [**Zhu13**]).

QUESTION 1.1. [**BGi14**, Question F] Does every Hopf algebra have either infinite or integral GK-dimension?

For further general questions on the GK-dimension of Hopf algebras, see [**BGi14**].

1.3. Artin-Schelter and Cohen-Macaulay properties. The following homological concepts are key to the analysis of the Hopf algebras studied in this paper. Recall that a Hopf $\Bbbk$-algebra H is *Artin-Schelter Gorenstein* (*AS-Gorenstein* for short) if

(AS1) $\operatorname{injdim}\, {}_H H = d < \infty$,
(AS2) $\dim_{\Bbbk} \operatorname{Ext}^d_H({}_H\Bbbk, {}_H H) = 1$, $\operatorname{Ext}^i_H({}_H\Bbbk, {}_H H) = 0$ for all $i \neq d$,
(AS3) the right H-module versions of the conditions (AS1), (AS2) hold.

We say H is *Artin-Schelter regular* (*AS-regular* for short) if it is AS-Gorenstein and has finite global dimension; in this case $\operatorname{gldim} H = d$. The *homological grade* of a non-zero H-module M is defined to be $j(M) := \min(\{j : \operatorname{Ext}^j_H(M, H) \neq 0\} \cup \{\infty\})$. Suppose that H has finite GK-dimension. Then H is *GK-Cohen-Macaulay* if, for all non-zero finitely generated H-modules M,

$$j(M) + \operatorname{GKdim} M = \operatorname{GKdim} H.$$

Notice that if H is AS-Gorenstein and GK-Cohen-Macaulay then

$$\operatorname{injdim} H = \operatorname{GKdim} H.$$

All known noetherian Hopf algebras are AS-Gorenstein [**BZ08**]. A well-known conjecture in the study of noetherian Hopf algebras is the following, which was posted in [**Br07**, Question E] and [**Go13**, Question 3.5].

CONJECTURE 1.2 (Brown-Goodearl Conjecture). *Every noetherian Hopf algebra is AS-Gorenstein.*

We refer to [**Br98, Br07, BZ08, Go13**] for further discussion of homological aspects of noetherian Hopf algebras.

1.4. Organization. The paper is organized as follows. In Section 1 we consider prime Hopf algebras of GK-dimension one, and in Section 2, we mainly consider Hopf algebra domains of GK-dimension two, with a digression to review Hopf Ore extensions. These two sections can be viewed as an expanded version of [**Go13**, Section 4]. Section 3 contains a review and brief commentary on the many questions stated in Sections 0, 1 and 2.

2. Hopf algebras of GK-dimension 1

In this section we study Hopf algebras which are prime of GK-dimension one. By a famous result of Small and Warfield [**SW84**], when such an algebra (whether Hopf or not) is affine it is a finite module over its affine center, so it is noetherian and satisfies a polynomial identity. Thus if H is a prime affine Hopf algebra of GK-dimension 1, it is AS-Gorenstein and GK-Cohen-Macaulay by [**WZ03**, Theorems 0.1 and 0.2]. In particular, the injective dimension of H (and its global dimension if H is regular) are both 1.

2.1. Preliminaries and key examples. Recall the classical result that the only connected algebraic groups of dimension one over an algebraically closed field $\Bbbk$ are the additive and multiplicative groups of $\Bbbk$, see for example, [**Hu75**, Theorem 20.5] or [**Bo91**, Theorem III.10.9]. Translating to the language of Hopf algebras, the only affine commutative Hopf $\Bbbk$-algebra domains of GK-dimension 1 are $\Bbbk[X]$ and $\Bbbk[X^{\pm 1}]$. But Tsen's theorem [**Co77**, p. 374] permits us to omit the commutative hypothesis, so combining [**LWZ07**, Corollary 7.8 (a)$\Rightarrow$(b)] with [**GZ10**, Proposition 2.1] yields the following.

LEMMA 2.1. [**GZ10**, Proposition 2.1(a,b)], [**BZ20**, Lemma 5.1] *The only affine Hopf $\Bbbk$-algebra domains of GK-dimension one are the coordinate rings $\Bbbk[X]$ and $\Bbbk[X^{\pm 1}]$ of $(\Bbbk, +)$ and $(\Bbbk^{\times}, \times)$ respectively.*

REMARK 2.2. In fact one can at least partly remove from Lemma 2.1 the hypothesis that H is affine: Tsen's theorem still ensures that the algebras involved are commutative, so $S^2 = \mathrm{Id}$ by [**Mo93**, Corollary 1.5.12] and hence H is a union of affine Hopf subalgebras. Let H be a non-affine Hopf $\Bbbk$-algebra domain of GK-dimension one. The following statements follow from Lemma 2.1.

(1) [**GZ10**, Proposition 2.1(c)] Assume that $\Bbbk$ has characteristic 0. Then H is a group algebra $\Bbbk G$ where G is a non-cyclic torsion-free abelian group of rank one, i.e., a non-cyclic subgroup of $\mathbb{Q}$.
(2) When $\Bbbk$ has positive characteristic and H contains a nontrivial grouplike element, then H is a group algebra $\Bbbk G$ where G is a non-cyclic torsion-free abelian group of rank one.
(3) If H does not contain a nontrivial grouplike element, then $\Bbbk$ has positive characteristic, say p. If further H is countably generated, then one can check that

$$H = \Bbbk\langle X_i : [X_i, X_j] = 0,\ X_i = F_i(X_{i+1}),\ (i, j \geq 1)\rangle,$$

with X_i primitive for all i, where $\{F_i(t)\}_{i\geq 1}$ is a sequence of polynomials of the form

$$F_i(t) = \sum_{s\geq 0} \alpha_{i,s} t^{p^s}$$

for some $\alpha_{i,s} \in \Bbbk$.

The following issue thus remains to be clarified.

QUESTION 2.3. Let $\Bbbk$ have positive characteristic, and let H be an uncountably generated Hopf domain over $\Bbbk$ of GK-dimension one. What is the structure of H?

For prime Hopf algebras of GK-dimension one which are not domains, the story is much more complicated and as yet incomplete even in characteristic 0. We begin by recalling some key examples.

EXAMPLE 2.4. Let $\mathbb{D}$ be the infinite Dihedral group $\langle g, x \mid g^2 = 1, gxg = x^{-1}\rangle$ and let H be the group algebra $\Bbbk\mathbb{D}$. Since $\mathbb{D}$ has no nontrivial finite normal subgroups, H is prime by Connell's theorem [**Pas77**, Theorem 4.2.10], and H is regular if and only if $\Bbbk$ does not have characteristic 2, see [**Pas77**, Theorem 10.3.13].

The second family is also well-known, sometimes called the *infinite Taft algebras.*

EXAMPLE 2.5. [**Ta71**], [**LWZ07**, Examples 2.7 and 7.3] Let n and t be positive integers with $n \geq 2$ and $1 \leq t \leq n$, and let $\xi \in \Bbbk$ be a primitive nth root of 1. Let $H := H(n,t,\xi)$ be the $\Bbbk$-algebra generated by x and g subject to the relations

$$g^n = 1, \quad \text{and} \quad xg = \xi gx,$$

so H is not commutative. The coalgebra structure of H is defined by

$$\Delta(g) = g \otimes g, \qquad \epsilon(g) = 1$$

and

$$\Delta(x) = x \otimes 1 + g^t \otimes x, \qquad \epsilon(x) = 0.$$

Thus H is cocommutative if and only if $t = n$. The antipode S of H is defined by

$$S(g) = g^{-1}, \quad \text{and} \quad S(x) = -g^{-t}x = -\xi^t x g^{-t}.$$

Then H is affine prime AS-regular of GK-dimension one, with center $Z(H) = \Bbbk[x^n]$.

In 2009 Liu discovered a new family of Hopf algebras of GK-dimension one:

EXAMPLE 2.6. [**Li09**, Proposition 2.1] Let $n \geq 2$ and $\xi \in \Bbbk$ be a primitive nth root of 1. Let $A(n,\xi)$ be generated by x, g and the inverse g^{-1}, with relations

$$xg = \xi gx, \quad \text{and} \quad x^n = 1 - g^n.$$

The coalgebra structure of $A(n,\xi)$ is defined by

$$\Delta(g) = g \otimes g, \qquad \epsilon(g) = 1$$

and

$$\Delta(x) = x \otimes 1 + g \otimes x, \qquad \epsilon(x) = 0.$$

Then $A(n,\xi)$ is affine prime AS-regular of GK-dimension one with center $\Bbbk[x^n]$.

2.2. Integral order and minor. The project to classify prime regular Hopf algebras was begun in [**LWZ07**]. The homological integral was defined there - this has proved crucial for further analysis of AS-Gorenstein Hopf algebras, including classification projects. In this subsection we review the basic definitions, generalising the classical integrals of finite dimensional Hopf algebras [**Mo93**, Chapter 2].

DEFINITION 2.7. [**LWZ07**, Definition 1.1] Let H be an AS-Gorenstein Hopf algebra of injective dimension d. Any nonzero element in $\mathrm{Ext}^d_H({}_H\Bbbk, {}_HH)$ is called a *left homological integral* of H. Write $\int^l := \mathrm{Ext}^d_H({}_H\Bbbk, {}_HH)$. A nonzero element in $\int^r := \mathrm{Ext}^d_{H^{op}}(\Bbbk_H, H_H)$ is called a *right homological integral* of H. By abuse of language we also call $\int^l$ and $\int^r$ the *left* and *right homological integrals* of H respectively.

Notice that $\int^l$ is an H-bimodule of $\Bbbk$-dimension 1, trivial as a left H-module but in general nontrivial on the right. For any 1-dimensional left (or right) H-module M, we can write it as H/I for an ideal I with $\dim_{\Bbbk} H/I = 1$, or equivalently view it as an algebra homomorphism $\phi_M : H \to H/I$. Then $\phi := \phi_M$ is a grouplike element in the dual Hopf algebra H°, whose order is denoted by $o(M) \in \mathbb{N} \cup \{\infty\}$. Equivalently, $o(M)$ is the least positive integer m such that $M^{\otimes m}$ is the trivial module $\Bbbk$; so $o(M) = 1$ if and only if $M = \Bbbk$.

Continuing with the above notation, the *left winding automorphism* of H associated to ϕ is defined to be

$$\Xi^l_\phi : h \to \sum \phi(h_1)h_2, \quad \forall\, h \in H,$$

and the *right winding automorphism* associated to ϕ is

$$\Xi^r_\phi : h \to \sum h_1\phi(h_2), \quad \forall\, h \in H.$$

Both Ξ^l_ϕ and Ξ^r_ϕ are in the group $\mathrm{Aut}_{alg}(H)$ of algebra automorphisms of H. Let G^l_ϕ [resp. G^r_ϕ] (or G^l_M [resp. G^r_M]) denote the subgroups of $\mathrm{Aut}_{alg}(H)$ generated by Ξ^l_ϕ [resp. Ξ^r_ϕ].

DEFINITION 2.8. Let H be an AS-Gorenstein Hopf algebra.

(1) [**LWZ07**, Definition 2.2] The *integral order* of H is

$$io(H) := o(\int^l) = |G^l_{\int^l}| \in \mathbb{N} \cup \{\infty\},$$

where $\int^l$ is considered as a right H-module.

(2) H is called *unimodular* if $io(H) = 1$.

(3) [**BZ10**, Definition 2.4] Suppose $io(H) < \infty$. The *integral minor* of H is

$$im(H) := |G^l_{\int^l}/(G^l_{\int^l} \cap G^r_{\int^l})|.$$

where the intersection is taken inside $\mathrm{Aut}_{alg}(H)$.

By [**LWZ07**, Lemma 2.1], $io(H)$ equals $o(\int^r)$, viewing $\int^r$ as a left H-module. So it is not hard to show [**LWZ07, BZ10**] that

$$io(H) = |G^r_{\int^l}| = |G^l_{\int^r}| = |G^r_{\int^r}|.$$

Clearly $im(H)$ divides $io(H)$ if $io(H)$ is finite. For the Taft algebras $H = H(n,t,\xi)$ of Example 2.5 it is not hard to check that $io(H) = n$ and $im(H) = n/\gcd(n,t)$, [**BZ10**, page 273].

2.3. Classification results. The connection between the concepts just defined and classification was made clear in [**LWZ07**], with the following result.

THEOREM 2.9. [**LWZ07**, Theorem 7.1] *Let H be an affine prime AS-regular Hopf algebra of GK-dimension one. Then $io(H) < \infty$ and it equals the PI-degree of H. As a consequence, H is unimodular if and only if H is commutative.*

The commutative algebras which occur in Theorem 2.9 are listed in Lemma 2.1. Progress was achieved towards a classification of the noncommutative ones in characteristic 0 by analysing their integral order and minor. Here are the two key theorems from [**BZ10**].

THEOREM 2.10. [**BZ10**, Theorem 4.1] *Assume that $\Bbbk$ has characteristic 0, and let H be an affine prime AS-regular Hopf $\Bbbk$-algebra of GK-dimension one. Suppose $io(H) > 1$ and $im(H) = 1$. Then H is one of the following.*

(1) *H is the Taft algebra $H(n,n,\xi)$ of Example 2.5. In this case $io(H) = n$.*

(2) *H is the Dihedral group algebra $\Bbbk\mathbb{D}$ of Example 2.4. In this case $io(H) = 2$.*

In both cases, H is cocommutative.

The next result classifies all Hopf algebras in Theorem 2.9 with $im(H) = io(H)$. To save space we will not give the definition of the generalized Liu algebras, denoted by $B(n, w, \xi)$, see [**BZ10**, Section 3.4] for details. Here ξ is a primitive nth root of 1 and w is a positive integer. When $\gcd(n, w) = 1$ one obtains precisely Liu's algebras from Example 2.6.

THEOREM 2.11. [**BZ10**, Theorem 6.1] *Assume that* $\Bbbk$ *has characteristic 0, and let* H *be an affine prime AS-regular Hopf* $\Bbbk$*-algebra of GK-dimension one. Suppose* $io(H) = im(H) > 1$. *Then* H *is one of the following.*

(1) *The Taft algebra* $H(n, 1, \xi)$ *of Example* 2.5.
(2) *The generalized Liu algebra* $B(n, w, \xi)$.

In both cases, $io(H) = n$ *and* H *is not cocommutative.*

In [**BZ10**, Question 7.1] we asked if, in characteristic 0, every Hopf algebra in Theorem 2.9 satisfies either $im(H) = 1$ or $im(H) = io(H)$. The remarkable construction and result of Wu, Liu and Ding [**WLD16**] answers this question in the negative, and in doing so provides the complete classification of the Hopf algebras in Theorem 2.9 in the characteristic 0 case. The new class of Hopf algebras $D(m, d, \xi)$ is defined in [**WLD16**, Subsection 4.1], requiring a couple of pages, and the proof that $D(m, d, \xi)$ is a Hopf algebra occupies 10 pages of [**WLD16**, Subsection 4.2]. Basic properties of $D(m, d, \xi)$ are described in [**WLD16**, Subsection 4.3]; in particular $io(D(m, d, \xi)) = 2m$ and $\mathrm{im}(D(m, d, \xi)) = m$ [**WLD16**, Lemma 4.5].

THEOREM 2.12. [**WLD16**, Theorem 8.3] *Asume that* $\Bbbk$ *has characteristic 0, and let* H *be an affine prime AS-regular Hopf* $\Bbbk$*-algebra of GK-dimension one. Suppose* $io(H) \neq im(H) \neq 1$. *Then* H *is isomorphic to* $D(m, d, \xi)$ *for some choice of* m, d, ξ.

From Theorems 2.9, 2.10, 2.11 and 2.12 one sees that, when $\Bbbk$ has characteristic 0 and the Hopf $\Bbbk$-algebra H is prime affine AS-regular of GK-dimension one,

$$io(H)/im(H) \text{ is either } 1,\ 2 \text{ or } io(H).$$

By [**WLD16**, Proposition 4.9], $D(m, d, \xi)$ is not pointed, giving a negative answer to [**Go13**, Question 2.16] and [**BZ10**, Section 0.5].

2.4. Further questions and comments on GK-dimension one. After the progress described in Subsection 2.3 it is natural to ask if one can achieve a classification extending that of Theorems 2.10-2.12 by removing one or more of the hypotheses that (A) $\Bbbk$ has characteristic 0, (B) H is AS-regular or (C) H is prime. We briefly consider each of these in turn.

(A) This may be the easiest of the three, but it has received minimal attention so far. For the sake of precision, we formulate a specific project:

PROJECT 2.13. Let $\Bbbk$ have characteristic $p > 0$. Classify the prime affine AS-regular Hopf $\Bbbk$-algebras H of GK-dimension 1 when (i) $p \nmid io(H)$ and (ii) $p | io(H)$.

It may be that - at least in case (i) - this requires nothing more than a careful check of the arguments used for the results of Subsection 2.3. Along similar lines, one should also consider what happens when $\Bbbk$ is not algebraically closed. Even in the commutative case, new examples occur, as noted in [**LWZ07**, Example 8.3].

(B) Consider now the AS-regularity hypothesis. It was known already from [**BZ10**, Example 7.3] that even in characteristic 0 there exist prime affine Hopf algebras of GK-dimension 1 which are not regular. Liu [**Li20**] started a program to classify all prime affine non-regular Hopf algebras H of GK-dimension one when $\Bbbk$ has characteristic 0. By [**WZ03**], despite the absence of regularity, such an algebra H is still AS-Gorenstein, so homological integrals are defined and homological invariants such as the integral order and integral minor can continue to be used. Liu's strategy is to impose two hypotheses on H which are known *consequences* of regularity. For example, one of the hypotheses is to assume the existence of a nontrivial one-dimensional H-module M, specifically a module M such that $o(M) = \mathrm{PIdeg} H$, where $o(M)$ is as defined in Subsection 2.2. Assuming these two extra hypotheses Liu is able to obtain a beautiful classification in which many new Hopf algebras occur, these being what he calls *fractional* versions of some of the regular algebras. This leaves an obvious project:

PROJECT 2.14. Classify prime affine Hopf $\Bbbk$-algebras of GK-dimension 1 in characteristic 0 without assuming Liu's two supplementary hypotheses.

(C) What happens when the primeness hypothesis is omitted? Recall that an affine algebra of GK-dimension one satisfies a polynomial identity (PI), thanks to [**SSW85**]. By [**LWZ07**, Lemma 5.3(b)], every noetherian affine AS-regular Hopf algebra satisfying a PI is a direct sum of prime rings. In view of this and considering known examples, Lu, Wu and Zhang propose the following conjectural structure of a noetherian affine AS-regular Hopf $\Bbbk$-algebra H of GK-dimension one. Such an algebra H should fit into a short exact sequence

$$0 \to H_{dis} \to H \to H_{conn} \to 0 \tag{E1.14.1}$$

where the *connected component* H_{conn} is an AS-regular prime factor Hopf algebra of H of GK-dimension 1, and the *discrete component* H_{dis} should be a finite dimensional subalgebra of H with a braided Hopf algebra structure. In [**LWZ07**, Theorem 6.5(b)], Lu, Wu and Zhang were able to prove much of this, but *not* the finite dimensionality of H_{dis}. We therefore ask:

QUESTION 2.15. Is the above conjecture correct?

The answer to this question might well be important for the structure of AS-regular noetherian Hopf algebras which are *not* of GK-dimension 1, nor even PI.

In parallel with trying to answer Question 2.15, one should try to construct all possible algebras provided by the recipe of Lu, Wu and Zhang. This philosophy motivates the following question.

PROJECT 2.16 (Gongxiang Liu [**Li20a**]). Let $\Bbbk$ have characteristic 0 and let H be an affine prime AS-regular Hopf algebra of GK-dimension 1, as listed in Lemma 2.1 and Theorems 2.10, 2.11 and 2.12. Classify finite dimensional semisimple braided Hopf algebras in the Yetter-Drinfeld category ${}^H_H\mathcal{YD}$. These objects should play the role of H_{dis} in (E1.14.1).

2.5. The bridge from GK-dimension 1 to GK-dimension 2. As a step towards looking at prime affine Hopf algebras of GK-dimension 2 in the next section, we first note that, when $\Bbbk$ has characteristic 0, many such algebras can be constructed as extensions of an (in general braided) Hopf algebra of GK-dimension 1 by a second such algebra. This is true in the commutative case, as is recalled in Subsection 3.1 below, and is also true in the cocommutative case, as can be deduced from the structure theorem of Cartier-Kostant-Gabriel [**Mo93**, 5.6.4, 5.6.5]. Many more examples can be constructed from the algebras featuring in Subsection 2.3 by use of the tensor product, as in part (3) of the following result.

PROPOSITION 2.17. *Let $\Bbbk$ be an algebraically closed field of characteristic 0, and let H and K be prime affine AS-regular Hopf $\Bbbk$-algebras of GK-dimension one. Let n and m be the integral orders of H and K respectively.*

(1) *The center $Z(H)$ of H is isomorphic to either $\Bbbk[t]$ or $\Bbbk[t^{\pm 1}]$.*
(2) *The Goldie quotient ring of H is isomorphic to $M_n(\Bbbk(t))$.*
(3) *The tensor product $H \otimes K$ is a noetherian prime affine AS-regular Hopf algebra of GK-dimension 2. Moreover $H \otimes K$ is a free module of rank nm over its center $Z(H) \otimes Z(K)$.*

PROOF. (1) This follows from the classification in Subsection 2.3. One can check the assertion directly for the algebras in Theorems 2.9, 2.10 and 2.11(1). For the generalized Liu algebra $B(n, w, \xi)$ in Theorem 2.11(2), the assertion follows from [**BZ10**, Theorem 3.4(b)]. For the algebras $D := D(m, d, \xi)$ in Theorem 2.12, we claim that $Z(D) = \Bbbk[t]$ where $t = x + x^{-1}$. We outline the argument, freely using the notation introduced in [**WLD16**]. Indeed, the algebra D has a bigrading $D = \bigoplus D_{ij}$ with details given in [**WLD16**, equation (4.7)]. From this bigrading it is not hard to see that $Z(D) \subseteq D_{00}$, where D_{00} is exactly the Hopf subalgebra $\Bbbk[x^{\pm 1}]$. Since $xu_i = u_i x^{-1}$ for $0 \leq i \leq m-1$ by the definition of $D(m, d, \xi)$, the center is contained in the subalgebra $\Bbbk[x + x^{-1}]$. Conversely, it is easy to check that $x + x^{-1}$ commutes with all generators of D.

(2) This follows from part (1), by [**SW84**] and Tsen's theorem [**Co77**, p. 374].

(3) First, H and K are finite over their affine centers. By the Künneth formula, we obtain that $H \otimes K$ has global dimension at most 2 (a similar argument was used in the proof of [**LWZ09**, Theorem 3.2]). As a consequence, $H \otimes K$ is a noetherian affine PI AS-regular Hopf algebra of GK-dimension two. By [**BG97**, Theorem A], $H \otimes K$ is a finite direct sum of prime rings.

We claim that $H \otimes K$ is prime. By (2), $Q(H) \cong M_n(\Bbbk(t))$ and $Q(K) \cong M_m(\Bbbk(s))$. Then $Q(H) \otimes Q(K) \cong M_{nm}(\Bbbk(s) \otimes \Bbbk(t))$, and hence

$$Q(H \otimes K) = Q(Q(H) \otimes Q(K)) \cong M_{nm}(\Bbbk(s,t)). \tag{E1.17.1}$$

This forces $H \otimes K$ to be prime by Goldie's theorem. Next, using (E1.17.1),

$$Z(H) \otimes Z(K) \subseteq Z(H \otimes K) \subseteq Z(Q(H \otimes K)) = \Bbbk(s,t) = Q(Z(H) \otimes Z(K)). \tag{E1.17.2}$$

Note that $Z(H \otimes K)$ is a finite $Z(H) \otimes Z(K)$-module and thus integral over $Z(H) \otimes Z(K)$. Since the latter algebra is integrally closed we deduce from (E1.17.2) that $Z(H \otimes K) = Z(H) \otimes Z(K)$. Finally, the freeness of $H \otimes K$ over this (possibly partly Laurent) polynomial subalgebra follows from the Cohen-Macaulay property [**WZ03**, Theorems 0.1 and 0.2] and [**BM17**, Theorem 3.7(ii)], and the rank from (E1.17.1) and (E1.17.2). □

The following question, though at first glance rather technical, is suggested for the algebras $D(m, d, \xi)$ by considering the pattern followed by the other GK-dimension one families, and is highly relevant to the classification program for GK-dimension two Hopf domains discussed in Section 2.

QUESTION 2.18. Let $D(m, d, \xi)$ be the Hopf algebra in Theorem 2.12. Is there a noetherian affine Hopf domain H of GK-dimension two containing either a grouplike element g or a skew-primitive element x with $H/\langle g-1\rangle \cong D(m,d,\xi)$ or $H/\langle x\rangle \cong D(m,d,\xi)$ as Hopf algebras?

If such a Hopf algebra exists, it could *not* be pointed (since $D(m,d,\xi)$ is not) and so would demonstrate the existence of non-pointed affine Hopf domains in dimension 2. At the moment, the project to classify noetherian affine Hopf domains of GK-dimension two only produces pointed Hopf algebras.

Note that the requirement that H be a domain is key in Question 2.18, since $H = D(m,d,\xi)[t]$ with t primitive or $H = D(m,d,\xi)[t^{\pm 1}]$ with t grouplike give easy prime examples as special cases of Proposition 2.17.

3. Hopf algebras of GK-dimension 2

In Section 2 we survey some results concerning Hopf algebra domains of GK-dimension two. We continue to assume that $\Bbbk$ is algebraically closed, and - except in Subsections 3.3 and 3.6 - we assume that $\Bbbk$ has characteristic 0.

3.1. Commutative examples and dimension of tangent spaces. The tensor product recipe of Proposition 2.17 applied to the commutative Hopf algebras of Lemma 2.1 yields the following three commutative Hopf domains of GK-dimension two:

(E2.1.0) $$\mathcal{O}(\Bbbk)^{\otimes 2}, \mathcal{O}(\Bbbk^{\times})^{\otimes 2}, \mathcal{O}(\Bbbk) \otimes \mathcal{O}(\Bbbk^{\times}).$$

The remaining affine commutative Hopf $\Bbbk$-algebra domains of GK-dimension two arise as follows.

EXAMPLE 3.1. [**GZ10**, Construction 1.1] Let n be a positive integer. Let $A(n,1)$ be the algebra generated by x, x^{-1}, y subject to the relation $xy = yx$. There is a unique Hopf algebra structure on A with comultiplication

$$\Delta(x) = x \otimes x, \quad \Delta(y) = y \otimes 1 + x^n \otimes y.$$

Expressed in alternative language, examples (E2.1.0) and 3.1 give a list (which is of course well-known) of all connected two-dimensional affine algebraic groups over $\Bbbk$, as noted in [**GZ10**, Lemma 2.2].

In trying to classify all the affine Hopf domains of GK-dimension two it turns out to be both useful and natural to first impose an additional hypothesis. Namely we consider first such Hopf algebras H for which

(E2.1.1) $$\mathrm{Ext}^1_H(\Bbbk, \Bbbk) \neq 0.$$

Recall that if A is *any* affine commutative $\Bbbk$-domain of GK-dimension two and $\mathfrak{m}$ is a maximal ideal of A, then

$$\dim_{\Bbbk} \mathfrak{m}/\mathfrak{m}^2 = \dim_{\Bbbk} \mathrm{Ext}^1_A(A/\mathfrak{m}, A/\mathfrak{m}) \geq 2.$$

Indeed $\dim_{\Bbbk} \mathrm{Ext}^1_A(A/\mathfrak{m}, A/\mathfrak{m})$ is equal to the dimension of the tangent space at the point $\mathfrak{m}$ in Spec A, so it seems reasonable to consider the class of quantum groups H for which the tangent space at any point of H is nontrivial. In fact, as we record in the next lemma, (E2.1.1) is equivalent to the condition that the corresponding quantum group contains a classical algebraic group of dimension 1; one might compare this with the analogous fact that every quantum projective plane in the sense of Artin and Schelter contains a classical commutative curve of dimension 1.

LEMMA 3.2. [**GZ10**, Theorem 3.9] *Assume that* $\mathbb{k}$ *has characteristic 0 and let* H *be a Hopf* $\mathbb{k}$*-algebra of GK-dimension at most two which is either affine or noetherian. Then* H *satisfies* (E2.1.1) *if and only if* H *has a Hopf quotient, denoted by* $\overline{H}$*, isomorphic to either* $\mathbb{k}[t^{\pm 1}]$ *with* t *grouplike or* $\mathbb{k}[t]$ *with* t *primitive.*

3.2. Noetherian Hopf domains of GK-dimension 2. Following Lemma 3.2 the classification of Hopf $\mathbb{k}$-algebras of GK-dimension 2 satisfying (E2.1.1) now splits into two cases: (i) $\overline{H} = \mathbb{k}[t^{\pm 1}]$ or (ii) $\overline{H} = \mathbb{k}[t]$. The following result is due to Goodearl and Zhang [**GZ10**]. In it, families (I), (III) and (IV) constitute case (i), (II) and (V) case (ii).

THEOREM 3.3. [**GZ10**, Theorem 0.1] *Let* $\mathbb{k}$ *be algebraically closed of characteristic 0, and let* H *be a Hopf* $\mathbb{k}$*-algebra domain of GK-dimension 2 satisfying* (E2.1.1)*. Then* H *is noetherian if and only if* H *is affine if and only if* H *is isomorphic to one of the following:*

- (I) *The group algebra* $\mathbb{k}\Gamma$*, where* Γ *is either*
  - (Ia) *the free abelian group* $\mathbb{Z}^2$*, or*
  - (Ib) *the nontrivial semidirect product* $\mathbb{Z} \rtimes \mathbb{Z}$*.*
- (II) *The enveloping algebra* $U(\mathfrak{g})$*, where* $\mathfrak{g}$ *is either*
  - (IIa) *the 2-dimensional abelian Lie algebra over* $\mathbb{k}$*, or*
  - (IIb) *the Lie algebra over* $\mathbb{k}$ *with basis* $\{x, y\}$ *and* $[x, y] = y$*.*
- (III) *The Hopf algebras* $A(n, q)$ *in Example* 3.4 *below.*
- (IV) *The Hopf algebras* $B(n, p_0, \cdots, p_s, q)$ *in Example* 3.5 *below.*
- (V) *The Hopf algebras* $C(n)$ *in Example* 3.6 *below.*

EXAMPLE 3.4. [**GZ10**, Construction 1.1] Let n be a positive integer and $q \in \mathbb{k}^{\times}$, and define $A(n, q)$ to be the algebra $\mathbb{k}\langle x^{\pm 1}, y \mid xy = qyx\rangle$. There is a unique Hopf algebra structure on $A(n, q)$ under which x is grouplike and y is skew primitive, with $\Delta(y) = y \otimes 1 + x^n \otimes y$. Clearly Example 3.1 is the case $q = 1$.

EXAMPLE 3.5. [**GZ10**, Construction 1.2] Let $n, p_0, p_1, \ldots, p_s$ be positive integers and $q \in \mathbb{k}^{\times}$ with the following properties:

- (a) $s \geq 2$ and $1 < p_1 < p_2 < \cdots < p_s$;
- (b) $p_0 \mid n$ and $p_0, p_1, \ldots, p_s$ are pairwise relatively prime;
- (c) q is a primitive ℓth root of unity, where $\ell = (n/p_0)p_1 p_2 \cdots p_s$.

Set $m = p_1 p_2 \cdots p_s$ and $m_i = \frac{m}{p_i}$ for $i = 1, \ldots, s$. Fix an indeterminate y and consider the subalgebra $A := \mathbb{k}\langle y_1, \ldots, y_s\rangle \subseteq \mathbb{k}[y]$, where $y_i := y^{m_i}$ for $i = 1, \ldots, s$. The $\mathbb{k}$-algebra automorphism of $\mathbb{k}[y]$ sending $y \to qy$ restricts to an automorphism σ of A. There is a unique Hopf algebra structure on the skew Laurent polynomial ring $B = A[x^{\pm 1}; \sigma]$ such that x is grouplike and the y_i are skew primitive, with

$$\Delta(y_i) = y_i \otimes 1 + x^{m_i n} \otimes y_i$$

for $i = 1, \ldots, s$. This Hopf algebra is denoted by $B(n, p_0, \ldots, p_s, q)$.

EXAMPLE 3.6. [**GZ10**, Construction 1.4] Let n be a positive integer, and set $C(n) := \mathbb{k}[y^{\pm 1}][x; (y^n - y)\frac{d}{dy}]$. That is, $C(n)$ is an Ore extension of $\mathbb{k}[y^{\pm 1}]$ with indeterminate x satisfying the equation

$$xf(y) = f(y)x + (y^n - y)\frac{df(y)}{dy}$$

for all $f(y) \in \Bbbk[y^{\pm 1}]$. There is a unique Hopf algebra structure on $C(n)$ such that y is grouplike and x is skew primitive, with

$$\Delta(x) = x \otimes y^{n-1} + 1 \otimes x.$$

Naturally, in the light of Theorem 3.3 Goodearl and Zhang asked the obvious question: does the classification remain true without assuming (E2.1.1)? In 2013 Wang, Zhang and Zhuang [**WZZ13**] showed that the answer is "No". We'll review their result in Subsection 3.5

Note that all the algebras listed in Theorem 3.3 are assembled from two one-dimensional domains, as discussed in Subsection 2.5. In particular, $A(n, q)$ and $C(n)$ are Ore extensions of the Hopf algebra $\Bbbk[t^{\pm 1}]$, special cases of the Hopf Ore construction which we review in the next subsection.

3.3. Hopf Ore extensions and IHOEs. Recall that, given an algebra automorphism σ of a $\Bbbk$-algebra A, a left σ-derivation δ of A is a $\Bbbk$-linear endomorphism of A such that

$$\delta(ab) = \sigma(a)\delta(b) + \delta(a)b$$

for all $a, b \in A$. The Ore extension of A associated to (σ, δ), denoted by $A[x; \sigma, \delta]$, is an algebra generated by A and x and subject to relation of the form

$$xa = \sigma(a)x + \delta(a)$$

for all $a \in A$. Hopf Ore extensions of a Hopf algebra A, also denoted by $A[x; \sigma, \delta]$, were defined and studied by Panov [**Pan03**] in 2003. His definition was refined and extended in [**BO15**] and then in [**Hu19**], to *iterated Hopf Ore extensions* (IHOEs) over a field $\Bbbk$. The following definition is significantly more general than the original one given in [**Pan03**, Definition 1.0] and its modification in [**BO15**, Definition 2.1]. The improvement here is due to recent work of Huang [**Hu19**].

DEFINITION 3.7. Let $\Bbbk$ be a field.

(1) Let A be a Hopf $\Bbbk$-algebra. A *Hopf Ore extension* (abbreviated to *HOE*) of A is an algebra H such that
  (1a) H is a Hopf $\Bbbk$-algebra with Hopf subalgebra A;
  (1b) there exist an algebra automorphism σ of A and a σ-derivation δ of A such that $H = A[x; \sigma, \delta]$.

(2) An *(n-step) iterated Hopf Ore extension of* $\Bbbk$ (abbreviated to *(n-step) IHOE (of* $\Bbbk$*)*) is a Hopf algebra

$$H = \Bbbk[X_1][X_2; \sigma_2, \delta_2] \cdots [X_n; \sigma_n, \delta_n],$$

where
  (2a) H is a Hopf algebra;
  (2b) $H_{(i)} := \Bbbk\langle X_1, \ldots, X_i\rangle$ is a Hopf subalgebra of H for $i = 1, \ldots, n$;
  (2c) σ_i is an algebra automorphism of $H_{(i-1)}$, and δ_i is a σ_i-derivation of $H_{(i-1)}$, for $i = 2, \ldots, n$.

The definition of an HOE in [**BO15**, Definition 2.1] required, in addition to Definition 3.7(1a,1b), that

(1c) there are $a, b \in A$ and $v, w \in A \otimes A$ such that

$$\Delta(x) = a \otimes x + x \otimes b + v(x \otimes x) + w.$$

However, by Huang's theorem [**Hu19**, Theorem 1.3], if $A \subseteq H$ are noetherian $\Bbbk$-algebras satisfying hypotheses (1a) and (1b) of Definition 3.7(1), and $A \otimes A$ is a domain, then, up to a change of variable,

(1d) there are $a \in A$ and $w \in A \otimes A$ such that

$$\Delta(x) = a \otimes x + x \otimes 1 + w.$$

This means that under the conditions in [**Hu19**, Theorem 1.3], namely that A is noetherian and that $A \otimes A$ is a domain, Definition 3.7(1) is equivalent to [**BO15**, Definition 2.1]. For the subalgebras $H_{(i)}$ appearing in Definition 3.7(2) it is easy to show that $H_{(i)} \otimes H_{(i)}$ is a noetherian domain, so a similar comment applies to the comultiplication of an IHOE over $\Bbbk$. There are moreover strong restrictions on the automorphisms σ_i and derivations δ_i which can occur - for details, see [**BO15**], [**Hu19**] or [**BZ20**].

In Examples 3.8 and 3.9 in the next subsection we construct Hopf Ore extensions that are generalizations of the Hopf Ore extensions $A(n, q)$ and $C(n)$ of Theorem 3.3.

3.4. Non-noetherian Hopf domains of GK-dimension 2. In this subsection we describe two families of Hopf domains of GK-dimension two which are not affine and not noetherian.

EXAMPLE 3.8. [**GZ17**, Example 1.2] Let $K = \Bbbk G$ where G is a group, let $\chi : G \to \Bbbk^{\times}$ be a character of G and choose any element e in the center of G, denoted by $Z(G)$. Define an algebra automorphism $\sigma_\chi : K \to K$ by

$$\sigma_\chi(g) = \chi(g)g, \quad \forall\, g \in G,$$

so σ_χ is the winding automorphism associated to the algebra map from K to $\Bbbk$ induced by χ. Let $\delta = 0$. By [**WZZ16**, Example 5.4], $K[z; \sigma_\chi]$ is an HOE of K with $\Delta(z) = z \otimes 1 + e \otimes z$. This HOE is denoted by $A_G(e, \chi)$.

Now carry out the above construction when G is a non-trivial non-cyclic subgroup of $(\mathbb{Q}, +)$. Then

$$\operatorname{GKdim} \Bbbk G = 1, \quad \text{and} \quad \operatorname{GKdim} A_G(e, \chi) = 2,$$

using [**HK96**, Lemma 2.2] for the second equality. By [**GZ17**, Theorem 0.1(4)], $A_G(e, \chi)$ satisfies (E2.1.1). When, for instance, $\Bbbk = \mathbb{C}$, there are many characters of G; for example, with $\lambda \in \mathbb{R}$, $\exp_\lambda : r \to \exp(2\pi i r \lambda)$ is a character from $(\mathbb{Q}, +)$ to $\mathbb{C}^{\times}$. Since $A_G(e, \chi)$ is a free $\Bbbk G$-module it is clear that $A_G(e, \chi)$ is not noetherian. Being a skew polynomial algebra with coefficient ring the group algebra of a torsion-free abelian group, $A_G(e, \chi)$ is a domain.

EXAMPLE 3.9. [**GZ17**, Example 1.3] Again, let $K = \Bbbk G$ where G is a group and fix $e \in Z(G)$. Let $\tau : G \to (\Bbbk, +)$ be an additive character of G. Define a $\Bbbk$-linear derivation $\delta : K \to K$ by

$$\delta(g) = \tau(g)g(e - 1), \quad \forall\, g \in G.$$

Then $C_G(e, \tau) := K[z; \delta]$ is an HOE of K when we define $\Delta(z) = z \otimes 1 + e \otimes z$.

As in the previous example, let G be a nontrivial non-cyclic subgroup of $(\mathbb{Q}, +)$. Again by [**HK96**, Lemma 2.2] (or [**Zha97**, Theorem 1.2(1)])

$$\operatorname{GKdim} \Bbbk G = 1, \quad \text{and} \quad \operatorname{GKdim} C_G(e, \tau) = 2.$$

Since we are assuming $\Bbbk$ has characteristic zero, there are many additive characters from G to $(\Bbbk, +)$. For example, let $\lambda \in \Bbbk$; then $i_\lambda : r \to r\lambda$ is an additive character from $G \to (\Bbbk, +)$.

Denote the augmentation ideal of K by K^+. Since $C_G(e,\tau)K^+$ is an ideal of $C_G(e,\tau)$ with $C_G(e,\tau)/C_G(e,\tau)K^+ \cong \Bbbk[z]$, $C_G(e,\chi)$ satisfies (E2.1.1). In the same way as Example 3.8, $C_G(e,\tau)$ is not noetherian, but it is a domain.

The next family of examples are *not* HOEs, although they are like the others assembled from two domains of GK-dimension 1. Since the details are quite technical, we only give a brief sketch here, referring the reader to [**GZ17**] for more information.

EXAMPLE 3.10. [**GZ17**, Construction 3.1] Let G be a nontrivial non-cyclic subgroup of $(\mathbb{Q}, +)$, let $\{p_i : i \in I\}$ be an infinite set of pairwise relatively prime integers, and let χ be a character of a certain group GM with $G \subseteq GM \subseteq (\mathbb{Q}, +)$. Given this data one can construct (over several pages) a non-affine Hopf domain of GK-dimension two, denoted by $B_G(\{p_i\}, \chi)$. By [**GZ17**, Construction 1.2], $B_G(\{p_i\}, \chi)$ satisfies (E2.1.1) and is not noetherian.

Here is the classification result parallel to Theorem 3.3. Let $\mathbb{Z}_{\langle 2\rangle}$ denote the localization of the ring $\mathbb{Z}$ at the maximal ideal $\langle 2\rangle$.

THEOREM 3.11. [**GZ17**, Theorem 0.1] *Assume that $\Bbbk$ has characteristic 0 and let H be a Hopf domain over $\Bbbk$ of GK-dimension two which is not noetherian and satisfies* (E2.1.1). *Then H is isomorphic as a Hopf algebra to one of the following.*

(1) *$\Bbbk G$ where G is a non-finitely generated subgroup of $\mathbb{Q}^2$ containing $\mathbb{Z}^2$.*
(2) *$\Bbbk G$ where $G = L \rtimes R$ for subgroups L of $\mathbb{Q}$ containing $\mathbb{Z}$ and R of $\mathbb{Z}_{(2)}$ containing $\mathbb{Z}$, with at least one of L and R not finitely generated.*
(3) *$A_G(e, \chi)$ in Example* 3.8.
(4) *$C_G(e, \tau)$ in Example* 3.9.
(5) *$B_G(\{p_i\}, \chi)$ in Example* 3.10.

Thus Theorems 3.3 and 3.11 together with Lemma 2.1 and Remark 2.2(1) give a complete classification when $\Bbbk$ has characteristic 0 of the Hopf $\Bbbk$-algebra domains of GK-dimension at most 2 which satisfy (E2.1.1). But not all Hopf domains of GK-dimension 2 satisfy (E2.1.1), as we explain in the next subsection.

3.5. Hopf domains of GK-dimension 2 with no infinite classical subgroup. The hope that the algebras of Subsection 3.2 constitute a complete list of affine Hopf domains of GK-dimension 2 is demolished by the following result. As the notation suggests, the new algebras are variants of those in Examples 3.5.

THEOREM 3.12. [**WZZ13**] *Assume that $\Bbbk$ has characteristic 0. Let $s, p_1, \dots, p_s$ be integers, all greater than 1 and with $\gcd(p_i, p_j) = 1$ when $i \neq j$. Fix scalars $\sigma_1, \dots, \sigma_s$ in $\Bbbk$, not all equal. For $\ell := \Pi_{i=1}^s p_i$, choose q, an ℓth root of 1 in $\Bbbk$. Then there is a family $B(n, \{p_i\}_1^s, q, \{\sigma_i\}_1^s)$ of Hopf $\Bbbk$-algebras with the following properties.*

(1) *$B(n, \{p_i\}_1^s, q, \{\sigma_i\}_1^s)$ is an affine domain of GK-dimension 2.*
(2) *$B(n, \{p_i\}_1^s, q, \{\sigma_i\}_1^s)$ does not satisfy* (E2.1.1).
(3) *$B(n, \{p_i\}_1^s, q, \{\sigma_i\}_1^s)$ is pointed, generated by grouplikes and skew primitives.*

(4) $B(n, \{p_i\}_1^s, q, \{\sigma_i\}_1^s)$ *is a finite module over its affine center, hence is noetherian and satisfies a polynomial identity.*
(5) $\mathrm{gldim} B(n, \{p_i\}_1^s, q, \{\sigma_i\}_1^s) < \infty$ *if and only if* $s = 2$ *and* $\sigma_1 \neq \sigma_2$. *In this case the global dimension is 2.*
(6) *Up to a permutation of* $\{1, \dots, s\}$ *and multiplication of* $\{\sigma_i\}_1^s$ *by a non-zero scalar, these Hopf algebras are mutually non-isomorphic.*

We refer to [**WZZ13**, Section 2] for more details about the algebras in the above theorem and some constraints concerning q and n. Wang, Zhang and Zhuang show in [**WZZ13**] that any as yet unknown example of an affine Hopf domain of GK-dimension 2 in characteristic 0, especially if generated by grouplikes and skew primitives, would have to exhibit very constrained features. They therefore ask:

QUESTION 3.13. [**WZZ13**, Remark 4.5] Assume $\Bbbk$ has characteristic 0. Are the algebras of Theorem 3.12 the only affine Hopf $\Bbbk$-algebra domains of GK-dimension 2 not satisfying (E2.1.1)?

It is tempting, given the classification in characteristic 0 of prime AS-regular algebras of dimension 1 and the partial classification of GK-dimension 2 Hopf domains, to consider prime AS-regular Hopf algebras of GK-dimension 2. Proposition 2.17 indicates that this may be a formidable task, so perhaps a preliminary step might be the following.

QUESTION 3.14 (Dingguo Wang [**Wa20**]). Find homological invariants (additional to, or generalising, the integral order and integral minor) which are defined for pointed Hopf algebras, and which may help to classify prime pointed Hopf algebras of GK-dimension two.

3.6. 2-step IHOEs in positive characteristic. In this subsection let p be a prime and let $\Bbbk$ be an algebraically closed field of characteristic p. Consider the project to classify the affine Hopf $\Bbbk$-algebra domains of GK-dimension at most 2. If H is such an algebra with $\mathrm{GKdim} H = 1$, then H is commutative by Tsen's theorem [**Co77**, p. 374], and so $H = \Bbbk[t]$ or $\Bbbk[t^{\pm 1}]$ by Lemma 2.1.

However when $\mathrm{GKdim} H = 2$ the situation appears much more complex than in characteristic 0. The *connected* Hopf $\Bbbk$-algebras, that is those whose coradical is $\Bbbk$, form a good starting point. In arbitrary characteristic every IHOE is connected by [**BO15**, Proposition 2.5]. The cocommutative connected Hopf algebras in characteristic 0 are the enveloping algebras of Lie algebras. Zhuang showed in [**Zhu13**, Proposition 7.6] that in characteristic 0, there are only two connected Hopf algebras of GK-dimension 2, and both are cocommutative IHOEs: namely the enveloping algebras of the two 2-dimensional Lie algebras. But in positive characteristic it's a different story, as we now explain.

Let $\mathbf{d_s} = \{\mathbf{d_s}\}_{s \geq 0}$, $\mathbf{b_s} = \{b_s\}_{s \geq 0}$ and $\mathbf{c_{s,t}} = \{c_{s,t}\}_{0 \leq s < t}$ be sequences of scalars in $\Bbbk$ with only finitely many nonzero terms. Let $H(\mathbf{d_s}, \mathbf{b_s}, \mathbf{c_{s,t}})$ denote the Ore extension $\Bbbk[X_1][X_2; \mathrm{Id}, \delta]$ where

$$\delta(X_1) = \sum_{s \geq 0} d_s X_1^{p^s},$$

and where the coalgebra structure is as follows (which is verified in [**BZ20**]). Define maps Δ, ϵ, and S on $\{X_1, X_2\}$ by setting

$$\begin{aligned}\Delta(X_1) &= X_1 \otimes 1 + 1 \otimes X_1,\\ \Delta(X_2) &= X_2 \otimes 1 + 1 \otimes X_2 + w,\\ \epsilon(X_1) &= \epsilon(X_2) = 0,\\ S(X_1) &= -X_1,\\ S(X_2) &= -X_2 - m(\mathrm{Id} \otimes S)(w),\end{aligned}$$

where

$$w = \sum_{s \geq 0} b_s \left(\sum_{i=1}^{p-1} \frac{(p-1)!}{i!(p-i)!} (X_1^{p^s})^i \otimes (X_1^{p^s})^{p-i} \right) + \sum_{0 \leq s < t} c_{s,t} \left(X_1^{p^s} \otimes X_1^{p^t} - X_1^{p^t} \otimes X_1^{p^s} \right).$$

THEOREM 3.15. [**BZ20**, Theorem 0.3(3)] *Let $\Bbbk$ be algebraically closed of characteristic $p > 0$ and let H be a 2-step IHOE over $\Bbbk$.*

(1) *The Hopf algebra H is isomorphic to $H(\mathbf{d_s}, \mathbf{b_s}, \mathbf{c_{s,t}})$ for some choice of scalars $\{d_s\}_{s \geq 0}$, $\{b_s\}_{s \geq 0}$ and $\{c_{s,t}\}_{0 \leq s < t}$.*
(2) *Two such Hopf algebras $H(\mathbf{d_s}, \mathbf{b_s}, \mathbf{c_{s,t}})$ and $H((\mathbf{d'_s}, \mathbf{b'_s}, \mathbf{c'_{s,t}})$ are isomorphic as Hopf algebras if and only if there are nonzero scalars $\alpha, \beta \in \Bbbk$ such that*

$$d'_s = d_s \alpha^{p^s - 1} \beta^{-1}, \quad b'_s = b_s \alpha^{p^{s+1}} \beta^{-1}, \quad c'_{s,t} = c_{s,t} \alpha^{p^s + p^t} \beta^{-1}$$

for all s, t.

The connected unipotent $\Bbbk$-groups U of dimension n are precisely those algebraic groups whose coordinate ring is an n-step commutative IHOE over $\Bbbk$, as explained in [**BZ20**, §6.1]. (Also see [**BGi14**, Theorem 6.1] when char $\Bbbk = 0$.) Thus Theorem 3.15 incorporates the classification of the 2-dimensional connected unipotent groups over $\Bbbk$ - namely, their coordinate rings are the algebras $H(\mathbf{0}, \mathbf{b_s}, \mathbf{c_{s,t}})$. The classification in terms of the groups rather than their coordinate rings can be found in [**KMT74**, §3.7]. For further information concerning the algebras $H(\mathbf{d_s}, \mathbf{b_s}, \mathbf{c_{s,t}})$, see [**BZ20**].

As suggested above, one should view Theorem 3.15 as the first step in a program to classify the connected affine Hopf domains of GK-dimension 2 over $\Bbbk$ when $\Bbbk$ has positive characteristic. From this perspective and in the light of Zhuang's classification in characteristic 0, [**Zhu13**, Proposition 7.6], it is natural to ask:

QUESTION 3.16. [**BZ20**, Question 6.3] Is every affine connected Hopf domain over $\Bbbk$ of GK-dimension 2 an IHOE?

4. Further comments, projects and open questions

In this final section $\Bbbk$ denotes an arbitrary algebraically closed field unless otherwise stated. We review the questions scattered throughout the previous sections and gather them into a few overarching projects, and then suggest a couple of further directions of research. All these questions and projects should be viewed as supplementing (and in some cases repeating) questions in other survey articles, for example [**And14, And17, Br98, Br07, BGi14, BGLZ14, Go13, KWZ19**].

First we note the familiar questions reappearing in this paper which concern the entire category of affine Hopf algebras of finite GK-dimension, namely Question 1.1 and Conjecture 1.2. To these we add the following, which has been implicit at several points in the foregoing pages.

QUESTION 4.1. Is every affine Hopf algebra of finite GK-dimension noetherian?

The only evidence in favor of a positive answer to Question 4.1 seems to be the absence of counterexamples. The converse is false - consider for example the group algebra of a polycyclic group which is not nilpotent-by-finite. It does however seem to be unknown whether a noetherian Hopf algebra has to be affine.

4.1. Projects in GK-dimensions one and two. The questions listed earlier concerning Hopf algebras of GK-dimension 1 are Projects 2.13, 2.14, 2.16, and Question 2.15. These should be regarded as key components on the road towards:

PROJECT 4.2. Classify semiprime noetherian Hopf $\Bbbk$-algebras of GK-dimension one.

Question 2.18 focuses on the possibility of a close link between GK-dimensions one and two. It would be answered as an immediate corollary of a classification result in GK-dimension two.

The target of classifying all affine or noetherian semiprime (or even prime) Hopf algebras of GK-dimension two seems rather too ambitious at present, so we propose:

PROJECT 4.3. Classify Hopf $\Bbbk$-algebra domains of GK-dimension two.

Additional hypotheses to impose as milestones on the route to Project 4.3 include those of Question 3.13 - namely that $\Bbbk$ has characteristic 0 and the algebra is affine; and those of Question 3.16, that $\Bbbk$ has positive characteristic and the Hopf algebra is affine and connected.

4.2. Beyond GK-dimension two. The following special case of Question 1.1 is currently open:

QUESTION 4.4. Does there exist a Hopf $\Bbbk$-algebra H with $2 < \mathrm{GKdim} H < 3$?

Strong evidence in support of a negative answer to this question is given by [**WZZ16**, Theorem 0.1], which confirms that the answer is "No" when $\Bbbk$ has characteristic 0, with H affine and pointed. In the light of their result Wang, Zhang and Zhuang proposed the following, suggesting that double Ore extensions may be an important tool for this purpose:

PROJECT 4.5. [**WZZ16**, Remark 7.5] In characteristic 0, classify affine pointed Hopf domains of GK-dimension three.

4.3. Connected (graded) Hopf algebras. Observe that if the pointed hypothesis in Project 4.5 is strengthened to require that the Hopf algebra is connected, the proposed classification has been achieved by Zhuang [**Zhu13**], and even in dimension four in [**WZZ15**]. As these results suggest, the class of connected Hopf algebras, and even the smaller classes of those which are graded as algebras and/or as coalgebras, form interesting testing grounds.

For example, in [**BGZ19**], Brown, Gilmartin and Zhang studied, for $\Bbbk$ of characteristic 0, Hopf $\Bbbk$-algebras of finite GK-dimension that are connected graded as an algebra. In particular, they proved that such a Hopf algebra is a noetherian domain which is GK-Cohen-Macaulay, AS regular and Auslander regular of global dimension equal to its GK-dimension, and is Calabi-Yau. As a consequence, it satisfies Poincaré duality.

Remaining in characteristic 0, but imposing the stronger hypothesis on the Hopf algebra H of finite GK-dimension, that it is connected graded both as an algebra *and* as a coalgebra, Zhou, Shen and Lu proved the striking result [**ZSL19**] that H is an IHOE [**ZSL19**, Theorem B]. Naturally, this leads to

QUESTION 4.6. [**ZSL19**, Question C] With $\Bbbk$ of characteristic 0, suppose the Hopf $\Bbbk$-algebra H has finite GK-dimension and is connected graded as an algebra. Is H an IHOE?

Acknowledgments

The authors thank the editors for the invitation to provide this submission, thank Professors Gongxiang Liu and Dingguo Wang for several conversations on the subject, for reading an early draft of the paper, for suggesting some projects, and thank the referee for his/her very careful reading and valuable comments.

References

[And14] Nicolás Andruskiewitsch, *On finite-dimensional Hopf algebras*, Proceedings of the International Congress of Mathematicians—Seoul 2014. Vol. II, Kyung Moon Sa, Seoul, 2014, pp. 117–141. MR3728608

[And17] Nicolás Andruskiewitsch, *An introduction to Nichols algebras*, Quantization, geometry and noncommutative structures in mathematics and physics, Math. Phys. Stud., Springer, Cham, 2017, pp. 135–195. MR3751453

[AA06] Nicolás Andruskiewitsch and Iván Ezequiel Angiono, *On Nichols algebras with generic braiding*, Modules and comodules, Trends Math., Birkhäuser Verlag, Basel, 2008, pp. 47–64, DOI 10.1007/978-3-7643-8742-6_3. MR2742620

[AAH19] Nicolás Andruskiewitsch, Iván Angiono, and István Heckenberger, *On finite GK-dimensional Nichols algebras of diagonal type*, Tensor categories and Hopf algebras, Contemp. Math., vol. 728, Amer. Math. Soc., Providence, RI, 2019, pp. 1–23, DOI 10.1090/conm/728/14653. MR3943743

[AS00] Nicolás Andruskiewitsch and Hans-Jürgen Schneider, *Finite quantum groups and Cartan matrices*, Adv. Math. **154** (2000), no. 1, 1–45, DOI 10.1006/aima.1999.1880. MR1780094

[AS04] Nicolás Andruskiewitsch and Hans-Jürgen Schneider, *A characterization of quantum groups*, J. Reine Angew. Math. **577** (2004), 81–104, DOI 10.1515/crll.2004.2004.577.81. MR2108213

[Ang13] Iván Angiono, *On Nichols algebras of diagonal type*, J. Reine Angew. Math. **683** (2013), 189–251, DOI 10.1515/crelle-2011-0008. MR3181554

[Bo91] Armand Borel, *Linear algebraic groups*, 2nd ed., Graduate Texts in Mathematics, vol. 126, Springer-Verlag, New York, 1991, DOI 10.1007/978-1-4612-0941-6. MR1102012

[Br98] Kenneth A. Brown, *Representation theory of Noetherian Hopf algebras satisfying a polynomial identity*, Trends in the representation theory of finite-dimensional algebras (Seattle, WA, 1997), Contemp. Math., vol. 229, Amer. Math. Soc., Providence, RI, 1998, pp. 49–79, DOI 10.1090/conm/229/03310. MR1676211

[Br07] Kenneth A. Brown, *Noetherian Hopf algebras*, Turkish J. Math. **31** (2007), no. suppl., 7–23. MR2368080

[BGi14] Kenneth A. Brown and Paul Gilmartin, *Hopf algebras under finiteness conditions*, Palest. J. Math. **3** (2014), no. Special issue, 356–365. MR3274613

[BGZ19] K. A. Brown, P. Gilmartin, and J. J. Zhang, *Connected (graded) Hopf algebras*, Trans. Amer. Math. Soc. **372** (2019), no. 5, 3283–3317, DOI 10.1090/tran/7686. MR3988611

[BG97] K. A. Brown and K. R. Goodearl, *Homological aspects of Noetherian PI Hopf algebras of irreducible modules and maximal dimension*, J. Algebra **198** (1997), no. 1, 240–265, DOI 10.1006/jabr.1997.7109. MR1482982

[BGLZ14] *Mini-Workshop: infinite dimensional Hopf algebras*, Oberwolfach Rep. **11** (2014), no. 2, 1111–1137, DOI 10.4171/OWR/2014/20. Abstracts from the mini-workshop held April 13–19, 2014; Organized by Ken Brown, Ken Goodearl, Tom Lenagan, and James Zhang. MR3379364

[BM17] Ken A. Brown and Marjory J. Macleod, *The Cohen Macaulay property for non-commutative rings*, Algebr. Represent. Theory **20** (2017), no. 6, 1433–1465, DOI 10.1007/s10468-017-9694-z. MR3735914

[BO15] K. A. Brown, S. O'Hagan, J. J. Zhang, and G. Zhuang, *Connected Hopf algebras and iterated Ore extensions*, J. Pure Appl. Algebra **219** (2015), no. 6, 2405–2433, DOI 10.1016/j.jpaa.2014.09.007. MR3299738

[BZ08] K. A. Brown and J. J. Zhang, *Dualising complexes and twisted Hochschild (co)homology for Noetherian Hopf algebras*, J. Algebra **320** (2008), no. 5, 1814–1850, DOI 10.1016/j.jalgebra.2007.03.050. MR2437632

[BZ10] K. A. Brown and J. J. Zhang, *Prime regular Hopf algebras of GK-dimension one*, Proc. Lond. Math. Soc. (3) **101** (2010), no. 1, 260–302, DOI 10.1112/plms/pdp060. MR2661247

[BZ20] K.A. Brown and J.J. Zhang, *Iterated Hopf Ore Extensions in positive characteristic*, preprint (2020) `arXiv:2004.14702`, Journal of Noncommutative Geometry, (accepted for publication).

[Co77] Paul Moritz Cohn, *Algebra. Vol. 2*, John Wiley & Sons, London-New York-Sydney, 1977. With errata to Vol. I. MR0530404

[EGNO15] Pavel Etingof, Shlomo Gelaki, Dmitri Nikshych, and Victor Ostrik, *Tensor categories*, Mathematical Surveys and Monographs, vol. 205, American Mathematical Society, Providence, RI, 2015, DOI 10.1090/surv/205. MR3242743

[Go13] K. R. Goodearl, *Noetherian Hopf algebras*, Glasg. Math. J. **55** (2013), no. A, 75–87, DOI 10.1017/S0017089513000517. MR3110805

[GZ10] K. R. Goodearl and J. J. Zhang, *Noetherian Hopf algebra domains of Gelfand-Kirillov dimension two*, J. Algebra **324** (2010), no. 11, 3131–3168, DOI 10.1016/j.jalgebra.2009.11.001. MR2732991

[GZ17] K. R. Goodearl and J. J. Zhang, *Non-affine Hopf algebra domains of Gelfand-Kirillov dimension two*, Glasg. Math. J. **59** (2017), no. 3, 563–593, DOI 10.1017/S0017089516000410. MR3682000

[Hu19] Hongdi Huang, *Hopf Ore extensions*, Algebr. Represent. Theory **23** (2020), no. 4, 1477–1486, DOI 10.1007/s10468-019-09901-8. MR4125587

[HK96] Chan Huh and Chol On Kim, *Gelfand-Kirillov dimension of skew polynomial rings of automorphism type*, Comm. Algebra **24** (1996), no. 7, 2317–2323, DOI 10.1080/00927879608825702. MR1390376

[Hu75] James E. Humphreys, *Linear algebraic groups*, Springer-Verlag, New York-Heidelberg, 1975. Graduate Texts in Mathematics, No. 21. MR0396773

[KMT74] Tatsuji Kambayashi, Masayoshi Miyanishi, and Mitsuhiro Takeuchi, *Unipotent algebraic groups*, Lecture Notes in Mathematics, Vol. 414, Springer-Verlag, Berlin-New York, 1974. MR0376696

[KL00] Günter R. Krause and Thomas H. Lenagan, *Growth of algebras and Gelfand-Kirillov dimension*, Revised edition, Graduate Studies in Mathematics, vol. 22, American Mathematical Society, Providence, RI, 2000, DOI 10.1090/gsm/022. MR1721834

[KWZ19] Henning Krause, Sarah Witherspoon, and James Zhang, *Mini-Workshop: Cohomology of Hopf Algebras and Tensor Categories*, Oberwolfach Rep. **16** (2019), no. 1, 663–693, DOI 10.4171/OWR/2019/11. MR4072199

[Li09] Gongxiang Liu, *On Noetherian affine prime regular Hopf algebras of Gelfand-Kirillov dimension 1*, Proc. Amer. Math. Soc. **137** (2009), no. 3, 777–785, DOI 10.1090/S0002-9939-08-09034-5. MR2457414

[Li20] Gongxiang Liu, *A classification result on prime Hopf algebras of GK-dimension one*, J. Algebra **547** (2020), 579–667, DOI 10.1016/j.jalgebra.2019.12.003. MR4042591

[Li20a] G. Liu, Oral communication, (2020).

[LWZ07] D.-M. Lu, Q.-S. Wu, and J. J. Zhang, *Homological integral of Hopf algebras*, Trans. Amer. Math. Soc. **359** (2007), no. 10, 4945–4975, DOI 10.1090/S0002-9947-07-04159-1. MR2320655

[LWZ09] D.-M. Lu, Q.-S. Wu, and J. J. Zhang, *Hopf algebras with rigid dualizing complexes*, Israel J. Math. **169** (2009), 89–108, DOI 10.1007/s11856-009-0005-1. MR2460900

[Mo93] Susan Montgomery, *Hopf algebras and their actions on rings*, CBMS Regional Conference Series in Mathematics, vol. 82, Published for the Conference Board of the Mathematical Sciences, Washington, DC; by the American Mathematical Society, Providence, RI, 1993, DOI 10.1090/cbms/082. MR1243637

[Pan03] A. N. Panov, *Ore extensions of Hopf algebras* (Russian, with Russian summary), Mat. Zametki **74** (2003), no. 3, 425–434, DOI 10.1023/A:1026115004357; English transl., Math. Notes **74** (2003), no. 3-4, 401–410. MR2022506

[Pas77] Donald S. Passman, *The algebraic structure of group rings*, Pure and Applied Mathematics, Wiley-Interscience [John Wiley & Sons], New York-London-Sydney, 1977. MR470211

[SW84] L. W. Small and R. B. Warfield Jr., *Prime affine algebras of Gel′fand-Kirillov dimension one*, J. Algebra **91** (1984), no. 2, 386–389, DOI 10.1016/0021-8693(84)90110-8. MR769581

[SSW85] L. W. Small, J. T. Stafford, and R. B. Warfield Jr., *Affine algebras of Gel′fand-Kirillov dimension one are PI*, Math. Proc. Cambridge Philos. Soc. **97** (1985), no. 3, 407–414, DOI 10.1017/S0305004100062976. MR778674

[Ta71] Earl J. Taft, *The order of the antipode of finite-dimensional Hopf algebra*, Proc. Nat. Acad. Sci. U.S.A. **68** (1971), 2631–2633, DOI 10.1073/pnas.68.11.2631. MR286868

[Wa20] D.-G. Wang, Oral communication, (2020).

[WZZ13] D.-G. Wang, J. J. Zhang, and G. Zhuang, *Hopf algebras of GK-dimension two with vanishing Ext-group*, J. Algebra **388** (2013), 219–247, DOI 10.1016/j.jalgebra.2013.03.032. MR3061686

[WZZ15] D.-G. Wang, J. J. Zhang, and G. Zhuang, *Connected Hopf algebras of Gelfand-Kirillov dimension four*, Trans. Amer. Math. Soc. **367** (2015), no. 8, 5597–5632, DOI 10.1090/S0002-9947-2015-06219-9. MR3347184

[WZZ16] D.-G. Wang, J. J. Zhang, and G. Zhuang, *Primitive cohomology of Hopf algebras*, J. Algebra **464** (2016), 36–96, DOI 10.1016/j.jalgebra.2016.07.003. MR3533424

[WLD16] Jinyong Wu, Gongxiang Liu, and Nanqing Ding, *Classification of affine prime regular Hopf algebras of GK-dimension one*, Adv. Math. **296** (2016), 1–54, DOI 10.1016/j.aim.2016.03.037. MR3490761

[WZ03] Q.-S. Wu and J. J. Zhang, *Noetherian PI Hopf algebras are Gorenstein*, Trans. Amer. Math. Soc. **355** (2003), no. 3, 1043–1066, DOI 10.1090/S0002-9947-02-03106-9. MR1938745

[ZSL19] G.-S. Zhou, Y. Shen, and D.-M. Lu, *The structure of connected (graded) Hopf algebras*, Adv. Math. **372** (2020), 107292, 31, DOI 10.1016/j.aim.2020.107292. MR4125517

[Zha97] James J. Zhang, *A note on GK dimension of skew polynomial extensions*, Proc. Amer. Math. Soc. **125** (1997), no. 2, 363–373, DOI 10.1090/S0002-9939-97-03602-2. MR1350966

[Zhu13] Guangbin Zhuang, *Properties of pointed and connected Hopf algebras of finite Gelfand-Kirillov dimension*, J. Lond. Math. Soc. (2) **87** (2013), no. 3, 877–898, DOI 10.1112/jlms/jds079. MR3073681

School of Mathematics and Statistics, University of Glasgow, Glasgow G12 8QW, Scotland

Email address: ken.brown@glasgow.ac.uk

Department of Mathematics, Box 354350, University of Washington, Seattle, Washington 98195

Email address: zhang@math.washington.edu

Contemporary Mathematics
Volume **771**, 2021
https://doi.org/10.1090/conm/771/15508

On braided double-biproduct Hopf algebras

D. Bulacu and B. Torrecillas

ABSTRACT. By using a categorical interpretation [**3**] for a double-biproduct Hopf algebra [**17**], we show that braided double-biproduct Hopf algebras can be identified to braided biproduct Hopf algebras. Furthermore, under some conditions, this remains true even after a deformation with an invertible 2-cocycle. Concrete examples that fit into this theory are presented, and the famous quantum group $U_q(\mathfrak{g})$ is among them.

Introduction

Quantum groups and their representation theory appeared for the first time in theoretical physics in connection to the theory of quantum inverse scattering method, theory which has as aim to construct and solve integrable quantum systems. Initially, quantum groups were called those algebras with generators and relations defined by an integrable system, by means of a certain "matrix of constants" depending on it. Independently, Drinfeld and Jimbo [**8, 9**] showed that these algebras are Hopf algebras, and that in most of the cases they are actually 1-parameter quantized enveloping algebras of a semisimple Lie algebra $\mathfrak{g}$. The further development of quantum group theory uncovered connections of this theory with the quantum theory of angular momentum, 2-dimensional conformal field theory, the topology of knots, the theory of 3-dimensional manifolds, etc.

From a purely Hopf algebraic point of view, the Drinfeld-Jimbo quantum group $U_q(\mathfrak{g})$ can be described in two equivalent ways. First, Majid [**17**] characterized $U_q(\mathfrak{q})$ as a sort of twist deformation of a double-bosonization Hopf algebra of two braided Hopf algebras $\overline{C}$, B living in categories of corepresentations of a certain coquasitriangular (CQT for short) Hopf algebra H. The word twist refers here to the existence of a sort of pairing of $\overline{C}$ and B, that contributes to the deformation of the algebra structure of the double-bosonization Hopf algebra. By taking B equals the Lusztig's algebra $\mathfrak{f}$ (see [**14**, Chapter 1]) and suitable $\overline{C}$ and H, modulo the q-Serre-relations the Majid's construction recovers the Lusztig's form of $U_q(\mathfrak{g})$. Secondly, Bespalov and Drabant [**3**] described $U_q(\mathfrak{g})$ as the deformation by an invertible 2-cocycle of a certain tensor product Hopf algebra in the category of Hopf bimodules over a Hopf algebra with bijective antipode. We will explain the

2020 *Mathematics Subject Classification.* Primary 16T05, 16S40, 18D10.
The first author thanks the University of Almeria (Spain) for its warm hospitality.
Work supported by the project MTM2017-86987-P "Anillos, modulos y algebra de Hopf."

connection between the Majid and Bespalov-Drabant approaches of $U_q(\mathfrak{g})$ below, and we will relate these constructions with our present work, too.

Hopf bimodules over a Hopf algebra H were introduced by Woronowicz [**25**] under the name of bicovariant bimodules. They form a braided category ${}^H_H\mathcal{M}^H_H$, and according to Schauenburg [**22**], ${}^H_H\mathcal{M}^H_H$ is braided equivalent to the category of left Yetter-Drinfeld modules over H, ${}^H_H\mathcal{YD}$ (considered braided with the braiding coming from its identification to the left centre of the monoidal category ${}_H\mathcal{M}$ of H-representations). These results were generalized to Hopf algebras H in braided categories $(\mathcal{C}, c)$ by Bespalov and Drabant in [**1, 2**]; as in the ordinary case, the braided equivalence is provided by structure theorems for Hopf bimodules. Furthermore, Bespalov and Drabant showed that Hopf algebras in ${}^H_H\mathcal{C}^H_H$ (which are obviously in a one to one correspondence with the Hopf algebras in ${}^H_H\mathcal{YD}(\mathcal{C})$) can be characterized as Hopf algebras A in $\mathcal{C}$ that admit a Hopf algebra projection $A \to H$ in $\mathcal{C}$ or, equivalently, as (left) biproduct Hopf algebras $\mathfrak{A} \times H$ (a construction previously introduced by Radford for ordinary Hopf algebras, which gives the description of Hopf algebras with a projection; see [**20**]). With these categorical results in mind, the Majid and Bespalov-Drabant constructions mentioned above lead to the same Hopf algebra structure. Philosophically speaking, both constructions start with a Hopf algebra H in $\mathcal{C}$ and two braided Hopf algebras A, B in ${}^H_H\mathcal{YD}(\mathcal{C})$ and $\mathcal{YD}(\mathcal{C})^H_H$, respectively, related through a "pairing" $\langle , \rangle : A \otimes B \to \underline{1}$ in the Majid case, and through a morphism $\rho : B \otimes A \to \underline{1}$ satisfying certain conditions in the Bespalov-Drabant case ($\underline{1}$ is the unit object of $\mathcal{C}$). Then Majid gives a concrete Hopf algebra structure on the object $A \otimes H \otimes B$, and denotes it by $U(A, H, B)$. The technique used by Bespalov and Drabant is categorical: they consider $A \otimes H$ and $H \otimes B$ as Hopf algebras in ${}^H_H\mathcal{C}^H_H$, and then take the tensor product algebra and coalgebra $\underline{Z} := (A \otimes H) \otimes_H (H \otimes B) \simeq A \otimes H \otimes B$ in ${}^H_H\mathcal{C}^H_H$. Under a natural condition, imposed by the fact that ${}^H_H\mathcal{C}^H_H$ is not a symmetric category, $\underline{Z}$ is a Hopf algebra in ${}^H_H\mathcal{C}^H_H$ and, according to [**2**], we can associate to it a Hopf algebra Z in $\mathcal{C}$ this time, built also on $A \otimes H \otimes B$. Furthermore, we have a convolution invertible 2-cocycle $\widehat{\rho}$ of Z (built with the help of ρ), and the so called twisted Hopf algebra $Z_{\widehat{\rho}}$ associated to this pair (Z, ρ) is nothing but the Majid's Hopf algebra $U(A, H, B)$. For reasons that we next explain, so far we avoid the use of terminology "double-biproduct".

The main goal of this paper is to study when is $U(A, H, B)$ a biproduct Hopf algebra or, equivalently, a braided Hopf algebra with a projection. To achieve this goal we split the problem in two parts. The first part refers to the study of when Z as above is a biproduct Hopf algebra, while the second part refers to the study of when a 2-cocycle deformation of a biproduct Hopf algebra remains a biproduct Hopf algebra. Again, philosophically speaking, Z comes from a braided Hopf algebra $\underline{Z}$ within ${}^H_H\mathcal{C}^H_H$, so by the results in [**1, 2**] mentioned above Z should be isomorphic to a biproduct Hopf algebra. Thus the first problem is reduced to finding the braided Hopf algebra $\mathcal{A}$ in ${}^H_H\mathcal{YD}(\mathcal{C})$ responsible for the Hopf algebra isomorphism $Z \simeq \mathcal{A} \times H$ in $\mathcal{C}$. To this end we need first some braided isomorphisms, and this is the topic of Section 2; more exactly, we establish that the braided equivalence functors $\mathcal{YD}(\mathcal{C})^H_H \xrightarrow{\mathcal{F}_r} {}^H_H\mathcal{C}^H_H \xleftarrow{\mathcal{F}_l} {}^H_H\mathcal{YD}$ are connected through a braided functor $\overline{\mathcal{G}} : \mathcal{YD}(\mathcal{C})^H_H \to {}^H_H\mathcal{YD}(\mathcal{C})$, in the sense that $\mathcal{F}_l\overline{\mathcal{G}}$ and $\mathcal{F}_r$ are braided, monoidally isomorphic functors; see Theorem 2.4 and Corollary 2.5. This result allows us to show in Section 3 that a two-sided smash product (co)algebra is isomorphic to a left smash product (co)algebra, and therefore that a so called double-biproduct Hopf algebra is isomorphic to a left biproduct Hopf algebra. Thus we get almost for free that the desired braided Hopf algebra $\mathcal{A}$, denoted in what follows by $\mathcal{A}_H(A,B)$, is $A\widetilde{\otimes}\overline{B}$, where $\overline{B} := \overline{\mathcal{G}}(B)$ considered as Hopf algebra in ${}_H^H\mathcal{YD}$ with structure determined by the Hopf algebra structure of B in $\mathcal{YD}_H^H$ and the monoidal structure of $\overline{\mathcal{G}}$, and $A\widetilde{\otimes}\overline{B}$ is the tensor product Hopf algebra of A and $\overline{B}$ within ${}_H^H\mathcal{YD}(\mathcal{C})$. Summing up, $Z = A \times H \times B$, i.e. Z is a double-biproduct Hopf algebra and, moreover, $Z \simeq \mathcal{A}_H(A,B) \times H$, i.e. Z can be seen as a left biproduct Hopf algebra. This solves the first problem.

Section 4 deals with the second problem, imposed now by the fact that the Majid's or, equivalently, the Bespalov-Drabant construction identifies to a 2-cocycle deformation of the left biproduct Hopf algebra $\mathcal{A}_H(A,\overline{B}) \times H$. This problem has an elegant solution if the 2-cocycle ϑ_1 of a biproduct Hopf algebra, say $A \times H$, is determined by a 2-cocycle ϑ of the Hopf algebra $A\otimes H$ in ${}_H^H\mathcal{C}_H^H$ as in Proposition 4.1: in this case there exists a 2-cocycle ϑ_0 of A in ${}_H^H\mathcal{YD}$ such that $(A\times H)_{\vartheta_1} = A_{\vartheta_0} \times H$ is a left biproduct Hopf algebra. Unfortunately, in general, not all the (invertible) 2-cocycles of $A \times H$ are of this form, and such a situation appears precisely in the case studied by us, that of a double-biproduct Hopf algebra. Nevertheless, we consider 2-cocycles $\widehat{\sigma} = \varepsilon_{A\rtimes H} \otimes \sigma \otimes \varepsilon_{H\ltimes B}$ defined by a morphism $\sigma : B \otimes A \to \underline{1}$ in $\mathcal{C}$. Necessary and sufficient conditions for σ that turn $\widehat{\sigma}$ into a 2-cocycle of $Z = A\times H\times B$ are uncovered in Proposition 4.2, and, in the classical case, modulo a permutation of the factors of the monomial tensors, these are exactly the conditions required by Majid in [**17**]. Furthermore, in the proof of Proposition 4.2 we see that these conditions on σ are equivalent to those required by Bespalov and Drabant for their ρ, and this explains once more why the two constructions lead to the same Hopf algebra object. On the other hand, by using a given Hopf algebra isomorphism $\chi : Z \to \mathcal{A}_H(A,B) \times H$, we find a unique 2-cocycle ω of $\mathcal{A}_H(A,B) \times H$ such that $\chi : Z_{\widehat{\sigma}} \to (\mathcal{A}_H(A,B) \times H)_\omega$ remains a Hopf algebra isomorphism. Furthermore, ω is completely determined by a morphism $\Sigma : \overline{B} \otimes A \to \underline{1}$, and Σ and σ determine each other. By Proposition 4.3, ω is a 2-cocycle of $\mathcal{A}_H(A,B) \times H$ if and only if Σ is a so called almost invertible skew pairing of $\overline{B}$, A in ${}_H^H\mathcal{YD}$, and we said almost because Σ has almost all the necessary properties for a skew pairing in ${}_H^H\mathcal{YD}$, with one exception: Σ is not left H-colinear. Thus giving a "pairing" in the sense of Majid or a morphism ρ as in [**3**] is equivalent to giving an almost skew-pairing Σ of $\overline{B}$, A in ${}_H^H\mathcal{YD}$. When Σ is a skew pairing is uncovered in Lemma 4.4: the morphism σ that defines Σ should be H-cobalanced, a property dual to the H-balanced property for modules. If this is the case, as any skew pairing does, Σ defines a 2-cocycle ϑ_0 of $\mathcal{A}_H(A,B)$ in ${}_H^H\mathcal{YD}$, and so a 2-cocycle ϑ of $\mathcal{A}_H(A,B)\otimes H$ in ${}_H^H\mathcal{C}_H^H$. By Proposition 4.1, ϑ yields a 2-cocycle ϑ_1 of $\mathcal{A}_H(A,B) \times H$ such that $(\mathcal{A}_H(A,B) \times H)_{\vartheta_1} = \mathcal{A}_H(A,B)_{\vartheta_0} \times H$ as Hopf algebras in $\mathcal{C}$. It turns out that $\vartheta_1 = \omega$, and so when σ is, moreover, H-cobalanced the deformation by the 2-cocycle $\widehat{\sigma}$ of the double-biproduct Hopf algebra $A \times H \times B$ is isomorphic to a left biproduct Hopf algebra, as wished. Whether or not this result holds true for σ not H-cobalanced remains unclear at this moment. Note also that in the H-cobalanced case $Z_{\widehat{\sigma}}$ identifies to the biproduct Hopf algebra of the generalized quantum double $D_\Sigma(A,\overline{B})$ and H, see Theorem 4.6 and its proof.

Section 5, the last one, is dedicated to the presentation of some general examples and of some concrete ones that derive from them. After we specialize our results to ordinary Hopf algebras, we produce two classes of examples of double-biproduct

Hopf algebras; both are inspired by some examples presented in [**17**]. Namely, we start with a CQT Hopf algebra $(H, \mathcal{R})$ and two free unital algebras $\mathfrak{e}$ and $\mathfrak{f}$ on the sets $\{e_j, j \in J\}$, and respectively $\{f_i, i \in I\}$, such that we have maps from I, J to $G(H)$, the set of grouplike elements of H. Depending on the choice of these maps, we have various Hopf algebra structures for $\mathfrak{e}$ and $\mathfrak{f}$ in ${}^H_H\mathcal{YD}$ and $\mathcal{YD}^H_H$, respectively. Roughly speaking, if the maps are of the form $j \mapsto K_j$ and $i \mapsto K_i$ we can construct an H-cobalanced σ, and so in this case $(\mathfrak{e} \times H \times \mathfrak{f})_{\widehat{\sigma}}$ is isomorphic to a left biproduct Hopf algebra. Surprisingly, $(\mathfrak{e} \times H \times \mathfrak{f})_{\widehat{\sigma}} = \mathfrak{e} \times H \times \mathfrak{f}$, although all the structures involved are not trivial. In the second case we take maps of the form $j \mapsto K_j^{-1}$ and $i \mapsto K_i$, and this time we have a morphism σ that contributes to the Hopf algebra structure of $(\mathfrak{e} \times H \times \mathfrak{f})_{\widehat{\sigma}}$; unlike to the first case, σ is not H-cobalanced, and so we cannot conclude from our general results that $(\mathfrak{e} \times H \times \mathfrak{f})_{\widehat{\sigma}}$ is isomorphic to a left biproduct Hopf algebra. Nevertheless, the Drinfeld-Jimbo quantum group fits in our second case, thus $U_q(\mathfrak{g})$ can be also characterized as a deformation by an invertible 2-cocycle of a biproduct Hopf algebra of the form $\mathcal{A}_H('e^-, \mathfrak{f}) \times H$, built for some H, $A = '\mathfrak{e}$ and $\mathfrak{f}$ as in the above; for more details we refer to Theorem 5.19. Finally, we exemplify our second construction for the Hopf algebra H_ν, considered in [**21**] and equipped with the CQT structure from [**7, 13**]; see Example 5.17.

1. Preliminaries

1.1. Braided Hopf algebras. We assume the reader familiar with the basic theory and graphical calculus of braided monoidal categories; see [**5, 12, 18**]. In what follows, $\mathcal{C}$ is a (pre)braided category with tensor product $\otimes : \mathcal{C} \times \mathcal{C} \to \mathcal{C}$, unit object $\underline{1}$ and braiding $c : \otimes \to \otimes \circ \tau$, where $\tau : \mathcal{C} \times \mathcal{C} \to \mathcal{C} \times \mathcal{C}$ is the twist functor. We denote $c_{X,Y}$ by 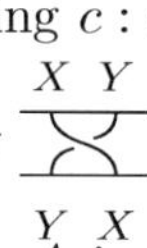

and its inverse $c^{-1}_{X,Y}$ by 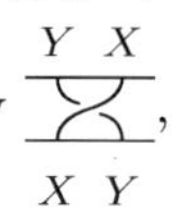

, for all $X, Y \in \mathcal{C}$.

For an algebra A in $\mathcal{C}$, the multiplication $m_A : A \otimes A \to A$ and the unit $\eta_A : \underline{1} \to A$ of A will be denoted by 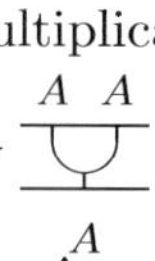

and 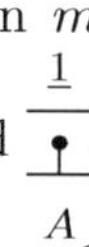

. Similarly, 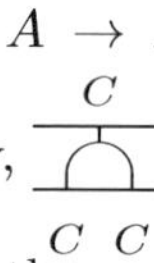

and 

will be our notation for the comultiplication $\Delta_C : C \to C \otimes C$ and the counit $\varepsilon_C : C \to \underline{1}$ of a coalgebra C in $\mathcal{C}$.

A bialgebra in a braided category $\mathcal{C}$ is a 5-tuple $(H, m_H, \eta_H, \Delta_H, \varepsilon_H)$, with (H, m_H, η_H) an algebra and $(H, \Delta_H, \varepsilon_H)$ a coalgebra in $\mathcal{C}$ such that $\Delta_H : H \to H \otimes H$ and $\varepsilon_H : H \to \underline{1}$ are algebra morphisms. Here $H \otimes H$ has the tensor product algebra structure and $\underline{1}$ is viewed as an algebra in $\mathcal{C}$ through the left or right unit constraints. For instance, the fact that Δ_H is multiplicative reads as

$$\Delta_H m_H = (m_H \otimes m_H)(\mathrm{Id}_H \otimes c_{H,H} \otimes \mathrm{Id}_H)(\Delta_H \otimes \Delta_H). \tag{1.1}$$

If H is a bialgebra in $\mathcal{C}$, the category ${}_H\mathcal{C}$ (resp. $\mathcal{C}_H$) of left (resp. right) H-modules in $\mathcal{C}$ is monoidal. The tensor product is the tensor product of $\mathcal{C}$ endowed with the left (resp. right) H-module structure given by (1.2), unit object $\underline{1}$ viewed as a left (resp. right) H-module via the counit ε_H of H, and the same constraint morphisms as those of $\mathcal{C}$;

$$\begin{array}{c} \mu^H_{X\otimes Y} = (\mu^H_X \otimes \mu^H_Y)((\mathrm{Id}_H \otimes c_{H,X})(\Delta_H \otimes \mathrm{Id}_X) \otimes \mathrm{Id}_Y) \\ (\text{resp. } \nu^H_{X\otimes Y} = (\nu^H_X \otimes \nu^H_Y)(\mathrm{Id}_X \otimes (c_{Y,H} \otimes \mathrm{Id}_H)(\mathrm{Id}_Y \otimes \Delta_H))). \end{array} \tag{1.2}$$

Here and in what follows $\mu_X^H =$ 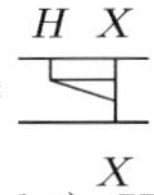 $: H \otimes X \to H$ (resp. $\nu_X^H =$ 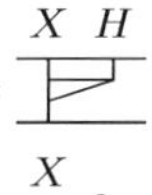 $: X \otimes H \to X$) stands for the left (resp right) H-module morphism structure of an object X of $\mathcal{C}$. By combining the structures in (1.2) we get that ${}_H\mathcal{C}_H$, the category of H-bimodules, is a monoidal category, too.

A Hopf algebra in a braided monoidal category $\mathcal{C}$ is a bialgebra H in $\mathcal{C}$ together with a morphism $S : H \to H$ in $\mathcal{C}$ (the antipode) satisfying the axioms

$$m_H(S \otimes \mathrm{Id}_H)\Delta_H = \eta_H\varepsilon_H = m_H(\mathrm{Id}_H \otimes S)\Delta_H. \tag{1.3}$$

It is well-known, see [**16**, Lemma 2.3], that the antipode S of a Hopf algebra H in a braided category $\mathcal{C}$ is an anti-morphism of algebras and coalgebras, that is

$$\begin{array}{ll}\text{(a)} & Sm_H = m_H(S \otimes S)c_{H,H}\ ,\ S\eta_H = \eta_H; \\ \text{(b)} & \Delta_H S = c_{H,H}(S \otimes S)\Delta_H\ ,\ \varepsilon_H S = \varepsilon_H.\end{array} \tag{1.4}$$

If the antipode S is an isomorphism in $\mathcal{C}$ we say that H is a Hopf algebra in $\mathcal{C}$ with skew antipode. In this case we will denote the composition inverse of S by $\overline{S}$.

1.2. Yetter-Drinfeld modules. One can introduce Yetter-Drinfeld modules by looking at the centre of a suitable monoidal category. More exactly, if H is a braided bialgebra then ${}_H\mathcal{C}$ is monoidal, and so we can consider $\mathcal{W}_l({}_H\mathcal{C})$, the left weak centre of ${}_H\mathcal{C}$; we refer to [**5**, Section 8.1] for the detailed construction of the left and right (weak) centre $(\mathcal{W}_{l/r}(\mathcal{D}))$ $\mathcal{Z}_{l/r}(\mathcal{D})$ of a monoidal category $\mathcal{D}$, previously introduced by Drinfeld (unpublished), Joyal and Street [**11**] and Majid [**15**].

An object in $\mathcal{W}_l({}_H\mathcal{C})$ consists of a left H-module V and a natural transformation $s_{V,-} = (s_{V,X} : V \otimes X \to X \otimes V)_{X \in {}_H\mathcal{C}}$, compatible with the monoidal structure of ${}_H\mathcal{C}$ in a natural way. In order to describe each component $s_{X,V}$ in terms of a certain left H-coaction $\lambda_V : V \to H \otimes V$, we need an extra assumption on $s_{V,-}$. Namely, any $X \in \mathcal{C}$ can be viewed as an object of ${}_H\mathcal{C}$ via the trivial H-action $\mu_X^{\mathrm{tr}} = \varepsilon_H \otimes \mathrm{Id}_X :\ H \otimes X \to X$; by X^{tr} we denote X equipped with this H-module structure in $\mathcal{C}$. Then the extra assumption we need is $s_{V,X^{\mathrm{tr}}} = c_{V,X}$, the braiding of $\mathcal{C}$, for all $X \in \mathcal{C}$. In what follows we denote by $\mathcal{WT}_l({}_H\mathcal{C})$ the full subcategory of $\mathcal{W}_l({}_H\mathcal{C})$ consisting of those objects $(V, s_{V,-})$ of $\mathcal{W}_l({}_H\mathcal{C})$ for which $s_{V,X^{\mathrm{tr}}} = c_{V,X}$, for all $X \in \mathcal{C}$. In a similar manner one can introduce $\mathcal{WT}_r(\mathcal{C}_H)$. Note that $\mathcal{WT}_l({}_H\mathcal{C})$ and $\mathcal{WT}_r(\mathcal{C}_H)$ are pre-braided monoidal categories (the braiding is not necessarily a natural isomorphism) with structure induced by that of $\mathcal{W}_l({}_H\mathcal{C})$ and $\mathcal{W}_r(\mathcal{C}_H)$.

By [**1**, Proposition 3.6.1], $\mathcal{WT}_l({}_H\mathcal{C})$ and $\mathcal{WT}_r(\mathcal{C}_H)$ identify to categories of H-Yetter-Drinfeld modules. In what follows, $\lambda_V^H =$ 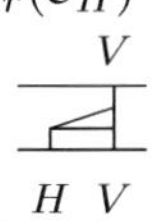 (resp. $\rho_W^H =$ 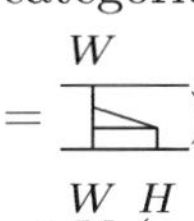) stands for the left (resp. right) H-comodule morphism structure of an object V (resp. W).

Definition 1.1. Let H be a bialgebra in $(\mathcal{C}, c)$. A left (resp. right) H-Yetter-Drinfeld module is a left (resp. right) H-module $\mathfrak{M}$ (resp. $\mathfrak{P}$) that is also a left (resp. right) H-comodule such that

$$\begin{array}{l}(m_H \otimes \mathrm{Id}_{\mathfrak{M}})(\mathrm{Id}_H \otimes c_{\mathfrak{M},H})(\lambda_{\mathfrak{M}}^H \otimes \mathrm{Id}_H)(\mu_{\mathfrak{M}}^H \otimes \mathrm{Id}_H)(\mathrm{Id}_H \otimes c_{H,\mathfrak{M}}) \\ \quad (\Delta_H \otimes \mathrm{Id}_{\mathfrak{M}}) = (m_H \otimes \mu_{\mathfrak{M}}^H)(\mathrm{Id}_H \otimes c_{H,H} \otimes \mathrm{Id}_{\mathfrak{M}})(\Delta_H \otimes \lambda_{\mathfrak{M}}^H) \\ (\text{resp. } (\mathrm{Id}_{\mathfrak{P}} \otimes m_H)(c_{H,\mathfrak{P}} \otimes \mathrm{Id}_H)(\mathrm{Id}_H \otimes \rho_{\mathfrak{P}}^H)(\mathrm{Id}_H \otimes \nu_{\mathfrak{P}}^H)(c_{\mathfrak{P},H} \otimes \mathrm{Id}_H) \\ \quad (\mathrm{Id}_{\mathfrak{P}} \otimes \Delta_H) = (\nu_{\mathfrak{P}}^H \otimes m_H)(\mathrm{Id}_{\mathfrak{P}} \otimes c_{H,H} \otimes \mathrm{Id}_H)(\rho_{\mathfrak{P}}^H \otimes \Delta_H)).\end{array} \tag{1.5}$$

The category of left (resp. right) H-Yetter-Drinfeld modules in $\mathcal{C}$ and left (resp. right) H-linear and colinear morphisms in $\mathcal{C}$ is denoted by ${}_H^H\mathcal{YD}(\mathcal{C})$ (resp. $\mathcal{YD}(\mathcal{C})_H^H$).

For $V \in {}_H^H\mathcal{YD}(H)$, define $s_{V,X} = (\mu_X^H \otimes \mathrm{Id}_V)(\mathrm{Id}_H \otimes c_{V,X})(\lambda_V \otimes \mathrm{Id}_X) : V \otimes X \to X \otimes V$, for all $X \in {}_H\mathcal{C}$. We then have that $(V, s_{V,-}) \in \mathcal{WT}_l({}_H\mathcal{C})$. Conversely, if $(V, s_{V,-}) \in \mathcal{WT}_l({}_H\mathcal{C})$ then $V \in {}_H\mathcal{C}$ and it is, moreover, a left H-Yetter-Drinfeld module with $\lambda_V^H = s_{V,H}(\mathrm{Id}_V \otimes \eta_H) : V \to H \otimes V$. Similarly, if $W \in \mathcal{YD}(H)_H^H$ then $(W, t_{-,W}) \in \mathcal{WT}_r(\mathcal{C}_H)$, where $t_{Y,W} = (\mathrm{Id}_W \otimes \nu_Y^H)(c_{Y,W} \otimes \mathrm{Id}_H)(\mathrm{Id}_Y \otimes \rho_W^H) : Y \otimes W \to W \otimes Y$, for all $Y \in \mathcal{C}_H$; conversely, if $(W, t_{-,W}) \in \mathcal{WT}_r(\mathcal{C}_H)$ then $W \in \mathcal{C}_H$ and it is a right H-Yetter-Drinfeld module with $\rho_W^H = t_{H,W}(\eta_H \otimes \mathrm{Id}_W) : W \to W \otimes H$.

As a consequence of the above facts, ${}_H^H\mathcal{YD}(H)$ (resp. $\mathcal{YD}(\mathcal{C})_H^H$) is a pre-braided monoidal category. The tensor product is the tensor product of $\mathcal{C}$, and if $V, V' \in {}_H^H\mathcal{YD}(\mathcal{C})$ (resp. $W, W' \in \mathcal{YD}(\mathcal{C})_H^H$) then $V \otimes V'$ (resp. $W \otimes W'$) is a left (resp. right) Yetter-Drinfeld module over H with structure determined by (1.2) and

$$
\begin{aligned}
&\lambda_{V\otimes V'}^H = (m_H \otimes \mathrm{Id}_{V\otimes V'})(\mathrm{Id}_H \otimes c_{V,H} \otimes \mathrm{Id}_{V'})(\lambda_V^H \otimes \lambda_{V'}^H)\\
&(\text{resp. } \rho_{W\otimes W'}^H = (\mathrm{Id}_{W\otimes W'} \otimes m_H)(\mathrm{Id}_W \otimes c_{H,W'} \otimes \mathrm{Id}_H)(\rho_W^H \otimes \rho_{W'}^H)).
\end{aligned}
\tag{1.6}
$$

All the other structures that define the monoidal structure of ${}_H^H\mathcal{YD}(\mathcal{C})$ (resp. $\mathcal{YD}(\mathcal{C})_H^H$) coincide with those of $\mathcal{C}$, and so the forgetful functor from ${}_H^H\mathcal{YD}(\mathcal{C})$ (resp. $\mathcal{YD}(\mathcal{C})_H^H$) to ${}_H\mathcal{C}$ (resp. $\mathcal{C}_H$) is a strong monoidal functor; we refer for instance to [**5**, Definition 1.22] for the meaning of this concept. Furthermore, ${}_H^H\mathcal{YD}(\mathcal{C})$ (resp. $\mathcal{YD}(\mathcal{C})_H^H$) is a pre-braided category with pre-braiding ${}^l\mathfrak{c}$ (resp. ${}^r\mathfrak{c}$) given, for all $V, V' \in {}_H^H\mathcal{YD}(\mathcal{C})$ (resp. $W, W' \in \mathcal{YD}(\mathcal{C})_H^H$), by

(1.7)

${}^l\mathfrak{c}_{V,V'} =$ 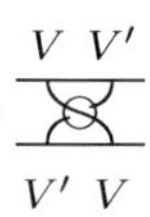 $:= (\mu_{V'}^H \otimes \mathrm{Id}_V)(\mathrm{Id}_H \otimes c_{V,V'})(\lambda_V^H \otimes \mathrm{Id}_{V'})$

(resp. ${}^r\mathfrak{c}_{W,W'} =$ 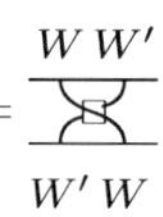 $:= (\mathrm{Id}_{W'} \otimes \nu_W^H)(c_{W,W'} \otimes \mathrm{Id}_H)(\mathrm{Id}_W \otimes \rho_{W'}^H)$).

If in the definition of $\mathcal{WT}_l({}_H\mathcal{C})$ we ask for an object $(V, s_{V,-})$ to be defined by $V \in {}_H\mathcal{C}$ and a natural isomorphism $s_{V,-} : V \otimes - \to - \otimes V$ then we get a new category, denoted by $\mathcal{ZT}_l({}_H\mathcal{C})$. In a similar way we introduce $\mathcal{ZT}_r(\mathcal{C}_H)$. It has been proved in [**1**] that ${}_H^H\mathcal{YD}(\mathcal{C}) = \mathcal{ZT}_l({}_H\mathcal{C})$ (resp. $\mathcal{YD}(\mathcal{C})_H^H = \mathcal{ZT}_r(\mathcal{C}_H)$), provided that H has skew antipode. Consequently, for H a Hopf algebra with skew antipode in $\mathcal{C}$, the category ${}_H^H\mathcal{YD}(\mathcal{C})$ (resp. $\mathcal{YD}(\mathcal{C})_H^H$) is braided, the inverse of the braiding being defined, for all $V, V' \in {}_H^H\mathcal{YD}(\mathcal{C})$ (resp. $W, W' \in \mathcal{YD}(\mathcal{C})_H^H$), by

(1.8)

${}^l\mathfrak{c}_{V,V'}^{-1} =$ 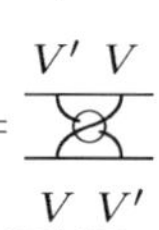 $:= c_{V,V'}^{-1}(\mu_{V'}^H c_{H,V'}^{-1}(\mathrm{Id}_{V'} \otimes \overline{S}) \otimes \mathrm{Id}_V)(\mathrm{Id}_{V'} \otimes \lambda_V^H)$

$({}^r\mathfrak{c}_{W,W'}^{-1} =$ 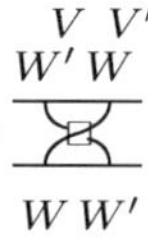 $:= c_{W,W'}^{-1}(\mathrm{Id}_{W'} \otimes \nu_W^H c_{W,H}^{-1}(\overline{S} \otimes \mathrm{Id}_W))(\rho_{W'}^H \otimes \mathrm{Id}_W))$.

Finally, if H is a Hopf algebra in $\mathcal{C}$ then (1.5) is equivalent to

$$
\begin{aligned}
\lambda_{\mathfrak{M}}^H \mu_{\mathfrak{M}}^H &= (m_H(\mathrm{Id}_H \otimes S) \otimes \mathrm{Id}_{\mathfrak{M}})(\mathrm{Id}_H \otimes c_{\mathfrak{M},H})\big((m_H \otimes \mu_{\mathfrak{M}}^H)\\
&\quad(\mathrm{Id}_H \otimes c_{H,H} \otimes \mathrm{Id}_{\mathfrak{M}})(\Delta_H \otimes \lambda_{\mathfrak{M}}^H) \otimes \mathrm{Id}_H\big)(\mathrm{Id}_H \otimes c_{H,\mathfrak{M}})(\Delta_H \otimes \mathrm{Id}_{\mathfrak{M}})\\
(\rho_{\mathfrak{P}}^H \nu_{\mathfrak{P}}^H &= (\mathrm{Id}_{\mathfrak{P}} \otimes m_H(S \otimes \mathrm{Id}_H))(c_{H,\mathfrak{P}} \otimes \mathrm{Id}_H)\big(\mathrm{Id}_H \otimes (\nu_{\mathfrak{P}}^H \otimes m_H)\\
&\quad(\mathrm{Id}_{\mathfrak{P}} \otimes c_{H,H} \otimes \mathrm{Id}_H)(\rho_{\mathfrak{P}}^H \otimes \Delta_H)\big)(c_{\mathfrak{P},H} \otimes \mathrm{Id}_H)(\mathrm{Id}_{\mathfrak{P}} \otimes \Delta_H)).
\end{aligned}
\tag{1.9}
$$

1.3. Hopf bimodules. Let H be a bialgebra in $(\mathcal{C}, c)$, so the category of H-bimodules ${}_H\mathcal{C}_H$ is monoidal and H is a coalgebra in ${}_H\mathcal{C}_H$. The category of H-Hopf bimodules in $\mathcal{C}$, denoted by ${}^H_H\mathcal{C}^H_H$, is the category of H-bicomodules in ${}_H\mathcal{C}_H$. Explicitly, an object of ${}^H_H\mathcal{C}^H_H$ is an H-bimodule which is at the same time an H-bicomodule such that the left and right coaction morphisms are left and right H-linear, where the H-module structures on a tensor product object are as in (1.2); a morphism in ${}^H_H\mathcal{C}^H_H$ is a left and right H-linear and colinear morphism in $\mathcal{C}$.

From now on, $\mathcal{C}$ is a braided category with (co)equalizers and such that any object of it is flat, coflat and robust. All these conditions are needed to endow canonically $M \otimes_H N$ with a Hopf H-bimodule structure, where for $M, N \in {}_H\mathcal{C}_H$ by $M \otimes_H N$ we denote the coequalizer of the morphisms $\nu^H_M \otimes \mathrm{Id}_N$, $\mathrm{Id}_M \otimes \mu^H_N$: $M \otimes H \otimes N \to M \otimes N$ in $\mathcal{C}$. We will explain this better in what follows.

Recall from [**4, 19, 23**] that an object X of a monoidal category $\mathcal{C}$ is called left (resp. right) coflat if the functor $- \otimes X : \mathcal{C} \to \mathcal{C}$ (resp. $X \otimes - : \mathcal{C} \to \mathcal{C}$) preserves coequalizers. Then X is called coflat if it left and right coflat. Furthermore, if A is an algebra in $\mathcal{C}$, $M \in \mathcal{C}_A$ is called right robust if and only if for any $P \in {}_A\mathcal{C}$ and $U \in \mathcal{C}$ the unique morphism $\theta_{M,P,U}$ closing commutatively the diagram

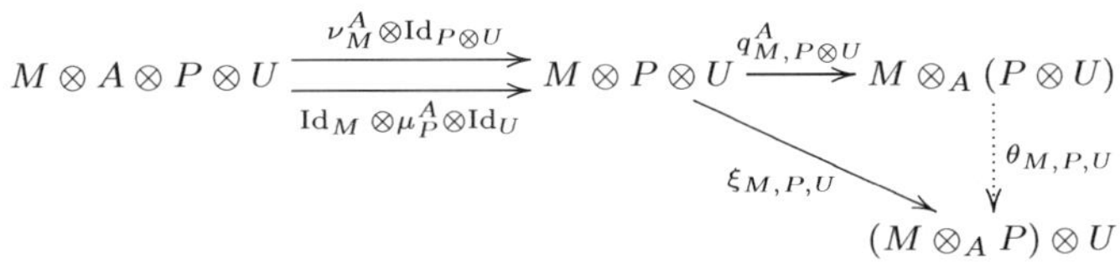

is an isomorphism in $\mathcal{C}$. Here $\xi_{M,P,U} := q^A_{M,P} \otimes \mathrm{Id}_U$ and $P \otimes U \in {}_A\mathcal{C}$ via $\mu^A_{P\otimes U} = \mu^A_P \otimes \mathrm{Id}_U$, and $\theta_{M,P,U}$ is given by the universality property of the coequalizer since $\xi_{M,P,U}(\nu^A_M \otimes \mathrm{Id}_{P\otimes U}) = \xi_{M,P,U}(\mathrm{Id}_M \otimes \mu^A_P \otimes \mathrm{Id}_U)$. Similarly, if A is an algebra in $\mathcal{C}$, we call $P \in {}_A\mathcal{C}$ left robust if for any $U \in \mathcal{C}$ and $M \in \mathcal{C}_A$ the morphism $\theta'_{U,M,P}$: $(U \otimes M) \otimes_A P \to U \otimes (M \otimes_A P)$ uniquely determined by $\theta'_{U,M,P} q^A_{U\otimes M,P} = \mathrm{Id}_U \otimes q^A_{M,P}$ is an isomorphism. We say that any object of $\mathcal{C}$ is robust if for any algebra A in $\mathcal{C}$, any $M \in \mathcal{C}_A$ and $P \in {}_A\mathcal{C}$ is right, and respectively left, robust.

The coflatness property allows to define left/right module structures on objects defined by $\otimes_H$. For instance, if H is left coflat, $M \in \mathcal{C}_H$ and $N \in {}_H\mathcal{C}_H$, then $\nu^H_{M\otimes_H N} : M \otimes_H N \otimes H \to M \otimes_H N$ determined by $\nu^H_{M\otimes_H N}(q^H_{M,N} \otimes \mathrm{Id}_H) = q^H_{M,N}(\mathrm{Id}_M \otimes \nu^H_N)$ turns $M \otimes_H N$ into an object of $\mathcal{C}_H$; likewise, $M \otimes_H N \in {}_H\mathcal{C}$ provided that H is right coflat and $M \in {}_H\mathcal{C}_H$, $N \in {}_H\mathcal{C}$. The robustness property yields left and right H-comodule structures. Namely, for $M \in \mathcal{C}_H$ and $N \in {}_H\mathcal{C}$, if $M \in {}^H\mathcal{C}$ then so is $M \otimes_H N$ with $\lambda^H_{M\otimes_H N} = \theta'_{H,M,N}(\lambda^H_M \otimes_H \mathrm{Id}_N)$ and if $N \in \mathcal{C}^H$ then so is $M \otimes_H N$ with $\rho^H_{M\otimes_H N} = \theta_{M,N,H}(\mathrm{Id}_M \otimes_H \rho^H_N)$. Here and in what follows, if $\zeta : P \to Q$ is in ${}_H\mathcal{C}$, $\tilde{\zeta} = \mathrm{Id}_M \otimes_A \zeta : M \otimes_H P \to M \otimes_H Q$ is the morphism satisfying $\tilde{\zeta} q^H_{M,P} = q^H_{M,Q}(\mathrm{Id}_M \otimes \zeta)$, and if $\phi : M \to P$ is in $\mathcal{C}_H$ then $\hat{\phi} = \phi \otimes_H \mathrm{Id}_N$: $M \otimes_H N \to P \otimes_H N$ is the morphism obeying $\hat{\phi} q^H_{M,N} = q^H_{P,N}(\phi \otimes \mathrm{Id}_N)$.

Under the coflatness and robustness conditions, ${}^H_H\mathcal{C}^H_H$ is monoidal with tensor product $\otimes_H$, unit object H, associativity constraint determined, for all $M, T, P \in {}_H\mathcal{C}_H$, by $\Sigma'_{M,T,P} : (M \otimes_H T) \otimes_H P \to M \otimes_H (T \otimes_H P)$ uniquely defined by $\Sigma'_{M,T,P} q^H_{M\otimes_H T,P}(q^H_{M,T} \otimes \mathrm{Id}_P) = q^H_{M,T\otimes_H P}(\mathrm{Id}_M \otimes q^H_{T,P})$, and left and right unit constraints given by the canonical isomorphisms $\Upsilon_M : M \otimes_A A \xrightarrow{\cong} M$ and Υ'_Y :

$A \otimes_A Y \xrightarrow{\cong} Y$ uniquely determined by $\Upsilon_M q^A_{M,A} = \nu^A_M$ and $\Upsilon'_Y q^A_{A,Y} = \mu^A_Y$, respectively.

A detailed proof for the next result can be found in [**1**].

PROPOSITION 1.2. *Let H be a Hopf algebra with skew antipode in a braided category $\mathcal{C}$. Then the categories $\mathcal{YD}(\mathcal{C})^H_H$ and ${}^H_H\mathcal{C}^H_H$ are monoidally equivalent.*

PROOF. Owing to [**1**, Theorem 3.9.6], the monoidal equivalence is produced by the following functors.

• If $\mathfrak{P} \in \mathcal{YD}(\mathcal{C})^H_H$, $H \otimes \mathfrak{P} \in {}^H_H\mathcal{C}^H_H$ with structure defined by

$\mu^H_{H\otimes\mathfrak{P}} =$ 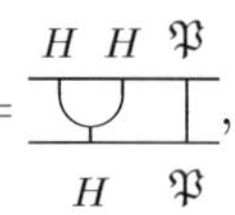, $\nu^H_{H\otimes\mathfrak{P}} =$ 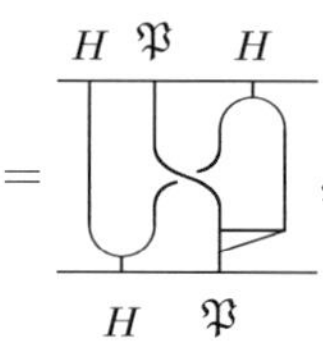, $\lambda^H_{H\otimes\mathfrak{P}} =$ 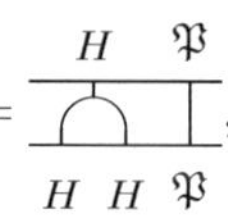, $\rho^H_{H\otimes\mathfrak{P}} =$ 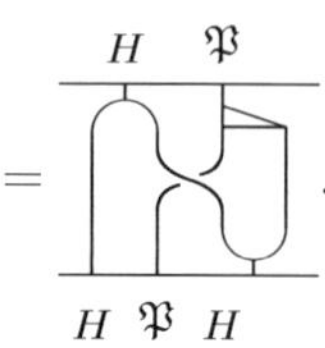.

So we have a functor $\mathcal{F}_r : \mathcal{YD}(\mathcal{C})^H_H \to {}^H_H\mathcal{C}^H_H$; $\mathcal{F}_r(\mathfrak{P}) = H \otimes \mathfrak{P}$ with structure as above, and $\mathcal{F}_r$ sends $f : \mathfrak{P} \to \mathfrak{Q}$ from $\mathcal{YD}(\mathcal{C})^H_H$ to $\mathrm{Id}_H \otimes f$, a morphism in ${}^H_H\mathcal{C}^H_H$.

• Set $M \in {}^H_H\mathcal{C}^H_H$ and consider the following equalizer in $\mathcal{C}$:

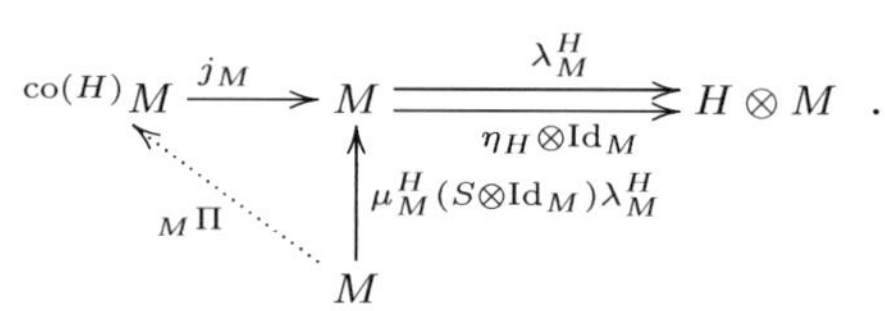

Then there is a unique morphism ${}_M\Pi : M \to {}^{\mathrm{co}(H)}M$ in $\mathcal{C}$ such that $j_M {}_M\Pi = \mu^H_M(S \otimes \mathrm{Id}_M)\lambda^H_M$. Thus ${}^{\mathrm{co}(H)}M \in \mathcal{YD}^H_H$ with right action and coaction defined by

$$j_M \circ \nu^H_{{}^{\mathrm{co}(H)}M} = \mu^H_M(S \otimes \nu^H_M)(c_{M,H} \otimes \mathrm{Id}_H)(\mathrm{Id}_M \otimes \Delta_H)(j_M \otimes \mathrm{Id}_H),$$
$$(j_M \otimes \mathrm{Id}_H) \circ \rho^H_{{}^{\mathrm{co}(H)}M} = \rho^H_M \circ j_M.$$

So we have a functor $\mathcal{G}_r : {}^H_H\mathcal{C}^H_H \to \mathcal{YD}(\mathcal{C})^H_H$; $\mathcal{G}_r$ sends $M \xrightarrow{g} N$ in $\mathcal{G}_r(g)$ defined by $j_N\mathcal{G}_r(g) = gj_M$.

The monoidal structure of $\mathcal{F}_r$ is given, for all $\mathfrak{P}, \mathfrak{Q} \in \mathcal{YD}(\mathcal{C})^H_H$, by $\varphi^r_{2,\mathfrak{P},\mathfrak{Q}} : (H \otimes \mathfrak{P}) \otimes_H (H \otimes \mathfrak{Q}) \to H \otimes \mathfrak{P} \otimes \mathfrak{Q}$ with

$$\varphi^r_{2,\mathfrak{P},\mathfrak{Q}} q^H_{H\otimes\mathfrak{P},H\otimes\mathfrak{Q}} = (m_H \otimes \nu^H_{\mathfrak{P}})(\mathrm{Id}_H \otimes (c_{\mathfrak{P},H} \otimes \mathrm{Id}_H)(\mathrm{Id}_{\mathfrak{P}} \otimes \Delta_H)) \otimes \mathrm{Id}_{\mathfrak{Q}},$$

and $\varphi^r_0 = \mathrm{Id}_H : H \to \mathcal{F}_r(\underline{1}) \simeq H$. The monoidal structure (ϕ^r_2, ϕ^r_0) of $\mathcal{G}_r$ is given by the canonical epimorphisms, $j_{M\otimes_H N}\phi^r_{2,M,N}(j_M \otimes j_N) = q^H_{M,N}$ for all $M, N \in {}^H_H\mathcal{C}^H_H$, and $\phi^r_0 = \mathrm{Id}_{\underline{1}} : \underline{1} \to \mathcal{G}_r(H) \simeq \underline{1}$. Moreover, for $M \in {}^H_H\mathcal{C}^H_H$, $\zeta_M := \mu^H_M(\mathrm{Id}_H \otimes j_M) : H \otimes {}^{\mathrm{co}(H)}M \to M$ is an isomorphism in ${}^H_H\mathcal{C}^H_H$ with inverse $\zeta^{-1}_M = (\mathrm{Id}_H \otimes {}_M\Pi)\lambda^H_M : M \to H \otimes {}^{\mathrm{co}(H)}M$. Thus $(\zeta_M)_{M \in {}^H_H\mathcal{C}^H_H}$ defines a natural strong isomorphism between $\mathcal{F}_r\mathcal{G}_r$ and the identity functor of ${}^H_H\mathcal{C}^H_H$. Likewise, the family $(\vartheta_{\mathfrak{P}} : {}^{\mathrm{co}(H)}(H \otimes \mathfrak{P}) \to \mathfrak{P})_{\mathfrak{P} \in \mathcal{YD}(\mathcal{C})^H_H}$ with $\vartheta_{\mathfrak{P}}$ given by $\vartheta_{\mathfrak{P}} j_{H\otimes\mathfrak{P}} = \varepsilon_H \otimes \mathrm{Id}_{\mathfrak{P}}$ yields a natural strong monoidal isomorphism ϑ between $\mathcal{G}_r\mathcal{F}_r$ and identity functor of $\mathcal{YD}(\mathcal{C})^H_H$. □

$\mathcal{F}_r : \mathcal{YD}(\mathcal{C})^H_H \to {}^H_H\mathcal{C}^H_H$ is a strong monoidal equivalence and $\mathcal{YD}(\mathcal{C})^H_H$ is braided. By [**5**, Proposition 1.59], there is a unique braiding d on ${}^H_H\mathcal{C}^H_H$ that turns $\mathcal{F}_r$ into a braided equivalence: for all $M, N \in {}^H_H\mathcal{C}^H_H$,

$$d_{M,N} = (\zeta_M \otimes_H \zeta_N)\varphi^{-1}_{2,M,N}\mathcal{F}_r(\mathfrak{c}_{{}^{\mathrm{co}(H)}M,{}^{\mathrm{co}(H)}N})\varphi^r_{2,{}^{\mathrm{co}(H)}M,{}^{\mathrm{co}(H)}N}(\zeta^{-1}_M \otimes_H \zeta^{-1}_N),$$

where the notation is as in the proof of Proposition 1.2. By [**2**, Theorem 4.3.1],

(1.10) $d_{M,N}q^H_{M,N} = q^H_{N,M}(\mathrm{Id}_N \otimes j_{M\,M}\Pi)(\mu^H_N \otimes \nu^H_M)(\mathrm{Id}_H \otimes c_{M,N} \otimes \mathrm{Id}_H)(\lambda^H_M \otimes \rho^H_N).$

2. Some braided isomorphisms

As it was pointed out in [**3**], we have a mirror-reversed version of Proposition 1.2. We will present it in detail in what follows, as well as the braided isomorphism ${}^H_H\mathcal{YD}(\mathcal{C}) \simeq \mathcal{YD}(\mathcal{C})^H_H$ induced by it. So we have a bijective correspondence between bialgebras/Hopf algebras in ${}^H_H\mathcal{YD}(\mathcal{C})$ and $\mathcal{YD}(\mathcal{C})^H_H$.

Let $\mathcal{C} = (\mathcal{C}, \otimes, \underline{1}, a, l, r)$ be a monoidal category. Then $\overline{\mathcal{C}} = (\mathcal{C}, \overline{\otimes} = \otimes\circ\tau, \underline{1}, \overline{a}, \overline{l} = r, \overline{r} = l)$ is as well a monoidal category, where $\overline{a}_{X,Y,Z} = a^{-1}_{Z,Y,X}$, for all $X, Y, Z \in \mathcal{C}$, and τ is the switch functor. $\overline{\mathcal{C}}$ is called the reverse monoidal category associated to $\mathcal{C}$. Note that $(\overline{\mathcal{C}}, \overline{c})$ is braided with $\overline{c}_{X,Y} = c_{Y,X}$, for all $X, Y \in \mathcal{C}$, provided that $(\mathcal{C}, c)$ is braided. Also, the identity functor of $\mathcal{C}$ is braided as a functor from $(\mathcal{C}, c)$ to $(\overline{\mathcal{C}}, \overline{c})$; its strong monoidal structure is given by $\varphi_0 = \mathrm{Id}_{\underline{1}}$ and $\varphi_2 = (c_{Y,X})_{X,Y\in\mathcal{C}}$ (see [**5**, Example 1.59]). Consequently, to any Hopf algebra H in $\mathcal{C}$ corresponds a Hopf algebra $\overline{H}$ in $\overline{\mathcal{C}}$. More exactly, the Hopf algebra structure of $\overline{H}$ in $\overline{\mathcal{C}}$ is determined by $m_{\overline{H}} = m_H \circ c_{H,H}$, $\eta_{\overline{H}} = \eta_H$, $\Delta_{\overline{H}} = c^{-1}_{H,H} \circ \Delta_H$, $\varepsilon_{\overline{H}} = \varepsilon_H$ and antipode S. Hence $\overline{H}$ identifies to the Hopf algebra $H^{\mathrm{op}+,\mathrm{cop}-}$ in $\mathcal{C}$, the object H with the algebra and coalgebra structure of $H^{\mathrm{op}+}$ and $H^{\mathrm{cop}-}$ as in [**5**, Proposition 2.48].

Lemma 2.1. *For $\mathcal{C}, \overline{\mathcal{C}}$, $H, \overline{H}$ as above, ${}^{\overline{H}}_{\overline{H}}\overline{\mathcal{C}}^{\overline{H}}_{\overline{H}} \simeq {}^H_H\mathcal{C}^H_H$ as braided categories.*

Proof. The left $\overline{H}$-modules in $\overline{\mathcal{C}}$ are actually the right $H^{\mathrm{op}+}$-modules in $\mathcal{C}$, so ${}_{\overline{H}}\overline{\mathcal{C}} \simeq \mathcal{C}_{H^{\mathrm{op}+}}$. Similarly, a right $\overline{H}$-module in $\overline{\mathcal{C}}$ is just a left $H^{\mathrm{op}+}$-module in $\mathcal{C}$, so $\overline{\mathcal{C}}_{\overline{H}} \simeq {}_{H^{\mathrm{op}+}}\mathcal{C}$. Dually, ${}^{\overline{H}}\overline{\mathcal{C}} \simeq \mathcal{C}^{H^{\mathrm{cop}-}}$ and $\overline{\mathcal{C}}^{\overline{H}} \simeq {}^{H^{\mathrm{cop}-}}\mathcal{C}$, and therefore ${}^{\overline{H}}_{\overline{H}}\overline{\mathcal{C}}^{\overline{H}}_{\overline{H}} \simeq {}^{H^{\mathrm{op}+,\mathrm{cop}-}}_{H^{\mathrm{op}+\mathrm{cop}-}}\mathcal{C}^{H^{\mathrm{op}+\mathrm{cop}-}}_{H^{\mathrm{op}+\mathrm{cop}-}}$ as ordinary categories.

By [**5**, Propositions 2.32 and 2.37], ${}_H\mathcal{C} \simeq \mathcal{C}_{H^{\mathrm{op}+}}$ and ${}^H\mathcal{C} \simeq \mathcal{C}^{H^{\mathrm{cop}-}}$ (for the second isomorphism we must consider in *loc. cit.* $\mathcal{C}^{\mathrm{in}}$ in place of $\mathcal{C}$, where $\mathcal{C}^{\mathrm{in}} = \mathcal{C}$ as monoidal category, viewed braided with $c^{\mathrm{in}}_{X,Y} = c^{-1}_{Y,X}$, for all $X, Y \in \mathcal{C}$). By replacing $\mathcal{C}$ with $\overline{\mathcal{C}}$, $\mathcal{C}_H \simeq {}_{H^{\mathrm{op}+}}\mathcal{C}$ and $\mathcal{C}^H \simeq {}^{H^{\mathrm{cop}-}}\mathcal{C}$. So we have correspondences,

$$
\begin{aligned}
{}_{\overline{H}}\overline{\mathcal{C}} \ni (M, \mu^{\overline{H}}_M) &\mapsto (M, \mu^H_M = \mu^{\overline{H}}_M c^{-1}_{M,H}) \in {}_H\mathcal{C};\\
\overline{\mathcal{C}}_{\overline{H}} \ni (M, \nu^{\overline{H}}_M) &\mapsto (M, \nu^H_M = \nu^{\overline{H}}_M c^{-1}_{H,M}) \in \mathcal{C}_H;\\
{}^{\overline{H}}\overline{\mathcal{C}} \ni (M, \lambda^{\overline{H}}_M) &\mapsto (M, \lambda^H_M = c_{M,H}\lambda^{\overline{H}}_M) \in {}^H\mathcal{C};\\
\overline{\mathcal{C}}^{\overline{H}} \ni (M, \rho^{\overline{H}}_M) &\mapsto (M, \rho^H_M = c_{H,M}\rho^{\overline{H}}_M) \in \mathcal{C}^H,
\end{aligned}
$$

which define an isomorphism ${}^{\overline{H}}_{\overline{H}}\overline{\mathcal{C}}^{\overline{H}}_{\overline{H}} \simeq {}^H_H\mathcal{C}^H_H$. For instance, $\lambda^{\overline{H}}_M \in {}^{\overline{H}}\overline{\mathcal{C}}$ reads in $\mathcal{C}$ as

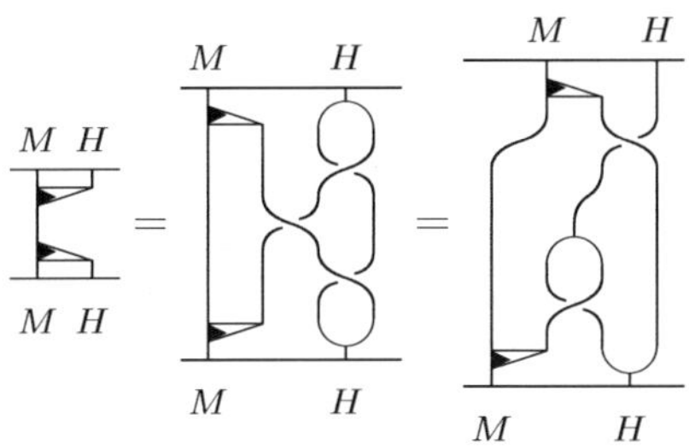

,

where we denoted $\mu_M^{\overline{H}} =$ 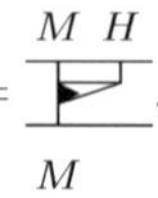, $\nu_M^{\overline{H}} =$ 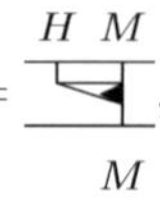, $\lambda_M^{\overline{H}} =$ 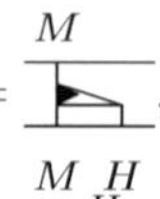, $\rho_M^{\overline{H}} =$ 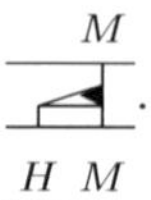.

With the help of the above equality we see that λ_M^H is left H-linear in $\mathcal{C}$:

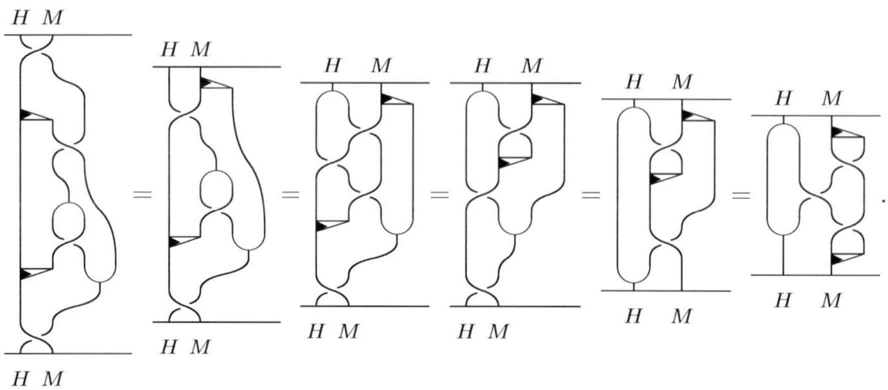

The other compatibilities can be proved similarly. In this way we have a functor $\mathcal{F} : {}^{\overline{H}}_{\overline{H}}\mathcal{C}^{\overline{H}}_{\overline{H}} \to {}^H_H\mathcal{C}^H_H$, and the composition ${}^{\overline{H}}_{\overline{H}}\mathcal{C}^{\overline{H}}_{\overline{H}} \simeq \overline{{}^{H^{\mathrm{op+,cop-}}}_{H^{\mathrm{op+cop-}}}\mathcal{C}^{H^{\mathrm{op+cop-}}}_{H^{\mathrm{op+cop-}}}} \simeq {}^H_H\mathcal{C}^H_H$ equals $\mathcal{F}$. So $\mathcal{F}$ is strong monoidal with $\varphi_0 = \mathrm{Id}_H$ and $\varphi_{2,M,N}$, $M, N \in {}^{\overline{H}}_{\overline{H}}\mathcal{C}^{\overline{H}}_{\overline{H}}$, defined by the identifications of the coequalizers in the diagram below

$$\begin{array}{ccccc} M\,\overline{\otimes}\,\overline{H}\,\overline{\otimes}\,N & \overset{\nu_M^{\overline{H}}\,\overline{\otimes}\,\mathrm{Id}_N}{\underset{\mathrm{Id}_M\,\overline{\otimes}\,\mu_N^{\overline{H}}}{\rightrightarrows}} & M\,\overline{\otimes}\,N & \xrightarrow{q^{\overline{H}}_{M,N}} & M\,\overline{\otimes}_{\overline{H}}\,N \\ \| & & \| & & \| \\ N\otimes H^{\mathrm{op+cop-}}\otimes M & \overset{\mu_N^{\overline{H}}\otimes \mathrm{Id}_M}{\underset{\mathrm{Id}_N\otimes\nu_M^{\overline{H}}}{\rightrightarrows}} & N\otimes M & \xrightarrow{q^{H^{\mathrm{op+cop-}}}_{N,M}} & N\otimes_{H^{\mathrm{op+cop-}}}M \\ {\scriptstyle (c_{H,M}\otimes \mathrm{Id}_N)(\mathrm{Id}_H\otimes c_{N,M})(c_{N,H}\otimes\mathrm{Id}_M)}\downarrow\ \simeq & & \simeq\ \downarrow{\scriptstyle c_{N,M}} & & \simeq\ \uparrow{\scriptstyle\varphi_{2,M,N}} \\ M\otimes H\otimes N & \overset{\nu_M^H\otimes\mathrm{Id}_N}{\underset{\mathrm{Id}_M\otimes\mu_N^H}{\rightrightarrows}} & M\otimes N & \xrightarrow{q^H_{M,N}} & M\otimes_H N \end{array}.$$

We show that $\mathcal{F}$ is a braided functor. Indeed, if Π_N is defined by the equalizer

$$\begin{array}{ccccc} N^{\mathrm{co}(H)} & \xrightarrow{\iota_N} & N & \overset{\rho_N^H}{\underset{\mathrm{Id}_N\otimes\eta_H}{\rightrightarrows}} & N\otimes H \\ & {\scriptstyle\Pi_N}\nwarrow & \uparrow{\scriptstyle \nu_N^H(\mathrm{Id}_N\otimes S)\rho_N^H} & & \\ & & N & & \end{array} \tag{2.1}$$

then $q^H_{M,N} \circ d_{M,N}$ equals

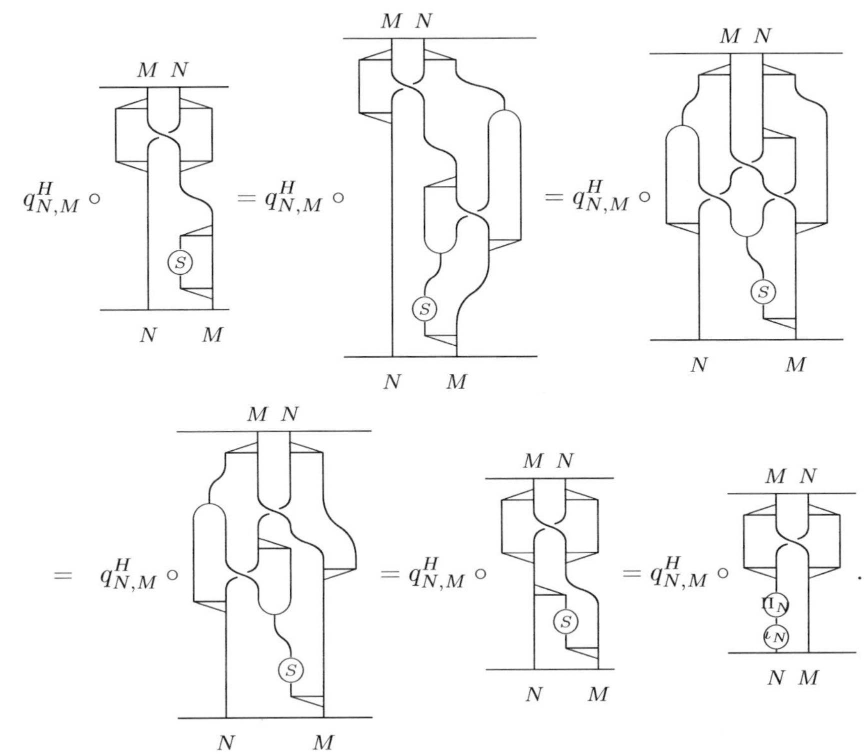

If $\overline{d}$ is the braiding of ${}^{\overline{H}}_{\overline{H}}\overline{\mathcal{C}}^{\overline{H}}_{\overline{H}}$, by applying the previous results to $\overline{H} \in \overline{\mathcal{C}}$ we get

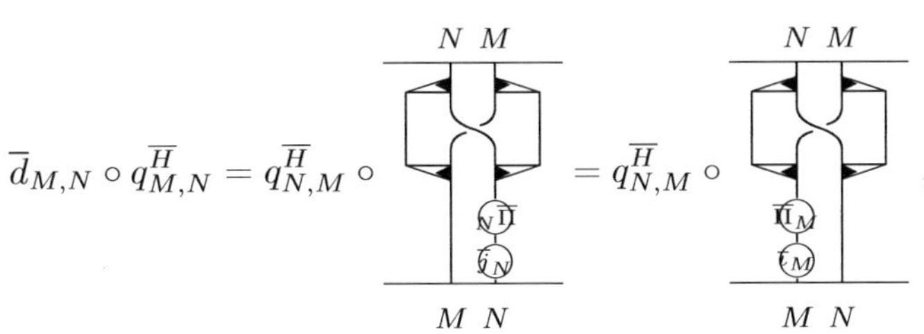

where the compositions are made in $\mathcal{C}$ rather than $\overline{\mathcal{C}}$; parallel to ${}_N\Pi$ and Π_M in $\mathcal{C}$ we have ${}_N\overline{\Pi}$ and $\overline{\Pi}_M$. Actually, a simple inspection shows that $\overline{\iota}_M \overline{\Pi}_M = j_M {}_M\Pi$, and therefore, for all $M, N \in {}_{H}^{\overline{H}}\overline{\mathcal{C}}_{\overline{H}}^{\overline{H}}$, $\varphi_{2,N,M} \circ d_{M,N} \circ q^{H}_{M,N}$ equals

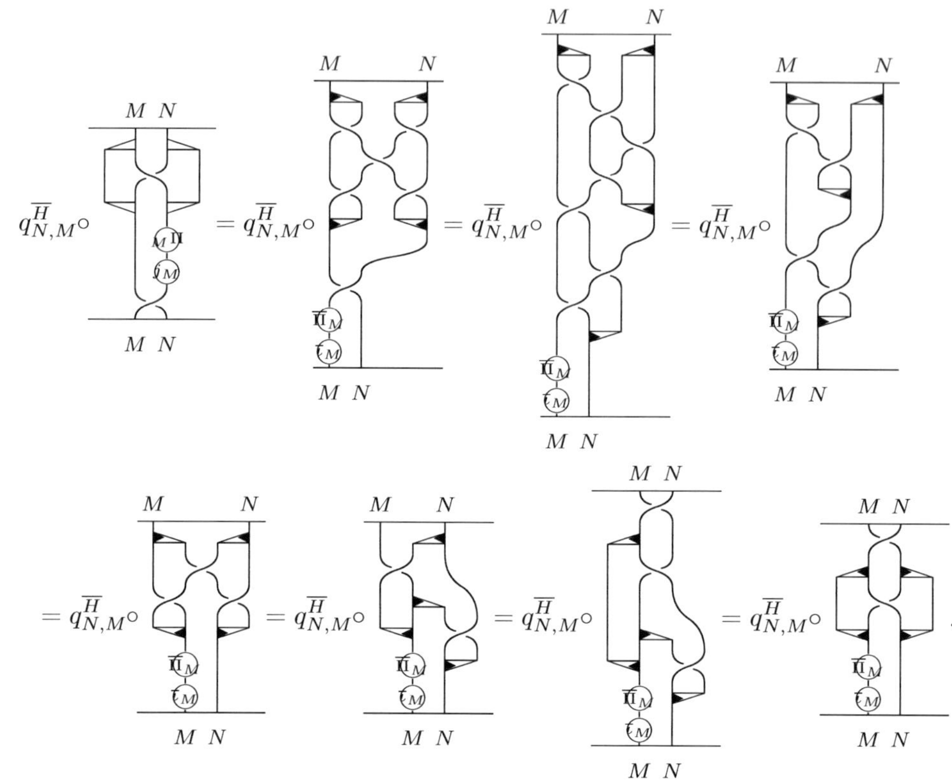

The latter is just $\overline{d}_{M,N} \circ \varphi_{2,M,N} \circ q^{H}_{M,N}$, as desired. □

The next results will be intensively used in what follows.

LEMMA 2.2. *If H has skew antipode, $\overline{{}_{H}^{H}\mathcal{YD}(\mathcal{C})} \simeq \mathcal{YD}(\mathcal{C})_{H}^{H}$ as (pre-)braided categories.*

PROOF. Owing to [**5**, Proposition 8.6], if $\mathcal{D}$ is a monoidal category then the reverse monoidal category associated to the left weak center of $\mathcal{D}$, $\overline{\mathcal{W}_l(\mathcal{D})}$, is pre-braided isomorphic to the right weak center of the reverse monoidal category associated to $\mathcal{D}$, $\mathcal{W}_r(\overline{\mathcal{D}})$. This isomorphism entails to a pre-braided isomorphism between $\overline{{}_{H}^{H}\mathcal{YD}(\mathcal{C})} = \overline{\mathcal{WT}_l({}_{H}\mathcal{C})}$ and $\mathcal{WT}_r(\overline{{}_{H}\mathcal{C}}) \cong \mathcal{WT}_r(\mathcal{C}_{H^{\rm op+cop-}}) = \mathcal{YD}(\mathcal{C})_{H^{\rm op+cop-}}^{H^{\rm op+cop-}}$, where the monoidal isomorphism between $\overline{{}_{H}\mathcal{C}}$ and $\mathcal{C}_{H^{\rm op+cop-}}$ is given by the identity functor, considered strong monoidal with $(c_{X,Y} : X \otimes Y \to X \overline{\otimes} Y = Y \otimes X)_{X,Y \in {}_{H}\mathcal{C}}$ and $\mathrm{Id}_{\underline{1}}$. Note that X is a right module over $H^{\rm op+cop-}$ in $\mathcal{C}$ with $\mu^{H}_{M} \circ c_{X,H}$, and so $c_{X,Y}$ is right $H^{\mathrm{op+cop-}}$-linear in $\mathcal{C}$ since

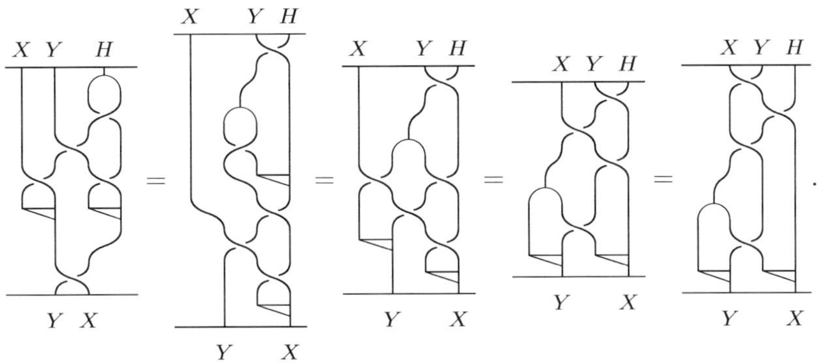

Thus we proved that $\overline{{}^H_H\mathcal{YD}(\mathcal{C})}$ and $\mathcal{YD}(\mathcal{C})^{H^{\mathrm{op+cop-}}}_{H^{\mathrm{op+cop-}}}$ are braided isomorphic. Now, the equalities in (1.4) express the fact that $S:\ H\to H^{\mathrm{op+cop-}}$ is a Hopf algebra isomorphism in $\mathcal{C}$. Consequently, by using [**5**, Propositions 2.34 and 2.39], we get that $\mathcal{YD}(\mathcal{C})^{H^{\mathrm{op+cop-}}}_{H^{\mathrm{op+cop-}}}$ and $\mathcal{YD}(\mathcal{C})^H_H$ are braided isomorphic, from where we conclude that $\overline{{}^H_H\mathcal{YD}(\mathcal{C})}$ and $\mathcal{YD}(\mathcal{C})^H_H$ are braided isomorphic, too.

Concretely, the functor $\mathcal{F}:\ \overline{{}^H_H\mathcal{YD}(\mathcal{C})}\to\mathcal{YD}(\mathcal{C})^H_H$ and its inverse $\mathcal{G}$ act as identity on morphisms, and on objects are defined by the correspondences

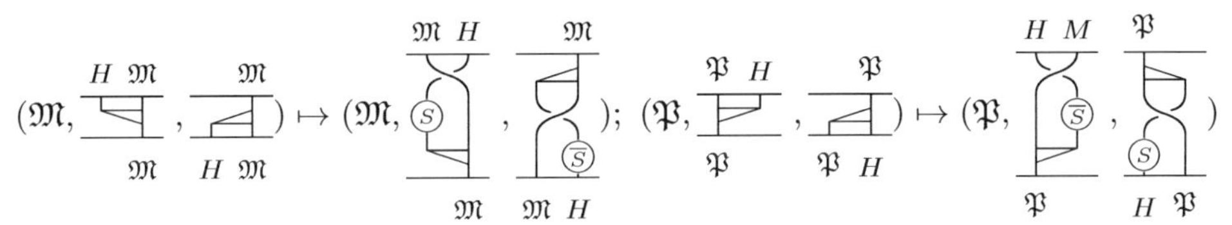

The monoidal structure of $\mathcal{F}$ is given by $(c_{\mathfrak{M},\mathfrak{N}})_{\mathfrak{M},\mathfrak{N}\in{}^H_H\mathcal{YD}(\mathcal{C})}$ and $\mathrm{Id}_{\underline{1}}$. □

PROPOSITION 2.3. *Let H be a Hopf algebra in $(\mathcal{C},c)$ with skew antipode. Then the categories ${}^H_H\mathcal{YD}(\mathcal{C})$ and ${}^H_H\mathcal{C}^H_H$ are braided equivalent.*

PROOF. We proved that $\overline{{}^H_H\mathcal{YD}(\mathcal{C})}\simeq\mathcal{YD}(\mathcal{C})^{H^{\mathrm{op+cop-}}}_{H^{\mathrm{op+cop-}}}$ and $\overline{{}^{H^{\mathrm{op+,cop-}}}_{H^{\mathrm{op+cop-}}}\mathcal{C}^{H^{\mathrm{op+cop-}}}_{H^{\mathrm{op+cop-}}}}\simeq{}^H_H\mathcal{C}^H_H$ as braided categories. We apply now Proposition 1.2 and the comments made after it to $(\mathcal{C},c)$ and $H^{\mathrm{op+cop-}}$ to deduce that $\mathcal{YD}(\mathcal{C})^{H^{\mathrm{op+cop-}}}_{H^{\mathrm{op+cop-}}}$ and ${}^{H^{\mathrm{op+cop-}}}_{H^{\mathrm{op+cop-}}}\mathcal{C}^{H^{\mathrm{op+cop-}}}_{H^{\mathrm{op+cop-}}}$ are braided equivalent categories. It follows now that $\overline{{}^H_H\mathcal{YD}(\mathcal{C})}$ and $\overline{{}^H_H\mathcal{C}^H_H}$ are braided equivalent, so the same are ${}^H_H\mathcal{YD}(\mathcal{C})$ and ${}^H_H\mathcal{C}^H_H$. Furthermore, this braided equivalence is produced by the following pair of functors $(\mathcal{F}_l,\mathcal{G}_l)$:

• the functor $\mathcal{F}_l:{}^H_H\mathcal{YD}(\mathcal{C})\to{}^H_H\mathcal{C}^H_H$ maps $f:\mathfrak{M}\to\mathfrak{N}$ in ${}^H_H\mathcal{YD}(\mathcal{C})$ to $f\otimes\mathrm{Id}_H$ and sends $\mathfrak{M}\in{}^H_H\mathcal{YD}(\mathcal{C})$ to $\mathfrak{M}\otimes H$, viewed as object in ${}^H_H\mathcal{C}^H_H$ with

$$\mu^H_{\mathfrak{M}\otimes H}=(\mu^H_{\mathfrak{M}}\otimes m_H)((\mathrm{Id}_H\otimes c_{H,\mathfrak{M}})(\Delta_H\otimes\mathrm{Id}_{\mathfrak{M}})\otimes\mathrm{Id}_H),\ \nu^H_{\mathfrak{M}\otimes H}=\mathrm{Id}_{\mathfrak{M}}\otimes m_H;$$
$$\lambda^H_{\mathfrak{M}\otimes H}=(m_H\otimes\mu^H_{\mathfrak{M}})(\mathrm{Id}_H\otimes c_{\mathfrak{M},H}\otimes\mathrm{Id}_H)(\lambda^H_{\mathfrak{M}}\otimes\Delta_H),\ \rho^H_{\mathfrak{M}\otimes H}=\mathrm{Id}_{\mathfrak{M}}\otimes\Delta_H.$$

• $\mathcal{G}_l:{}^H_H\mathcal{C}^H_H\to{}^H_H\mathcal{YD}(\mathcal{C})$ sends $N\in{}^H_H\mathcal{C}^H_H$ to $N^{\mathrm{co}(H)}$ and $g:M\to N$ in ${}^H_H\mathcal{C}^H_H$ to $\mathcal{G}_l(g):M^{\mathrm{co}(H)}\to N^{\mathrm{co}(H)}$ given by $\iota_N\mathcal{G}_l(g)=g\iota_M$; $N^{\mathrm{co}(H)}\in{}^H_H\mathcal{YD}(\mathcal{C})$ via $\iota_N\mu^H_{\mathcal{G}_l(N)}=\nu^H_N(\mu^H_N\otimes S)(\mathrm{Id}_N\otimes c_{H,N})(\Delta_H\otimes\mathrm{Id}_N)(\mathrm{Id}_H\otimes\iota_N)$, $(\mathrm{Id}_H\otimes\iota_N)\circ\lambda^H_{\mathcal{G}_l(N)}=\lambda^H_N\circ\iota_N$.

The natural isomorphism between $\mathcal{F}_l\mathcal{G}_l$ and the identity functor of ${}^H_H\mathcal{C}^H_H$ is given by $\overline{\zeta}_M = \nu^H_M(\iota_M \otimes \mathrm{Id}_H) : M^{\mathrm{co}(H)} \otimes H \to M$, $M \in {}^H_H\mathcal{C}^H_H$; $\overline{\zeta}_M$ is an isomorphism in ${}^H_H\mathcal{C}^H_H$ with inverse $\overline{\zeta}_M^{-1} : M \to M^{\mathrm{co}(H)} \otimes H$ determined by $\overline{\zeta}_M^{-1} = (\Pi_M \otimes \mathrm{Id}_H)\rho^H_M$. Likewise, the family $(\overline{\vartheta}_{\mathfrak{M}} : (\mathfrak{M} \otimes H)^{\mathrm{co}(H)} \to \mathfrak{M})_{\mathfrak{M}\in {}^H_H\mathcal{YD}(\mathcal{C})}$ determined by $\overline{\vartheta}_{\mathfrak{M}}\iota_{\mathfrak{M}\otimes H} = \mathrm{Id}_{\mathfrak{M}} \otimes \varepsilon_H$ yields a natural strong monoidal isomorphism between $\mathcal{G}_l\mathcal{F}_l$ and the identity functor of ${}^H_H\mathcal{YD}(\mathcal{C})$. Here $\mathcal{F}_l$ is strong monoidal with $\varphi^l_{2,\mathfrak{M},\mathfrak{N}} : (\mathfrak{M} \otimes H) \otimes_H (\mathfrak{N} \otimes H) \to (\mathfrak{M} \otimes \mathfrak{N}) \otimes H$ given, for all $\mathfrak{M}, \mathfrak{N} \in {}^H_H\mathcal{YD}(\mathcal{C})$, by $q^H_{\mathfrak{M}\otimes H,\mathfrak{N}\otimes H}\varphi^l_{2,\mathfrak{M},\mathfrak{N}} = \mathrm{Id}_{\mathfrak{M}} \otimes (\mu^H_{\mathfrak{N}} \otimes m_H)((\mathrm{Id}_H \otimes c_{H,\mathfrak{N}})(\Delta_H \otimes \mathrm{Id}_{\mathfrak{N}}) \otimes \mathrm{Id}_H)$ and $\varphi^l_0 = \mathrm{Id}_H$. The verification of all these details is left to the reader. □

The functors $\mathcal{F}_l$ and $\mathcal{F}_r$ can be related.

THEOREM 2.4. *Let $(\mathcal{C}, c)$ be a braided category and H a Hopf algebra in $\mathcal{C}$ with skew antipode. If $\mathcal{G} : \mathcal{YD}(\mathcal{C})^H_H \to {}^H_H\mathcal{YD}(\mathcal{C})$ is the functor built in Lemma 2.2 and $\mathcal{YD}(\mathcal{C})^H_H \xrightarrow{\mathcal{F}_r} {}^H_H\mathcal{C}^H_H \xleftarrow{\mathcal{F}_l} {}^H_H\mathcal{YD}(\mathcal{C})$ are the braided equivalence functors defined in Propositions 1.2 and 2.3 then $\mathcal{F}_r$ and $\mathcal{F}_l\mathcal{G}$ are isomorphic.*

PROOF. For $\mathfrak{P} \in \mathcal{YD}(\mathcal{C})^H_H$ and $H \otimes \mathfrak{P} = \mathcal{F}_r(\mathfrak{P}) \in {}^H_H\mathcal{C}^H_H$, consider the equalizer

$$(H \otimes \mathfrak{P})^{\mathrm{co}(H)} \xrightarrow{\iota_{H\otimes\mathfrak{P}}} H \otimes \mathfrak{P} \underset{\mathrm{Id}_{H\otimes\mathfrak{P}} \otimes \eta_H}{\overset{\rho^H_{H\otimes\mathfrak{P}}}{\rightrightarrows}} H \otimes \mathfrak{P} \otimes H \tag{2.2}$$

and $\psi_{\mathfrak{P}} = (\mathrm{Id}_H \otimes \nu^H_{\mathfrak{P}})(c_{\mathfrak{P},H} \otimes \mathrm{Id}_H)(\mathrm{Id}_{\mathfrak{P}} \otimes \Delta_H \circ S)\rho^H_{\mathfrak{P}} : \mathfrak{P} \to H \otimes \mathfrak{P}$. We have

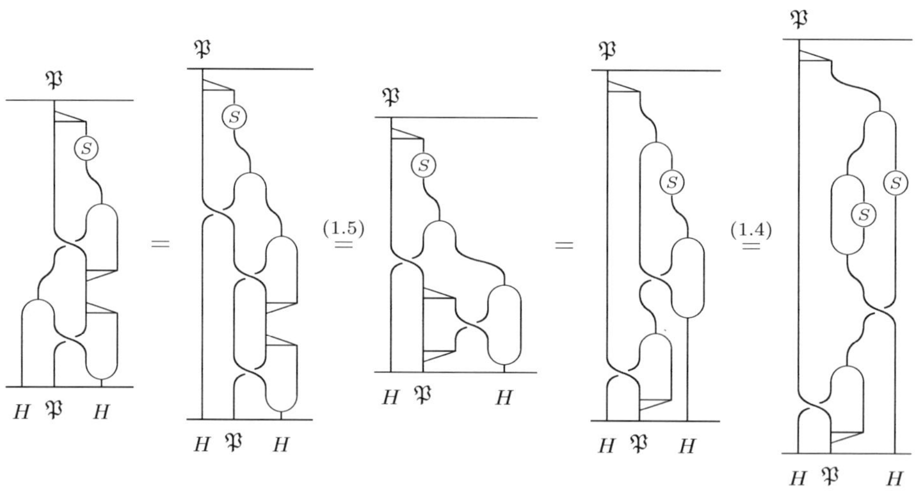

and so by (1.3) we get $\rho^H_{H\otimes\mathfrak{P}}\psi_{\mathfrak{P}} = (\mathrm{Id}_{H\otimes\mathfrak{P}} \otimes \eta_H)\psi_{\mathfrak{P}}$. The equalizer in (2.2) yields now $\Psi_{\mathfrak{P}} : \mathfrak{P} \to \mathcal{G}_r(\mathfrak{P})$ in $\mathcal{C}$, defined by $\iota_{H\otimes\mathfrak{P}}\Psi_{\mathfrak{P}} = \psi_{\mathfrak{P}}$. We claim that $\Psi^{-1}_{\mathfrak{P}} : (H \otimes \mathfrak{P})^{\mathrm{co}(H)} \to \mathfrak{P}$ given by $\Psi^{-1}_{\mathfrak{P}} = \nu^H_{\mathfrak{P}}(\mathrm{Id}_{\mathfrak{P}} \otimes \overline{S})c^{-1}_{H,\mathfrak{P}}\iota_{H\otimes\mathfrak{P}}$ is inverse for $\Psi_{\mathfrak{P}}$.

In the computation below we use: the naturality of c applied to $\Delta_H \circ \overline{S}$ in the fourth equality, applied to $\rho^H_{\mathfrak{P}}$ and Δ_H in the fifth equality, applied to $\nu^H_{\mathfrak{P}}$ and $c_{H,\mathfrak{P}}$ in the sixth equality, applied to Δ_H and to m_H after we used that it is associative in the seventh equality, and in the tenth equality the coassociativity of Δ_H followed by the naturality of c applied to it. So we have:

$\iota_{H\otimes\mathfrak{P}}\Psi_{\mathfrak{P}}\Psi_{\mathfrak{P}}^{-1} = \psi_{\mathfrak{P}}\nu_{\mathfrak{P}}^{H}(\mathrm{Id}_{\mathfrak{P}}\otimes\overline{S})c_{H,\mathfrak{P}}^{-1}\iota_{H\otimes\mathfrak{P}}$

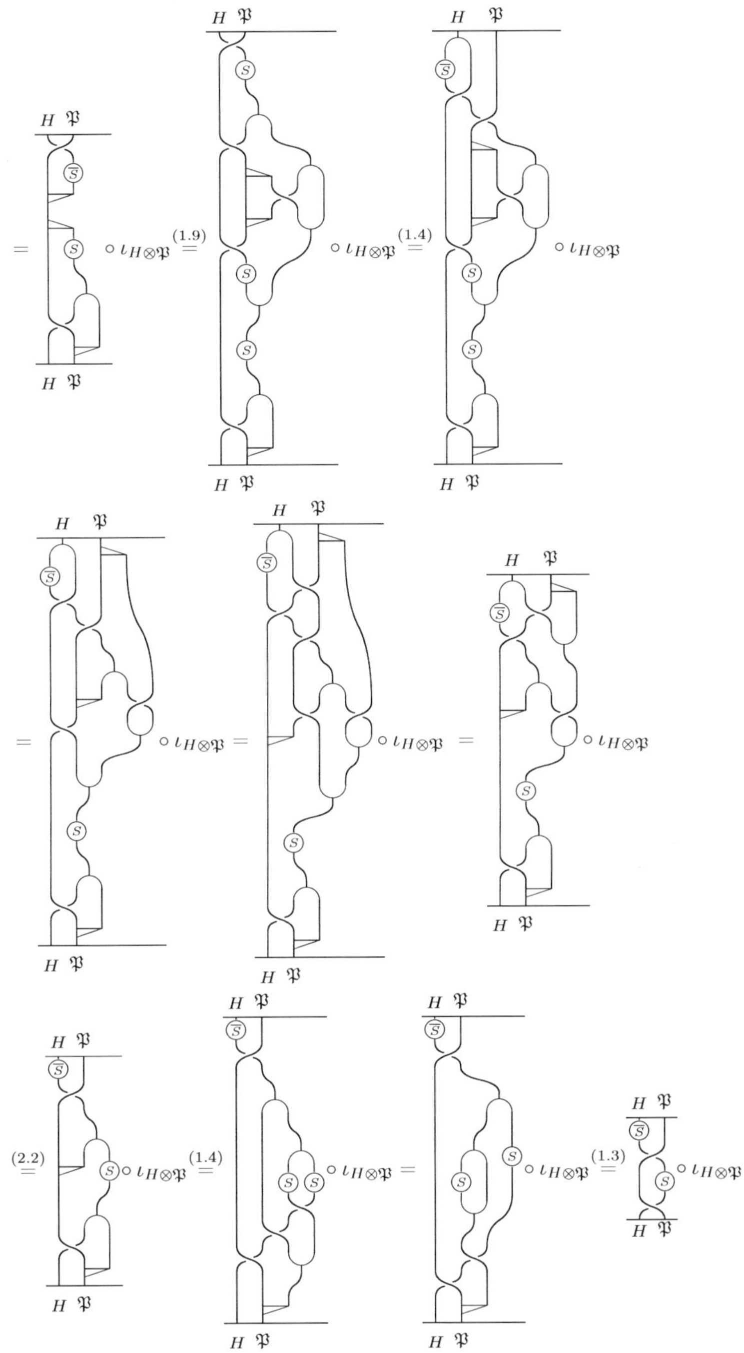

and the latter is equal to $\iota_{H\otimes\mathfrak{P}}$, as needed. That $\Psi_{\mathfrak{P}}^{-1}$ is as well a left inverse of $\Psi_{\mathfrak{P}}$ follows more easily by applying the naturality of c to $\nu_{\mathfrak{P}}^{H}$ and then (1.3).

It is clear at this point that there is a unique structure on $\mathfrak{P}$ that turns the isomorphism $\Psi_{\mathfrak{P}} : \mathfrak{P} \to \mathcal{G}_l\mathcal{F}_r(\mathfrak{P})$ in $\mathcal{C}$ into an isomorphism in ${}^{H}_{H}\mathcal{YD}(\mathcal{C})$. If we denote by $\underline{\mathfrak{P}}$ the object $\mathfrak{P}$ endowed with this left Yetter-Drinfeld module structure over H, we state that $\underline{\mathfrak{P}} = \mathcal{G}(\mathfrak{P})$. This will finish the proof since the results proved so far entail to $\mathcal{F}_l\mathcal{G}(\mathfrak{P}) \simeq \mathcal{F}_l\mathcal{G}_l\mathcal{F}_r(\mathfrak{P}) \simeq \mathcal{F}_r(\mathfrak{P})$ in ${}^H_H\mathcal{C}^H_H$.

Keeping in mind the structure of $\mathcal{G}(\mathfrak{P})$ in ${}^H_H\mathcal{YD}(\mathcal{C})$ coming from that of $\mathfrak{P}$ in $\mathcal{YD}(\mathcal{C})^H_H$ we deduce that $\Psi_{\mathfrak{P}} : \mathcal{G}(\mathfrak{P}) \to (H \otimes \mathfrak{P})^{\mathrm{co}(H)}$ is left H-linear in $\mathcal{C}$.

Indeed, by applying for several times the naturality of the braiding c and the (co)associativity of the morphisms involved in the computation below, we see that $\iota_{H\otimes\mathfrak{P}}\Psi_{\mathfrak{P}}\mu^H_{\underline{\mathfrak{P}}}$ is equal to

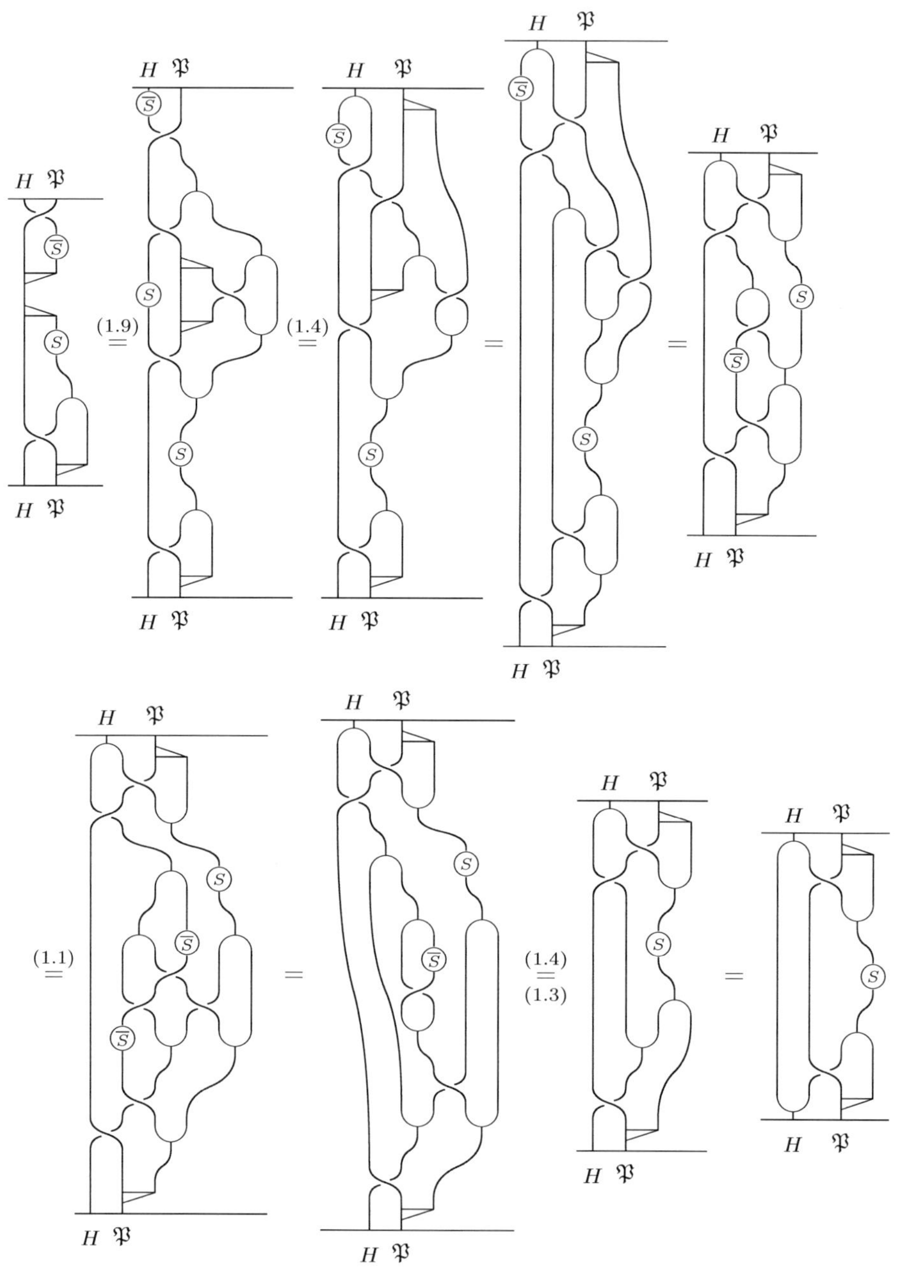

On the other hand, we use in $(*_1)$ and $(*_2)$ below the naturality of c applied twice to $\nu_{\mathfrak{P}}^H$ and m_H, respectively, to see that

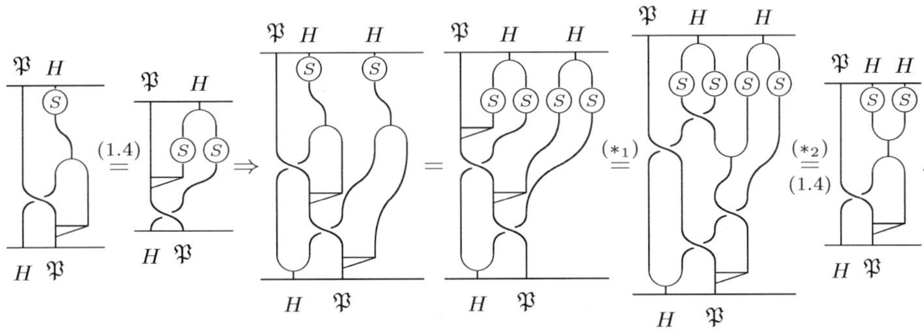

With the help of the above equality and (1.4) we compute that (the second equality uses the naturality of c applied to $(\mathrm{Id}_H \otimes \nu_{\mathfrak{P}}^H)(c_{\mathfrak{P},H} \otimes \mathrm{Id}_H)(\mathrm{Id}_{\mathfrak{P}} \otimes \Delta_H \circ S)$)

$$\iota_{H\otimes\mathfrak{P}}\mu^H_{\mathcal{G}_l(\mathfrak{P})}(\mathrm{Id}_H \otimes \Psi_{\mathfrak{P}}) =$$

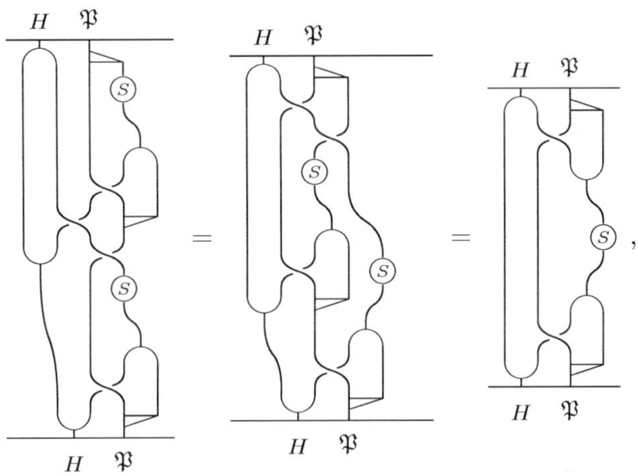

as needed. Finally, $(\mathrm{Id}_H \otimes \iota_{H\otimes\mathfrak{P}}\Psi_{\mathfrak{P}})\lambda^H_{\underline{\mathfrak{P}}}$ equals

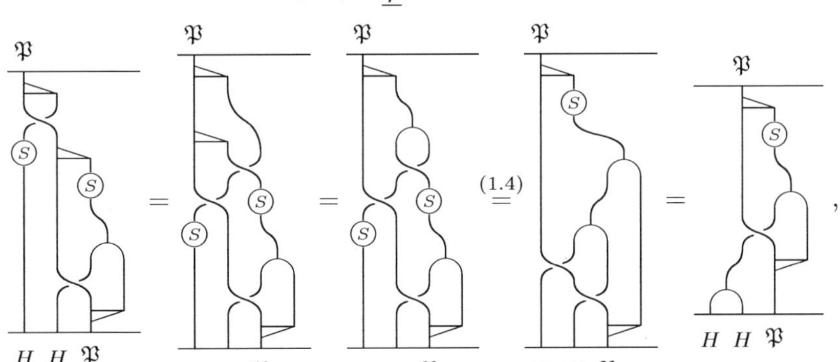

and therefore $\Psi_{\mathfrak{P}} : \underline{\mathfrak{P}} \to \mathcal{G}_l(H \otimes \mathfrak{P})$ is left H-colinear. So the proof is finished. $\square$

The functors $\mathcal{F}_l\mathcal{G}$, $\mathcal{F}_r$ are not strong monoidally isomorphic, because $\mathcal{G}$ is strong monoidal as functor from $\mathcal{YD}(\mathcal{C})^H_H$ to $\overline{{}^H_H\mathcal{YD}(\mathcal{C})}$ and not to ${}^H_H\mathcal{YD}(\mathcal{C})$. This is why we replace $\mathcal{G}$ with $\overline{\mathcal{G}} : \mathcal{YD}(\mathcal{C})^H_H \xrightarrow{\mathcal{G}} \overline{{}^H_H\mathcal{YD}(\mathcal{C})} \xrightarrow{\mathcal{J}} {}^H_H\mathcal{YD}(\mathcal{C})$. $\mathcal{J}$ is the identity functor, considered braided with $(c_{\mathfrak{M},\mathfrak{N}})_{\mathfrak{M},\mathfrak{N}\in {}^H_H\mathcal{YD}(\mathcal{C})}$ and $\mathrm{Id}_{\underline{1}}$.

COROLLARY 2.5. *The functors $\mathcal{F}_l\overline{\mathcal{G}}$ and $\mathcal{F}_r$ are monoidally isomorphic.*

PROOF. Let $\omega := (\omega_{\mathfrak{P}} := \overline{\zeta}_{H\otimes\mathfrak{P}}(\Psi_{\mathfrak{P}} \otimes \mathrm{Id}_H) : \underline{\mathfrak{P}} \otimes H \to H \otimes \mathfrak{P})_{\mathfrak{P}\in\mathcal{YD}(\mathcal{C})^H_H}$. So

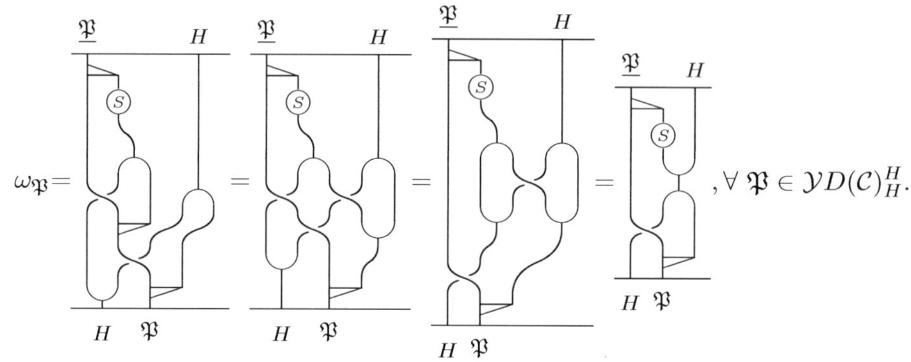

We show that ω is a monoidal natural transformation from $\mathcal{F}_l\overline{\mathcal{G}}$ to $\mathcal{F}_r$, i.e. $\omega_{\underline{1}}\phi_0 = \varphi_0^r$ and $\omega_{\mathfrak{P}\otimes\mathfrak{Q}}\phi_{2,\underline{\mathfrak{P}},\underline{\mathfrak{Q}}} = \varphi^r_{2,\mathfrak{P},\mathfrak{Q}}(\omega_{\mathfrak{P}} \otimes \omega_{\mathfrak{Q}})$, for $\mathfrak{P}, \mathfrak{Q} \in \mathcal{YD}(\mathcal{C})^H_H$; by (ϕ_2, ϕ_0) we denoted the monoidal structure of $\mathcal{F}_l\overline{\mathcal{G}}$: $\phi_0 = \mathrm{Id}_{\underline{1}}$ and $\phi_{2,\underline{\mathfrak{P}},\underline{\mathfrak{Q}}}q^H_{\underline{\mathfrak{P}}\otimes H,\underline{\mathfrak{Q}}\otimes H}$ equals

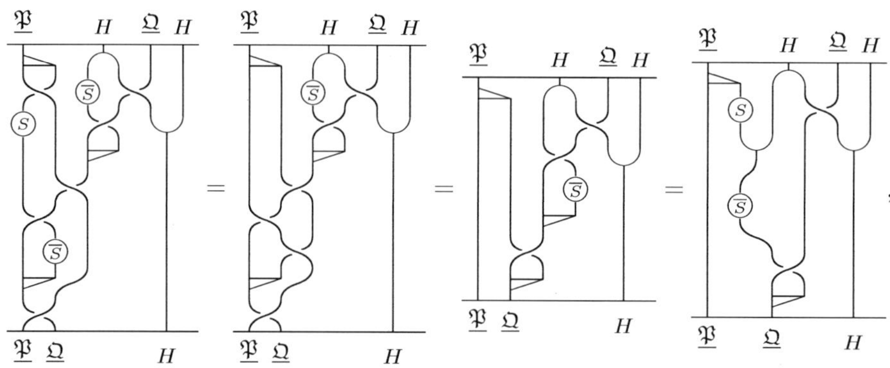

for all $\mathfrak{P}, \mathfrak{Q} \in \mathcal{YD}(\mathcal{C})^H_H$. Together with the equality

(2.3)

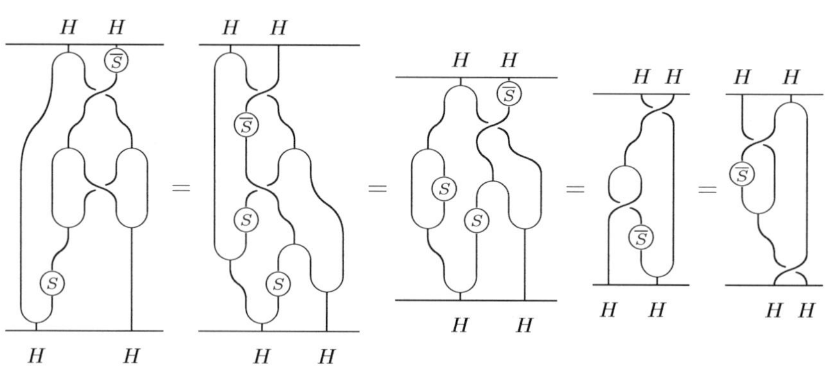

allows to see that $\omega_{\mathfrak{P}\otimes\mathfrak{Q}}\phi_{2,\underline{\mathfrak{P}},\underline{\mathfrak{Q}}} = \varphi^r_{2,\mathfrak{P},\mathfrak{Q}}(\omega_{\mathfrak{P}} \otimes \omega_{\mathfrak{Q}})$, for all $\mathfrak{P}, \mathfrak{Q} \in \mathcal{YD}(\mathcal{C})^H_H$ (we will not perform all the computations but we will point out the relations that are used at each step). Indeed, we have

$\omega_{\mathfrak{P}\otimes\mathfrak{Q}}\phi_{2,\underline{\mathfrak{P}},\underline{\mathfrak{Q}}}q^{H}_{\mathfrak{P}\otimes H,\mathfrak{Q}\otimes H}$

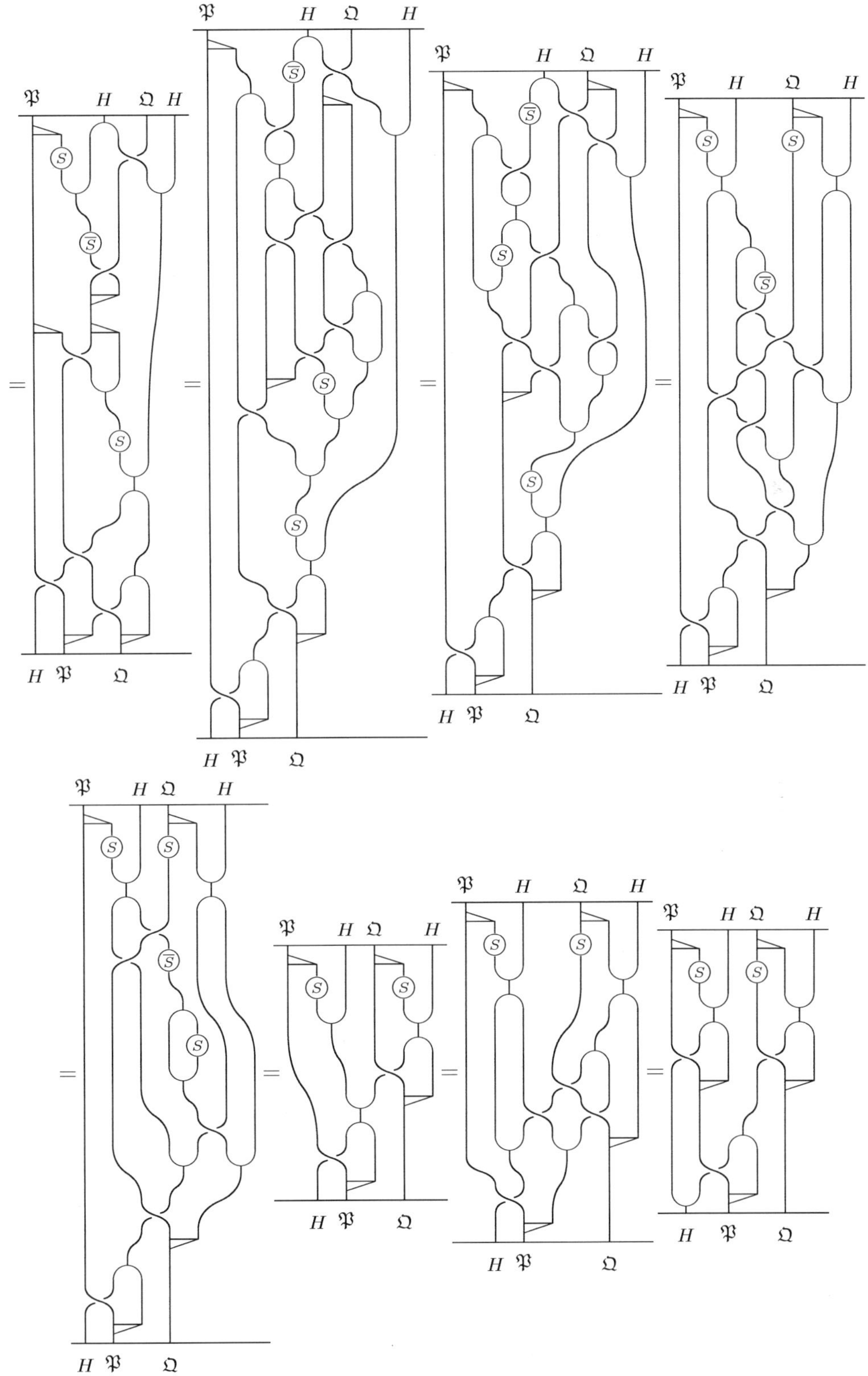

and the latter is just $\varphi^r_{2,\mathfrak{P},\mathfrak{Q}}(\omega_{\mathfrak{P}}\otimes\omega_{\mathfrak{Q}})$. We used: in the first equality (1.9), the coassociativity of $\rho^H_{\mathfrak{P}}$ and the naturality of c applied to Δ_H and $\rho^H_{\mathfrak{Q}}$, in the second equality the naturality of c applied once to Δ_H and twice to m_H (together with its associativity) and $\rho^H_{\mathfrak{Q}}$, in the third equality (2.3), the naturality of c^{-1} applied to $c_{H,\mathfrak{Q}}$, Δ_H and $c^{-1}_{H,H}$ and the coassociativity of Δ_H, in the fourth one the naturality of c^{-1} applied to $c^{-1}_{H,H}(\mathrm{Id}_H\otimes S)\Delta_H$ and m_H and the naturality of c applied to the same morphism $c^{-1}_{H,H}(\mathrm{Id}_H\otimes S)\Delta_H$, in the fifth equality (1.3) and the naturality of c applied to m_H, in the sixth equality (1.1) and the naturality of c applied to Δ_H, and in the last but one equality the naturality of c applied to m_H and Δ_H. $\square$

3. Double-biproduct Hopf algebras

As before, $(\mathcal{C},c)$ is a (strict) braided category and $H\in\mathcal{C}$ is a Hopf algebra with skew antipode.

By [**2**], the Hopf algebras in $({}^H_H\mathcal{C}^H_H,\otimes_H,H,d)$ are determined by Hopf algebras in ${}^H_H\mathcal{YD}(\mathcal{C})$ or, alternatively, by the so called biproduct Hopf algebras in $\mathcal{C}$. We will use these characterizations to show that a double-biproduct Hopf algebra [**17**] is isomorphic to a biproduct Hopf algebra.

The braided equivalence between ${}^H_H\mathcal{C}^H_H$ and ${}^H_H\mathcal{YD}(\mathcal{C})$ gives us a first description for the algebras (resp. coalgebras, etc.) in ${}^H_H\mathcal{C}^H_H$.

3.1. Left and right smash product algebras. If $\mathbb{A}$ is an algebra in ${}^H_H\mathcal{C}^H_H$, then $A=\mathcal{G}_l(\mathbb{A})$ is an algebra in ${}^H_H\mathcal{YD}(\mathcal{C})$ and $\overline{\zeta}_{\mathbb{A}}:A\otimes H\to\mathbb{A}$ is an isomorphism of algebras in ${}^H_H\mathcal{C}^H_H$. In general, for an algebra A in ${}^H_H\mathcal{YD}(\mathcal{C})$, $A\otimes H=\mathcal{F}_l(A)$ is an algebra in ${}^H_H\mathcal{C}^H_H$ via the strong monoidal structure of $\mathcal{F}_l$; explicitly, the multiplication $m^H_{A\otimes H}$ of $A\otimes H$ in ${}^H_H\mathcal{C}^H_H$ is $m^H_{A\otimes H}=m_{A\otimes H}q^H_{A\otimes H,A\otimes H}$ with

$$m_{A\otimes H}=(m_A\otimes m_H)(\mathrm{Id}_A\otimes(\mu^H_A\otimes\mathrm{Id}_H)(\mathrm{Id}_H\otimes c_{H,A})(\Delta_H\otimes\mathrm{Id}_A)\otimes\mathrm{Id}_H)$$

and the unit of $A\otimes H$ in ${}^H_H\mathcal{C}^H_H$ is $\eta_A\otimes\mathrm{Id}_H$. Thus the algebra structure of $\mathcal{F}_l(A)$ in ${}^H_H\mathcal{C}^H_H$ is determined by an algebra structure $(m_{A\otimes H},\eta_A\otimes\eta_H)$ of $A\otimes H$ in $\mathcal{C}$, and the latter is nothing but the left smash product algebra $A\#H$ of A and H. We should point out that the construction of the algebra $A\#H$ in $\mathcal{C}$ makes sense for A an algebra in ${}_H\mathcal{C}$ only, but here we need A to be an algebra in ${}^H_H\mathcal{YD}(\mathcal{C})$ in order to obtain $A\#H$ an algebra in ${}^H_H\mathcal{C}^H_H$; for otherwise $A\#H$ is an algebra only in ${}_H\mathcal{C}^H_H$, the category of right H-comodules in ${}_H\mathcal{C}_H$.

If we consider the pair $(\mathcal{F}_r,\mathcal{G}_r)$ instead of $(\mathcal{F}_l,\mathcal{G}_l)$ then $\zeta^{-1}_{\mathbb{A}}:\ \mathbb{A}\simeq\mathcal{F}_r\mathcal{G}_r(\mathbb{A})=H\otimes\mathfrak{A}$ is an isomorphism of algebras in ${}^H_H\mathcal{C}^H_H$, where $\mathfrak{A}=\mathcal{G}_r(\mathbb{A})$ is an algebra in $\mathcal{YD}(\mathcal{C})^H_H$. As above, if $\mathfrak{A}$ is an algebra in $\mathcal{YD}(\mathcal{C})^H_H$ then $\mathcal{F}_r(\mathfrak{A})=H\otimes\mathfrak{A}$ is an algebra in ${}^H_H\mathcal{C}^H_H$ with multiplication $m^H_{H\otimes\mathfrak{A}}=q^H_{H\otimes\mathfrak{A},H\otimes\mathfrak{A}}m_{H\otimes A}$ and unit $\mathrm{Id}_H\otimes\eta_{\mathfrak{A}}$, where

$$m_{H\otimes A}=(m_H\otimes m_{\mathfrak{A}})(\mathrm{Id}_H\otimes(\mathrm{Id}_H\otimes\nu^H_{\mathfrak{A}})(c_{\mathfrak{A},H}\otimes\mathrm{Id}_H)(\mathrm{Id}_{\mathfrak{A}}\otimes\Delta_H)\otimes\mathrm{Id}_{\mathfrak{A}}).$$

Concluding, algebras $\mathbb{A}$ in ${}^H_H\mathcal{C}^H_H$ are also characterized by the right smash product algebras $H\#\mathfrak{A}=(H\otimes\mathfrak{A},m_{H\otimes A},\eta_H\otimes\eta_{\mathfrak{A}})$ of H and algebras $\mathfrak{A}$ in $\mathcal{YD}(\mathcal{C})^H_H$.

LEMMA 3.1. *If $\mathcal{H},\mathcal{K}:\mathcal{D}\to\mathcal{E}$ are isomorphic monoidal functors then for any algebra $\mathcal{A}$ in $\mathcal{D}$ the algebras $\mathcal{H}(\mathcal{A})$ and $\mathcal{K}(\mathcal{A})$ defined by the monoidal structures of $\mathcal{H}$ and $\mathcal{K}$ are isomorphic in $\mathcal{E}$. Consequently, if $\mathfrak{A}$ is an algebra in $\mathcal{YD}(\mathcal{C})^H_H$ then $H\#\mathfrak{A}\simeq\overline{\mathfrak{A}}\#H$ as algebras in $\mathcal{C}$, where $\overline{\mathfrak{A}}=\overline{\mathcal{G}}(\mathfrak{A})$ and $\overline{\mathcal{G}}$ is as in Corollary 2.5.*

PROOF. The proof of the general result in the statement is completely obvious. The particular one follows by specializing it to the isomorphic strong monoidal functors $\mathcal{H} = \mathcal{F}_l\overline{\mathcal{G}}$ and $\mathcal{K} = \mathcal{F}_r$; see Corollary 2.5. So, for $\mathfrak{A}$ an algebra in $\mathcal{YD}(\mathcal{C})_H^H$, $\mathcal{F}_l\overline{\mathcal{G}}(\mathfrak{A}) = \overline{\mathfrak{A}}\#H$ and $\mathcal{F}_r(\mathfrak{A}) = H\#\mathfrak{A}$ are isomorphic both as algebras in ${}_H^H\mathcal{C}_H^H$ and $\mathcal{C}$. Note that $\overline{\mathfrak{A}}$ is an algebra in ${}_H^H\mathcal{YD}(\mathcal{C})$ with structure defined by

(3.1) $\mu_{\overline{\mathfrak{A}}}^H =$ 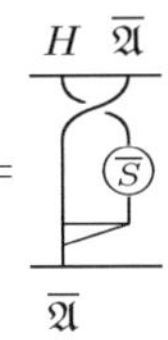 , $\lambda_{\overline{\mathfrak{A}}}^H =$ 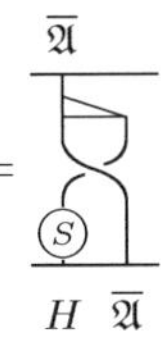 , $m_{\overline{\mathfrak{A}}} =$ 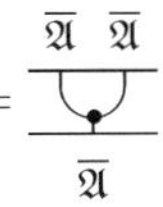 $=$ 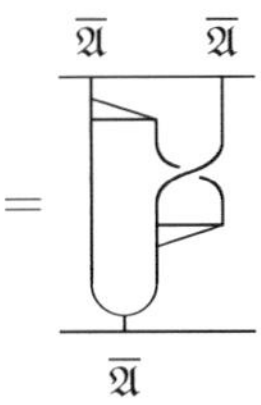 , $\eta_{\overline{\mathfrak{A}}} = \eta_{\mathfrak{A}}$.

The algebra isomorphism $\overline{\mathfrak{A}}\#H \simeq H\#\mathfrak{A}$ is produced by $\omega_{\mathfrak{A}}$; see Corollary 2.5. □

Let $\mathfrak{A}$, $\mathfrak{B}$ be algebras in $\mathcal{YD}(\mathcal{C})_H^H$ and ${}_H^H\mathcal{YD}(\mathcal{C})$ and $\mathcal{F}_r(\mathfrak{A})$, $\mathcal{F}_l(\mathfrak{B})$ the algebras in ${}_H^H\mathcal{C}_H^H$ corresponding to them. Denote by $\mathcal{U}(\mathfrak{A}, H, \mathfrak{B}) := \mathcal{F}_l(\mathfrak{B}) \otimes_H \mathcal{F}_r(\mathfrak{A})$ the tensor product algebra of $\mathcal{F}_l(\mathfrak{B})$ and $\mathcal{F}_r(\mathfrak{A})$ in ${}_H^H\mathcal{C}_H^H$, and identify $\mathcal{U}(\mathfrak{A}, H, \mathfrak{B})$ with $\mathfrak{B} \otimes H \otimes \mathfrak{A}$ canonically.

PROPOSITION 3.2. *Under the above hypothesis, $\mathcal{U}(\mathfrak{A}, H, \mathfrak{B}) \simeq (\mathfrak{B}\widetilde{\otimes}\overline{\mathfrak{A}})\#H$ as algebras in $\mathcal{C}$, where $\mathfrak{B}\widetilde{\otimes}\overline{\mathfrak{A}}$ is the tensor product algebra of $\mathfrak{B}$, $\overline{\mathfrak{A}}$ in ${}_H^H\mathcal{YD}(\mathcal{C})$.*

PROOF. For $(\mathcal{F}, \varphi_0, \varphi_2) : \mathcal{D} \to \mathcal{E}$ a strong monoidal functor between two braided categories and A an algebra in $\mathcal{D}$, the algebra structure on $\mathcal{F}(A)$ in $\mathcal{E}$ defined by the monoidal structure of $\mathcal{F}$ and algebra structure of A is designed in such a way that $\varphi_{2,A,A} : \mathcal{F}(A) \otimes \mathcal{F}(A) \to \mathcal{F}(A \otimes A)$ is an algebra isomorphism, where $A \otimes A$ has the tensor product algebra structure in $\mathcal{D}$.

The above result implies $\mathcal{U}(\mathfrak{A}, H, \mathfrak{B}) = \mathcal{F}_l(\mathfrak{B}) \otimes_H \mathcal{F}_r(\mathfrak{A}) \simeq \mathcal{F}_l(\mathfrak{B}) \otimes_H \mathcal{F}_l\overline{\mathcal{G}}(\mathfrak{A}) \simeq \mathcal{F}_l(\mathfrak{B}\widetilde{\otimes}\overline{\mathfrak{A}}) = (\mathfrak{B}\widetilde{\otimes}\overline{\mathfrak{A}}) \otimes H$ as algebras in ${}_H^H\mathcal{C}_H^H$ and $\mathcal{C}$. The multiplication $m_{\mathcal{U}(\mathfrak{A},H,\mathfrak{B})}^H$ of $\mathfrak{B} \otimes H \otimes \mathfrak{A}$ given by the tensor product is determined by

(3.2) $m_{\mathcal{U}(\mathfrak{A},H,\mathfrak{B})} =$ 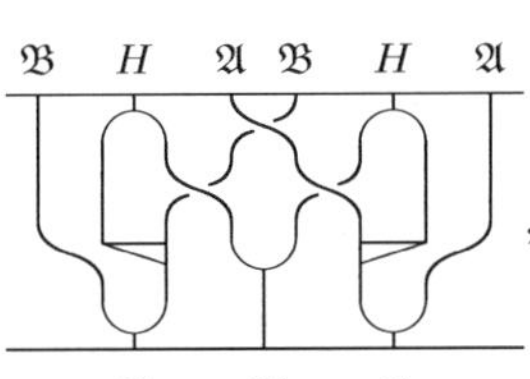 ,

i.e. $m_{\mathcal{U}(\mathfrak{A},H,\mathfrak{B})}^H q_{\mathcal{U}(\mathfrak{A},H,\mathfrak{B}),\mathcal{A}_H(\mathfrak{A},\mathfrak{B})}^H = m_{\mathcal{U}(\mathfrak{A},H,\mathfrak{B})}$. Indeed, since $\Pi_{\mathfrak{B}\otimes H} = \mathrm{Id}_{\mathfrak{B}} \otimes \eta_H\varepsilon_H$,

$$d_{H\otimes\mathfrak{A},\mathfrak{B}\otimes H}\, q_{H\otimes\mathfrak{A},\mathfrak{B}\otimes H}^H = q_{\mathfrak{B}\otimes H,H\otimes\mathfrak{A}}^H \circ$$ 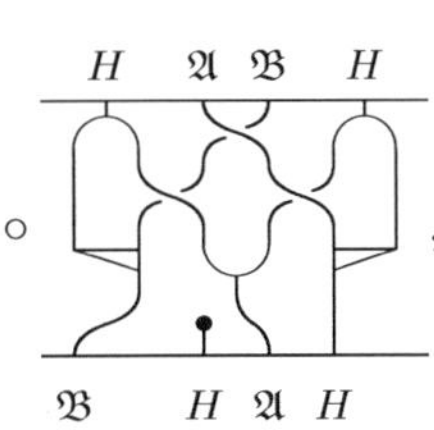 ,

so $m_{\mathcal{U}(\mathfrak{A},H,\mathfrak{B})}^H q_{\mathcal{U}(\mathfrak{A},H,\mathfrak{B}),\mathcal{U}(\mathfrak{A},H,\mathfrak{B})}^H = m_{\mathcal{U}(\mathfrak{A},H,\mathfrak{B})}$. The unit of $\mathcal{U}(\mathfrak{A}, H, \mathfrak{B})$ is $\eta_{\mathfrak{B}} \otimes \mathrm{Id}_H \otimes \eta_{\mathfrak{A}}$ (resp. $\eta_{\mathfrak{B}} \otimes \eta_H \otimes \eta_{\mathfrak{A}}$) if it is regarded as an algebra in ${}_H^H\mathcal{C}_H^H$ (resp. in $\mathcal{C}$).

The algebra structure of $\mathfrak{B}\widetilde{\otimes}\overline{\mathfrak{A}}$ in ${}^H_H\mathcal{YD}(\mathcal{C})$ is given by $\eta_{\mathfrak{B}\widetilde{\otimes}\overline{\mathfrak{A}}} = \eta_{\mathfrak{B}} \otimes \eta_{\mathfrak{A}}$,

$\mu^H_{\mathfrak{B}\widetilde{\otimes}\overline{\mathfrak{A}}} =$ 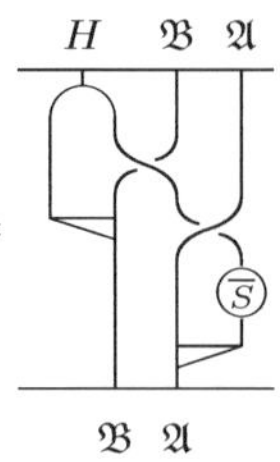

, $\lambda^H_{\mathfrak{B}\widetilde{\otimes}\overline{\mathfrak{A}}} =$ 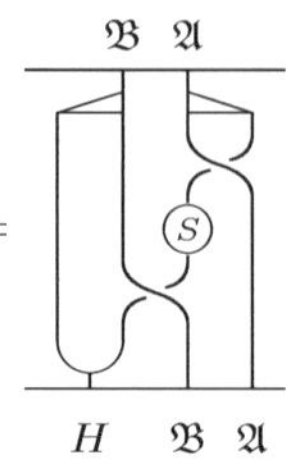

, $m_{\mathfrak{B}\widetilde{\otimes}\overline{\mathfrak{A}}} =$ 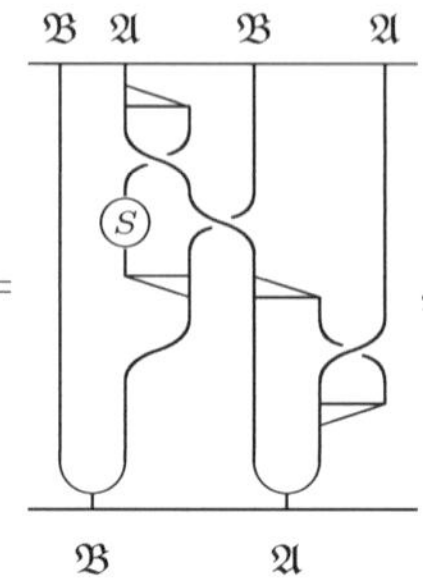

,

and $\chi := \varphi^l_{2,\mathfrak{B},\underline{\mathfrak{A}}}(\mathrm{Id}_{\mathfrak{B}\otimes H} \otimes_H \omega_{\mathfrak{A}}^{-1}) : \mathcal{U}(\mathfrak{A}, H, \mathfrak{B}) \to (\mathfrak{B}\widetilde{\otimes}\overline{\mathfrak{A}})\#H$ is invertible. As

$\omega_{\mathfrak{P}}^{-1} =$ 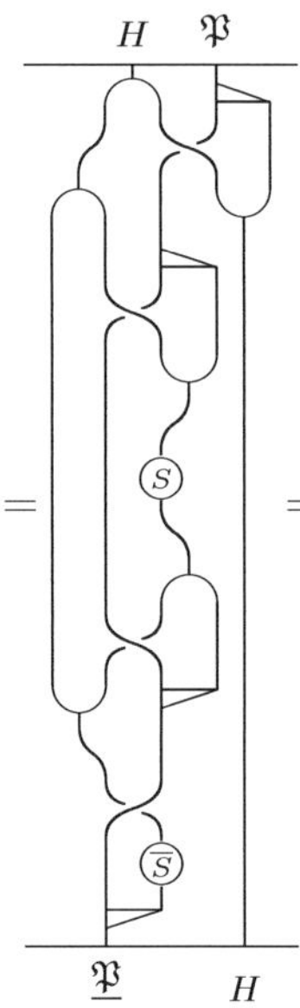

$=$ 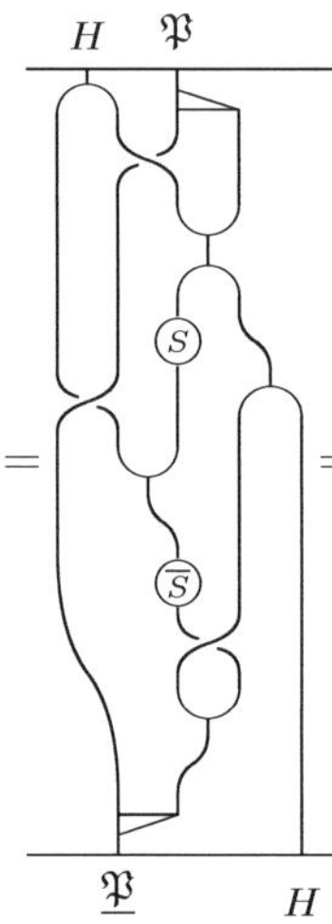

$=$ 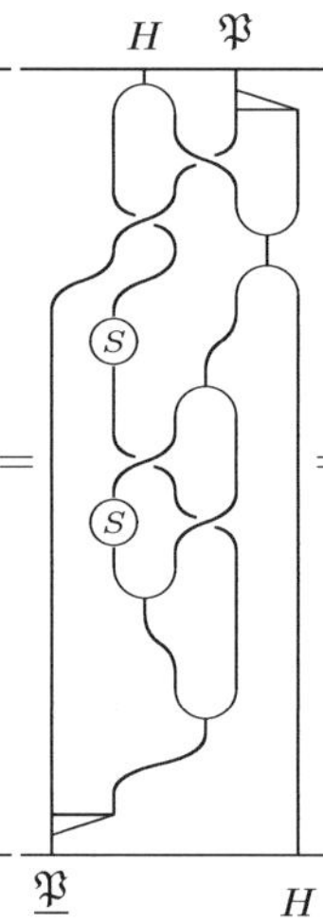

$=$ 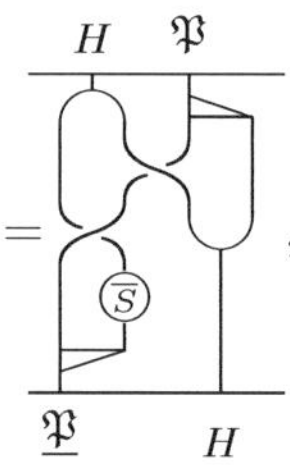

, $\chi =$ 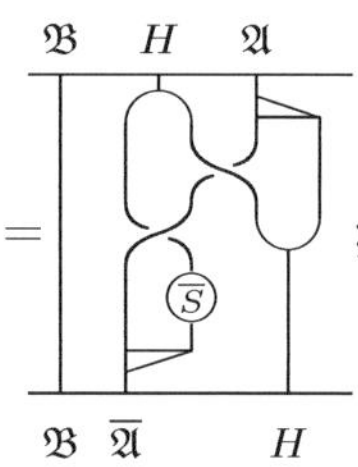

;

$\mathfrak{P} \in \mathcal{YD}(\mathcal{C})^H_H$. From here we deduce that $\chi = \mathrm{Id}_{\mathfrak{B}} \otimes \omega_{\mathfrak{A}}^{-1}$. □

The algebra $\mathfrak{B} \otimes H \otimes \mathfrak{A}$ in $\mathcal{C}$ contains both $\mathfrak{B}\#H$ and $H\#\mathfrak{A}$ as subalgebras. This is why it will be called the two-sided smash product algebra of $\mathfrak{B}$, H and $\mathfrak{A}$, and will be denoted by $\mathfrak{B}\#H\#\mathfrak{A}$. Thus $\mathfrak{B}\#H\#\mathfrak{A} \simeq (\mathfrak{B}\widetilde{\otimes}\overline{\mathfrak{A}})\#H$ as algebras in $\mathcal{C}$. A similar result holds for coalgebras, too.

3.2. Left and right smash product coalgebras. Owing to the braided equivalence between ${}^H_H\mathcal{C}^H_H$ and $\mathcal{YD}(\mathcal{C})^H_H$, any coalgebra $\mathfrak{C}$ in ${}^H_H\mathcal{C}^H_H$ is determined by a coalgebra $C \in \mathcal{YD}(\mathcal{C})^H_H$: $\zeta_{\mathfrak{C}}^{-1} : \mathfrak{C} \to H \otimes C$ is an isomorphism of coalgebras in ${}^H_H\mathcal{C}^H_H$, where $C = {}^{\mathrm{co}(H)}\mathfrak{C} = \mathcal{G}_r(\mathfrak{C})$. The coalgebra structure on $H \otimes C = \mathcal{F}_r(C)$ is given by the strong monoidal structure of $\mathcal{F}_r$ and the coalgebra structure of C in $\mathcal{YD}(\mathcal{C})^H_H$: $\Delta_{H\otimes C} = q^H_{H\otimes C, H\otimes C}(\mathrm{Id}_{H\otimes C} \otimes \eta_H \otimes \mathrm{Id}_C)(\mathrm{Id}_H \otimes \Delta_C)$ and $\varepsilon_{H\otimes C} = \mathrm{Id}_H \otimes \varepsilon_C$.

Likewise, if $\mathfrak{C}$ is a coalgebra in ${}^H_H\mathcal{C}^H_H$ then $\overline{\zeta}_{\mathfrak{C}}^{-1} : \mathfrak{C} \to D \otimes H$ is an isomorphism of coalgebras in ${}^H_H\mathcal{C}^H_H$, where $D = \mathcal{G}_l(\mathfrak{C})$ is a coalgebra in ${}^H_H\mathcal{YD}(\mathcal{C})$. The object $D \otimes H = \mathcal{F}_l(D)$ of ${}^H_H\mathcal{C}^H_H$ is considered coalgebra with $\Delta_{D\otimes H} = q^H_{D\otimes H, D\otimes H}(\mathrm{Id}_D \otimes \eta_H \otimes \mathrm{Id}_{D\otimes H})(\Delta_D \otimes \mathrm{Id}_H)$ and $\varepsilon_{D\otimes H} = \varepsilon_D \otimes \mathrm{Id}_H$.

LEMMA 3.3. *If C is a coalgebra in $\mathcal{YD}(\mathcal{C})_H^H$ and $\overline{C} = \overline{\mathcal{G}}(C)$ with $\overline{\mathcal{G}} : \mathcal{YD}(\mathcal{C})_H^H \to {}_H^H\mathcal{YD}(\mathcal{C})$ as in Corollary 2.5, then $H \otimes C \simeq \overline{C} \otimes H$ as coalgebras in ${}_H^H\mathcal{C}_H^H$.*

PROOF. Parallel to Lemma 3.1, if $\mathcal{H}, \mathcal{K} : \mathcal{D} \to \mathcal{E}$ are isomorphic strong monoidal functors and $D \in \mathcal{D}$ is a coalgebra then $\mathcal{H}(D) \simeq \mathcal{K}(D)$ as coalgebras in $\mathcal{E}$. If we apply this result to the isomorphic strong monoidal functors $\mathcal{F}_l\overline{\mathcal{G}}$ and $\mathcal{F}_r$ and coalgebra C we deduce that $\overline{C} \otimes H$ and $H \otimes C$ are isomorphic as coalgebras in ${}_H^H\mathcal{C}_H^H$, the isomorphism being produced by the morphism ω_C; see Corollary 2.5.

For further use, note that $\overline{C}$ is a coalgebra in ${}_H^H\mathcal{YD}(\mathcal{C})$ with:

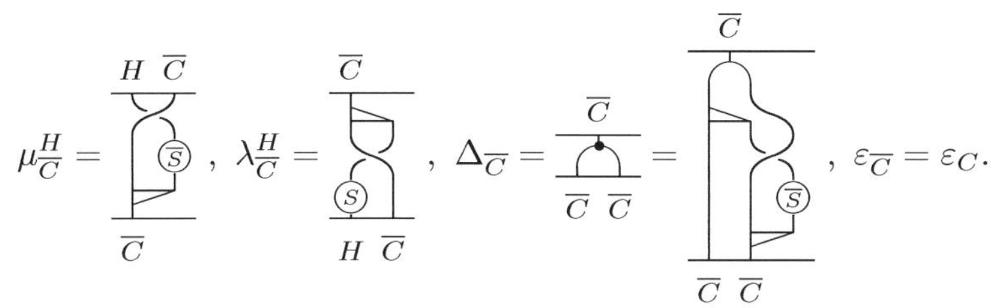

We skip the details to the reader. □

By [**2**], coalgebras in ${}_H^H\mathcal{C}_H^H$ can be also characterized as coalgebras $C \in {}_H\mathcal{C}_H$, equipped with a morphism of coalgebras $\pi : C \to H$ in ${}_H\mathcal{C}_H$. Here ${}_H\mathcal{C}_H$ is considered monoidal via the tensor product $\otimes$ of $\mathcal{C}$ and Hopf algebra structure of H, so the unit object is $\underline{1}$ and H has a canonical coalgebra structure in ${}_H\mathcal{C}_H$. To avoid any confusion we denote ${}_H\mathcal{C}_H$ with this monoidal structure by $\widetilde{{}_H\mathcal{C}_H}$ and keep the notation ${}_H\mathcal{C}_H$ for ${}_H\mathcal{C}_H$ when it is considered monoidal via $\otimes_H$, and unit object H.

A coalgebra $\mathfrak{C}$ in ${}_H^H\mathcal{C}_H^H$ can be seen as a coalgebra in $\widetilde{{}_H\mathcal{C}_H}$ with the newly defined structure ($\underline{\Delta}_{\mathfrak{C}} = \upsilon_{\mathfrak{C},\mathfrak{C}}\Delta_{\mathfrak{C}}, \underline{\varepsilon}_{\mathfrak{C}} = \varepsilon_H\varepsilon_{\mathfrak{C}}$); for $M, N \in {}_H^H\mathcal{C}_H^H$, the coequalizer $(M \otimes_H N, q_{M,N}^H)$ gives a unique morphism $\upsilon_{M,N} : M \otimes_H N \to M \otimes N$ satisfying

$$\upsilon_{M,N}q_{M,N}^H = (\nu_M^H \otimes \mu_N^H)(\mathrm{Id}_M \otimes c_{H,H} \otimes \mathrm{Id}_N)(\rho_M^H \otimes \lambda_N^H).$$

Then $\pi = \varepsilon_C : C \to H$ becomes a coalgebra morphism in $\widetilde{{}_H\mathcal{C}_H}$. Conversely, if $(\mathfrak{C}, \pi)$ consists of a coalgebra $(\mathfrak{C}, \underline{\Delta}_{\mathfrak{C}}, \underline{\varepsilon}_{\mathfrak{C}})$ in $\widetilde{{}_H\mathcal{C}_H}$ and a coalgebra morphism $\pi : C \to H$ in $\widetilde{{}_H\mathcal{C}_H}$ then $\mathfrak{C}$ is a coalgebra in ${}_H^H\mathcal{C}_H^H$ with the initial H-bimodule structure and

$$\lambda_{\mathfrak{C}}^H = (\pi \otimes \mathrm{Id}_{\mathfrak{C}})\underline{\Delta}_{\mathfrak{C}},\ \rho_{\mathfrak{C}}^H = (\mathrm{Id}_{\mathfrak{C}} \otimes \pi)\underline{\Delta}_{\mathfrak{C}},\ \Delta_{\mathfrak{C}} = q_{\mathfrak{C},\mathfrak{C}}^H(\Pi_{\mathfrak{C}} \otimes \mathrm{Id}_{\mathfrak{C}})\underline{\Delta}_{\mathfrak{C}} \text{ and } \varepsilon_{\mathfrak{C}} = \pi,$$

where $\Pi_{\mathfrak{C}}$ is as in (2.1), defined for $\mathfrak{C} \in {}_H^H\mathcal{C}_H^H$ with structure we just described.

Hence, for $\mathfrak{C} = H \otimes C$, $C \in \mathcal{YD}(\mathcal{C})_H^H$ a coalgebra, the morphisms

$$\underline{\Delta}_{H\otimes C} = (\mathrm{Id}_H \otimes(\mathrm{Id}_C \otimes m_H)(c_{H,C} \otimes \mathrm{Id}_H)(\mathrm{Id}_H \otimes \rho_C^H) \otimes \mathrm{Id}_C)(\Delta_H \otimes \Delta_C)$$

and $\underline{\varepsilon}_{H\otimes C} = \varepsilon_H \otimes \varepsilon_C$ turn $\mathfrak{C}$ into a coalgebra in $\widetilde{{}_H\mathcal{C}_H}$. Equivalently, $\mathfrak{C} = H \ltimes C$, the right smash product coalgebra of C and H.

For $\mathfrak{C} = D \otimes H$, $D \in {}_H^H\mathcal{YD}(\mathcal{C})$ a coalgebra, $\underline{\varepsilon}_{D\otimes H} = \varepsilon_D \otimes \varepsilon_H$ and $\underline{\Delta}_{D\otimes H} = (\mathrm{Id}_D \otimes(m_H \otimes \mathrm{Id}_D)(\mathrm{Id}_H \otimes c_{D,H})(\lambda_D^H \otimes \mathrm{Id}_H) \otimes \mathrm{Id}_H)(\Delta_D \otimes \Delta_H)$ make $D \otimes H \in \widetilde{{}_H\mathcal{C}_H}$ a coalgebra, equals $D \ltimes H$ (the left smash product coalgebra of D and H).

COROLLARY 3.4. *Let C be a coalgebra in $\mathcal{YD}(\mathcal{C})_H^H$ and $\overline{C}$ the coalgebra in ${}_H^H\mathcal{YD}(\mathcal{C})$ defined by C and the braided functor $\overline{\mathcal{G}} : \mathcal{YD}(\mathcal{C})_H^H \to {}_H^H\mathcal{YD}(\mathcal{C})$. Then $\overline{C} \rtimes H \simeq H \ltimes C$ as coalgebras in $\widetilde{{}_H\mathcal{C}_H}$.*

PROOF. $\omega_C : \overline{C} \rtimes H \to H \ltimes C$ is a coalgebra isomorphism in $\widetilde{{}_H\mathcal{C}_H}$. □

Consider $C \in \mathcal{YD}(\mathcal{C})^H_H$, $D \in {}^H_H\mathcal{YD}(\mathcal{C})$ coalgebras and $\mathcal{F}_r(C)$, $\mathcal{F}_l(D)$ the coalgebras in ${}^H_H\mathcal{C}^H_H$ corresponding to them. As in the algebra case, denote by $\mathcal{U}(C, H, D) := (D \otimes H) \otimes_H (H \otimes C) \cong D \otimes H \otimes C$ the tensor product coalgebra in ${}^H_H\mathcal{C}^H_H$ of $\mathcal{F}_l(D)$ and $\mathcal{F}_r(C)$, and by $D\widetilde{\otimes}\overline{C}$ the tensor product coalgebra in ${}^H_H\mathcal{YD}(\mathcal{C})$ of D and $\overline{C} = \overline{\mathcal{G}}(C)$.

PROPOSITION 3.5. *Let C, D be as in the above. Then $\mathcal{U}(C, H, D)$ with the H-bimodule coalgebra structure induced by that in ${}^H_H\mathcal{C}^H_H$ is isomorphic to $(D\widetilde{\otimes}\overline{C}) \rtimes H$.*

PROOF. If $(\mathcal{F}, \varphi_0, \varphi_2) : \mathcal{D} \to \mathcal{E}$ is a strong monoidal functor between two braided categories and $D \in \mathcal{D}$ is a coalgebra, then $\varphi_{2,C,C} : \mathcal{F}(C) \otimes \mathcal{F}(C) \to \mathcal{F}(C)$ is a coalgebra isomorphism. Also, if ω is a natural monoidal transformation between the strong monoidal functors $\mathcal{H}, \mathcal{K} : \mathcal{D} \to \mathcal{E}$ and $D \in \mathcal{D}$ is a coalgebra then $\omega_D : \mathcal{H}(D) \to \mathcal{K}(D)$ is an isomorphism of coalgebras in $\mathcal{E}$. These two results entail to the following sentence of coalgebra isomorphisms in ${}^H_H\mathcal{C}^H_H$: $\mathcal{U}(C, H, D) \cong \mathcal{F}_l(D) \otimes_H \mathcal{F}_r(C) \simeq \mathcal{F}_l(D) \otimes_H \mathcal{F}_l\mathcal{G}(C) \simeq \mathcal{F}_l(D\widetilde{\otimes}\overline{C})$. As before, we still denote by $\mathcal{U}(C, H, D)$ the coalgebra in $\widetilde{{}_H\mathcal{C}_H}$ corresponding to the coalgebra structure $(\Delta_{\mathcal{U}(C,H,D)}, \varepsilon_{\mathcal{U}(C,H,D)})$ of $\mathcal{U}(C, H, D)$ in ${}^H_H\mathcal{C}^H_H$, that is $\mathcal{U}(C, H, D) = D \otimes H \otimes C$, $\underline{\Delta}_{\mathcal{U}(C,H,D)} = \upsilon_{\mathcal{U}(C,H,D),\mathcal{U}(C,H,D)} \Delta_{\mathcal{U}(C,H,D)}$ and $\underline{\varepsilon}_{\mathcal{U}(C,H,D)} = \varepsilon_H \varepsilon_{\mathcal{U}(C,H,D)}$.

The difficult part of the proof (omitted in [**3**]) consists in giving a simple description for $\underline{\Delta}_{\mathcal{U}(C,H,D)}$. For this, we notice first that $d_{D\otimes H, H\otimes C} \circ q^H_{D\otimes H, H\otimes C}$ equals

$q^H_{H\otimes C, D\otimes H} \circ$ 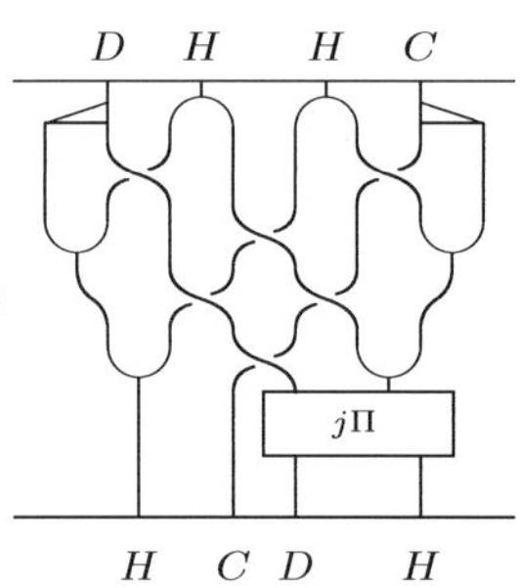

, where $j\Pi = j_{D\otimes H\, D\otimes H}\Pi =$ 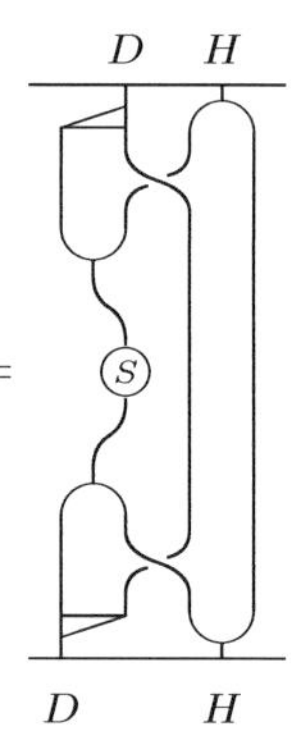

.

This gives the comultiplication of $\mathcal{U} := \mathcal{U}(C, H, D) \equiv D \otimes H \otimes C$ in ${}^H_H\mathcal{C}^H_H$:

$\Delta_{\mathcal{U}} = q^H_{\mathcal{U},\mathcal{U}} \circ$ 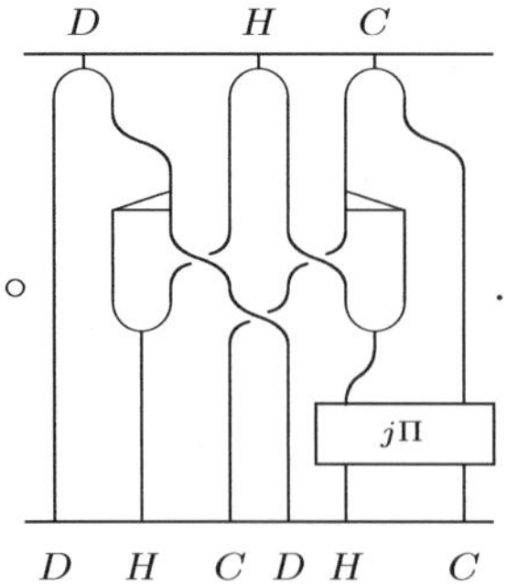

. As $\upsilon_{\mathcal{U},\mathcal{U}} q^H_{\mathcal{U},\mathcal{U}} =$ 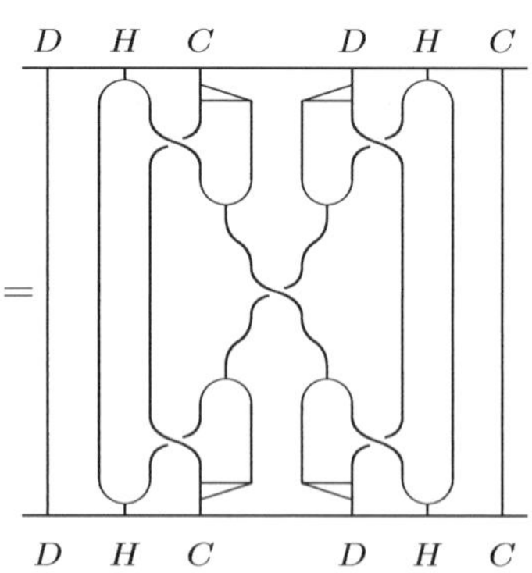

we see that $\underline{\Delta}_{\mathcal{U}} = \upsilon_{\mathcal{U},\mathcal{U}}\Delta_{\mathcal{U}}$ is equal to

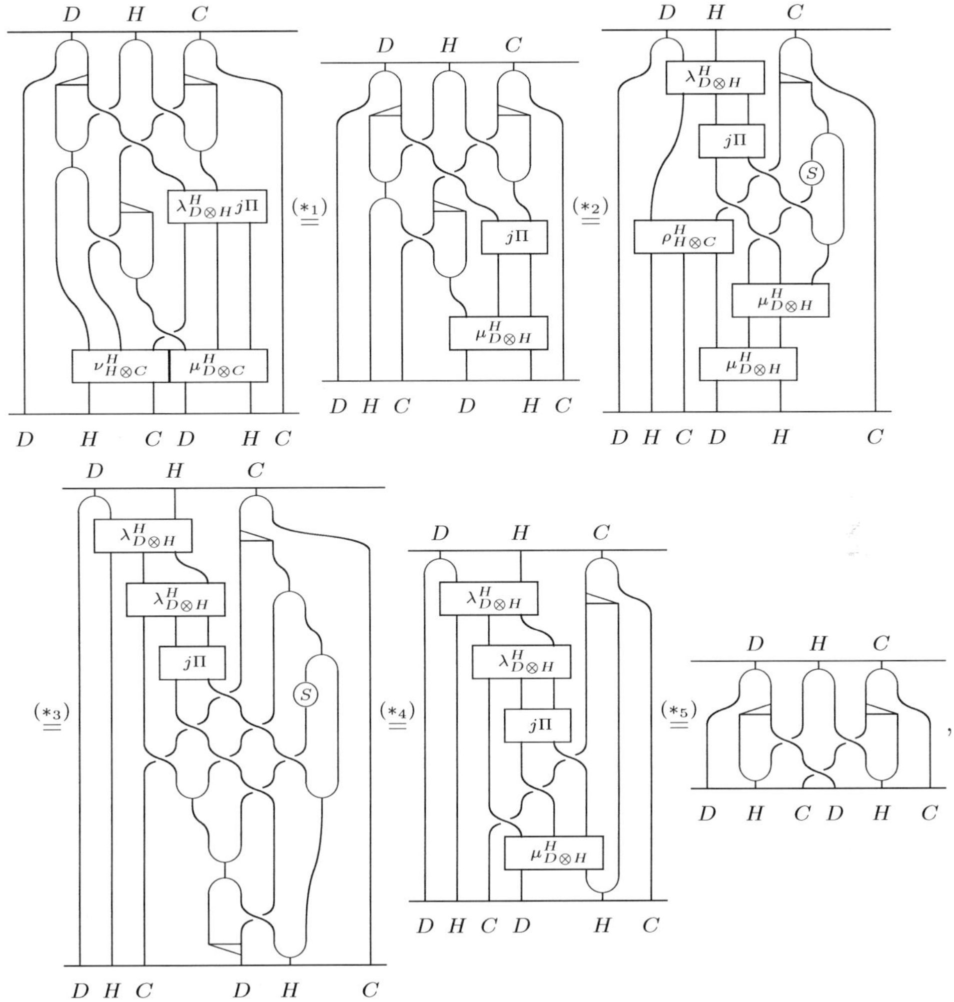

where $(*_1)$ refers to the equality, valid for all $M \in {}^H_H\mathcal{C}^H_H$,

$\lambda^H_M j_{M\,M}\Pi$ =

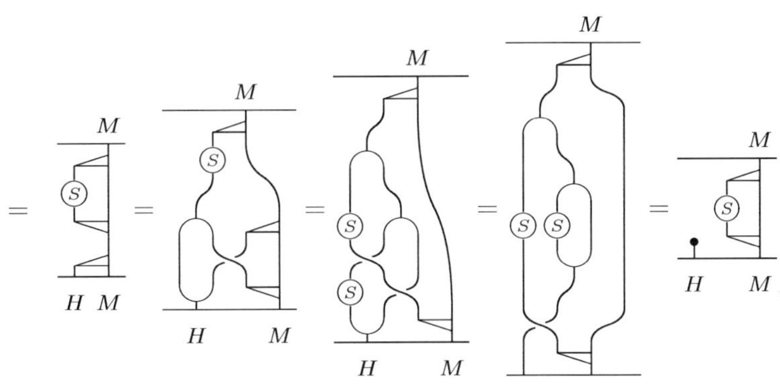

$= \quad \eta_H \otimes j_{M\,M}\Pi,$

specialized for $M = \mathcal{F}_l(D)$, $(*_2)$ refers to the fact that $\Pi : M \to {}^{\mathrm{co}(H)}M$ is right H-linear for all $M \in {}^H_H\mathcal{C}^H_H$ (which follows easily from (1.4); then we specialize it for $M = \mathcal{F}_l(D)$) and to the naturality of c applied to $j\Pi$, $(*_3)$ refers to the naturality of c applied twice to ρ^H_C, the associativity of $\mu^H_{D\otimes H}$ and coassociativity of Δ_C, $(*_4)$ refers to the associativity of m_H, the naturality of c applied twice to m_H, the coassociativity of Δ_C and (1.3), while $(*_5)$ refers to the naturality of c applied to $\mu^H_{D\otimes H}$ and the fact that $\zeta_{D\otimes H} = \mu^H_{D\otimes H}(\mathrm{Id}_H \otimes j)$ is an isomorphism with inverse $\zeta^{-1}_{D\otimes H} = (\mathrm{Id}_H \otimes \Pi)\lambda^H_{D\otimes H}$. Clearly, $\underline{\varepsilon}_{\mathcal{A}} = \varepsilon_D \otimes \varepsilon_H \otimes \varepsilon_C$. Note also that in the computations above me made use of the structure of $\mathcal{U}$ in ${}^H_H\mathcal{C}^H_H$, given by:

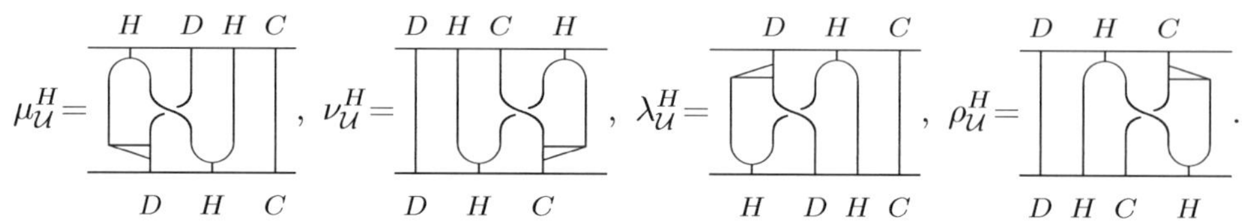

We end the proof by recording the coalgebra structure of $D\widetilde{\otimes}\overline{C}$ in ${}^H_H\mathcal{Y}D$,

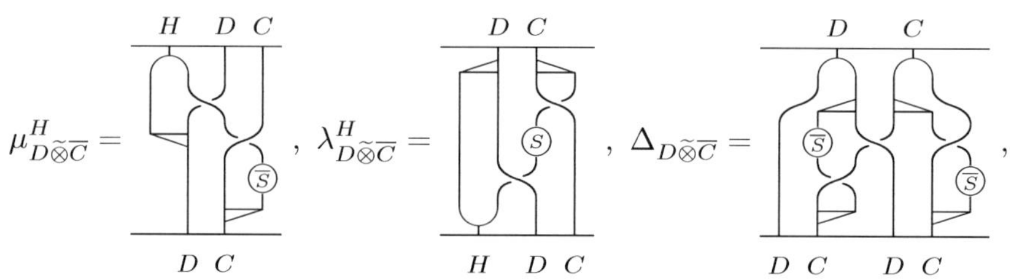

$\varepsilon_{D\widetilde{\otimes}\overline{C}} = \varepsilon_D \otimes \varepsilon_C$, and that $(D\widetilde{\otimes}\overline{C})\#H$ is a coalgebra in $\widetilde{{}_H\mathcal{C}_H}$ isomorphic to $\mathcal{U}$. □

We denote by $D \bowtie H \ltimes C$ the coalgebra structure of $\mathcal{A} = D \otimes H \otimes C$ in ${}_H\mathcal{C}_H$ described in the proof of Proposition 3.5, and call it the double-smash product coalgebra of D, C and H. The notation and terminology are justified by the fact that $D \bowtie H \ltimes C$ contains both $D \bowtie H$ and $H \ltimes C$ as H-bimodule subcoalgebras. As we have seen, $D \bowtie H \ltimes C \simeq (D\widetilde{\otimes}\overline{C}) \bowtie H$ as H-bimodule coalgebras in $\mathcal{C}$, the isomorphism being produced by $\chi = \mathrm{Id}_D \otimes \omega_C^{-1}$; see the proof of Proposition 3.2.

3.3. Left and right biproduct Hopf algebras. Throughout this subsection $A \in {}^H_H\mathcal{Y}D(\mathcal{C})$ and $B \in \mathcal{Y}D(\mathcal{C})^H_H$ are bialgebras or Hopf algebras. From the previous results we get two bialgebras (resp. Hopf algebras) $A\otimes H$ and $B\otimes H$ in ${}^H_H\mathcal{C}^H_H$, and the two correspondences defined by them can be identified since $\omega_B : \overline{B}\otimes H \to H\otimes B$ is an isomorphism of bialgebras (resp. Hopf algebras) in $\mathcal{C}$; as before, $\overline{B}$ is the algebra (resp. coalgebra, etc.) obtained from B and the braided structure of the functor $\overline{\mathcal{G}} : \mathcal{Y}D(\mathcal{C})^H_H \to {}^H_H\mathcal{Y}D(\mathcal{C})$.

To $A \otimes H$ and $B \otimes H$ one can associate bialgebras (resp. Hopf algebras) in $\mathcal{C}$, denoted by $A \times H$ and $H \times B$. Namely, $A \times H$ (resp. $H \times B$) equals $A\#H$ (resp. $H\#B$) as an algebra and $A \bowtie H$ (resp. $H \ltimes B$) as a coalgebra. In the Hopf algebra case $A \times H$, $H \times B$ are Hopf algebras in with antipode defined by (see [**2**])

$$S_{A\times H} = (\mu^H_A \otimes \mathrm{Id}_H)(\mathrm{Id}_H \otimes c_{H,A})(\Delta_H \otimes \mathrm{Id}_A)(Sm_H \otimes S_A)(\mathrm{Id}_H \otimes c_{A,H})(\lambda^H_A \otimes \mathrm{Id}_H),$$

and respectively by

$$S_{H\times B} = (\mathrm{Id}_H \otimes \nu^H_B)(c_{B,H} \otimes \mathrm{Id}_H)(\mathrm{Id}_B \otimes \Delta_H)(S_B \otimes Sm_H)(c_{H,B} \otimes \mathrm{Id}_H)(\mathrm{Id}_H \otimes \rho^H_B).$$

This is the (left) braided version of the result of Radford [**20**] which asserts that the (left) smash product algebra $A\#H$ and coalgebra $A \rtimes H$ afford a bialgebra (resp. Hopf algebra) structure on $A \otimes H$ if and only if A is a bialgebra (resp. Hopf algebra) within ${}^H_H\mathcal{YD}(\mathcal{C})$. This is why we call $A \times H$ the left biproduct bialgebra (resp. Hopf algebra) of A and H, and $H \times B$ the right biproduct bialgebra (resp. Hopf algebra) of B and H. Thus giving a bialgebra (resp. Hopf algebra) in ${}^H_H\mathcal{C}^H_H$ is equivalent to giving a left or right biproduct bialgebra (resp. Hopf algebra) between H and a suitable bialgebra (resp. Hopf algebra) in the category ${}^H_H\mathcal{YD}(\mathcal{C})/\mathcal{YD}(\mathcal{C})^H_H$.

By Lemma 3.1 and Corollary 3.3, we obtain the following.

PROPOSITION 3.6. *Let $B \in \mathcal{YD}^H_H$ be a bialgebra (resp. Hopf algebra). Then $\omega_B : \overline{B} \times H \to H \times B$ is an isomorphism of bialgebras (resp. Hopf algebras) in $\mathcal{C}$.*

Consider $\mathcal{U}(B,H,A) = (A \otimes H) \otimes_H (H \otimes B) \equiv A \otimes H \otimes B$, the tensor product algebra and coalgebra of $\mathcal{F}_l(A) = A \otimes H$, $\mathcal{F}_r(B) = H \otimes B$ in ${}^H_H\mathcal{C}^H_H$. We know that $\mathcal{U}(B,H,A) \simeq A\#H\#B \simeq (A\widetilde{\otimes}B)\#H$ as algebras in $\mathcal{C}$ and $\mathcal{U}(B,H,A) \simeq A \rtimes H \ltimes B \simeq (A\widetilde{\otimes}B) \rtimes H$ as coalgebras in $\widetilde{{}_H\mathcal{C}_H}$ (and so in $\mathcal{C}$). Hence $\mathcal{U}(B,H,A)$ is a bialgebra (resp. Hopf algebra) in ${}^H_H\mathcal{C}^H_H$ if and only if so is $A\widetilde{\otimes}\overline{B}$ in ${}^H_H\mathcal{YD}(\mathcal{C})$.

LEMMA 3.7. *Under the above notation, the following assertions are equivalent:*
(i) $\mathcal{U}(B,H,A)$ is a bialgebra (resp. Hopf algebra) in ${}^H_H\mathcal{C}^H_H$;
(ii) $A\widetilde{\otimes}\overline{B}$ is a bialgebra (resp. Hopf algebra) in ${}^H_H\mathcal{YD}(\mathcal{C})$;
(iii) the following compatibility relation holds:

(3.3) 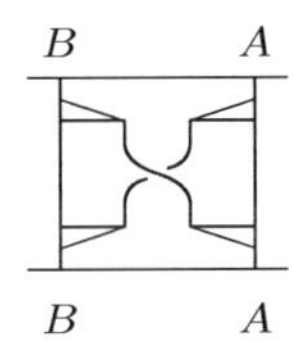 $= \mathrm{Id}_{B\otimes A}$.

PROOF. Owing to [**3**, Corollary 2.17], if U, V are bialgebras (resp. Hopf algebras) in a braided category $(\mathcal{E}, b)$ then the tensor product algebra and colgebra of U and V afford a bialgebra (resp. coalgebra) structure on $U \otimes V$ if and only if $b_{U,V} \circ b_{V,U} = \mathrm{Id}_{V\otimes U}$.

$(i) \Leftrightarrow (ii)$. It follows from the comments made above.

$(i) \Leftrightarrow (iii)$. $\mathcal{U}(B,H,A) = (A \otimes H) \otimes_H (H \otimes B)$ is a bialgebra/Hopf algebra in ${}^H_H\mathcal{C}^H_H$ if and only if $d_{\mathcal{F}_l(A),\mathcal{F}_r(B)} \circ d_{\mathcal{F}_r(B),\mathcal{F}_l(A)} = \mathrm{Id}_{\mathcal{F}_r(B),\mathcal{F}_l(A)}$, if and only if (3.3) holds; see [**3**, Lemma 3.1]. □

If, moreover, A, B are Hopf algebras then $\mathcal{U}(B,H,A)$ is a Hopf algebra;

$$S_{A\otimes H\otimes B} = ((\mathrm{Id}_A \otimes m_H)(\mathrm{Id}_{A\otimes H} \otimes m_H) \otimes \mathrm{Id}_B)(S_{A\times H} \otimes \mathrm{Id}_H \otimes S_{H\times B})$$
$$((\mathrm{Id}_{A\otimes H} \otimes \Delta_H)(\mathrm{Id}_A \otimes \Delta_H) \otimes \mathrm{Id}_B)$$

is its antipode. Therefore, $\mathcal{U}(B,H,A)$ contains both $A \times H$ and $H \times B$ as Hopf subalgebras. In analogy with the (co)algebra case we call $\mathcal{U}(B,H,A)$ the double-biproduct bialgebra (resp. Hopf algebra) of A, B and H, and denote it by $A\times H\times B$. Thus the construction of $A \times H \times B$ makes sense only if A, B fulfill (3.3), in which case $A \times H \times B \simeq (A\widetilde{\otimes}\overline{B}) \times H$ as bialgebras (resp. Hopf algebras) in $\mathcal{C}$.

4. 2-cocycle deformations on double-biproduct Hopf algebras

The Majid's double-bosonization process [**17**] involves the deformation of the algebra structure of $A \times H \times B$ by a kind of "pairing" between A, B. A braided version for this deformation was proposed in [**3**, Proposition 3.4] and it leads to the theory of convolution invertible 2-cocycles in a braided category. In this section, we uncover the categorical meaning of the construction presented in [**3**] and connect it with the notion of generalized quantum double as it was proposed in [**17**].

Recall from [**3**] that a 2-cocycle of a braided bialgebra H in $(\mathcal{C}, c)$ is a morphism $\tau : H \otimes H \to \underline{1}$ in $\mathcal{C}$ obeying $\tau(\eta_H \otimes \mathrm{Id}_H) = \varepsilon_H = \tau(\mathrm{Id}_H \otimes \eta_H)$ and

(4.1) 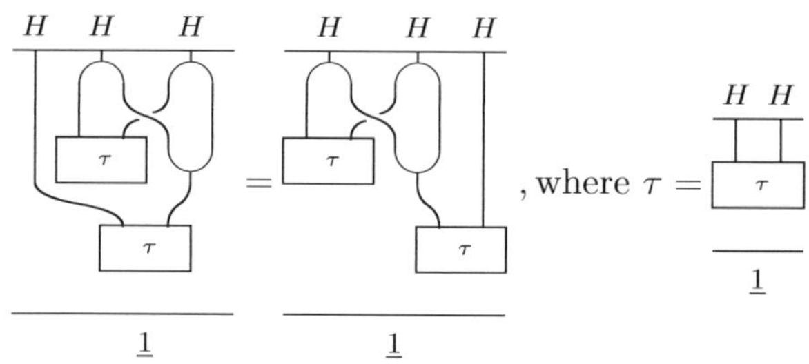

and for simplicity we assumed $\mathcal{C}$ strict monoidal. The 2-cocycle τ is called invertible if it is convolution invertible: for $\mathfrak{C}$ a coalgebra in $\mathcal{C}$ denote by $*$ the convolution product on the monoid $\mathrm{Hom}_{\mathcal{C}}(\mathfrak{C}, \underline{1})$ given by $\zeta * \zeta' = (\zeta \otimes \zeta')\Delta_{\mathfrak{C}}$, for all $\zeta, \zeta' : \mathfrak{C} \to \underline{1}$ in $\mathcal{C}$; then $\zeta : \mathfrak{C} \to \underline{1}$ is convolution invertible if it is an invertible element of this monoid, in which case we denote its inverse by $\overline{\zeta}$.

Let τ be an invertible 2-cocycle of H. We denote by H_τ the coalgebra H equipped with the newly defined multiplication

(4.2) $m_H^\tau :=$ 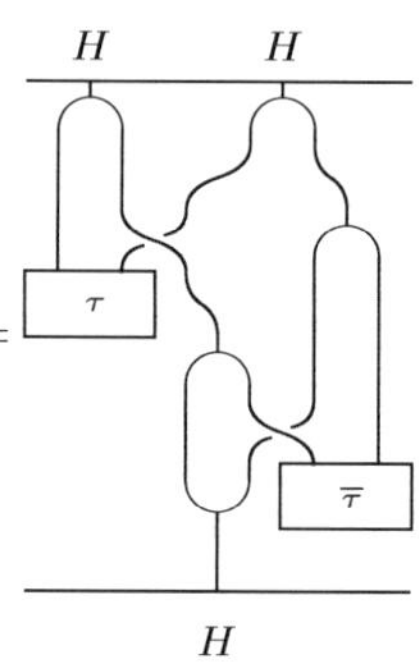 ,

and unit η_H. Then H_τ is as well a bialgebra in $\mathcal{C}$ and, moreover, a Hopf algebra with antipode $S^\tau := (u_\tau \otimes S \otimes u_{\overline{\tau}})(\mathrm{Id}_H \otimes \Delta_H)\Delta_H$, provided that H is a Hopf algebra in $\mathcal{C}$ with antipode S; here $u_\tau := \tau(\mathrm{Id}_H \otimes S)\Delta_H : H \to \underline{1}$, and similar for $u_{\overline{\tau}}$. For more details we refer to [**3**, Definition and Proposition 1.6].

According to [**6**], if $(\mathcal{F}, \varphi_0, \varphi_2) : \mathcal{C} \to \mathcal{D}$ is a braided equivalence and H is a bialgebra in $\mathcal{C}$ then the assignment ($\underline{I}$ is the unit object of $\mathcal{D}$)

$$(4.3)\quad (\tau : H \otimes H \to \underline{1}) \mapsto (\tau_{\mathcal{F}} : \mathcal{F}(H) \otimes \mathcal{F}(H) \xrightarrow{\varphi_{2,H,H}} \mathcal{F}(H \otimes H) \xrightarrow{\mathcal{F}(\tau)} \mathcal{F}(\underline{1}) \xrightarrow{\varphi_0} \underline{I})$$

provides a one to one correspondence between (invertible) 2-cocycles of H and (invertible) 2-cocycles of $\mathcal{F}(H)$, and $\mathcal{F}(H_\tau) = \mathcal{F}(H)_{\tau_{\mathcal{F}}}$ as bialgebras or Hopf algebras.

PROPOSITION 4.1. *Let H be a Hopf algebra in a braided category $\mathcal{C}$ and $A \in {}^H_H\mathcal{YD}(\mathcal{C})$ a bialgebra/Hopf algebra. If $\vartheta : \mathcal{F}_l(A) \otimes_H \mathcal{F}_l(A) \to H$ is an invertible 2-cocycle of $\mathcal{F}_l(A)$ in ${}^H_H\mathcal{C}^H_H$ then there exists an invertible 2-cocycle ϑ_0 of A in ${}^H_H\mathcal{YD}(\mathcal{C})$ such that $\mathcal{F}_l(A)_\vartheta = \mathcal{F}_l(A_{\vartheta_0})$ as bialgebras/Hopf algebras in ${}^H_H\mathcal{C}^H_H$. Furthermore, ϑ*

defines an invertible 2*-cocycle* ϑ_1 *on* $A \times H$ *in such a way that the bialgebra/Hopf algebra in* $\mathcal{C}$ *corresponding to the bialgebra/Hopf algebra* $(A \otimes H)_\vartheta$ *in* ${}^H_H\mathcal{C}^H_H$ *is precisely* $(A \times H)_{\vartheta_1}$*, hence* $(A \times H)_{\vartheta_1} = A_{\vartheta_0} \times H$ *is a biproduct bialgebra/Hopf algebra.*

PROOF. We specialize (4.3) for $\mathcal{F}_l$ and see that any (invertible) 2-cocycle ϑ of $\mathcal{F}_l(A)$ in ${}^H_H\mathcal{C}^H_H$ is of the form $\vartheta = (\vartheta_0 \otimes \mathrm{Id}_H)\varphi^l_{2,A,A}$ for a certain (invertible) 2-cocycle ϑ_0 of A in ${}^H_H\mathcal{YD}$. The latter means that $\vartheta_0 : A \otimes A \to \underline{1}$ is a (convolution invertible) morphism in $\mathcal{C}$ that satisfies the conditions $\vartheta_0(\eta_A \otimes \mathrm{Id}_A) = \varepsilon_A = \vartheta_0(\mathrm{Id}_A \otimes \eta_A)$ and (4.4)

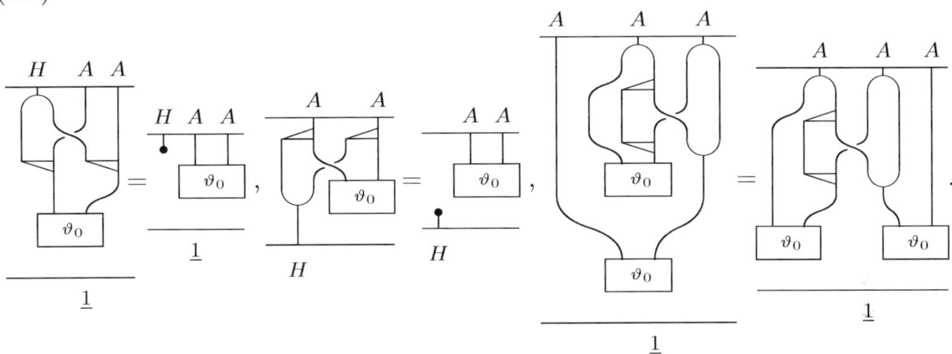

In addition, $\mathcal{F}_l(A)_\vartheta = \mathcal{F}_l(A_{\vartheta_0})$ as bialgebras (resp. Hopf algebras) in $\mathcal{C}$, provided that ϑ is invertible. Thus in this case $(A \otimes H)_\vartheta = A_{\vartheta_0} \otimes H$ as bialgebras/Hopf algebras in ${}^H_H\mathcal{C}^H_H$. Define now

$\vartheta_1 =$ 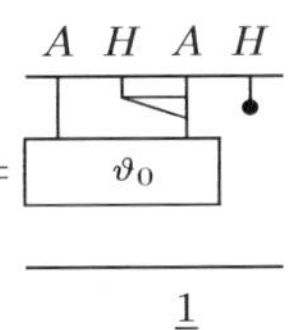 and $\overline{\vartheta_1} =$ 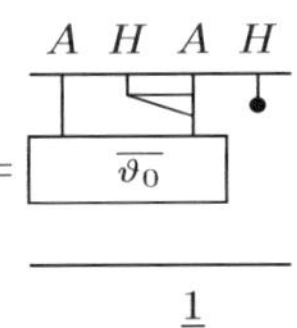 .

It can be easily seen that $\overline{\vartheta_1}$ is the convolution inverse of $\vartheta_1 : A \times H \otimes A \times H \to \underline{1}$, that is $\vartheta_1 * \overline{\vartheta_1} = \varepsilon_{A\times H} \otimes \varepsilon_{A \times H} = \overline{\vartheta_1} * \vartheta_1$; the details are left to the reader. Furthermore, we show that ϑ_1 verifies the 2-cocycle condition for $A \times H$, which in this case is equivalent to

(4.5)

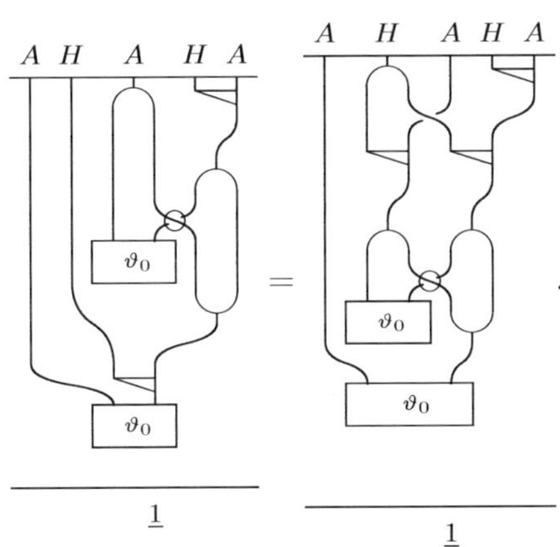

Now, to show that the equivalent condition (4.5) holds it suffices to see that

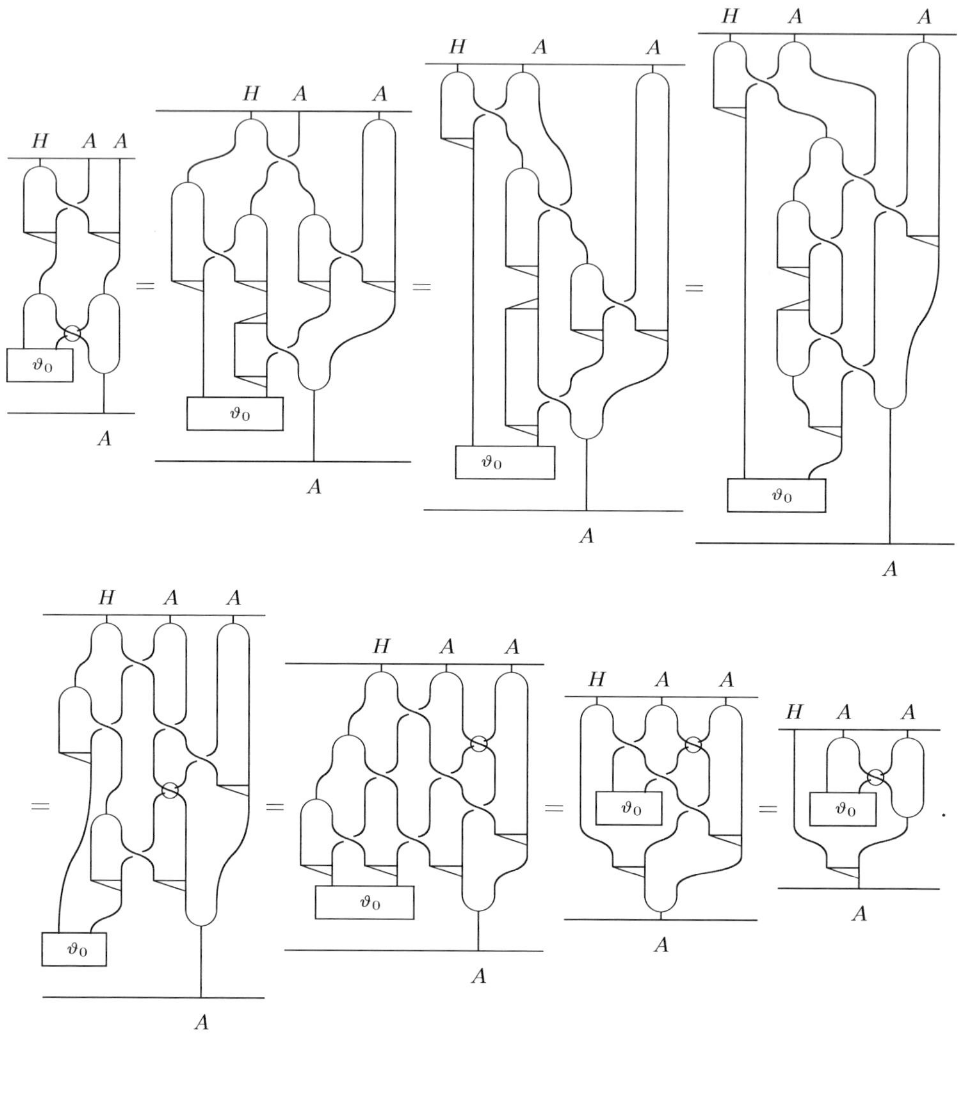

So ϑ_1 is a 2-cocycle of $A \times H$ (clearly ϑ_1 behaves well with respect to the unit of $A \times H$), and we show that the bialgebra (resp. Hopf algebra) in $\mathcal{C}$ corresponding

to $(A \otimes H)_\vartheta$ in ${}^H_H\mathcal{C}^H_H$ equals $(A \times H)_{\vartheta_1}$. If $m^{\vartheta_1} = m_{(A\times H)_{\vartheta_1}}$ and

$m^\vartheta =$ 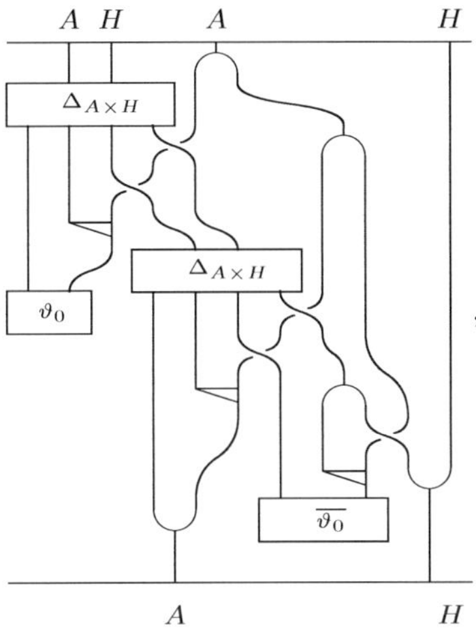 , $m^{\vartheta_1} =$ 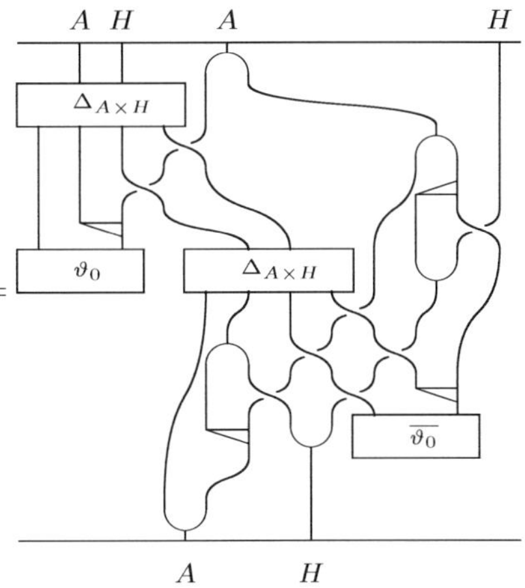

and $m_{(A\otimes H)_\vartheta} q^H_{A\otimes H, A\otimes H} = m^\vartheta$. So we have to show that $m^{\vartheta_1} = m^\vartheta$. Indeed,

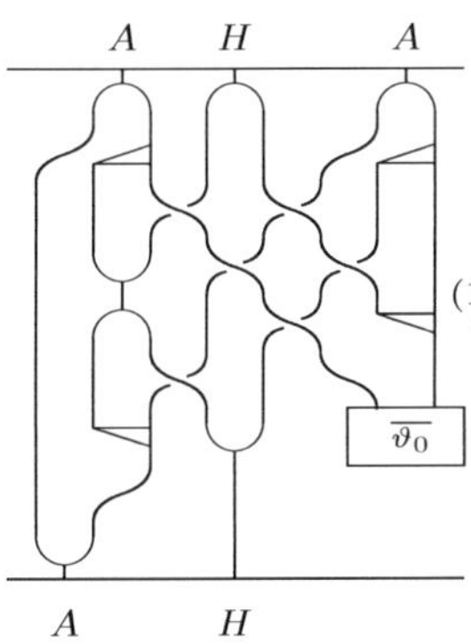 $\overset{(1.1)}{=}$ 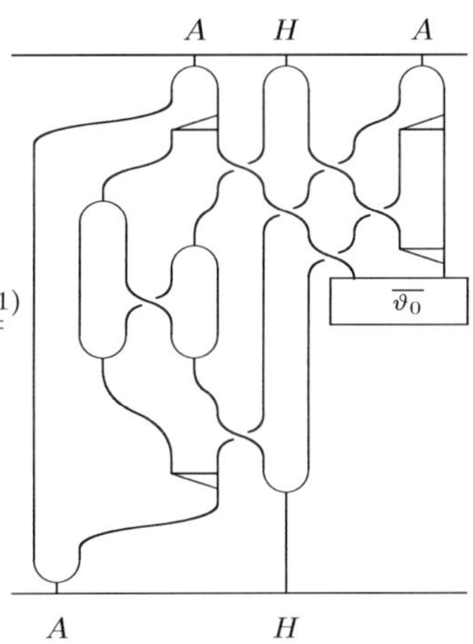 $=$ 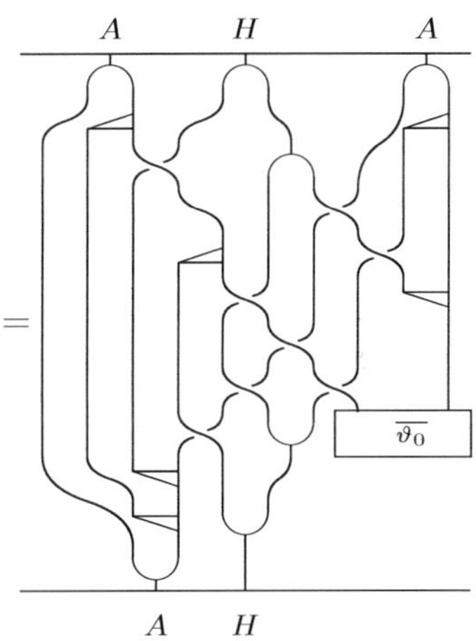

$=$ 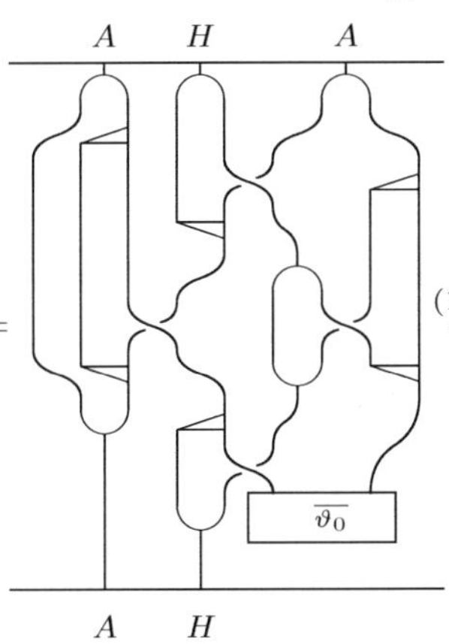 $\overset{(1.5)}{=}$ 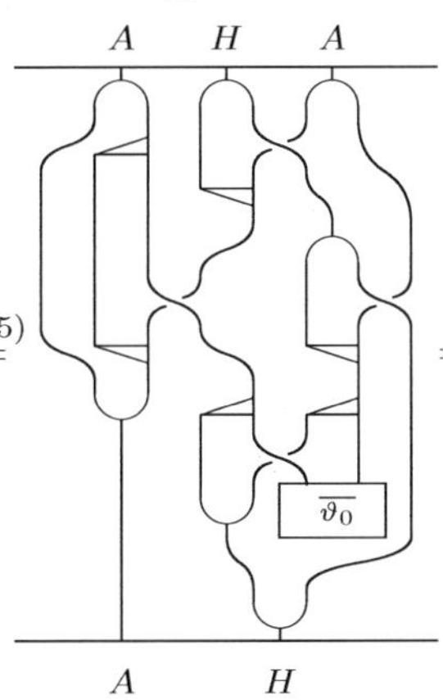 $=$ 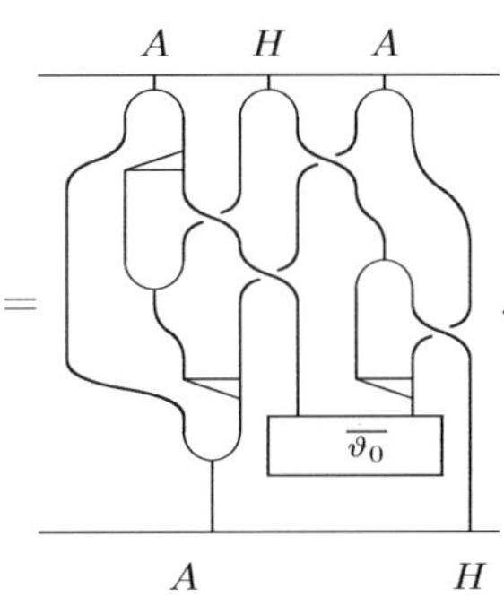 .

The equality we just proved can be rewritten as

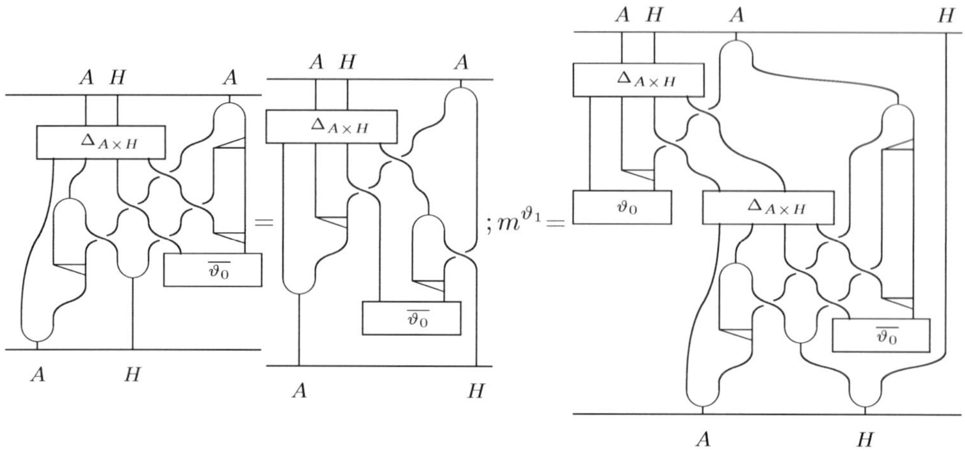

and from here we deduce that $m_{(A\times H)_{\vartheta_1}} = m^{\vartheta_1} = m^{\vartheta}$, as claimed. $\square$

We cannot provide a converse for Proposition 4.1: if ϑ_1 is an invertible 2-cocycle of $A \times H$ there is no guarantee that there is an invertible 2-cocycle ϑ of $A \otimes H$ in ${}^H_H\mathcal{C}^H_H$ such that the Hopf algebra in $\mathcal{C}$ corresponding to $(A \otimes H)_\vartheta \in {}^H_H\mathcal{C}^H_H$ is $(A \times H)_{\vartheta_1}$. For instance, such a situation occurs when we consider 2-cocycles of $A \times H \times B \simeq (A\widetilde{\otimes}\overline{B}) \times H$ of the form $\widehat{\sigma} = \varepsilon_{A\rtimes H} \otimes \sigma \otimes \varepsilon_{H\ltimes B}$, with $\sigma : B \otimes A \to \underline{1}$.

PROPOSITION 4.2. *$\widehat{\sigma}$ is an invertible 2-cocycle of $A \times H \times B$ if and only if*
(4.6)

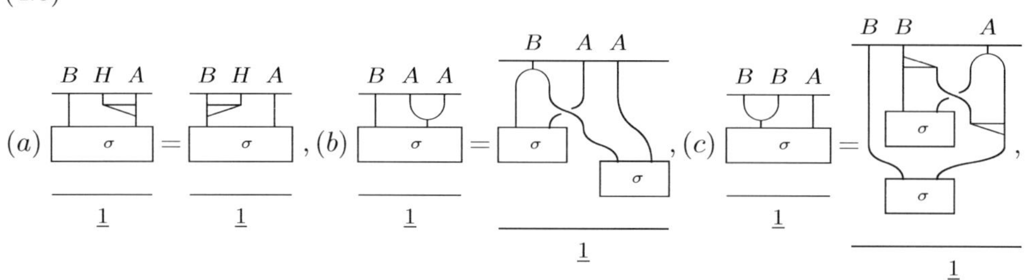

$\sigma(\eta_B \otimes \mathrm{Id}_A) = \varepsilon_A$, $\sigma(\mathrm{Id}_B \otimes \eta_A) = \varepsilon_B$ *and σ is convolution invertible.*

PROOF. The 2-cocycle condition for $\widehat{\sigma}$ reads as follows:

(4.7)

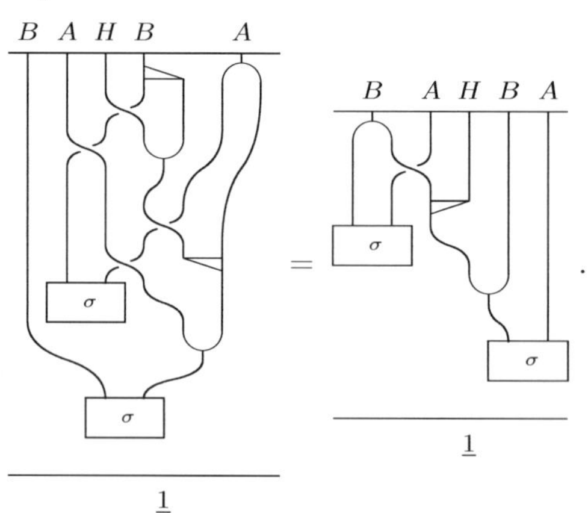

It is clear that $\widehat{\sigma}$ obeys the unital conditions if and only if $\sigma(\eta_B \otimes \mathrm{Id}_A) = \varepsilon_A$ and $\sigma(\mathrm{Id}_B \otimes \eta_A) = \varepsilon_B$. Now, if we compose both sides of (4.7) to the right with $\mathrm{Id}_B \otimes \eta_A \otimes \mathrm{Id}_H \otimes \eta_B \otimes \mathrm{Id}_A$ we get (4.6.a). Likewise, if we compose both sides of (4.7) to the right with $\mathrm{Id}_{B\otimes A} \otimes \eta_H \otimes \eta_B \otimes \mathrm{Id}_A$ we obtain (4.6.b), and by composing both sides of (4.7) to the right with $\mathrm{Id}_B \otimes \otimes\, \eta_A \otimes \eta_H \otimes \mathrm{Id}_{B\otimes A}$ we deduce (4.6.c). Observe that (4.6.a, b, c) are equivalent to (4.6.a, b) and (4.8) below,

(4.8)

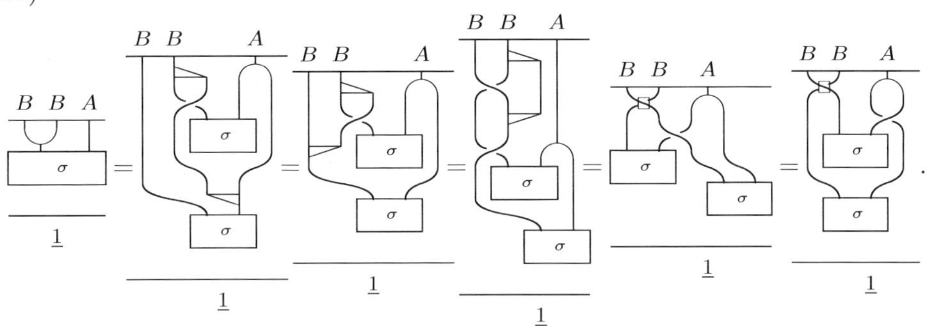

Thus our conditions (4.6.a, b, c) are equivalent to those in [**3**, Proposition 3.4], and everything follows now from the quoted result. Note that the inverse of $\widehat{\sigma}$ is $\varepsilon_{A\bowtie H} \otimes \overline{\sigma} \otimes \varepsilon_{H\bowtie B}$. □

Thus $\widehat{\sigma}$ is an invertible 2-cocycle of $A \times H \times B$ if and only if σ satisfies the conditions (4.6.a-c) or, equivalently, (4.6.a, b) and (4.8). Parallel to the classical Hopf algebra case, we call such a morphism σ an H-balanced mixed skew pairing of B and A; H-balanced because of (4.6.a), mixed because of (4.8) and skew pairing because of (4.6.b) and (4.8).

Owing to the Hopf algebra isomorphism $\chi : A \times H \times B \to (A\widetilde{\otimes}\overline{B})\#H$, there exists a unique invertible 2-cocycle of $(A\widetilde{\otimes}\overline{B}) \times H$, denoted by ω, that turns $\chi : (A \times H \times B)_{\widehat{\sigma}} \to ((A\widetilde{\otimes}\overline{B}) \times H)_\omega$ into a Hopf algebra isomorphism. Explicitly, ω is given by

$\omega = \widehat{\sigma}(\chi^{-1} \otimes \chi^{-1}) =$ 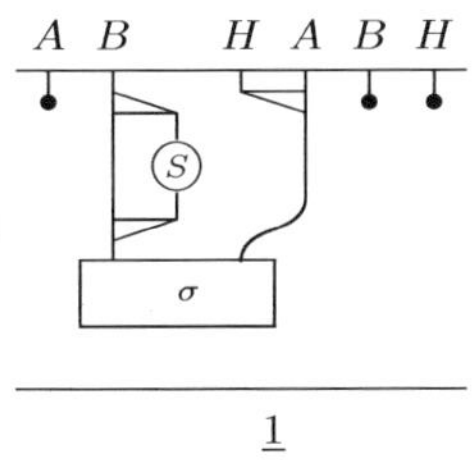 $=$ 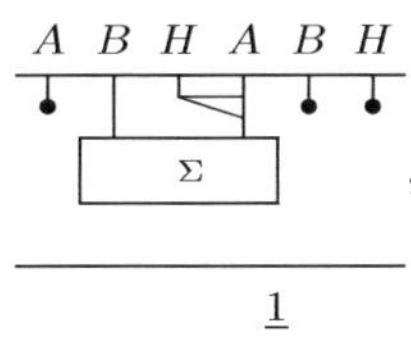 , where $\Sigma =$ 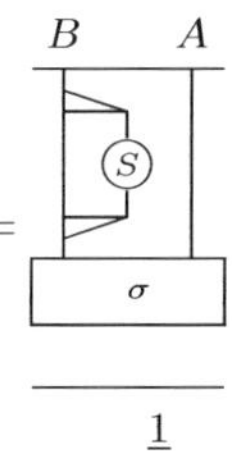 .

Note that when σ is invertible, $\widehat{\sigma}$ invertible implies ω invertible and its inverse is defined by a morphism $\overline{\Sigma} : B \otimes A \to \underline{1}$ in $\mathcal{C}$ determined by $\overline{\sigma}$ in the same way as Σ is determined by σ.

For simplicity, we replace $\overline{B}$ by an arbitrary Hopf algebra in ${}^H_H\mathcal{YD}(\mathcal{C})$ and study when an ω defined by a Σ as in the above is a 2-cocycle of $(A\widetilde{\otimes}C) \times H$.

PROPOSITION 4.3. *Let A, C be bialgebras in ${}^H_H\mathcal{YD}(\mathcal{C})$ and denote by $A\widetilde{\otimes}C$ their tensor product bialgebra in ${}^H_H\mathcal{YD}(\mathcal{C})$, so it is implicitly understood that $c_{A,C} \circ c_{C,A} = \mathrm{Id}_{C\otimes A}$. If $\Sigma : C \otimes A \to \underline{1}$ is a morphism in $\mathcal{C}$ then $\omega := \varepsilon_A \otimes \Sigma(\mathrm{Id}_C \otimes \mu^H_A) \otimes \varepsilon_B \otimes \varepsilon_H$ is a 2-cocycle of $(A\widetilde{\otimes}C) \times H$ if and only if $\Sigma(\eta_C \otimes \mathrm{Id}_A) = \varepsilon_A$, $\Sigma(\mathrm{Id}_C \otimes \eta_A) = \varepsilon_C$,* *and the following equalities hold:*

(4.9)

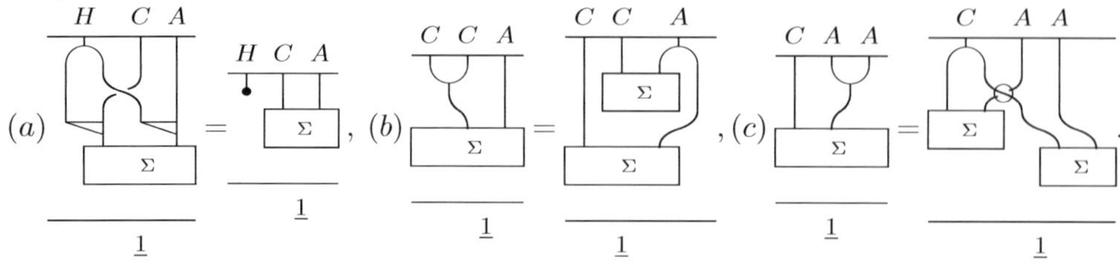

Furthermore, $\overline{\omega} := \varepsilon_A \otimes \overline{\Sigma}(\mathrm{Id}_C \otimes \mu_A^H) \otimes \varepsilon_B \otimes \varepsilon_H$ *is inverse for* ω *if and only if*

(4.10)

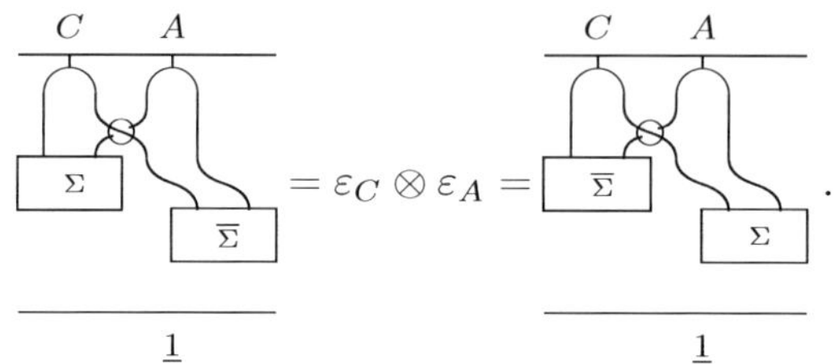

PROOF. Clearly, ω behaves well with respect to the unit of $(A\tilde{\otimes}C) \times H$ if and only if $\Sigma(\eta_C \otimes \mathrm{Id}_A) = \varepsilon_A$ and $\Sigma(\mathrm{Id}_C \otimes \eta_A) = \varepsilon_C$. Also, ω obeys the 2-cocycle condition for $(A\tilde{\otimes}C) \times H$ if and only if

(4.11)

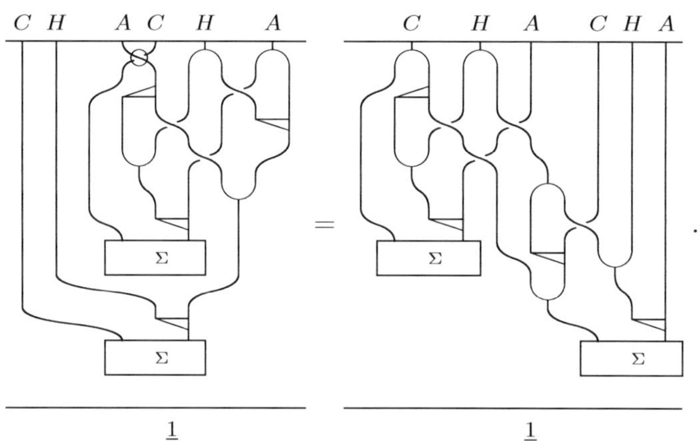

If we compose both sides of (4.11) to the right with $\eta_C \otimes \mathrm{Id}_{H\otimes A\otimes C} \otimes \eta_H \otimes \mathrm{Id}_A$ we land at (4.9.a), i.e. at the fact that Σ is left H-linear in $\mathcal{C}$. If we compose both sides of (4.11) to the right with $\mathrm{Id}_C \otimes \eta_H \otimes \mathrm{Id}_{A\otimes C} \otimes \eta_H \otimes \mathrm{Id}_A$ we get that

(4.12)

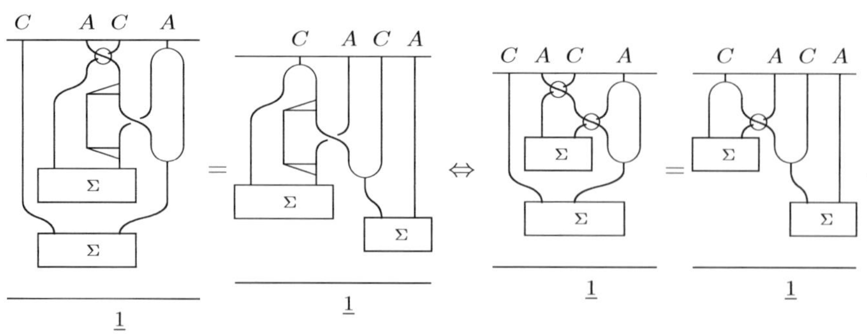

By composing the both sides of (4.12) to the right with $\mathrm{Id}_C \otimes \eta_A \otimes \mathrm{Id}_{C\otimes A}$ and $\mathrm{Id}_{C\otimes A} \otimes \eta_C \otimes \mathrm{Id}_A$ we obtain (4.9.b) and (4.9.c), respectively.

Conversely, although Σ is not a morphism in ${}^H_H\mathcal{YD}(\mathcal{C})$ (and so the naturality of the braiding ${}^l\mathfrak{c}$ does not apply to Σ), the fact that Σ is left H-linear implies

(4.13) 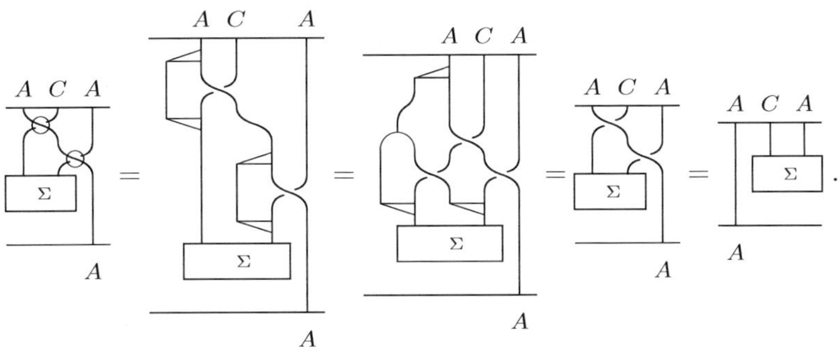

We can compute now that

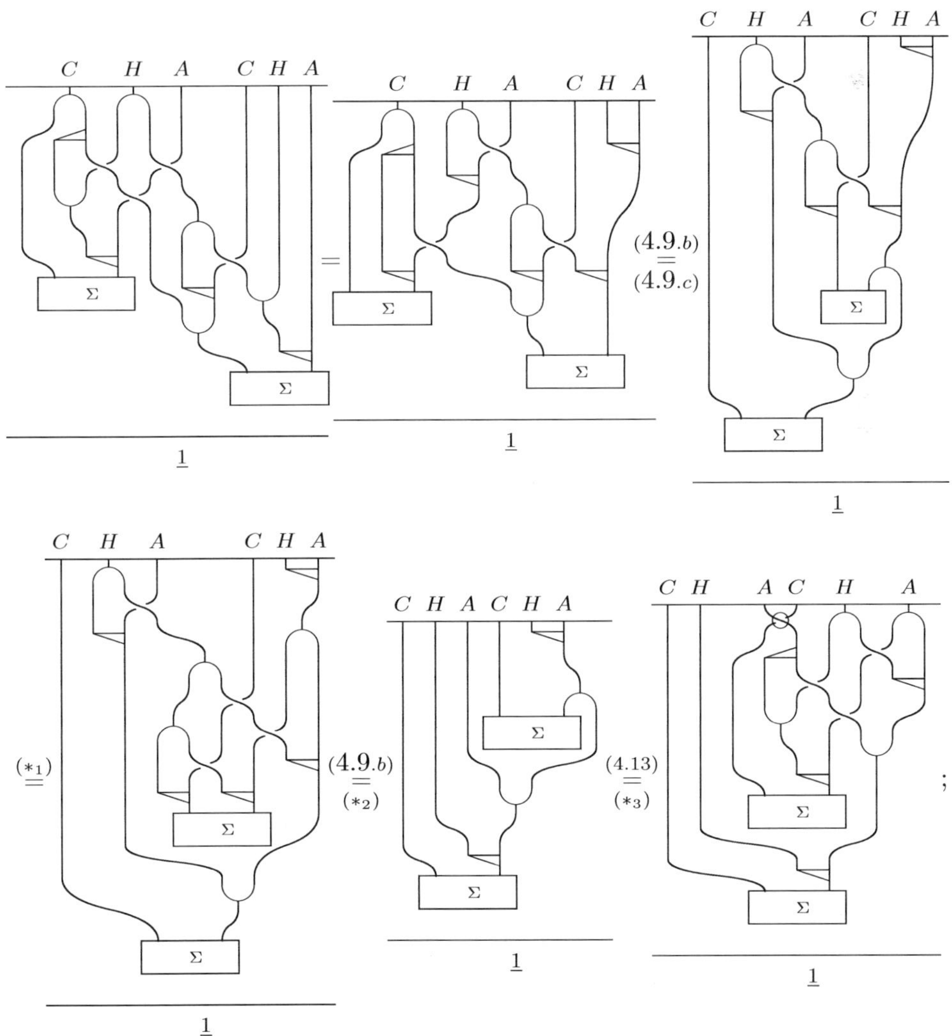

$(*_1)$ refers to the fact that Δ_A is in ${}_H\mathcal{C}$ and to the naturality of c applied to Δ_H, $(*_2)$ refers to the naturality of c applied to Σ and to the fact that m_A is in ${}_H\mathcal{C}$, and $(*_3)$ refers to the naturality of the braiding c applied to μ_A^H and the associativity of μ_A^H. Finally, ω, $\overline{\omega}$ are convolution inverses if and only if (4.10) holds. □

So $\widehat{\sigma}$ is an invertible 2-cocycle of $A \times H \times B$ if and only if ω is an invertible 2-cocycle of $(A\widetilde{\otimes}\overline{B}) \times H$, if and only if $\Sigma : \overline{B}\widetilde{\otimes}A \to \underline{1}$ is an almost invertible skew pairing of $\overline{B}$, A in ${}_H^H\mathcal{YD}(H)$; almost since Σ is not in ${}^H\mathcal{C}$. Note that σ can be recovered from Σ as $\Sigma(\mathrm{Id}_B \otimes \mu_A^H)(\rho_B^H \otimes \mathrm{Id}_A)$, and that these correspondences $\sigma \leftrightarrow \Sigma$ are inverses of each other. Hence, H-balanced invertible mixed skew pairings of B, A in $\mathcal{C}$ are in bijection with invertible skew pairings of $\overline{B}$, A in ${}_H^H\mathcal{YD}(\mathcal{C})$.

We next see when σ is defined by an invertible skew pairing of $\overline{B}$, A in ${}_H^H\mathcal{YD}(\mathcal{C})$.

LEMMA 4.4. *$\Sigma : \overline{B}\widetilde{\otimes}A \to \underline{1}$ is left H-colinear if and only if σ satisfies*

$$(\mathrm{Id}_H \otimes \sigma)(c_{B,H} \otimes \mathrm{Id}_A)(\mathrm{Id}_B \otimes \lambda_A^H) = (\sigma \otimes \mathrm{Id}_H)(\mathrm{Id}_B \otimes c_{H,A})(\rho_B^H \otimes \mathrm{Id}_A). \tag{4.14}$$

PROOF. $\Sigma = \sigma(\mathrm{Id}_B \otimes \mu_A^H)(\mathrm{Id}_B \otimes S \otimes \mathrm{Id}_A)(\rho_B^H \otimes \mathrm{Id}_A)$ left H-colinear reads as

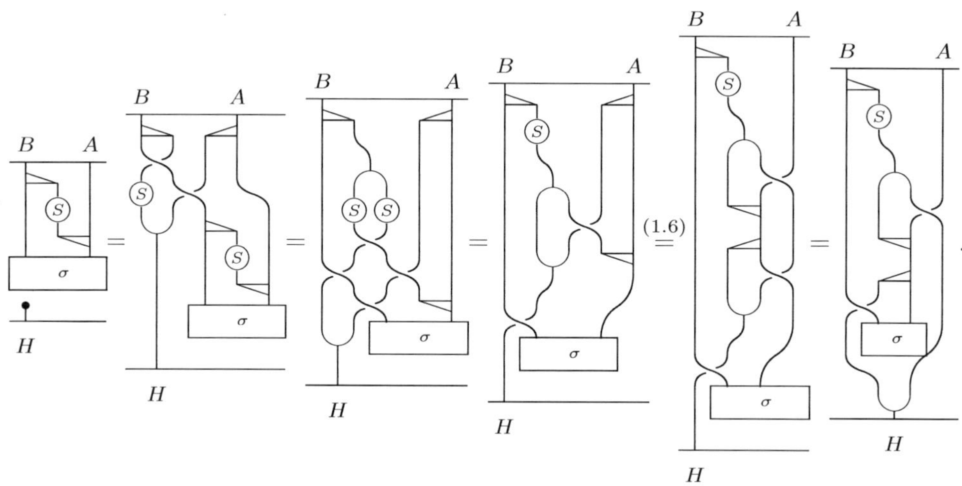

Now, the relation (4.14) implies that Σ is left H-colinear since

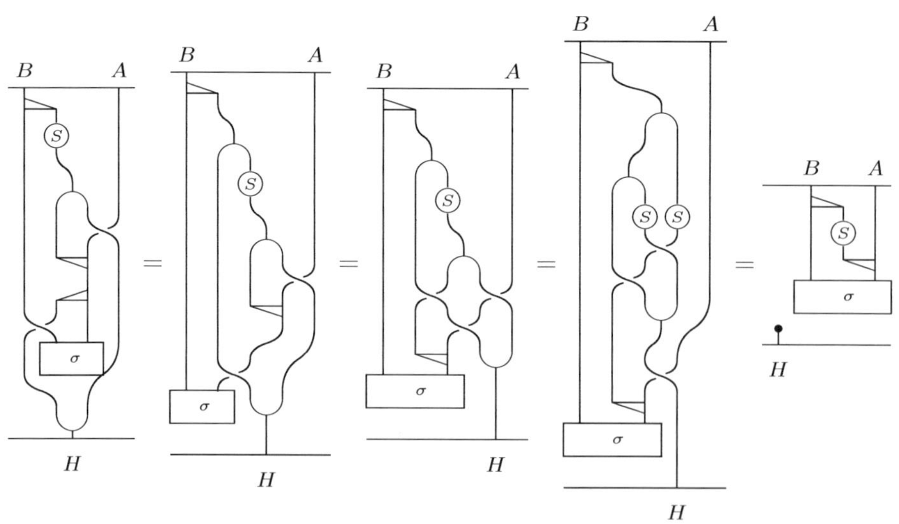

(In the second equality we applied the naturality of c to μ_A^H, in the third equality the naturality of c applied to m_H, (1.4) and the coassociativity of Δ_H, and in the last equality the naturality of c applied to m_H and (1.3).)

To show that Σ in ${}^H\mathcal{C}$ implies (4.14), we first notice that Σ in ${}^H\mathcal{C}$ leads to

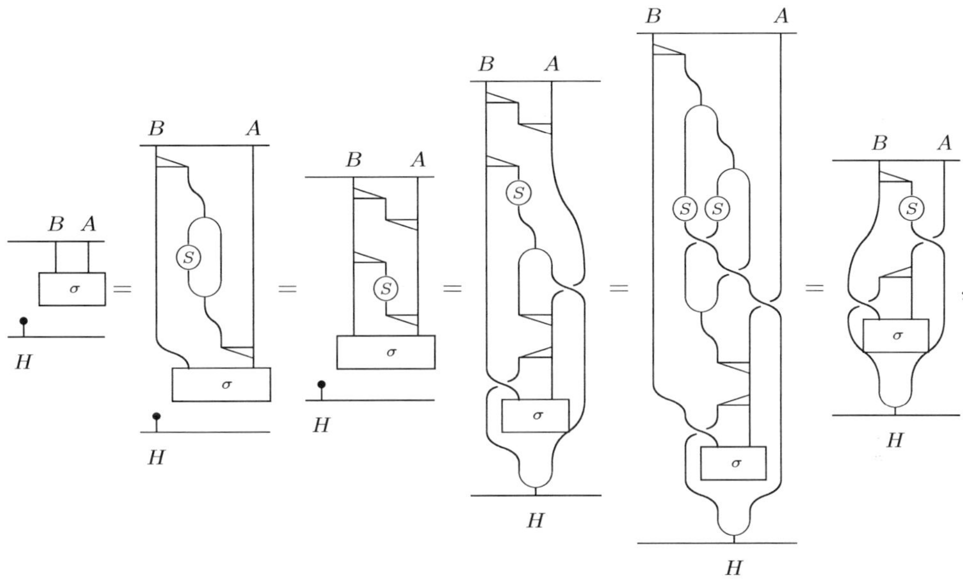

and with the help of this relation we compute that

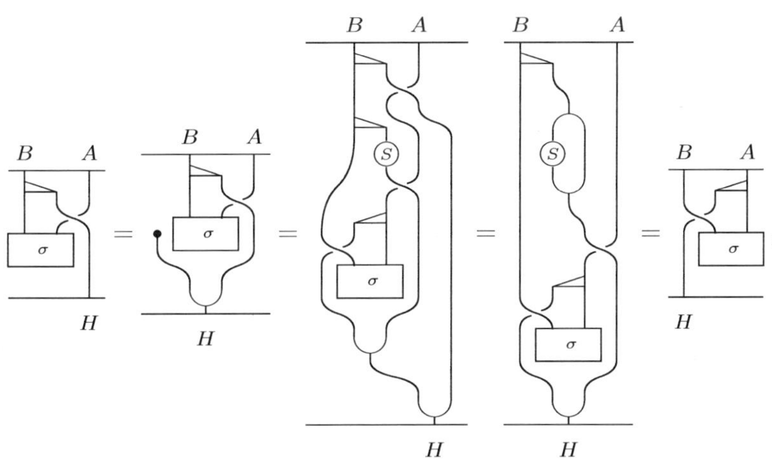

as desired. So our proof is finished. $\square$

We say that $\sigma : B \otimes A \to \underline{1}$ is H-cobalanced if it obeys (4.14).

REMARKS 4.5. In Hopf algebra theory, the connection between skew pairings and some 2-cocycles was established in [**10**]. To be able to state and prove the main result of this section we need a similar connection, but now in the braided monoidal setting.

(i) Let $(\mathcal{C}, c)$ be a braided category and A, A' bialgebras in $\mathcal{C}$ such that $A\otimes A'$ is a bialgebra in $\mathcal{C}$, too. Consider $\tau : A\otimes A'\otimes A\otimes A' \to \underline{1}$ of the form $\tau = \eta_A \otimes \Sigma \otimes \varepsilon_{A'}$, where $\Sigma : A' \otimes A \to \underline{1}$ is a given morphism in $\mathcal{C}$. Then τ is a 2-cocycle of $A \otimes A'$ if and only if $\Sigma(\eta_{A'} \otimes \mathrm{Id}_A) = \varepsilon_A$, $\Sigma(\mathrm{Id}_{A'} \otimes \varepsilon_A) = \varepsilon_{A'}$,

$$(4.15) \qquad \begin{aligned} \Sigma(m_{A'} \otimes \mathrm{Id}_A) &= \Sigma(\mathrm{Id}_{A'} \otimes \Sigma \otimes \mathrm{Id}_A)(\mathrm{Id}_{A' \otimes A'} \otimes \Delta_A), \\ \Sigma(\mathrm{Id}_{A'} \otimes m_A) &= (\Sigma \otimes \Sigma)(\mathrm{Id}_{A'} \otimes c_{A',A} \otimes \mathrm{Id}_A)(\Delta_{A'} \otimes \mathrm{Id}_{A \otimes A}). \end{aligned}$$

Comparing to (4.9), τ is a 2-cocycle if and only if Σ is a skew paring of A', A.

Again by analogy with the classical Hopf algebra case, perhaps it is more adequate to call a skew pairing of A, A' a morphism $\theta : A \otimes A' \to \underline{1}$ obeying

$$(4.16) \qquad \begin{aligned} \theta(m_A \otimes \mathrm{Id}_{A'}) &= \theta(\mathrm{Id}_A \otimes (\mathrm{Id}_{A'} \otimes \theta)(c_{A,A'} \otimes \mathrm{Id}_{A'})(\mathrm{Id}_A \otimes \Delta_{A'})), \\ \theta(\mathrm{Id}_A \otimes m_{A'}) &= (\theta \otimes \theta)((\mathrm{Id}_A \otimes c_{A,A'})(c_{A,A}\Delta_A \otimes \mathrm{Id}_{A'}) \otimes \mathrm{Id}_{A'}) \end{aligned}$$

and the unital relations. Nevertheless, these two definitions for a skew paring are equivalent since Σ is a skew pairing of A', A in our sense if and only if $\theta := \Sigma c^{-1}_{A',A}$ is a skew pairing, let's say in the spirit of Doi and Takeuchi [**10**]. Note that another definition for pairing was proposed in [**3**, Proposition 3.6].

(ii) If $\Sigma : A' \otimes A \to \underline{1}$ is a skew pairing (in our sense) and A is a Hopf algebra in $\mathcal{C}$ then it is automatic that Σ is convolution invertible with inverse $\overline{\Sigma} := \Sigma(\mathrm{Id}_{A'} \otimes S_A)$, where S_A is the antipode of A. Likewise, if A' is a Hopf algebra with skew antipode in $\mathcal{C}$ then Σ is invertible and $\overline{\Sigma} = \Sigma(\overline{S}_{A'} \otimes \mathrm{Id}_A)$. Hence $\Sigma(S_{A'} \otimes S_A) = \Sigma$, provided that A, A' are Hopf algebras and $S_{A'}$ is invertible.

For H, A, B as above, let $\sigma : B \otimes A \to \underline{1}$ be an H-balanced invertible mixed skew pairing of B and A, so that $\widehat{\sigma} = \varepsilon_{A \rtimes H} \otimes \sigma \otimes \varepsilon_{H \ltimes B}$ is an invertible 2-cocycle of $A \times H \times B$. Then the unique invertible 2-cocycle ω that turns $\chi : (A \times H \times B)_{\widehat{\sigma}} \to ((A \widetilde{\otimes} \overline{B}) \times H)_\omega$ into a Hopf algebra isomorphism is completely determined by an almost skew pairing Σ of $\overline{B}$ and A in ${}^H_H\mathcal{YD}(\mathcal{C})$; recall that Σ is defined by σ and that Σ is an invertible skew pairing of $\overline{B}$ and A in ${}^H_H\mathcal{YD}(\mathcal{C})$ if and only if σ is, moreover, H-cobalanced in $\mathcal{C}$. If all these conditions are satisfied, we call (H, A, B, σ) a biproduct Drinfeld data. The terminology we use is justified by the following.

THEOREM 4.6. *Let (H, A, B, σ) be a biproduct Drinfeld data in a braided category $(\mathcal{C}, c)$. Then $(A \times H \times H)_{\widehat{\sigma}}$, the Hopf algebra obtained from the double-biproduct Hopf algebra $A \times H \times B$ and the 2-cocycle $\widehat{\sigma}$, is isomorphic as a Hopf algebra to a biproduct Hopf algebra in $\mathcal{C}$.*

PROOF. We are in the situation when $\Sigma : \overline{B} \otimes A \to \underline{1}$ is a skew pairing of $\overline{B}$, A in ${}^H_H\mathcal{YD}(\mathcal{C})$. By Remarks 4.5, Σ is convolution invertible with $\overline{\Sigma} = \Sigma(\mathrm{Id}_B \otimes S_A)$, where S_A is the antipode of A in ${}^H_H\mathcal{YD}(\mathcal{C})$. Thus, by the same Remarks 4.5, we see that $\vartheta_0 := \varepsilon_A \otimes \Sigma \otimes \varepsilon_B : (A \widetilde{\otimes} \overline{B}) \otimes (A \widetilde{\otimes} \overline{B}) \to \underline{1}$ is an invertible 2-cocycle of $A \widetilde{\otimes} B$ in ${}^H_H\mathcal{YD}(\mathcal{C})$. Hence, by Proposition 4.1, we have a 2-cocycle ϑ of $(A \widetilde{\otimes} \overline{B}) \otimes H$ in ${}^H_H\mathcal{C}^H_H$, given by $\vartheta = (\vartheta_0 \otimes \mathrm{Id}_H)\varphi^l_{2, A\widetilde{\otimes}\overline{B}, A\widetilde{\otimes}\overline{B}}$. So ϑ is completely determined by

$\vartheta q^H_{A\widetilde{\otimes}\overline{B}, A\widetilde{\otimes}\overline{B}} =$ 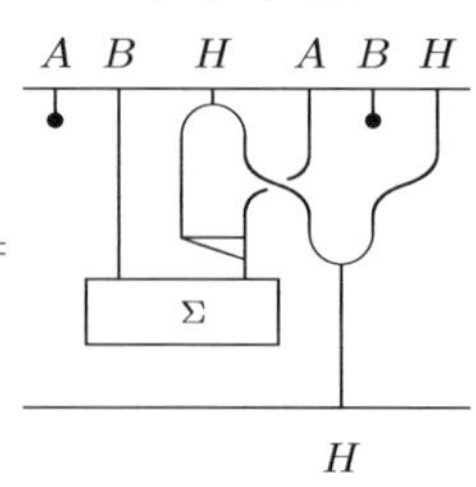

and defines $\vartheta_1 =$ 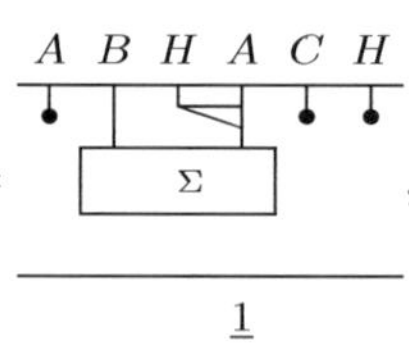

,

an invertible 2-cocycle of $(A \widetilde{\otimes} B) \times H$, see Proposition 4.1. But $\vartheta_1 = \omega$, and the same Proposition 4.1 yields $(A \times H \times B)_{\widehat{\sigma}} \simeq ((A \widetilde{\otimes} \overline{B}) \times H)_\omega = (A \widetilde{\otimes} \overline{B})_{\vartheta_0} \times H$, a biproduct Hopf algebra in $\mathcal{C}$. We end by noting that $(A\widetilde{\otimes}\overline{B})_{\vartheta_0}$ is called a generalized quantum double and is usually denoted by $D_\Sigma(A,\overline{B})$. We leave it to the reader to check that the multiplication and the antipode of $D_\Sigma(A,\overline{B})$ are given by

(4.17)
$$(m_A\otimes m_{\overline{B}})(\mathrm{Id}_A\otimes(\Sigma\otimes\mathrm{Id}_{A\otimes\overline{B}}\otimes\overline{\Sigma})(\mathrm{Id}_{\overline{B}}\otimes(\Delta_A\otimes\Delta_{\overline{B}})^l\mathfrak{c}_{\overline{B},A}\otimes\mathrm{Id}_A)(\Delta_{\overline{B}}\otimes\Delta_A)\otimes\mathrm{Id}_{\overline{B}});$$
$$(S_A\otimes S_B\otimes\overline{\Sigma})(\mathrm{Id}_A\otimes(\Delta_{\overline{B}}\otimes S_A)^l\mathfrak{c}_{A,\overline{B}})(\Delta_A(\overline{\Sigma}\otimes\mathrm{Id}_A)(\mathrm{Id}_{\overline{B}}\otimes\Delta_A)^l\mathfrak{c}_{A,\overline{B}}\otimes\mathrm{Id}_{\overline{B}})(\mathrm{Id}_A\otimes\Delta_{\overline{B}}).$$

As coalgebra, $D_\Sigma(A,\overline{B})=A\widetilde{\otimes}\overline{B}$, and the unit is $\eta_A\otimes\eta_B$. □

5. Some examples

Our first goal is to characterize double-biproduct Hopf algebras as biproduct Hopf algebras; then we present the case when the deformation by a 2-cocycle of a double-biproduct Hopf algebra still remains a biproduct Hopf algebra. The second aim is to give various concrete examples.

5.1. Double-biproduct Hopf algebras. Let k be a field. We will specialize the results obtained in this paper for $\mathcal{C}={}_k\mathcal{M}$, the category of k-vector spaces. For a Hopf algebra H in ${}_k\mathcal{M}$ we make use of the Sweedler notation for the comultiplication Δ_H: $\Delta_H(h)=h_1\otimes h_2$, for all $h\in H$. A left (resp. right) H-coaction on a k-vector space V will be denoted by $\lambda_V^H: V\ni v\mapsto v_{\{-1\}}\otimes v_{\{0\}}\in H\otimes V$ (resp. $\rho_V^H: V\ni v\mapsto v_{(0)}\otimes v_{(1)}\in V\otimes H$). Also, the categories of left and right Yetter-Drinfeld modules over H in ${}_k\mathcal{M}$ will be denoted by ${}_H^H\mathcal{YD}$ and $\mathcal{YD}_H^H$. If A is a bialgebra in ${}_H^H\mathcal{YD}$, for the comultiplication Δ_A of A in ${}_H^H\mathcal{YD}$ we use the Sweedler type notation $\Delta_A(a)=a_{\underline{1}}\otimes a_{\underline{2}}$, for all $a\in A$; a similar notation we use for the comultiplication Δ_B of a bialgebra B in $\mathcal{YD}_H^H$: $\Delta_B(b)=b_{\underline{1}}\otimes b_{\underline{2}}$, for all $b\in B$. When A (resp. B) is, moreover, a braided Hopf algebra, by S_A (resp. S_B) we denote the antipode of A (resp. B); the antipode of H will be denoted by S.

For A an algebra in the monoidal category of left H-representations ${}_H\mathcal{M}$ ($H\otimes A\ni h\otimes a\mapsto h\cdot a\in A$ will be our notation for the left H-action on A), one can define on $A\otimes H$ a k-algebra structure (denoted by $A\#H$ and called the left smash product of A and H; the elements of $A\#H$ will be denoted by $a\#h$). The multiplication of $A\#H$ is given, for all $a,a'\in A$ and $h,h'\in H$, by $(a\#h)(a'\#h')=a(h_1\cdot a')\#h_2h'$, and its unit is $1_A\#1_H$. Likewise, to B an algebra in the monoidal category $\mathcal{M}_H$ of right H-representations (the right H-coaction on B is denoted by $B\otimes H\ni b\otimes h\mapsto b\cdot h\in B$) one can define a k-algebra structure on $B\otimes H$ as follows: $(b\#h)(b'\#h')=hh'_1\#(b\cdot h'_2)b'$, for all $h,h'\in H$ and $b,b'\in B$. $H\#B$ is unital with unit $1_H\#1_B$ and is called the right smash product algebra of H and B.

Left and right smash product algebras identify when we work with algebras in categories of Yetter-Drinfeld modules over a Hopf algebra; see Proposition 3.2.

Corollary 5.1. *Let H be a Hopf algebra with bijective antipode and B an algebra within $\mathcal{YD}_H^H$. We denote by $\overline{B}$ the k-vector space B regarded as an algebra in ${}_H^H\mathcal{YD}$ with structure given, for all $h\in H$ and $b,b'\in B$, by*

$$h\triangleright b=b\cdot S^{-1}(h),\ \lambda_{\overline{B}}^H:\overline{B}\ni b\mapsto S(b_{(1)})\otimes b_{(0)}\in H\otimes\overline{B},\ b\bullet b'=b_{(0)}(b'\cdot b_{(1)}),\ 1_{\overline{B}}=1_B,$$

where the juxtaposition stands for the multiplication of B in $\mathcal{YD}_H^H$. Then $\omega_B:\overline{B}\#H\to H\#B$ determined by $\omega_B(b\#h)=S(b_{(1)_2})h_1\#b_{(0)}\cdot S(b_{(1)_1})h_2$, for all $b\in B$ and $h\in H$, is an isomorphism of k-algebras. Its inverse is $\omega_B^{-1}:H\#B\to\overline{B}\#H$ given by $\omega_B^{-1}(h\#b)=b_{(0)}\cdot S^{-1}(h_1)\#h_2b_{(1)}$, for all $h\in H$ and $b\in B$.

A similar result holds for coalgebras. If A is a coalgebra in the monoidal category of left H-corepresentations ${}^H\mathcal{M}$, $A\otimes H$ is a k-coalgebra with comultiplication $\Delta(a\bowtie h)=a_{\underline{1}}\bowtie (a_{\underline{2}})_{\{-1\}}h_1\otimes (a_{\underline{2}})_{\{0\}}\bowtie h_2$ and counit $\varepsilon(a\bowtie h)=\varepsilon_A(a)\varepsilon_H(h)$, for all $a\in A$, $h\in H$. $A\otimes H$ equipped with this structure is called the left smash product coalgebra of A and H and is denoted by $A\bowtie H$ (this justifies why we write $a\bowtie h$ in place of $a\otimes h$). For the right version of a smash product coalgebra, we start with B a coalgebra in $\mathcal{M}^H$, the category of right H-corepresentations. Then $H\ltimes B$ is $H\otimes B$ equipped with the coalgebra structure, $h\in H$, $b\in B$,

$$\Delta(h\ltimes b)=h_1\ltimes (b_{\underline{1}})_{(0)}\otimes h_2(b_{\underline{1}})_{(1)}\ltimes b_{\underline{2}} \text{ and } \varepsilon(h\ltimes b)=\varepsilon_H(h)\varepsilon_B(b).$$

We wrote $b\ltimes h$ in place of $b\otimes h$ to distinguish this k-coalgebra structure on $H\otimes B$.

By specializing Proposition 3.5 to ${}_k\mathcal{M}$ we get the following.

Corollary 5.2. *Let H be a Hopf algebra with bijective antipode and $B\in\mathcal{YD}^H_H$ a coalgebra. Denote by $\overline{B}$ the k-vector space equipped with the following coalgebra structure in ${}^H_H\mathcal{YD}$: $h\triangleright b=b\cdot S^{-1}(h)$, $\lambda^H_{\overline{B}}:\overline{B}\ni b\mapsto S(b_{(1)})\otimes b_{(0)}\in H\otimes B$,*

$$\Delta_{\overline{B}}(b)=(b_{\underline{1}})_{(0)}\otimes b_{\underline{2}}\cdot S^{-1}((b_{\underline{1}})_{(1)}),\ \varepsilon_{\overline{B}}=\varepsilon_B.$$

Then $\overline{B}\bowtie H$ and $H\ltimes B$ are isomorphic k-coalgebras, the isomorphism being produced by the same linear map ω_B as in Corollary 5.1.

A famous result of Radford [**20**] asserts that a left (resp. right) smash product algebra and coalgebra structure afford on $A\otimes H$ (resp. $H\otimes B$) a bialgebra/Hopf algebra structure if and only if A (resp. B) is a bialgebra/Hopf algebra in ${}^H_H\mathcal{YD}$ (resp. $\mathcal{YD}^H_H$). If this is the case, we denote by $A\times H$ (resp. $H\times B$) the bialgebra/Hopf algebra equals $A\#H$ and $A\bowtie H$ (resp. $H\#B$ and $H\ltimes B$) as an algebra and coalgebra. Note that in the Hopf case the antipode of $A\times H$ (resp. $B\times H$) is defined, for all $a\in A$ and $h\in H$ (resp. for all $h\in H$ and $b\in B$), by

$$\begin{aligned} {}^l s(a\times h) &= (1_A\times S(a_{\{-1\}}h))(S_A(a_{\{0\}})\times 1_H)\\ (\text{resp. } {}^r s(h\times b) &= (1_H\times S_B(b_{(0)}))(S(hb_{(1)})\times 1_B)). \end{aligned}$$

By Corollaries 5.1 and 5.2 we have the following result; see also the comments made after the proof of Lemma 3.7. As before, H has bijective antipode.

Corollary 5.3. *Let $B\in\mathcal{YD}^H_H$ be a bialgebra (resp. Hopf algebra) and denote by $\overline{B}$ the k-vector space B endowed with the algebra and coalgebra structure in ${}^H_H\mathcal{YD}$ presented in Corollaries 5.1 and 5.2. Then $\overline{B}$ is a bialgebra (resp. Hopf algebra with antipode equals S_B) in ${}^H_H\mathcal{YD}$, and ω_B considered as a morphism from $\overline{B}\times H$ to $H\times B$ is a bialgebra (and so automatically a Hopf algebra) isomorphism.*

The next result completes [**17**, Theorem A1]; again, H has bijective antipode.

Theorem 5.4. *Let H be a Hopf algebra and $A\in{}^H_H\mathcal{YD}$, $B\in\mathcal{YD}^H_H$.*

(i) If A, B are monoidal algebras, $A\otimes H\otimes B$ is a k-algebra with multiplication

$$(a\#h\#b)(a'\#h'\#b')=a(h_1\cdot a')\#h_2h'_1\#(b\cdot h'_2)b',$$

for all $a,a'\in A$, $h,h'\in H$ and $b,b'\in B$, and unit $1_A\#1_H\#1_B$, where we wrote $a\#h\#b$ in place of $a\otimes h\otimes b$ in order to distinguish this algebra structure on $A\otimes H\otimes B$ (denoted by $A\#H\#B$ and called the double-smash product of A,H,B).

(ii) If A,B are monoidal coalgebras, $A\otimes H\otimes B$ is a k-coalgebra with

$$\Delta(a\bowtie h\ltimes b)=a_{\underline{1}}\bowtie (a_{\underline{2}})_{\{-1\}}h_1\ltimes (b_{\underline{1}})_{(0)}\otimes (a_{\underline{2}})_{\{0\}}\bowtie h_2(b_{\underline{1}})_{(1)}\ltimes b_{\underline{2}}$$

and $\varepsilon(a \rtimes h \ltimes b) = \varepsilon_A(a)\varepsilon(h)\varepsilon_B(b)$, $a \in A$, $h \in H$, $b \in B$; *we wrote* $a \rtimes h \ltimes b$ *in place of* $a \otimes h \otimes b$ *to distinguish this coalgebra structure on* $A \otimes H \otimes B$ *(denoted by* $A \rtimes H \ltimes B$ *and called the double-smash coproduct coalgebra of* A, H, B*).*

(iii) Suppose that A, B *are braided bialgebras such that, for all* $a \in A$, $b \in B$,

$$b_{(0)} \cdot a_{\{-1\}} \otimes b_{(1)} \cdot a_{\{0\}} = b \otimes a. \tag{5.1}$$

Then the double-smash product algebra and coalgebra structures afford on $A \otimes H \otimes B$ *a* k*-bialgebra structure, denoted by* $A \times H \times B$ *and called the double-biproduct bialgebra of* A, H, B*. Furthermore, if* A, B *are braided Hopf algebras then* $A \times H \times B$ *is a Hopf algebra with antipode* s *defined, for all* $a \in A$, $h \in H$ *and* $b \in B$, *by*

$$s(a \times h \times b) = (1_A \times S(a_{\{-1\}}h_1) \times 1_B)(S_A(a_{\{0\}}) \times h_2 \times S_B(b_{(0)}))(1_A \times S(h_3 b_{(1)}) \times 1_B).$$

(iv) Let $\overline{B}$ *be the algebra/coalgebra in* ${}^H_H\mathcal{YD}$ *associated to the algebra/coalgebra* B *in* $\mathcal{YD}^H_H$ *as in Corollary 5.3, and* $A\widetilde{\otimes}\overline{B}$ *the tensor product algebra/coalgebra of* A, $\overline{B}$ *in* ${}^H_H\mathcal{YD}$*. Then* $\chi := \mathrm{Id}_A \otimes \omega_B : A \otimes B \otimes H \to A \otimes H \otimes B$ *yields*

- *a* k*-algebra/coalgebra isomorphism* $(A\widetilde{\otimes}\overline{B})\#H \simeq A \times H \times B$, *respectively* $(A\widetilde{\otimes}\overline{B}) \rtimes H \simeq A \rtimes H \ltimes B$;
- *a* k*-bialgebra/Hopf algebra isomorphism between* $(A\widetilde{\otimes}\overline{B}) \times H$ *and* $A \times H \times B$, *provided that* A, B *are braided bialgebras/Hopf algebras satisfying* (5.1).

PROOF. (i) follows from the proof of Proposition 3.2 (see also (3.2)), (ii) follows from the proof of Proposition 3.5 and (iii) from Lemma 3.7 and the comments made after its proof. Finally, (iv) is a consequence of all these quoted results. □

The terminology used in Theorem 5.4 is based on the fact that $A\#H\#B$ (resp. $A \rtimes H \ltimes B$, $A \times H \times B$) contains both $A\#H$, $H\#B$ ($A \rtimes H$, $H \ltimes B$ and, respectively, $A \times H$, $H \times B$) as subalgebras (resp. subcoalgebras, subbialgebras/Hopf subalgebras). Note that the multiplication of $(A\widetilde{\otimes}B)\#H$ is given by

$$((a\widetilde{\otimes}b)\#h)((a'\widetilde{\otimes}b')\#h') = (a(S(b_{(2)})h_1 \cdot a')\widetilde{\otimes}b_{(0)}(b' \cdot S^{-1}(h_2)b_{(1)}))\#h_3h',$$

the comultiplication of $(A\widetilde{\otimes}\overline{B}) \rtimes H$ is, for all $a, a' \in A$, $b, b' \in B$, $h, h' \in H$,

$$\Delta((a\widetilde{\otimes}b) \rtimes h) = (a_{\underline{1}}\widetilde{\otimes}(b_{\underline{1}})_{(0)} \cdot S^{-1}((a_{\underline{2}})_{\{-2\}})) \rtimes (a_{\underline{2}})_{\{-1\}}(b_{\underline{1}})_{(1)}S((b_{\underline{1}})_{(3)}(b_{\underline{2}})_{(1)})h_1 \\ \otimes((a_{\underline{2}})_{\{0\}}\widetilde{\otimes}(b_{\underline{2}})_{(0)} \cdot S^{-1}((b_{\underline{1}})_{(2)})) \rtimes h_2,$$

and the antipode ${}^l s$ of $(A\widetilde{\otimes}\overline{B}) \times H$ is determined by

$${}^l s((a\widetilde{\otimes}b) \times h) = (1_A\widetilde{\otimes}1_B \times S(a_{\{-1\}}S(b_{(1)})))((S_A(a_{\langle 0\rangle})\widetilde{\otimes}S_B(b_{(0)})) \times 1_H),$$

where we used the standard sigma notation for comodules:

$$b_{(0)} \otimes b_{(1)} \otimes b_{(2)} := b_{(0)_{(0)}} \otimes b_{(0)_{(1)}} \otimes b_{(1)} = b_{(0)} \otimes b_{(1)_1} \otimes b_{(1)_2}, \text{ etc.}$$
$$a_{\{-2\}} \otimes a_{\{-1\}} \otimes a_{\{0\}} := a_{\{-1\}} \otimes a_{\{0\}_{\{-1\}}} \otimes a_{\{0\}_{\{0\}}} = a_{\{-1\}_1} \otimes a_{\{-1\}_2} \otimes a_{\{0\}}, \text{ etc.}$$

All the above structure is necessary in order to give an explicit description for the deformation of $A \times H \times B$ by a convolution invertible 2-cocycle $\widehat{\sigma}$, defined in terms of a suitable k-linear map $\sigma : B \otimes A \to k$. Namely, according to Proposition 4.2, $\widehat{\sigma} = \varepsilon_{A \rtimes H} \otimes \sigma \otimes \varepsilon_{H \ltimes B}$ is an invertible 2-cocycle of $A \times H \times B$ if and only if $\sigma(1_B \otimes a) = \varepsilon_A(a)$, $\sigma(b \otimes 1_A) = \varepsilon_B(b)$ and σ obeys the conditions

$$\sigma(b \cdot h \otimes a) = \sigma(b \otimes h \cdot a), \tag{5.2}$$

$$\sigma(b \otimes aa') = \sigma(b_{\underline{1}} \otimes a)\sigma(b_{\underline{2}} \otimes a'), \tag{5.3}$$

$$\sigma(bb' \otimes a) = \sigma(b \otimes b'_{(1)} \cdot a_{\underline{2}})\sigma(b'_{(0)} \otimes a_{\underline{1}}), \tag{5.4}$$

for all $a, a' \in A$ and $b, b' \in B$. We should point out that, after a permutation of tensor components, (5.2)-(5.4) are precisely the first three conditions of (56) in [**17**], required for a k-linear map $\langle , \rangle : A \otimes B \to k$ to be a "pairing". The fourth (which is the last one) reads in our case as $\sigma(b \otimes S_A(a)) = \sigma(S_B^{-1}(b) \otimes a)$, for all $a \in A$ and $b \in B$, and is needed to deduce that the convolution inverse of σ is $\overline{\sigma}$ given by $\overline{\sigma}(b \otimes a) = \sigma(b \otimes S_A(a)) = \sigma(S_B^{-1}(b) \otimes a)$, for all $a \in A$ and $b \in B$. From our general results, and also from the above (5.2)-(5.4), one can see easily that both $\overline{\sigma}_1, \overline{\sigma}_2 : B \otimes A \to k$ defined by $\overline{\sigma}_1(b \otimes a) = \sigma(b \otimes S_A(a))$ and $\overline{\sigma}_2(b \otimes a) = \sigma(S_B^{-1}(b) \otimes a)$, for all $a \in A$ and $b \in B$, are convolution inverses for σ, and therefore equal. So the fourth condition of (56) in [**17**] is unnecessary, as it is implied by the previous three ones. To strengthen our claim, let's show that $\sigma(b_{\underline{1}} \otimes a_{\underline{1}})\overline{\sigma}_2(b_{\underline{2}} \otimes a_{\underline{2}}) = \varepsilon_B(b)\varepsilon_A(a)$, for all $a \in A$ and $b \in B$. To this end, observe that (5.4) and (5.2) imply

$$\sigma((b \cdot S^{-1}(b'_{(1)}))b'_{(0)} \otimes a) = \sigma(b \otimes a_{\underline{2}})\sigma(b' \otimes a_{\underline{1}}),$$

for all $a \in A$ and $b, b' \in B$. Now, the fact that S_B^{-1} is an anti-morphism of coalgebras in $\mathcal{YD}_H^H$ gives that $S_B^{-1}(bb') = (S_B^{-1}(b') \cdot S^{-1}(b_{(1)}))S_B^{-1}(b_{(0)})$, for all $b, b' \in B$, and together with the fact that S_B is right H-colinear entails to

$$\varepsilon_B(h)1_B = S_B^{-1}(S(b_{\underline{1}})b_{\underline{2}}) = (S^{-1}(b_{\underline{2}}) \cdot S^{-1}((b_{\underline{1}})_{(1)}))(b_{\underline{1}})_{(0)},$$

for all $b \in B$. One can compute now that

$$\sigma(b_{\underline{1}} \otimes a_{\underline{1}})\overline{\sigma}_2(b_{\underline{2}} \otimes a_{\underline{2}}) = \sigma((S^{-1}(b_{\underline{2}}) \cdot S^{-1}((b_{\underline{1}})_{(1)}))(b_{\underline{1}})_{(0)} \otimes a) = \varepsilon_B(b)\varepsilon_A(a),$$

as stated. The verification of the remaining details is left to the reader.

For $\widehat{\sigma}$ defined by σ as in the above, we have the new Hopf algebra $(A \times H \times B)_{\widehat{\sigma}}$, which is equal to $A \times H \times B$ as a coalgebra, has the multiplication defined by

$$(5.5)\quad (a \times h \times b) \cdot_\sigma (a' \times h' \times b') = \sigma((b_{\underline{1}})_{(0)}, a'_{\underline{1}})\sigma(b_{\underline{3}}, S_A((a'_{\underline{3}})_{(0)}))a(h_1(b_{\underline{1}})_{(1)} \cdot a'_{\underline{2}}) \\ \times h_2(b_{\underline{1}})_{(2)}(a'_{\underline{3}})_{\{-2\}}h'_1 \times (b_{\underline{2}} \cdot (a'_{\underline{3}})_{\{-1\}}h'_2)b',$$

unit $1_A \times 1_H \times 1_B$ and antipode s^σ given by

$$s^\sigma(a \times h \times b) = \sigma((b_{\underline{1}})_{(0)} \otimes S(a_{\{-1\}}h_1(b_{\underline{1}})_{(1)}) \cdot S_A(a_{\{0\}_{\underline{1}}}))a_{\{0\}_{\underline{2}}} \rtimes a_{\{0\}_{\underline{3}\{-2\}}}h_2 \\ (b_{\underline{1}})_{(2)} \ltimes (b_{\underline{2}})_{(0)}\overline{\sigma}((b_{\underline{3}})_{(0)} \otimes S(a_{\{0\}_{\underline{3}\{-1\}}}h_3(b_{\underline{1}})_{(3)}(b_{\underline{2}})_{(1)}(b_{\underline{3}})_{(1)}) \cdot S_A(a_{\{0\}_{\underline{3}\{0\}}}));$$

here $a, a' \in A$, $h, h' \in H$ and $b, b' \in B$.

PROPOSITION 5.5. *Let H be a Hopf algebra with bijective antipode and $A \in {}_H^H\mathcal{YD}$, $B \in \mathcal{YD}_H^H$ braided Hopf algebras related through a k-linear map $\sigma : B \otimes A \to k$ fulfilling the natural unital conditions and (5.2)-(5.4). Define $\widehat{\sigma} := \varepsilon_{A \rtimes H} \otimes \sigma \otimes \varepsilon_{H \ltimes B} : (A \times H \times B) \otimes (A \times H \otimes B) \to H$, $\Sigma : B \otimes A \to k$ given by $\Sigma(b \otimes a) = \sigma(b_{(0)} \otimes S(b_{(1)}) \cdot a)$, for all $a \in A$ and $b \in B$, and $\omega = \varepsilon_A \otimes \Sigma(\mathrm{Id}_B \otimes \mu_A^H) \otimes \varepsilon_B \otimes \varepsilon_H : (A \widetilde{\otimes} \overline{B}) \times H \otimes (A \widetilde{\otimes} \overline{B}) \times H \to k$. Then:*

(i) $\widehat{\sigma}/\omega$ are invertible 2-cocycles of the Hopf algebras $A \times H \times B/(A \widetilde{\otimes} \overline{B}) \times H$.

(ii) $\chi = \mathrm{Id}_A \otimes \omega_B$, regarded as a map from $(A \times H \times B)_{\widehat{\sigma}}$ to $((A \widetilde{\otimes} \overline{B}) \times H)_\omega$, is a Hopf algebra isomorphism.

PROOF. It follows from the comments preceding Theorem 4.6. Note that ω is determined by an almost skew-pairing Σ of $\overline{B}$ and A in ${}_H^H\mathcal{YD}$. This means that Σ is left H-linear, satisfies the unital conditions, and the following relations hold:

$$(5.6)\qquad \Sigma(b \bullet b' \otimes a) = \Sigma(b \otimes a_{\underline{2}})\Sigma(b' \otimes a_{\underline{1}}),$$

$$(5.7)\qquad \Sigma(b \otimes aa') = \Sigma(b_{\overline{1}} \otimes S((b_{\overline{2}})_{(1)}) \cdot a)\Sigma((b_{\overline{2}})_{(0)} \otimes a'),$$

for all $a, a' \in A$ and $b, b' \in B$. Here $b \bullet b' = b_{(0)}(b' \cdot b_{(1)})$ is the multiplication of $b, b' \in \overline{B}$ in ${}^H_H\mathcal{YD}$, while $\Delta_{\overline{B}} = b_{\overline{1}} \otimes b_{\overline{2}}$ is our sigma notation for the comultiplication of $\overline{B}$ in ${}^H_H\mathcal{YD}$: $b_{\overline{1}} \otimes b_{\overline{2}} = (b_{\underline{1}})_{(0)} \otimes b_{\underline{2}} \cdot S^{-1}((b_{\underline{1}})_{(1)})$, for all $b \in B$. □

To turn Σ into a skew pairing of $\overline{B}$ and A in ${}^H_H\mathcal{YD}$ we need σ to be H-cobalanced, see Lemma 4.4. This means that (4.14) holds, which reads now as

$$\sigma(b \otimes a_{\{0\}})a_{\{-1\}} = \sigma(b_{(0)} \otimes a)b_{(1)}, \ \forall\, a \in A,\ b \in B. \tag{5.8}$$

When this happens, (H, A, B, σ) is called a biproduct Drinfeld data.

THEOREM 5.6. *Let H be a Hopf algebra with bijective antipode and (H, A, B, σ) a biproduct Drinfeld data. Then:*

(i) $\Sigma : B \otimes A \to k$ given by $\Sigma(b \otimes a) = \sigma(b_{(0)} \otimes S(b_{(1)}) \cdot a)$, for all $a \in A$ and $b \in B$, is a convolution invertible skew pairing of $\overline{B}$ and A in ${}^H_H\mathcal{YD}$.

(ii) $\vartheta_0 := \varepsilon_A \otimes \Sigma \otimes \varepsilon_B$ is an invertible 2-cocycle of $A\widetilde{\otimes}\overline{B}$ in ${}^H_H\mathcal{YD}$.

(iii) $\chi = \mathrm{Id}_A \otimes \omega_B : (A \times H \times B)_{\widehat{\sigma}} \to (A\widetilde{\otimes}\overline{B})_{\vartheta_0} \times H$ is a Hopf algebra isomorphism, hence $(A \times H \times B)_{\widehat{\sigma}}$ identifies in this case to a left biproduct Hopf algebra.

PROOF. Everything follows from Theorem 4.6, specialized for $\mathcal{C} = {}_k\mathcal{M}$. □

5.2. The coquasitriangular case I. Throughout this subsection $(H, \mathcal{R})$ is a coquasitriangular (CQT for short) Hopf algebra, I, J are non-empty sets and $I \ni i \mapsto K_i \in G(H)$, $J \ni j \mapsto K_j \in G(H)$ are given maps, where $G(H) = \{g \in G \mid \Delta_H(g) = g \otimes g,\ \varepsilon_H(g) = 1\}$ is the set of grouplike elements of H. Recall that H is called CQT if the k-bilinear morphism $\mathcal{R} : H \otimes H \to k$ satisfies $\mathcal{R}(1_H \otimes h) = \mathcal{R}(h \otimes 1_H) = \varepsilon_H(h)$, $\mathcal{R}(h_1 \otimes h'_1)h_2h'_2 = \mathcal{R}(h_2 \otimes h'_2)h'_1h_1$,

$$\mathcal{R}(hh' \otimes \hbar) = \mathcal{R}(h \otimes \hbar_1)\mathcal{R}(h' \otimes \hbar_2) \text{ and } \mathcal{R}(\hbar \otimes hh') = \mathcal{R}(\hbar_2 \otimes h)\mathcal{R}(\hbar_1 \otimes h'),$$

for all $h, h', \hbar \in H$. Equivalently, H is CQT if $\mathcal{M}^H$ is braided such that the forgetful functor to ${}_k\mathcal{M}$ is braided: the braidings on $\mathcal{M}^H$ are in this case in a one to one correspondence to the CQT structures on H. For instance, $\mathcal{R}$ as above turns $\mathcal{M}^H$ into a braided category with braiding ${}^r\mathfrak{b}$ determined by ${}^r\mathfrak{b}_{W,W'} : W \otimes W' \ni w \otimes w' \mapsto \mathcal{R}(w_{(1)} \otimes w'_{(1)})w'_{(0)} \otimes w_{(0)} \in W' \otimes W$, for all $W, W' \in \mathcal{M}^H$.

Any braided category embeds into its left/right centre as a braided category; see [**5**]. Thus, for $(H, \mathcal{R})$ CQT, we have a braided functor $\mathcal{M}^H \to \mathcal{YD}^H_H$. Hence, $(W, \rho_W) \in \mathcal{M}^H$ belongs to $\mathcal{YD}^H_H$ via ρ_W and $w \cdot h = \mathcal{R}(w_{(1)}, h)w_{(0)}$; $w \in W$, $h \in H$.

Let $\mathfrak{f}$ be the unital free algebra with unit $1_\mathfrak{f}$, and letters $(f_i)_{i \in I}$. As the K_i's are in $G(H)$, $\rho_\mathfrak{f} : \mathfrak{f} \to \mathfrak{f} \otimes H$ sending f_i to $f_i \otimes K_i$, for all $i \in I$, extended as an algebra morphism from $\mathfrak{f}$ to $\mathfrak{f} \otimes H$, defines a right H-coaction on $\mathfrak{f}$. With this $\rho_\mathfrak{f}$ and H-action defined by $1_\mathfrak{f} \cdot h = \varepsilon_H(h)1_\mathfrak{f}$ and $f_i \cdot h = \mathcal{R}(K_i, h)f_i$, for all $h \in H$ and $i \in I$, $\mathfrak{f}$ becomes an algebra in $\mathcal{YD}^H_H$. Note that $\cdot$ extends to the whole H by using the equalities $(uv) \cdot h = (u \cdot h_1)(v \cdot h_2)$, valid for all $u, v \in \mathfrak{f}$, $h \in H$.

PROPOSITION 5.7. *The algebra $\mathfrak{f}$ is a Hopf algebra in $\mathcal{YD}^H_H$ with*

$$\Delta_\mathfrak{f}(1_\mathfrak{f}) = 1_\mathfrak{f}\overline{\otimes}1_\mathfrak{f},\ \Delta_\mathfrak{f}(f_i) = f_i\overline{\otimes}1_\mathfrak{f} + 1_\mathfrak{f}\overline{\otimes}f_i,\ \varepsilon_\mathfrak{f}(1_\mathfrak{f}) = 1_k \text{ and } \varepsilon_\mathfrak{f}(f_i) = 0,\ \forall\, i \in I,$$

extended as algebra morphisms from $\mathfrak{f}$ to $\mathfrak{f}\overline{\otimes}\mathfrak{f}$ (the tensor product algebra of $\mathfrak{f}$ and itself within $\mathcal{YD}^H_H$), and respectively from $\mathfrak{f}$ to k. The antipode is given by $S_\mathfrak{f}(1_\mathfrak{f}) = 1_\mathfrak{f}$ and $S_\mathfrak{f}(f_i) = -f_i$, extended as an anti-morphism of algebras in $\mathcal{YD}^H_H$.

PROOF. We apply the universal property of the free algebra $\mathfrak{f}$ to the k-algebra $\mathfrak{f}\overline{\otimes}\mathfrak{f}$ and the map $f_i \mapsto f_i\overline{\otimes}1_{\mathfrak{f}} + 1_{\mathfrak{f}}\overline{\otimes}f_i \in \mathfrak{f}\overline{\otimes}\mathfrak{f}$, $i \in I$, to obtain a unique algebra morphism $\Delta_{\mathfrak{f}} : \mathfrak{f} \to \mathfrak{f}\overline{\otimes}\mathfrak{f}$ obeying $\Delta_{\mathfrak{f}}(f_i) = f_i\overline{\otimes}1_{\mathfrak{f}} + 1_{\mathfrak{f}}\overline{\otimes}f_i$, for all $i \in I$. So, to extend $\Delta_{\mathfrak{f}}$ to the whole $\mathfrak{f}$ we should make use of the braiding ${}^r\mathfrak{c}$ of $\mathcal{YD}_H^H$; for instance,

$$\Delta_{\mathfrak{f}}(f_{i_1}f_{i_2}) = f_{i_1}f_{i_2}\overline{\otimes}1_{\mathfrak{f}} + f_{i_1}\overline{\otimes}f_{i_2} + \mathcal{R}(K_{i_1}, K_{i_2})f_{i_2}\overline{\otimes}f_{i_1} + 1_{\mathfrak{f}}\overline{\otimes}f_{i_1}f_{i_2},\ \forall\, i_1,\ i_2 \in I.$$

Similarly, $\varepsilon_{\mathfrak{f}} : \mathfrak{f} \to k$ is the algebra morphism defined by the null morphism, $f_i \mapsto 0_k$.

To show that $\Delta_{\mathfrak{f}}$ is coassociative, we apply again the universal property of the free algebra $\mathfrak{f}$ to $\mathfrak{f}\overline{\otimes}\mathfrak{f}\overline{\otimes}\mathfrak{f}$ and the map $\varrho : f_i \mapsto f_i\overline{\otimes}1_{\mathfrak{f}}\overline{\otimes}1_{\mathfrak{f}} + 1_{\mathfrak{f}}\overline{\otimes}f_i\overline{\otimes}1_{\mathfrak{f}} + 1_{\mathfrak{f}}\overline{\otimes}1_{\mathfrak{f}}\overline{\otimes}f_i$, $i \in I$. This yields a unique algebra morphism from $\mathfrak{f}$ to $\mathfrak{f}\overline{\otimes}\mathfrak{f}\overline{\otimes}\mathfrak{f}$ which on each f_i is equal to $\varrho(f_i)$. But one can see easily that $(\Delta_{\mathfrak{f}} \otimes \mathrm{Id}_{\mathfrak{f}})\Delta_{\mathfrak{f}}(f_i) = (\mathrm{Id}_{\mathfrak{f}} \otimes \Delta_{\mathfrak{f}})\Delta_{\mathfrak{f}}(f_i) = \varrho(f_i)$, for all $i \in I$, and this forces $(\Delta_{\mathfrak{f}} \otimes \mathrm{Id}_{\mathfrak{f}})\Delta_{\mathfrak{f}} = (\mathrm{Id}_{\mathfrak{f}} \otimes \Delta_{\mathfrak{f}})\Delta_{\mathfrak{f}}$, i.e. $\Delta_{\mathfrak{f}}$ is coassociative.

As $(\varepsilon_{\mathfrak{f}} \otimes \mathrm{Id}_{\mathfrak{f}})\Delta_{\mathfrak{f}}(f_i) = (\mathrm{Id}_{\mathfrak{f}} \otimes \varepsilon_{\mathfrak{f}})\Delta_{\mathfrak{f}}(f_i) = f_i$ $(i \in I)$ and $(\varepsilon_{\mathfrak{f}} \otimes \mathrm{Id}_{\mathfrak{f}})\Delta_{\mathfrak{f}}$, $(\mathrm{Id}_{\mathfrak{f}} \otimes \varepsilon_{\mathfrak{f}})\Delta_{\mathfrak{f}} : \mathfrak{f} \to \mathfrak{f}$ are algebra morphisms, we get $(\varepsilon_{\mathfrak{f}} \otimes \mathrm{Id}_{\mathfrak{f}})\Delta_{\mathfrak{f}} = (\mathrm{Id}_{\mathfrak{f}} \otimes \varepsilon_{\mathfrak{f}})\Delta_{\mathfrak{f}} = \mathrm{Id}_{\mathfrak{f}}$.

The antipode $S_{\mathfrak{f}}$ is defined by the universal property of $\mathfrak{f}$ applied to the opposite algebra $\mathfrak{f}^{\mathrm{op}}$ of $\mathfrak{f}$ in $\mathcal{YD}_H^H$ and the map $f_i \mapsto -f_i$, $i \in I$. The multiplication $\circ$ of $\mathfrak{f}^{\mathrm{op}}$ is $u \circ v = v_{(0)}(u \cdot v_{(1)})$, for all $u, v \in \mathfrak{f}$; so $S_{\mathfrak{f}}$ extends to $\mathfrak{f}$ by using $S_{\mathfrak{f}}(uv) = S_{\mathfrak{f}}(u) \circ S_{\mathfrak{f}}(v) = S_{\mathfrak{f}}(v_{(0)})(S_{\mathfrak{f}}(u) \cdot v_{(1)})$, $u, v \in \mathfrak{f}$ (clearly, $S_{\mathfrak{f}}$ is a morphism in $\mathcal{YD}_H^H$).

Under the notation used for a graphical computation, we have that

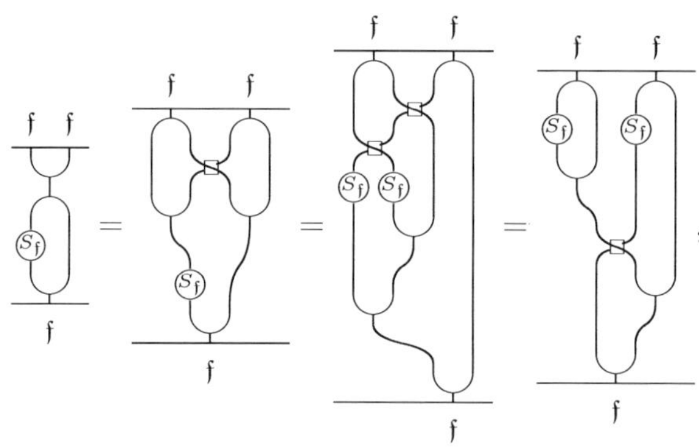

,

where the equalities take place in $\mathcal{YD}_H^H$. The above computation tells that $S_f * \mathrm{Id}_{\mathfrak{f}} = \eta_{\mathfrak{f}}\varepsilon_{\mathfrak{f}}$ holds on the product uv of two elements u, v of $\mathfrak{f}$, provided that it holds on u and v. A similar statement is true for $\mathrm{Id}_{\mathfrak{f}} * S_{\mathfrak{f}} = \eta_{\mathfrak{f}}\varepsilon_{\mathfrak{f}}$, where $*$ is the convolution product in $\mathrm{Hom}_{\mathcal{YD}_H^H}(\mathfrak{f}, \mathfrak{f})$. We end the proof by noting that

$$S_{\mathfrak{f}}(f_{i_1} \cdots f_{i_r}) = (-1)^r \left(\prod_{l=2}^{r} \mathcal{R}(K_{i_{l-1}} \cdots K_{i_1}, K_{i_l}) \right) f_{i_r} \cdots f_{i_1},$$

for any natural number $r \geq 2$ and $i_1, \cdots, i_r \in I$. □

REMARK 5.8. Any k-algebra B is $\mathfrak{f}/\mathfrak{I}$ for a certain $\mathfrak{f}$ and $\mathfrak{I} \subseteq \mathfrak{f}$ ideal. B can be seen as a Hopf algebra in $\mathcal{YD}_H^H$ such that the canonical projection $\pi : \mathfrak{f} \to B$ is a Hopf algebra morphism in $\mathcal{YH}_H^H$, provided that $\mathfrak{I}$ is a right H-submodule and subcomodule of $\mathfrak{f}$, $\Delta_{\mathfrak{f}}(\mathfrak{I}) \subseteq \mathfrak{I} \otimes \mathfrak{f} + \mathfrak{f} \otimes \mathfrak{I}$, $\varepsilon_{\mathfrak{f}}(\mathfrak{I}) = 0$ and $S_{\mathfrak{f}}(\mathfrak{I}) \subseteq \mathfrak{I}$ ($\mathfrak{I}$ is a Hopf ideal).

We produce now a left version of Proposition 5.7. By $\mathfrak{e}$ we denote the free algebra with unit $1_{\mathfrak{e}}$ and letters $(e_j)_{j \in J}$, considered in ${}^H\mathcal{M}$ with $e_j \mapsto K_j \otimes e_j$, $j \in J$, extended as an algebra morphism from $\mathfrak{e}$ to $H \otimes \mathfrak{e}$. This left coaction makes $\mathfrak{e}$ an algebra in the monoidal category ${}^H\mathcal{M}$ of left H-corepresentations. As

$(H,\mathcal{R})$ is CQT, ${}^{H}\mathcal{M}$ is braided with ${}^{l}\mathfrak{b}$ given by ${}^{l}\mathfrak{b}_{V,V'}: V\otimes V' \ni v\otimes v' \mapsto \overline{\mathcal{R}}(v'_{\{-1\}}\otimes v_{\{-1\}})v'_{\{0\}}\otimes v_{\{0\}} \in V'\otimes V$, for all $V,V'\in {}^{H}\mathcal{M}$, where $\overline{\mathcal{R}}$ is the convolution inverse of $\mathcal{R}$: $\overline{\mathcal{R}}(h\otimes h') = \mathcal{R}(S(h)\otimes h') = \mathcal{R}(h\otimes S^{-1}(h'))$, for all $h,h'\in H$. Furthermore, we have a braided functor from ${}^{H}\mathcal{M}$ to ${}_{H}^{H}\mathcal{YD}$, so any $V\in {}^{H}\mathcal{M}$ becomes an object in ${}_{H}^{H}\mathcal{YD}$ with the initial left H-comodule structure and left H-action $h\triangleright v := \overline{\mathcal{R}}(h, v_{\{-1\}})v_{\{0\}}$; $h\in H$, $v\in V$. Consequently, $\mathfrak{e}$ is an algebra in ${}_{H}^{H}\mathcal{YD}$ with the above left H-coaction and left H-action determined by $h\triangleright 1_{\mathfrak{e}} = \varepsilon(h)1_{\mathfrak{e}}$ and $h\triangleright e_j = \overline{\mathcal{R}}(h,K_j)e_j = \mathcal{R}(h,K_j^{-1})e_j$, for all $j\in J$. Again, $\triangleright$ extends to the whole $\mathfrak{e}$ with the help of $h\triangleright(ab) = (h_1\triangleright a)(h_2\triangleright b)$; $h\in H$, $a,b\in\mathfrak{e}$.

PROPOSITION 5.9. *The algebra $\mathfrak{e}$ is a Hopf algebra in ${}_{H}^{H}\mathcal{YD}$ with*

- coalgebra structure defined by $\Delta_{\mathfrak{e}}(1_{\mathfrak{e}}) = 1_{\mathfrak{e}}\widetilde{\otimes}1_{\mathfrak{e}}$, $\Delta_{\mathfrak{e}}(e_j) = e_j\widetilde{\otimes}1_{\mathfrak{e}} + 1_{\mathfrak{e}}\widetilde{\otimes}e_j$, $\varepsilon_{\mathfrak{e}}(1_{\mathfrak{e}}) = 1_k$, $\varepsilon_{\mathfrak{e}}(e_j) = 0$, for all $j\in J$, extended as algebra morphisms from $\mathfrak{e}$ to $\mathfrak{e}\widetilde{\otimes}\mathfrak{e}$ (the tensor product algebra of $\mathfrak{e}$ and itself within ${}_{H}^{H}\mathcal{YD}$), respectively from $\mathfrak{e}$ to k;

- antipode determined by $S_{\mathfrak{e}}(1_{\mathfrak{e}}) = 1_{\mathfrak{e}}$ and $S_{\mathfrak{e}}(e_j) = -e_j$, for all $j\in J$, extended as an anti-morphisms of algebras in ${}_{H}^{H}\mathcal{YD}$ between $\mathfrak{e}$ and itself.

PROOF. It is similar to the one of Proposition 5.7, so we omit it. □

We next see that Theorem 5.4 applies to the braided Hopf algebras $\mathfrak{f}$ and $\mathfrak{e}$.

EXAMPLE 5.10. Let $(H,\mathcal{R})$ be a CQT Hopf algebra, I,J non-empty sets and $I\ni i\mapsto K_i\in G(H)$ and $J\ni j\mapsto K_j\in G(H)$ two maps. Denote by $\mathcal{A}_H(\mathfrak{e},\mathfrak{f})$ the unital algebra generated by the families $(E_j)_{j\in J}$ and $(F_i)_{i\in I}$ with relations $F_iE_j = \mathcal{R}(K_i,K_j)E_jF_i$, for all $(i,j)\in I\times J$. Then:

(i) $\mathcal{A}_H(\mathfrak{e},\mathfrak{f})$ is an algebra in ${}_{H}^{H}\mathcal{YD}$ with structure given by

$$h\cdot E_j = \overline{\mathcal{R}}(h,K_j)E_j,\ h\cdot F_i = \overline{\mathcal{R}}(K_i,h)F_i,\ E_j\mapsto K_j\otimes E_j,\ F_i\mapsto K_i^{-1}\otimes F_i,$$

for all $(i,j)\in I\times J$. As before, the left H-comodule structure is extended to $\mathcal{A}_H(\mathfrak{e},\mathfrak{f})$ as an algebra morphism, while the left H-action is extended to $\mathcal{A}_H(\mathfrak{e},\mathfrak{f})$ by using the rule $h\cdot(uv) = (h_1\cdot u)(h_2\cdot v)$, for all $h\in H$ and $u,v\in\mathcal{A}_H(\mathfrak{e},\mathfrak{f})$.

$\mathcal{A}_H(\mathfrak{e},\mathfrak{f})$ is a Hopf algebra in ${}_{H}^{H}\mathcal{YD}$ with comultiplication, counit and antipode

$$\Delta(E_j) = E_j\otimes 1 + 1\otimes E_j,\ \Delta(F_i) = F_i\otimes 1 + 1\otimes F_i,$$

$\varepsilon(E_j) = \varepsilon(F_i) = 0$, $S(E_j) = -E_j$, $S(F_i) = -F_i$, for all $(i,j)\in I\times J$. As usual, Δ is extended as an algebra morphism from $\mathcal{A}_H(\mathfrak{e},\mathfrak{f})$ to $\mathcal{A}_H(\mathfrak{e},\mathfrak{f})\widetilde{\otimes}\mathcal{A}_H(\mathfrak{e},\mathfrak{f})$ (the tensor product algebra of $\mathcal{A}_H(\mathfrak{e},\mathfrak{f})$ and itself within ${}_{H}^{H}\mathcal{YD}$), ε as an algebra morphism from $\mathcal{A}_H(\mathfrak{e},\mathfrak{f})$ to k and S as an anti-morphism of algebras in ${}_{H}^{H}\mathcal{YD}$ from $\mathcal{A}_H(\mathfrak{e},\mathfrak{f})$ to itself.

(ii) If $\mathfrak{e},\mathfrak{f}$ are the free unital algebras on the sets $\{e_j, j\in J\}$ and $\{f_i, i\in I\}$, respectively, viewed as Hopf algebras in ${}_{H}^{H}\mathcal{YD}$ and $\mathcal{YD}_{H}^{H}$ as in Propositions 5.9 and 5.7, then $\mathfrak{e}$ and $\mathfrak{f}$ satisfy (5.1) and the associated double-biproduct Hopf algebra $\mathfrak{e}\times H\times f$ is isomorphic to the left biproduct Hopf algebra $\mathcal{A}_H(\mathfrak{e},\mathfrak{f})\times H$.

PROOF. We show that $\mathcal{A}_H(\mathfrak{e},\mathfrak{f}) = \mathfrak{e}\widetilde{\widetilde{\otimes}}\mathfrak{f}$ as algebras and coalgebras in ${}_{H}^{H}\mathcal{YD}$. Clearly, $\mathcal{A}_H(\mathfrak{e},\mathfrak{f})$ is generated as a unital algebra by $\{E_j := e_j\widetilde{\otimes}1_{\mathfrak{f}}, j\in J\}$ and $\{F_i := 1_{\mathfrak{e}}\widetilde{\otimes}f_i, i\in I\}$ with relations $F_iE_j = (1_{\mathfrak{e}}\widetilde{\otimes}f_i)(e_j\widetilde{\otimes}1_{\mathfrak{f}}) = K_i^{-1}\triangleright e_j\widetilde{\otimes}f_i(1_{\mathfrak{f}}\cdot K_i) = \mathcal{R}(K_i,K_j)E_jF_i$, for all $(i,j)\in I\times J$. Furthermore, keeping in mind the structure that makes $\mathfrak{e}\widetilde{\otimes}\mathfrak{f}$ an object in ${}_H^H\mathcal{Y}D$, we see that, for all $(i,j) \in I \times J$,

$$h \cdot E_j = h_1 \triangleright e_j \widetilde{\otimes} 1_\mathfrak{f} \cdot S^{-1}(h_2) = \overline{\mathcal{R}}(h, K_j)E_j,$$
$$h \cdot F_i = h \cdot (1_\mathfrak{e} \widetilde{\otimes} f_i) = 1_\mathfrak{e} \widetilde{\otimes} f_i \cdot S^{-1}(h) = \mathcal{R}(K_i, S^{-1}(h))f_i = \overline{\mathcal{R}}(K_i, h)F_i,$$
$$E_j = e_j \widetilde{\otimes} 1_\mathfrak{f} \mapsto K_j \otimes e_j \widetilde{\otimes} 1_\mathfrak{f} = K_j \otimes E_j, F_i = 1_\mathfrak{e} \widetilde{\otimes} f_i \mapsto S(K_i) \otimes 1_\mathfrak{e} \widetilde{\otimes} f_i = K_i^{-1} \otimes F_i.$$

In a similar way we deduce that the coalgebra structure of $\mathcal{A}_H(\mathfrak{e}, \mathfrak{f})$ induced by that of $\mathfrak{e}\widetilde{\otimes}\mathfrak{f}$ in ${}_H^H\mathcal{Y}D$ is just the one mentioned in the statement (i) above.

Clearly (5.1) is satisfied if b or c is $1_\mathfrak{f}$. It is satisfied for all $b \in \mathfrak{f}$, $c \in \mathfrak{e}$ since

$$(f_{i_1} \cdots f_{i_r}) \cdot K_{j_1} \cdots K_{j_t} \otimes K_{i_1} \cdots K_{i_r} \triangleright (e_{j_1} \cdots e_{j_t}) = f_{i_1} \cdots f_{i_r} \otimes e_{j_1} \cdots e_{j_t},$$

for all $r, t \geq 1$, $i_1, \cdots, i_r \in I$ and $j_1, \cdots, j_t \in J$. Thus $\mathfrak{e}\widetilde{\otimes}\mathfrak{f}$ is a Hopf algebra in ${}_H^H\mathcal{Y}D$ with antipode $S_\mathfrak{e} \otimes S_\mathfrak{f}$, and therefore so is $\mathcal{A}_H(\mathfrak{e}, \mathfrak{f})$. A simple inspection shows that the antipode S is determined by $S(E_j) = -E_j$, $S(F_i) = -F_i$, $(i,j) \in I \times J$.

Finally, the Hopf algebra isomorphism between $\mathfrak{e} \times H \times \mathfrak{f}$ and $\mathcal{A}_H(\mathfrak{e}, \mathfrak{f}) \times H$ is produced by χ defined, in general, in (iv) of Theorem 5.4. In this case,

$$\chi(e_j \times h \times f_i) = \mathcal{R}(K_i \otimes S^{-1}(h_1))E_jF_i \times h_2K_i = \overline{\mathcal{R}}(K_i \otimes h_1)E_jF_i \times h_2K_i,$$

for all $(i,j) \in I \times J$ and $h \in H$. Also, $\mathfrak{e} \times H \times \mathfrak{f}$ is the unital k-algebra generated by $\{e_j, j \in J\}$, $\{f_i, i \in I\}$ and the elements h of a CQT Hopf algebra $(H, \mathcal{R})$ with

$$he_j = \overline{\mathcal{R}}(h_1 \otimes K_j)e_jh_2 \;\Leftrightarrow\; e_jh = \mathcal{R}(h_1 \otimes K_j)h_2e_j,$$
$$f_ih = \mathcal{R}(K_i, h_2)h_1f_i \;\Leftrightarrow\; hf_i = \overline{\mathcal{R}}(K_i \otimes h_2)f_ih_1,$$

$e_jf_i = f_ie_j$, for $(i,j) \in I \times J$ and $h \in H$; $(K_i)_{i\in I}$, $(K_j)_{j\in J}$ are families of grouplike elements of H. The coalgebra structure of $\mathfrak{e} \times H \times \mathfrak{f}$ is determined by

$$\Delta(e_j) = e_j \otimes 1 + K_j \otimes e_j,\ \varepsilon(e_j) = 1\ ;\ \Delta(f_i) = f_i \otimes K_i + 1 \otimes f_i,\ \varepsilon(f_i) = 0, \tag{5.9}$$

extended as algebra morphisms. The antipode s is defined by $s(e_j) = -K_j^{-1}e_j = -\mathcal{R}(K_j \otimes K_j)e_jK_j^{-1}$, $s(h) = S(h)$, $s(f_i) = -f_iK_i^{-1} = -\overline{\mathcal{R}}(K_i, K_i)K_i^{-1}f_i$, for $(i,j) \in I \times J$ and $h \in H$, extended as an anti-morphism of algebras. □

The structures of $\mathfrak{e}$, $\mathfrak{f}$ as Yetter-Drinfeld modules were chosen such that (5.1) is satisfied, as well as (5.8) to work for a canonical map $\sigma : \mathfrak{f} \otimes \mathfrak{e} \to k$ that we next describe. The price we must pay for all these is the condition $\overline{\mathcal{R}}(h \otimes h') = \mathcal{R}(h' \otimes h)$, for all $h, h' \in H$, i.e. $(H, \mathcal{R})$ is a cotriangular Hopf algebra (or, equivalently, the category ${}^H\mathcal{M}$ is symmetric monoidal), required to make σ an H-balanced map.

LEMMA 5.11. *Let $(H, \mathcal{R})$ be a cotriangular Hopf algebra and $\mathfrak{e}$, $\mathfrak{f}$ the braided Hopf algebras constructed above. Define $\sigma : \mathfrak{f} \otimes \mathfrak{e} \to k$ by*

$$\sigma(1_\mathfrak{f} \otimes a) = \varepsilon_\mathfrak{e}(a),\ \sigma(b \otimes 1_\mathfrak{e}) = \varepsilon_\mathfrak{f}(b),\ \sigma(f_i \otimes e_j) = \varpi_{i,j}\delta_{i,j},$$

for all $a \in \mathfrak{e}$, $b \in \mathfrak{f}$, $(i,j) \in I \times J$, and extend it to $\mathfrak{e} \otimes \mathfrak{f}$ by using (5.3), (5.4). Here $\varpi_{i,j}$ is a family of scalars indexed by $I \times J$ and $\delta_{i,j}$ is the Kronecker's delta.

Then σ is H-balanced and cobalanced, that is σ verifies (5.2) and (5.8). So σ is an invertible H-balanced and cobalanced mixed skew pairing of $\mathfrak{f}$ and $\mathfrak{e}$.

PROOF. We show that (5.2) is satisfied by σ in two steps. First, we show that $\sigma(b \cdot h \otimes e_j) = \sigma(b \otimes h \triangleright e_j)$, for all $b \in \mathfrak{f}$, $h \in H$ and $j \in J$. Towards this end, we compute that $\sigma(1_{\mathfrak{f}}\cdot h\otimes e_j)=\varepsilon_H(h)\varepsilon_{\mathfrak{e}}(e_j)=0=\sigma(1_{\mathfrak{f}}\otimes h\triangleright e_j)$ and

$$\begin{aligned}\sigma(f_i\cdot h\otimes e_j)&=\mathcal{R}(K_i,h)\varpi_{i,j}\delta_{i,j}=\overline{\mathcal{R}}(h,K_j)\sigma(f_i\otimes e_j)=\sigma(f_i\otimes h\triangleright e_j),\\ \sigma(f_{i_1}f_{i_2}\cdot h\otimes e_j)&=\mathcal{R}(K_{i_1}K_{i_2}\otimes h)\sigma(f_{i_1}\otimes K_{i_2}\triangleright (e_j)_{\underline{1}})\sigma(f_{i_2}\otimes (e_j)_{\underline{2}})=0,\\ \sigma(f_{i_1}f_{i_2},h\triangleright e_j)&=\overline{\mathcal{R}}(h,K_j)\sigma(f_{i_1},K_{i_2}\triangleright (e_j)_{\underline{2}})\sigma(f_{i_2},(e_j)_{\underline{1}})=0.\end{aligned}$$

With the help of (5.4), a computation similar to the one above gives that

$$\sigma((f_{i_1}\cdots f_{i_r})\cdot h,e_j)=0=\sigma(f_{i_1}\cdots f_{i_r}\otimes h\triangleright e_j),$$

for all $r\geq 2$, $h\in H$ and $i_1,\cdots,i_r\in I$. This ends the proof of the first step.

Secondly, we prove that if $a,a'\in\mathfrak{e}$ verify (5.4) for all $b\in\mathfrak{f}$ and $h\in H$ then the same property has the element aa' of $\mathfrak{e}$. Indeed, for all $b\in B$ and $h\in H$ we have

$$\sigma(b\cdot h\otimes aa')=\sigma(b_{\underline{1}}\otimes h_1\triangleright a)\sigma(b_{\underline{2}}\otimes h_2\triangleright a')=\sigma(b\otimes h\triangleright (aa')),$$

as desired. All these facts, together with $\sigma(b\cdot h\otimes 1_{\mathfrak{e}})=\varepsilon_{\mathfrak{f}}(b)\varepsilon_H(h)=\sigma(b\otimes h\triangleright 1_{\mathfrak{e}})$, valid for all $b\in\mathfrak{f}$ and $h\in H$, entail to σ is H-balanced.

The proof of the fact that σ is H-cobalanced is similar. Remark that for $a=1_{\mathfrak{e}}$ the relation (5.8) reduces to $\varepsilon_{\mathfrak{f}}(b)1_H=\varepsilon_{\mathfrak{f}}(b)1_H$, for all $b\in\mathfrak{f}$. For $a=e_j$, $j\in J$, (5.8) reduces to $\sigma(b\otimes e_j)K_j=\sigma(b_{(0)}\otimes e_j)b_{(1)}$, for all $b\in\mathfrak{f}$, and this is true since $\sigma(1_{\mathfrak{f}}\otimes e_j)K_j=0=\sigma(1_{\mathfrak{f}}\otimes e_j)1_H$ and

$$\begin{aligned}\sigma(f_i\otimes e_j)K_j&=\varpi_{i,j}\delta_{i,j}K_j=\varpi_{i,j}\delta_{i,j}K_i=\sigma(f_i\otimes e_j)K_i,\\ \sigma(f_{i_1}\cdots f_{i_r}\otimes e_j)K_j&=0=\sigma(f_{i_1}\cdots f_{i_r}\otimes e_j)K_{i_1}\cdots K_{i_r},\end{aligned}$$

for all $r\geq 2$ and $i,i_1,\cdots,i_r\in I$. Note also that, if for $a,a'\in\mathfrak{e}$ the relation (5.8) holds for all $b\in B$ then aa' also verifies (5.8) for all $b\in B$ since

$$\begin{aligned}\sigma(b\otimes a_{\{0\}}a'_{\{0\}})a_{\{-1\}}a'_{\{-1\}}&=\sigma(b_{\underline{1}}\otimes a_{\{0\}})\sigma(b_{\underline{2}}\otimes a'_{\{0\}})a_{\{-1\}}a'_{\{-1\}}=\sigma((b_{\underline{1}})_{(0)}\otimes a)\\ \sigma((b_{\underline{2}})_{(0)}\otimes a')(b_{\underline{1}})_{(1)}(b_{\underline{2}})_{(1)}&=\sigma(b_{(0)_{\underline{1}}}\otimes a)\sigma(b_{(0)_{\underline{2}}}\otimes a')b_{(1)}=\sigma(b_{(0)}\otimes aa')b_{(1)},\end{aligned}$$

and our proof ends. □

Surprisingly, nothing new appears when we deform the Hopf algebra $\mathfrak{e}\times H\times\mathfrak{f}$ by the 2-cocycle $\widehat{\sigma}$ defined by σ built in Lemma 5.11.

PROPOSITION 5.12. *Consider $(H,\mathcal{R})$, $\mathfrak{e}$, $\mathfrak{f}$ and $\mathcal{A}_H(\mathfrak{e},\mathfrak{f})$ as in Example 5.10, $\widehat{\sigma}$ defined by σ in Lemma 5.11 as in Proposition 4.2, and $D_\Sigma(\mathfrak{e},\mathfrak{f})$ the generalized quantum double of $\mathfrak{e}$, $\mathfrak{f}$ and Σ as it was defined in the proof of Theorem 4.6. Then $(\mathfrak{e}\times H\times\mathfrak{f})_{\widehat{\sigma}}=\mathfrak{e}\times H\times\mathfrak{f}$ and $D_\Sigma(\mathfrak{e},\mathfrak{f})=\mathcal{A}_H(\mathfrak{e},\mathfrak{f})$ as Hopf algebras, and therefore $(\mathfrak{e}\times H\times\mathfrak{f})_{\widehat{\sigma}}=\mathcal{A}_H(\mathfrak{e},\mathfrak{f})\times H$ as Hopf algebras.*

PROOF. According to (5.5), the multiplication of $(\mathfrak{e}\times H\times\mathfrak{f})_{\widehat{\sigma}}$ is given by

$$\begin{aligned}(e_{j_1}hf_{i_1})(e_{j_2}h'f_{i_2})&=\varpi_{i_1,j_2}\delta_{i_1,j_2}e_{i_1}hK_{i_1}h'f_{i_2}+\mathcal{R}(K_{j_2},h_1)\mathcal{R}(K_{i_1},h'_2)e_{j_1}e_{j_2}h_2\\ &\quad h'_1f_{i_1}f_{i_2}-\varpi_{i_1,j_2}\delta_{i_1,j_2}e_{i_1}hK_{j_2}h'f_{i_2}=\mathcal{R}(K_{j_2},h_1)\mathcal{R}(K_{i_1},h'_2)e_{j_1}e_{j_2}h_2h'_1f_{i_1}f_{i_2},\end{aligned}$$

for all $i_1,i_2\in I$, $j_1,j_2\in J$ and $h,h'\in H$, where we wrote e_jhf_i in place of $e_j\times h\times f_i$. Since this is the multiplication of $\mathfrak{e}\times H\times\mathfrak{f}$, it follows that $(\mathfrak{e}\times H\times\mathfrak{f})_{\widehat{\sigma}}=\mathfrak{e}\times H\times\mathfrak{f}$.

Similarly, by specializing (4.17) to our context, we see that the multiplication $\diamond$ of $D_\Sigma(\mathfrak{e},\mathfrak{f})$ is determined, for all $i_1,i_2\in I$ and $j_1,j_2\in J$, by

$$\begin{aligned}(f_{i_1}\otimes e_{j_1})\diamond(f_{i_2}\otimes e_{j_2})&=\Sigma(f_{i_1}\otimes e_{j_2})e_{j_1}\otimes f_{i_2}+\Sigma(f_{i_1}\otimes 1_{\mathfrak{e}})e_{j_1}\otimes\overline{\Sigma}(1_{\mathfrak{f}}\otimes e_{j_2})f_{i_2}\\ &\quad+\mathcal{R}(K_{i_1},K_{j_2})e_{j_1}e_{j_2}\otimes f_{i_1}\bullet f_{i_2}+\overline{\Sigma}(f_{i_1}\otimes e_{j_2})e_{j_1}\otimes f_{i_2},\end{aligned}$$

where $\bullet$ is the multiplication of $\bar{\mathfrak{f}}$ in ${}_H^H\mathcal{YD}$. But $\Sigma(f_i \otimes e_j) = \sigma(f_i \otimes K_i^{-1} \triangleright e_j) = \overline{\mathcal{R}}(K_i^{-1}, K_j)\sigma(f_i \otimes e_j) = \mathcal{R}(K_i \otimes K_j)\varpi_{i,j}\delta_{i,j}$ and $\overline{\Sigma}(f_i \otimes e_j) = -\Sigma(f_i \otimes e_j)$, for all $i \in I$ and $j \in J$. We get from here that $(f_{i_1} \otimes e_{j_1}) \diamond (f_{i_2} \otimes e_{j_2}) = \mathcal{R}(K_{i_1}, K_{j_2}) e_{j_1} e_{j_2} \otimes f_{i_1} \bullet f_{i_2}$ or, equivalently, that $D_\Sigma(\mathfrak{e}, \mathfrak{f}) = \mathcal{A}_H(\mathfrak{e}, \mathfrak{f})$ as algebras. This finishes the proof. □

Concluding, although we have a non-trivial σ it does not contribute to the algebra structure of $(\mathfrak{e} \times H \times \mathfrak{f})_{\widehat{\sigma}}$. To circumvent this disappearance of σ in computations we make another choice among the grouplike elements of H. As we will see, the price we must pay in this case is the lost of the H-cobalance property of σ.

5.3. The coquasitriangular case II. We fix the same context as in the case I: $(H, \mathcal{R})$ is a CQT Hopf algebra, and $\mathfrak{f}$ and $\mathfrak{e}$ are the unital free algebras on the sets $\{f_i, i \in I\}$ and $\{e_j, j \in J\}$; furthermore, we have maps $I \ni i \mapsto K_i \in G(H)$ and $J \ni j \mapsto K_j \in G(H)$. The difference from the case I consists in the fact that in what follows we will work with the map $j' := S \circ j : J \ni j \mapsto K_j^{-1} \in G(H)$ instead of j. Thus, $\mathfrak{f}$ will be regarded as a Hopf algebra in $\mathcal{YD}_H^H$ as before (see Proposition 5.7) and $\mathfrak{e}$ will be considered an algebra in ${}_H^H\mathcal{YD}$ with a new structure:

$$e_j \mapsto K_j^{-1} \otimes e_j,\ h \blacktriangleright e_j = \overline{\mathcal{R}}(h \otimes K_j^{-1}) e_j = \mathcal{R}(h \otimes K_j) e_j, \forall\ j \in J,\ h \in H,$$

extended to the whole $\mathfrak{e}$ in a way that turns $\mathfrak{e}$ into an algebra in ${}_H^H\mathcal{YD}$. To avoid any possible confusion we denote $\mathfrak{e}$ with this new algebra structure in ${}_H^H\mathcal{YD}$ by $\mathfrak{e}^-$. Then the coalgebra structure and the antipode of $\mathfrak{e}^-$ in ${}_H^H\mathcal{YD}$ coincide with the comultiplication and the antipode of $\mathfrak{e}$, respectively (see Proposition 5.9). Also, (5.1) is satisfied by $A = \mathfrak{e}^-$, $B = \mathfrak{f}$, so we can consider $\mathfrak{e}^- \times H \times \mathfrak{f}$.

PROPOSITION 5.13. *(i) The biproduct Hopf algebra $\mathfrak{e}^- \times H \times \mathfrak{f}$ can be characterized as the unital k-algebra generated by the families $\{e_j, j \in J\}$, $\{f_i, i \in I\}$ and the elements of H, with relations $e_j f_i = f_i e_j$ and*

$$he_j = \mathcal{R}(h_1 \otimes K_j) e_j h_2 \ \Leftrightarrow\ e_j h = \overline{\mathcal{R}}(h_1 \otimes K_j) h_2 e_j,$$
$$f_i h = \mathcal{R}(K_i \otimes h_2) h_1 f_i \ \Leftrightarrow\ h f_i = \overline{\mathcal{R}}(K_i \otimes h_2) f_i h_1,$$

for all $(i, j) \in I \times J$ and $h \in H$. The coalgebra structure of $\mathfrak{e}^- \otimes H \times \mathfrak{f}$ is defined by

$$\Delta(e_j) = e_j \otimes 1 + K_j^{-1} \otimes e_j,\ \varepsilon(e_j) = 0,\ \Delta(f_i) = f_i \otimes K_i + 1 \otimes f_i,\ \varepsilon(f_i) = 0,$$

for all $(i, j) \in I \times J$, extended such that Δ and ε become algebra morphisms in ${}_k\mathcal{M}$. Its antipode s is determined by $s(e_j) = -\mathcal{R}(K_j \otimes K_j) e_j K_j$, $s(h) = S(h)$ and $s(f_i) = -\overline{\mathcal{R}}(K_i, K_i) K_i^{-1} f_i$, extended as an anti-morphism of algebras in ${}_k\mathcal{M}$.

(ii) If $\mathcal{A}_H(\mathfrak{e}^-, \mathfrak{f})$ is the unital k-algebra generated by the families $(E_j)_{j \in J}$ and $(F_i)_{i \in I}$ then $\mathcal{A}_H(\mathfrak{e}^-, \mathfrak{f})$ is an algebra in ${}_H^H\mathcal{YD}$ with left H-action and coaction

$$h \cdot E_j = \mathcal{R}(h \otimes K_j) E_j,\ h \cdot F_i = \overline{\mathcal{R}}(K_i \otimes h) F_i,\ E_j \mapsto K_j^{-1} \otimes E_j,\ F_i \mapsto K_i^{-1} \otimes F_i,$$

extended such that $\mathcal{A}_H(\mathfrak{e}^-, \mathfrak{f}) \in {}_H^H\mathcal{YD}$ is an algebra; it is also a coalgebra here with

$$\Delta(E_j) = E_j \otimes 1 + 1 \otimes E_j,\ \varepsilon(E_j) = 0,\ \Delta(F_i) = F_i \otimes 1 + 1 \otimes F_i,\ \varepsilon(F_i) = 0,$$

extended as algebra morphisms in ${}_H^H\mathcal{YD}$ and ${}_k\mathcal{M}$, and a braided Hopf algebra with $S(E_j) = -E_j$, $S(F_i) = -F_i$, extended as an anti-morphism of algebras in ${}_H^H\mathcal{YD}$.

(iii) The map $\chi : \mathfrak{e}^- \times H \times \mathfrak{f} \ni e_j h f_i \mapsto \overline{\mathcal{R}}(K_i \otimes h_1) E_j F_i \times h_2 K_i \in \mathcal{A}_H(\mathfrak{e}^-, \mathfrak{f}) \times H$ is a Hopf algebra isomorphism.

PROOF. Follows from Example 5.10, by replacing K_j with K_j^{-1} $(j \in J)$. □

For the next result we need $\mathcal{R}$ to be symmetric.

LEMMA 5.14. *Let $(H,\mathcal{R})$ be a CQT Hopf algebra with $\mathcal{R}$ symmetric. For $\mathfrak{e}^-$, $\mathfrak{f}$ as in Proposition 5.13, define $\sigma : \mathfrak{f} \otimes \mathfrak{e}^- \to k$ by*

$$\sigma(1_{\mathfrak{f}} \otimes a) = \varepsilon_{\mathfrak{e}}(a),\ \sigma(b \otimes 1_{\mathfrak{e}}) = \varepsilon_{\mathfrak{f}}(b),\ \sigma(f_i \otimes e_j) = \varpi_{i,j}\delta_{i,j},$$

for all $a \in \mathfrak{e}^-$, $b \in \mathfrak{f}$, $(i,j) \in I \times J$, and extend it to $\mathfrak{e}^- \otimes \mathfrak{f}$ by using (5.3), (5.4); $\varpi_{i,j}$ is a given family of scalars indexed by $I \times J$.

Then σ is H-balanced, and so an invertible H-balanced skew-paring of $\mathfrak{f}$, $\mathfrak{e}^-$.

PROOF. The only thing we must check is (5.3), specialized for $b = f_i$ and $a = e_j$; then the proof continues along the line of the first part of the proof of Lemma 5.11. We compute, for all $h \in H$,

$$\sigma(f_i \cdot h \otimes e_j) = \mathcal{R}(K_i \otimes h)\varpi_{i,j}\delta_{i,j} = \mathcal{R}(h \otimes K_j)\sigma(f_i \otimes e_j) = \sigma(f_i \otimes h \blacktriangleright e_j),$$

as desired. Note that we need $\mathcal{R}$ symmetric just to make this equality to work. □

Clearly σ is not H-cobalanced. Nevertheless, it defines an invertible 2-cocycle $\widehat{\sigma}$ for $\mathfrak{e}^- \times H \times \mathfrak{f}$, and this time contributes to the algebra structure of $(\mathfrak{e}^- \times H \times \mathfrak{f})_{\widehat{\sigma}}$.

PROPOSITION 5.15. *For $(H,\mathcal{R})$ a CQT Hopf algebra with $\mathcal{R}$ symmetric, the Hopf algebra $\mathbb{H} := (\mathfrak{e}^- \times H \times \mathfrak{f})_{\widehat{\sigma}}$ can be described as follows:*

- as an algebra, $\mathbb{H}$ is generated by the families $\{e_j, j \in J\}$, $\{f_i, i \in I\}$ and the elements of H, with relations, for all $(i,j) \in I \times J$ and $h \in H$,

$$\begin{aligned} he_j &= \mathcal{R}(h_1 \otimes K_j)e_jh_2 \ \Leftrightarrow\ e_jh = \mathcal{R}(S^{-1}(h_1) \otimes K_j)h_2e_j, \\ f_ih &= \mathcal{R}(K_i \otimes h_2)h_1f_i \ \Leftrightarrow\ hf_i = \overline{\mathcal{R}}(K_i \otimes h_2)f_ih_1, \\ [f_i, e_j] &= \varpi_{i,j}\delta_{i,j}(K_i - K_i^{-1}),\ \forall\ (i,j) \in I \times J; \end{aligned}$$

- the coalgebra structure of $\mathbb{H}$ is determined by $\Delta\mid_H = \Delta_H$, $\varepsilon\mid_H = \varepsilon_H$ and

$$\Delta(e_j) = e_j \otimes 1 + K_j^{-1} \otimes e_j,\ \varepsilon(e_j) = 0,\ \Delta(f_i) = f_i \otimes K_i + 1 \otimes f_i,\ \varepsilon(f_i) = 0,$$

for all $(i,j) \in I \times J$, extended as algebra morphisms;

- the antipode s^σ of $\mathbb{H}$ is defined, for all $(i,j) \in I \times J$ and $h \in H$, by

$$s^\sigma(e_j) = -\mathcal{R}(K_j \otimes K_j)e_jK_j,\ s^\sigma(h) = S(h) \text{ and } s^\sigma(f_i) = -\overline{\mathcal{R}}(K_i \otimes K_i)K_i^{-1}f_i,$$

and extended as an anti-morphism of k-algebras.

PROOF. The formula (5.5) gives the following rules for the multiplication of $\mathbb{H}$:

$$\begin{aligned} he_j &\equiv (1_{\mathfrak{e}^-} \times h \times 1_{\mathfrak{f}}) \cdot_\sigma (e_j \times 1_H \times 1_{\mathfrak{f}}) = (h_1 \blacktriangleright e_j) \times h_2 \times 1_{\mathfrak{f}} \equiv \mathcal{R}(h_1 \otimes K_j)e_jh_2, \\ f_ih &\equiv (1_{\mathfrak{e}^-} \times 1_H \times f_i) \cdot_\sigma (1_{\mathfrak{e}^-} \times h \times 1_{\mathfrak{f}}) = 1_{\mathfrak{e}^-} \times h_1 \times f_i \cdot h_2 \equiv \mathcal{R}(K_i \otimes h_2)h_1f_i, \\ f_ie_j &\equiv (1_{\mathfrak{e}^-} \times 1_H \times f_i) \cdot_\sigma (e_j \times 1_H \times 1_{\mathfrak{f}}) = \sigma(f_i \otimes e_j)1_{\mathfrak{e}^-} \times K_i \times 1_{\mathfrak{f}} + e_j \times 1_H \times f_i \\ &\qquad + \overline{\sigma}(f_i \otimes e_j)1_{\mathfrak{e}^-} \times K_j^{-1} \times 1_{\mathfrak{f}} \equiv e_jf_i + \varpi_{i,j}\delta_{i,j}(K_i - K_i^{-1}), \end{aligned}$$

for all $(i,j) \in I \times J$ and $h \in H$. As $\mathbb{H}$ and $\mathfrak{e}^- \times H \times \mathfrak{f}$ are equal as coalgebras, the coalgebra structure of $\mathbb{H}$ is obtained from Proposition 5.13. Then either the definition of $s^{\widehat{\sigma}}$ or a direct computation entails to the above description of s^σ. □

REMARK 5.16. Since σ is not H-cobalanced (unless K_i, K_j are all either equal to 1_H or of order 2), there is no warranty that $\mathbb{H}$ is a biproduct Hopf algebra. Nevertheless, $\mathbb{H}$ can be seen as a deformation by an invertible 2-cocycle ω of $\mathcal{A}_H(\mathfrak{e}^-, \mathfrak{f}) \times H$ described in Proposition 5.13; ω can be derived from σ as in Proposition 4.3.

A concrete situation when Proposition 5.15 applies is the following.

EXAMPLE 5.17. Let ν be an odd natural number and H_ν the unital k-algebra generated by g, x with relations $g^{2\nu} = 1$, $x^2 = 0$, $xg = -gx$. Then H_ν is a Hopf algebra with $g \in G(H_\nu)$, $\Delta(x) = x \otimes g^\nu + 1 \otimes x$, $\varepsilon(x) = 0$, $S(g) = g^{-1}$, $S(x) = g^\nu x$.

Following [**7, 13**], the CQT structures $\mathcal{R}$ of H_ν are parametrized by (s, β) consisting of odd natural numbers $1 \leq s \leq 2\nu - 1$ and $\beta \in k$, in the sense that

$$\mathcal{R} = \mathcal{R}_{s,\beta} = \sum_{n,m=0}^{2\nu-1} \zeta^{snm}(g^n)^* \otimes (g^m)^* + \beta \sum_{n,m=0}^{2\nu-1} (-1)^m \omega^{snm}(g^n x)^* \otimes (g^m x)^*,$$

where ζ is a primitive 2ν-root of unity in k and $\{(g^n x^i)^* \mid 0 \leq n \leq 2\nu-1,\ 0 \leq i \leq 1\}$ is the basis of H_ν^* dual to the basis $\{g^n x^i \mid 0 \leq n \leq 2\nu - 1,\ 0 \leq i \leq 1\}$ of H_ν. As we need $\mathcal{R}_{s,\beta}$ symmetric, this forces $\beta = 0$, and so we must work with $\mathcal{R}_s := \mathcal{R}_{s,0} = \sum_{n,m=0}^{2\nu-1} \omega^{snm}(g^n)^* \otimes (g^m)^*$.

Consider the families $(n_i)_{i \in I}$ and $(m_j)_{j \in J}$ with $n_i, m_j \in \{0, 1, \cdots, 2\nu - 1\}$, and the grouplike elements $K_i = g^{n_i}$ and $K_j = g^{m_j}$. Then the Hopf algebra $\mathbb{H}$ in Proposition 5.15 specializes as follows:

- as unital algebra, $\mathbb{H}$ is generated by the families $\{e_j, j \in J\}$, $\{f_i, i \in I\}$ and elements g, x, with relations (valid for all $i \in I$ and $j \in J$)

$$\begin{gathered} ge_j = \zeta^{sm_j} e_j g,\ f_i g = \zeta^{sn_i} g f_i,\ xe_j = e_j x,\ f_i x = \zeta^{s\nu n_i} x f_i, \\ xg = -gx,\ [f_i, e_j] = \varpi_{i,j}\delta_{i,j}(g^{n_i} - g^{-n_i}); \end{gathered}$$

- the coalgebra structure of $\mathbb{H}$ is defined, for all $i \in I$ and $j \in J$, by

$$\Delta(g) = g \otimes g,\ \Delta(x) = x \otimes g^\nu + 1 \otimes x,\ \Delta(e_j) = e_j \otimes 1 + g^{-m_j} \otimes e_j,\ \Delta(f_i) = f_i \otimes g^{n_i} + 1 \otimes f_i$$

and $\varepsilon(g) = 1$, $\varepsilon(x) = \varepsilon(e_j) = \varepsilon(f_i) = 0$, extended as algebra morphisms;

- the antipode s is given by $s(g) = g^{-1}$, $s(x) = g^\nu x$, $s(e_j) = -\zeta^{sm_j^2} e_j g^{m_j}$, $s(f_i) = -\zeta^{-sn_i^2} g_i^{-n_i} f_i$, extended as an anti-morphism of algebras.

The Hopf algebra $\mathbb{H}$ in Proposition 5.15 generalizes the Hopf algebra $U(\overline{C}, H, B)$ built and related to the Lusztig's construction of $U_q(\mathfrak{g})$ in [**17**, Proposition 4.3]. We next see that $U_q(\mathfrak{g})$, as it is presented in [**12**], is a quotient of $\mathbb{H}$ and can be seen as a deformation by an invertible 2-cocycle of a certain left biproduct Hopf algebra.

5.4. The Drinfeld-Jimbo quantum groups $U_q(\mathfrak{g})$. Let $(I, \cdot)$ be a Cartan datum as in [**14**, Chapter 1]. So we have a symmetric bilinear form $\cdot : \mathbb{Z}[I] \times \mathbb{Z}[I] \to \mathbb{Z}$ satisfying appropriate conditions. Here $(\mathbb{Z}[I], +)$ is the free abelian group with basis $\{i, i \in I\}$. The elements of $\mathbb{Z}[I]$ are denoted by $\{K_\nu, \nu \in \mathbb{Z}[I]\}$; then $K_\mu K_\nu = K_{\mu+\nu}$, so K_0 is the neutral element of $\mathbb{Z}[I]$, and $K_\mu^{-1} = K_{-\mu}$, $\mu, \nu \in \mathbb{Z}[I]$.

Let H be equal to the group Hopf algebra of $\mathbb{Z}[I]$ over k; thus any element K_ν of $\mathbb{Z}[I]$ is a grouplike element of H. Furthermore, for q a non-zero element of k, define $\mathcal{R} : H \otimes H \to k$ by $\mathcal{R}(K_\mu \otimes K_\nu) = q^{\mu \cdot \nu}$, for all $\mu, \nu \in \mathbb{Z}[I]$, extended by k-linearity. Clearly, $\mathcal{R}$ provides a CQT structure for H; the convolution inverse of $\mathcal{R}$ is $\overline{\mathcal{R}}$ given by $\overline{\mathcal{R}}(K_\mu \otimes K_\nu) = q^{-\mu \cdot \nu}$, for all $\mu, \nu \in \mathbb{Z}[I]$, extended by k-linearity.

Consider $\mathfrak{e}^-$ equals the free k-algebra on the set $\{e_i, i \in I\}$, and regard it as a Hopf algebra in ${}_H^H\mathcal{YD}$ with the structure defined at the beginning of Subsection 5.3 (with the mention that the chosen family of grouplike elements is $(K_i^{-1})_{i \in I}$, with $K_i \in \mathbb{Z}[I]$ as in the above). Likewise, consider $\mathfrak{f}$ equals the free k-algebra on the set $\{f_i, i \in I\}$ and regard it as a Hopf algebra in $\mathcal{YD}_H^H$ with structure as in the statement of Proposition 5.7; so this time the chosen family of grouplike elements is $(K_i)_{i\in I}$. Finally, denote by $\mathfrak{I}_{\mathfrak{e}}$ (resp. $\mathfrak{I}_{\mathfrak{f}}$) the ideal of $\mathfrak{e}^-$ (resp. $\mathfrak{f}$) generated by

$$\{e_ie_j - e_je_i \mid i \neq j \text{ s.t. } i\cdot j = 0\} \cup \{e_i^2e_j - [2]e_ie_je_i + e_je_i^2 \mid i \neq j \text{ s.t. } i\cdot i = 2,\ i\cdot j = -1\}$$
$$(\{f_if_j - f_jf_i \mid i \neq j \text{ s.t. } i\cdot j = 0\} \cup \{f_i^2f_j - [2]f_if_jf_i + f_jf_i^2 \mid i \neq j \text{ s.t. } i\cdot i = 2, i\cdot j = -1\}),$$

where $[2] = q + q^{-1}$. We next see that $'\mathfrak{e}^- = \mathfrak{e}^-/\mathfrak{I}_{\mathfrak{e}}$ and $'\mathfrak{f} := \mathfrak{f}/\mathfrak{I}_{\mathfrak{f}}$ remain braided Hopf algebras, and so we can consider the double-biproduct Hopf algebra $'\mathfrak{e} \times H \times '\mathfrak{f}$.

PROPOSITION 5.18. *Under the above context and notation, $\mathfrak{I}_{\mathfrak{e}}$ is a left (resp. right) H-module and subcomodule and at the same a Hopf ideal of $\mathfrak{e}^-$ (resp. $\mathfrak{f}$). Consequently, $'\mathfrak{e}^-$ (resp. $'\mathfrak{f}$) is a braided Hopf algebra within ${}^H_H\mathcal{YD}$ (resp. $\mathcal{YD}^H_H$).*

PROOF. We prove only the statement for $\mathfrak{I}_{\mathfrak{e}}$; the proof for $\mathfrak{I}_{\mathfrak{f}}$ is similar. Let $i, j \in I$ be such that $i \neq j$, and denote $\mathfrak{z} := e_i^2e_j - [2]e_ie_je_i + e_je_i^2$. We have $h \blacktriangleright (e_ie_j) = (h_1 \blacktriangleright e_i)(h_2 \blacktriangleright e_j) = \mathcal{R}(h_1 \otimes K_i)\mathcal{R}(h_2 \otimes K_j)e_ie_j = \mathcal{R}(h \otimes K_{i+j})e_ie_j$,

$$h \blacktriangleright (e_ie_j - e_je_i) = \mathcal{R}(h \otimes K_{i+j})(e_ie_j - e_je_i) \text{ and } h \blacktriangleright \mathfrak{z} = \mathcal{R}(h \otimes K_{2i+j})\mathfrak{z},$$

for all $h \in H$, proving that $\mathfrak{I}_{\mathfrak{e}}$ is a left H-submodule of $\mathfrak{e}^-$. By definition, $e_ie_j \mapsto K_i^{-1}K_j^{-1} \otimes e_ie_j = K_{i+j}^{-1} \otimes e_ie_j$, so

$$e_ie_j - e_je_i \mapsto K_{i+j}^{-1} \otimes (e_ie_j - e_je_i),\ \mathfrak{z} \mapsto K_{2i+j}^{-1} \otimes \mathfrak{z}.$$

Thus, so far $'\mathfrak{e}^- \in {}^H_H\mathcal{YD}$ such that the canonical projection $\mathfrak{e}^- \to {}'\mathfrak{e}^-$ is a morphism of algebras in ${}^H_H\mathcal{YD}$ (whether or not $i \cdot j = 0$). Clearly, $\varepsilon_{\mathfrak{e}^-}(\mathfrak{I}_{\mathfrak{e}}) = 0$. Hence, to show that $\mathfrak{I}_{\mathfrak{e}}$ is a coideal of $\mathfrak{e}^-$ it suffices to show that, for all $i \neq j$, we have

$$\Delta(e_ie_j - e_je_i)/\Delta(\mathfrak{z}) \in \mathfrak{I}_{\mathfrak{e}} \otimes \mathfrak{e}^- + \mathfrak{e}^- \otimes \mathfrak{I}_{\mathfrak{e}}, \text{ if } i\cdot j = 0/i\cdot i = 2 \text{ and } i\cdot j = -1.$$

Recall that the comultiplication of $\mathfrak{e}^-$ is defined by $\Delta(e_i) = e_i\widehat{\otimes}1_{\mathfrak{e}^-} + 1_{\mathfrak{e}^-}\widehat{\otimes}e_i$, for all $i \in I$, extended as an algebra morphism from $\mathfrak{e}^-$ to $\mathfrak{e}^-\widehat{\otimes}\mathfrak{e}^-$, the tensor product algebra in ${}^H_H\mathcal{YD}$ of $\mathfrak{e}^-$ and itself. So

$$\Delta(e_ie_j) \quad = \quad e_ie_j\widehat{\otimes}1_{\mathfrak{e}^-} + e_i\widehat{\otimes}e_j + q^{-i\cdot j}e_j\widehat{\otimes}e_i + 1_{\mathfrak{e}^-}\widehat{\otimes}e_ie_j,$$

hence $\Delta(e_ie_j - e_je_i) = (e_ie_j - e_je_i)\widehat{\otimes}1_{\mathfrak{e}^-} + (1 - q^{-i\cdot j})e_i\widehat{\otimes}e_j - (1 - q^{-i\cdot j})e_j\widehat{\otimes}e_i + 1_{\mathfrak{e}^-}\widehat{\otimes}(e_ie_j - e_je_i)$. It follows that for $i \cdot j = 0$, $\Delta(e_ie_j - e_je_i) \in \mathfrak{I}_{\mathfrak{e}} \otimes \mathfrak{e}^- + \mathfrak{e}^- \otimes \mathfrak{I}_{\mathfrak{e}}$.

In a similar manner we compute that

$$\begin{aligned}\Delta(e_ie_je_s) = {}& e_ie_je_s\widehat{\otimes}1_{\mathfrak{e}^-} + e_ie_j\widehat{\otimes}e_s + q^{-j\cdot s}e_ie_s\widehat{\otimes}e_j + e_i\widehat{\otimes}e_je_s \\ &+ q^{-i\cdot(j+s)}e_je_s\widehat{\otimes}e_i + q^{-i\cdot j}e_j\widehat{\otimes}e_ie_s + q^{-(i+j)\cdot s}e_s\widehat{\otimes}e_ie_j + 1_{\mathfrak{e}^-}\widehat{\otimes}e_ie_je_s.\end{aligned}$$

With the help of the above formula we see that

$$\begin{aligned}\Delta(\mathfrak{z}) = {}& \mathfrak{z}\widehat{\otimes}1_{\mathfrak{e}^-} + ((q^{-i\cdot j} + q^{-i\cdot(i+j)} - [2])e_ie_j + (1 + q^{-i\cdot i} - [2]q^{-i\cdot(i+j)})e_je_i)\widehat{\otimes}e_i \\ &+ 1_{\mathfrak{e}^-}\widehat{\otimes}\mathfrak{z} + (1 - [2]q^{-i\cdot j} + q^{-2i\cdot j})e_i^2\widehat{\otimes}e_j + (1 + q^{-i\cdot i} - [2]q^{-i\cdot(i+j)})e_i\widehat{\otimes}e_ie_j \\ &+ (q^{-2i\cdot j} - [2]q^{-i\cdot j} + 1)e_j\widehat{\otimes}e_i^2 + (q^{-i\cdot j} + q^{-i\cdot(i+j)} - [2])e_i \otimes e_je_i.\end{aligned}$$

When $i \cdot j = -1$ and $i \cdot i = 2$ (and so $i \cdot (i + j) = 1$) the above equality reduces to $\Delta(\mathfrak{z}) = \mathfrak{z}\widehat{\otimes}1_{\mathfrak{e}^-} + 1_{\mathfrak{e}^-}\widehat{\otimes}\mathfrak{z}$, and so $\mathfrak{I}_{\mathfrak{e}}$ is a coideal of $\mathfrak{e}^-$; it remains to show that $S_{\mathfrak{e}^-}(\mathfrak{I}_{\mathfrak{e}}) \subseteq \mathfrak{I}_{\mathfrak{e}}$. Since $S_{\mathfrak{e}^-}(e_i) = -e_i$ and extends as an anti-algebra morphism in ${}^H_H\mathcal{YD}$, we see that $S_{\mathfrak{e}^-}(e_ie_j) = S_{\mathfrak{e}^-}(K_i^{-1} \blacktriangleright e_j)S_{\mathfrak{e}^-}(e_i) = q^{-i\cdot j}e_je_i$. Thus $S_{\mathfrak{e}^-}(e_ie_j - e_je_i) = -q^{-i\cdot j}(e_ie_j - e_je_i) \in \mathfrak{I}_{\mathfrak{e}}$. Likewise, $S_{\mathfrak{e}^-}(e_ie_je_s) = -q^{-i\cdot j-(i+j)\cdot s}e_se_je_i$, from where we compute that $S_{\mathfrak{e}^-}(e_i^2e_j) = -e_je_i^2$, $S_{\mathfrak{e}^-}(e_ie_je_i) = -e_ie_je_i$ and $S_{\mathfrak{e}^-}(e_je_i^2) = -e_i^2e_j$, provided that $i \cdot j = -1$ and $i \cdot i = 2$. Thus $S_{\mathfrak{e}^-}(\mathfrak{z}) = -\mathfrak{z}$, provided that $i \cdot j = -1$ and $i \cdot i = 2$, and so $\mathfrak{I}_{\mathfrak{e}}$ is a Hopf ideal of $\mathfrak{e}^-$.

By the right version of Remark 5.8, $'\mathfrak{e}^-$ is a Hopf algebra in ${}^H_H\mathcal{YD}$ in a way that makes the canonical projection $\mathfrak{e}^- \to {}'\mathfrak{e}^-$ a Hopf algebra morphism in ${}^H_H\mathcal{YD}$. □

By Proposition 5.15 we get the following description for $U_q(\mathfrak{g})$.

THEOREM 5.19. *Let $(I, \cdot)$ be a Cartan datum, $(H, \mathcal{R})$ the group Hopf algebra of $\mathbb{Z}[I]$ over k considered CQT with $\mathcal{R}(K_\mu, K_\nu) = q^{\mu\cdot\nu}$ for a fixed $q \in k \setminus \{0\}$, and $'\mathfrak{e}^-$ and $'\mathfrak{f}$ the braided Hopf algebras from Proposition 5.18. Define $'\mathbb{H}$ as being the unital k-algebra generated by $\{e_i, f_i, K_i; i \in I\}$ with the K_i's invertible and such that $K_iK_j = K_jK_i$, and relations*

$$K_ie_jK_i^{-1} = q^{i\cdot j}e_j,\ K_if_jK_i^{-1} = q^{-i\cdot j}f_jK_i,\ [f_i, e_j] = \varpi_{i,j}\delta_{i,j}(K_i - K_i^{-1}),$$

$e_ie_j = e_je_i$ and $f_if_j = f_jf_i$ if $i \neq j$ such that $i\cdot j = 0$, and $e_i^2e_j - [2]e_ie_je_i + e_je_i^2 = 0$ and $f_i^2f_j - [2]f_if_jf_i + f_jf_i^2 = 0$ if $i \neq j$ such that $i \cdot i = 2$ and $i \cdot j = -1$; here $(\varpi_{i,j})_{i,j\in I}$ is a family of parameters from k. Then $'\mathbb{H}$ is a Hopf algebra with structure given by

$$\Delta(e_i) = e_i \otimes 1 + K_i^{-1} \otimes e_i,\ \varepsilon(e_i) = 0,\ \Delta(f_i) = f_i \otimes K_i + 1 \otimes f_i,\ \varepsilon(f_i) = 0,$$

for all $i \in I$, extended as algebra morphisms; the antipode $'s$ of $'\mathbb{H}$ is defined by

$$'s(e_i) = -q^{i\cdot i}e_iK_i = -K_ie_i,\ 's(f_i) = -q^{-i\cdot i}K_i^{-1}f_i = -f_iK_i^{-1},\ 's(K_i) = K_i^{-1},$$

for all $i \in I$, extended as an anti-morphism of algebras. Furthermore, $'\mathbb{H}$ is isomorphic to a 2-cocycle deformation of a left biproduct Hopf algebra.

PROOF. σ defined in Lemma 5.14 induces an invertible, H-balanced mixed skew pairing (still denoted by σ) of $'\mathfrak{f}$ and $'\mathfrak{e}^-$. Otherwise stated, $\sigma(\mathfrak{f}\otimes\mathfrak{I}_\mathfrak{e}) = 0 = \sigma(\mathfrak{I}_\mathfrak{f}\otimes\mathfrak{e})$. For this, we only prove that $\sigma(\mathfrak{I}_\mathfrak{f} \otimes \mathfrak{e}) = 0$; that $\sigma(\mathfrak{f} \otimes \mathfrak{I}_\mathfrak{e}) = 0$ is omitted.

For simplicity, denote $x := f_if_j - f_jf_i$ for $i \neq j$ such that $i \cdot j = 0$, and $y = f_i^2f_j - [2]f_if_jf_i + f_jf_i^2$ for $i \neq j$ such that $i \cdot i = 2$ and $i \cdot j = -1$. Clearly, $\sigma(x \otimes 1_{\mathfrak{e}^-}) = \varepsilon_\mathfrak{f}(x) = 0 = \sigma(y \otimes 1_{\mathfrak{e}^-})$ and

$$\sigma(f_if_j \otimes e_s) = \sigma(f_i \otimes 1_\mathfrak{e}^-)\sigma(f_j \otimes e_s) + q^{j\cdot s}\sigma(f_i \otimes e_s)\sigma(f_j \otimes 1_{\mathfrak{e}^-}) = 0,$$
$$\sigma(f_if_jf_l \otimes e_s) = q^{l\cdot s}\sigma(f_if_j \otimes e_s)\sigma(f_l \otimes 1_{\mathfrak{e}^-}) + \sigma(f_if_j \otimes 1_{\mathfrak{e}^-})\sigma(f_l \otimes e_s) = 0,$$

for all $i, j, l, s \in I$. It is immediate now that $\sigma(x \otimes e_s) = \sigma(y \otimes e_s) = 0$, for all $s \in I$.

The elements x, y are primitive, i.e. $\Delta(x) = x \otimes 1_\mathfrak{f} + 1_\mathfrak{f} \otimes x$, and similar for y. This guarantees that $\sigma(x \otimes uv) = \sigma(x \otimes u)\sigma(1_\mathfrak{f} \otimes v) + \sigma(1_\mathfrak{f} \otimes u)\sigma(x \otimes v)$, for all $u, v \in \mathfrak{e}^-$, and similar for y. As the e_s's generate the k-algebra $\mathfrak{e}^-$, it follows now that $\sigma(z \otimes \mathfrak{e}^-) = 0$, for all $z \in \mathfrak{I}_\mathfrak{f}$.

We have an invertible 2-cocycle $\widehat{\sigma}$ of $'\mathfrak{e}^- \times H \times {}'\mathfrak{f}$, and $'\mathbb{H} = ({}'\mathfrak{e}^- \times H \times {}'\mathfrak{f})_{\widehat{\sigma}}$. Otherwise stated, $'\mathbb{H}$ can be seen as the Hopf algebra $\mathbb{H}$ in Proposition 5.15, with the algebra structure "enriched" by the relations obtained through the factorizations in Proposition 5.18. For instance, $K_ie_j = \mathcal{R}(K_i \otimes K_j)e_jK_i = q^{i\cdot j}e_jK_i$, etc.

We know that $'\mathfrak{e}^- \times H \times {}'\mathfrak{f}$ is isomorphic to $\mathcal{A}_H({}'\mathfrak{e}^-, {}'\mathfrak{f}) \times H$ as Hopf algebras, where $\mathcal{A}_H({}'\mathfrak{e}^-, {}'\mathfrak{f})$ is the tensor product Hopf algebra of $'\mathfrak{e}^-$ and $\overline{'\mathfrak{f}}$ in ${}^H_H\mathcal{YD}$. It can be described as the unital algebra generated by $\{E_i, F_i; i \in I\}$ with relations $E_iE_j = E_jE_i$ and $F_iF_j = F_jF_i$ if $i \neq j$ s.t. $i \cdot j = 0$, and $E_i^2E_j - [2]E_iE_jE_i + E_jE_i^2 = 0$ and $F_i^2F_j - [2]F_iF_jF_i + F_jF_i^2 = 0$ if $i \neq j$ such that $i \cdot i = 2$ and $i \cdot j = -1$.

$\mathcal{A}_H({}'\mathfrak{e}^-, {}'\mathfrak{f})$ is a Hopf algebra in ${}^H_H\mathcal{YD}$ with structure determined by

$$K_\mu \cdot E_i = q^{\mu\cdot i}E_i,\ K_\mu \cdot F_i = q^{-i\cdot\mu}F_i,\ E_i \mapsto K_i^{-1} \otimes E_i,\ F_i \mapsto K_i^{-1} \otimes F_i;$$
$$\Delta(E_i) = E_i \otimes 1 + 1 \otimes E_i,\ \varepsilon(E_i) = 0,\ \Delta(F_i) = F_i \otimes 1 + 1 \otimes F_i,\ \varepsilon(F_i) = 0;$$

$S(E_i) = -E_i$, $S(F_i) = -F_i$. Δ, ε are extended as algebra morphisms in ${}^H_H\mathcal{YD}$ and ${}_k\mathcal{M}$, and S as an anti-morphism of algebras in ${}^H_H\mathcal{YD}$. Then $\chi : {}'\mathfrak{e}^- \times H \times {}'\mathfrak{f} \ni e_jK_sf_i \mapsto q^{-i\cdot s}E_jF_i \times K_sK_i \in \mathcal{A}_H({}'\mathfrak{e}^-, {}'\mathfrak{f}) \times H$ is a Hopf algebra isomorphism. Consequently, ${}'\mathbb{H}$ is isomorphic as a Hopf algebra with a deformation by a 2-cocycle ω of the left biproduct Hopf algebra $\mathcal{A}_H({}'\mathfrak{e}^-, {}'\mathfrak{f}) \times H$; see Proposition 5.13. □

REMARK 5.20. Let $(a_{ij})_{1\le i,j\le l} \in M_l(\mathbb{Z})$ be a symmetric, positive definite matrix such that $a_{ii} = 2$ and $a_{ij} \in \{0, -1\}$ if $i \neq j$; in other words, $(a_{ij})_{1\le i,j\le l}$ is the Cartan matrix of a complex semisimple Lie algebra $\mathfrak{g}$ of type A, D or E. If we take $i \cdot j = a_{ij}$, after we interchange e_i and f_i and take $\varpi_{i,j} = (q - q^{-1})^{-1}$, the resulting Hopf algebra ${}'\mathbb{H}$ in Theorem 5.19 is just the Drinfeld-Jimbo quantum group $U_q(\mathfrak{g})$; see [**12**, Notes VI.7 and VII.9]. Thus, up to isomorphism, $U_q(\mathfrak{g})$ is a deformation by an invertible 2-cocycle of a left biproduct Hopf algebra.

Acknowledgments

The authors are very grateful to the referee for the careful reading of the paper and valuable suggestions and comments.

The authors also thank Bodo Pareigis for sharing his "diagrams" program.

References

[1] Yu. N. Bespalov, *Crossed modules and quantum groups in braided categories*, Appl. Categ. Structures **5** (1997), no. 2, 155–204, DOI 10.1023/A:1008674524341. MR1456522

[2] Yuri Bespalov and Bernhard Drabant, *Hopf (bi-)modules and crossed modules in braided monoidal categories*, J. Pure Appl. Algebra **123** (1998), no. 1-3, 105–129, DOI 10.1016/S0022-4049(96)00105-3. MR1492897

[3] Yuri Bespalov and Bernhard Drabant, *Cross product bialgebras. I*, J. Algebra **219** (1999), no. 2, 466–505, DOI 10.1006/jabr.1999.7896. MR1706813

[4] D. Bulacu, S. Caenepeel, and B. Torrecillas, *Frobenius and separable functors for the category of entwined modules over cowreaths, I: General theory*, Algebr. Represent. Theory **23** (2020), no. 3, 1119–1157, DOI 10.1007/s10468-019-09884-6. MR4109150

[5] Daniel Bulacu, Stefaan Caenepeel, Florin Panaite, and Freddy Van Oystaeyen, *Quasi-Hopf algebras*, Encyclopedia of Mathematics and its Applications, vol. 171, Cambridge University Press, Cambridge, 2019. A categorical approach, DOI 10.1017/9781108582780. MR3929714

[6] Daniel Bulacu and Blas Torrecillas, *On sovereign, balanced and ribbon quasi-Hopf algebras*, J. Pure Appl. Algebra **224** (2020), no. 3, 1064–1091, DOI 10.1016/j.jpaa.2019.07.003. MR4009567

[7] G. Carnovale and J. Cuadra, *The Brauer group of some quasitriangular Hopf algebras*, J. Algebra **259** (2003), no. 2, 512–532, DOI 10.1016/S0021-8693(02)00563-X. MR1955530

[8] V. G. Drinfel'd, *Hopf algebras and the quantum Yang-Baxter equation* (Russian), Dokl. Akad. Nauk SSSR **283** (1985), no. 5, 1060–1064. MR802128

[9] V. G. Drinfel'd, *Quantum groups*, Proceedings of the International Congress of Mathematicians, Vol. 1, 2 (Berkeley, Calif., 1986), Amer. Math. Soc., Providence, RI, 1987, pp. 798–820. MR934283

[10] Yukio Doi and Mitsuhiro Takeuchi, *Multiplication alteration by two-cocycles—the quantum version*, Comm. Algebra **22** (1994), no. 14, 5715–5732, DOI 10.1080/00927879408825158. MR1298746

[11] André Joyal and Ross Street, *Tortile Yang-Baxter operators in tensor categories*, J. Pure Appl. Algebra **71** (1991), no. 1, 43–51, DOI 10.1016/0022-4049(91)90039-5. MR1107651

[12] Christian Kassel, *Quantum groups*, Graduate Texts in Mathematics, vol. 155, Springer-Verlag, New York, 1995, DOI 10.1007/978-1-4612-0783-2. MR1321145

[13] Larry A. Lambe and David E. Radford, *Introduction to the quantum Yang-Baxter equation and quantum groups: an algebraic approach*, Mathematics and its Applications, vol. 423, Kluwer Academic Publishers, Dordrecht, 1997, DOI 10.1007/978-1-4615-4109-7. MR1483071

[14] George Lusztig, *Introduction to quantum groups*, Progress in Mathematics, vol. 110, Birkhäuser Boston, Inc., Boston, MA, 1993. MR1227098

[15] Shahn Majid, *Representations, duals and quantum doubles of monoidal categories*, Proceedings of the Winter School on Geometry and Physics (Srní, 1990), Rend. Circ. Mat. Palermo (2) Suppl. **26** (1991), 197–206. MR1151906

[16] Shahn Majid, *Algebras and Hopf algebras in braided categories*, Advances in Hopf algebras (Chicago, IL, 1992), Lecture Notes in Pure and Appl. Math., vol. 158, Dekker, New York, 1994, pp. 55–105. MR1289422

[17] S. Majid, *Double-bosonization of braided groups and the construction of* $U_q(\mathfrak{g})$, Math. Proc. Cambridge Philos. Soc. **125** (1999), no. 1, 151–192, DOI 10.1017/S0305004198002576. MR1645545

[18] Shahn Majid, *Foundations of quantum group theory*, Cambridge University Press, Cambridge, 1995, DOI 10.1017/CBO9780511613104. MR1381692

[19] B. Pareigis, *Non-additive ring and module theory V. Projective and coflat objects*, Algebra Ber. **40** (1980).

[20] David E. Radford, *The structure of Hopf algebras with a projection*, J. Algebra **92** (1985), no. 2, 322–347, DOI 10.1016/0021-8693(85)90124-3. MR778452

[21] David E. Radford, *On Kauffman's knot invariants arising from finite-dimensional Hopf algebras*, Advances in Hopf algebras (Chicago, IL, 1992), Lecture Notes in Pure and Appl. Math., vol. 158, Dekker, New York, 1994, pp. 205–266. MR1289427

[22] Peter Schauenburg, *Hopf modules and Yetter-Drinfel′d modules*, J. Algebra **169** (1994), no. 3, 874–890, DOI 10.1006/jabr.1994.1314. MR1302122

[23] Peter Schauenburg, *Actions of monoidal categories and generalized Hopf smash products*, J. Algebra **270** (2003), no. 2, 521–563, DOI 10.1016/S0021-8693(03)00403-4. MR2019629

[24] Mitsuhiro Takeuchi, *Some topics on* $\mathrm{GL}_q(n)$, J. Algebra **147** (1992), no. 2, 379–410, DOI 10.1016/0021-8693(92)90212-5. MR1161300

[25] S. L. Woronowicz, *Differential calculus on compact matrix pseudogroups (quantum groups)*, Comm. Math. Phys. **122** (1989), no. 1, 125–170. MR994499

Faculty of Mathematics and Informatics, University of Bucharest, Str. Academiei 14, RO-010014 Bucharest 1, Romania

Email address: `daniel.bulacu@fmi.unibuc.ro`

Department of Algebra and Analysis, Universidad de Almería, E-04071 Almería, Spain

Email address: `btorreci@ual.es`

Contemporary Mathematics
Volume **771**, 2021
https://doi.org/10.1090/conm/771/15509

Module Eilenberg-Watts calculus

Jürgen Fuchs, Gregor Schaumann, and Christoph Schweigert

Abstract. The categorical formulation of the Eilenberg-Watts calculus relates, for any pair of finite categories $\mathcal{M}$ and $\mathcal{N}$, the finite categories $\mathcal{L}ex(\mathcal{N},\mathcal{M})$ and $\mathcal{R}ex(\mathcal{N},\mathcal{M})$ of linear left or right exact functors and the Deligne product $\overline{\mathcal{N}}\boxtimes\mathcal{M}$ by adjoint equivalences. We establish a variant of this calculus for the case that the finite categories $\mathcal{M}$ and $\mathcal{N}$ are module categories over a finite tensor category. This provides in particular canonical and explicitly computable equivalences between categories of left or right exact module functors (or, more generally, balanced functors) and certain twisted centers of bimodule categories.

1. Introduction

It is textbook knowledge (see e.g. [**BW**, Ch. 39]) that any right exact linear functor between categories of finite-dimensional modules over finite-dimensional algebras over a field is isomorphic to the functor given by tensoring with a bimodule, while any left exact such functor is isomorphic to a Hom functor involving a bimodule. These isomorphisms, which date back to Eilenberg [**Ei**] and Watts [**Wa**], are extremely useful, which partly accounts for the prominence of bimodules. Still, from a modern perspective they are unsatisfactory – they hide the fact that the underlying algebras only play the role of 'coordinates'. Indeed, a Morita invariant version exists. It is achieved by the following purely categorical formulation: For any pair of finite linear categories $\mathcal{A}$ and $\mathcal{B}$ there is a commuting triangle

(1.1)

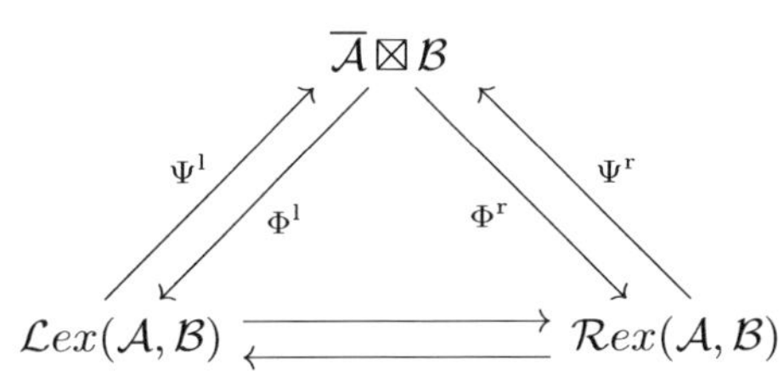

of two-sided adjoint equivalences between finite linear categories that can be explicitly expressed in categorical terms. Here $\mathcal{L}ex(\mathcal{A},\mathcal{B})$ and $\mathcal{R}ex(\mathcal{A},\mathcal{B})$ are categories of

---

2020 *Mathematics Subject Classification.* Primary 18D10.
The first author was supported by VR under project no. 2017-03836.
The third author was partially supported by the RTG 1670 "Mathematics inspired by String theory and Quantum Field Theory" and by the Deutsche Forschungsgemeinschaft (DFG, German Research Foundation) under Germany's Excellence Strategy - EXC 2121 "Quantum Universe"-QT.2.

linear left and right exact functors, respectively, and $\overline{\mathcal{A}}\boxtimes\mathcal{B}$ is the Deligne product of the category $\overline{\mathcal{A}}$ opposite to $\mathcal{A}$ with $\mathcal{B}$. In particular, the role of the bimodule in the classical setting is taken over by an object of the category $\overline{\mathcal{A}}\boxtimes\mathcal{B}$. These equivalences are given by [**Sh1**, **FSS2**]

$$
\begin{aligned}
\Phi^{\mathrm{l}} \equiv \Phi^{\mathrm{l}}_{\mathcal{A},\mathcal{B}} :\quad & \overline{\mathcal{A}}\boxtimes\mathcal{B} \xrightarrow{\simeq} \mathcal{L}ex(\mathcal{A},\mathcal{B})\,, \\
& \overline{a}\boxtimes b \longmapsto \mathrm{Hom}_{\mathcal{A}}(a,-)\otimes b\,, \\
\Psi^{\mathrm{l}} \equiv \Psi^{\mathrm{l}}_{\mathcal{A},\mathcal{B}} :\quad & \mathcal{L}ex(\mathcal{A},\mathcal{B}) \xrightarrow{\simeq} \overline{\mathcal{A}}\boxtimes\mathcal{B}\,, \\
& F \longmapsto \int^{a\in\mathcal{A}} \overline{a}\boxtimes F(a)\,,
\end{aligned} \tag{1.2}
$$

and

$$
\begin{aligned}
\Phi^{\mathrm{r}} \equiv \Phi^{\mathrm{r}}_{\mathcal{A},\mathcal{B}} :\quad & \overline{\mathcal{A}}\boxtimes\mathcal{B} \xrightarrow{\simeq} \mathcal{R}ex(\mathcal{A},\mathcal{B})\,, \\
& \overline{a}\boxtimes b \longmapsto \mathrm{Hom}_{\mathcal{A}}(-,a)^{*}\otimes b\,, \\
\Psi^{\mathrm{r}} \equiv \Psi^{\mathrm{r}}_{\mathcal{A},\mathcal{B}} :\quad & \mathcal{R}ex(\mathcal{A},\mathcal{B}) \xrightarrow{\simeq} \overline{\mathcal{A}}\boxtimes\mathcal{B}\,, \\
& G \longmapsto \int_{a\in\mathcal{A}} \overline{a}\boxtimes G(b)
\end{aligned} \tag{1.3}
$$

respectively.

We refer to these equivalences as *Eilenberg-Watts functors*. They can be used to set up an 'Eilenberg-Watts calculus' which provides, at the level of linear categories, e.g. a Morita invariant formulation of Nakayama functors [**FSS2**, Sect. 3.5] and embeds them into a Grothendieck-Verdier like picture by which they are related to monoidal units. The Eilenberg-Watts calculus has also been a very useful tool in the state-sum construction of a modular functor in [**FSS3**].

In [**FSS2**, Sect. 4] the Eilenberg-Watts calculus was already applied to the situation that the linear categories in question have the additional structure of finite module and bimodule categories over finite monoidal categories. It turned out that the Eilenberg-Watts functors can be combined with the structure of a module category to obtain a transparent setting for several further results. Specifically, a link between the Eilenberg-Watts calculus and Radford's theorem on the fourth power of the antipode of a Hopf algebra was established, and an extension of the calculus to bimodule categories was discussed. This also led to the theory of relative Serre functors. These exist if and only if the module category is exact (in the sense of [**EtGNO**, Ch. 7.5]), and in that case they are related to Nakayama functors via the distinguished invertible object of the finite tensor category. Relative Serre functors, in turn, give rise to the notion of an inner-product structure on a module category over a finite tensor category [**Sc2**, Def. 5.2], also known as a pivotal structure on an exact module category [**Sh2**, Def. 5.6]. In the presence of a pivotal structure e.g. the inner Ends of the module category are not just algebras, but even symmetric Frobenius algebras (see [**Sh3**, Thm. 3.15], and [**Sc1**, Thm. 6.6] for the result in the semisimple case).

In the present contribution we further develop the interaction between module categories over monoidal categories and Eilenberg-Watts functors. We work over general finite tensor categories, and in particular do not assume that they are endowed with a pivotal structure. As a consequence, the bidual functor and its powers constitute monoidal automorphisms of the finite tensor category which generically are non-trivial. In the module setting, instead of functors and the

Deligne product, the players in the game are now module functors and the relative Deligne product [**EtNOM**] of module categories. The latter is conveniently described [**FSS1**, Sect. 2.3] in terms of balancings, whereby also balanced functors and twisted centers, with the twisting by powers of the bidual functor (see Definitions 3.1 – 3.3), come into the game.

To study these structures, it is convenient to place them into a larger context involving (co)monads. To a comonad on a finite category one can associate induced comonads on functor categories as well as (see Proposition 6.1) 'transferred' comonads on other finite categories that are obtained by transporting the comonad along two-sided adjoint equivalences, such as the Eilenberg-Watts equivalences. Our main results are stated in Theorem 6.4: Based on canonical and explicitly computable equivalences between the categories of comodules over different comonads that are obtained by such a transfer and on the resulting universal properties (see Proposition 6.2), we obtain the *module Eilenberg-Watts equivalences* (6.14), as well as the isomorphisms (6.16) and (6.17) between coends over the twisted center $\mathcal{Z}^{\kappa}(\mathcal{M})$ of a module category and coends over $\mathcal{M}$ itself. As a special case, this gives a module version of the Eilenberg-Watts equivalences: two-sided adjoint equivalences between certain twisted centers and categories of left and right exact module functors, respectively. We present these in Proposition 4.1 and Corollary 4.2. We study this special situation separately, and before the general case, because the reader might be more interested in module functors than in general balanced functors, and also because we can approach it in a more direct manner that does not require constructions involving comonads. Along the way we also obtain a few other results that round the picture off. As an example, we show in Proposition 3.5 that certain twisted centers have a natural structure of right and left exact relative Deligne products, respectively.

In the sequel we freely use the framework of finite tensor categories and their module and bimodule categories, as described e.g. in Chapters 6 and 7 of [**EtGNO**], as well as properties of ends and coends. The latter can e.g. be found in [**FuS**] and in Section 2.2 of [**FSS2**]. As one crucial relation let us mention explicitly that, as a version of the (co-)Yoneda lemma, for F a linear functor between finite categories $\mathcal{A}$ and $\mathcal{B}$ there are natural isomorphisms [**Ri**, Cor. 1.4.5 & Ex. 1.4.6]

$$\int^{a\in\mathcal{A}} \mathrm{Hom}_{\mathcal{A}}(a,-)\otimes F(a) \,\cong\, F \,\cong \int_{a\in\mathcal{A}} \mathrm{Hom}_{\mathcal{A}}(-,a)^{*}\otimes F(a) \tag{1.4}$$

of linear functors. We also assume that the reader is familiar with the Deligne product $\boxtimes$ of finite linear categories [**De**, Sect. 5], which is universal for left exact as well as for right exact functors, with the relative Deligne product (see e.g. [**FSS1**, Sect. 2.5]) $\boxtimes_{\mathcal{A}}$ of right and left modules over a finite tensor category $\mathcal{A}$, and with basic aspects of monads and modules over monads (see e.g. [**BV**]).

Throughout, we fix an algebraically closed field $\Bbbk$ and denote by V^* the dual of a $\Bbbk$-vector space V. All vector spaces are finite-dimensional $\Bbbk$-vector spaces; by vect we denote the symmetric monoidal category of such vector spaces and linear maps. All categories are taken to be abelian and finite $\Bbbk$-linear, and all functors and natural transformations are taken to be $\Bbbk$-linear. The linear category opposite to $\mathcal{A}$ is denoted by $\overline{\mathcal{A}}$. Our conventions for dualities are as follows: The left and right dual of an object x are denoted by ${}^{\vee}x$ and $x^{\vee}$, respectively; the corresponding evaluation and coevaluation morphisms lie in the morphism spaces $\mathrm{Hom}(x\otimes {}^{\vee}x,\mathbf{1})$

and $\mathrm{Hom}(\mathbf{1}, {}^\vee x \otimes x)$ for the left duality, and in $\mathrm{Hom}(x^\vee \otimes x, \mathbf{1})$ and $\mathrm{Hom}(\mathbf{1}, x \otimes x^\vee)$ for the right duality.

2. More about the Eilenberg-Watts calculus

Before turning to the context of module categories, we make a few further observations that are easy consequences of the treatment in [**FSS2**].

Let us first point out that $\Psi^{\mathrm{l}}_{\mathcal{A},\mathcal{B}}$ as defined in (1.2) maps a left exact functor F to a coend $\int^{a\in\mathcal{A}} \overline{a} \boxtimes F(a)$, that is, to an object C in $\overline{\mathcal{A}} \boxtimes \mathcal{B}$ endowed with dinatural structure morphisms $\overline{a} \boxtimes F(a) \to C$ for every $a \in \mathcal{A}$. Since Ψ^{l} is an equivalence of categories, it follows conversely that every left exact functor $F\colon \mathcal{A} \to \mathcal{B}$ between finite categories must come with such canonical structural morphisms (that is, morphisms $\Phi^{\mathrm{l}}_{\mathcal{A},\mathcal{B}}(\overline{a} \boxtimes F(a)) = \mathrm{Hom}_{\mathcal{A}}(a,-) \otimes F(a) \to F$ for $a \boxtimes F(a) \in \overline{\mathcal{A}} \boxtimes \mathcal{B}$). This is indeed the case, even for any linear functor, and one recovers the Yoneda isomorphisms (1.4). Analogously, the equivalence Ψ^{r} endows every right exact functor between finite categories with the canonical structure of an end, given by the second Yoneda isomorphism in (1.4).

Next we note the following parameter version of the Eilenberg-Watts functors, from which the equivalences (1.2) are recovered by taking $\mathcal{P}$ to be vect:

LEMMA 2.1. [**FSS3**, Lemma B.4] *Let $\mathcal{A}$, $\mathcal{B}$ and $\mathcal{P}$ be finite categories. There are adjoint equivalences*

$$\begin{aligned} \mathcal{L}ex(\mathcal{P} \boxtimes \mathcal{A}, \mathcal{B}) &\simeq \mathcal{L}ex(\mathcal{P}, \overline{\mathcal{A}} \boxtimes \mathcal{B}) \qquad \text{and} \\ \mathcal{R}ex(\mathcal{P} \boxtimes \mathcal{A}, \mathcal{B}) &\simeq \mathcal{R}ex(\mathcal{P}, \overline{\mathcal{A}} \boxtimes \mathcal{B}) . \end{aligned} \tag{2.1}$$

PROOF. This follows immediately by observing that

$$\mathcal{L}ex(\mathcal{P} \boxtimes \mathcal{A}, \mathcal{B}) \simeq \overline{\mathcal{P} \boxtimes \mathcal{A}} \boxtimes \mathcal{B} \simeq \overline{\mathcal{P}} \boxtimes (\overline{\mathcal{A}} \boxtimes \mathcal{B}) \simeq \mathcal{L}ex(\mathcal{P}, \overline{\mathcal{A}} \boxtimes \mathcal{B}) , \tag{2.2}$$

where the first and third equivalences are the ordinary Eilenberg-Watts functors (1.2), while the one in the middle comes from the associativity property of the Deligne product. □

One virtue of the Eilenberg-Watts calculus is that it allows one to manipulate ends and coends in ways that are not covered by the Yoneda isomorphisms (1.4). Specifically, according to Proposition 3.4 of [**FSS2**] for any left exact functor $F \in \mathcal{L}ex(\mathcal{A}, \mathcal{B})$, one has

$$\mathrm{Hom}_{\overline{\mathcal{A}}\boxtimes\mathcal{B}} \big(- \,, \int^{a\in\mathcal{A}} \overline{a} \boxtimes F(a)\big) \cong \int^{a\in\mathcal{A}} \mathrm{Hom}_{\overline{\mathcal{A}}\boxtimes\mathcal{B}} \big(- \,, \overline{a} \boxtimes F(a)\big) , \tag{2.3}$$

and likewise a similar isomorphism for right exact functors. As a consequence, while left exact functors always commute with ends and right exact functors commute with coends, in our situation we have in addition

LEMMA 2.2. *Let $\mathcal{A}, \mathcal{B}, \mathcal{K}$ and $\mathcal{X}$ be finite linear categories.*
(i) *For any left exact functor $F \in \mathcal{L}ex(\mathcal{A} \boxtimes \overline{\mathcal{A}} \boxtimes \mathcal{K}, \mathcal{X})$ we have*

$$\int^{a\in\mathcal{A}} F(a \boxtimes \overline{a} \boxtimes k) \cong F\big(\int^{a\in\mathcal{A}} a \boxtimes \overline{a} \boxtimes k\big) \tag{2.4}$$

for all $k \in \mathcal{K}$. Analogously, for any right exact functor $G \in \mathcal{R}ex(\mathcal{A} \boxtimes \overline{\mathcal{A}} \boxtimes \mathcal{K}, \mathcal{X})$ there is an isomorphism $\int_a G(a \boxtimes \overline{a} \boxtimes k) \cong G(\int_a a \boxtimes \overline{a} \boxtimes k)$.

(ii) *Let* $F\colon \mathcal{A}\boxtimes\overline{\mathcal{A}}\boxtimes\mathcal{B}\boxtimes\overline{\mathcal{B}}\boxtimes\mathcal{K}\to\mathcal{X}$ *be a left exact or a right exact functor. Then there is a canonical natural isomorphism*

$$\int^{a\in\mathcal{A}}\!\int_{b\in\mathcal{B}} F(a\boxtimes\overline{a}\boxtimes b\boxtimes\bar{b}\boxtimes -) \cong \int_{b\in\mathcal{B}}\!\int^{a\in\mathcal{A}} F(a\boxtimes\overline{a}\boxtimes b\boxtimes\bar{b}\boxtimes -) \tag{2.5}$$

of functors from $\mathcal{K}$ *to* $\mathcal{X}$.

PROOF. (i) By combining the Yoneda isomorphism (1.4), the Fubini theorem for coends and the isomorphism (2.3) we obtain

$$\begin{aligned}\int^{a\in\mathcal{A}} F(a\boxtimes\overline{a}\boxtimes k) &\cong \int^{a\in\mathcal{A}}\!\int^{y\in\mathcal{A}\boxtimes\overline{\mathcal{A}}\boxtimes\mathcal{K}} \mathrm{Hom}(y,a\boxtimes\overline{a}\boxtimes k)\otimes F(y)\\ &\cong \int^{y\in\mathcal{A}\boxtimes\overline{\mathcal{A}}\boxtimes\mathcal{K}}\!\int^{a\in\mathcal{A}} \mathrm{Hom}(y,a\boxtimes\overline{a}\boxtimes k)\otimes F(y)\\ &\cong \int^{y\in\mathcal{A}\boxtimes\overline{\mathcal{A}}\boxtimes\mathcal{K}} \mathrm{Hom}(y,{\textstyle\int^{a\in\mathcal{A}}} a\boxtimes\overline{a}\boxtimes k)\otimes F(y)\\ &\cong F({\textstyle\int^{a\in\mathcal{A}}} a\boxtimes\overline{a}\boxtimes k)\,.\end{aligned} \tag{2.6}$$

The case of a right exact functor is treated dually.

(ii) If F is left exact, then it commutes with ends in the following sense [**FSS2**, Lemma 2.6]: for any $H\colon \overline{\mathcal{A}}\times\mathcal{A}\to\mathcal{A}\boxtimes\overline{\mathcal{A}}\boxtimes\mathcal{B}\boxtimes\overline{\mathcal{B}}\boxtimes\mathcal{K}$ such that the end $\int_{a\in\mathcal{A}}H(\overline{a},a)$ exists, there is a canonical isomorphism $F\big(\int_{a\in\mathcal{A}}H(\overline{a},a)\big)\cong\int_{a\in\mathcal{A}}F(H(\overline{a},a))$. Hence by invoking (i) we obtain

$$\begin{aligned}\int^{a\in\mathcal{A}}\!\int_{b\in\mathcal{B}} F(a\boxtimes\overline{a}\boxtimes b\boxtimes\bar{b}\boxtimes k) &\cong F\big(({\textstyle\int^{a\in\mathcal{A}}} a\boxtimes\overline{a})\boxtimes{\textstyle\int_{b\in\mathcal{B}}}(b\boxtimes\bar{b})\boxtimes k\big)\\ &\cong \int_{b\in\mathcal{B}}\!\int^{a\in\mathcal{A}} F(a\boxtimes\overline{a}\boxtimes b\boxtimes\bar{b}\boxtimes k)\end{aligned} \tag{2.7}$$

for any $k\in\mathcal{K}$. Similarly, if the functor F is right exact, it commutes with coends, and in the present situation, by (i), with ends as well; thus the argument works analogously. □

Since the triangle (1.1) relates left and right exact functors, and since the left adjoint of a left exact functor is right exact and vice versa, the Eilenberg-Watts functors are particularly useful in combination with adjoints of functors. For instance, as seen in Lemma 3.8 of [**FSS2**], for $\mathcal{A}$ and $\mathcal{B}$ finite linear categories and $F\in\mathcal{L}ex(\mathcal{A},\mathcal{B})$ we have an isomorphism

$$\Psi^{\mathrm{l}}(F)\equiv\int^{a\in\mathcal{A}}\overline{a}\boxtimes F(a)\cong\int^{b\in\mathcal{B}}\overline{F^{\mathrm{l.a.}}(b)}\boxtimes b \tag{2.8}$$

of coends, as well as a similar isomorphism of ends for right exact functors. Below we will need the following modest generalization of this result:

LEMMA 2.3. *Let* $\mathcal{A}$ *and* $\mathcal{B}$ *be finite linear categories. For any two left exact functors* $F_1\colon\mathcal{A}\to\mathcal{B}$ *and* $F_2\colon\mathcal{B}\to\mathcal{A}$ *there is an isomorphism*

$$\int^{b\in\mathcal{B}}\overline{F_1^{\mathrm{l.a.}}(b)}\boxtimes F_2(b)\cong\int^{a\in\mathcal{A}}\overline{a}\boxtimes F_2{\circ}F_1(a) \tag{2.9}$$

of coends. In the case of right exact functors there is a similar isomorphism of ends.

3. Balanced functors and twistings

We are now interested in the situation that the finite categories involved in the Eilenberg-Watts equivalences of Lemma 2.1 come equipped with actions of finite tensor categories. The resulting module versions of the Eilenberg-Watts equivalences involve module functors, balanced functors and (twisted) centers. In the present section we recall these and related pertinent notions. We formulate the categories that will appear in the equivalences in the convenient language of monads and comonads.

Let thus now $\mathcal{A}$ and $\mathcal{B}$ be finite tensor categories, and let $\mathcal{M} \equiv (\mathcal{M}, \triangleright, \triangleleft)$ be an $\mathcal{A}$-$\mathcal{B}$-bimodule category. We do not require the monoidal categories $\mathcal{A}$ and $\mathcal{B}$ to be endowed with a pivotal structure. As a consequence, by acting with powers of the monoidal functor of taking biduals we get generically different bimodules structures from a given one. To handle these different bimodules, we introduce the notation

$$a^{[\kappa]} := a^{\vee\cdots\vee} \tag{3.1}$$

for the κ-fold right dual of $a \in \mathcal{A}$, and ${}^{[\kappa]}a := {}^{\vee\cdots\vee}a$ for the κ-fold left dual. (Via the identification $a^{[-\kappa]} = {}^{[\kappa]}a$ these notations make sense also for negative values of κ.) Then for any pair $(\kappa, \kappa') \in 2\mathbb{Z} \times 2\mathbb{Z}$ we denote by ${}^{\kappa}\mathcal{M}^{\kappa'}$ the $\mathcal{A}$-$\mathcal{B}$-bimodule bimodule for which the left and right actions on $\mathcal{M}$ are twisted by the κ-fold left and κ'-fold right dual, respectively, that is, $\mathcal{A}$ acts as $m \mapsto {}^{[\kappa]}a \triangleright m$ and $\mathcal{B}$ as $m \mapsto m \triangleleft b^{[\kappa']}$. Thus in particular ${}^{0}\mathcal{M}^{0} = \mathcal{M}$; we also abbreviate ${}^{0}\mathcal{M}^{\kappa} \equiv \mathcal{M}^{\kappa}$ and ${}^{\kappa}\mathcal{M}^{0} \equiv {}^{\kappa}\mathcal{M}$. In the same vein we endow, for any pair $(\kappa, \kappa') \in 2\mathbb{Z}{+}1 \times 2\mathbb{Z}{+}1$, the opposite category $\overline{\mathcal{M}}$ with the structure of a $\mathcal{B}$-$\mathcal{A}$-bimodule by defining the actions of $a \in \mathcal{A}$ and $b \in \mathcal{B}$ to be

$$\overline{m} \triangleleft a := \overline{a^{[\kappa']} \triangleright m} \qquad \text{and} \qquad b \triangleright \overline{m} := \overline{m \triangleleft {}^{[\kappa]}b}\,, \tag{3.2}$$

respectively. We denote the so obtained bimodule categories by ${}^{\kappa}\overline{\mathcal{M}}^{\kappa'}$. By omitting one of the actions, analogous definitions apply to left and right module categories. In the common situation that we want to pass from a left module $\mathcal{N}$ to a right module with underlying category being the opposite $\overline{\mathcal{N}}$, we also use the special notation

$$\overline{\mathcal{N}}^{1} =: \mathcal{N}^{\#} \qquad \text{and} \qquad \overline{\mathcal{N}}^{-1} =: {}^{\#}\mathcal{N}\,. \tag{3.3}$$

In the context of (bi)module categories, it is natural to consider two types of extra structure on a functor. The first type is given by the notion of a module functor: For $\mathcal{M}$ and $\mathcal{N}$ left $\mathcal{A}$-modules, the category of *module functors* from $\mathcal{N}$ to $\mathcal{M}$, denoted by $\mathcal{F}un_{\mathcal{A}}(\mathcal{N}, \mathcal{M})$, has as objects functors $F\colon \mathcal{N} \to \mathcal{M}$ that come with a family of natural isomorphisms $F(a \triangleright n) \cong a \triangleright F(n)$ for $a \in \mathcal{A}$ and $n \in \mathcal{N}$ that is coherent with respect to the tensor product of $\mathcal{A}$ [**EtGNO**, Def. 7.2.1]. In our context we are particularly interested in the finite subcategories of left and right exact module functors, which we denote by $\mathcal{L}ex_{\mathcal{A}}(\mathcal{N}, \mathcal{M})$ and $\mathcal{R}ex_{\mathcal{A}}(\mathcal{N}, \mathcal{M})$, respectively.

The second type of functors with extra structure are balanced functors:

DEFINITION 3.1. [**EtNOM**, Def. 3.1]

(i) For a right module $\mathcal{N} = (\mathcal{N}, \triangleleft)$ and a left module $\mathcal{M} = (\mathcal{M}, \triangleright)$ over a finite tensor category $\mathcal{A}$, a *balanced functor* from $\mathcal{N} \boxtimes \mathcal{M}$ to a finite category $\mathcal{X}$ is a linear functor $F\colon \mathcal{N} \boxtimes \mathcal{M} \to \mathcal{X}$ together with a natural family of isomorphisms

$$F(n \triangleleft a \boxtimes m) \xrightarrow{\;\cong\;} F(n \boxtimes a \triangleright m) \tag{3.4}$$

for $a \in \mathcal{A}$, $m \in \mathcal{M}$ and $n \in \mathcal{N}$.

(ii) The category $\mathcal{F}un^{\mathrm{bal}}(\mathcal{N}\boxtimes\mathcal{M},\mathcal{X})$ of balanced functors has as objects the balanced functors from $\mathcal{N}\boxtimes\mathcal{M}$ to $\mathcal{X}$ and as morphisms those natural transformations between balanced functors which are compatible with the balancings.

Again there are left and right exact versions. A natural generalization is to consider balanced functors from an arbitrary $\mathcal{A}$-bimodule category $\mathcal{M}$ to $\mathcal{X}$. Still more generally, we set:

DEFINITION 3.2. Let $\mathcal{M}$ be a bimodule over a finite tensor category $\mathcal{A}$, and $\mathcal{X}$ a finite category. For $\kappa\in 2\mathbb{Z}$, the category of *κ-twisted balanced functors* from $\mathcal{M}$ to $\mathcal{X}$, denoted $\mathcal{F}un^{\kappa}(\mathcal{M},\mathcal{X})$, has as objects functors F that come with coherent isomorphisms

$$F(a\triangleright m) \xrightarrow{\cong} F(m\triangleleft a^{[\kappa]}) \tag{3.5}$$

for $a\in\mathcal{A}$.

The module Eilenberg-Watts calculus to be developed below relates module functors and balanced functors with (twisted) centers. We first recall

DEFINITION 3.3. [**FSS3**, Def. 3.2,3.4] Let $\mathcal{A}$ be a finite tensor category and $\mathcal{M}$ a finite $\mathcal{A}$-bimodule, and let $m\in\mathcal{M}$.

(i) A *balancing* on m is a natural family $(\sigma=(\sigma_a\colon a\triangleright m\to m\triangleleft a)_{a\in\mathcal{A}})$ of isomorphisms in $\mathcal{M}$ that is coherent with respect to the tensor product of $\mathcal{A}$, i.e. (taking the action of the monoidal unit to be strict) satisfies $\sigma_1=\mathrm{id}_m$ and

$$\sigma_{a\otimes a'} = (\sigma_a\triangleleft a')\circ(a\triangleright\sigma_{a'}): \quad (a\otimes a')\triangleright m\to m\triangleleft(a\otimes a') \tag{3.6}$$

for all $a,a'\in\mathcal{A}$ (here the constraint morphisms of the bimodule $\mathcal{M}$ are suppressed).

(ii) For $\kappa\in 2\mathbb{Z}$, a *κ-twisted balancing* on m is a natural family of isomorphisms

$$^{[\kappa-2]}a\triangleright m \xrightarrow{\cong} m\triangleleft a \tag{3.7}$$

in $\mathcal{M}$, for $a\in\mathcal{A}$, that is coherent with respect to the tensor product of $\mathcal{A}$.

(iii) The *κ-twisted center* of $\mathcal{M}$ is the category of objects of $\mathcal{M}$ endowed with κ-twisted balancings. We denote this category by $\mathcal{Z}^{\kappa}(\mathcal{M})\equiv\mathcal{Z}^{\kappa}_{\mathcal{A}}(\mathcal{M})$ (in [**FSS3**] this category is denoted by $\mathcal{M}\overset{\kappa}{\boxtimes}$).

The convention of having a shift by -2 in Equation (3.7) is chosen in such a way that the 0-twisted center is universal for balanced left exact functors, see Proposition 3.5 below, and that κ-balanced left exact functors correspond to functors from the κ-twisted center, as will be seen in Theorem 6.4. That our conventions favor left exact over right exact functors can be traced back to the more common use of Hom-spaces as compared to dual Hom-spaces. Also note that it is sufficient to define balancings just as a family of morphisms; owing to the dualities of $\mathcal{A}$ these morphisms are automatically invertible [**FSS3**, Lemma 3.3]. Balanced functors and twisted centers are related via the relative Deligne product [**EtNOM**], which comes in two variants.

DEFINITION 3.4. Let $\mathcal{M}$ and $\mathcal{N}$ be a right and left $\mathcal{A}$-module, respectively. The *right exact relative Deligne product* $\mathcal{M}\boxtimes^{\mathrm{r.e.}}_{\mathcal{A}}\mathcal{N}$ is the finite category characterized by the following universal property: It is equipped with a balanced right exact functor $B_{\mathcal{M},\mathcal{N}}\colon\mathcal{M}\boxtimes\mathcal{N}\to\mathcal{M}\boxtimes^{\mathrm{r.e.}}_{\mathcal{A}}\mathcal{N}$, such that for every finite category $\mathcal{X}$, pre-composition with $B_{\mathcal{M},\mathcal{N}}$ furnishes an adjoint equivalence

$$\mathcal{R}ex^{\mathrm{bal}}(\mathcal{M}\boxtimes\mathcal{N},\mathcal{X}) \simeq \mathcal{R}ex(\mathcal{M}\boxtimes^{\mathrm{r.e.}}_{\mathcal{A}}\mathcal{N},\mathcal{X})\,. \tag{3.8}$$

Analogously, the *left exact relative Deligne product* $\mathcal{M} \boxtimes^{\text{l.e.}}_{\mathcal{A}} \mathcal{N}$ is universal for left exact balanced functors.

In [**EtNOM**, Prop. 3.8] the right exact relative Deligne product is shown to be related to the twisted center. [1]

PROPOSITION 3.5. *The 0-twisted center* $\mathcal{Z}^0_{\mathcal{A}}(\mathcal{M}\boxtimes\mathcal{N})$ *has a natural structure of a left exact relative Deligne product of* $\mathcal{M}$ *and* $\mathcal{N}$. *The 4-twisted center* $\mathcal{Z}^4_{\mathcal{A}}(\mathcal{M}\boxtimes\mathcal{N})$ *has a natural structure of a right exact relative Deligne product of* $\mathcal{M}$ *and* $\mathcal{N}$. *The universal balanced functors are provided by the left adjoint, respectively right adjoint, of the forgetful functor from* $\mathcal{Z}^0_{\mathcal{A}}(\mathcal{M}\boxtimes\mathcal{N})$ *and from* $\mathcal{Z}^4_{\mathcal{A}}(\mathcal{M}\boxtimes\mathcal{N})$ *to* $\mathcal{M}\boxtimes\mathcal{N}$, *respectively.*

PROOF. We only recall the main idea of the proof. Let $F\colon \mathcal{M}\boxtimes\mathcal{N}\to\mathcal{X}$ be balanced right exact. Then it admits a right adjoint $F^{\text{r.a.}}$, and for all $x\in\mathcal{X}$, the object $F^{\text{r.a.}}(x)$ in $\mathcal{M}\boxtimes\mathcal{N}$ naturally has the structure of an object in $\mathcal{Z}^4_{\mathcal{A}}(\mathcal{M}\boxtimes\mathcal{N})$. Indeed, for all $m\boxtimes n\in\mathcal{M}\boxtimes\mathcal{N}$ there are natural isomorphisms

$$
\begin{aligned}
\mathrm{Hom}(m\boxtimes n, a\triangleright F^{\text{r.a.}}(x)) &\cong \mathrm{Hom}(m\boxtimes a^\vee\triangleright n, F^{\text{r.a.}}(x)) \\
&\cong \mathrm{Hom}(F(m\boxtimes a^\vee\triangleright n), x) \\
&\overset{(3.5)}{\cong} \mathrm{Hom}(F(m\triangleleft a^\vee\boxtimes n), x) \\
&\cong \mathrm{Hom}(m\triangleleft a^\vee\boxtimes n, F^{\text{r.a.}}(x)) \\
&\cong \mathrm{Hom}(m\boxtimes n, F^{\text{r.a.}}(x)\triangleleft a^{\vee\vee})\,.
\end{aligned}
\tag{3.9}
$$

These isomorphisms endow $F^{\text{r.a.}}(x)$ with the structure of an object in $\mathcal{Z}^4_{\mathcal{A}}(\mathcal{M}\boxtimes\mathcal{N})$ and give one part of the adjoint equivalence from Definition 3.4.

Analogously, if F is left exact and balanced, we obtain objects $F^{\text{l.a.}}(x)\in\mathcal{Z}^0_{\mathcal{A}}(\mathcal{M}\boxtimes\mathcal{N})$. The universal property follows in the same way as in the proof of [**EtNOM**, Prop. 3.8], and the characterization of the universal balanced functors as in [**FSS1**, Prop. 2.18]. □

REMARK 3.6. For any $\mathcal{A}$-bimodule $\mathcal{M}$ there is an equivalence

$$
F_{\mathcal{M}}:\quad \mathcal{Z}^0_{\mathcal{A}}(\mathcal{M}) \xrightarrow{\simeq} \mathcal{Z}^4_{\mathcal{A}}(\mathcal{M}) \tag{3.10}
$$

of linear categories which makes use of the distinguished invertible object $D\in\mathcal{A}$. Recall from [**EtNO**] that the object D comes with coherent natural isomorphisms $D\otimes{}^{\vee\vee}a\cong a^{\vee\vee}\otimes D$ for all $a\in\mathcal{A}$. It follows that by setting $F_{\mathcal{M}}(m):=D\triangleright m$ we get an equivalence between the two twisted centers as finite categories.

We now show that the categories of module functors, balanced functors, and centers can all be expressed as (co)modules over suitable (co)monads. Recall [**TV2**, Ch. 9] that the *central comonad* of a finite tensor category $\mathcal{A}$ is the endofunctor $Z\colon\mathcal{A}\to\mathcal{A}$ that maps objects as

$$
Z:\quad b\longmapsto \int_{a\in\mathcal{A}} a\otimes b\otimes a^\vee, \tag{3.11}
$$

and that the category of comodules over this comonad is equivalent to the center $\mathcal{Z}(\mathcal{A})$ of $\mathcal{A}$. We consider variants of the central comonad that are twisted by powers of the bidual. Let $\mathcal{M}$ be a bimodule category over finite tensor category $\mathcal{A}$. There

[1] The twist is omitted in [**EtNOM**], see [**FSS1**].

is then a comonad analogous to (3.11) in which the tensor products in the end (3.11) are replaced by the left and right actions of $\mathcal{A}$ on $\mathcal{M}$; for any $\kappa \in 2\mathbb{Z}$ this can be generalized [**FSS3**] to a comonad

$$Z_{[\kappa]}: \quad m \longmapsto \int_{a\in\mathcal{A}} a \triangleright m \triangleleft a^{[\kappa-1]} \tag{3.12}$$

on $\mathcal{M}$.

Let now $\mathcal{M}$ be an $\mathcal{A}$-bimodule, and $\mathcal{K}$ and $\mathcal{N}$ left $\mathcal{A}$ modules. Then the category $\mathcal{F}un(\mathcal{K},\mathcal{N})$ becomes an $\mathcal{A}$-bimodule by setting

$$a_1 \triangleright F \triangleleft a_2(k) := a_1 \triangleright F(a_2 \triangleright k) \tag{3.13}$$

for $a_1, a_2 \in \mathcal{A}$. Likewise, for any finite category $\mathcal{X}$, $\mathcal{F}un(\mathcal{M},\mathcal{X})$ is an $\mathcal{A}$-bimodule, with

$$(a_1 \triangleright G \triangleleft a_2)(m) := G(a_2 \triangleright m \triangleleft a_1)\,. \tag{3.14}$$

We then obtain the following categories of comodules over twisted central comonads on these $\mathcal{A}$-bimodules:

LEMMA 3.7. *Let $\kappa \in 2\mathbb{Z}$.*

(i) The category of comodules over $Z_{[\kappa]}$ on an $\mathcal{A}$-bimodule $\mathcal{M}$ is equivalent to the κ-twisted center $\mathcal{Z}^{\kappa}(\mathcal{M})$.

(ii) For $\mathcal{K}$ and $\mathcal{N}$ left $\mathcal{A}$-modules, the category of comodules over $Z_{[\kappa]}$ on the functor category $\mathcal{F}un(\mathcal{K},\mathcal{N})$ is equivalent to the category $\mathcal{F}un_{\mathcal{A}}({}_{\mathcal{A}}\mathcal{K}, {}^{\kappa-2}_{\mathcal{A}}\mathcal{N})$ of module functors.

(iii) For $\mathcal{M}$ an $\mathcal{A}$-bimodule, the category of comodules over $Z_{[\kappa]}$ on $\mathcal{F}un(\mathcal{M},\mathcal{X})$ is the category of twisted balanced functors $\mathcal{F}un^{2-\kappa}(\mathcal{M},\mathcal{X})$.

PROOF. The structure of a comodule over $Z_{[\kappa]}$ on m is equivalent to a coherent family of morphisms $m \to a \triangleright m \triangleleft a^{[\kappa-1]}$. By adjunction this is, in turn, equivalent to a coherent family ${}^{[\kappa-2]}a \triangleright m \to m \triangleleft a$. This shows claim (i) (see also [**FSS3**, Cor. B.3]).

As for claim (ii), notice that according to the formulas (3.12) and (3.13) the structure of a $Z_{[\kappa]}$-comodule on a functor $F \in \mathcal{F}un(\mathcal{K},\mathcal{N})$ gives rise to a coherent family $F(k) \to a \triangleright F(a^{[\kappa-1]} \triangleright k)$. This gives coherent isomorphisms $F(a \triangleright k) \cong {}^{[\kappa-2]}a \triangleright F(k)$, as required for a module functor from ${}_{\mathcal{A}}\mathcal{K}$ to ${}^{\kappa-2}_{\mathcal{A}}\mathcal{N}$.

Concerning claim (iii), we note that by (3.12) and (3.14) a functor $F \in \mathcal{F}un(\mathcal{M},\mathcal{X})$ with comodule structure over the comonad $Z_{[\kappa]}$ comes with coherent morphisms $F(m) \to F(a^{[\kappa-1]} \triangleright m \triangleleft a)$. Using the evaluation morphism, this provides us with a sequence of morphisms $F(m \triangleleft a^{\vee}) \to F(a^{[\kappa-1]} \triangleright (m \triangleleft a^{\vee}) \triangleleft a) \to F(a^{[\kappa-1]} \triangleright m)$. The composite of these morphisms is an isomorphism and a balancing, as required. The converse follows analogously. □

4. Eilenberg-Watts calculus for module functors

To proceed we invoke Corollary 4.3 of [**FSS2**], which states that for any $\mathcal{A}$-$\mathcal{B}$-bimodule category $\mathcal{M} = (\mathcal{M}, \triangleright, \triangleleft)$ over finite tensor categories there are, coherently with respect to the monoidal structures of $\mathcal{A}$ and $\mathcal{B}$, natural isomorphisms

$$\begin{aligned}\int_{m\in\mathcal{M}} \overline{m}\boxtimes a\triangleright m\triangleleft b &\cong \int_{m\in\mathcal{M}} \overline{{}^{\vee}a\triangleright m\triangleleft b^{\vee}}\boxtimes m \qquad \text{and} \\ \int^{m\in\mathcal{M}} \overline{m}\boxtimes a\triangleright m\triangleleft b &\cong \int^{m\in\mathcal{M}} \overline{a^{\vee}\triangleright m\triangleleft {}^{\vee}b}\boxtimes m\end{aligned} \tag{4.1}$$

(be aware of the different appearance of left and right duals in the two formulas). Let now F and G be left and right exact module functors, respectively, between finite left $\mathcal{A}$-modules $\mathcal{N}\equiv{}_{\mathcal{A}}\mathcal{N}$ and $\mathcal{M}\equiv{}_{\mathcal{A}}\mathcal{M}$, with $\mathcal{A}$ a finite tensor category. We use the notations $\mathcal{N}^{\#}$ and ${}^{\#}\!\mathcal{N}$ introduced in (3.3). By applying the result (4.1) to the $\mathcal{A}$-bimodules ${}^{\#}\!\mathcal{N}\boxtimes\mathcal{M}$ and $\mathcal{N}^{\#}\boxtimes\mathcal{M}$, respectively, we get isomorphisms

$$\begin{aligned}\Psi^{\mathrm{l}}_{\mathcal{N},\mathcal{M}}(F(a\triangleright -)) &= \int^{n\in\mathcal{N}} \overline{n}\boxtimes F(a\triangleright n) \cong \int^{n\in\mathcal{N}} \overline{n}\boxtimes a\triangleright F(n) \\ &\overset{(4.1)}{\cong} \int^{n\in\mathcal{N}} \overline{a^{\vee}\triangleright n}\boxtimes F(n) \cong \int^{n\in\mathcal{N}} n^{\#}\triangleleft a\boxtimes F(n)\end{aligned} \tag{4.2}$$

and

$$\begin{aligned}\Psi^{\mathrm{r}}_{\mathcal{N},\mathcal{M}}(G(a\triangleright -)) &= \int_{n\in\mathcal{N}} \overline{n}\boxtimes G(a\triangleright n) \cong \int_{n\in\mathcal{N}} \overline{n}\boxtimes a\triangleright G(n) \\ &\overset{(4.1)}{\cong} \int_{n\in\mathcal{N}} \overline{{}^{\vee}a\triangleright n}\boxtimes G(n) \cong \int_{n\in\mathcal{N}} {}^{\#}n\triangleleft a\boxtimes G(n)\,,\end{aligned} \tag{4.3}$$

respectively, coherent in $a\in\mathcal{A}$. Together with the exactness of the action functor, which allows one to exchange it with (co)ends, this implies that the object $\Psi^{\mathrm{l}}(F)\in\overline{\mathcal{N}}\boxtimes\mathcal{M}$ has a natural structure of an object in the center $\mathcal{Z}(\mathcal{N}^{\#}\boxtimes\mathcal{M})$, while $\Psi^{\mathrm{r}}(G)$ has a natural structure of an object in $\mathcal{Z}({}^{\#}\!\mathcal{N}\boxtimes\mathcal{M})$.

Conversely, given an object $n^{\#}\boxtimes m$ in the center $\mathcal{Z}(\mathcal{N}^{\#}\boxtimes\mathcal{M})$, there are coherent balancing isomorphisms $\overline{a^{\vee}\triangleright n}\boxtimes m\cong\overline{n}\boxtimes a\triangleright m$, and hence we have

$$\begin{aligned}\mathrm{Hom}_{\mathcal{N}}(n,a\triangleright -)\otimes m &\cong \mathrm{Hom}_{\mathcal{N}}(a^{\vee}\triangleright n,-)\otimes m \\ &= \Phi^{\mathrm{l}}_{\mathcal{N},\mathcal{M}}(\overline{a^{\vee}\triangleright n}\boxtimes m) \cong \Phi^{\mathrm{l}}_{\mathcal{N},\mathcal{M}}(\overline{n}\boxtimes a\triangleright m) \\ &= \mathrm{Hom}_{\mathcal{N}}(n,-)\otimes a\triangleright m\end{aligned} \tag{4.4}$$

for any $a\in\mathcal{A}$. Thus the left exact functor $\mathrm{Hom}_{\mathcal{N}}(n,-)\otimes m$ is endowed with the structure of a module functor from ${}_{\mathcal{A}}\mathcal{N}$ to ${}_{\mathcal{A}}\mathcal{M}$. Analogously one obtains, invoking the definition of $\Phi^{\mathrm{r}}_{\mathcal{N},\mathcal{M}}$, that

$$\mathrm{Hom}_{\mathcal{N}}(a\triangleright -,n)^{*}\otimes m \cong \mathrm{Hom}_{\mathcal{N}}(-,n)^{*}\otimes a\triangleright m \tag{4.5}$$

for ${}^{\#}n\boxtimes m\in\mathcal{Z}({}^{\#}\!\mathcal{N}\boxtimes\mathcal{M})$, showing that the right exact functor $\mathrm{Hom}_{\mathcal{N}}(-,n)^{*}\otimes m$ comes with the structure of a module functor from ${}_{\mathcal{A}}\mathcal{N}$ to ${}_{\mathcal{A}}\mathcal{M}$, too.

Taken together, we have the following module version of the Eilenberg-Watts equivalences:

PROPOSITION 4.1. *For finite left modules $\mathcal{N}$ and $\mathcal{M}$ over a finite tensor category $\mathcal{A}$ the equivalences* (1.2) *induce two-sided adjoint equivalences*

$$\begin{array}{ccccc} & \mathcal{Z}(\mathcal{N}^{\#}\boxtimes\mathcal{M}) & & \mathcal{Z}({}^{\#}\!\mathcal{N}\boxtimes\mathcal{M}) & \\ {\scriptstyle\Psi^{\mathrm{l}}}\nearrow\swarrow{\scriptstyle\Phi^{\mathrm{l}}} & & \textit{and} & & {\scriptstyle\Phi^{\mathrm{r}}}\searrow\nwarrow{\scriptstyle\Psi^{\mathrm{r}}} \\ \mathcal{L}ex_{\mathcal{A}}(\mathcal{N},\mathcal{M}) & & & & \mathcal{R}ex_{\mathcal{A}}(\mathcal{N},\mathcal{M}) \end{array} \tag{4.6}$$

of finite categories.

In case the finite tensor category $\mathcal{A}$ is pivotal, the pivotal structure gives rise to a distinguished equivalence between the right $\mathcal{A}$-modules $\mathcal{N}^{\#}$ and ${}^{\#}\!\mathcal{N}$, and hence also between the finite categories $\mathcal{Z}(\mathcal{N}^{\#}\boxtimes\mathcal{M})$ and $\mathcal{Z}({}^{\#}\!\mathcal{N}\boxtimes\mathcal{M})$. Thus we have

COROLLARY 4.2. *For finite left modules $\mathcal{N}$ and $\mathcal{M}$ over a* pivotal *finite tensor category $\mathcal{A}$ the Eilenberg-Watts equivalences restrict to a commuting triangle*

$$\begin{array}{ccc} & \mathcal{Z}(\mathcal{N}^{\#}\boxtimes\mathcal{M}) \simeq \mathcal{Z}({}^{\#}\!\mathcal{N}\boxtimes\mathcal{M}) & \\ {\scriptstyle\Psi^{\mathrm{l}}}\nearrow\swarrow{\scriptstyle\Phi^{\mathrm{l}}} & & {\scriptstyle\Phi^{\mathrm{r}}}\searrow\nwarrow{\scriptstyle\Psi^{\mathrm{r}}} \\ \mathcal{L}ex_{\mathcal{A}}(\mathcal{N},\mathcal{M}) & \rightleftarrows & \mathcal{R}ex_{\mathcal{A}}(\mathcal{N},\mathcal{M}) \end{array} \tag{4.7}$$

of two-sided adjoint equivalences of finite categories.

5. Eilenberg-Watts calculus for balanced functors

For left $\mathcal{A}$-modules $\mathcal{M}$ and $\mathcal{N}$, besides the category of module functors there is also another pertinent category of functors with structures related to the $\mathcal{A}$-actions: The category $\mathcal{F}un^{\mathrm{bal}}(\mathcal{N}_{\mathcal{A}}\boxtimes{}_{\mathcal{A}}\mathcal{M}^{\#},\mathrm{vect})$ of balanced functors, as described in Definition 3.1.

As it turns out, the relation between module functors and balanced functors fits well with the Eilenberg-Watts calculus. Indeed, notice that by Lemma 2.1 it follows that for any two finite linear categories $\mathcal{B}$ and $\mathcal{C}$ there is an equivalence

$$\begin{aligned} \Xi_{\mathcal{B},\mathcal{C}}:\quad \mathcal{L}ex(\mathcal{B},\mathcal{C}) &\xrightarrow{\;\simeq\;} \mathcal{L}ex(\mathcal{B}\boxtimes\overline{\mathcal{C}},\mathrm{vect}) \\ F &\longmapsto \Xi_{\mathcal{B},\mathcal{C}}(F) := \mathrm{Hom}_{\mathcal{C}}(-,F(-)) \end{aligned} \tag{5.1}$$

of categories [**FSS3**, Eq. (B.5)]. Let now $\mathcal{M}$ and $\mathcal{N}$ be left $\mathcal{A}$-modules and endow the opposite category $\overline{\mathcal{M}}$ with the right $\mathcal{A}$-module structure given by $\mathcal{M}^{\#}$. Then for any left exact $\mathcal{A}$-module functor $F\colon\mathcal{N}\to\mathcal{M}$ we get

$$\begin{aligned} \Xi_{\mathcal{N},\mathcal{M}}(F)(a\triangleright n\boxtimes\overline{m}) &\equiv \mathrm{Hom}_{\mathcal{M}}(m,F(a\triangleright n)) \\ &\cong \mathrm{Hom}_{\mathcal{M}}(m,a\triangleright F(n)) \\ &\cong \mathrm{Hom}_{\mathcal{M}}(a^{\vee}\triangleright m,F(n)) \\ &= \Xi_{\mathcal{N},\mathcal{M}}(F)(n\boxtimes\overline{m}\triangleleft a)\,, \end{aligned} \tag{5.2}$$

where we first use the module functor structure, then the duality of $\mathcal{A}$ and finally the definition of the action on $\mathcal{M}^{\#}$. It follows that the functor $\Xi_{\mathcal{M},\mathcal{N}}(F)$ is a balanced functor. Since $\Xi_{\mathcal{N},\mathcal{M}}$ is an equivalence, it follows as well that, conversely, a balancing on $\Xi_{\mathcal{N},\mathcal{M}}(F)$ is the same as the structure of a module functor on F. We conclude that $\Xi_{\mathcal{N},\mathcal{M}}$ induces an equivalence

$$\mathcal{L}ex_{\mathcal{A}}(\mathcal{N},\mathcal{M}) \simeq \mathcal{L}ex^{\mathrm{bal}}(\mathcal{N}\boxtimes\mathcal{M}^{\#},\mathrm{vect})\,. \tag{5.3}$$

Together with the results of the previous section we thus arrive at the following relations between module structures and the Eilenberg-Watts equivalences:

PROPOSITION 5.1. *Let $\mathcal{N}$ and $\mathcal{M}$ be left $\mathcal{A}$-modules. The Eilenberg-Watts equivalences provide the following sequence of equivalences:*

$$\begin{array}{ccc} \mathcal{L}ex_{\mathcal{A}}(\mathcal{N},\mathcal{M}) & \xrightarrow[\simeq]{(4.6)} & \mathcal{Z}(\mathcal{N}^{\#}\boxtimes\mathcal{M}) \\ {\scriptstyle (5.3)}\Big\downarrow{\scriptstyle\simeq} & & {\scriptstyle\simeq}\Big\downarrow{\scriptstyle Y_{\mathcal{Z}(\mathcal{N}^{\#}\boxtimes\mathcal{M})}} \\ \mathcal{L}ex^{\mathrm{bal}}(\mathcal{N}\boxtimes\mathcal{M}^{\#},\mathrm{vect}) & & \mathcal{L}ex(\overline{\mathcal{Z}(\mathcal{N}^{\#}\boxtimes\mathcal{M})},\mathrm{vect})\,. \end{array} \tag{5.4}$$

The equivalence in the right column of the diagram (5.4) is the Yoneda embedding $Y_{\mathcal{X}}\colon \mathcal{X} \xrightarrow{\simeq} \overline{\overline{\mathcal{X}}}\boxtimes\mathrm{vect} \xrightarrow{(1.2)} \mathcal{L}ex(\overline{\mathcal{X}},\mathrm{vect})$ for $\mathcal{X}=\mathcal{Z}(\mathcal{N}^{\#}\boxtimes\mathcal{M})$.

It is tempting to expect that the so obtained equivalence $\mathcal{L}ex^{\mathrm{bal}}(\mathcal{N}\boxtimes\mathcal{M}^{\#},\mathrm{vect}) \simeq \mathcal{L}ex(\overline{\mathcal{Z}(\mathcal{N}^{\#}\boxtimes\mathcal{M})},\mathrm{vect})$ is part of a family

$$\mathcal{L}ex^{\mathrm{bal}}(\mathcal{N}\boxtimes\mathcal{M}^{\#},\mathcal{X}) \simeq \mathcal{L}ex(\overline{\mathcal{Z}(\mathcal{N}^{\#}\boxtimes\mathcal{M})},\mathcal{X}) \tag{5.5}$$

of equivalences that satisfies the universal property of the (left exact) relative Deligne product. To prove that this is indeed the case, it is useful to express all categories in question as categories of comodules over suitable comonads. This will be discussed in the next section.

REMARK 5.2. In [**Sh3**, Def. 2.1] the related notion of a *module profunctor* between left $\mathcal{A}$-module categories $\mathcal{N}$ and $\mathcal{M}$ is introduced: A module profunctor F from $\mathcal{N}$ to $\mathcal{M}$ is a functor $F\colon \mathcal{N}\boxtimes\overline{\mathcal{M}}\to\mathrm{vect}$ together with a collection of coherent dinatural morphisms $F(n\boxtimes\overline{m})\to F(a\triangleright n\boxtimes\overline{a\triangleright m})$ for all $n\in\mathcal{N}$, $m\in\mathcal{M}$ and $a\in\mathcal{A}$. In our setting, the category of module profunctors is equivalent to the category of module functors. Indeed, such a collection of dinatural morphisms is the same as a morphism $F\to Z_{[2]}(F)$, with $Z_{[2]}$ as in (3.12) acting on $\mathcal{F}un(\mathcal{N}\boxtimes\mathcal{M}^{\#},\mathrm{vect})$. The coherence of the dinatural family is thereby the same as requiring this morphism $F\to Z_{[2]}(F)$ to be a comodule action. Moreover, this correspondence extends to the morphisms between module profunctors. Thus we conclude that the category of module profunctors from $\mathcal{N}$ to $\mathcal{M}$ is equivalent to the category $\mathrm{Comod}_{Z_{[2]}}(\mathcal{F}un(\mathcal{N}\boxtimes\mathcal{M}^{\#},\mathrm{vect}))$, which by Lemma 3.7 is equivalent to the category $\mathcal{F}un^{\mathrm{bal}}(\mathcal{N}\boxtimes\mathcal{M}^{\#},\mathrm{vect})$. With Proposition 5.1 we finally conclude that $\mathcal{L}ex_{\mathcal{A}}(\mathcal{N},\mathcal{M})$ is equivalent to the category of left exact module profunctors from $\mathcal{N}$ to $\mathcal{M}$.

6. Transporting (co)monads along Eilenberg-Watts

The situations considered in the previous sections may be thought of as particular instances of applying the Eilenberg-Watts equivalence

$$\mathcal{R}ex(\mathcal{N}\boxtimes\mathcal{K},\mathcal{X}) \simeq \overline{\mathcal{N}\boxtimes\mathcal{K}}\boxtimes\mathcal{X} \simeq \mathcal{L}ex(\mathcal{N}\boxtimes\mathcal{K},\mathcal{X}) \tag{6.1}$$

to situations in which a finite tensor category $\mathcal{A}$ has two actions on two of the three finite categories $\mathcal{N}$, $\mathcal{K}$ and $\mathcal{X}$: In Section 4, $\mathcal{A}$ acts from the left on $\mathcal{N}$ and on $\mathcal{X} = \mathcal{M}$, while $\mathcal{K}$ is just vect (which, being a unit for the Deligne product, can be omitted) and does not carry an $\mathcal{A}$-action; and in Section 5, $\mathcal{A}$ acts on $\mathcal{K} = \mathcal{M}$ and on $\mathcal{N}$, but does not act on $\mathcal{X} = \mathrm{vect}$. For treating these different cases uniformly, the language of monads and comonads turns out to be convenient; in particular it allows for a clear description of the relative Deligne product. Indeed, by invoking Lemma 3.7, the twisted center, module functors, and balanced functors can all be seen as comodules over suitable comonads that are transported via Eilenberg-Watts functors.

We are now going to show that, when suitable finiteness conditions are satisfied, the Eilenberg-Watts calculus yields a convenient description of the categories of modules over a monad on a functor category. As a preparation, we note the following transfer principle:

PROPOSITION 6.1. *Let $\Phi\colon \mathcal{X} \leftrightarrows \mathcal{Y} : \Psi$ be an adjoint equivalence between categories $\mathcal{X}$ and $\mathcal{Y}$.*

(i) *Given a (co)monad $T^{\mathcal{X}}$ on $\mathcal{X}$, the endofunctor*

$$\Phi \circ T^{\mathcal{X}} \circ \Psi =: T^{\mathcal{Y}} \tag{6.2}$$

has a canonical structure of a (co)monad on $\mathcal{Y}$.

We refer to the (co)monad $T^{\mathcal{Y}}$ as the one transferred *along the adjoint equivalence.*

(ii) *The equivalence between $\mathcal{X}$ and $\mathcal{Y}$ can be lifted in a unique way to an adjoint equivalence*

$$\widehat{\Phi} : \mathit{(Co)mod}_{T^{\mathcal{X}}}(\mathcal{X}) \rightleftarrows \mathit{(Co)mod}_{T^{\mathcal{Y}}}(\mathcal{Y}) : \widehat{\Psi} \tag{6.3}$$

between the categories of (co)modules over the transferred (co)monads such that the diagrams

$$\begin{array}{ccc} \mathcal{X} & \xrightarrow{\quad\Phi\quad} & \mathcal{Y} \\ {\scriptstyle \mathrm{Ind}_{T^{\mathcal{X}}}}\big\downarrow & & \big\downarrow{\scriptstyle \mathrm{Ind}_{T^{\mathcal{Y}}}} \\ \mathit{Mod}_{T^{\mathcal{X}}}(\mathcal{X}) & \xrightarrow{\ \widehat{\Phi}\ } & \mathit{Mod}_{T^{\mathcal{Y}}}(\mathcal{Y}) \end{array} \tag{6.4}$$

and

$$\begin{array}{ccc} \mathcal{X} & \xrightarrow{\quad\Phi\quad} & \mathcal{Y} \\ {\scriptstyle \mathrm{coInd}_{T^{\mathcal{X}}}}\big\downarrow & & \big\downarrow{\scriptstyle \mathrm{coInd}_{T^{\mathcal{Y}}}} \\ \mathit{Comod}_{T^{\mathcal{X}}}(\mathcal{X}) & \xrightarrow{\ \widehat{\Phi}\ } & \mathit{Comod}_{T^{\mathcal{Y}}}(\mathcal{Y}) \end{array} \tag{6.5}$$

commute. Here for a monad T on $\mathcal{X}$, the induction functor $\mathrm{Ind}_T\colon \mathcal{X} \to \mathit{Mod}_T(\mathcal{X})$ has T as underlying functor, and an analogous statement holds for the coinduction $\mathrm{coInd}_T\colon \mathcal{X} \to \mathit{Comod}_T(\mathcal{X})$.

PROOF. The statement is shown by direct computation. Note that for the unit axiom of the induced (co)monad $T^{\mathcal{Y}}$ on $\mathcal{Y}$ it is necessary that the equivalence is an adjoint equivalence. The uniqueness of the functor $\widehat{\Phi}$ follows by considering the adjoints of the diagrams (6.4) and (6.5), which are commuting diagrams involving the forgetful functor. □

In the sequel we denote the (co)induction functor for a (co)monad T just by its underlying functor T. Given a (co)monad $T\colon \mathcal{M} \to \mathcal{M}$, for any category $\mathcal{X}$ the functor category $\mathcal{F}un(\mathcal{M}, \mathcal{X})$ inherits a (co)monad T^* by pre-composition with T. We call a (co)monad *left exact* iff its underlying endofunctor is left exact.

PROPOSITION 6.2. *Let $T\colon \mathcal{M} \to \mathcal{M}$ be a left exact comonad on a finite category $\mathcal{M}$, let $\mathcal{X}$ be a finite category, and let T^* be the induced comonad on $\mathcal{L}ex(\mathcal{M}, \mathcal{X})$.*

(i) *The comonad on the category $\overline{\mathcal{M}} \boxtimes \mathcal{X}$ that is obtained by transferring the comonad T^* on the functor category $\mathcal{L}ex(\mathcal{M}, \mathcal{X})$ along the Eilenberg-Watts equivalence $\Psi^{\rm l}\colon \mathcal{L}ex(\mathcal{M}, \mathcal{X}) \to \overline{\mathcal{M}} \boxtimes \mathcal{X}$ is*

$$\overline{T^{\rm l.a.}} \boxtimes \mathrm{Id}\colon \quad \overline{\mathcal{M}} \boxtimes \mathcal{X} \to \overline{\mathcal{M}} \boxtimes \mathcal{X}\,. \tag{6.6}$$

Moreover, there is a canonical equivalence

$$Comod_{\overline{T^{\rm l.a.}}}(\overline{\mathcal{M}}) \simeq \overline{Comod_T(\mathcal{M})} \tag{6.7}$$

between the category of $\overline{T^{\rm l.a.}}$-comodules and the opposite of the category of T-comodules.

(ii) *Using implicitly the canonical equivalence* (6.7)*, the Eilenberg-Watts equivalences* (1.2) *induce an equivalence $\Theta^{\rm l}$ between the category of T^*-comodules and the category $\mathcal{L}ex(Comod_T(\mathcal{M}), \mathcal{X})$ of left exact functors, giving rise to a commuting triangle*

$$\begin{array}{ccc} Comod_{T^*}(\mathcal{L}ex(\mathcal{M},\mathcal{X})) & \xrightarrow[\simeq]{\ \Theta^{\rm l}\ } & \mathcal{L}ex(Comod_T(\mathcal{M}),\mathcal{X}) \\ & {\scriptstyle\widehat{\Psi}^{\rm l}}\searrow \qquad \nearrow{\scriptstyle\Phi^{\rm l}} & \\ & \overline{Comod_T(\mathcal{M})} \boxtimes \mathcal{X} & \end{array} \tag{6.8}$$

A quasi-inverse of the equivalence $\Theta^{\rm l}$ is given by

$$\mathcal{L}ex(Comod_T(\mathcal{M}), \mathcal{X}) \ni G \mapsto G \circ T \in Comod_{T^*}(\mathcal{L}ex(\mathcal{M}, \mathcal{X}))\,. \tag{6.9}$$

(iii) *Concretely, for any $F \in Comod_{T^*}(\mathcal{L}ex(\mathcal{M}, \mathcal{X}))$ we have a canonical isomorphism*

$$\int^{z \in Comod_T(\mathcal{M})} \overline{z} \boxtimes \Theta^{\rm l}(F)(z) \cong \int^{m \in \mathcal{M}} \overline{m} \boxtimes F(m) \tag{6.10}$$

of objects in $\overline{Comod_T(\mathcal{M})} \boxtimes \mathcal{X}$.

PROOF. (i) Consider the transfer of the comonad T^* on $\mathcal{L}ex(\mathcal{M}, \mathcal{X})$ to $\overline{\mathcal{M}} \boxtimes \mathcal{X}$ along the Eilenberg-Watts equivalence $\Psi^{\rm l}$. It follows from (2.8) that this transferred comonad is $\overline{T^{\rm l.a.}} \boxtimes \mathrm{Id}$. By considering, for a comodule m with coaction $m \to Tm$ in $\mathcal{M}$, the morphism in $\overline{\mathcal{M}}$ that corresponds to $T^{\rm l.a.} m \to m$, one then sees that there is an equivalence $\mathrm{Comod}_{\overline{T^{\rm l.a.}}}(\overline{\mathcal{M}}) \simeq \overline{\mathrm{Comod}_T(\mathcal{M})}$.

(ii) The existence of an induced equivalence $\Theta^{\rm l}$ as in (6.8) readily follows from Proposition 6.1 with the help of the Eilenberg-Watts equivalences. To show the statement about the quasi-inverse, we first apply the quasi-inverse Ψ^{l} of Φ^{l} in (6.8) to $G \in \mathcal{L}ex(\mathrm{Comod}_T(\mathcal{M}), \mathcal{X})$ and then use the quasi-inverse $\widehat{\Phi}^{\mathrm{l}}$ to $\widehat{\Psi}^{\mathrm{l}}$ to arrive at

$$\begin{aligned} \widehat{\Phi}^{\mathrm{l}}(\Psi^{\mathrm{l}}(G)) &= \int^{z\in\mathrm{Comod}_T(\mathcal{M})} \mathrm{Hom}_{\mathcal{M}}(U(z),-)\otimes G(z) \\ &\cong \int^{m\in\mathcal{M}} \mathrm{Hom}_{\mathcal{M}}(m,-)\otimes G(T(m)) \,\cong\, G\circ T\,. \end{aligned} \tag{6.11}$$

Here the second step uses that $T^{\mathrm{l.a.}} = U\colon \mathrm{Comod}_T(\mathcal{M}) \to \mathcal{M}$ is the forgetful functor together with Lemma 2.3.

Part (iii) is a direct consequence of part (ii). □

REMARK 6.3. (i) Note that as compared with the ordinary Eilenberg-Watts situation in (1.2), in the diagram (6.8) we deal with three instead of two categories – the adjoint pair of quasi-inverse functors Φ and Ψ splits up into a triangle of equivalences. (Also, in (6.8) we use the same notation $\widehat{\Psi}$ as in (6.3) even though we actually deal with the composition of the functor $\widehat{\Psi}$ from (6.3) with the canonical equivalence (6.7).)

(ii) It is worth to note that the functor $\Theta^{\mathrm{l}}(F)$ in (6.10) can be defined point-wise through the equalizer

$$\Theta^{\mathrm{l}}(F)(z) \to F(U(z)) \underset{\varphi_2}{\overset{\varphi_1}{\rightrightarrows}} F(T(z))\,, \tag{6.12}$$

where the morphisms φ_1 and φ_2 are defined via the T_*-comodule structure of F and by the T-comodule structure of z, respectively.

Dually, for a right exact monad T on $\mathcal{M}$ we obtain an equivalence involving categories of modules over monads:

$$\Theta^{\mathrm{r}}: \quad \mathrm{Mod}_{T^*}(\mathcal{R}ex(\mathcal{M},\mathcal{X})) \simeq \mathcal{R}ex(\mathrm{Mod}_T(\mathcal{M}),\mathcal{X})\,. \tag{6.13}$$

We are now in a position to show that the Eilenberg-Watts calculus can be lifted to twisted centers. To this end we apply Proposition 6.2 to the case that T is the central comonad $Z_{[\kappa]}$ for an $\mathcal{A}$-bimodule $\mathcal{M}$. For the category $\mathcal{L}ex^{\kappa}(\mathcal{M},\mathcal{X})$ of κ-balanced functors, we then get

THEOREM 6.4. *Let $\mathcal{M}$ be a finite bimodule category over a finite tensor category $\mathcal{A}$, $\mathcal{X}$ a finite category, and $\kappa \in 2\mathbb{Z}$.*

(i) *The Eilenberg-Watts calculus provides explicit equivalences*

$$\begin{array}{ccc} \mathcal{L}ex^{\kappa}(\mathcal{M},\mathcal{X}) & \xrightarrow{\quad\Theta^{\mathrm{l}}\quad} & \mathcal{L}ex(\mathcal{Z}^{\kappa}(\mathcal{M}),\mathcal{X}) \\ & {\scriptstyle\widehat{\Psi}^{\mathrm{l}}}\searrow \qquad \nearrow{\scriptstyle\Phi^{\mathrm{l}}} & \\ & \overline{\mathcal{Z}^{\kappa}(\mathcal{M})}\boxtimes\mathcal{X} & \end{array} \tag{6.14}$$

of linear categories for all $\mathcal{X}$. A quasi-inverse of the equivalence Θ^{l} is given by

$$\mathcal{L}ex(\mathcal{Z}^{\kappa}(\mathcal{M}),\mathcal{X}) \ni G \mapsto G\circ Z_{[\kappa]} \in \mathcal{L}ex^{\kappa}(\mathcal{M},\mathcal{X})\,. \tag{6.15}$$

In particular, for $\mathcal{M} = \mathcal{K}_{\mathcal{A}}\boxtimes_{\mathcal{A}}\mathcal{N}$ with $\mathcal{K}$ and $\mathcal{N}$ finite right and left $\mathcal{A}$-modules, respectively, and $\kappa = 0$, the category $\mathcal{Z}^0(\mathcal{K}\boxtimes\mathcal{N})$ becomes a left exact relative Deligne product of $\mathcal{K}$ and $\mathcal{N}$ according to Definition 3.4.

(ii) *For any left exact κ-balanced functor $F\colon \mathcal{M}\to\mathcal{X}$ there is an isomorphism*

$$\int^{z\in\mathcal{Z}^\kappa(\mathcal{M})} \overline{z}\boxtimes\Theta^{\mathrm{l}}(F)(z) \cong \int^{m\in\mathcal{M}} \overline{m}\boxtimes F(m) \tag{6.16}$$

of objects in $\overline{\mathcal{Z}^\kappa(\mathcal{M})}\boxtimes\mathcal{X}$.

(iii) *Specifically, for the co-induction functor $Z_{[\kappa]}\colon \mathcal{M}\to\mathcal{Z}^\kappa(\mathcal{M})$ that corresponds to the comonad $Z_{[\kappa]}$, the corresponding functor $\Theta^{\mathrm{l}}(Z_{[\kappa]})\colon \mathcal{Z}^\kappa(\mathcal{M})\to\mathcal{Z}^\kappa(\mathcal{M})$ is the identity functor, whereby Equation* (6.16) *reduces to*

$$\begin{aligned}\int^{z\in\mathcal{Z}^\kappa(\mathcal{M})} \overline{z}\boxtimes\mathrm{Id}(z) &\cong \int^{m\in\mathcal{M}} \overline{m}\boxtimes Z_{[\kappa]}(m)\\ &\cong \int^{m\in\mathcal{M}} \overline{m}\boxtimes\int_{a\in\mathcal{A}} a\triangleright m\triangleleft a^{[\kappa-1]}.\end{aligned} \tag{6.17}$$

Proof. In view of Proposition 6.2, it only remains to be shown that the category $\mathrm{Comod}_{(Z_{[\kappa]})^*}(\mathcal{L}ex(\mathcal{M},\mathcal{X}))$ is equivalent to the category $\mathcal{L}ex^\kappa(\mathcal{M},\mathcal{X})$. By Lemma 3.7, the latter category is equivalent to the category of comodules over the comonad $Z_{[2-\kappa]}$ on $\mathcal{L}ex(\mathcal{M},\mathcal{X})$. Now by direct computation one verifies that $(Z_{[\kappa]})^* = Z_{[2-\kappa]}$ as comonads. Thus the statement follows. □

We refer to the functors $\widehat{\Psi}^{\mathrm{l}}$ and Φ^{l} in the commuting triangle (6.14) as the (*left exact*) *module Eilenberg-Watts equivalences*.

By applying this result to the opposite category, an isomorphism analogous to (6.17) holds for ends instead of coends: We have an isomorphism

$$\int_{z\in\mathcal{Z}^\kappa(\mathcal{M})} \overline{z}\boxtimes z \cong \int_{m\in\mathcal{M}} \overline{m}\boxtimes Z^{[\kappa]}(m) \tag{6.18}$$

of objects in $\overline{\mathcal{Z}^\kappa(\mathcal{M})}\boxtimes\mathcal{Z}^\kappa(\mathcal{M})$, where $Z^{[\kappa]}$ is the κ-twisted central monad, which has $m\mapsto\int^{a\in\mathcal{A}} a\triangleright m\triangleleft a^{[\kappa-3]}$ as underlying endofunctor [**FSS3**, Eq. (3.17)].

Remark 6.5. Recall that the category of comodules over the central comonad (3.11) of a finite tensor category $\mathcal{A}$ is equivalent to the center $\mathcal{Z}(\mathcal{A})$. Thus in this case the result (6.17) gives the isomorphism

$$\int^{z\in\mathcal{Z}(\mathcal{A})} \overline{z}\boxtimes z \cong \int^{b\in\mathcal{A}} \overline{b}\boxtimes\int_{a\in\mathcal{A}} a\triangleright b\triangleleft a^\vee \tag{6.19}$$

of objects in $\overline{\mathcal{Z}(\mathcal{A})}\boxtimes\mathcal{Z}(\mathcal{A})$, where the two balancings on each the two tensor factors on the right hand side are provided by the isomorphisms (4.1). Note that the balancing on the first factor involves both the b- and the a-variables. This reproduces the description of the coend of the center of a fusion category, as given in [**BV**, Sect. 9.3] and recalled in [**TV2**, Sect. 9.2.3]. Indeed, in that work (note their different convention for duals in $\mathcal{A}$) it is shown that

$$\int^{z\in\mathcal{Z}(\mathcal{A})} z^\vee\otimes z \cong \int^{b\in\mathcal{A}} T(b)^\vee\otimes b \tag{6.20}$$

as objects in $\mathcal{Z}(\mathcal{A})$, where $T(b)$ appearing on the right hand side is the coend $T(b) = \int^a a^\vee\otimes b\otimes a$, so that $T(b)^\vee \cong \int_a a^\vee\otimes b^\vee\otimes a^{\vee\vee}$. If we combine (6.19) with the isomorphism (4.1) and then apply the exact functor from $\overline{\mathcal{A}}\boxtimes\mathcal{A}$ to $\mathcal{A}$ that is given by $\overline{a}\boxtimes b\mapsto a^\vee\otimes b$, we obtain

$$\begin{aligned}\int^{z\in\mathcal{Z}(\mathcal{A})} z^\vee\otimes z \;&\cong \int^{b\in\mathcal{A}}\!\!\int_{a\in\mathcal{A}} (a^\vee\otimes b\otimes a)^\vee\otimes b \\ &\cong \int^{b\in\mathcal{A}}\!\!\int_{a\in\mathcal{A}} a^\vee\otimes b^\vee\otimes a^{\vee\vee}\otimes b\,,\end{aligned}\tag{6.21}$$

thus reproducing (6.20).

REMARK 6.6. In the semisimple case, i.e. if $\mathcal{A}$ is a fusion category, ends and coends coincide, and (6.19) reduces to an isomorphism

$$\bigoplus_\alpha \overline{z_\alpha}\boxtimes z_\alpha \;\cong\; \bigoplus_{i,j}\overline{b_i}\boxtimes a_j\triangleright b_i\triangleleft a_j^\vee \tag{6.22}$$

of objects in $\mathcal{Z}(\mathcal{A})$, where α ranges over a set of representatives for the isomorphism classes of simple objects of $\mathcal{Z}(\mathcal{A})$, while i and j range over a set representing the isomorphism classes of simples in $\mathcal{A}$. The isomorphism (6.22) relates $\mathcal{A}$ and its Drinfeld center. It leads to Lemma 7.4 of [**BaK**] after applying the functor $\overline{\mathcal{Z}(\mathcal{A})}\boxtimes\mathcal{Z}(\mathcal{A})\to\mathcal{Z}(\mathcal{A})$ that is given by $\overline{x}\boxtimes y\mapsto x^\vee\otimes y$ to both sides. In case $\mathcal{A}$ has a spherical structure, the isomorphism (6.22) plays a central role in approaches [**BaK, TV1**] to relate the Turaev-Viro TFT for a spherical fusion category $\mathcal{A}$ with the Reshetikhin-Turaev TFT for $\mathcal{Z}(\mathcal{A})$, in which it is responsible for the factorization property of the state spaces: It is the algebraic counterpart of factoring a surface that has been obtained by gluing two surfaces along a common circular boundary, compare [**FSS3**, Thm. 5.2].

REMARK 6.7. Statements analogous to those of Theorem 6.4 hold for right exact functors. In particular we have explicit equivalences

$$\mathcal{R}ex^{\kappa-4}(\mathcal{M},\mathcal{X}) \simeq \mathcal{R}ex(\mathcal{Z}^\kappa(\mathcal{M}),\mathcal{X})\,. \tag{6.23}$$

By considering the $\mathcal{A}$-bimodule $\mathcal{M}=\mathcal{N}_\mathcal{A}\boxtimes{}_\mathcal{A}\mathcal{N}'$, where $\mathcal{N}$ and $\mathcal{N}'$ are a right and a left $\mathcal{A}$-module, respectively, with Proposition 3.5 we then recover for $\kappa=+4$ the statement that the right exact relative Deligne product is equivalent to the twisted center. Recalling from (1.1) that the Eilenberg-Watts equivalences provide an equivalence of the categories of left and right exact functors between finite categories, we arrive at the following module version of that picture:

$$\begin{array}{ccccc} \mathcal{L}ex^\kappa(\mathcal{M},\mathcal{X}) & & \xrightarrow{\quad\Theta^{\mathrm{l}}\quad} & & \mathcal{L}ex(\mathcal{Z}^\kappa(\mathcal{M}),\mathcal{X}) \\ \downarrow & \searrow^{\widehat{\Psi}^{\mathrm{l}}} & & \nearrow^{\Phi^{\mathrm{l}}} & \downarrow \\ & & \overline{\mathcal{Z}^\kappa(\mathcal{M})}\boxtimes\mathcal{X} & & \\ & \nearrow^{\widehat{\Psi}^{\mathrm{r}}} & & \searrow^{\Phi^{\mathrm{r}}} & \\ \mathcal{R}ex^{\kappa-4}(\mathcal{M},\mathcal{X}) & & \xrightarrow{\quad\Theta^{\mathrm{r}}\quad} & & \mathcal{R}ex(\mathcal{Z}^\kappa(\mathcal{M}),\mathcal{X}) \end{array} \tag{6.24}$$
$$\text{(diagonal arrows: } \widehat{\Psi}^{\mathrm{l}}, \widehat{\Psi}^{\mathrm{r}} \text{ into } \overline{\mathcal{Z}^\kappa(\mathcal{M})}\boxtimes\mathcal{X}\text{;} \ \Phi^{\mathrm{l}}, \Phi^{\mathrm{r}} \text{ out of it)}$$

where the unlabeled vertical arrows are defined by the commutativity of the diagram.

Combining the module Eilenberg-Watts equivalences (6.14) with Lemma 2.1 we obtain the following generalization of Corollary 4.2, which is a module version of Lemma 2.1:

LEMMA 6.8. *Let $\mathcal{M}$ and $\mathcal{N}$ be finite left modules over a finite tensor category $\mathcal{A}$. For any finite category $\mathcal{X}$ the equivalences from Lemma* 2.1 *induce equivalences*

$$\mathcal{L}ex_{\mathcal{A}}(\mathcal{N},\mathcal{M}\boxtimes\mathcal{X}) \simeq \mathcal{L}ex^{\mathrm{bal}}(\mathcal{N}\boxtimes\mathcal{M}^{\#},\mathcal{X}) \simeq \mathcal{L}ex(\mathcal{Z}^{0}(\mathcal{N}\boxtimes\mathcal{M}^{\#}),\mathcal{X})\,. \tag{6.25}$$

PROOF. The first equivalence is a straightforward generalization of Proposition 5.1. The second is Theorem 6.4(i). □

The case of bimodule functors is treated as follows. For bimodules ${}_{\mathcal{A}}\mathcal{M}_{\mathcal{B}}$ and ${}_{\mathcal{A}}\mathcal{N}_{\mathcal{B}}$, there are two corresponding comonads on $\mathcal{L}ex(\mathcal{N},\mathcal{M})$, with obvious distributive law, whose comodules are the corresponding categories of bimodule functors. For the $\mathcal{A}$-action this is $Z_{[2],\mathcal{A}}$, see Lemma 3.7(ii). For the $\mathcal{B}$-action, the relevant comonad $Z_{[2],\mathcal{B}}$ on $F\colon \mathcal{N}\to\mathcal{M}$ is given by $Z_{[2],\mathcal{B}}(F)(n) = \int_{b\in\mathcal{B}} F(n\triangleleft b)\triangleleft b^{\vee}$. To summarize, $Z := Z_{[2],\mathcal{A}}\circ Z_{[2],\mathcal{B}}$ is canonically a comonad on the category $\mathcal{L}ex(\mathcal{N},\mathcal{M})$ whose category of comodules is equivalent to the category $\mathcal{L}ex_{\mathcal{A},\mathcal{B}}(\mathcal{N},\mathcal{M})$ of bimodule functors.

COROLLARY 6.9. *Let ${}_{\mathcal{A}}\mathcal{M}_{\mathcal{B}}$ and ${}_{\mathcal{A}}\mathcal{N}_{\mathcal{B}}$ be finite bimodules over finite tensor categories $\mathcal{A}$ and $\mathcal{B}$. The Eilenberg-Watts equivalences induce an equivalence*

$$\mathcal{L}ex_{\mathcal{A},\mathcal{B}}(\mathcal{N},\mathcal{M}) \simeq \mathcal{Z}_{\mathcal{A},\mathcal{B}}(\mathcal{N}^{\#}\boxtimes\mathcal{M}) \tag{6.26}$$

of categories, where the category $\mathcal{N}^{\#}$ is equipped with the left $\mathcal{B}$-action $b\triangleright\overline{m} = \overline{m\triangleleft{}^{\vee}b}$ and $\mathcal{Z}_{\mathcal{A},\mathcal{B}}$ denotes the center with respect to the $\mathcal{A}$ and $\mathcal{B}$-actions.

PROOF. According to Proposition 6.2 we need to transfer the comonad Z along the Eilenberg-Watts equivalence $\mathcal{L}ex(\mathcal{N},\mathcal{M})\simeq\overline{\mathcal{N}}\boxtimes\mathcal{M}$. A straightforward computation yields the comonad on $\overline{\mathcal{N}}\boxtimes\mathcal{M}$ that is given by

$$\overline{n}\boxtimes m \longmapsto \int_{a\in\mathcal{A}}\int_{b\in\mathcal{B}} \overline{a^{\vee\vee}\triangleright n\triangleleft{}^{\vee\vee}b}\boxtimes a\triangleright n\triangleleft b\,. \tag{6.27}$$

The category of comodules over this comonad is $\mathcal{Z}_{\mathcal{A},\mathcal{B}}(\mathcal{N}^{\#}\boxtimes\mathcal{M})$. □

REMARK 6.10. Given any three $\mathcal{A}$-$\mathcal{B}$-bimodules ${}_{\mathcal{A}}\mathcal{M}_{\mathcal{B}}$, ${}_{\mathcal{A}}\mathcal{N}_{\mathcal{B}}$ and ${}_{\mathcal{A}}\mathcal{K}_{\mathcal{B}}$, the equivalence (6.26) induces a composition operation on the centers: The composition

$$\mathcal{L}ex_{\mathcal{A},\mathcal{B}}(\mathcal{N},\mathcal{M})\times\mathcal{L}ex_{\mathcal{A},\mathcal{B}}(\mathcal{M},\mathcal{K}) \to \mathcal{L}ex_{\mathcal{A},\mathcal{B}}(\mathcal{N},\mathcal{K}) \tag{6.28}$$

of bimodules functors yields a functor

$$\mathcal{Z}_{\mathcal{A},\mathcal{B}}(\mathcal{N}^{\#}\boxtimes\mathcal{M})\boxtimes\mathcal{Z}_{\mathcal{A},\mathcal{B}}(\mathcal{M}^{\#}\boxtimes\mathcal{K}) \to \mathcal{Z}_{\mathcal{A},\mathcal{B}}(\mathcal{N}^{\#}\boxtimes\mathcal{K})\,. \tag{6.29}$$

This is given by contraction with the Hom functor:

$$(\overline{n}\boxtimes m)\boxtimes(\overline{m'}\boxtimes k) \mapsto \mathrm{Hom}_{\mathcal{M}}(m',m)\otimes(\overline{n}\boxtimes k)\,. \tag{6.30}$$

This is seen as follows. In [**FSS2**, Cor. 3.7] the corresponding functor in the absence of bimodule structures is shown to be given by the Hom-functor. Since the composition of functors induces a composition of bimodule functors, the induced composition (6.29) is given by the Hom-functor as well. The balancing on the right hand side follows by combining the balancings on the left and the dualities of $\mathcal{A}$ and $\mathcal{B}$ inside the Hom space.

References

[BaK] B. Balsam and A.N. Kirillov, *Turaev-Viro invariants as an extended TQFT*, preprint `math.GT/1004.1533`.

[BV] Alain Bruguières and Alexis Virelizier, *Quantum double of Hopf monads and categorical centers*, Trans. Amer. Math. Soc. **364** (2012), no. 3, 1225–1279, DOI 10.1090/S0002-9947-2011-05342-0. MR2869176

[BW] Tomasz Brzezinski and Robert Wisbauer, *Corings and comodules*, London Mathematical Society Lecture Note Series, vol. 309, Cambridge University Press, Cambridge, 2003, DOI 10.1017/CBO9780511546495. MR2012570

[De] P. Deligne, *Catégories tannakiennes* (French), The Grothendieck Festschrift, Vol. II, Progr. Math., vol. 87, Birkhäuser Boston, Boston, MA, 1990, pp. 111–195. MR1106898

[Ei] Samuel Eilenberg, *Abstract description of some basic functors*, J. Indian Math. Soc. (N.S.) **24** (1960), 231–234 (1961). MR125148

[EtGNO] Pavel Etingof, Shlomo Gelaki, Dmitri Nikshych, and Victor Ostrik, *Tensor categories*, Mathematical Surveys and Monographs, vol. 205, American Mathematical Society, Providence, RI, 2015, DOI 10.1090/surv/205. MR3242743

[EtNO] Pavel Etingof, Dmitri Nikshych, and Viktor Ostrik, *An analogue of Radford's S^4 formula for finite tensor categories*, Int. Math. Res. Not. **54** (2004), 2915–2933, DOI 10.1155/S1073792804141445. MR2097289

[EtNOM] Pavel Etingof, Dmitri Nikshych, and Victor Ostrik, *Fusion categories and homotopy theory*, Quantum Topol. **1** (2010), no. 3, 209–273, DOI 10.4171/QT/6. With an appendix by Ehud Meir. MR2677836

[FSS1] Jürgen Fuchs, Gregor Schaumann, and Christoph Schweigert, *A trace for bimodule categories*, Appl. Categ. Structures **25** (2017), no. 2, 227–268, DOI 10.1007/s10485-016-9425-3. MR3638361

[FSS2] Jürgen Fuchs, Gregor Schaumann, and Christoph Schweigert, *Eilenberg-Watts calculus for finite categories and a bimodule Radford S^4 theorem*, Trans. Amer. Math. Soc. **373** (2020), no. 1, 1–40, DOI 10.1090/tran/7838. MR4042867

[FSS3] J. Fuchs, G. Schaumann, and C. Schweigert, *A modular functor from state sums for finite tensor categories and their bimodules*, preprint `math.QA/1911.06214`.

[FuS] Jürgen Fuchs and Christoph Schweigert, *Coends in conformal field theory*, Lie algebras, vertex operator algebras, and related topics, Contemp. Math., vol. 695, Amer. Math. Soc., Providence, RI, 2017, pp. 65–81, DOI 10.1090/conm/695/13996. MR3709706

[Ri] Emily Riehl, *Categorical homotopy theory*, New Mathematical Monographs, vol. 24, Cambridge University Press, Cambridge, 2014, DOI 10.1017/CBO9781107261457. MR3221774

[Sc1] Gregor Schaumann, *Traces on module categories over fusion categories*, J. Algebra **379** (2013), 382–425, DOI 10.1016/j.jalgebra.2013.01.013. MR3019263

[Sc2] Gregor Schaumann, *Pivotal tricategories and a categorification of inner-product modules*, Algebr. Represent. Theory **18** (2015), no. 6, 1407–1479, DOI 10.1007/s10468-015-9547-6. MR3435098

[Sh1] Kenichi Shimizu, *On unimodular finite tensor categories*, Int. Math. Res. Not. IMRN **1** (2017), 277–322, DOI 10.1063/1.436405. MR3632104

[Sh2] Kenichi Shimizu, *Further results on the structure of (co)ends in finite tensor categories*, Appl. Categ. Structures **28** (2020), no. 2, 237–286, DOI 10.1007/s10485-019-09577-7. MR4077241

[Sh3] K. Shimizu, *Relative Serre functor for comodule algebras*, preprint `math.CT/1904.00376`.

[TV1] V.G. Turaev and A. Virelizier, *On two approaches to 3-dimensional TQFTs*, preprint `math.GT/1006.3501`.

[TV2] Vladimir Turaev and Alexis Virelizier, *Monoidal categories and topological field theory*, Progress in Mathematics, vol. 322, Birkhäuser/Springer, Cham, 2017, DOI 10.1007/978-3-319-49834-8. MR3674995

[Wa] Charles E. Watts, *Intrinsic characterizations of some additive functors*, Proc. Amer. Math. Soc. **11** (1960), 5–8, DOI 10.2307/2032707. MR118757

Teoretisk fysik, Karlstads Universitet, Universitetsgatan 21, S – 651 88 Karlstad
Email address: juergen.fuchs@kau.se

Institut für Mathematik, Universität Würzburg, Mathematische Physik, Emil-Fischer-Strasse 31, D – 97 074 Würzburg
Email address: gregor.schaumann@uni-wuerzburg.de

Fachbereich Mathematik, Universität Hamburg, Bereich Algebra und Zahlentheorie, Bundesstrasse 55, D – 20 146 Hamburg
Email address: christoph.schweigert@uni-hamburg.de

Contemporary Mathematics
Volume **771**, 2021
https://doi.org/10.1090/conm/771/15510

Coherent unit actions on braided operads and Hopf algebras

Li Guo and Yunnan Li

ABSTRACT. The notion of a coherent unit action on algebraic operads was first introduced by Loday for binary quadratic regular operads and generalized by Holtkamp, to ensure that the free objects of the operads carry a Hopf algebra structure. There were also criteria for such operads in the binary quadratic regular case. We generalize the notion of coherent unit action to braided operads and show that free objects of braided operads with such an action carries a braided Hopf algebra structure. Under the conditions of being binary, quadratic and regular, we give criteria for the braided operads that allow a coherent unit action and thus carry a braided Hopf algebra structure on their free objects.

1. Introduction

This paper explores braided Hopf algebra structures on free objects from braided operads by the method of coherent unit actions.

As the importance of Hopf algebras continued to show in more and more areas of mathematics, ad hoc instances of Hopf algebras sometimes turned out to be just special cases of a general construction. One such general construction is the free objects of various (nonassociative) algebras (operads). The first instance is the dendriform algebra introduced by Loday [**17**] who showed that, even though a dendriform algebra is not an associative algebra, free dendriform algebras has a natural Hopf algebra structure[**20**]. This Hopf algebra is realized on planar binary trees and was found to be isomorphic to the noncommutative analog (by Foissy and Holtkamp [**5**, **10**]) of the Hopf algebra of rooted trees in the Connes-Kreimer theory of renormalization of quantum field theory [**2**]. After Hopf algebra structures were discovered for several related algebraic structures, such as tridendriform algebras and quadri-algebras [**1**, **21**], Loday showed in [**17**] that a nonsymmetric (also called regular) operad with a so-called coherent unit action endows a natural Hopf algebra structure on its free objects. Similar results hold for the corresponding algebraic structures with certain commutativity [**18**, **19**]. This uniform approach not only included as special cases many Hopf algebras obtained case by case during that period of time, it also recovered the classical Hopf algebra structures on the shuffle

2020 *Mathematics Subject Classification.* Primary 16T05, 16T25, 16W99, 05C05.

Key words and phrases. Operad, unit action, braided Hopf algebra, braided dendriform algebra, braided tridendriform algebra.

The second author was supported by Natural Science Foundation of China (Grant Nos. 11501214 and 11771142).

algebra and quasi-shuffle algebra, as the algebras are shown to be the free objects of commutative dendriform and tridendriform algebras. This work was further generalized to regular operads by Holtkamp [**11**]. Given the great interest in such Hopf algebras, criteria were achieved for binary quadratic regular operads which have a coherent unit action, and hence allow for Hopf algebra structures on their free objects [**3**].

Operads have been extended to various broader contexts. We are particularly interested in braided operads [**4**, **7**] for their connection with Yang-Baxter equations and quantum theory[**15**]. In particular, in this context quantum shuffle algebras, quantum quasi-shuffle algebras and generally quantum multi-brace algebras have been obtained by Rosso, Jian-Rosso-Zhang and Jian-Rosso [**12**–**14**, **23**] respectively. Recently, with motivation from braided construction of rooted trees from the work of Connes-Kreimer on renormalization of quantum field theory, braided structures for dendriform algebras and tridendriform algebras have attracted attention [**6**, **8**, **9**]. There again the free objects can be equipped with braided Hopf algebra structures.

Thus it appears to be the time to provide a uniform approach in the context of braided operads, in the direction of coherent unit actions on algebraic operads developed by Loday and Holtkamp [**11**, **18**] as noted above. In particular, it is interesting to study the coherent unit actions such that the free objects of the braided operads are braided Hopf algebras. This is the purpose of this article. We also extend the criteria for coherent unit actions on binary quadratic regular operads in [**3**] to the braided context, thereby determining such braided operads that give rise to braided Hopf algebras from their free objects.

Here is the layout of the paper. In Section 2, we provide background on braided operads with some details on the dendriform and tridendriform cases. In Section 3, we extend the notion of coherent unit actions on regular operads to braided operads and show that such braided operads have a braided Hopf algebra structure on their free objects. We consider the two cases of braided regular and completely commutative operads, so that we can cover for instance both the braided dendriform operad and braided commutative dendriform (Zinbiel) operad. In Section 4, we give criteria for braided binary quadratic regular operads with coherent unit actions.

NOTATIONS. In this paper, we fix a ground field $\mathbf{k}$ of characteristic 0. All the objects under discussion, including vector spaces, algebras and tensor products, are taken over $\mathbf{k}$ by default. Also denote $\mathbb{S}_n$ (resp. $\mathbb{B}_n$) the n-th symmetric group (resp. braid group) for any $n \geq 0$ with $\mathbb{S}_0 = \emptyset$ (resp. $\mathbb{B}_0 = \emptyset$). For the convenience of discussion, we also fix a braided tensor category $\mathcal{C}$ over $\mathbf{k}$ to involve all the forthcoming braided objects.

2. Braided algebraic operads

We start with a basic notion.

DEFINITION 2.1. A **braided vector space** is a vector space V together with a linear operator σ on $V^{\otimes 2}$, called the **Yang-Baxter operator** characterized by the equation

$$(\sigma \otimes \mathrm{id}_V)(\mathrm{id}_V \otimes \sigma)(\sigma \otimes \mathrm{id}_V) = (\mathrm{id}_V \otimes \sigma)(\sigma \otimes \mathrm{id}_V)(\mathrm{id}_V \otimes \sigma). \tag{2.1}$$

For any braided vector space (V, σ) and $n \geq 1$, the tensor space $V^{\otimes n}$ becomes a representation of $\mathbb{B}_n$ with its usual generators b_i acting as $\sigma_i := \mathrm{id}^{\otimes(i-1)} \otimes \sigma \otimes$

$\mathrm{id}^{\otimes(n-1-i)}$ for $1 \leq i \leq n-1$. For braided vector spaces $(V,\sigma),(V',\sigma')$, a linear map $f : V \to V'$ is called a **homomorphism of braided vector spaces** if $(f \otimes f)\sigma = \sigma'(f \otimes f)$. Let **VS** (resp. **BV**) denote the category of (braided) vector spaces.

Let $\pi_n : \mathbb{B}_n \to \mathbb{S}_n$ be the natural projection mapping b_i to transpositions s_i, $i = 1, \ldots, n-1$. It is well known that there exists an injective map $\iota_n : \mathbb{S}_n \to \mathbb{B}_n$, satisfying $\pi_n \iota_n = \mathrm{id}_{\mathbb{S}_n}$ and called the **Matsumoto-Tits section**. If $w = s_{i_1} \cdots s_{i_r}$ is any reduced expression of $w \in \mathbb{S}_n$, then $\iota_n(w) = b_{i_1} \cdots b_{i_r}$, which is uniquely determined by w. Given any braided vector space (V,σ), we denote the action of $\iota_n(w)$ on $V^{\otimes n}$ by $T_w^\sigma := \sigma_{i_1} \cdots \sigma_{i_r}$. In particular, for the usual flip map τ of V, we have

$$T_w^\tau(v_1 \otimes \cdots \otimes v_n) = v_{w^{-1}(1)} \otimes \cdots \otimes v_{w^{-1}(n)}.$$

Definition 2.2. Let A be an algebra with product μ, and σ be a braiding on A. We call the triple (A,μ,σ) a **braided algebra** if it satisfies the conditions

$$(\mu \otimes \mathrm{Id}_A)\sigma_2\sigma_1 = \sigma(\mathrm{Id}_A \otimes \mu), \quad (\mathrm{Id}_A \otimes \mu)\sigma_1\sigma_2 = \sigma(\mu \otimes \mathrm{Id}_A). \tag{2.2}$$

Moreover, if A is unital with unit 1_A and satisfies

$$\sigma(a \otimes 1_A) = 1_A \otimes a, \quad \sigma(1_A \otimes a) = a \otimes 1_A \quad \text{for all } a \in A, \tag{2.3}$$

then A is called a **unital braided algebra**.

Dually, we call $(C,\Delta,\varepsilon,\sigma)$ a **braided coalgebra** if (C,Δ,ε) is a coalgebra with braiding σ and satisfies

$$\sigma_1\sigma_2(\Delta \otimes \mathrm{Id}_C) = (\mathrm{Id}_C \otimes \Delta)\sigma, \ \sigma_2\sigma_1(\mathrm{Id}_C \otimes \Delta) = (\Delta \otimes \mathrm{Id}_C)\sigma, \tag{2.4}$$

$$(\varepsilon \otimes \mathrm{Id}_C)\sigma = \mathrm{Id}_C = (\mathrm{Id}_C \otimes \varepsilon)\sigma. \tag{2.5}$$

Also, a quintuple $(H,\mu,\Delta,\varepsilon,\sigma)$ is called a **braided bialgebra**, if (H,μ,σ) (resp. $(H,\Delta,\varepsilon,\sigma)$) is a braided algebra (resp. coalgebra) such that

$$\Delta\,\mu = (\mu \otimes \mu)\sigma_2(\Delta \otimes \Delta). \tag{2.6}$$

When H has unit 1_H and antipode $S : H \to H$ such that $\mu(S \otimes \mathrm{Id}_H)\Delta = \varepsilon 1_H = \mu(\mathrm{Id}_H \otimes S)\Delta$, then the septuple $(H,\mu,\Delta,\varepsilon,S,\sigma)$ is a **braided Hopf algebra**. Similar to the case of braided vector spaces, the homomorphisms for these braided objects are the usual homomorphisms commuting with braidings.

The notion of braided operads was originally introduced by Fiedorowicz in [**4**]. Especially when the Yang-Baxter operators are chosen to be the flip maps, the framework recovers the usual definition of operads. For the study of algebraic operads and braided operads, we refer the reader to [**7, 22**].

Definition 2.3. A **braided (algebraic) operad** over $\mathbf{k}$ is an analytic functor $\mathcal{P} : \mathbf{BV} \to \mathbf{BV}$ such that $\mathcal{P}(0) = 0$ and is equipped with an associative natural transformation of functors $\gamma : \mathcal{P} \circ \mathcal{P} \to \mathcal{P}$ and a unit $\eta : \mathrm{id}_{\mathbf{BV}} \to \mathcal{P}$ such that the following diagrams commute:

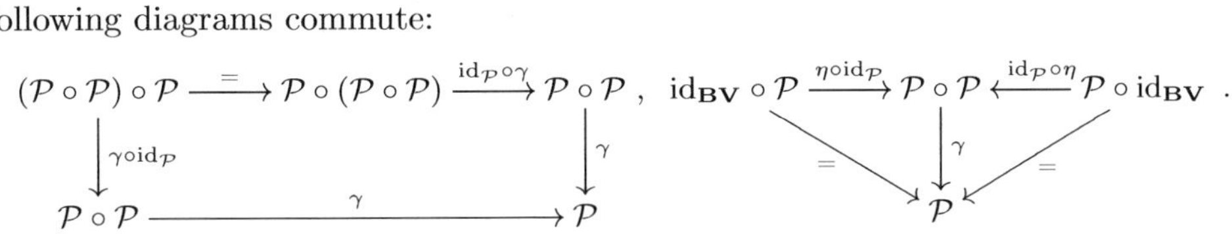

Namely,

$$\gamma_V\gamma_{\mathcal{P}(V)} = \gamma_V\mathcal{P}(\gamma_V), \quad \gamma_V\eta_{\mathcal{P}(V)} = \mathrm{id}_{\mathcal{P}(V)} = \gamma_V\mathcal{P}(\eta_V) \quad \text{for any } (V,\sigma) \in \mathbf{BV}.$$

For two braided operads $(\mathcal{P},\gamma_{\mathcal{P}},\eta_{\mathcal{P}})$, $(\mathcal{Q},\gamma_{\mathcal{Q}},\eta_{\mathcal{Q}})$, a natural transformation $\alpha:\mathcal{P}\to\mathcal{Q}$ is called a **morphism of braided operads**, if $\gamma_{\mathcal{Q}}\circ(\alpha,\alpha)=\alpha\circ\gamma_{\mathcal{P}}$ and $\eta_{\mathcal{Q}}=\alpha\circ\eta_{\mathcal{P}}$.

According to [**4**, Definition 3.2] or the construction in [**7**, §5.1], a braided operad $\mathcal{P}$ consists of a $\mathbb{B}$-module $\{\mathcal{P}(n)\}_{n\geq 0}$, with the composition maps

$$\gamma(i_1,\ldots,i_k):\mathcal{P}(k)\otimes\mathcal{P}(i_1)\otimes\cdots\otimes\mathcal{P}(i_k)\to\mathcal{P}(i_1+\cdots+i_k)$$

and the unit map η satisfying the associativity and the unity conditions, and also the following equivalent conditions:

$$\begin{aligned}\gamma(\mu b;\mu_1,\ldots,\mu_k)&=\gamma(\mu;\mu_{b^{-1}(1)},\ldots,\mu_{b^{-1}(k)})b(i_1,\ldots,i_k),\\ \gamma(\mu;\mu_1b_1,\ldots,\mu_kb_k)&=\gamma(\mu;\mu_1,\ldots,\mu_k)(b_1\times\cdots\times b_k),\end{aligned}$$

for any $\mu\in\mathcal{P}(k),\mu_1\in\mathcal{P}(i_1),\ldots,\mu_k\in\mathcal{P}(i_k)$, where $b\in\mathbb{B}_k$ acts on $\{1,\ldots,k\}$ by permutations via the natural projection π_k, $b(i_1,\ldots,i_k)$ is the braid obtained from b by replacing its j-th strand with i_j parallel strands for any $j=1,\ldots,k$, and $b_1\times\cdots\times b_k$ denotes the direct product of braids $b_1,\ldots,b_k$. It is easy to see that $\gamma(i_1,\ldots,i_k)$ factors through

$$\mathcal{P}(k)\otimes_{\mathbb{B}_k}\operatorname{Ind}^{\mathbb{B}_{i_1+\cdots+i_k}}_{\mathbb{B}_{i_1}\times\cdots\times\mathbb{B}_{i_k}}\mathcal{P}(i_1)\otimes\cdots\otimes\mathcal{P}(i_k)=\mathcal{P}(k)\otimes_{\mathbb{B}_k}(\mathcal{P}(i_1)\otimes\cdots\otimes\mathcal{P}(i_k))\otimes_{\mathbb{B}_{i_1}\times\cdots\times\mathbb{B}_{i_k}}\mathbf{k}[\mathbb{B}_{i_1+\cdots+i_k}]$$

by the homomorphism of $\mathbb{B}_{i_1+\cdots+i_k}$-modules sending any $\mu\otimes_{\mathbb{B}_k}(\mu_1\otimes\cdots\otimes\mu_k)\otimes_{\mathbb{B}_{i_1}\times\cdots\times\mathbb{B}_{i_k}}b$ to $\gamma(\mu;\mu_1,\ldots,\mu_k)b$, as we have $b(i_1,\ldots,i_k)(b_1\times\cdots\times b_k)=(b_{b^{-1}(1)}\times\cdots\times b_{b^{-1}(k)})b(i_1,\ldots,i_k)$.

DEFINITION 2.4. A **$\mathbb{B}$-module** over $\mathbf{k}$ is a family $M=\{M(n)\}_{n\geq 0}$ of right $\mathbb{B}_n$-modules $M(n)$. A morphism of $\mathbb{B}$-modules $f:M\to N$ is a family of homomorphisms of $\mathbb{B}_n$-modules $f_n:M(n)\to N(n)$.

Given a $\mathbb{B}$-module M and a braided vector space (V,σ), we define a functor $\widetilde{M}:\mathbf{BV}\to\mathbf{BV}$ by

$$\widetilde{M}(V):=\bigoplus_{n\geq 0}M(n)\otimes_{\mathbb{B}_n}V^{\otimes n},$$

whose Yang-Baxter operator, denoted by σ_M, is determined by the following equalities,

$$\sigma_M((\mu,u_1,\ldots,u_i)\otimes(\nu,v_1,\ldots,v_j)):=\sum(\nu,v_1',\ldots,v_j')\otimes(\mu,u_1',\ldots,u_i'),\tag{2.7}$$

for any $(\mu,u_1,\ldots,u_i)\in M(i)\otimes_{\mathbb{B}_i}V^{\otimes i}$ and $(\nu,v_1,\ldots,v_j)\in M(j)\otimes_{\mathbb{B}_j}V^{\otimes j}$, where we denote

$$T^{\sigma}_{\beta_{ij}}(u_1\otimes\cdots\otimes u_i\underline{\otimes}v_1\otimes\cdots\otimes v_j):=\sum v_1'\otimes\cdots\otimes v_j'\underline{\otimes}u_1'\otimes\cdots\otimes u_i',$$

with $\beta_{ij}\in\mathbb{S}_{i+j}$ such that $\beta_{ij}(k)=j+k$ if $1\leq k\leq i$, and $k-i$ if $i+1\leq k\leq i+j$.

Define the **tensor product** of $\mathbb{B}$-modules M and N to be the $\mathbb{B}$-module $M\otimes N$ with

$$(M\otimes N)(n):=\bigoplus_{i+j=n}\operatorname{Ind}^{\mathbb{B}_n}_{\mathbb{B}_i\times\mathbb{B}_j}M(i)\otimes N(j)=\bigoplus_{i+j=n}(M(i)\otimes N(j))\otimes_{\mathbb{B}_i\times\mathbb{B}_j}\mathbf{k}[\mathbb{B}_n],\quad n\geq 0.$$

Define the **composite** of $\mathbb{B}$-modules M and N to be the $\mathbb{B}$-module

$$M \circ N := \bigoplus_{k\geq 0} M(k) \otimes_{\mathbb{B}_k} N^{\otimes k}.$$

More precisely,

$$(M\circ N)(n) := \bigoplus_{k\geq 0} M(k)\otimes_{\mathbb{B}_k}\left(\bigoplus_{i_1+\cdots+i_k=n} \mathrm{Ind}^{\mathbb{B}_n}_{\mathbb{B}_{i_1}\times\cdots\times\mathbb{B}_{i_k}} N(i_1)\otimes\cdots\otimes N(i_k)\right),\ n\geq 0,$$

where the left module action of $\mathbb{B}_k$ on $N^{\otimes k}$ is defined as follows. Any braid $b\in\mathbb{B}_k$ sends

$$(\mu_1,\ldots,\mu_k,c)\in (N(i_1)\otimes\cdots\otimes N(i_k))\otimes_{\mathbb{B}_{i_1}\times\cdots\times\mathbb{B}_{i_k}}\mathbf{k}[\mathbb{B}_n]$$

to

$$\begin{aligned}(\mu_{b^{-1}(1)},\ldots,\mu_{b^{-1}(k)},b(i_1,\ldots,i_k)c)\in (N(i_{b^{-1}(1)})\otimes\cdots\otimes N(i_{b^{-1}(k)}))\\ \otimes_{\mathbb{B}_{i_{b^{-1}(1)}}\times\cdots\times\mathbb{B}_{i_{b^{-1}(k)}}}\mathbf{k}[\mathbb{B}_n],\end{aligned}$$

where $b\in\mathbb{B}_k$ acts on $\{1,\ldots,k\}$ by permutations via the natural projection π_k, and $b(i_1,\ldots,i_k)$ is the braid obtained from b by replacing its j-th strand with i_j parallel strands for any $j=1,\ldots,k$.

PROPOSITION 2.5. *For any $\mathbb{B}$-modules M, N and braided vector space V, we have*

$$(\widetilde{M\otimes N})(V)=\widetilde{M}(V)\otimes\widetilde{N}(V),\ (\widetilde{M\circ N})(V)=\widetilde{M}(\widetilde{N}(V)).$$

PROOF. By the definition of $\widetilde{M}$ and $M\otimes N$, we find

$$\begin{aligned}(\widetilde{M\otimes N})(V) &= \bigoplus_{n\geq 0}(M\otimes N)(n)\otimes_{\mathbb{B}_n} V^{\otimes n}\\ &= \bigoplus_{n\geq 0}\left(\bigoplus_{i+j=n}(M(i)\otimes N(j))\otimes_{\mathbb{B}_i\times\mathbb{B}_j}\mathbf{k}[\mathbb{B}_n]\right)\otimes_{\mathbb{B}_n} V^{\otimes n}\\ &= \bigoplus_{i,j\geq 0}(M(i)\otimes N(j))\otimes_{\mathbb{B}_i\times\mathbb{B}_j} V^{\otimes(i+j)}\\ &= \left(\bigoplus_{i\geq 0} M(i)\otimes_{\mathbb{B}_i} V^{\otimes i}\right)\otimes\left(\bigoplus_{j\geq 0} N(j)\otimes_{\mathbb{B}_j} V^{\otimes j}\right)\\ &= \widetilde{M}(V)\otimes\widetilde{N}(V).\end{aligned}$$

On the other hand, by the definition of $\widetilde{M}$ and $M \circ N$, we have

$$\begin{aligned}
&(\widetilde{M \circ N})(V) \\
&\quad = \bigoplus_{n\geq 0} (M \circ N)(n) \otimes_{\mathbb{B}_n} V^{\otimes n} \\
&\quad = \bigoplus_{n\geq 0} \left(\bigoplus_{k\geq 0} M(k) \otimes_{\mathbb{B}_k} \left(\bigoplus_{i_1+\cdots+i_k=n} (N(i_1)\otimes\cdots\otimes N(i_k)) \otimes_{\mathbb{B}_{i_1}\times\cdots\times\mathbb{B}_{i_k}} \mathbf{k}[\mathbb{B}_n] \right)\right) \\
&\qquad\qquad \otimes_{\mathbb{B}_n} V^{\otimes n} \\
&\quad = \bigoplus_{k\geq 0} M(k) \otimes_{\mathbb{B}_k} \left(\bigoplus_{i_1,\dots,i_k\geq 0} (N(i_1)\otimes\cdots\otimes N(i_k)) \otimes_{\mathbb{B}_{i_1}\times\cdots\times\mathbb{B}_{i_k}} V^{\otimes(i_1+\cdots+i_k)} \right) \\
&\quad = \bigoplus_{k\geq 0} M(k) \otimes_{\mathbb{B}_k} \left(\left(\bigoplus_{i_1\geq 0} N(i_1) \otimes_{\mathbb{B}_{i_1}} V^{\otimes i_1} \right) \otimes\cdots\otimes \left(\bigoplus_{i_k\geq 0} N(i_k) \otimes_{\mathbb{B}_{i_k}} V^{\otimes i_k} \right)\right) \\
&\quad = \bigoplus_{k\geq 0} M(k) \otimes_{\mathbb{B}_k} \widetilde{N}(V)^{\otimes k} \\
&\quad = \widetilde{M}(\widetilde{N}(V)),
\end{aligned}$$

where the third equality is due to the fact that the left action of $\mathbb{B}_k$ on $\mathrm{Ind}_{\mathbb{B}_{i_1}\times\cdots\times\mathbb{B}_{i_k}}^{\mathbb{B}_n} N(i_1)\otimes\cdots\otimes N(i_k)$ commutes with the right one of $\mathbb{B}_n$, and the fifth equality is based on the representation $\widetilde{N}(V)^{\otimes k}$ of $\mathbb{B}_k$ via the Yang-Baxter operator σ_N of $\widetilde{N}(V)$. $\square$

Given any $(V,\sigma) \in\mathbf{BV}$, we have

$$\mathcal{P}(V) := \bigoplus_{n\geq 0} \mathcal{P}(n) \otimes_{\mathbb{B}_n} V^{\otimes n},$$

with its Yang-Baxter operator $\sigma_{\mathcal{P}}$ defined as in Eq. (2.7). If $\mathcal{P}$ is **regular**, then $\mathcal{P}(n) = \mathcal{P}_n \otimes \mathbf{k}[\mathbb{B}_n]$, $n \geq 0$, where $\bigoplus_{n\geq 0}\mathcal{P}_n$ is a graded vector space. Hence, $\mathcal{P}(V) = \bigoplus_{n\geq 0}\mathcal{P}_n \otimes V^{\otimes n}$.

DEFINITION 2.6. Given any (braided) operad $\mathcal{P}$, a (braided) vector space A is called an **algebra over** $\mathcal{P}$ or $\mathcal{P}$**-algebra**, if it is equipped with a homomorphism $\theta_A : \mathcal{P}(A) \to A$ of (braided) vector spaces such that the following diagrams commute:

$$\begin{array}{ccccc}
(\mathcal{P}\circ\mathcal{P})(A) & \xrightarrow{=} & \mathcal{P}(\mathcal{P}(A)) & \xrightarrow{\mathcal{P}(\theta_A)} & \mathcal{P}(A) \\
\downarrow{\scriptstyle \gamma_A} & & & & \downarrow{\scriptstyle \theta_A} \\
\mathcal{P}(A) & & \xrightarrow{\theta_A} & & A
\end{array}, \qquad
\begin{array}{ccc}
A & \xrightarrow{\eta_A} & \mathcal{P}(A) \\
 & {\scriptstyle =}\searrow & \downarrow{\scriptstyle \theta_A} \\
 & & A
\end{array}.$$

For two $\mathcal{P}$-algebras A and B, a homomorphism $\varphi : A \to B$ of (braided) vector spaces is called a **morphism of** $\mathcal{P}$**-algebras** if $\varphi\theta_A = \theta_B\mathcal{P}(\varphi)$.

In particular, the **free** $\mathcal{P}$**-algebra** generated by a (braided) vector space V is given by $(\mathcal{P}(V), \gamma_V)$ with $\eta_V : V \to \mathcal{P}(V)$, such that for any $\mathcal{P}$-algebra A and homomorphism $f : V \to A$ of (braided) vector spaces, there exists a unique morphism $\tilde{f} : \mathcal{P}(V) \to A$ of $\mathcal{P}$-algebras satisfying $f = \tilde{f}\eta_V$.

EXAMPLE 2.7. A dendriform algebra D is a vector space with two binary operators $\prec, \succ$ such that

$$\begin{aligned}(a \prec b) \prec c &= a \prec (b \prec c + b \succ c),\\ (a \succ b) \prec c &= a \succ (b \prec c),\\ a \succ (b \succ c) &= (a \prec b + a \succ b) \succ c,\end{aligned}$$

for any $a, b, c \in D$. Intrinsically, any dendriform algebra is an algebra over the regular operad $(\mathcal{P}_{\mathcal{D}}, \gamma, I)$ generated by

$$\mathcal{P}_{\mathcal{D},1} = \mathbf{k}I,\ \mathcal{P}_{\mathcal{D},2} = \mathbf{k} \prec \oplus \mathbf{k} \succ,$$

with relations

$$\begin{aligned}&\gamma(\prec; \prec, I) = \gamma(\prec; I, \prec + \succ),\\ &\gamma(\prec; \succ, I) = \gamma(\succ; I, \prec),\\ &\gamma(\succ; I, \succ) = \gamma(\succ; \prec + \succ, I),\\ &\gamma(\prec; I, I) = \prec,\ \gamma(\succ; I, I) = \succ,\\ &\gamma(I; \prec) = \prec,\ \gamma(I; \succ) = \succ .\end{aligned}$$

As a braided analogue, we interpret the braided dendriform algebras in [**8**] as the braided $\mathcal{P}_{\mathcal{D}}$-algebras. A **braided dendriform algebra** $(D, \prec, \succ, \sigma)$ is a braided vector space (D, σ) endowed with the dendriform algebra structure $(\prec, \succ)$ such that

$$\sigma(\mathrm{Id}_D \otimes \prec) = (\prec \otimes \mathrm{Id}_D)\sigma_2\sigma_1, \quad \sigma(\prec \otimes \mathrm{Id}_D) = (\mathrm{Id}_D \otimes \prec)\sigma_1\sigma_2, \tag{2.8}$$

$$\sigma(\mathrm{Id}_D \otimes \succ) = (\succ \otimes \mathrm{Id}_D)\sigma_2\sigma_1, \quad \sigma(\succ \otimes \mathrm{Id}_D) = (\mathrm{Id}_D \otimes \succ)\sigma_1\sigma_2. \tag{2.9}$$

Since the map $\theta_D : \mathcal{P}_{\mathcal{D}}(D) \to D$ is a homomorphism of braided vector spaces, we clearly have

$$(\theta_D \otimes \theta_D)\sigma_{\mathcal{P}_{\mathcal{D}}} = \sigma(\theta_D \otimes \theta_D).$$

Then for any $(I, a) \in \mathcal{P}_{\mathcal{D},1} \otimes D$ and $(\prec, b, c) \in \mathcal{P}_{\mathcal{D},2} \otimes D^{\otimes 2}$, we have

$$(\theta_D \otimes \theta_D)\sigma_{\mathcal{P}_{\mathcal{D}}}((I, a) \otimes (\prec, b, c)) = (\theta_D \otimes \theta_D)\left(\sum (\prec, b', c') \otimes (I, a')\right) = \sum (b' \prec c') \otimes a',$$

where we set $T^{\sigma}_{\beta_{12}}(a \otimes b \otimes c) = \sigma_2\sigma_1(a \otimes b \otimes c) = \sum b' \otimes c' \otimes a'$. On the other hand,

$$\sigma(\theta_D \otimes \theta_D)((I, a) \otimes (\prec, b, c)) = \sigma(a \otimes (b \prec c)).$$

Hence,

$$\sigma(a \otimes (b \prec c)) = \sum (b' \prec c') \otimes a',$$

that is, $\sigma(\mathrm{Id}_D \otimes \prec) = (\prec \otimes \mathrm{Id}_D)\sigma_2\sigma_1$. The other conditions in Eq. (2.8), (2.9) can be verified similarly.

In particular, the free $\mathcal{P}_{\mathcal{D}}$-algebra over a braided vector space (V, σ) is the free braided dendriform algebra, realized as the braided analogue of the Loday-Ronco algebra of planar binary rooted trees; see [**9**, Theorem 2.8].

EXAMPLE 2.8. A tridendriform algebra T is a vector space with three binary operators $\prec, \succ, *$ such that

$$\begin{aligned}
&(a \prec b) \prec c = a \prec (b \prec c + b \succ c + b * c),\\
&(a \succ b) \prec c = a \succ (b \prec c),\\
&a \succ (b \succ c) = (a \prec b + a \succ b + a * b) \succ c,\\
&(a * b) \prec c = a * (b \prec c),\\
&(a \prec b) * c = a * (b \succ c),\\
&(a \succ b) * c = a \succ (b * c),\\
&(a * b) * c = a * (b * c),
\end{aligned}$$

for $a, b, c \in T$. Intrinsically, any tridendriform algebra is an algebra over the regular operad $(\mathcal{P}_{\mathcal{T}}, \gamma, I)$ generated by

$$\mathcal{P}_{\mathcal{T},1} = \mathbf{k}I,\ \mathcal{P}_{\mathcal{T},2} = \mathbf{k} \prec \oplus \mathbf{k} \succ \oplus \mathbf{k} *,$$

with relations

$$\begin{aligned}
&\gamma(\prec; \prec, I) = \gamma(\prec; I, \prec + \succ + *),\\
&\gamma(\prec; \succ, I) = \gamma(\succ; I, \prec),\\
&\gamma(\succ; I, \succ) = \gamma(\succ; \prec + \succ + *, I),\\
&\gamma(\prec; *, I) = \gamma(*; I, \prec),\\
&\gamma(*; \prec, I) = \gamma(*; I, \succ),\\
&\gamma(*; \succ, I) = \gamma(\succ; I, *),\\
&\gamma(*; *, I) = \gamma(*; I, *),\\
&\gamma(\prec; I, I) = \prec,\ \gamma(\succ; I, I) = \succ,\ \gamma(*; I, I) = *,\\
&\gamma(I; \prec) = \prec,\ \gamma(I; \succ) = \succ,\ \gamma(I; *) = *.
\end{aligned}$$

A **braided tridendriform algebra** [**9**], denoted $(T, \prec, \succ, *, \sigma)$ is a braided vector space (T, σ) endowed with a tridendriform algebra structure $(\prec, \succ, *)$ such that

$$\sigma(\mathrm{Id}_T \otimes \prec) = (\prec \otimes \mathrm{Id}_T)\sigma_2\sigma_1, \quad \sigma(\prec \otimes \mathrm{Id}_T) = (\mathrm{Id}_T \otimes \prec)\sigma_1\sigma_2, \tag{2.10}$$

$$\sigma(\mathrm{Id}_T \otimes \succ) = (\succ \otimes \mathrm{Id}_T)\sigma_2\sigma_1, \quad \sigma(\succ \otimes \mathrm{Id}_T) = (\mathrm{Id}_T \otimes \succ)\sigma_1\sigma_2, \tag{2.11}$$

$$\sigma(\mathrm{Id}_T \otimes *) = (* \otimes \mathrm{Id}_T)\sigma_2\sigma_1, \quad \sigma(* \otimes \mathrm{Id}_T) = (\mathrm{Id}_T \otimes *)\sigma_1\sigma_2. \tag{2.12}$$

Then the braided tridendriform algebras are exactly the braided $\mathcal{P}_{\mathcal{T}}$-algebras, as the conditions in (2.10)–(2.12) can be recovered by the fact that the map $\theta_T : \mathcal{P}_{\mathcal{T}}(T) \to T$ is a homomorphism of braided vector spaces. In particular, the free $\mathcal{P}_{\mathcal{T}}$-algebra over a braided vector space (V, σ) is the free braided tridendriform algebra, constructed as the braided analogue of the Loday-Ronco algebra of planar rooted trees; see [**9**, Theorem 4.5].

3. Coherent unit actions and braided $\mathcal{P}$-Hopf algebras

Let $\mathcal{P}$ be any (braided) operad such that $\mathcal{P}(0) = 0$ and $\mathcal{P}(1) = \mathbf{k} = \mathbf{k}I$. Let $\mathcal{I}$ be a 0-ary element adjoined to $\mathcal{P}$ by

$$\mathcal{P}'(i) := \begin{cases} \mathcal{P}(i), & i \geq 1, \\ \mathbf{k}\mathcal{I}, & i = 0. \end{cases}$$

A **unit action** on the operad $(\mathcal{P}, \gamma, I)$ is a partial extension of the composition map γ on $\mathcal{P}'$ with

$$\gamma(i_1, \dots, i_k) : \mathcal{P}'(k) \otimes \mathcal{P}'(i_1) \otimes \cdots \otimes \mathcal{P}'(i_k) \to \mathcal{P}'(i_1 + \cdots + i_k)$$

defined for all $i_j \geq 0$ with $j = 1, \dots, k$ and $i_1 + \cdots + i_k > 0$ if $k \geq 2$. Note that $\gamma(i_1, \cdots, i_k)$ is not defined when $i_1 = \cdots = i_k = 0$ and $k \geq 2$.

In particular, for $\gamma(1,0) : \mathcal{P}_2 \otimes \mathbf{k}I \otimes \mathbf{k}\mathcal{I} \to \mathbf{k}I$ and $\gamma(0,1) : \mathcal{P}_2 \otimes \mathbf{k}\mathcal{I} \otimes \mathbf{k}I \to \mathbf{k}I$ in the regular case, there exist linear maps $\alpha, \beta : \mathcal{P}_2 \to \mathbf{k}$ such that

$$\gamma(\mu; I, \mathcal{I}) = \alpha(\mu)I,\ \gamma(\mu; \mathcal{I}, I) = \beta(\mu)I \quad \text{for any } \mu \in \mathcal{P}_2. \tag{3.1}$$

Given a $\mathcal{P}$-algebra A, we define $A^+ := \mathbf{k} \oplus A$, called a **unitary** $\mathcal{P}$-algebra, with structure map $\theta_{A^+} : \mathcal{P}'(A^+) \to A^+$ extending θ_A by sending $\mathcal{I} \in \mathcal{P}'(0)$ to $1 \in A^+$.

DEFINITION 3.1. For any (braided) regular operad $(\mathcal{P}, \gamma, I)$ generated by operation sets $M_k \subseteq \mathcal{P}_k$, $k \geq 2$, with a unit action, if there are operations $\star_n \in \mathcal{P}_n$ for all $n \geq 0$ satisfying

$$\gamma(\star_n; I, \dots, \overset{i\text{ th}}{\mathcal{I}}, \dots, I) = \star_{n-1},\ 1 \leq i \leq n, \tag{3.2}$$

$$\gamma(\star_2; \mathcal{I}, I) = \gamma(\star_2; I, \mathcal{I}) = I, \tag{3.3}$$

then we say that the unit action respects the operations $\star_n$. Particularly $\star_0 = \mathcal{I}$, $\star_1 = I$ and $\alpha(\star_2) = \beta(\star_2) = 1$. Then one can further extend γ by requiring

$$\gamma(\star_n; \overbrace{\mathcal{I}, \dots, \mathcal{I}}^{n \text{ times}}) = \mathcal{I}.$$

In this case, such a unit action is called **coherent**, if for any two $\mathcal{P}$-algebras A and B,

$$A \boxtimes B := (A \otimes \mathbf{k}) \oplus (\mathbf{k} \otimes B) \oplus (A \otimes B)$$

is again a $\mathcal{P}$-algebra defined as follows. Let c be the braiding in the braided tensor category $\mathcal{C}$; see [**15**, Ch. XIII]. We let

$$\sigma := \sigma_{A,B} := c_{A^+ \oplus B^+, A^+ \oplus B^+}$$

denote the corresponding braiding on $A^+ \oplus B^+$ when $A, B \in \mathcal{C}$. For any $p \in M_n$, $a_i \in A^+$ and $b_i \in B^+$, $1 \leq i \leq n$ with $n \geq 2$, let

(3.4)

$$p(a_1 \otimes b_1, \dots, a_n \otimes b_n)$$

$$:= \begin{cases} \sum \star_n(a'_1, \dots, a'_n) \otimes p(b'_1, \dots, b'_n), & \text{if at least one } b_i \in B, \\ p(a_1, \dots, a_n) \otimes 1, & \text{if all } b_i = 1,\ p(a_1, \dots, a_n) \text{ is defined}, \\ \text{undefined}, & \text{otherwise}, \end{cases}$$

where we denote $T^{\sigma}_{w_n}(a_1 \otimes b_1 \underline{\otimes} \cdots \underline{\otimes} a_n \otimes b_n) = \sum a'_1 \otimes \cdots \otimes a'_n \otimes b'_1 \otimes \cdots \otimes b'_n$ with $w_n \in \mathbb{S}_{2n}$ such that $w_n(2i-1) = i$, $w_n(2i) = n + i$ for $i = 1, \dots, n$. Then Eq. (3.4) actually defines the structure map $\theta_{A \boxtimes B}$ of $A \boxtimes B$ as a $\mathcal{P}$-algebra.

LEMMA 3.2. *For a unit action on a braided regular operad $(\mathcal{P}, \gamma, I)$, let $\star_0 := \mathcal{I}$, $\star_1 := I$, and $\star_2 \in \mathcal{P}_2$ satisfying Eq. (3.3). If $\star_2$ is associative, i.e. $\gamma(\star_2; \star_2, I) = \gamma(\star_2; I, \star_2)$, then one can recursively define a sequence of operations $\star_n$, $n \geq 0$, by*

$$\star_n := \gamma(\star_2; \star_{n-1}, I) \in \mathcal{P}_n,$$

satisfying Eq. (3.2)*. Moreover,*

$$\gamma(\star_2; \star_i, \star_{n-i}) = \star_n, \tag{3.5}$$

for all $i = 0, 1, \dots, n$ *with* $n \geq 0$*. Then we call an operad* $(\mathcal{P}, \gamma, I)$ **associative**, *if it has such an associative operation* $\star := \star_2 \in \mathcal{P}_2$.

PROOF. First we inductively prove that $\star_n$, $n \geq 0$, satisfy Eq. (3.2). It is clear when $n = 1, 2$. If $n > 2$, then by the definition of $\star_n$ and the induction hypothesis we have

$$\begin{aligned}
\gamma(\star_n; I, \dots, \overset{i\text{ th}}{\mathcal{I}}, \dots, I) &= \gamma(\gamma(\star_2; \star_{n-1}, I); I, \dots, \overset{i\text{ th}}{\mathcal{I}}, \dots, I) \\
&= \begin{cases} \gamma(\star_2; \gamma(\star_{n-1}; I, \dots, \mathcal{I}, \dots, I), I), & 1 \leq i < n, \\ \gamma(\star_2; \gamma(\star_{n-1}; I, \dots, I), \mathcal{I}), & i = n, \end{cases} \\
&= \begin{cases} \gamma(\star_2; \star_{n-2}, I), & 1 \leq i < n, \\ \gamma(\star_2; \star_{n-1}, \mathcal{I}), & i = n, \end{cases} \\
&= \begin{cases} \gamma(\star_2; \star_{n-2}, I), & 1 \leq i < n, \\ \gamma(\gamma(\star_2; I, \mathcal{I}); \star_{n-1}), & i = n, \end{cases} \\
&= \star_{n-1}.
\end{aligned}$$

Further, Eq. (3.5) can also be proved by induction on n. Indeed, it is obviously true when $n = 0, 1, 2$. If $n > 2$, then

$$\begin{aligned}
\gamma(\star_2; \star_i, \star_{n-i}) &= \gamma(\star_2; \gamma(\star_2; \star_{i-1}, I), \star_{n-i}) \\
&= \gamma(\gamma(\star_2; \star_2, I); \star_{i-1}, I, \star_{n-i}) \\
&= \gamma(\gamma(\star_2; I, \star_2); \star_{i-1}, I, \star_{n-i}) \\
&= \gamma(\star_2; \star_{i-1}, \gamma(\star_2; I, \star_{n-i})) \\
&= \gamma(\star_2; \star_{i-1}, \star_{n+1-i}),
\end{aligned}$$

for any $i = 2, \dots, n-1$. It then follows that

$$\star_n = \gamma(\star_2; \star_n, \mathcal{I}) = \gamma(\star_2; \star_{n-1}, I) = \cdots = \gamma(\star_2; \mathcal{I}, \star_n).$$

Hence, Eq. (3.5) holds for any $n \geq 2$. □

EXAMPLE 3.3. Let $\mathcal{P}_\mathcal{D}$ be the braided dendriform operad given in Example 2.7. There is a unit action on $\mathcal{P}_\mathcal{D}$ defined as follows. First extend $\mathcal{P}_\mathcal{D}$ to be $\mathcal{P}'_\mathcal{D}$ such that

$$\mathcal{P}'_\mathcal{D}(0) = \mathbf{k}\mathcal{I},\ \mathcal{P}'_{\mathcal{D},1} = \mathbf{k}I,\ \mathcal{P}'_{\mathcal{D},2} = \mathbf{k} \prec \oplus\, \mathbf{k} \succ,$$

and then extend γ to $\mathcal{P}'_\mathcal{D}$ such that

$$\begin{aligned}
&\gamma(\prec; I, \mathcal{I}) = I,\ \gamma(\prec; \mathcal{I}, I) = 0, \\
&\gamma(\succ; I, \mathcal{I}) = 0,\ \gamma(\succ; \mathcal{I}, I) = I.
\end{aligned}$$

Let $\star_0 := \mathcal{I}$, $\star_1 := I$, $\star_2 := \prec + \succ$ and $\star_n := \gamma(\star_2; \star_{n-1}, I)$ inductively for any $n \geq 3$. Then $\star_2$ clearly satisfies Eq. (3.3). By the relations of $\mathcal{P}'_\mathcal{D}$, we know that $\star_2$ is also associative. According to Lemma 3.2, the sequence of operations $\{\star_n\}_{n \geq 0}$ satisfies Eqs. (3.2) and (3.5).

Now for the braided dendriform operad case, we further have the following property.

PROPOSITION 3.4. *The stated unit action on $\mathcal{P}_{\mathcal{D}}$ generated by $M_2 = \{\prec, \succ\}$ is coherent.*

PROOF. It is enough to check that the operations $\prec$ and $\succ$ on $A \boxtimes B$ defined by Eq. (3.4) satisfy the dendriform relations and the compatibility conditions (2.8), (2.9) for any $\mathcal{P}_{\mathcal{D}}$-algebras A, B. The proof is the same as the case of the algebraic operads treated in [**9**, Proposition 4.9] to which we refer the reader for details. □

Let $\mathcal{P}$ be a (braided) regular operad equipped with a coherent unit action and $A^+ := \mathbf{k} \oplus A$ be a unitary $\mathcal{P}$-algebra. Do the same for $A^+ \otimes A^+ \cong (A \boxtimes A)^+$ via Eq. (3.4). Let $\Delta : A^+ \to (A \boxtimes A)^+ \cong A^+ \otimes A^+$ be a coassociative linear map, such that $\Delta(1) = 1 \otimes 1$ and $\Delta'(a) := \Delta(a) - a \otimes 1 - 1 \otimes a \in A \otimes A$ for any $a \in A$.

If Δ is a morphism of unitary $\mathcal{P}$-algebras, we call A^+ together with Δ a **$\mathcal{P}$-bialgebra**. Moreover, if there is a direct sum $A^+ = \bigoplus_{n \geq 0} A^{(n)}$ of vector spaces, such that $A^{(0)} = \mathbf{k}$ and

$$\Delta(A^{(n)}) \subseteq \sum_{i=0}^{n} A^{(i)} \otimes A^{(n-i)}$$

for all $n \geq 0$, we call A^+ a **connected graded $\mathcal{P}$-Hopf algebra**.

As a braided analogue of [**11**, Lemma 6], we provide the following theorem.

THEOREM 3.5. *Let $\mathcal{P}$ be a braided regular operad equipped with a coherent unit action and let $A^+ := \mathbf{k} \oplus A$ with $A := \mathcal{P}(V)$ be the free unitary $\mathcal{P}$-algebra generated by a braided vector space V. Then there is a coassociative morphism $\Delta_A : A^+ \to A^+ \otimes A^+$ of $\mathcal{P}$-algebras defined by*

$$\Delta_A(x) = x \otimes 1 + 1 \otimes x,\ x \in V,$$

which provides A^+ with a connected graded $\mathcal{P}$-bialgebra structure. It further provides A^+ with a connected braided bialgebra structure and hence a braided Hopf algebra structure.

Furthermore, if the Yang-Baxter operator σ_A on A^+ becomes a $\mathcal{P}$-algebra isomorphism, then Δ_A is twisted cocommutative, that is, $\sigma_A \Delta_A = \Delta_A$.

PROOF. Since $\mathcal{P}(V)$ is the free $\mathcal{P}$-algebra on V and $\mathcal{P}(V)^{\boxtimes n}, n = 2, 3$, are $\mathcal{P}$-algebras by the coherent unit action, there is a unique $\mathcal{P}$-algebra morphism

$$\Delta_A : \mathcal{P}(V) \to \mathcal{P}(V)^{\boxtimes 2}, v \mapsto 1 \otimes v + v \otimes 1, v \in V,$$

identifying V with $\mathcal{P}_1 \otimes V$. It then induces a braided (associative) algebra homomorphism $\Delta : (\mathcal{P}(V), \star) \to (\mathcal{P}(V)^{\boxtimes 2}, \star)$. By the same argument, we obtain the compatibility requirement

$$\Delta_A \theta_A = \theta_{A \boxtimes A} \mathcal{P}(\Delta_A),$$

where $\theta_A = \gamma_V$ and $\theta_{A \boxtimes A}$ are defined in Eq. (3.4). Then it also guarantees the commutativity between Δ_A and the braidings,

$$(\Delta_A \otimes \Delta_A)\sigma_A = \beta_{22}(\Delta_A \otimes \Delta_A).$$

On the other hand, since

$$(\Delta_A \otimes \mathrm{id}_{A^+})\Delta_A(x) = x \otimes 1 \otimes 1 + 1 \otimes x \otimes 1 + 1 \otimes 1 \otimes x = (\mathrm{id}_{A^+} \otimes \Delta_A)\Delta_A(x)$$

for any $x \in V$, by the universal property of $\mathcal{P}(V)$ again, Δ_A as a morphism of $\mathcal{P}$-algebras is coassociative. Also, $\Delta_A(1) = 1 \otimes 1$ and $\Delta'_A(a) := \Delta_A(a) - a \otimes 1 - 1 \otimes a \in A \otimes A$ for any $a \in A$. Furthermore, the free unitary $\mathcal{P}$-algebra A^+ allows a grading $A^{(n)} := \mathcal{P}_n \otimes V^{\otimes n}$ which is respected by all $\mathcal{P}$-operations. The same holds for $A^+ \otimes A^+$. Thus Δ_A makes A^+ into a connected graded $\mathcal{P}$-bialgebra. Thus $(A^+, \star, \Delta_A)$ is a connected graded braided bialgebra and hence a braided Hopf algebra; see [**16, 24**].

Finally, the stated cocommutativity of Δ_A follows from its definition. □

EXAMPLE 3.6. For the braided dendriform operad, by Proposition 3.4 and Theorem 3.5, we know that the free unitary $\mathcal{P}_{\mathcal{D}}$-algebra over a braided vector space is a connected graded $\mathcal{P}_{\mathcal{D}}$-Hopf algebra. In particular, we recover the braided Hopf algebra structures of free braided dendriform algebras, namely, the braided analogue of the Loday-Ronco Hopf algebra of planar binary rooted trees, directly verified in [**9**].

PROPOSITION 3.7. *The conclusion of Theorem* 3.5 *remains true for any braided vector space V with involutive braiding σ and braided operad $(\mathcal{P}, \gamma, I)$ with a unit action, provided the associativity condition stated in Lemma* 3.2 *is replaced by a* **completely commutativity condition** *for the braided operad $(\mathcal{P}, \gamma, I)$, in the sense that there are operations $\star_n \in \mathcal{P}(n)$, $n \geq 0$, invariant under the right $\mathbb{B}_n$-module action and satisfying Eqs.* (3.2) *and* (3.3).

PROOF. Under this condition, if the braided tensor category $\mathcal{C}$ is symmetric, then $\mathcal{P}$-operations on $A \boxtimes B$ are still well-defined by Eq. (3.4), thus again endow $A \boxtimes B$ with a $\mathcal{P}$-algebra structure given by the same $\theta_{A\boxtimes B}$ from $\mathcal{P}(A \boxtimes B) = \bigoplus_{n\geq 0} \mathcal{P}(n) \otimes_{\mathbb{B}_n} (A \boxtimes B)^{\otimes n}$ to $A \boxtimes B$ in Eq. (3.4). In short, there is again a coherent unit action on $\mathcal{P}$. Then Proposition 3.7 follows from the same argument. □

EXAMPLE 3.8. Typical examples of completely commutative (braided) operads with coherent unit actions are the (braided) commutative dendriform (identified with Zinbiel) operad and the braided commutative tridendriform (CTD) operad. Thus Proposition 3.7 in particular implies the braided Hopf algebra structures on free braided commutative (tri)dendriform algebras. Free braided commutative dendriform algebras are shown [**8**] to be isomorphic to the quantum shuffle algebras defined in [**23**] for involutive braidings. Thus the coherent unit action approach of braided Hopf algebras in Proposition 3.7 puts the braided Hopf algebra structure on quantum shuffle algebras in a broader context. A similar approach can be given to the quantum quasi-shuffle algebra considered in [**14**].

4. Criteria for braided binary quadratic regular operads with coherent unit actions

In [**3**] the authors provided defining equations and criteria for all the associative, binary, quadratic and regular (ABQR) operads with coherent unit actions. In this section we extend these criteria to the braided context.

We first characterize braided ABQR operad with coherent unit actions by a set of equations.

THEOREM 4.1. *Let $(\mathcal{P}, \gamma, I, \star)$ be a braided ABQR operad. A unit action on $\mathcal{P}$ is coherent if and only if, for every $(\sum_i \odot_i^{(1)} \otimes \odot_i^{(2)}, \sum_j \odot_j^{(3)} \otimes \odot_j^{(4)}) \in \mathcal{P}_2^{\otimes 2} \oplus \mathcal{P}_2^{\otimes 2}$ from the quadratic relations of $\mathcal{P}$, the following* **coherence equations** *hold in terms of the linear maps α, β defined in Eq.* (3.1).

(C1) $\sum_i \beta(\odot_i^{(1)})\odot_i^{(2)} = \sum_j \beta(\odot_j^{(3)})\odot_j^{(4)}$,

(C2) $\sum_i \alpha(\odot_i^{(1)})\odot_i^{(2)} = \sum_j \beta(\odot_j^{(4)})\odot_j^{(3)}$,
(C3) $\sum_i \alpha(\odot_i^{(2)})\odot_i^{(1)} = \sum_j \alpha(\odot_j^{(4)})\odot_j^{(3)}$,
(C4) $\sum_i \beta(\odot_i^{(2)})\odot_i^{(1)} = \sum_j \beta(\odot_j^{(3)})\beta(\odot_j^{(4)})\star$,
(C5) $\sum_i \alpha(\odot_i^{(1)})\alpha(\odot_i^{(2)})\star = \sum_j \alpha(\odot_j^{(3)})\,\odot_j^{(4)}$.

PROOF. A braided ABQR operad $\mathcal{P}$ is generated by $\mathcal{P}_1 = \mathbf{k}I$, $\mathcal{P}_2 = \mathbf{k}\Omega$ with relations

$$\sum_i \gamma(\odot_i^{(1)};\odot_i^{(2)},I) = \sum_j \gamma(\odot_j^{(3)};I,\odot_j^{(4)}),$$

where $(\sum_i \odot_i^{(1)}\otimes\odot_i^{(2)}, \sum_j \odot_j^{(3)}\otimes\odot_j^{(4)}) \in \mathcal{P}_2^{\otimes 2}\oplus\mathcal{P}_2^{\otimes 2}$.

A unit action on $\mathcal{P}$ is coherent if and only if, for any $\mathcal{P}$-algebra A and B and for any $a, a', a'' \in A^+$ and $b, b', b'' \in B^+$, we have the equation

$$\sum_i \Big((a\otimes b)\odot_i^{(1)}(a'\otimes b')\Big)\odot_i^{(2)}(a''\otimes b'') = \sum_j (a\otimes b)\odot_j^{(3)}\Big((a'\otimes b')\odot_j^{(4)}(a''\otimes b'')\Big). \tag{4.1}$$

Then there are eight mutually disjoint cases for eight subsets of b, b', b'' which are in $\mathbf{k}$ (and hence not in B). Note that when all of b, b', b'' are in $\mathbf{k}$ or all of b, b', b'' are in B, Eq. (4.1) just means that $(\sum_i \odot_i^{(1)}\otimes\odot_i^{(2)}, \sum_j \odot_j^{(3)}\otimes\odot_j^{(4)})$ is a relation for $\mathcal{P}$, so is automatically true. Thus to prove the theorem we only need to prove

Case 1. Eq. (4.1) holds for $b\in\mathbf{k}$, $b', b''\in B$ if and only if **(C1)** is true;
Case 2. Eq. (4.1) holds for $b'\in\mathbf{k}$, $b, b''\in B$ if and only if **(C2)** is true;
Case 3. Eq. (4.1) holds for $b''\in\mathbf{k}$, $b, b'\in B$ if and only if **(C3)** is true;
Case 4. Eq. (4.1) holds for $b, b'\in\mathbf{k}$, $b''\in B$ if and only if **(C4)** is true;
Case 5. Eq. (4.1) holds for $b', b''\in\mathbf{k}$, $b\in B$ if and only if **(C5)** is true;
Case 6. Eq. (4.1) holds for $b, b''\in\mathbf{k}$, $b'\in B$ if **(C1)** is true.

All these cases are routinely checked as in the original paper [**3**] using the binary operations on $A\boxtimes B$ defined in Eq. (3.4). □

We now apply Theorem 4.1 to give a simple criterion for all braided ABQR operads with coherent unit actions.

THEOREM 4.2. *Let $(\mathcal{P},\gamma,I,\star)$ be a braided ABQR operad of dimension n (that is, $\dim\mathcal{P}_2 = n$).*

(1) *There is a coherent unit action (α,β) on $\mathcal{P}$ with $\alpha\neq\beta$ if and only if there is a basis $\{\odot_i\}$ of $\mathcal{P}_2$ with $\star = \sum_i\odot_i$ such that the space Λ of quadratic relations for $\mathcal{P}$ are contained in the subspace $\Lambda'_{n,\,\mathrm{coh}}$ of $\mathcal{P}_2^{\otimes 2}\oplus\mathcal{P}_2^{\otimes 2}$ with basis*

$$\lambda'_{n,\,\mathrm{coh}} := \left\{\begin{array}{l} (\star\otimes\odot_2,\odot_2\otimes\odot_2),\\ (\odot_1\otimes\odot_1,\odot_1\otimes\star),\\ (\odot_i\otimes\odot_1,\odot_i\otimes\odot_1),\ 2\le i\le n,\\ (\odot_2\otimes\odot_j,\odot_2\otimes\odot_j),\ 3\le j\le n,\\ (\odot_1\otimes\odot_i,\odot_i\otimes\odot_2),\ 3\le i\le n,\\ (\odot_i\otimes\odot_j,0),\ (0,\odot_i\otimes\odot_j),\ 3\le i,j\le n. \end{array}\right\} \tag{4.2}$$

(2) *There is a coherent unit action (α,β) on $\mathcal{P}$ with $\alpha=\beta$ if and only if there is a basis $\{\odot_i\}$ of $\mathcal{P}_2$ with $\star = \sum_i\odot_i$ such that the space Λ of quadratic*

relations for $\mathcal{P}$ are contained in the subspace $\Lambda''_{n,\mathrm{coh}}$ of $\mathcal{P}_2^{\otimes 2} \oplus \mathcal{P}_2^{\otimes 2}$ with basis

$$\lambda''_{n,\mathrm{coh}} := \left\{ \begin{array}{l} (\odot_1 \otimes \star, \odot_1 \otimes \star) + (\star \otimes \odot_1, \star \otimes \odot_1) - (\odot_1 \otimes \odot_1, \odot_1 \otimes \odot_1), \\ (\odot_i \otimes \odot_j, 0), \ (0, \odot_i \otimes \odot_j), \ 2 \leq i, j \leq n. \end{array} \right\}. \tag{4.3}$$

PROOF. The proof is the same as the one for the algebraic operads in [**3**, Theorem 4.9], so we refer the reader there for details. □

By inspection, the braided dendriform algebra (see Proposition 3.4) and braided tridendriform algebras have coherent unit actions.

Acknowledgments

We thank the organizers for hospitality during the International Workshop on Hopf Algebras and Tensor Categories held in Nanjing University, September 2019, and thank the anonymous referees for helpful suggestions.

References

[1] Marcelo Aguiar and Jean-Louis Loday, *Quadri-algebras*, J. Pure Appl. Algebra **191** (2004), no. 3, 205–221, DOI 10.1016/j.jpaa.2004.01.002. MR2059613

[2] Alain Connes and Dirk Kreimer, *Hopf algebras, renormalization and noncommutative geometry*, Comm. Math. Phys. **199** (1998), no. 1, 203–242, DOI 10.1007/s002200050499. MR1660199

[3] Kurusch Ebrahimi-Fard and Li Guo, *Coherent unit actions on regular operads and Hopf algebras*, Theory Appl. Categ. **18** (2007), No. 13, 348–371. MR2342165

[4] Zbigniew Fiedorowicz, *The symmetric bar construction*, Preprint, `http://www.math.ohio-state.edu/~fiedorow/`, 1992.

[5] Loïc Foissy, *Les algèbres de Hopf des arbres enracinés décorés. I* (French, with English and French summaries), Bull. Sci. Math. **126** (2002), no. 3, 193–239, DOI 10.1016/S0007-4497(02)01108-9. MR1905177

[6] Loïc Foissy, *Les algèbres de Hopf des arbres enracinés décorés. II* (French, with English and French summaries), Bull. Sci. Math. **126** (2002), no. 4, 249–288, DOI 10.1016/S0007-4497(02)01113-2. MR1909461

[7] Benoit Fresse, *Homotopy of operads and Grothendieck-Teichmüller groups. Part 2*, Mathematical Surveys and Monographs, vol. 217, American Mathematical Society, Providence, RI, 2017. The applications of (rational) homotopy theory methods, DOI 10.1090/surv/217.2. MR3616816

[8] Li Guo and Yunnan Li, *Braided Rota-Baxter algebras, quantum quasi-shuffle algebras and braided dendriform algebras*, `arXiv:1901.02843`, published online at J. Algebra and Its Applications, `https://doi.org/10.1142/S0219498822501341`.

[9] Li Guo and Yunnan Li, *Braided dendriform and tridendriform algebras and braided Hopf algebras of planar trees*, J. Algebraic Combin. 53 (2021), no. 4, 1147–1185. MR4263646

[10] Ralf Holtkamp, *Comparison of Hopf algebras on trees*, Arch. Math. (Basel) **80** (2003), no. 4, 368–383, DOI 10.1007/s00013-003-0796-y. MR1982837

[11] Ralf Holtkamp, *On Hopf algebra structures over free operads*, Adv. Math. **207** (2006), no. 2, 544–565, DOI 10.1016/j.aim.2005.12.004. MR2271016

[12] Run-Qiang Jian, *Quantum quasi-shuffle algebras II*, J. Algebra **472** (2017), 480–506, DOI 10.1016/j.jalgebra.2016.10.024. MR3584887

[13] Run-Qiang Jian and Marc Rosso, *Braided cofree Hopf algebras and quantum multi-brace algebras*, J. Reine Angew. Math. **667** (2012), 193–220, DOI 10.1515/crelle.2011.126. MR2929677

[14] Run-Qiang Jian, Marc Rosso, and Jiao Zhang, *Quantum quasi-shuffle algebras*, Lett. Math. Phys. **92** (2010), no. 1, 1–16, DOI 10.1007/s11005-010-0382-8. MR2607915

[15] Christian Kassel, *Quantum groups*, Graduate Texts in Mathematics, vol. 155, Springer-Verlag, New York, 1995, DOI 10.1007/978-1-4612-0783-2. MR1321145

[16] Vladislav K. Kharchenko, *Connected braided Hopf algebras*, J. Algebra **307** (2007), no. 1, 24–48, DOI 10.1016/j.jalgebra.2006.02.037. MR2278041

[17] Jean-Louis Loday, *Dialgebras*, Dialgebras and related operads, Lecture Notes in Math., vol. 1763, Springer, Berlin, 2001, pp. 7–66, DOI 10.1007/3-540-45328-8_2. MR1860994

[18] Jean-Louis Loday, *Scindement d'associativité et algèbres de Hopf* (French, with English and French summaries), Actes des Journées Mathématiques à la Mémoire de Jean Leray, Sémin. Congr., vol. 9, Soc. Math. France, Paris, 2004, pp. 155–172. MR2145941

[19] Jean-Louis Loday, *On the algebra of quasi-shuffles*, Manuscripta Math. **123** (2007), no. 1, 79–93, DOI 10.1007/s00229-007-0086-2. MR2300061

[20] Jean-Louis Loday and María O. Ronco, *Hopf algebra of the planar binary trees*, Adv. Math. **139** (1998), no. 2, 293–309, DOI 10.1006/aima.1998.1759. MR1654173

[21] Jean-Louis Loday and María Ronco, *Trialgebras and families of polytopes*, Homotopy theory: relations with algebraic geometry, group cohomology, and algebraic K-theory, Contemp. Math., vol. 346, Amer. Math. Soc., Providence, RI, 2004, pp. 369–398, DOI 10.1090/conm/346/06296. MR2066507

[22] Jean-Louis Loday and Bruno Vallette, *Algebraic operads*, Grundlehren der Mathematischen Wissenschaften [Fundamental Principles of Mathematical Sciences], vol. 346, Springer, Heidelberg, 2012, DOI 10.1007/978-3-642-30362-3. MR2954392

[23] Marc Rosso, *Quantum groups and quantum shuffles*, Invent. Math. **133** (1998), no. 2, 399–416, DOI 10.1007/s002220050249. MR1632802

[24] Mitsuhiro Takeuchi, *Survey of braided Hopf algebras*, New trends in Hopf algebra theory (La Falda, 1999), Contemp. Math., vol. 267, Amer. Math. Soc., Providence, RI, 2000, pp. 301–323, DOI 10.1090/conm/267/04277. MR1800719

Department of Mathematics and Computer Science, Rutgers University, Newark, New Jersey 07102

Email address: liguo@rutgers.edu

School of Mathematics and Information Science, Guangzhou University, Waihuan Road West 230, Guangzhou 510006, People's Republic of China

Email address: ynli@gzhu.edu.cn

Contemporary Mathematics
Volume **771**, 2021
https://doi.org/10.1090/conm/771/15511

A novel quantum affine algebra of type $A_1^{(1)}$ and its PBW basis

Naihong Hu and Rushu Zhuang

ABSTRACT. In this paper, we construct a novel quantum affine algebra $\mathcal{U}_q(\widehat{\mathfrak{sl}_2})$ associated to the quantum structure constant matrix $\begin{pmatrix} -q^2 & q^{-2} \\ q^{-2} & -q^2 \end{pmatrix}$. We prove that it is neither isomorphic to the usual quantum affine algebra of type $A_1^{(1)}$ nor to its co-opposite as Hopf algebras. We also introduce a quantum Weyl group as a subgroup of the group of algebra automorphisms of $\mathcal{U}_q(\widehat{\mathfrak{sl}_2})$. Using this quantum Weyl group, we give an explicit description of a Poincaré-Birkhoff-Witt basis of $\mathcal{U}_q(\widehat{\mathfrak{sl}_2})$.

1. Introduction

To a symmetrizable generalized Cartan matrix $A = (a_{ij})_{1\le i,j\le n}$ (i.e., DA is symmetric, $D = \mathrm{diag}(d_1, \cdots, d_n)$ with co-prime positive integers d_i), one associates a quantized universal enveloping algebra $U_{\mathbf{q}}(\mathfrak{g}_A)$ of a Kac-Moody algebra $\mathfrak{g}_A$ [**16**]. Here $\mathbf{q} = (q_{ij})_{1\le i,j\le n}$ satisfies the Cartan condition: $q_{ij}q_{ji} = q_{ii}^{a_{ij}}$. This $\mathbf{q}$ is called the structure constant matrix or braiding matrix of $U_{\mathbf{q}}(\mathfrak{g}_A)$. When $\mathbf{q}$ is positive (i.e., all $q_{ij} > 0$), this, as a certain classification condition, appeared in several papers. For instance, Fronsdal [**7**] classified those generalizations of quantum groups equipped with universal R-matrices that are the so-called "coboundary" Hopf algebras. These $U_{\mathbf{q}}(\mathfrak{g}_A)$ occurred in general positions on algebraic surfaces (the Cartan condition as defining algebraic equations) in parameter space, cf. condition (1.8) in [**7**], also see [**8**], [**9**]. The positive $\mathbf{q}$ is just what Rosso obtained as a classification result if Gelfand-Kirillov dimension of $U_{\mathbf{q}}(\mathfrak{g}_A)^+$ over complex numbers $\mathbb{C}$ (the quantum symmetric or Nichols algebra) is assumed to be finite, see Theorem 21 [**18**]. Furthermore, under the assumption of positive braidings, Andruskiewitsch-Schneider gave a characterization to a class of pointed Hopf algebras (with a finitely generated abelian group algebra as the coradical) to be domains of finite GK dimensions. Later on, the Cartan-Rosso condition became a starting point of the classification work on finite-dimensional pointed Hopf algebras of diagonal type with finite datum of Cartan type, e.g. [**1**] and [**18**]. This is also used to define a

2020 *Mathematics Subject Classification.* Primary 16T05; Secondary 17B37.
Key words and phrases. Admissible quantum affine algebra, quanutm Weyl group, PBW basis.
The work was supported by the NNSFC (Grants No. 11771142 and 11801394) and in part by Science and Technology Commission of Shanghai Municipality (No. 18dz2271000).

class of multiparameter quantum groups or two-parameter quantum (affine) algebras, e.g. [**16**], [**5**], [**10**], [**11**] and references therein.

It is well-known that $\mathbf{p} = (p_{ij}) = (q^{d_i a_{ij}})$ is a standard structure constant matrix of Cartan type for a quantum group $U_q(\mathfrak{g}_A)$ of Drinfeld-Jimbo type, where $q \in \mathbb{C}^*$ is not a root of unity or an indeterminate. The class of the usual quantum affine algebras for A being affine types has been extensively investigated in the literature, e.g. [**3**], [**4**], [**6**], [**13**] and many others. Let us say that $\mathbf{q}' = (q'_{ij})_{1\le i,j\le n}$ is equivalent by twist to $\mathbf{q}$ if $q'_{ii} = q_{ii}$ and $q'_{ij}q'_{ji} = q_{ij}q_{ji}$.

In this paper, to start with, we define a class of pointed Hopf algebras from $\mathbf{q} = \begin{pmatrix} \varepsilon_1 q^2 & tq^{-2} \\ t^{-1}q^{-2} & \varepsilon_2 q^2 \end{pmatrix}$ of Cartan type (where $\varepsilon_i \in \{\pm 1\}$ and $t \in \mathbb{Q}(q)^*$) of affine type $A_1^{(1)}$, but not equivalent by twist to $\mathbf{p}$ of the same type. Precisely, in Section 2, we define a class of admissible quantum affine algebras of type $A_1^{(1)}$, which are the Drinfeld doubles of their Borel subalgebras. Secondly, in Section 3, we specialize our $\mathbf{q}$ at $\varepsilon_1 = \varepsilon_2 = -1$ and $t = 1$ and denote by $\mathcal{U}_q(\widehat{\mathfrak{sl}_2})$ the newly defined quantum affine algebra. We prove that it is neither isomorphic to the usual quantum affine algebra $U_q(\widehat{\mathfrak{sl}_2})$ (see [**6**]), nor its co-opposite $U_q^{\mathrm{cop}}(\widehat{\mathfrak{sl}_2})$, as pointed Hopf algebras (see Theorem 3.4). Thirdly, we introduce the quantum Weyl group as the subgroup of algebra automorphisms of $\mathcal{U}_q(\widehat{\mathfrak{sl}_2})$ generated by T_i, $i = 1, 2$ that can be thought of as liftings of the generators s_i of the affine Weyl group of type $A_1^{(1)}$; see Theorem 3.5, Lemmas 3.15, 3.16, Remark 3.17 and Proposition 3.18. In Section 4, we define the imaginary root vectors $E_{n\delta}$ and real root vectors $E_{m\delta+\alpha_i}$ by means of the quantum Weyl group, and calculate their commutation rules following the ideas of [**6**]. Finally, in Section 5, we construct a PBW basis of $\mathcal{U}_q(\widehat{\mathfrak{sl}_2})$, adapting an argument from [**17**] applied there to quantum groups of finite type A_n.

Since the admissible quantum affine algebra $\mathcal{U}_q(\widehat{\mathfrak{sl}_2})$ of type $A_1^{(1)}$ considered here has different q-Serre relations (B5) (B6) than the usual one, together with Theorem 3.5, we conjecture that its vertex operator representation theory will be different than the standard one. Also we ask whether its nilpotent part is a Nichols algebra of affine type.

In what follows, let us write $\mathbb{Z}$, $\mathbb{Z}_+$, $\mathbb{N}$ the set of integers, nonnegative integers, positive integers, respectively.

2. Admissible quantum affine algebras $\mathbf{U_q}(\widehat{\mathfrak{sl}_2})$ of type $A_1^{(1)}$

2.1. Definition of $\mathbf{U_q}(\widehat{\mathfrak{sl}_2})$. Let $\mathbb{Q}(q)$ denote a field of rational functions with an indeterminate q. Next, we define a class of admissible quantum affine algebras $\mathbf{U_q}(\widehat{\mathfrak{sl}_2})$, for $t \in \mathbb{Q}(q)^*$ and $\varepsilon_1, \varepsilon_2 \in \{\pm 1\}$. Their quantum structure constant matrices are of forms: $\mathbf{q} = \begin{pmatrix} \varepsilon_1 q^2 & tq^{-2} \\ t^{-1}q^{-2} & \varepsilon_2 q^2 \end{pmatrix}$, which still satisfy the Cartan condition, see [**18**], where $A = (a_{ij})$ with $a_{ii} = 2$, $a_{ij} = -2$ $(i \ne j)$ is the Cartan matrix of affine Lie algebra $\widehat{\mathfrak{sl}_2}$ of type $A_1^{(1)}$. We also prove that these admissible quantum affine algebras have Drinfel'd double structure. Following Definition 7 [**16**], we have

DEFINITION 2.1. Let $\mathbf{U_q}(\widehat{\mathfrak{sl}_2})$ be the unital associative algebra over $\mathbb{Q}(q)$ generated by e_i, f_i, $w_i^{\pm 1}$, $w_i'^{\pm 1}$ $(i = 1, 2)$, such that $w_i^{\pm 1}$, $w_i'^{\pm 1}$ commute with each other, and the following relations $(B1)$—$(B6)$ are satisfied.

(B1) $$w_iw_i^{-1} = w_i'w_i'^{-1} = 1.$$

(B2) $$\begin{aligned} w_1e_1 &= \varepsilon_1 q^2 e_1 w_1, & w_1 f_1 &= \varepsilon_1 q^{-2} f_1 w_1,\\ w_1e_2 &= t^{-1}q^{-2}e_2w_1, & w_1f_2 &= tq^2 f_2 w_1,\\ w_2e_1 &= tq^{-2}e_1w_2, & w_2f_1 &= t^{-1}q^2 f_1 w_2,\\ w_2e_2 &= \varepsilon_2 q^2 e_2 w_2, & w_2 f_2 &= \varepsilon_2 q^{-2} f_2 w_2. \end{aligned}$$

(B3) $$\begin{aligned} w_1'e_1 &= \varepsilon_1 q^{-2} e_1 w_1', & w_1' f_1 &= \varepsilon_1 q^{2} f_1 w_1',\\ w_1'e_2 &= t^{-1}q^{2}e_2w_1', & w_1'f_2 &= tq^{-2} f_2 w_1',\\ w_2'e_1 &= tq^{2}e_1w_2', & w_2'f_1 &= t^{-1}q^{-2} f_1 w_2',\\ w_2'e_2 &= \varepsilon_2 q^{-2} e_2 w_2', & w_2' f_2 &= \varepsilon_2 q^{2} f_2 w_2'. \end{aligned}$$

(B4) $$[e_i, f_j] = \delta_{ij}\frac{w_i - w_i'}{q - q^{-1}}, \qquad i, j \in \{1, 2\}.$$

The q-Serre relations for e_i's:

(B5)
$$e_1^3e_2 - t^{-1}(\varepsilon_1 + q^2 + q^{-2})e_1^2e_2e_1 + t^{-2}\varepsilon_1(\varepsilon_1 + q^2 + q^{-2})e_1e_2e_1^2 - t^{-3}\varepsilon_1e_2e_1^3 = 0,$$
$$e_2^3e_1 - t(\varepsilon_2 + q^2 + q^{-2})e_2^2e_1e_2 + t^{2}\varepsilon_2(\varepsilon_2 + q^2 + q^{-2})e_2e_1e_2^2 - t^{3}\varepsilon_2e_1e_2^3 = 0.$$

The q-Serre relations for f_i's:

(B6)
$$f_1^3f_2 - t(\varepsilon_1 + q^2 + q^{-2})f_1^2f_2f_1 + t^{2}\varepsilon_1(\varepsilon_1 + q^2 + q^{-2})f_1f_2f_1^2 - t^{3}\varepsilon_1f_2f_1^3 = 0,$$
$$f_2^3f_1 - t^{-1}(\varepsilon_2 + q^2 + q^{-2})f_2^2f_1f_2 + t^{-2}\varepsilon_2(\varepsilon_2 + q^2 + q^{-2})f_2f_1f_2^2 - t^{-3}\varepsilon_2f_1f_2^3 = 0.$$

PROPOSITION 2.2. *$\mathbf{U_q}(\widehat{\mathfrak{sl}_2})$ is a Hopf algebra with the comultiplication:*

$$\begin{aligned} \Delta(w_i^{\pm1}) &= w_i^{\pm1}\otimes w_i^{\pm1}, & \Delta(w_i'^{\pm1}) &= w_i'^{\pm1}\otimes w_i'^{\pm1},\\ \Delta(e_i) &= e_i\otimes 1 + w_i\otimes e_i, & \Delta(f_i) &= 1\otimes f_i + f_i\otimes w_i'. \end{aligned}$$

2.2. A Drinfel'd (quantum) double. Let $\mathcal{B}$ (resp., $\mathcal{B}'$) denote the Hopf Borel subalgebra of $\mathbf{U_q}(\widehat{\mathfrak{sl}_2})$ generated by e_i, $w_i^{\pm1}$ (resp., f_i, $w_i'^{\pm1}$), $i = 1, 2$. Similar to Proposition 2.3 [**5**], we have

PROPOSITION 2.3. *There exists a unique skew-dual pairing $\langle\, ,\, \rangle : \mathcal{B}'\times\mathcal{B} \to \mathbb{Q}(q)$ of the Hopf subalgebras $\mathcal{B}$ and $\mathcal{B}'$ such that*

$$\langle f_i, e_j\rangle = \delta_{ij}\frac{1}{q^{-1}-q},$$
$$\langle w_i', w_i\rangle = \langle w_i'^{-1}, w_i^{-1}\rangle = \varepsilon_i q^2, \qquad i = 1, 2$$
$$\langle w_i', w_i^{-1}\rangle = \langle w_i'^{-1}, w_i\rangle = \varepsilon_i q^{-2}, \qquad i = 1, 2$$
$$\langle w_1', w_2\rangle = \langle w_1'^{-1}, w_2^{-1}\rangle = tq^{-2}, \qquad \langle w_1'^{-1}, w_2\rangle = \langle w_1', w_2^{-1}\rangle = t^{-1}q^2,$$
$$\langle w_2', w_1\rangle = \langle w_2'^{-1}, w_1^{-1}\rangle = t^{-1}q^{-2}, \qquad \langle w_2'^{-1}, w_1\rangle = \langle w_2', w_1^{-1}\rangle = tq^2,$$

and all the other pairs of generators assume 0. Moreover, we have $\langle S(a), S(b)\rangle = \langle a, b\rangle$ for $a\in\mathcal{B}'$, $b\in\mathcal{B}$.

Based on an argument going back to Drinfeld, the reader can check the following

PROPOSITION 2.4. *The admissible quantum affine algebra* $\mathbf{U_q}(\widehat{\mathfrak{sl}_2})$ *is isomorphic to the Drinfel'd quantum double* $\mathcal{D}(\mathcal{B},\mathcal{B}')$.

3. New quantum affine algebra $\mathcal{U}_q(\widehat{\mathfrak{sl}_2})$ of type $A_1^{(1)}$ and its quantum Weyl group

3.1. The distinct $\mathcal{U}_q(\widehat{\mathfrak{sl}_2})$.

DEFINITION 3.1. Let $\mathcal{U}_q(\widehat{\mathfrak{sl}_2})$ be the Hopf algebra generated by $E_i, F_i, K_i^{\pm 1}$ ($i = 1, 2$) over $\mathbb{Q}(q)$ with the multiplication, comultiplication satisfying the relations:

$$K_iK_j = K_jK_i, \qquad K_iK_i^{-1} = 1 = K_i^{-1}K_i,$$

$$K_iE_i = -q^2E_iK_i, \qquad K_iF_i = -q^{-2}F_iK_i,$$

$$K_iE_j = q^{-2}E_jK_i, \qquad K_iF_j = q^2F_jK_i, \qquad (i \neq j),$$

$$[\, E_i, F_j \,] = \delta_{ij}\frac{K_i - K_i^{-1}}{q - q^{-1}},$$

$$E_i^3E_j + (1 - q^2 - q^{-2})E_i^2E_jE_i + (1 - q^2 - q^{-2})E_iE_jE_i^2 + E_jE_i^3 = 0, \quad (i \neq j)$$

$$F_i^3F_j + (1 - q^2 - q^{-2})F_i^2F_jF_i + (1 - q^2 - q^{-2})F_iF_jF_i^2 + F_jF_i^3 = 0, \quad (i \neq j)$$

$$\Delta(K_i^{\pm 1}) = K_i^{\pm 1} \otimes K_i^{\pm 1},$$

$$\Delta(E_i) = E_i \otimes 1 + K_i \otimes E_i, \qquad \Delta(F_i) = 1 \otimes F_i + F_i \otimes K_i^{-1}.$$

REMARK 3.2. This 'new' quantum affine algebra $\mathcal{U}_q(\widehat{\mathfrak{sl}_2})$ of type $A_1^{(1)}$ is isomorphic to $\mathbf{U_q}(\widehat{\mathfrak{sl}_2})/(w_i' - w_i^{-1})$, where $(w_i' - w_i^{-1})$ is the Hopf ideal generated by the elements $w_i' - w_i^{-1}$, and the $\mathbf{q}$ here is a specialization of the generic $\mathbf{q}$ in Section 2 at $\varepsilon_1 = \varepsilon_2 = -1$ and $t = 1$.

LEMMA 3.3. *For* $i = 1,\ 2$, *in* $\mathcal{U}_q(\widehat{\mathfrak{sl}_2})$, *we have*

(i) $P_{1,K_i} = \mathbb{Q}(q)(1 - K_i) + \mathbb{Q}(q)E_i + \mathbb{Q}(q)F_iK_i$, *and* $P_{1,K_i^{-1}} = \mathbb{Q}(q)(1 - K_i^{-1})$;

(ii) $P_{K_i^{-1},1} = \mathbb{Q}(q)(1-K_i^{-1})+\mathbb{Q}(q)F_i+\mathbb{Q}(q)E_iK_i^{-1}$, *and* $P_{K_i,1} = \mathbb{Q}(q)(1-K_i)$;

(iii) $P_{1,\sigma} = P_{\sigma,1} = \mathbb{Q}(q)(1 - \sigma)$, *for any group-like element* $\sigma \notin \{K_1^{\pm 1}, K_2^{\pm 1}\}$,

where $P_{g,h} = \{x \in \mathcal{U}_q(\widehat{\mathfrak{sl}_2}) \mid \Delta(x) = x \otimes g + h \otimes x\}$, *for some group-likes* g, h *in* $\mathcal{U}_q(\widehat{\mathfrak{sl}_2})$.

THEOREM 3.4. $\mathcal{U}_q(\widehat{\mathfrak{sl}_2})$ *is neither isomorphic to the standard quantum affine algebra* $U_q(\widehat{\mathfrak{sl}_2})$ *of type* $A_1^{(1)}$ *nor to its co-opposite object* $U_q^{cop}(\widehat{\mathfrak{sl}_2})$ *as pointed Hopf algebras.*

PROOF. By definition, we know both $U_q(\widehat{\mathfrak{sl}_2})$ (resp. $U_q^{cop}(\widehat{\mathfrak{sl}_2})$) and $\mathcal{U}_q(\widehat{\mathfrak{sl}_2})$ are pointed Hopf algebras generated by group-like and skew-primitive elements. Suppose that $f : U_q(\widehat{\mathfrak{sl}_2}) \longrightarrow \mathcal{U}_q(\widehat{\mathfrak{sl}_2})$ (resp., $g : U_q^{cop}(\widehat{\mathfrak{sl}_2}) \longrightarrow \mathcal{U}_q(\widehat{\mathfrak{sl}_2})$) is a Hopf algebra isomorphism. Let $K_i^{*\,\pm 1}$, E_i^*, F_i^* denote the generators of $U_q(\widehat{\mathfrak{sl}_2})$, respectively. Since f, g preserve the group-likes (resp., the skew-primitives), and

$$(f \otimes f)(\Delta(E_1^*)) = \Delta(f(E_1^*)), \quad (\textit{resp.,} \quad (g \otimes g)(\Delta^{cop}(E_1^*)) = \Delta(g(E_1^*))\,)$$

where $f(E_1^*) \in P_{1,f(K_1^*)}$ (resp., $g(E_1^*) \in P_{g(K_1^*),1}$). Since it is impossible for $f(K_1^*) = K_i^{-1}$ (resp., $g(K_1^*) = K_i$), otherwise, $f(E_1^*) = a(1 - K_i^{-1}) = f(a(1 - K_1^*))$

(resp., $g(E_1^*) = b(1 - K_i) = g(b(1 - K_1^*)))$ for some $a, b \in \mathbb{Q}(q)^*$, this contradicts the injectivity of f (resp., g). Thereby, we have two possibilities:

Case (i):

$$f(K_1^*) = K_1, \quad f(E_1^*) = a_{11}E_1 + a_{12}F_1K_1 + a_{13}(1 - K_1),$$

respectively,

$$g(K_1^*) = K_1^{-1}, \quad g(E_1^*) = b_{11}F_1 + b_{12}E_1K_1^{-1} + b_{13}(1 - K_1^{-1}),$$

Case (ii):

$$f(K_1^*) = K_2, \quad f(E_1^*) = a_{21}E_2 + a_{22}F_2K_2 + a_{23}(1 - K_2),$$

respectively,

$$g(K_1^*) = K_2^{-1}, \quad g(E_1^*) = b_{21}F_2 + b_{22}E_2K_2^{-1} + b_{23}(1 - K_2^{-1}),$$

where $a_{ij}, b_{ij} \in \mathbb{Q}(q)$. Applying f (resp., g) to the relation:

$$K_1^*E_1^* = q^2E_1^*K_1^*,$$

we get

$$\begin{aligned}
&K_1\Big(a_{11}E_1{+}a_{12}F_1K_1{+}a_{13}(1 - K_1)\Big) = q^2\Big(a_{11}E_1{+}a_{12}F_1K_1{+}a_{13}(1{-}K_1)\Big)K_1,\\
&\Rightarrow 2q^2a_{11}E_1{+}(q^2{+}q^{-2})a_{12}F_1{+}(q^2{-}1)a_{13}(1{-}K_1) = 0,\\
&K_2\Big(a_{21}E_2{+}a_{22}F_2K_2{+}a_{23}(1{-}K_2)\Big) = q^2\Big(a_{21}E_2{+}a_{22}F_2K_2{+}a_{23}(1{-}K_2)\Big)K_2,\\
&\Rightarrow 2q^2a_{21}E_2{+}(q^2{+}q^{-2})a_{22}F_2{+}(q^2{-}1)a_{23}(1{-}K_2) = 0,
\end{aligned}$$

respectively,

$$\begin{aligned}
&K_1\Big(b_{11}F_1{+}b_{12}E_1K_1^{-1}{+}b_{13}(1{-}K_1^{-1})\Big) = q^2\Big(b_{11}F_1{+}b_{12}E_1K_1^{-1}{+}b_{13}(1{-}K_1^{-1})\Big)K_1,\\
&\Rightarrow (q^2{+}q^{-2})b_{11}F_1{+}2q^2b_{12}E_1K_1^{-1}{+}(q^2{-}1)b_{13}(1{-}K_1^{-1}) = 0,\\
&K_2\Big(b_{21}F_2{+}b_{22}E_2K_2^{-1}{+}b_{23}(1{-}K_2^{-1})\Big) = q^2\Big(b_{21}F_2{+}b_{22}E_2K_2^{-1}{+}b_{23}(1{-}K_2^{-1})\Big)K_2,\\
&\Rightarrow (q^2{+}q^{-2})b_{21}F_2{+}2q^2b_{22}E_2K_2^{-1}{+}(q^2{-}1)b_{23}(1{-}K_2^{-1}) = 0.
\end{aligned}$$

We get $a_{ij},\ b_{ij} = 0$ ($i \in \{1, 2\},\ j \in \{1, 2, 3\}$). This contradicts the hypothesis. □

3.2. Quantum Weyl group. As we know, the affine Weyl group of $\widehat{\mathfrak{sl}}_2$ is generated by s_1, s_2, satisfying $s_1^2 = s_2^2 = 1$. As the lifting of the affine Weyl group, we shall define the 'quantum Weyl group' (that will take the same role as the braid group in [**6**]) to be the subgroup of algebra automorphisms of $\mathcal{U}_q(\widehat{\mathfrak{sl}}_2)$ generated by

T_1 and T_2, where the actions of T_i $(i=1,2)$ on K_j, E_j, F_j $(j=1,2)$ are given by

$$\begin{aligned}
&T_i(K_j)=K_jK_i^{-a_{ij}}, \qquad T_i(E_i)=-F_iK_i, \qquad T_i(F_i)=-K_i^{-1}E_i,\\
&T_i(E_j)=E_i^{(2)}E_j-q^{-2}E_jE_i^{(2)}+\frac{1-q^{-2}}{[2]_q}E_iE_jE_i, \qquad i\neq j,\\
&T_i^{-1}(E_j)=-q^{-2}E_i^{(2)}E_j+E_jE_i^{(2)}+\frac{1-q^{-2}}{[2]_q}E_iE_jE_i, \qquad i\neq j,\\
&T_i(F_j)=-q^2F_i^{(2)}F_j+F_jF_i^{(2)}+\frac{1-q^2}{[2]_q}F_iF_jF_i, \qquad i\neq j,\\
&T_i^{-1}(F_j)=F_i^{(2)}F_j-q^2F_jF_i^{(2)}+\frac{1-q^2}{[2]_q}F_iF_jF_i, \qquad i\neq j,
\end{aligned}$$

where

$$\begin{aligned}
E_i^{(n)}&=\frac{E_i^n}{[n]_q!}, \qquad F_i^{(n)}=\frac{F_i^n}{[n]_q!}, \quad n\in\mathbb{N},\\
[n]_q&=\frac{q^n-q^{-n}}{q-q^{-1}}, \qquad [n]_q!=[1]_q\cdots[n]_q, \quad n\in\mathbb{N}.
\end{aligned}$$

There exists a (Hopf) $\mathbb{Q}(q)$-algebra automorphism Φ of $\mathcal{U}_q(\widehat{\mathfrak{sl}_2})$ (which interchanges the indices 1 and 2):

$$\Phi(E_i)=E_j, \quad \Phi(F_i)=F_j, \quad \Phi(K_i)=K_j, \quad (i\neq j)$$

where $\Phi^2=\mathrm{id}$, $T_1\Phi=\Phi T_2$ and $\Phi T_1^{-1}=T_2^{-1}\Phi$.

Moreover, there exists a $\mathbb{Q}$-antiautomorphism Ω of $\mathcal{U}_q(\widehat{\mathfrak{sl}_2})$ defined by

$$\Omega(E_i)=F_i, \quad \Omega(F_i)=E_i, \quad \Omega(K_i)=K_i^{-1}, \quad \Omega(q)=q^{-1}, \qquad i=1,2.$$

Obviously, $\Omega^2=\mathrm{id}$, $\Omega T_i=T_i\Omega$ $(i=1,2)$ and $\Phi\Omega=\Omega\Phi$.

In what follows, we will show that this new quantum affine algebra $\mathcal{U}_q(\widehat{\mathfrak{sl}_2})$ of type $A_1^{(1)}$ also has the quantum Weyl group as its algebra automorphism subgroup.

To this end, let us introduce the following notations.

$$\begin{aligned}
&E_\delta=E_1E_2-q^{-2}E_2E_1, && F_\delta=F_2F_1-q^2F_1F_2,\\
&E_{\delta+\alpha_1}=\frac{E_\delta E_1+E_1E_\delta}{[2]_q}, && F_{\delta+\alpha_1}=\frac{F_\delta F_1+F_1F_\delta}{[2]_q}.\\
&E_{2\delta+\alpha_1}\triangleq\frac{1}{[2]_q}(E_\delta E_{\delta+\alpha_1}+E_{\delta+\alpha_1}E_\delta) &&\\
&E_{3\delta+\alpha_1}\triangleq\frac{1}{[2]_q}(E_\delta E_{2\delta+\alpha_1}+E_{2\delta+\alpha_1}E_\delta), && E_{\delta+\alpha_2}\triangleq\frac{1}{[2]_q}(E_\delta E_2+E_2E_\delta),\\
&E_{2\delta}\triangleq E_1E_{\delta+\alpha_2}+q^{-2}E_{\delta+\alpha_2}E_1. &&
\end{aligned}$$

then we have

$$[E_1, F_\delta] = -q^2[2]_q F_2 K_1, \qquad [E_2, F_\delta] = [2]_q F_1 K_2^{-1},$$
$$[E_\delta, F_1] = -[2]_q E_2 K_1^{-1}, \qquad [E_\delta, F_2] = q^{-2}[2]_q E_1 K_2,$$
$$[E_\delta, F_\delta] = [2]_q \frac{K_1K_2 - K_1^{-1}K_2^{-1}}{q-q^{-1}},$$
$$[E_1, F_{\delta+\alpha_1}] = F_\delta K_1, \qquad [E_2, F_{\delta+\alpha_1}] = q^{-1}[2]_q F_1^2 K_2^{-1},$$
$$[E_{\delta+\alpha_1}, F_1] = -E_\delta K_1^{-1}, \qquad [E_{\delta+\alpha_1}, F_2] = q^{-3}[2]_q E_1^2 K_2,$$
$$[E_{\delta+\alpha_1}, F_\delta] = -[2]_q E_1 K_1 K_2, \qquad [E_\delta, F_{\delta+\alpha_1}] = [2]_q F_1 K_1^{-1} K_2^{-1},$$
$$[E_{\delta+\alpha_1}, F_{\delta+\alpha_1}] = \frac{K_1^2K_2 - K_1^{-2}K_2^{-1}}{q-q^{-1}},$$
$$E_{\delta+\alpha_1} = T_1(E_2), \qquad F_{\delta+\alpha_1} = T_1(F_2).$$

THEOREM 3.5. *T_k $(k = 1, 2)$ are well-defined algebra endomorphisms of $\mathcal{U}_q(\widehat{\mathfrak{sl}}_2)$.*

PROOF. We need to check that T_k preserves the following algebra relations.

(1) $T_k(K_i)T_k(K_i^{-1}) = T_k(K_iK_i^{-1}) = 1$, $T_k(K_i)T_k(K_j) = T_k(K_iK_j)$. This is clear.

(2) $T_k(K_i)T_k(E_i) = -q^2T_k(E_i)T_k(K_i)$, $T_k(K_i)T_k(F_i) = -q^{-2}T_k(F_i)T_k(K_i)$. We need to consider the cases k, i, $j \in \{1, 2\}$:

$$T_1(K_1)T_1(E_1)T_1(K_1)^{-1} = -{K_1}^{-1}F_1{K_1}^2 = q^2F_1K_1 = -q^2T_1(E_1),$$
$$T_1(K_2)T_1(E_2)T_1(K_2)^{-1} = K_2{K_1}^2T_1(E_2){K_2}^{-1}{K_1}^{-2} = -q^2T_1(E_2),$$
$$T_2(K_1)T_2(E_1)T_2(K_1)^{-1} = K_1{K_2}^2T_2(E_1){K_1}^{-1}{K_2}^{-2} = -q^2T_2(E_1),$$
$$T_2(K_2)T_2(E_2)T_2(K_2)^{-1} = -{K_2}^{-1}F_2{K_2}^2 = q^2F_2K_2 = -q^2T_2(E_2).$$

Similarly, one can check the second relation in (2).

(3) $T_k(K_i)T_k(E_j) = q^{-2}T_k(E_j)T_k(K_i)$, $T_k(K_i)T_k(F_j) = q^2T_k(F_j)T_k(K_i)$, $i \neq j$. It is similar to (2), we need to consider the following cases:

$$T_1(K_1)T_1(E_2)T_1(K_1)^{-1} = {K_1}^{-1}T_1(E_2)K_1 = q^{-2}T_1(E_2),$$
$$T_1(K_2)T_1(E_1)T_1(K_2)^{-1} = K_2{K_1}^2T_1(E_1){K_2}^{-1}{K_1}^{-2} = q^{-2}T_1(E_1),$$
$$T_2(K_1)T_2(E_2)T_2(K_1)^{-1} = K_1{K_2}^2T_2(E_2){K_1}^{-1}{K_2}^{-2} = q^{-2}T_2(E_2),$$
$$T_2(K_2)T_2(E_1)T_2(K_2)^{-1} = {K_2}^{-1}T_2(E_1)K_2 = q^{-2}T_2(E_1).$$

Similarly, one can check the second relation in (3).

(4) $[T_k(E_i), T_k(F_j)] = T_k[E_i, F_j]$.

We need do it in some steps: When $k = 1, i \neq j \in \{1, 2\}$, we have

$$\begin{aligned}
&[T_1(E_1), T_1(F_2)] = -F_1K_1T_1(F_2) + T_1(F_2)F_1K_1\\
&= -F_1K_1(-q^2F_1^{(2)}F_2 + F_2F_1^{(2)} + \frac{1-q^2}{[2]_q}F_1F_2F_1) + T_1(F_2)F_1K_1\\
&= -\frac{q^{-2}}{[2]_q}F_1(-q^2F_1^2F_2 + F_2F_1^2 + (1-q^2)F_1F_2F_1)K_1 + T_1(F_2)F_1K_1\\
&= -\frac{q^{-2}}{[2]_q}(-q^2F_1^3F_2 + F_1F_2F_1^2 + (1-q^2){F_1}^2F_2F_1)K_1 + T_1(F_2)F_1K_1
\end{aligned}$$

$$
\begin{aligned}
&= -\frac{q^{-2}}{[2]_q}\Big\{-q^2[(-1+q^{-2}+q^2)(F_1^2F_2F_1+F_1F_2F_1^2)-F_2F_1^3]\\
&\quad +F_1F_2F_1^2+(1-q^2){F_1}^2F_2F_1\Big\}K_1+T_1(F_2)F_1K_1\\
&= -\frac{q^{-2}}{[2]_q}\Big\{-q^4F_1^2F_2+(q^2-q^4)F_1F_2F_1+q^2F_2F_1^2\Big\}F_1K_1+T_1(F_2)F_1K_1\\
&= \Big\{q^2F_1^{(2)}F_2+\frac{(-1+q^2)}{[2]_q}F_1F_2F_1-F_2F_1^{(2)}\Big\}F_1K_1+T_1(F_2)F_1K_1\\
&= 0 = T_1[E_1,F_2].
\end{aligned}
$$

$$
\begin{aligned}
&[T_1(E_2),T_1(F_1)] = -q^2K_1^{-1}T_1(E_2)E_1+K_1^{-1}E_1T_1(E_2)\\
&= -\frac{q^2}{[2]_q}K_1^{-1}\Big\{E_1^2E_2E_1-q^{-2}E_2E_1^3+(1-q^{-2})E_1E_2{E_1}^2\Big\}+K_1^{-1}E_1T_1(E_2)\\
&= -\frac{q^2}{[2]_q}K_1^{-1}\Big\{E_1^2E_2E_1-q^{-2}\Big[-{E_1}^3E_2+(q^2+q^{-2}-1){E_1}^2E_2E_1\\
&\quad +(q^2+q^{-2}-1)E_1E_2{E_1}^2\Big]+(1-q^{-2})E_1E_2{E_1}^2\Big\}+K_1^{-1}E_1T_1(E_2)\\
&= 0 = T_1[E_2,F_1].
\end{aligned}
$$

It is easy to check

$$
\begin{aligned}
&[T_1(E_2),T_1(F_2)] = \frac{K_1^2K_2-K_1^{-2}K_2^{-1}}{q-q^{-1}} = T_1[E_1,F_2],\\
&[T_1(E_1),T_1(F_1)] = F_1E_1-E_1F_1 = [F_1,E_1] = \frac{-K_1+K_1^{-1}}{q-q^{-1}} = T_1\Big(\frac{K_1-K_1^{-1}}{q-q^{-1}}\Big).
\end{aligned}
$$

(5) We need to check that T_1 preserves the q-Serre relations

$$
E_i^3E_j+(1-q^2-q^{-2})E_i^2E_jE_i+(1-q^2-q^{-2})E_iE_jE_i^2+E_jE_i^3=0,\quad (i\neq j).
$$

It suffices to verify:

$$
\begin{aligned}
\text{(i)}\ & T_1(E_1)^3T_1(E_2)+(1-q^2-q^{-2})\big(T_1(E_1)^2T_1(E_2)T_1(E_1)+T_1(E_1)T_1(E_2)T_1(E_1)^2\big)\\
&\qquad +T_1(E_2)T_1(E_1)^3=0,\\
\text{(ii)}\ & T_1(E_2)^3T_1(E_1)+(1-q^2-q^{-2})\big(T_1(E_2)^2T_1(E_1)T_1(E_2)+T_1(E_2)T_1(E_1)T_1(E_2)^2\big)\\
&\qquad +T_1(E_1)T_1(E_2)^3=0.
\end{aligned}
$$

For (i), we calculate the following

$$
\begin{aligned}
&T_1(E_1)^3T_1(E_2) = (-F_1K_1)^3T_1(E_2) = -q^2F_1K_1F_1K_1(T_1(E_2)F_1-{K_1}^{-1}E_\delta)K_1\\
&= -q^4F_1K_1(T_1(E_2)F_1-{K_1}^{-1}E_\delta)K_1F_1K_1+q^2F_1K_1(E_\delta F_1+[2]_qE_2{K_1}^{-1})K_1\\
&= -q^6F_1T_1(E_2)K_1F_1K_1F_1K_1+q^4F_1E_\delta K_1F_1K_1-q^2F_1E_\delta K_1F_1K_1+q^2[2]_qF_1K_1E_2\\
&= -q^6(T_1(E_2)F_1-{K_1}^{-1}E_\delta)K_1F_1K_1F_1K_1\\
&\quad +(q^4-q^2)(E_\delta F_1+[2]_qE_2{K_1}^{-1})K_1F_1K_1-q^4[2]_qK_1F_1E_2\\
&= -q^6T_1(E_2)(F_1K_1)^3+(-q^6+q^4-q^2)E_\delta(F_1K_1)^2+[2]_q(q^4-q^2+1)E_2F_1K_1,
\end{aligned}
$$

$$T_1(E_1)^2T_1(E_2)T_1(E_1) = -q^2F_1K_1(T_1(E_2)F_1 - K_1{}^{-1}E_\delta)K_1F_1K_1$$
$$= -q^4(T_1(E_2)F_1 - K_1{}^{-1}E_\delta)K_1F_1K_1F_1K_1 + q^2E_\delta(F_1K_1)^2 + q^2[2]_qE_2F_1K_1$$
$$= -q^4T_1(E_2)(F_1K_1)^3 + (q^2 - q^4)E_\delta(F_1K_1)^2 + q^2[2]_qE_2F_1K_1,$$
$$T_1(E_1)T_1(E_2)T_1(E_1)^2 = -q^2(T_1(E_2)F_1 - K_1{}^{-1}E_\delta)K_1F_1K_1F_1K_1$$
$$= -q^2T_1(E_2)(F_1K_1)^3 - q^2T_1(E_2)(F_1K_1)^3E_\delta F_1K_1F_1K_1,$$
$$- T_1(E_2)(F_1K_1)^3.$$

So, the left-hand side of (i) equals 0.

For (ii), we have

$$T_1(E_2)^3T_1(E_1) = -T_1(E_2)^3F_1K_1,$$
$$T_1(E_2)^2T_1(E_1)T_1(E_2) = -q^2T_1(E_2)^2(T_1(E_2)F_1 - K_1{}^{-1}E_\delta)K_1$$
$$= -q^2T_1(E_2)^3F_1K_1 - q^2T_1(E_2)^2E_\delta,$$
$$T_1(E_2)T_1(E_1)T_1(E_2)^2 = -q^4T_1(E_2)(T_1(E_2)F_1 - K_1{}^{-1}E_\delta)T_1(E_2)K_1$$
$$= -q^4T_1(E_2)^3F_1K_1 - q^4T_1(E_2)^2E_\delta - q^2T_1(E_2)E_\delta T_1(E_2),$$
$$T_1(E_1)T_1(E_2)^3 = -q^6(T_1(E_2)F_1 - K_1{}^{-1}E_\delta)T_1(E_2)^2K_1$$
$$= -q^6T_1(E_2)(T_1(E_2)F_1 - K_1{}^{-1}E_{12})T_1(E_2)K_1 - q^2E_\delta T_1(E_2)^2$$
$$= -q^6T_1(E_2)^2(T_1(E_2)F_1 - K_1{}^{-1}E_\delta)K_1 - q^4T_1(E_2)E_\delta T_1(E_2) - q^2E_\delta T_1(E_2)^2.$$

So, the left-hand side of (ii) equals

$$T_1(E_2)^2E_\delta + (1-q^2)T_1(E_2)E_\delta T_1(E_2) - q^2E_\delta T_1(E_\delta)^2$$
$$= T_1(E_2)(T_1(E_2)E_\delta + E_\delta T_1(E_2)) - q^2(T_1(E_2)E_\delta + E_\delta T_1(E_2))T_1(E_2)$$
$$= E_{\delta+\alpha_1}E_{2\delta+\alpha_1} - q^2E_{2\delta+\alpha_1}E_{\delta+\alpha_1}.$$

To complete (ii), it suffices to verify:

$$E_{2\delta+\alpha_1}E_{\delta+\alpha_1} = q^{-2}E_{\delta+\alpha_1}E_{2\delta+\alpha_1}. \tag{3.1}$$

This will be done in Proposition 3.13.

(6) Applying the anti-automorphism Ω to identities in (5), we can verify that T_k preserves the q-Serre relations of $\mathcal{U}_q(\widehat{\mathfrak{sl}}_2)^-$. □

LEMMA 3.6. *The following hold:*

(1) $E_{\delta+\alpha_1}E_1 = q^{-2}E_1E_{\delta+\alpha_1}$,

(2) $E_{2\delta+\alpha_1}E_1 + q^{-2}E_1E_{2\delta+\alpha_1} = (1+q^{-2})E_{\delta+\alpha_1}{}^2$,

(3) $E_{3\delta+\alpha_1}E_1 - q^{-2}E_1E_{3\delta+\alpha_1} = (2+q^{-2})E_{2\delta+\alpha_1}E_{\delta+\alpha_1} - (1+2q^{-2})E_{\delta+\alpha_1}E_{2\delta+\alpha_1}$.

PROPOSITION 3.7. $E_{2\delta} = E_{\delta+\alpha_1}E_2 + q^{-2}E_2E_{\delta+\alpha_1}$.

LEMMA 3.8. *The following hold:*

(1) $E_{2\delta}E_1 - E_1E_{2\delta} = (q^{-2} - q^2)E_{\delta+\alpha_1}E_\delta - q^{-2}[2]_qE_{2\delta+\alpha_1}$,

(2) $E_{2\delta}F_1 - F_1E_{2\delta} = [2]_q(q^{-2} - q^2)E_\delta E_2K_1{}^{-1} - q^{-2}[2]_q{}^2E_{\delta+\alpha_2}K_1{}^{-1}$,

(3) $E_{2\delta}F_2 - F_2E_{2\delta} = (-q^{-1} + q^{-3})[2]_qE_1E_\delta K_2 + (q^{-2}[2]_q - q^{-3}[2]_q{}^2)E_{\delta+\alpha_1}K_2$.

It is easy to check the following

PROPOSITION 3.9. *$[[E_{2\delta}, E_\delta], F_i] = 0$, for $i = 1, 2$.*

LEMMA 3.10. *If $x \in \mathcal{U}^+_{q,\alpha}(\alpha \in R_+)$ satisfies $[x, F_i] = 0$, $\forall\, i = 1, 2$, then x must be zero.*

PROOF. See the proof of Lemma 1 of [**6**]. □

THEOREM 3.11. *$[E_\delta, E_{2\delta}] = 0$.*

PROOF. It follows from Proposition 3.9 and Lemma 3.10. □

LEMMA 3.12. *$E_{2\delta}E_{\delta+\alpha_1} - E_{\delta+\alpha_1}E_{2\delta} = (q^{-2} - q^2)E_{2\delta+\alpha_1}E_\delta - q^{-2}[2]_q E_{3\delta+\alpha_1}$.*

PROPOSITION 3.13. *$E_{2\delta+\alpha_1}E_{\delta+\alpha_1} = q^{-2}E_{\delta+\alpha_1}E_{2\delta+\alpha_1}$.*

LEMMA 3.14. *$T_2(E_\delta) = {T_1}^{-1}(E_\delta) = \Phi(E_\delta)$. This means $T_1T_2(E_\delta) = E_\delta$.*

PROOF. By definition, we have

$$\begin{aligned}
T_2(E_\delta) &= T_2(E_1E_2 - q^{-2}E_2E_1)\\
&= \frac{1}{[2]_q}\Big\{\Big(\Phi(E_\delta)E_2 + E_2\Phi(E_\delta)\Big)T_2(E_2) - q^{-2}T_2(E_2)\Big(\Phi(E_\delta)E_2 + E_2\Phi(E_\delta)\Big)\Big\}\\
&= -\frac{1}{[2]_q}\Big\{\Big(\Phi(E_\delta)E_2 + E_2\Phi(E_\delta)\Big)F_2 - F_2\Big(\Phi(E_\delta)E_2 + E_2\Phi(E_\delta)\Big)\Big\}K_2\\
&= -\frac{1}{[2]_q}\Phi\Big(E_\delta E_1F_1 + E_1E_\delta F_1 - F_1E_\delta E_1 - F_1E_1E_\delta\Big)K_2\\
&= -\Phi\Big(E_{\delta+\alpha_1}F_1 - F_1E_{\delta+\alpha_1}\Big)K_2 = \Phi(E_\delta {K_1}^{-1})K_2 = \Phi(E_\delta).
\end{aligned}$$

Similarly, we calculate ${T_1}^{-1}(E_\delta) = \Phi(E_\delta)$. □

LEMMA 3.15. *$T_i{T_i}^{-1}(E_j) = {T_i}^{-1}T_i(E_j) = E_j$, $\forall\, i, j \in \{1, 2\}$. This means $T_i \in \mathrm{Aut}(\mathcal{U}_q(\widehat{\mathfrak{sl}}_2))$.*

PROOF. Thanks to $T_1\Phi = \Phi T_2$, we only need to prove the case for $i = 1$. By definition, it is clear that $T_1{T_1}^{-1}(E_1) = {T_1}^{-1}T_1(E_1) = E_1$. For $j = 2$, we have

$$\begin{aligned}
T_1{T_1}^{-1}(E_2) &= \frac{1}{[2]_q}T_1\Big(\Phi(E_\delta)E_1 + E_1\Phi(E_\delta)\Big) = \frac{1}{[2]_q}\Big(E_\delta T_1(E_1) + T_1(E_1)E_\delta\Big)\\
&= -\frac{1}{[2]_q}\Big(E_\delta F_1 - F_1E_\delta\Big)K_1 = E_2,\\
{T_1}^{-1}T_1(E_2) &= \frac{1}{[2]_q}{T_1}^{-1}\Big(E_\delta E_1 + E_1E_\delta\Big) = -\frac{1}{[2]_q}{K_1}^{-1}\Big(-\Phi(E_\delta)F_1 + F_1\Phi(E_\delta)\Big)\\
&= \frac{1}{[2]_q}{K_1}^{-1}\Phi\Big(E_\delta F_2 - F_2E_\delta\Big) = {K_1}^{-1}\Phi\Big(q^{-2}E_1K_2\Big) = E_2.
\end{aligned}$$

Thereby, T_i $(i = 1, 2)$ are algebra automorphisms of $\mathcal{U}_q(\widehat{\mathfrak{sl}}_2)$. □

LEMMA 3.16. *For any $n \in \mathbb{Z}_+ \setminus \{0\}$, we have the identities as follows*

(1) $T_i^n(E_i) = (-1)^{\frac{n+1}{2}}q^{-n+1}F_iK_i^n$, *when n is odd.*
(2) $T_i^n(E_i) = (-1)^{\frac{n}{2}}q^{-n}E_iK_i^{-n}$, *when n is even.*
(3) $(T_1^{\pm1}T_2^{\pm1})^n(K_1) = K_1^{2n+1}K_2^{2n}$, $(T_2^{\pm1}T_1^{\pm1})^n(K_1) = K_1^{1-2n}K_2^{-2n}$.
(4) $T_1^{\pm1}T_2^{\pm1}T_1^{\pm1}(K_1) = K_1^{-3}K_2^{-2}$, $T_2^{\pm1}T_1^{\pm1}T_2^{\pm1}(K_1) = K_1^3K_2^4$.

PROOF. By definition and induction on n. □

REMARK 3.17. From Lemma 3.16, we have $T_1^{\pm 1}T_2^{\pm 1} \neq T_2^{\pm 1}T_1^{\pm 1}$, $T_1^{\pm 1}T_2^{\pm 1}T_1^{\pm 1} \neq T_2^{\pm 1}T_1^{\pm 1}T_2^{\pm 1}$.

Since K_i is of infinite order by definition, Lemma 3.16 deduces the following

PROPOSITION 3.18. *The quantum Weyl group is a non-abelian infinite group with T_i $(i = 1, 2)$, $T_1T_2^{\pm 1}$ being of infinite order.*

4. Definition of imaginary and real root vectors

Denote by $R = \{n\alpha_1 + m\alpha_2 \mid |n - m| \leq 1, nm \geq 0, n, m \in \mathbb{Z}\} \setminus \{0\}$ the root system of $\widehat{\mathfrak{sl}}_2$, with the imaginary root system $R^{\mathrm{im}} = \{n\delta \in R \mid n \in \mathbb{Z}^*\}$ (where $\delta = \alpha_1 + \alpha_2$), the real root system $R^{\mathrm{re}} = \{n\alpha_1 + m\alpha_2 \in R \mid n \neq m\}$. Denote by $R_+ = \{n\alpha_1 + m\alpha_2 \in R \mid n, m \geqslant 0\}$, the positive root system; $R_+^{\mathrm{re}} = R_+ \cap R^{\mathrm{re}}$, the real positive root system; and $R_+^{\mathrm{im}} = R_+ \cap R^{\mathrm{im}}$, the imaginary positive roots, respectively.

According to the root system of $\widehat{\mathfrak{sl}}_2$, we have

$$\underbrace{s_1 s_2 \cdots s_i}_{n}(\alpha_j) = n\delta + \alpha_1, \qquad \underbrace{s_2 s_1 \cdots s_i}_{n}(\alpha_j) = n\delta + \alpha_2, \quad (i \neq j).$$

Recall that the action of quantum Weyl group on $\mathcal{U}_q(\widehat{\mathfrak{sl}}_2)$ is a natural generalization of the induced action of Weyl group on $\widehat{\mathfrak{sl}}_2$ defined by

$$E_{n\delta+\alpha_1} = \underbrace{T_1^{\pm 1}T_2^{\pm 1} \cdots T_i^{\pm 1}}_{n}(E_j), \qquad E_{n\delta+\alpha_2} = \underbrace{T_2^{\pm 1}T_1^{\pm 1} \cdots T_i^{\pm 1}}_{n}(E_j), \quad (i \neq j).$$

We are left to choose the suitable signs of the exponents and prove that the forms of E_α $(\alpha \in R_+^{\mathrm{re}})$ act as the root vectors in $\mathcal{U}_q(\widehat{\mathfrak{sl}}_2)^+$.

4.1. Imaginary positive root vectors $E_{n\delta}$. We can define inductively the possible form of the imaginary root vectors $E_{n\delta}$ $(n \in \mathbb{N})$: once we have $E_{(n-1)\delta+\alpha_2}$ in hand, we can define

$$E_{n\delta} = E_1 E_{(n-1)\delta+\alpha_2} + (-1)^n q^{-2} E_{(n-1)\delta+\alpha_2} E_1,$$
$$\widetilde{E}_{n\delta} = E_2 E_{(n-1)\delta+\alpha_1} + (-1)^n q^{-2} E_{(n-1)\delta+\alpha_1} E_2.$$

As for the base step, we can define

$$E_\delta := E_1 E_2 - q^{-2} E_2 E_1, \tag{4.1}$$
$$\widetilde{E}_\delta := E_2 E_1 - q^{-2} E_1 E_2. \tag{4.2}$$

Then we have

$$E_\delta E_1 + E_1 E_\delta = [2]_q T_1(E_2), \tag{4.3}$$
$$E_\delta E_2 + E_2 E_\delta = [2]_q T_2^{-1}(E_1), \tag{4.4}$$
$$\widetilde{E}_\delta E_1 + E_1 \widetilde{E}_\delta = [2]_q T_1^{-1}(E_2), \tag{4.5}$$
$$\widetilde{E}_\delta E_2 + E_2 \widetilde{E}_\delta = [2]_q T_2(E_1), \tag{4.6}$$

where $\Phi(E_{n\delta}) = \widetilde{E}_{n\delta}$.

Note that

$$\Phi(E_{n\delta+\alpha_1}) = \widetilde{E}_{n\delta+\alpha_2}, \quad \Phi(E_{n\delta+\alpha_2}) = \widetilde{E}_{n\delta+\alpha_1}, \qquad \forall n \in \mathbb{Z}_+.$$

By (4.3) and (4.4), we set

$$E_{\delta+\alpha_1} = T_1(E_2), \qquad E_{\delta+\alpha_2} = T_2^{-1}(E_1),$$

and

$$E_\delta E_1 + E_1 E_\delta = [2]_q E_{\delta+\alpha_1}, \qquad E_\delta E_2 + E_2 E_\delta = [2]_q E_{\delta+\alpha_2}.$$

Similarly, by (4.5) and (4.6), we set

$$\widetilde{E}_{\delta+\alpha_1} = T_1^{-1}(E_2), \qquad \widetilde{E}_{\delta+\alpha_2} = T_2(E_1),$$

and

$$\widetilde{E}_\delta E_1 + E_1 \widetilde{E}_\delta = [2]_q \widetilde{E}_{\delta+\alpha_1}, \qquad \widetilde{E}_\delta E_2 + E_2 \widetilde{E}_\delta = [2]_q \widetilde{E}_{\delta+\alpha_2}.$$

4.2. The real positive root vectors. To give a natural definition of $E_\alpha{}'s$ $(\alpha \in R_+^{\mathrm{re}})$, let us prove the following result once we fix our definition of E_δ.

THEOREM 4.1. *$T_1T_2(E_\delta) = E_\delta$. Moreover, $\forall\, n \in \mathbb{Z}_+$ $(i \neq j)$,*

$$E_\delta \underbrace{T_1T_2\cdots T_i}_{n}(E_j) + \underbrace{T_1T_2\cdots T_i}_{n}(E_j)E_\delta = [2]_q \underbrace{T_1T_2\cdots T_j}_{n+1}(E_i),$$

$$E_\delta \underbrace{T_2^{-1}T_1^{-1}\cdots T_i^{-1}}_{n}(E_j) + \underbrace{T_2^{-1}T_1^{-1}\cdots T_i^{-1}}_{n}(E_j)E_\delta = [2]_q \underbrace{T_2^{-1}T_1^{-1}\cdots T_j^{-1}}_{n+1}(E_i).$$

PROOF. Similar to that of Lemma 1 (Section 3.2) of [**6**]. □

Noting that $T_1\Phi = \Phi T_2$ and $\Phi T_1{}^{-1} = T_2{}^{-1}\Phi$, and according to the Theorem above, we can give a natural definition: for any $n \in \mathbb{Z}_+$,

$$E_{n\delta+\alpha_1} = \underbrace{T_1T_2\cdots T_i}_{n}(E_j) = (T_1\Phi)^n(E_1),$$

$$E_{n\delta+\alpha_2} = \underbrace{T_2^{-1}T_1^{-1}\cdots T_i^{-1}}_{n}(E_j) = (T_2^{-1}\Phi)^n(E_2).$$

Therefore, we have: for any $n \in \mathbb{Z}_+$,

$$E_\delta E_{n\delta+\alpha_1} + E_{n\delta+\alpha_1}E_\delta = [2]_q E_{(n+1)\delta+\alpha_1}, \tag{4.7}$$

$$E_\delta E_{n\delta+\alpha_2} + E_{n\delta+\alpha_2}E_\delta = [2]_q E_{(n+1)\delta+\alpha_2}. \tag{4.8}$$

This shows that $E_\alpha{}'s \in \mathcal{U}_q(\widehat{\mathfrak{sl}}_2)^+$ $(\forall\, \alpha \in R_+^{\mathrm{re}})$, and also defines the concrete forms of $E_{n\delta+\alpha_1}$ and $E_{n\delta+\alpha_2}$.

4.3. The commutation rules of $E_{n\delta+\alpha_i}E_{m\delta+\alpha_i}$. In this subsection, we must exhibit the commutation rules among the real positive root vectors with respect to an ordering of R_+^{re}.

The Proposition below solves the ordering of elements in sets $\{E_{n\delta+\alpha_1} \mid n \in \mathbb{Z}_+\}$ and $\{E_{n\delta+\alpha_2} \mid n \in \mathbb{Z}_+\}$, respectively.

PROPOSITION 4.2. *For any $n, m \in \mathbb{Z}_+$ with $n > m$, there exist $a_h^{(r)}, a_h'^{(r)} \in \mathbb{Z}[q, q^{-1}]$, $r \in \mathbb{Z}_+$, $h \in \{0, ..., [r/2]\}$, such that*

$$E_{n\delta+\alpha_1}E_{m\delta+\alpha_1} = \sum_{h=0}^{[(n-m)/2]} a_h^{(n-m)} E_{(m+h)\delta+\alpha_1}E_{(n-h)\delta+\alpha_1},$$

$$E_{m\delta+\alpha_2}E_{n\delta+\alpha_2} = \sum_{h=0}^{[(n-m)/2]} a_h'^{(n-m)} E_{(n-h)\delta+\alpha_2}E_{(m+h)\delta+\alpha_2}.$$

Particularly, $a_0^{(r)} = a_0'^{(r)} = (-1)^{r-1}q^{-2}, \quad \forall\, r \in \mathbb{Z}_+$.

PROOF. Similar to that of Proposition 2 (Section 4.1) of [**6**]. □

4.4. The other commutation rules. The remaining problem is to calculate:
(1) the commutation rule between $E_{n\delta}$ and $E_{m\delta}$ $(n,\ m \in \mathbb{N})$;
(2) the commutation rule between $E_{n\delta}$ $(n \in \mathbb{N})$ and $E_{m\delta+\alpha_i}$ $(m \in \mathbb{Z}_+,\ i = 1,\ 2)$;
(3) the commutation rule between $E_{n\delta+\alpha_1}$ and $E_{m\delta+\alpha_2}$ $(n,\ m \in \mathbb{Z}_+)$.

PROPOSITION 4.3. *The following statements are equivalent:*
(1) $[E_\delta, E_{n\delta}] = 0, \quad \forall\, n \in \mathbb{N}$,
(2) $E_{r\delta+\alpha_1}E_{s\delta+\alpha_2}+(-1)^{r+s+1}q^{-2}E_{s\delta+\alpha_2}E_{r\delta+\alpha_1}=E_{(r+s+1)\delta}$, $r,\ s \in \mathbb{Z}_+, r+s = n$,
(3) $T_2(E_{(n+1)\delta}) = \Phi(E_{(n+1)\delta})$ $(\Rightarrow T_1T_2(E_{(n+1)\delta}) = E_{(n+1)\delta})$, $\quad \forall\, n \in \mathbb{N}$.

PROOF. The proof is similar to that of Proposition 1 (Section 4.2) in [**6**]. □

Next, we will prove that $T_2(E_{n\delta}) = \Phi(E_{n\delta})$ $(\forall\, n \in \mathbb{N})$ holds by induction on n. For $r \leq n$, Suppose that we have

$$T_2(E_{r\delta}) = \Phi(E_{r\delta}), \qquad r \leq n. \tag{4.9}$$

In what follows, we will prove that $T_2(E_{(n+1)\delta}) = \Phi(E_{(n+1)\delta})$ holds for any $n \in \mathbb{N}$.

To this end, we need first to study the commutation rules between $E_{n\delta}$ and E_α $(\alpha \in R_+^{\mathrm{re}})$.

PROPOSITION 4.4. *Suppose that $n \in \mathbb{N}$ satisfies (4.9). Then there exist ${b'}_r^{(n)}$ and $b_r^{(n)} \in \mathbb{Q}(q)$ $(r \leq n)$ so that*

$$E_{n\delta}E_1 + (-1)^{n-1}E_1E_{n\delta} = \sum_{r=1}^{n-1} {b'}_r^{(n)} E_{r\delta+\alpha_1}E_{(n-r)\delta} + {b'}_n^{(n)} E_{n\delta+\alpha_1},$$

$$E_{n\delta}E_2 + (-1)^{n-1}E_2E_{n\delta} = \sum_{r=1}^{n-1} b_r^{(n)} E_{(n-r)\delta}E_{r\delta+\alpha_2} + b_n^{(n)} E_{n\delta+\alpha_2},$$

where ${b'}_r^{(n)}$ and $b_r^{(n)}$ can be precisely calculated.

PROOF. Similar to those of Proposition 1 (Section 4.3) of [**6**]. □

REMARK 4.5. (1) If $T_2(E_{n\delta}) = \Phi(E_{n\delta})$ is valid, then the Proposition holds. Moreover, we have

$$E_{n\delta}E_{m\delta+\alpha_1}+(-1)^{n-1}E_{m\delta+\alpha_1}E_{n\delta} = \sum_{r=1}^{n-1} {b'}_r^{(n)} E_{(m+r)\delta+\alpha_1}E_{(n-r)\delta}+{b'}_n^{(n)} E_{(n+m)\delta+\alpha_1},$$

$$E_{n\delta}E_{m\delta+\alpha_2}+(-1)^{n-1}E_{m\delta+\alpha_2}E_{n\delta} = \sum_{r=1}^{n-1} b_r^{(n)} E_{(n-r)\delta}E_{(m+r)\delta+\alpha_2}+b_n^{(n)} E_{(n+m)\delta+\alpha_2}.$$

(2) Assume that $n \in \mathbb{N}$ satisfies (4.9). Applying $(T_1\Phi)^{-1}$ and ΦT_2 to the identities in Proposition 4.4, we obtain

$$[E_{n\delta}, F_1] = (-1)^{n-1}\Big(\sum_{r=1}^{n-1} b_r^{(n)} E_{(n-r)\delta}E_{(r-1)\delta+\alpha_2} + b_n^{(n)} E_{(n-1)\delta+\alpha_2}\Big)K_1^{-1},$$

$$[E_{n\delta}, F_2] = (-1)^n K_2\Big(\sum_{r=1}^{n-1} {b'}_r^{(n)} E_{(r-1)\delta+\alpha_1}E_{(n-r)\delta} + {b'}_n^{(n)} E_{(n-1)\delta+\alpha_1}\Big).$$

4.5. The remaining commutation rules. Now we still need to study the remaining commutation rules and shall prove that the hypothesis condition (4.9) is correct.

LEMMA 4.6. *Assume that* $n \in \mathbb{N}$, $\forall r < n$, $s < n-1$, $r+s>0$, *and* $\forall m < n$ *such that* (4.9) *holds. Then there exist* $X_{r,s}^{(i)}$, $Y_{r,s}^{(i)} \in \mathcal{U}_q^{\geq 0}$, $i \in \{1, 2\}$ (*where* $\mathcal{U}_q^{\geq 0}$ *is the* $\mathbb{Q}(q)$*-subalgebra of* $\mathcal{U}_q$ *generated by* $\mathcal{U}_q^+$ *and* $\mathcal{U}_q^0$) *such that*

$$\begin{aligned}
[[E_{n\delta}, E_{m\delta}], F_1] &= \sum_{s=1}^{n-1} X_{0,s}^{(1)}[E_{m\delta}, E_{(n-s)\delta}] + \sum_{r=1}^{m-1} X_{r,0}^{(2)}[E_{(m-r)\delta}, E_{n\delta}] \\
&\quad + \sum_{s=1}^{n-1}\sum_{r=1}^{m-1} X_{r,s}^{(3)}[E_{(m-r)\delta}, E_{(n-s)\delta}], \\
[[E_{n\delta}, E_{m\delta}], F_2] &= \sum_{s=1}^{n-1} Y_{0,s}^{(1)}[E_{m\delta}, E_{(n-s)\delta}] + \sum_{r=1}^{m-1} Y_{r,0}^{(2)}[E_{(m-r)\delta}, E_{n\delta}] \\
&\quad + \sum_{s=1}^{n-1}\sum_{r=1}^{m-1} Y_{r,s}^{(3)}[E_{(m-r)\delta}, E_{(n-s)\delta}].
\end{aligned}$$

PROOF. Similar to those of Lemma 1 (Section 4.4) of [**6**]. □

REMARK 4.7. If $T_2(E_{n\delta}) = \Phi(E_{n\delta})$ is valid, then the Lemma holds.

PROPOSITION 4.8. $[E_{n\delta}, E_{m\delta}] = 0$, *for any* $n, m \in \mathbb{N}$.

PROOF. It follows by the induction hypothesis on (m, n). □

REMARK 4.9. Up to now, we have proved that $T_2(E_{(n+1)\delta}) = \Phi(E_{(n+1)\delta})$ in Proposition 4.3 (3) holds, as required.

From Propositions 4.8 & 4.3, we have

COROLLARY 4.10. $\forall n$, $m \in \mathbb{Z}_+$, *we have*

$$E_{n\delta+\alpha_1}E_{m\delta+\alpha_2} + (-1)^{n+m+1}q^{-2}E_{m\delta+\alpha_2}E_{n\delta+\alpha_1} = E_{(n+m+1)\delta}.$$

5. A Poincaré-Birkhoff-Witt basis of $\mathcal{U}_q(\widehat{\mathfrak{sl}_2})$

Recall the ordering $<$ on the set of positive roots R_+ of type $A_1^{(1)}$:

$$n\delta+\alpha_1 < (n+1)\delta+\alpha_1 < \cdots < (m+1)\delta < m\delta < \cdots < (l+1)\delta+\alpha_2 < l\delta+\alpha_2,$$

where n, m, $l \in \mathbb{Z}_+$.

Let us define a set

$$\mathrm{B}_+ := \{E_{\beta_1}^{r_1}\cdots E_{\beta_N}^{r_N} \mid N \in \mathbb{N}, \beta_1 < \cdots < \beta_N \in \mathrm{R}_+, r_1\cdots r_N \in \mathbb{Z}_+\}.$$

PROPOSITION 5.1. *Any element of the subalgebra* $\mathcal{U}_q^+$ *generated by* E_i $(i = 1, 2)$ *in* $\mathcal{U}_q(\widehat{\mathfrak{sl}_2})$ *can be written as a linear combination of elements in* B_+.

PROOF. It suffices to consider any word $E_{\beta_{i_1}}\cdots E_{\beta_{i_N}} \in \mathcal{U}_q^+$ of length N ($\beta_{i_j} \in \mathrm{R}_+$). Using the double induction on (N, β_{i_1}): for $N = 1$, $\forall \beta_{i_1} \in \mathrm{R}_+$, we show that the claim is true. By induction on N, $\forall \beta_{i_j} \in \mathrm{R}_+$, $E_{\beta_{i_1}}\cdots E_{\beta_{i_N}}$ can be written as a linear combination of elements in B_+ with length less than and equal to N. Let us consider any word $E_{\beta_{i_1}}\cdots E_{\beta_{i_{N+1}}}$ of length $N{+}1$. Now we will prove it in three cases.

(1) If $\beta_{i_1}=m\delta+\alpha_1$ $(m\in\mathbb{Z}_+)$, we use the induction hypothesis on m. For $m=0$, $E_{\beta_{i_1}}\cdots E_{\beta_{i_{N+1}}}=E_{\alpha_1}\cdots E_{\beta_{i_{N+1}}}$. By induction, $E_{\beta_{i_2}}\cdots E_{\beta_{i_{N+1}}}$ is a linear combination of $E_{\beta_{l_1}}\cdots E_{\beta_{l_M}}$'s with $M\le N$ and $\beta_{l_1}\le\cdots\le\beta_{l_M}$. As $\alpha_1\le\beta_{l_1}$, $E_{\alpha_1}E_{\beta_{l_1}}\cdots E_{\beta_{l_M}}\in \mathrm{B}_+$, the claim holds. Suppose that it is true for $m\le n$. Now consider any word $E_{\beta_{i_1}}\cdots E_{\beta_{i_{N+1}}}=E_{(n+1)\delta+\alpha_1}E_{\beta_{i_2}}\cdots E_{\beta_{i_{N+1}}}$ of length $N{+}1$. Use induction on subword $E_{\beta_{i_2}}\cdots E_{\beta_{i_{N+1}}}$, and it suffices to consider those words: $E_{(n+1)\delta+\alpha_1}E_{\beta_{l_1}}\cdots E_{\beta_{l_M}}$ with $\beta_{l_1}\le\cdots\le\beta_{l_M}$.

If $(n{+}1)\delta+\alpha_1\le\beta_{l_1}$, there is nothing to prove.

If $(n{+}1)\delta+\alpha_1>\beta_{l_1}$, then by convention on the ordering $<$, $\exists\ k<n{+}1$, such that $\beta_{l_1}=k\delta+\alpha_1$. By Proposition 4.2, we have

$$E_{\beta_{i_1}}E_{\beta_{l_1}}\cdots E_{\beta_{l_M}}=\sum_{h=0}^{[\frac{n-k}{2}]}a_h^{(n-k)}E_{\beta_{l_1}+h\delta}E_{\beta_{i_1}-h\delta}E_{\beta_{l_2}}\cdots E_{\beta_{l_M}},\tag{5.1}$$

where $(k{+}h)\delta{+}\alpha_1=\beta_{l_1}+h\delta\le\beta_{i_1}-h\delta\le\beta_{i_1}$. Note that $k+h<n+1$, so by induction on m, this gives us the result as required.

(2) If $\beta_{i_1}=m\delta$, we adopt induction on m. For $m=1$, $E_{\beta_{i_1}}\cdots E_{\beta_{i_{N+1}}}=E_\delta E_{\beta_{i_2}}\cdots E_{\beta_{i_{N+1}}}$, by induction, $E_{\beta_{i_2}}\cdots E_{\beta_{i_{N+1}}}$ is a linear combination of $E_{\beta_{j_1}}\cdots E_{\beta_{j_M}}$'s with $M\le N$ and $\beta_{j_1}\le\cdots\le\beta_{j_M}$. If $\delta\le\beta_{j_1}$, $E_\delta E_{\beta_{j_1}}\cdots E_{\beta_{j_M}}\in\mathrm{B}_+$. If $\delta>\beta_{j_1}$, by definition, $E_{\beta_{j_1}}=E_{b\delta+\alpha_1}$ $(b\in\mathbb{Z}_+)$, by (4.7), we have that

$$E_{\beta_{i_1}}E_{\beta_{j_1}}\cdots E_{\beta_{j_M}}=-E_{\beta_{j_1}}E_{\beta_{i_1}}E_{\beta_{j_2}}\cdots E_{\beta_{j_M}}+[2]_qE_{\beta_{i_1}+\beta_{j_1}}E_{\beta_{j_2}}\cdots E_{\beta_{j_M}}.$$

Since $E_{\beta_{j_1}}=E_{b\delta+\alpha_1}$ and $E_{\beta_{i_1}+\beta_{j_1}}=E_{(b+m)\delta+\alpha_1}$, the two summands have been reduced to the case (1) above.

Suppose that it is true for $m\le n$. Now consider the word $E_{(n+1)\delta}E_{\beta_{i_2}}\cdots E_{\beta_{i_{N+1}}}$. By induction, $E_{\beta_{i_2}}\cdots E_{\beta_{i_{N+1}}}$ is a linear combination of $E_{\beta_{j_1}}\cdots E_{\beta_{j_M}}$'s with $M\le N$ $(\beta_{j_1}\le\cdots\le\beta_{j_M})$.

If $(n{+}1)\delta\le\beta_{j_1}$, then $E_{(n+1)\delta}E_{\beta_{j_1}}\cdots E_{\beta_{j_M}}\in\mathrm{B}_+$.

If $(n{+}1)\delta>\beta_{j_1}$, then $\beta_{j_1}\in\{\,k\delta+\alpha_1,s\delta\,\}$ with $s>n{+}1$. When $\beta_{j_1}=s\delta$, Proposition 4.8 gives $E_{\beta_{j_1}}E_{(n+1)\delta}\cdots E_{\beta_{j_M}}\in B_+$. When $\beta_{j_1}=k\delta+\alpha_1$, by the 1st formula of Remark 4.5 (1), we have

$$\begin{aligned}E_{(n+1)\delta}E_{k\delta+\alpha_1}E_{\beta_{j_2}}\cdots E_{\beta_{j_M}}&=\sum_{r=0}^{n}{b'_r}^{(n+1)}E_{(k+r)\delta+\alpha_1}E_{(n+1-r)\delta}E_{j_2}\cdots E_{\beta_{j_M}}\\&\quad+{b'_{n+1}}^{(n+1)}E_{(n+k+1)\delta+\alpha_1}E_{\beta_{j_2}}\cdots E_{\beta_{j_M}}.\end{aligned}\tag{5.2}$$

These summands have already been reduced to the case (1) above, so the claim is true.

(3) For $\beta_{i_1}=m\delta+\alpha_2$: we proceed (3) as in cases (1) and (2), using the 2nd formula of Remark 4.5 (1), or Corollary 4.10. Thus the claim is valid.

This completes the proof. □

Let us define an associative $\mathbb{Q}(q)$-subalgebra A of $\mathcal{U}_q^+$ generated by $E_{n\delta}$'s $(\forall\, n\in\mathbb{N})$, and define a set $A_+:=\{{E_\delta}^{\ell_1}\cdots {E_{s\delta}}^{\ell_s}\mid(\ell_1,\cdots,\ell_s)\in\mathbb{Z}_+^s,\ \forall\ s\in\mathbb{N}\}$.

Proposition 5.2. *The elements of A_+ form a basis of A.*

PROOF. It is obvious that each element of A is a linear combination among the elements in A_+. We now prove that the elements of A_+ are linearly independent. The polynomial ring $\mathbb{Q}(q)[x_i \mid i \in \mathbb{N}]$ is an A-module in the following way:

$$E_{s\delta} \circ 1 = x_s, \quad E_{s\delta} \circ x_{i_1} \cdots x_{i_r} = x_s x_{i_1} \cdots x_{i_r}.$$

If we have $\forall (\ell_1, \cdots, \ell_s) \in \mathbb{Z}_+^s, a_{\ell_1,\cdots,\ell_s} \in \mathbb{Q}(q)$ such that

$$\sum_{(\ell_1,\cdots,\ell_s)} a_{\ell_1,\cdots,\ell_s} E_\delta{}^{\ell_1} \cdots E_{s\delta}{}^{\ell_s} = 0,$$

then we can obtain

$$\sum_{(\ell_1,\cdots,\ell_s)} a_{\ell_1,\cdots,\ell_s} E_\delta{}^{\ell_1} \cdots E_{s\delta}{}^{\ell_s} \circ 1 = \sum_{(\ell_1,\cdots,\ell_s)} a_{\ell_1,\cdots,\ell_s} x_1{}^{\ell_1} \cdots x_s{}^{\ell_s} = 0.$$

We have already known that $\{x_1{}^{\ell_1} \cdots x_s{}^{\ell_s} \mid (\ell_1, \cdots, \ell_s) \in \mathbb{Z}_+^s, \forall s \in \mathbb{N}\}$ is a basis of $\mathbb{Q}(q)[x_i \mid i \in \mathbb{N}]$, and so all $a_{\ell_1,\cdots,\ell_s} = 0$. The claim is valid. □

LEMMA 5.3. *In each summand of $\Delta(E_{n_1\delta+\alpha_1})$, the root of the left-hand element of $\otimes$ is not $m_1\delta + \alpha_2$, and the root of the right-hand element of $\otimes$ is not $m_2\delta + \alpha_1$ except $E_{n_1\delta+\alpha_1}$, for $m_1, m_2 < n_1, \forall n_1, m_1, m_2 \in \mathbb{Z}_+$.*

PROOF. We can prove the claim by induction on n_1. For $n_1 = 0$, $\Delta(E_1) = E_1 \otimes 1 + K_1 \otimes E_1$ is true. For $n_1 = n$, the claim is also valid by the induction hypothesis. We now prove the claim for $n_1 = n + 1$. Using (4.7), we obtain

$$\Delta([2]_q E_{(n+1)\delta+\alpha_1}) = \Delta(E_{n\delta+\alpha_1})\Delta(E_\delta) + \Delta(E_\delta)\Delta(E_{n\delta+\alpha_1}),$$

where

$$\Delta(E_\delta) = E_\delta \otimes 1 + K_1K_2 \otimes E_\delta + (1 - q^{-4})E_1K_2 \otimes E_2,$$
$$\Delta(E_{n\delta+\alpha_1}) = E_{n\delta+\alpha_1} \otimes 1 + \cdots + K_1{}^{n+1}K_2{}^n \otimes E_{n\delta+\alpha_1}.$$

In each summand of $\Delta(E_{(n+1)\delta+\alpha_1})$, the root of the left-hand element of $\otimes$ is not $m_1\delta + \alpha_2$, and the root of the right-hand element of $\otimes$ is not $m_2\delta + \alpha_1$ except $E_\delta K_1{}^{n+1}K_2{}^n \otimes E_{n\delta+\alpha_1} + K_1{}^{n+1}K_2{}^n E_\delta \otimes E_{n\delta+\alpha_1} + K_1{}^{n+2}K_2{}^{n+1} \otimes (E_\delta E_{n\delta+\alpha_1} + E_{n\delta+\alpha_1}E_\delta)$, and for $E_\delta K_1{}^{n+1}K_2{}^n + K_1{}^{n+1}K_2{}^n E_\delta = 0$, the claim is valid. □

LEMMA 5.4. *In each summand of $\Delta(E_{n_2\delta+\alpha_2})$, the root of the left-hand element of $\otimes$ is not $l_1\delta + \alpha_2$ except $E_{n_2\delta+\alpha_2}$, and the root of the right-hand element of $\otimes$ is not $l_2\delta + \alpha_1$, for $l_1, l_2 < n_2, \forall n_2, l_1, l_2 \in \mathbb{Z}_+$.*

PROOF. It follows from (4.8) and a similar argument of the above Lemma. □

LEMMA 5.5. *In each summand of $\Delta(E_{n_3\delta})$, the root of the left-hand element of $\otimes$ is not $k_1\delta + \alpha_2$, and the root of the right-hand element of $\otimes$ is not $k_2\delta + \alpha_1$, for $k_1, k_2 < n_3, \forall n_3 \in \mathbb{N}, k_1, k_2 \in \mathbb{Z}_+$.*

PROOF. We can prove the claim by induction on n_3. For $n_3 = 0$, $\Delta(E_\delta) = E_\delta \otimes 1 + K_1K_2 \otimes E_\delta + (1 - q^{-4})E_1K_2 \otimes E_2$ is true. For $n_3 = n$, the claim is also valid by the induction hypothesis. We now prove the claim for $n_3 = n + 1$. Using Proposition 4.3 (2), we obtain

$$\Delta(E_{(n+1)\delta}) = \Delta(E_1)\Delta(E_{n\delta+\alpha_2}) + (-1)^{n+1}q^{-2}\Delta(E_{n\delta+\alpha_2})\Delta(E_1),$$

where

$$\Delta(E_1) = E_1 \otimes 1 + K_1 \otimes E_1,$$
$$\Delta(E_{n\delta+\alpha_2}) = E_{n\delta+\alpha_2} \otimes 1 + \cdots + K_1{}^nK_2{}^{n+1} \otimes E_{n\delta+\alpha_2}.$$

In each summand of $\Delta(E_{(n+1)\delta})$, the root of the right-hand element of $\otimes$ is not $k_2\delta+\alpha_1$, and the root of the left-hand element of $\otimes$ is not $l_1\delta+\alpha_2$ except $K_1E_{n\delta+\alpha_2}\otimes E_1+(-1)^{n+1}q^{-2}E_{n\delta+\alpha_2}K_1\otimes E_1$, and for $K_1E_{n\delta+\alpha_2}+(-1)^{n+1}q^{-2}E_{n\delta+\alpha_2}K_1=0$, the claim is valid. □

THEOREM 5.6. *The elements of* B_+ *form a basis of* $\mathcal{U}_q^+$.

PROOF. It follows from a similar argument to the one in [**17**] by using the comultiplication to separate linearly independent terms. □

Let us define the elements of B_+ such as $E_{\beta_1}^{r_1}\cdots E_{\beta_N}^{r_N}\triangleq E_\eta$, where $\eta=r_1\beta_1+\cdots+r_N\beta_N\in Q_+,\beta_1<\cdots<\beta_N$.

COROLLARY 5.7. $\{E_\eta K_1^nK_2^m\mid \eta\in Q_+,\, n,\, m\in\mathbb{Z}\}$ *is a basis of* $\mathcal{U}_q^{\geq 0}$.

THEOREM 5.8. *Set* $F_\alpha:=\Omega(E_\alpha)$ $(\forall\,\alpha\in Q_+)$. *Then the monomials* $F_\varsigma K_1^nK_2^mE_\eta$ *form a Poincaré-Birkhoff-Witt basis of* $\mathcal{U}_q(\widehat{\mathfrak{sl}_2})$.

Acknowledgments

The authors would like to express their many thanks to the referees for the valuable comments and improvements for the manuscript, as well as to the organizers of the International Workshop on Hopf Algebras and Tensor Categories (Sept. 9–13, 2019) held at Nanjing for having the opportunity to publish the results here.

References

[1] Nicolás Andruskiewitsch and Hans-Jürgen Schneider, *Pointed Hopf algebras*, New directions in Hopf algebras, Math. Sci. Res. Inst. Publ., vol. 43, Cambridge Univ. Press, Cambridge, 2002, pp. 1–68, DOI 10.2977/prims/1199403805. MR1913436

[2] Nicolás Andruskiewitsch and Hans-Jürgen Schneider, *A characterization of quantum groups*, J. Reine Angew. Math. **577** (2004), 81–104, DOI 10.1515/crll.2004.2004.577.81. MR2108213

[3] Jonathan Beck, *Convex bases of PBW type for quantum affine algebras*, Comm. Math. Phys. **165** (1994), no. 1, 193–199. MR1298947

[4] Jonathan Beck, *Braid group action and quantum affine algebras*, Comm. Math. Phys. **165** (1994), no. 3, 555–568. MR1301623

[5] Nantel Bergeron, Yun Gao, and Naihong Hu, *Drinfel′d doubles and Lusztig's symmetries of two-parameter quantum groups*, J. Algebra **301** (2006), no. 1, 378–405, DOI 10.1016/j.jalgebra.2005.08.030. MR2230338

[6] Ilaria Damiani, *A basis of type Poincaré-Birkhoff-Witt for the quantum algebra of* $\widehat{\mathrm{sl}}(2)$, J. Algebra **161** (1993), no. 2, 291–310, DOI 10.1006/jabr.1993.1220. MR1247357

[7] Christian Frønsdal, *Generalization and exact deformations of quantum groups*, Publ. Res. Inst. Math. Sci. **33** (1997), no. 1, 91–149, DOI 10.2977/prims/1195145535. MR1442494

[8] Christian Frønsdal, *On the classification of q-algebras*, Lett. Math. Phys. **53** (2000), no. 2, 105–120, DOI 10.1023/A:1026726700677. MR1804186

[9] Christian Frønsdal, *q-algebras and arrangements of hyperplanes*, J. Algebra **278** (2004), no. 2, 433–455, DOI 10.1016/j.jalgebra.2004.03.024. MR2070845

[10] Naihong Hu, Marc Rosso, and Honglian Zhang, *Two-parameter quantum affine algebra* $U_{r,s}(\widehat{\mathfrak{sl}_n})$, *Drinfel′d realization and quantum affine Lyndon basis*, Comm. Math. Phys. **278** (2008), no. 2, 453–486, DOI 10.1007/s00220-007-0405-1. MR2372766

[11] Naihong Hu and Xiuling Wang, *Convex PBW-type Lyndon bases and restricted two-parameter quantum groups of type B*, J. Geom. Phys. **60** (2010), no. 3, 430–453, DOI 10.1016/j.geomphys.2009.11.005. MR2600005

[12] Victor G. Kac, *Infinite-dimensional Lie algebras*, 3rd ed., Cambridge University Press, Cambridge, 1990, DOI 10.1017/CBO9780511626234. MR1104219

[13] Serge Levendorskii, Yan Soibelman, and Vladimir Stukopin, *The quantum Weyl group and the universal quantum R-matrix for affine Lie algebra* $A_1^{(1)}$, Lett. Math. Phys. **27** (1993), no. 4, 253–264, DOI 10.1007/BF00777372. MR1219498

[14] G. Lusztig, *Quantum deformations of certain simple modules over enveloping algebras*, Adv. in Math. **70** (1988), no. 2, 237–249, DOI 10.1016/0001-8708(88)90056-4. MR954661

[15] George Lusztig, *Quantum groups at roots of* 1, Geom. Dedicata **35** (1990), no. 1-3, 89–113, DOI 10.1007/BF00147341. MR1066560

[16] Yufeng Pei, Naihong Hu, and Marc Rosso, *Multi-parameter quantum groups and quantum shuffles. I*, Quantum affine algebras, extended affine Lie algebras, and their applications, Contemp. Math., vol. 506, Amer. Math. Soc., Providence, RI, 2010, pp. 145–171, DOI 10.1090/conm/506/09939. MR2642565

[17] Marc Rosso, *An analogue of P.B.W. theorem and the universal R-matrix for* $U_h\mathrm{sl}(N+1)$, Comm. Math. Phys. **124** (1989), no. 2, 307–318. MR1012870

[18] Marc Rosso, *Quantum groups and quantum shuffles*, Invent. Math. **133** (1998), no. 2, 399–416, DOI 10.1007/s002220050249. MR1632802

School of Mathematical Sciences, Shanghai Key Laboratory of PMMP, East China Normal University, Shanghai 200241, People's Republic of China
Email address: nhhu@math.ecnu.edu.cn

School of Mathematical Sciences, Shanghai Key Laboratory of PMMP, East China Normal University, Shanghai 200241, People's Republic of China
Email address: 51170601031@stu.ecnu.edu.cn

Contemporary Mathematics
Volume **771**, 2021
https://doi.org/10.1090/conm/771/15512

Finite quasi-quantum groups over abelian groups

Hua-Lin Huang, Gongxiang Liu, Yuping Yang, and Yu Ye

Abstract. This is a survey of some recent results on finite quasi-quantum groups over abelian groups. The works rely heavily on a unified and explicit formula of 3-cocycles on finite abelian groups. The formula reveals a useful separation of the 3-cocycles into two types, namely the abelian and the nonabelian ones. We give a brief summary to the classification of finite quasi-quantum groups of diagonal type induced by abelian 3-cocycles, and mention some initial advances of finite quasi-quantum groups of nondiagonal type induced by nonabelian 3-cocycles. Some problems for further research are also proposed.

1. Introduction

The classification problem of finite quasi-quantum groups (i.e., quasi-Hopf and coquasi-Hopf algebras) is motivated mainly by the theory of finite tensor categories [**17**] initiated by Etingof and Ostrik in the beginning of the century. As is clear that the general classification problem is far out of reach, it is necessary to narrow the scope and focus first on some interesting classes of tensor categories and quasi-quantum groups. In their pioneering work [**13**], Etingof and Gelaki proposed to classify nonsemisimple pointed finite tensor categories. By pointed it is meant that the simple objects are invertible. There are multifold reasons for this restriction: first, this kind of reduction is standard and powerful in representation theory; second, this class of tensor categories are essentially concrete, i.e., they admit quasi-fiber functors and they can be realized as the module categories of finite-dimensional elementary quasi-Hopf algebras by the Tannakian formalism [**17**]; third, this theory is a natural generalization of the deep and beautiful theory of finite-dimensional pointed Hopf algebras, see for example [**1–3, 9, 19, 22**] among many other works.

Similar to the Hopf situation, the familar reduction procedures of degeneration and deformation (or lifting), factorization and bosonization (or biproduct) are useful in the quasi-Hopf situation as well. Let H be a finite-dimensional pointed coquasi-Hopf algebra with associator Φ. By $\{H_n\}_{n\geq 0}$ we denote its coradical filtration, and

$$\operatorname{gr} H = H_0 \oplus H_1/H_0 \oplus H_2/H_1 \oplus \cdots$$

2020 *Mathematics Subject Classification.* Primary 16T05, 18D10, 17B37.
Key words and phrases. Quasi-quantum group, tensor category.
Supported by NSFC 11701468, 11722016, 11971181 and 11971449.
The third author is the corresponding author.

the corresponding coradically graded coalgebra. Then naturally $\operatorname{gr} H$ inherits from H a graded coquasi-Hopf algebra structure and is called a degeneration of H. The corresponding graded associator $\operatorname{gr}\Phi$ satisfies $\operatorname{gr}\Phi(\bar{a},\bar{b},\bar{c})=0$ for all homogeneous $\bar{a},\bar{b},\bar{c}\in \operatorname{gr} H$ unless they all lie in H_0. In particular, H_0 is a coquasi-Hopf subalgebra and it turns out to be $(kG, \operatorname{gr}\Phi)$ for $G=G(H)$, the set of group-like elements of H. We call a pointed coquasi-Hopf algebra H coradically graded if $H\cong \operatorname{gr} H$ as coquasi-Hopf algebras. Thus, any finite-dimensional pointed coquasi-Hopf algebra has a coradically graded degeneration. Conversely, one may lift or deform coradically graded pointed coquasi-Hopf algebras to get general ones. Now let $H=\oplus_{i\geq 0}H_i$ be a coradically graded pointed coquasi-Hopf algebra. Then H can be factorized as the form $R\#\Bbbk G$ where R is a connected graded Hopf algebra in a twisted Yetter-Drinfeld category of G. This reduces the study of coradically graded pointed coquasi-Hopf algebras to that of connected graded Hopf algebras in twisted Yetter-Drinfeld categories of groups. One can recover the former by the latter via the bosonization procedure.

To the best of our knowledge, so far most of the classification results on finite-dimensional quasi-Hopf algebras are in the coradically graded case. In [**13, 14**], Etingof and Gelaki obtained a series of classification results about graded elementary quasi-Hopf algebras over cyclic groups of prime order; in [**15, 18**], they studied graded elementary quasi-Hopf algebras over general cyclic groups and their liftings. One main achievement of this series of works is a complete classification of elementary quasi-Hopf algebras of rank 1. More importantly, a novel method of constructing genuine quasi-Hopf algebras (i.e., not twist equivalent to ordinary Hopf algebras) from known pointed Hopf algebras is invented. Along the same vein, Angiono classified in [**5**] finite-dimensional elementary quasi-Hopf algebras over cyclic groups whose orders have no small prime divisors. The basic idea of Etingof and Gelaki in [**13–15**] is embedding a genuine elementary quasi-Hopf algebras into an elementary quasi-Hopf algebra, possibly up to twist equivalence. The crux of these constructions is that there is a resolution for any given 3-cocycle on a cyclic group, namely, for any 3-cocycle σ on $\mathbb{Z}_n=\langle g|g^n=1\rangle$, the pull-back $\pi^*(\sigma)$ along the natural projection $\pi:\mathbb{Z}_{n^2}\to\mathbb{Z}_n$ is a 3-coboundary on $\mathbb{Z}_{n^2}$.

The previous idea can be easily extended to a full generality theoretically. However, there are several serious problems on the way of an explicit fulfillment. Some results in this direction are included in [**27, 28, 33**].

(1) The first problem is to find a resolution, if possible, for any normalized 3-cocycle on a finite group. This essentially lies in the cohomology theory of finite groups and is naturally split into two separated, namely the abelian and the nonabelian, cases. It is not surprised that the abelian case is relatively easier to handle. By extending the idea of [**30**], we are able to give a unified and explicit formula for a complete set of representatives of normalized 3-cocycles on any finite abelian groups. These 3-cocycles are further split into two types, that is the abelian and the nonabelian cocycles, according to whether or not the simple objects of the associated twisted Yetter-Drinfeld categories are all 1-dimensional. Moreover, we show that a 3-cocycle is resolvable by a finite abelian group if and only if it is abelian and we give an explicit resolution if this is indeed the case. For nonabelian 3-cocycles, there is no chance to resolve them and the associated twisted Yetter-Drinfeld categories are much more complicated.

For nonabelian groups, even the formulas for normalized 3-cocycles are not available yet.

(2) The second problem is to give a clear description of the Nichols algebras in a twisted Yetter-Drinfeld category ${}^{\mathbb{G}}_{\mathbb{G}}\mathcal{YD}^{\Phi}$. In case $\mathbb{G}$ and Φ are abelian, we can transform them to those in the usual Yetter-Drinfeld category ${}^{G}_{G}\mathcal{YD}$ by a delicate manipulation, where G is a finite abelian group with canonical projection $\pi : G \to \mathbb{G}$ such that $\pi^*(\Phi)$ is a 3-coboundary on G. The possibility of such a transformation is guaranteed since all the abelian 3-cocycles are resolvable. Then by combining Heckenberger's classification of arithmetic root systems [**22**], we achieve a complete classification of diagonal Nichols algebras with arithmetic root systems in ${}^{\mathbb{G}}_{\mathbb{G}}\mathcal{YD}^{\Phi}$. When the 3-cocycle Φ is nonabelian, the associated Nichols algebras become very complicated. Though we know in most cases they are infinite-dimensional, for the moment it is still far from getting a complete classification of the finite-dimensional ones.

(3) The third is the generation problem. In the diagonal case, with the transformation in (2) we can also reduce the problem of generation into that of Nichols algebras in the usual Yetter-Drinfeld categories of finite abelian groups. With a help of Angiono's result [**6**], we prove that finite-dimensional pointed coquasi-Hopf algebras of diagonal type are generated by group-likes and skew-primitive elements. Then we obtain a complete classification of finite-dimensional coradically graded pointed coquasi-Hopf algebras of diagonal type in a conceptual way. In nondiagonal case, a satisfactory theory of Nichols algebras is not available yet. So far only those Nichols algebras of semisimple twisted Yetter-Drinfeld modules with few summands are touched. It turns out that if the number of summands is less than or equal to 2, then we are able to make a connection from this to the diagonal case.

(4) The fourth problem is find a method to turn the conceptual classification into an operable procedure of constructions. In the diagonal case, for any given finite abelian group with fixed 3-cocycle and a compatible arithmetic root system, the construction is essentially a computational problem of linear congruence equations. We find two efficient ways, for most cases, to generate series of new genuine finite-dimensional pointed coquasi-Hopf algebras. In the nondiagonal case, basically we know nothing except very few sporadic examples.

There are also some other approaches to finite quasi-quantum groups and pointed finite tensor categories. For example, in [**7, 8**] Angiono, Galindo and Pereira studied the de-equivariantization of the category of comodules over a Hopf algebra and characterized pointed finite tensor categories over abelian groups constructed as de-equivariantizations of tensor categories of comodules over finite-dimensional pointed Hopf algebras. Some results obtained in *ibid.* are close to ours in [**27, 28**]. Specifically, abelian cocycles are called trivializable in the paper [**7**] by Angiono-Galindo, where the same description on the commutativity of the twisted Drinfeld double, and also an explicit characterization using results by Breen on cohomology [**11**] are found. In the same work the result about generation in degree one is included as well as a classification result in terms of de-equivariantizations when the 3-cocycle is abelian. In our earlier works [**25, 26**], we proposed an approach via quivers and representations. In particular, a Drozd trichotomy of graded pointed finite tensor categories was obtained in [**29**, **31**] by the well known theory of representation types of algebras. Due to the lack of space, we do not include any more details of these aspects.

The rest of the paper is organized as follows. In Section 2, we recall the formulas of normalized 3-cocycles on finite abelian groups and the application to a classification of braided Gr-categories. Section 3 is devoted to a brief summary of finite quasi-quantum groups of diagonal type. In Section 4, we present a first attempt to the classification problem of finite quasi-quantum groups of nondiagonal type. Finally, in Section 5 we provide some further research problems. Throughout the paper, $\Bbbk$ is an algebraically closed field with characteristic zero and all linear spaces are over $\Bbbk$. In accordance with our previous works [**25**, **26**, **29**, **31**], we only work on pointed coquasi-Hopf algebras. By taking linear dual, one has the version for elementary quasi-Hopf algebras. All the results mentioned in this paper can be found in [**27**, **28**, **32**, **33**]. In most cases, technical computations and long proofs are omitted. The interested reader is referred to *ibid.* for details.

2. Normalized cocycles on finite abelian groups and braided Gr-categories

In this section, we will present unified formulas for normalized cocycles of all degrees on finite abelian groups. As applications of the unified formulas of normalized 3-cocycles, we provide an explicit description of braided Gr-categories which are a simplest class of pointed finite tensor categories. For later applications, we also recall the definition of abelian 3-cocycles and their resolutions. The main results are from [**28**, **30**, **32**]. Throughout this section, we will use freely the concepts and notations about group cohomology in the book [**35**].

2.1. Normalized cocycles on finite abelian groups. Let G be a group and $(B_\bullet, \partial_\bullet)$ be its normalized bar resolution. Applying $\mathrm{Hom}_{\mathbb{Z}G}(-, \Bbbk^*)$ one gets a complex $(B_\bullet^*, \partial_\bullet^*)$. Denote the group of normalized k-cocycles by $Z^k(G, \Bbbk^*)$, which is $\mathrm{Ker}\,\partial_k^*$. In general, it is hard to determine $Z^n(G, \Bbbk^*)$ directly as the normalized bar resolution is far too large.

Our approach of formulating the normalized cocycles is straightforward and elementary. First we construct a Koszul-like resolution of a finite abelian group G by tensoring the minimal resolutions of cyclic factors of G and give a complete set of representatives of cocycles for this resolution. Then we construct a chain map from the normalized bar resolution to this Koszul-like resolution. Finally we get the desired explicit and unified formulas of normalized cocycles on G by pulling back those on the Koszul-like resolution along the chain map.

In the following we record only the formulas of normalized cocycles. The interested reader is referred to [**32**] for more details. Without loss of generality, we can assume that $G = \mathbb{Z}_{m_1} \times \cdots \times \mathbb{Z}_{m_n}$ where $m_i | m_{i+1}$ for all $1 \le i \le n-1$. Let g_i be a fixed generator of $\mathbb{Z}_{m_i}$, $1 \le i \le n$. Let $\alpha = (\alpha_1, \dots, \alpha_k)$ and $\alpha_u = (\alpha_{u1}, \dots, \alpha_{un})$ for $1 \le u \le k$, where each $\alpha_{ij} \in [0, m_j)$ and is viewed as an integer modulo m_j for all $1 \le i \le k$. For α_u, $1 \le u \le k$, we denote

$$g^{\alpha_u} = g_1^{\alpha_{u1}} \cdots g_n^{\alpha_{un}}.$$

For each positive integer m, fix a primitive m-th root of unity ζ_m. Let $1 \leq a < b \leq k, 1 \leq r \leq n$ be integers. Then for a fixed α we can define

$$\eta^{\alpha}_{r,[a,b]} := \begin{cases} [\frac{\alpha_{br}+\alpha_{b-1,r}}{m_r}]\cdots[\frac{\alpha_{a+1,r}+\alpha_{ar}}{m_r}], & b-a \text{ odd;} \\ [\frac{\alpha_{br}+\alpha_{b-1,r}}{m_r}]\cdots[\frac{\alpha_{a+2,r}+\alpha_{a+1,r}}{m_r}]\alpha_{ar}, & b-a \text{ even.} \end{cases}$$

Here by $[\frac{s}{t}]$ we denote the integer part of $\frac{s}{t}$ for any positive integers s and t. Let

$$\mathcal{R}_l := \{(r_1, r_2, \cdots, r_l) | 1 \leq r_1 < r_2 < \cdots < r_l \leq n\},$$

$$\Lambda_l := \{(\lambda_1, \lambda_2, \cdots, \lambda_l) | \lambda_1 + \lambda_2 + \cdots \lambda_l = k, \ \lambda_1 \text{ odd}, \ \lambda_i \geq 1 \text{ for } 1 \leq i \leq l\}.$$

Let Γ be the set of all sequences of the form

$$(a_{r_1^{\lambda_1}\cdots r_l^{\lambda_l}})_{(r_1,\cdots,r_l)\in\mathcal{R}_l, \ (\lambda_1,\cdots,\lambda_l)\in\Lambda_l, \ 1\leq l\leq k}$$

such that $0 \leq a_{r_1^{\lambda_1}\cdots r_l^{\lambda_l}} \leq m_{r_1}$ for $(r_1, \cdots, r_l) \in \mathcal{R}_l, (\lambda_1, \cdots, \lambda_l) \in \Lambda_l, 1 \leq l \leq k$.

Then we have the following

PROPOSITION 2.1. [**32**, Corollary 2.5] *For each sequence* $a = (a_{r_1^{\lambda_1}\cdots r_l^{\lambda_l}})_{(r_1,\cdots,r_l)\in\mathcal{R}_l, \ (\lambda_1,\cdots,\lambda_l)\in\Lambda_l, \ 1\leq l\leq k}$ *in* Γ, *we can define a k-cochains* $\omega \in \mathrm{Hom}_{\mathbb{Z}G}(B_k, \mathbb{k}^*)$ *by*

$$(2.1) \quad \omega_a([g^{\alpha_1}, \ldots, g^{\alpha_k}]) = \prod_{l=1}^{k} \prod_{\substack{(r_1,\cdots,r_l)\in\mathcal{R}_l \\ (\lambda_1,\cdots,\lambda_l)\in\Lambda_l}} \zeta_{m_{r_1}}^{(-1)^{\sum_{1\leq i<j\leq l}\lambda_i\lambda_j}\eta^{\alpha}_{r_1,[a_1,b_1]}\cdots\eta^{\alpha}_{r_l,[a_l,b_l]}a_{r_1^{\lambda_1}\cdots r_l^{\lambda_l}}}$$

where $a_u = \sum_{i=u+1}^{l} \lambda_i + 1, \quad b_u = \sum_{i=u}^{l} \lambda_i$ *for* $(\lambda_1, \cdots, \lambda_l) \in \Lambda_l, \ 1 \leq u \leq l$. *The set* $\{\omega_a | a \in \Gamma\}$ *is a complete set of representatives of k-cocycles of the complex* $(B^*_\bullet, \partial^*_\bullet)$.

2.2. Braided Gr-categories. The monoidal category of finite dimensional vector spaces graded by a group G, with the usual tensor product and associativity constraint given by a 3-cocycle ω is denoted by Vec^{ω}_G. Such a monoidal category is called a linear Gr-category. The terminology goes back to Hoàng Xuân Sính [**24**], a student of Grothendieck. The aim of this subsection is to give a complete description to all braided linear Gr-categories with a help of the explicit formulas of normalized 3-cocycles.

Recall that the category Vec_G of finite-dimensional G-graded vector spaces has simple objects $\{S_g | g \in G\}$ where $(S_g)_h = \delta_{g,h}\mathbb{k}, \ \forall h \in G$. The tensor product is given by $S_g \otimes S_h = S_{gh}$, and S_1 (1 is the identity of G) is the unit object. Without loss of generality we may assume that the left and right unit constraints are identities. If a is an associativity constraint on Vec_G, then it is given by $a_{S_f,S_g,S_h} = \omega(f,g,h)\,\mathrm{id}$, where $\omega : G \times G \times G \to \mathbb{k}^*$ is a function. The pentagon axiom and the triangle axiom give

$$\omega(ef, g, h)\omega(e, f, gh) = \omega(e, f, g)\omega(e, fg, h)\omega(f, g, h),$$
$$\omega(f, 1, g) = 1,$$

which say exactly that ω is a normalized 3-cocycle on G. Note that cohomologous cocycles define equivalent monoidal structures, therefore the classification of monoidal structures on Vec_G is equivalent to determining a complete set of representatives of normalized 3-cocycles on G.

Keep the notations of Subsection 2.1. One can easily give the explicit formulas of normalized 3-cocycle on G by (2.1). Define $\mathcal{A}$ to be the set of all sequences like

$$(a_1, \ldots, a_l, \ldots, a_n, a_{12}, \ldots, a_{ij}, \ldots, a_{n-1,n}, a_{123}, \ldots, a_{rst}, \ldots, a_{n-2,n-1,n}) \tag{2.2}$$

such that $0 \leq a_l < m_l$, $0 \leq a_{ij} < m_i$, $0 \leq a_{rst} < m_r$ for $1 \leq l \leq n$, $1 \leq i < j \leq n$, $1 \leq r < s < t \leq n$ where a_{ij} and a_{rst} are ordered by the lexicographic order. In the following, the sequence (2.2) is denoted by $\underline{\mathbf{a}}$ for short. In the special case $k = 3$, if we abbreviate a_{r^3}, $a_{r^1s^2}$ and $a_{r^1s^1t^1}$ in formula (2.1) by a_r, a_{rs} and a_{rst} respectively, then (2.1) becomes

$$\begin{aligned} &\Phi_{\underline{\mathbf{a}}} : B_3 \longrightarrow \Bbbk^* \\ &[g_1^{i_1} \cdots g_n^{i_n}, g_1^{j_1} \cdots g_n^{j_n}, g_1^{k_1} \cdots g_n^{k_n}] \\ &\mapsto \prod_{r=1}^{n} \zeta_{m_r}^{a_r i_r [\frac{j_r + k_r}{m_r}]} \prod_{1 \leq r < s \leq n} \zeta_{m_r}^{a_{rs} k_r [\frac{i_s + j_s}{m_s}]} \prod_{1 \leq r < s < t \leq n} \zeta_{m_r}^{-a_{rst} k_r j_s i_t} \end{aligned} \tag{2.3}$$

where $0 \leq a_r, a_{rs}, a_{rst} < m_r$. This is a complete set of representatives of normalized 3-cocycles on G.

Now we consider the braided structures on a linear Gr-category Vec_G^ω. Recall that a braiding in Vec_G^ω is a commutativity constraint $c : \otimes \to \otimes^{\mathrm{op}}$ satisfying the hexagon axiom. Note that c is given by $c_{S_x, S_y} = \mathcal{R}(x, y)\,\mathrm{id}$, where $\mathcal{R} : G \times G \to \Bbbk^*$ is a function, and the hexagon axiom of c says that

$$\frac{\mathcal{R}(xy, z)}{\mathcal{R}(x, z)\mathcal{R}(y, z)} \frac{\omega(x, z, y)}{\omega(x, y, z)\omega(z, x, y)} = 1 = \frac{\mathcal{R}(x, yz)}{\mathcal{R}(x, y)\mathcal{R}(x, z)} \frac{\omega(x, y, z)\omega(y, z, x)}{\omega(y, x, z)} \tag{2.4}$$

for all $x, y, z \in G$.

In other words, $\mathcal{R}$ is a quasi-bicharacter of G with respect to ω. Therefore, the classification of braidings in Vec_G^ω is equivalent to determining all the quasi-bicharacters of G with respect to ω. Clearly, any quasi-bicharacter $\mathcal{R}$ is uniquely determined by the following values:

$$r_{ij} := \mathcal{R}(g_i, g_j), \qquad \text{for all } 1 \leq i,\ j \leq n.$$

PROPOSITION 2.2. [**32**, Proposition 3.2] *Let $r_{ij} \in \Bbbk^*$ for $1 \leq i,\ j \leq n$. Then there is a quasi-bicharacter $\mathcal{R}$ with respect to ω satisfying $\mathcal{R}(g_i, g_j) = r_{ij}$ if and only if the following equations are satisfied:*

$$\begin{aligned} r_{ii}^{m_i} &= \zeta_{m_i}^{a_i} = \zeta_{m_i}^{-a_i}, && \text{for } 1 \leq i \leq n, \\ r_{ij}^{m_i} = r_{ji}^{m_i} &= 1, \quad a_{ij} = 0, && \text{for } 1 \leq i < j \leq n, \\ a_{rst} &= 0, && \text{for } 1 \leq r < s < t \leq n. \end{aligned}$$

2.3. Abelian 3-cocycles. In order to define abelian 3-cocycles, we need to recall first twisted quantum doubles [**12**]. By definition, the twisted quantum double $D^\Phi(G)$ of G with respect to the 3-cocycle Φ over G is the semisimple quasi-Hopf algebra with underlying vector space $(\Bbbk G)^* \otimes \Bbbk G$ in which multiplication, comultiplication Δ, associator ϕ, counit ε, antipode $\mathcal{S}$, α and β are given by

$$\begin{aligned}
&(e(g)\otimes x)(e(h)\otimes y) = \theta_g(x,y)\delta_{g,h}e(g)\otimes xy,\\
&\Delta(e(g)\otimes x) = \sum_{hk=g}\gamma_x(h,k)e(h)\otimes x\otimes e(k)\otimes x,\\
&\phi = \sum_{g,h,k\in G}\Phi(g,h,k)^{-1}e(g)\otimes 1\otimes e(h)\otimes 1\otimes e(k)\otimes 1,\\
&\mathcal{S}(e(g)\otimes x) = \theta_{g^{-1}}(x,x^{-1})^{-1}\gamma_x(g,g^{-1})^{-1}e(g^{-1})\otimes x^{-1},\\
&\varepsilon(e(g)\otimes x) = \delta_{g,1},\quad \alpha = 1,\quad \beta = \sum_{g\in G}\Phi(g,g^{-1},g)e(g)\otimes 1,
\end{aligned}$$

where $\{e(g)|g\in G\}$ is the dual basis of $\{g|g\in G\}$, $\delta_{x,y}$ is the Kronecker delta, and

$$\begin{aligned}
\theta_g(x,y) &= \frac{\Phi(g,x,y)\Phi(x,y,g)}{\Phi(x,g,y)},\\
\gamma_g(x,y) &= \frac{\Phi(x,y,g)\Phi(g,x,y)}{\Phi(x,g,y)}
\end{aligned}$$

for any $x,y,g\in G$.

DEFINITION 2.3. A 3-cocycle Φ over G is called abelian if $D^{\Phi}(G)$ is a commutative algebra.

With a help of our explicit formulas of normalized 3-cocycles, one can easily single out the abelian ones. Let $\mathbb{G}$ be a finite abelian group. So $\mathbb{G}\cong \mathbb{Z}_{\mathbb{m}_1}\times\cdots\times\mathbb{Z}_{\mathbb{m}_n}$ with $\mathbb{m}_j\in\mathbb{N}$ for $1\le j\le n$ and $\mathbb{m}_i|\mathbb{m}_{i+1}$ for all $1\le i\le n-1$. Let $\mathbb{g}_i$ be a generator of $\mathbb{Z}_{\mathbb{m}_i}$. Suppose Φ is a normalized 3-cocycle on $\mathbb{G}$. Thanks to (2.3), we may assume that $\Phi=\Phi_{\underline{\mathbf{a}}}$ for some $\mathbf{a}\in A$. Then we have the following numerical description of abelian 3-cocycles.

PROPOSITION 2.4. *The 3-cocycle $\Phi_{\underline{\mathbf{a}}}$ is abelian if and only if $a_{rst}=0$ for all $1\le r<s<t\le n$.*

PROOF. "$\Leftarrow$:" If all $a_{rst}=0$, then by (2.3) it is not hard to find that

$$\Phi_{\underline{\mathbf{a}}}(x,y,z)=\Phi_{\underline{\mathbf{a}}}(x,z,y)$$

for $x,y,z\in\mathbb{G}$. From this, we can find that

$$\theta_g(x,y)=\theta_g(y,x)$$

for $g,x,y\in\mathbb{G}$, which implies that $D^{\Phi_{\underline{\mathbf{a}}}}(\mathbb{G})$ is commutative.

"$\Rightarrow$:" If $a_{rst}\neq 0$ for some $r<s<t$. Through direct computations, we have

$$\theta_{\mathbb{g}_r}(\mathbb{g}_s,\mathbb{g}_t)=1,\qquad \theta_{\mathbb{g}_r}(\mathbb{g}_t,\mathbb{g}_s)=\zeta_{\mathbb{m}_r}^{-a_{rst}}.$$

This implies that

$$(e(\mathbb{g}_r)\otimes\mathbb{g}_s)(e(\mathbb{g}_r)\otimes\mathbb{g}_t)\neq(e(\mathbb{g}_r)\otimes\mathbb{g}_t)(e(\mathbb{g}_r)\otimes\mathbb{g}_s).$$

$\square$

2.4. Resolution. Let $\mathbb{G} = \mathbb{Z}_{\mathbbm{m}_1} \times \cdots \times \mathbb{Z}_{\mathbbm{m}_n}$ as before and $\Phi_{\underline{\mathbf{a}}}$ be an abelian 3-cocycle of $\mathbb{G}$. Then $\Phi_{\underline{\mathbf{a}}}$ can be "resolved" in a slightly bigger abelian group G. More precisely, take $G = \mathbb{Z}_{m_1} \times \cdots \times \mathbb{Z}_{m_n}$ for $m_i = \mathbbm{m}_i^2$ $(1 \le i \le n)$. As before, let $\mathbbm{g}_i$ (resp. g_i) be a generator of $\mathbb{Z}_{\mathbbm{m}_i}$ (resp. $\mathbb{Z}_{m_i}$) for $1 \le i \le n$. Using such notations, we have a canonical group epimorphism:

$$\pi : G \to \mathbb{G}, \quad g_i \mapsto \mathbbm{g}_i \ (1 \le i \le n).$$

By pulling back the 3-cocycles on $\mathbb{G}$ along π, we get 3-cocycles

$$\pi^*(\Phi_{\underline{\mathbf{a}}}) : G \times G \times G \to k^*, \quad (g, h, z) \mapsto \Phi_{\underline{\mathbf{a}}}(\pi(g), \pi(h), \pi(z)), \quad g, h, z \in G$$

on G. Then $\pi^*(\Phi_{\underline{\mathbf{a}}})$ is in fact a coboundary. To see this clearly, consider the following map

(2.5)

$$J_{\underline{\mathbf{a}}} : G \times G \to k^*; \quad (g_1^{x_1} \cdots g_n^{x_n}, g_1^{y_1} \cdots g_n^{y_n}) \mapsto \prod_{l=1}^{n} \zeta_{m_l}^{a_l x_l (y_l - y_l')} \prod_{1 \le s < t \le n} \zeta_{\mathbbm{m}_s \mathbbm{m}_t}^{a_{st} x_t (y_s - y_s')}$$

where y_i' is the remainder of y_i divided by $\mathbbm{m}_i$ for $1 \le i \le n$. Here for simple, we just take $\zeta_t = e^{\frac{2\pi i}{t}}$ for $t \in \mathbb{N}$. Thus, we have

PROPOSITION 2.5. *The differential of $J_{\underline{\mathbf{a}}}$ equals to $\pi^*(\Phi_{\underline{\mathbf{a}}})$, that is*

$$\partial(J_{\underline{\mathbf{a}}}) = \pi^*(\Phi_{\underline{\mathbf{a}}}).$$

PROOF.

$$\begin{aligned}
&\partial(J_{\underline{\mathbf{a}}})(g_1^{i_1} \cdots g_n^{i_n}, g_1^{j_1} \cdots g_n^{j_n}, g_1^{k_1} \cdots g_n^{k_n}) \\
&= \frac{J_{\underline{\mathbf{a}}}(g_1^{j_1} \cdots g_n^{j_n}, g_1^{k_1} \cdots g_n^{k_n}) J_{\underline{\mathbf{a}}}(g_1^{i_1} \cdots g_n^{i_n}, g_1^{j_1+k_1} \cdots g_n^{j_n+k_n})}{J_{\underline{\mathbf{a}}}(g_1^{i_1+j_1} \cdots g_n^{i_n+j_n}, g_1^{k_1} \cdots g_n^{k_n}) J_{\underline{\mathbf{a}}}(g_1^{i_1} \cdots g_n^{i_n}, g_1^{j_1} \cdots g_n^{j_n})} \\
&= \frac{\prod_{l=1}^n \zeta_{m_l}^{a_l j_l (k_l - k_l')} \prod_{1 \le s < t \le n} \zeta_{\mathbbm{m}_s \mathbbm{m}_t}^{a_{st} j_t (k_s - k_s')} \prod_{l=1}^n \zeta_{m_l}^{a_l i_l (j_l + k_l - (j_l + k_l)')} \prod_{1 \le s < t \le n} \zeta_{\mathbbm{m}_s \mathbbm{m}_t}^{a_{st} i_t (j_s + k_s - (j_s + k_s)')}}{\prod_{l=1}^n \zeta_{m_l}^{a_l (i_l + j_l)(k_l - k_l')} \prod_{1 \le s < t \le n} \zeta_{\mathbbm{m}_s \mathbbm{m}_t}^{a_{st} (i_t + j_t)(k_s - k_s')} \prod_{l=1}^n \zeta_{m_l}^{a_l i_l (j_l - j_l')} \prod_{1 \le s < t \le n} \zeta_{\mathbbm{m}_s \mathbbm{m}_t}^{a_{st} i_t (j_s - j_s')}} \\
&= \prod_{l=1}^{n} \zeta_{m_l}^{a_l i_l (j_l' + k_l' - (j_l + k_l)')} \prod_{1 \le s < t \le n} \zeta_{\mathbbm{m}_s \mathbbm{m}_t}^{a_{st} i_t (j_s' + k_s' - (j_s + k_s)')} \\
&= \prod_{l=1}^{n} \zeta_{\mathbbm{m}_l}^{a_l i_l' [\frac{j_l' + k_l'}{\mathbbm{m}_l}]} \prod_{1 \le s < t \le n} \zeta_{\mathbbm{m}_t}^{a_{st} i_t' [\frac{j_s' + k_s'}{\mathbbm{m}_s}]} \\
&= \pi^*(\Phi_{\underline{\mathbf{a}}})(g_1^{i_1} \cdots g_n^{i_n}, g_1^{j_1} \cdots g_n^{j_n}, g_1^{k_1} \cdots g_n^{k_n}).
\end{aligned}$$

□

However, there is no chance to resolve non-abelian 3-cocycles on bigger abelian groups.

PROPOSITION 2.6. [**27**, Proposition 3.17] *Let $\Phi_{\underline{\mathbf{a}}}$ be a non-abelian 3-cocycle on $\mathbb{G}$ and G be an arbitrary finite abelian group with a group epimorphism $\pi : G \twoheadrightarrow \mathbb{G}$. Then $\pi^*(\Phi_{\underline{\mathbf{a}}})$ is not a coboundary on G.*

In other words, for a 3-cocycle on a finite abelian group, it is abelian if and only if it is resolvable. We remark that abelian 3-cocycles are called trivializable in [**7, 11**] and some similar equivalent conditions are provided therein.

3. Finite quasi-quantum groups of diagonal type: A brief summary

In this section, we give a brief summary on a classification of finite-dimensional coradically graded pointed coquasi-Hopf algebras of diagonal type. The results presented here are included in [**27**, **28**].

Let $(\mathbb{M}, \Phi)$ be a finite-dimensional coradically graded pointed coquasi-Hopf algebra. Recall from Introduction that $\mathbb{M}$ can be factorized as $R\#\Bbbk\mathbb{G}$ with $\mathbb{G}$ a finite group and R a Hopf algebra in the twisted Yetter-Drinfeld category ${}^{\mathbb{G}}_{\mathbb{G}}\mathcal{YD}^{\Phi}$. If $\mathbb{G}$ is abelian, then $(\Bbbk\mathbb{G}, \Phi)$ is completely clear by Section 2. So the study of $\mathbb{M}$ is reduced to that of R. In accordance with the theory of finite-dimensional pointed Hopf algebras, the key here is the Nichols algebras in twisted Yetter-Drinfeld categories of finite abelian groups. We call the Nichols algebra $\mathcal{B}(V)$ of an object $V \in {}^{\mathbb{G}}_{\mathbb{G}}\mathcal{YD}^{\Phi}$ is of diagonal type, if V itself is of diagonal type (that is, a direct sum of 1-dimensional simple objects in ${}^{\mathbb{G}}_{\mathbb{G}}\mathcal{YD}^{\Phi}$.) Similarly, we call $\mathbb{M}$ diagonal if R is diagonal in ${}^{\mathbb{G}}_{\mathbb{G}}\mathcal{YD}^{\Phi}$. Thanks to Subsection 2.3, all simple objects of ${}^{\mathbb{G}}_{\mathbb{G}}\mathcal{YD}^{\Phi}$ are 1-dimensional if and only if both $\mathbb{G}$ and Φ are abelian. Moreover, in this case the Nichols algebra $\mathcal{B}(V)$ of $V \in {}^{\mathbb{G}}_{\mathbb{G}}\mathcal{YD}^{\Phi}$ can be transformed to a usual Nichols algebra as the abelian 3-cocycle Φ is resolvable. With this definite connection, the theory of finite-dimensional pointed Hopf algebras over abelian groups can be applied to our quasi situation.

Thus, in order to classify finite-dimensional coradically graded pointed coquasi-Hopf algebras of diagonal type, the main task is to give a classification of the Nichols algebras of diagonal type with arithmetic root system in ${}^{\mathbb{G}}_{\mathbb{G}}\mathcal{YD}^{\Phi}$. The idea to realize our purpose consists of five steps. Firstly, we can assume that the support group of $\mathcal{B}(V)$ is $\mathbb{G}$, and from this assumption we can prove that Φ must be an abelian 3-cocycle over $\mathbb{G}$. Secondly, we will develop a technique to change the base group from $\mathbb{G}$ to a bigger one G together with a group epimorphism $\pi : G \to \mathbb{G}$. Thirdly, we will show that any Nichols algebra $\mathcal{B}(V)$ in ${}^{\mathbb{G}}_{\mathbb{G}}\mathcal{YD}^{\Phi}$ is isomorphic to a Nichols algebra in ${}^{G}_{G}\mathcal{YD}^{\pi^*(\Phi)}$, which is thus twist equivalent to a usual Nichols algebra by Proposition 2.5. Fourthly, we want to get a return ticket, that is, we will give a sufficient and necessary condition to determine when a Nichols algebra in ${}^{\mathbb{G}}_{\mathbb{G}}\mathcal{YD}^{\pi^*(\Phi)}$ is isomorphic to one in ${}^{\mathbb{G}}_{\mathbb{G}}\mathcal{YD}^{\Phi}$. Finally, combining these results and Heckenberger's classification of arithmetic root systems, we obtain the classification of Nichols algebras of diagonal type with arithmetic root system in ${}^{\mathbb{G}}_{\mathbb{G}}\mathcal{YD}^{\Phi}$.

3.1. Yetter-Drinfeld modules over $(\Bbbk\mathbb{G}, \Phi)$ and Nichols algebras. There are general definitions of Yetter-Drinfeld modules over coquasi-Hopf algebras and Nichols algebras of braided tensor categories. For our purpose, it is enough to recall only the definitions of Yetter-Drinfeld modules over coquasi-Hopf algebras of form $(\Bbbk\mathbb{G}, \Phi)$ with $\mathbb{G}$ an abelian group and Nichols algebra in ${}^{\mathbb{G}}_{\mathbb{G}}\mathcal{YD}^{\Phi}$.

Assume that V is a left $\Bbbk\mathbb{G}$-comudule with comodule structure map $\delta_L : V \to \Bbbk\mathbb{G} \otimes V$. Define ${}^gV := \{v \in V | \delta_L(v) = g \otimes v\}$ and thus $V = \bigoplus_{g\in G} {}^gV$. Here we call g the degree of the elements in ${}^gV$ and denote by $\deg v = g$ for $v \in {}^gV$. For the 3-cocycle Φ on $\mathbb{G}$ and any $g \in \mathbb{G}$, define

$$\widetilde{\Phi}_g : \mathbb{G} \times \mathbb{G} \to \Bbbk^*, \quad (e, f) \mapsto \frac{\Phi(g, e, f)\Phi(e, f, g)}{\Phi(e, g, f)}. \tag{3.1}$$

Direct computation shows that

$$\widetilde{\Phi}_g \in \mathbb{Z}^2(\mathbb{G}, \Bbbk^*).$$

DEFINITION 3.1. A left $\Bbbk\mathbb{G}$-comudule V is a left-left Yetter-Drinfeld module over $(\Bbbk\mathbb{G}, \Phi)$ if each ${}^gV$ is a projective $\mathbb{G}$-representation with respect to the 2-cocycle $\widetilde{\Phi}_g$, namely the $\mathbb{G}$-action $\triangleright$ on ${}^gV$ satisfies

$$e \triangleright (f \triangleright v) = \widetilde{\Phi}_g(e, f)(ef) \triangleright v, \quad \forall e, f \in G, \ v \in {}^gV. \tag{3.2}$$

The category of left-left Yetter-Drinfeld modules is denoted by ${}^{\mathbb{G}}_{\mathbb{G}}\mathcal{YD}^{\Phi}$. Similarly, one can define left-right, right-left and right-right Yetter-Drinfeld modules over $(\Bbbk\mathbb{G}, \Phi)$. As the familiar Hopf case, ${}^{\mathbb{G}}_{\mathbb{G}}\mathcal{YD}^{\Phi}$ is a braided tensor category. More precisely, for any $M, N \in {}^{\mathbb{G}}_{\mathbb{G}}\mathcal{YD}^{\Phi}$, the structure maps of $M \otimes N$ as a left-left Yetter-Drinfeld module are given by

$$\delta_L(m_g \otimes n_h) := gh \otimes m_g \otimes n_h, \quad x \triangleright (m_g \otimes n_h) := \widetilde{\Phi}_x(g, h) x \triangleright m_g \otimes x \triangleright n_h \tag{3.3}$$

for all $x, g, h \in \mathbb{G}$ and $m_g \in {}^gM, \ n_h \in {}^hN$. The associativity constraint a and the braiding c of ${}^{\mathbb{G}}_{\mathbb{G}}\mathcal{YD}^{\Phi}$ are given respectively by

$$a((u_e \otimes v_f) \otimes w_g) = \Phi(e, f, g)^{-1} u_e \otimes (v_f \otimes w_g) \tag{3.4}$$

$$c(u_e \otimes v_f) = e \triangleright v_f \otimes u_e \tag{3.5}$$

for all $e, f, g \in \mathbb{G}, \ u_e \in {}^eU, \ v_f \in {}^fV, \ w_g \in {}^gW$ and $U, V, W \in {}^{\mathbb{G}}_{\mathbb{G}}\mathcal{YD}^{\Phi}$.

Nichols algebras can be defined by various equivalent ways, see for example [**2**]. Here we adopt the defining method in terms of the universal property. Roughly, Nichols algebras are the analogue of the usual symmetric algebras in more general braided tensor categories.

Let V be a nonzero object in ${}^{\mathbb{G}}_{\mathbb{G}}\mathcal{YD}^{\Phi}$. By $T_\Phi(V)$ we denote the tensor algebra in ${}^{\mathbb{G}}_{\mathbb{G}}\mathcal{YD}^{\Phi}$ generated freely by V. It is clear that $T_\Phi(V)$ is isomorphic to $\bigoplus_{n \geq 0} V^{\otimes \vec{n}}$ as a linear space, where $V^{\otimes \vec{n}}$ means

$$\underbrace{(\cdots((}_{n-1} V \otimes V) \otimes V) \cdots \otimes V).$$

This induces a natural $\mathbb{N}$-graded structure on $T_\Phi(V)$. Define a comultiplication on $T_\Phi(V)$ by $\Delta(X) = X \otimes 1 + 1 \otimes X$, $\forall X \in V$, a counit by $\varepsilon(X) = 0$, and an antipode by $S(X) = -X$. These provide a graded Hopf algebra structure on $T_\Phi(V)$ in the braided tensor category ${}^{\mathbb{G}}_{\mathbb{G}}\mathcal{YD}^{\Phi}$.

DEFINITION 3.2. The Nichols algebra $\mathcal{B}(V)$ of V is defined to be the quotient Hopf algebra $T_\Phi(V)/I$ in ${}^{\mathbb{G}}_{\mathbb{G}}\mathcal{YD}^{\Phi}$, where I is the unique maximal graded Hopf ideal generated by homogeneous elements of degree greater than or equal to 2.

To stress that our Nichols algebras may be nonassociative in some occasions, we will call an associative Nichols algebra, e.g. $\mathcal{B}(V) \in {}^{\mathbb{G}}_{\mathbb{G}}\mathcal{YD}$, a usual Nichols algebra. The following definition is used widely in the rest of this paper.

DEFINITION 3.3. Let $V \in {}^{\mathbb{G}}_{\mathbb{G}}\mathcal{YD}^{\Phi}$ be a Yetter-Drinfeld module of diagonal type and $\{X_i | 1 \leq i \leq n\}$ be a standard basis of V, and $\{g_i | 1 \leq i \leq n\}$ be the corresponding degrees, that is $\delta_L(X_i) = g_i \otimes X_i$ for $1 \leq i \leq n$. Then we call the subgroup $\mathbb{G}' = \langle g_1, \cdots, g_n \rangle$ generated by $g_1, \ldots, g_n$ the support group of V, which is denoted by $\mathbb{G}_V$.

It is obvious that the definition does not depend on the choices of the standard bases and

$$\mathbb{G}_V = \mathbb{G}_{\mathcal{B}(V)} = \mathbb{G}_{T_\Phi(V)}.$$

The twisting process for coquasi-Hopf algebras can be transferred to Nichols algebras directly. In fact, let $(V, \triangleright, \delta_L) \in {}^{\mathbb{G}}_{\mathbb{G}}\mathcal{YD}^{\Phi}$, and J a 2-cochain of $\mathbb{G}$. Then we can define a new action $\triangleright_J$ of $\mathbb{G}$ over V by

$$g \triangleright_J X = \frac{J(g,x)}{J(x,g)} g \triangleright X \tag{3.6}$$

for $X \in V$ and $g \in \mathbb{G}$. We denote $(V, \triangleright_J, \delta_L)$ by V^J and by definition we have $V^J \in {}^{\mathbb{G}}_{\mathbb{G}}\mathcal{YD}^{\Phi * \partial(J)}$. Moreover there is a tensor equivalence $(F_J, \varphi_0, \varphi_2) : {}^{\mathbb{G}}_{\mathbb{G}}\mathcal{YD}^{\Phi} \to {}^{\mathbb{G}}_{\mathbb{G}}\mathcal{YD}^{\Phi * \partial(J)}$ which takes V to V^J and

$$\varphi_2(U, V) : (U \otimes V)^J \to U^J \otimes V^J, \quad Y \otimes Z \mapsto J(y,z)^{-1} Y \otimes Z$$

for $Y \in U, \ Z \in V$.

Let $\mathcal{B}(V)$ be a usual Nichols algebra in ${}^{\mathbb{G}}_{\mathbb{G}}\mathcal{YD}$, then it is clear that $\mathcal{B}(V)^J$ is a Hopf algebra in ${}^{\mathbb{G}}_{\mathbb{G}}\mathcal{YD}^{\partial J}$ with multiplication $\circ$ determined by

$$X \circ Y = J(x,y) XY \tag{3.7}$$

for all homogenous elements $X, Y \in \mathcal{B}(V)$, here $x = \deg X, y = \deg Y$. Using the same terminology as (co)quasi-Hopf algebras, we call $\mathcal{B}(V)$ and $\mathcal{B}(V)^J$ are twist equivalent. The following fact is obvious, but important for our exposition.

LEMMA 3.4. *The twisting $\mathcal{B}(V)^J$ of $\mathcal{B}(V)$ is a Nichols algebra in ${}^{\mathbb{G}}_{\mathbb{G}}\mathcal{YD}^{\partial J}$ and $\mathcal{B}(V)^J \cong \mathcal{B}(V^J)$.*

3.2. Pointed coquasi-Hopf algebras and related notions. Recall that a coquasi-Hopf algebra $\mathbb{M}$ is said to be pointed if the underlying coalgebra is so. Given a pointed coquasi-Hopf algebra $(\mathbb{M}, \Delta, \varepsilon, \mathrm{M}, \mu, \Phi, \mathcal{S}, \alpha, \beta)$, let $\{\mathbb{M}_n\}_{n \geq 0}$ be its coradical filtration, and

$$\operatorname{gr} \mathbb{M} = \mathbb{M}_0 \oplus \mathbb{M}_1/\mathbb{M}_0 \oplus \mathbb{M}_2/\mathbb{M}_1 \oplus \cdots$$

the corresponding coradically graded coalgebra. Then $\operatorname{gr} \mathbb{M}$ is a coradically graded coquasi-Hopf algebra. The corresponding graded associator $\operatorname{gr} \Phi$ satisfies $\operatorname{gr} \Phi(\bar{a}, \bar{b}, \bar{c}) = 0$ for all homogeneous $\bar{a}, \bar{b}, \bar{c} \in \operatorname{gr} \mathbb{M}$ unless they all lie in $\mathbb{M}_0$. Similar condition holds for $\operatorname{gr} \alpha$ and $\operatorname{gr} \beta$. In particular, $\mathbb{M}_0$ is a coquasi-Hopf subalgebra and it turns out to be the coquasi-Hopf algebra $(\mathbb{k}\mathbb{G}, \operatorname{gr} \Phi)$ for $\mathbb{G} = \mathbb{G}(\mathbb{M})$, the set of group-like elements of $\mathbb{M}$. We call a pointed Majid algebra $\mathbb{M}$ *graded* if $\mathbb{M} \cong \operatorname{gr} \mathbb{M}$ as Majid algebras.

DEFINITION 3.5. Let $(\mathbb{M}, \Delta, \varepsilon, \mathrm{M}, \mu, \Phi, \mathcal{S}, \alpha, \beta)$ be a coquasi-Hopf algebra. A convolution-invertible linear map

$$J : \ \mathbb{M} \otimes \mathbb{M} \to \mathbb{k}$$

is called a twisting (or gauge transformation) on $\mathbb{M}$ if

$$J(h, 1) = \varepsilon(h) = J(1, h)$$

for all $h \in \mathbb{M}$.

Given a coquasi-Hopf algebra $\mathbb{M}$ and a twisting J, then one can construct a new coquasi-Hopf algebra $\mathbb{M}^J$ as follows: $\mathbb{M}^J = \mathbb{M}$ as a coalgebra and the multiplication "$\circ$" on $\mathbb{M}^J$ is given by

$$a \circ b := J(a_1, b_1) a_2 b_2 J^{-1}(a_3, b_3) \tag{3.8}$$

for all $a, b \in \mathbb{M}$. The associator Φ^J and the quasi-antipode $(\mathcal{S}^J, \alpha^J, \beta^J)$ are given as:

$$\Phi^J(a,b,c) = J(b_1,c_1)J(a_1,b_2c_2)\Phi(a_2,b_3,c_3)J^{-1}(a_3b_4,c_4)J^{-1}(a_4,b_5),$$
$$\mathcal{S}^J = \mathcal{S}, \quad \alpha^J(a) = J^{-1}(\mathcal{S}(a_1),a_3)\alpha(a_2), \quad \beta^J(a) = J(a_1,\mathcal{S}(a_3))\beta(a_2)$$

for all $a, b, c \in \mathbb{M}$.

DEFINITION 3.6. Two coquasi-Hopf algebras $\mathbb{M}_1$ and $\mathbb{M}_2$ are called twist equivalent if there is a twisting J on $\mathbb{M}_1$ such that $\mathbb{M}_1^J \cong \mathbb{M}_2$ as coquasi-Hopf algebras. Denote $\mathbb{M}_1 \sim \mathbb{M}_2$ if $\mathbb{M}_1$ is twist equivalent to $\mathbb{M}_2$. We call a coquasi-Hopf algebra $\mathbb{M}$ genuine if it is not twist equivalent to a Hopf algebra.

Now suppose $\mathbb{M} = \sum_{i\geq 0} \mathbb{M}_i$ is a coradically graded pointed coquasi-Hopf algebra. Let $\pi: \ \mathbb{M} \to \mathbb{M}_0$ be the canonical projection. Then $\mathbb{M}$ is a $k\mathbb{G}$-bicomodule naturally via

$$\delta_L := (\pi \otimes \mathrm{id})\Delta, \quad \delta_R := (\mathrm{id} \otimes \pi)\Delta.$$

Thus there is a $\mathbb{G}$-bigrading on $\mathbb{M}$, that is,

$$\mathbb{M} = \bigoplus_{g,h\in\mathbb{G}} {}^g\mathbb{M}^h$$

where ${}^g\mathbb{M}^h = \{m \in \mathbb{M} | \delta_L(m) = g \otimes m, \ \delta_R(m) = m \otimes h\}$. As a convention, we only deal with homogeneous elements with respect to this $\mathbb{G}$-bigrading in this subsection. For example, whenever we write $\Delta(X) = X_1 \otimes X_2$, all X, X_1, X_2 are assumed homogeneous, and for any capital $X \in {}^g\mathbb{M}^h$, we use its lowercase x to denote gh^{-1}.

Define the coinvariant subalgebra of $\mathbb{M}$ by

$$\mathcal{R} := \{m \in \mathbb{M} | (\mathrm{id} \otimes \pi)\Delta(m) = m \otimes 1\}.$$

Clearly $1 \in \mathcal{R}$. There is a $(\mathbb{k}\mathbb{G}, \Phi)$-action on $\mathcal{R}$ via

$$f \triangleright X := \frac{\Phi(fg, f^{-1}, f)}{\Phi(f, f^{-1}, f)}(f \cdot X) \cdot f^{-1} \tag{3.9}$$

for all $f, g \in \mathbb{G}$ and $X \in {}^g\mathcal{R}$. Here $\cdot$ is the multiplication in $\mathbb{M}$. Then $(\mathcal{R}, \delta_L, \triangleright)$ is a left-left Yetter-Drinfeld module over $(\mathbb{k}\mathbb{G}, \Phi)$.

$$\mathrm{M}: \ \mathcal{R} \otimes \mathcal{R} \to \mathcal{R}, \quad (X,Y) \mapsto XY := X \cdot Y;$$
$$u\colon \mathbb{k} \to \mathcal{R}, \quad \lambda \mapsto \lambda 1;$$
$$\Delta_{\mathcal{R}}: \ \mathcal{R} \to \mathcal{R} \otimes \mathcal{R}, \quad X \mapsto \Phi(x_1, x_2, x_2^{-1}) X_1 \cdot x_2^{-1} \otimes X_2;$$
$$\varepsilon_{\mathcal{R}}: \ \mathcal{R} \to \mathbb{k}, \quad \varepsilon_{\mathcal{R}} := \varepsilon|_{\mathcal{R}};$$
$$\mathcal{S}_{\mathcal{R}}: \ \mathcal{R} \to \mathcal{R}, \quad X \mapsto \tfrac{1}{\Phi(x,x^{-1},x)} x \cdot \mathcal{S}(X).$$

Then it is routine to verify that $(\mathcal{R}, \mathrm{M}, u, \Delta_{\mathcal{R}}, \varepsilon_{\mathcal{R}}, \mathcal{S}_{\mathcal{R}})$ is a Hopf algebra in ${}^{\mathbb{G}}_{\mathbb{G}}\mathcal{YD}^{\Phi}$.

Conversely, let H be a Hopf algebra in ${}^{\mathbb{G}}_{\mathbb{G}}\mathcal{YD}^{\Phi}$. Since H is a left $\mathbb{G}$-comodule, there is a $\mathbb{G}$-grading on H:

$$H = \bigoplus_{x\in G} {}^xH$$

where ${}^xH = \{X \in H | \delta_L(X) = x \otimes X\}$. As before, we only need to deal with $\mathbb{G}$-homogeneous elements. As a convention, homogeneous elements in H are denoted by capital letters, say $X, Y, Z, \ldots$, and the associated degrees are denoted by their lower cases, say $x, y, z, \ldots$.

For our purpose, we also assume that H is $\mathbb{N}$-graded with $H_0 = \mathbb{k}$. If $X \in H_n$, then we say that X has length n. Moreover, we assume that both gradings are compatible in the sense that

$$H = \bigoplus_{g\in\mathbb{G}} {}^{g}H = \bigoplus_{g\in\mathbb{G}} \bigoplus_{n\in\mathbb{N}} {}^{g}H_n.$$

For example, the Hopf algebra $\mathcal{R}$ in ${}_{\mathbb{G}}^{\mathbb{G}}\mathcal{YD}^{\Phi}$ considered above satisfies these assumptions as $\mathcal{R} = \bigoplus_{i\in\mathbb{N}} \mathcal{R}_i$ is coradically graded. In this case, we call $\dim\mathcal{R}_1$ the rank of $\mathcal{R}$ and $\mathbb{M}$. For any $X \in H$, we write its comultiplication as

$$\Delta_H(X) = X_{(1)} \otimes X_{(2)}.$$

The following bosonization formulas come from [**34**], a more general version for coquasi-bialgebras can be found in [**10**].

LEMMA 3.7. [**34**, Proposition 3.3] *Keep the assumptions on H as above. Define on $H \otimes \mathbb{k}\mathbb{G}$ a product by*

$$(X \otimes g)(Y \otimes h) = \frac{\Phi(xg, y, h)\Phi(x, y, g)}{\Phi(x, g, y)\Phi(xy, g, h)} X(g \triangleright Y) \otimes gh, \tag{3.10}$$

and a coproduct by

$$\Delta(X \otimes g) = \Phi(x_{(1)}, x_{(2)}, g)^{-1}(X_{(1)} \otimes x_{(2)}g) \otimes (X_{(2)} \otimes g). \tag{3.11}$$

Then $H \otimes \mathbb{k}\mathbb{G}$ becomes a graded coquasi-Hopf algebra with a quasi-antipode $(\mathcal{S}, \alpha, \beta)$ given by

$$\mathcal{S}(X \otimes g) = \frac{\Phi(g^{-1},g,g^{-1})}{\Phi(x^{-1}g^{-1},xg,g^{-1})\Phi(x,g,g^{-1})}(1 \otimes x^{-1}g^{-1})(\mathcal{S}_H(X) \otimes 1), \tag{3.12}$$

$$\alpha(1 \otimes g) = 1, \quad \alpha(X \otimes g) = 0, \tag{3.13}$$

$$\beta(1 \otimes g) = \Phi(g, g^{-1}, g)^{-1}, \quad \beta(X \otimes g) = 0, \tag{3.14}$$

here $g, h \in \mathbb{G}$ and X, Y are homogeneous elements of length ≥ 1.

Finally, we give the definition of connected pointed coquasi-Hopf algebras.

DEFINITION 3.8. Suppose $\mathbb{M}$ is a pointed coquasi-Hopf algebra and R is the coinvariant subalgebra of $\mathrm{gr}(\mathbb{M})$, then we say that $\mathbb{M}$ is connected if $\mathbb{G}(\mathbb{M}) = \mathbb{G}_R$, where $\mathbb{G}_R$ is the support group of R.

3.3. Arithmetic root systems and generalized Dynkin diagrams.

Arithmetic root systems are invariants of Nichols algebras of diagonal type with certain finiteness property. A complete classification of arithmetic root systems was given by Heckenberger [**22**]. This is a crucial ingredient for the classification program of finite-dimensional pointed Hopf algebras, and turns out to be equally important in the broader situation of pointed coquasi-Hopf algebras.

Suppose $\mathcal{B}(V)$ is a usual Nichols algebra of diagonal type in ${}_{\mathbb{G}}^{\mathbb{G}}\mathcal{YD}$. Let $\{X_i | 1 \leq i \leq n\}$ be a standard basis of V with $\delta_l(X_i) = h_i \otimes X_i$. The structure constants of $\mathcal{B}(V)$ are $\{q_{ij} | 1 \leq i, j \leq n\}$ such that $h_i \triangleright X_j = q_{ij}X_j$. Let $E = \{e_i | 1 \leq i \leq d\}$ be a canonical basis of $\mathbb{Z}^n$, and χ be a bicharacter of $\mathbb{Z}^n$ determined by $\chi(e_i, e_j) = q_{ij}$. As defined in [**19**, Sec.3], $\triangle^+(\mathcal{B}(V))$ is the set of degrees of the (restricted) Poincare-Birkhoff-Witt generators counted with multiplicities and $\triangle(\mathcal{B}(V)) := \triangle^+(\mathcal{B}(V)) \bigcup - \triangle^+ (\mathcal{B}(V))$, which is called the root system of $\mathcal{B}(V)$.

Moreover, the triple $(\triangle(\mathcal{B}(V)), \chi, E))$ is called an arithmetic root system of $\mathcal{B}(V)$ if the corresponding Weyl groupoid $W_{\chi,E}$ is full and finite. In this case, we denote this arithmetic root system by $\triangle(\mathcal{B}(V))_{\chi,E}$ for brevity. If there is another arithmetic root system $\triangle_{\chi',E'}$, and an isomorphism $\tau : \mathbb{Z}^n \to \mathbb{Z}^n$ such that

$$\tau(E) = E', \qquad \chi'(\tau(e), \tau(e)) = \chi(e, e),$$

$$\chi'(\tau(e_1), \tau(e_2))\chi'(\tau(e_2), \tau(e_1)) = \chi(e_1, e_2)\chi'(e_2, e_1),$$

then we say that $\triangle_{\chi,E}$ and $\triangle_{\chi',E'}$ are twist equivalent.

Generalized Dynkin diagrams are invariants of arithmetic root systems, and they can determine arithmetic root systems up to twist equivalence.

DEFINITION 3.9. The generalized Dynkin diagram of an arithmetic root system $\triangle_{\chi,E}$ is a nondirected graph $\mathcal{D}_{\chi,E}$ with the following properties:

1) There is a bijective map ϕ from $I = \{1, 2, \ldots, d\}$ to the set of vertices of $\mathcal{D}_{\chi,E}$.
2) For all $1 \le i \le d$, the vertex $\phi(i)$ is labelled by q_{ii}.
3) For all $1 \le i, j \le d$, the number n_{ij} of edges between $\phi(i)$ and $\phi(j)$ is either 0 or 1. If $i = j$ or $q_{ij}q_{ji} = 1$ then $n_{ij} = 0$, otherwise $n_{ij} = 1$ and the edge is labelled by $\widetilde{q_{ij}} = q_{ij}q_{ji}$ for $1 \le i < j \le n$.

An arithmetic root system is called connected provided the corresponding generalized Dynkin diagram $\mathcal{D}_{\chi,E}$ is connected. All the connected arithmetic root systems are classified and the corresponding generalized Dynkin diagrams are listed in [**20**–**22**].

3.4. Classification results. In this subsection, we provide a classification of finite-dimensional Nichols algebras of diagonal type in ${}^{\mathbb{G}}_{\mathbb{G}}\mathcal{YD}^{\Phi}$. This also leads to a complete classification of connected finite-dimensional pointed coquasi-Hopf algebras of diagonal type.

Since Nichols algebras in twisted Yetter-Drinfeld categories ${}^{\mathbb{G}}_{\mathbb{G}}\mathcal{YD}^{\Phi}$ are nonassociative algebras, the structures of these algebras depend on $\mathbb{G}$ and the 3-cocycle Φ on $\mathbb{G}$. We will call $\mathbb{G}$ the base group of $\mathcal{B}(V)$. An important method is to change the base groups of Nichols algebras.

DEFINITION 3.10. Let $\mathcal{B}(V)$ and $\mathcal{B}(U)$ be Nichols algebras in ${}^{G}_{G}\mathcal{YD}^{\Phi}$ and ${}^{H}_{H}\mathcal{YD}^{\Psi}$ respectively with $\dim V = \dim U = l$. We say $\mathcal{B}(V)$ is isomorphic to $\mathcal{B}(U)$ if there is a $\mathbb{Z}^l$-graded linear isomorphism $\mathcal{F} : \mathcal{B}(V) \to \mathcal{B}(U)$ which preserves the multiplication and comultiplication.

LEMMA 3.11. [**28**, Lemma 4.4] *Suppose $V \in {}^{G}_{G}\mathcal{YD}^{\Phi}$ and $U \in {}^{H}_{H}\mathcal{YD}^{\Psi}$. Let G' and H' be support groups of V and U respectively. If there are a linear isomorphism $F : V \to U$ and a group epimorphism $f : G' \to H'$ such that:*

$$\delta \circ F = (f \otimes F) \circ \delta, \tag{3.15}$$

$$F(g \triangleright v) = f(g) \triangleright F(v), \tag{3.16}$$

$$\Phi|_{G'} = f^* \Psi|_{H'} \tag{3.17}$$

for any $g \in G'$, $v \in V$. Then $\mathcal{B}(V)$ is isomorphic to $\mathcal{B}(U)$.

If (F, f) is an isomorphism from $\mathcal{B}(V)$ to $\mathcal{B}(U)$ as in Lemma 3.11, then we say $\mathcal{B}(V)$ is isomorphic to $\mathcal{B}(U)$ through the group morphism f.

In the following of this subsection, suppose $\mathbb{G} = \mathbb{Z}_{\mathbb{m}_1} \times \cdots \times \mathbb{Z}_{\mathbb{m}_n} = \langle \mathbb{g}_1 \rangle \times \cdots \langle \mathbb{g}_n \rangle$ and $G = \mathbb{Z}_{m_1} \times \cdots \times \mathbb{Z}_{m_n} = \langle g_1 \rangle \times \cdots \langle g_n \rangle$ where $m_i = \mathbb{m}_i^2$ for $1 \leq i \leq n$. Let

$$\pi : \Bbbk G \to \Bbbk \mathbb{G}, \quad g_i \mapsto \mathbb{g}_i, \quad 1 \leq i \leq n \tag{3.18}$$

be the canonical epimorphism. Observe that π has a section

$$\iota : \Bbbk \mathbb{G} \to \Bbbk G, \quad \prod_{i=1}^{n} \mathbb{g}_i^{i_j} \mapsto \prod_{i=1}^{n} g_i^{i_j} \tag{3.19}$$

which is not a group morphism. Let δ_L and $\triangleright$ be the comodule and module structure maps of $V \in {}^{\mathbb{G}}_{\mathbb{G}}\mathcal{YD}^{\Phi}$. Define

$$\rho_L : V \to \Bbbk G \otimes V, \quad \rho_L = (\iota \otimes \mathrm{id})\delta_L$$
$$\blacktriangleright : \Bbbk G \otimes V \to V, \quad g \blacktriangleright Z = \pi(g) \triangleright Z$$

for all $g \in G$ and $Z \in V$. Then

LEMMA 3.12. *$\widetilde{V} := (V, \rho_L, \blacktriangleright)$ is an object in ${}^{G}_{G}\mathcal{YD}^{\pi^*(\Phi)}$.*

Moreover we have

PROPOSITION 3.13. *For any Nichols algebra $\mathcal{B}(V) \in {}^{\mathbb{G}}_{\mathbb{G}}\mathcal{YD}^{\Phi}$, the Nichols algebra $\mathcal{B}(\widetilde{V}) \in {}^{G}_{G}\mathcal{YD}^{\pi^*(\Phi)}$ is isomorphic to $\mathcal{B}(V)$. Moreover, if $B(V)$ is of diagonal type, then $\mathcal{B}(\widetilde{V})$ is twist equivalent to a usual Nichols algebra.*

To summarize so far, we have found the route in Figure I of transforming a *nonassociative* Nichols algebra to a *usual* one:

$\mathcal{B}(V) \in {}^{\mathbb{G}}_{\mathbb{G}}\mathcal{YD}^{\Phi}$ Original diagonal Nichols algebra

↓

$\mathcal{B}(V) \cong \mathcal{B}(\widetilde{V}) \in {}^{G}_{G}\mathcal{YD}^{\pi^*(\Phi)}$ Lemma 3.11+Lemma 3.12

↓ Proposition 2.5

$\mathcal{B}(\widetilde{V})$ is twisted equivalent to a usual Nichols algebra $\mathcal{B}(V)'$

FIGURE 1

According to this diagram, every diagonal Nichols algebra $\mathcal{B}(V) \in {}^{\mathbb{G}}_{\mathbb{G}}\mathcal{YD}^{\Phi}$ corresponds to a usual diagonal Nichols algebra, denoted by $\mathcal{B}(V)'$ for convenience, in a canonical way.

DEFINITION 3.14. The arithmetic root system of $\mathcal{B}(V)$ is defined to be that of $\mathcal{B}(V)'$. That is, $\triangle(\mathcal{B}(V))_{\chi,E} := \triangle(\mathcal{B}(V'))_{\chi,E}$ by the prescribed notations in Subsection 3.3. In particular, the root system $\triangle(\mathcal{B}(V))$ of $\mathcal{B}(V)$ equals to $\triangle(\mathcal{B}(V)')$.

In order to complete the classification of diagonal Nichols algebras, we need to find a return trip of the above diagram. To this end, we shall answer this question: For a usual diagonal type Nichols algebra $\mathcal{B}$ with arithmetic root system, when is $\mathcal{B}$ gotten from a Nichols algebra $\mathcal{B}(V) \in {}^{\mathbb{G}}_{\mathbb{G}}\mathcal{YD}^{\Phi}$?

Firstly we observe the following fact.

LEMMA 3.15. [**28**, Lemma 4.9] *Let $\mathcal{B}(\widetilde{V}) \in {}^{G}_{G}\mathcal{YD}^{\pi^*(\Phi)}$ be a Nichols algebra of diagonal type and $\{Y_i|1 \le i \le m\}$ be a standard basis of $\widetilde{V}$. Then $\mathcal{B}(\widetilde{V})$ is isomorphic to a Nichols algebra in ${}^{\mathbb{G}}_{\mathbb{G}}\mathcal{YD}^{\Phi}$ through π if and only if*

$$g_i^{\mathbb{m}_i} \blacktriangleright Y_j = Y_j, \quad 1 \le i \le n, 1 \le j \le m. \tag{3.20}$$

We point out that the result of Lemma 3.15 can also be deduced from [**8**, Proposition 4.3]. Now fix an usual Nichols algebra of diagonal type $\mathcal{B}(V)' \in {}^{G}_{G}\mathcal{YD}$ with support group G. According to Figure I, we need to answer the following question:

When is $\mathcal{B}(V)'^{J_{\underline{\mathbf{a}}}}$ isomorphic to a Nichols algebra in ${}^{\mathbb{G}}_{\mathbb{G}}\mathcal{YD}^{\Phi_{\underline{\mathbf{a}}}}$ through π? (Question ($\diamond$))

Let $\{X_i|1 \le i \le m\}$ be a standard basis of V. Assume that

$$\delta'_L(X_i) = h_i \otimes X_i, \quad g_k \triangleright' X_j = q_{kj} X_j$$

for $1 \le i,j \le m$, $1 \le k \le n$, $h_i \in G$ and $q_{kj} \in \mathbb{k}^*$, where δ'_L (resp. $\triangleright'$) is the comodule (resp. module) structure map of $\mathcal{B}(V)' \in {}^{G}_{G}\mathcal{YD}$. So there are $0 \le x_{kj}, s_{ik} < m_k$ such that

$$q_{kj} = \zeta_{m_k}^{x_{kj}}, \quad h_i = \prod_{k=1}^{n} g_k^{s_{ik}}$$

for $1 \le i,j \le m$ and $1 \le k \le n$. Let $\mathbf{X} = (x_{ij})_{n \times m}$. By assumption, the support group $G_{\mathcal{B}(V)'} = G$, $\{h_i|1 \le i \le m\}$ generate the group G, so there are $t_{jl} \in \mathbb{N}$ such that

$$g_j = \prod_{l=1}^{m} h_l^{t_{jl}}, \quad 1 \le j \le n.$$

By $\mathbf{S}$ and $\mathbf{T}$, we denote the matrices $(s_{ik})_{m \times n}$ and $(t_{jl})_{n \times m}$. It is obvious that

$$\mathbf{TS} \equiv \begin{pmatrix} 1 \pmod{m_1} & 0 \pmod{m_1} & \cdots & 0 \pmod{m_1} \\ 0 \pmod{m_2} & 1 \pmod{m_2} & \cdots & 0 \pmod{m_2} \\ \cdots & \cdots & \cdots & \cdots \\ 0 \pmod{m_n} & 0 \pmod{m_n} & \cdots & 1 \pmod{m_n} \end{pmatrix}. \tag{3.21}$$

With these notations, we can give the answer to Question ($\diamond$) now.

PROPOSITION 3.16. [**28**, Proposition 4.10] *The twisting $\mathcal{B}(V)'^{J_{\underline{\mathbf{a}}}}$ is isomorphic to a Nichols algebra in ${}^{\mathbb{G}}_{\mathbb{G}}\mathcal{YD}^{\Phi_{\underline{\mathbf{a}}}}$ through π if and only if the following congruence equalities hold:*

$$\sum_{j=1}^{m} x_{ij} t_{lj} \equiv 0 \pmod{\mathbb{m}_i}, \ 1 \le l < i \le n, \tag{3.22}$$

$$\sum_{j=1}^{m} x_{ij} t_{ij} \equiv a_i \pmod{\mathbb{m}_i}, \ 1 \le i \le n, \tag{3.23}$$

$$(\sum_{j=1}^{m} x_{ij} t_{lj})\mathbb{m}_l \equiv \mathbb{m}_i a_{il} \pmod{\mathbb{m}_i \mathbb{m}_l}, \ 1 \le i < l \le n. \tag{3.24}$$

PROPOSITION 3.17. [**28**, Corollary 4.12] *For the Nichols algebra $\mathcal{B}(V)' \in {}^{G}_{G}\mathcal{YD}$, there is at most one $\underline{\mathbf{a}} \in \mathcal{A}$ such that $\mathcal{B}(V)'^{J_{\underline{\mathbf{a}}}}$ is isomorphic to a Nichols algebra in*

${}^{\mathbb{G}}_{\mathbb{G}}\mathcal{YD}^{\Phi_{\underline{\mathbf{a}}}}$ *through* π. *Moreover, this* $\underline{\mathbf{a}}$ *exists if and only if the Equations* (3.22) *hold and in this case* $\underline{\mathbf{a}}$ *can be taken in the following way:*

$$a_i \equiv \sum_{j=1}^{m} x_{ij} t_{ij} \ (\mathrm{mod}\ \mathbb{m}_i); \quad a_{il} \equiv \frac{\mathbb{m}_l}{\mathbb{m}_i} \sum_{j=1}^{m} x_{ij} t_{lj} \ (\mathrm{mod}\ \mathbb{m}_l); \quad a_{ilt} = 0 \tag{3.25}$$

for $1 \leq i \leq n$, $1 \leq i < l \leq n$ *and* $1 \leq i < l < t \leq n$.

Now we are in the position to find the "return trip" in Figure II.

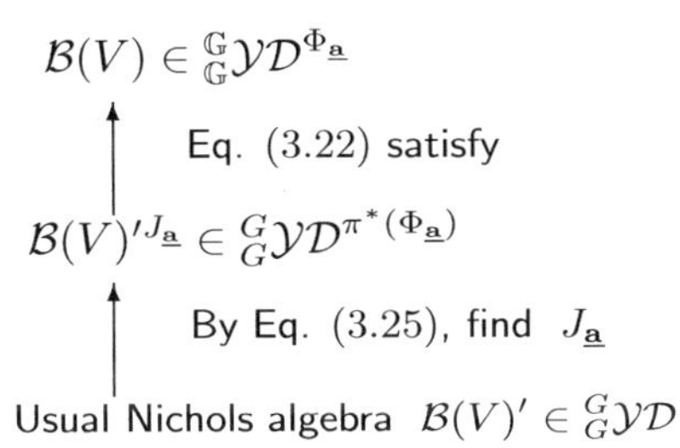

Figure II

Next we can give a complete classification of Nichols algebras of diagonal type with arithmetic root systems in ${}^{\mathbb{G}}_{\mathbb{G}}\mathcal{YD}^{\Phi}$. Suppose $(\triangle, \chi, E)$ is an arithmetic root system, $\mathcal{D}_{\chi,E}$ is the Dynkin diagram of $(\triangle, \chi, E)$. Up to twist equivalence, $(\triangle, \chi, E)$ is uniquely determined by $\mathcal{D}_{\chi,E}$. Fix a Dynkin diagram with m vertices, we call

$$\{q_{ii} = \chi(e_i, e_i),\ \widetilde{q_{ij}} = \chi(e_i, e_j)\chi(e_j, e_i) | 1 \leq i, j \leq m\}$$

the structure constants of $\mathcal{D}_{\chi,E}$.

DEFINITION 3.18. Let $\mathbb{G} = \mathbb{Z}_{\mathbb{m}_1} \times \cdots \times \mathbb{Z}_{\mathbb{m}_n}$ be the abelian group defined as above and set $m_i := \mathbb{m}_i^2$ for $1 \leq i \leq n$. Suppose $\mathcal{D}_{\chi,E}$ is a Dynkin diagram of an arithmetic root system $\triangle_{\chi,E}$ and $(q_{ii}, \widetilde{q_{ij}})$ is the set of structure constants. Assume that there exist parameter matrices $\mathbf{S}$ and $\mathbf{X}$ satisfying

1. $\mathbf{S} = (s_{ij})_{m \times n}$ is a matrix with integer entries $0 \leq s_{ij} < m_i$ for all $1 \leq i \leq m, 1 \leq j \leq n$ such that there exists a matrix $\mathbf{T} = (t_{ij})_{n \times m}$ satisfying (3.21).
2. $\mathbf{X} = (x_{ij})_{n \times m}$ with integer entries $0 \leq x_{ij} < m_i$ for all $1 \leq i \leq m, 1 \leq j \leq n$ such that $q_{ii} = \prod_{k=1}^{n} \zeta_{m_k}^{s_{ik} x_{ki}}$, $\widetilde{q_{ij}} = \prod_{k=1}^{n} \zeta_{m_k}^{s_{ik} x_{kj} + s_{jk} x_{ki}}$, and satisfy Equations (3.22).

Then we call $\mathfrak{D} = \mathfrak{D}(\mathcal{D}_{\chi,E}, \mathbf{S}, \mathbf{X})$ a root datum over $\mathbb{G}$, and $\triangle_{\chi,E}$ the arithmetic root system of $\mathfrak{D}$.

For a fixed root datum $\mathfrak{D} = \mathfrak{D}(\mathcal{D}_{\chi,E}, \mathbf{S}, \mathbf{X})$ over $\mathbb{G}$, define a sequence $\underline{\mathbf{a}} \in A$ through Equations (3.25). Now we can define a Nichols algebra $\mathcal{B}(V_{\mathfrak{D}}) \in {}^{G}_{G}\mathcal{YD}^{\pi^*(\Phi_{\underline{\mathbf{a}}})}$ in the following way: Let $V_{\mathfrak{D}}$ be the Yetter-Drinfeld module in ${}^{G}_{G}\mathcal{YD}^{\pi^*(\Phi_{\underline{\mathbf{a}}})}$ with a standard basis $\{X_i | 1 \leq i \leq m\}$ such that

$$\delta_L(X_i) = \prod_{k=1}^{n} g_k^{s_{ik}} \otimes X_i, \quad g_i \blacktriangleright X_j = \zeta_{m_i}^{x_{ij}} \frac{J_{\underline{\mathbf{a}}}(g_i, \prod_{k=1}^{n} g_k^{s_{ik}})}{J_{\underline{\mathbf{a}}}(\prod_{k=1}^{n} g_k^{s_{ik}}, g_i)} X_j.$$

THEOREM 3.19. [**28**, Theorem 4.14]

(1) *The Nichols algebra* $\mathcal{B}(V_{\mathfrak{D}})$ *is isomorphic to a Nichols algebra of diagonal type with arithmetic root system in* ${}_{\mathbb{G}}^{\mathbb{G}}\mathcal{YD}^{\Phi_{\underline{\mathbf{a}}}}$ *through the group epimorphism* $\pi : G \to \mathbb{G}$.

(2) *Suppose* $\mathcal{B}(V)$ *is a Nichols algebra of diagonal type with arithmetic root system in* ${}_{\mathbb{G}}^{\mathbb{G}}\mathcal{YD}^{\Phi_{\underline{\mathbf{a}}}}$ *and the support group is* $\mathbb{G}$, *then there exists a root datum* $\mathfrak{D}$ *over* $\mathbb{G}$ *such that* $\mathcal{B}(V_{\mathfrak{D}}) \cong \mathcal{B}(V)$ *through the group epimorphism* $\pi : G \to \mathbb{G}$.

By this theorem, we know that the Nichols algebra $\mathcal{B}(V_{\mathfrak{D}})$ is isomorphic to a unique Nichols algebra in ${}_{\mathbb{G}}^{\mathbb{G}}\mathcal{YD}^{\Phi_{\underline{\mathbf{a}}}}$. For convenience, this Nichols algebra is denoted by $\mathcal{B}(\mathfrak{D})$ and $\mathbb{M}(\mathfrak{D}) = \mathcal{B}(\mathfrak{D})\#\Bbbk\mathbb{G}$.

With the previous preparation, now we are in the position to give the classification result of connected finite-dimensional pointed coquasi-Hopf algebras of diagonal type. Suppose that $\mathbb{M}$ is a finite-dimensional connected pointed coquasi-Hopf algebras of diagonal type with associator $\Phi_{\underline{\mathbf{a}}}$, $\mathbb{G} = G(\mathbb{M})$ and $\mathcal{R}$ the coinvariant subalgebra of $\mathbb{M}$. Note that $\mathcal{R} = \sum_{i\geq 0} \mathcal{R}_i$ is a graded Hopf algebras in ${}_{\mathbb{G}}^{\mathbb{G}}\mathcal{YD}^{\Phi_{\underline{\mathbf{a}}}}$. In fact we have

PROPOSITION 3.20. [**28**, Lemma 4.1, Proposition 5.1] $\Phi_{\underline{\mathbf{a}}}$ *is an abelian* 3*-cocycle on* $\mathbb{G} = G(\mathbb{M})$, *and* $\mathcal{R} \cong \mathcal{B}(\mathcal{R}_1)$ *is a Nichols algebra in* ${}_{\mathbb{G}}^{\mathbb{G}}\mathcal{YD}^{\Phi_{\underline{\mathbf{a}}}}$.

Combining Theorem 3.19 and Proposition 3.20, we obtain our main classification result.

THEOREM 3.21. [**28**, Theorem 5.7] *Keep the notations as before. We have*

(1). *The coquasi-Hopf algebra* $\mathbb{M}(\mathfrak{D})$ *is a connected coradical graded pointed coquasi-Hopf algebra of diagonal type over the group* $\mathbb{G}$. *Moreover,* $\mathbb{M}(\mathfrak{D})$ *is finite-dimensional if and only if the heights of all restricted Poincare-Birkhoff-Witt generators of* $\mathcal{B}(\mathfrak{D})$ *are finite.*

(2). *Any finite-dimensional connected coradical graded pointed coquasi-Hopf algebra of diagonal type over* $\mathbb{G}$ *is isomorphic to a* $\mathbb{M}(\mathfrak{D})$ *for some* $\mathfrak{D}$.

In [**3**, Conjecture 1.4], Andruskiewitch-Schneider conjectured that every finite-dimensional pointed Hopf algebras over $\Bbbk$ is generated by group-like and skew-primitive elements. This is the so called generation in degree one problem, which plays an important role on the classification of pointed Hopf algebras. It is true in many cases, see [**6**]. This conjecture was generalized to finite-dimensional pointed coquasi-Hopf algebras or even to pointed finite tensor categories [**16**].

COROLLARY 3.22. [**28**, Corollary 5.8] *Suppose* $\mathbb{M}$ *is a finite-dimensional pointed coquasi-Hopf algebras of diagonal type, then* $\mathbb{M}$ *is generated by group-like and skew-primitive elements.*

To conclude this section, we remark that there are many new classes of genuine finite-dimensional coradically graded pointed coquasi-Hopf algebras constructed in [**27, 28**] explicitly.

4. Finite quasi-quantum groups of nondiagonal type: A first attempt

In this section, we will present some classification results of nondiagonal finite quasi-quantum groups over abelian groups obtained in [**33**]. This is an initial step to go further beyond [**27, 28**].

In the rest of this section, let G be a finite abelian group and Φ be a nonabelian 3-cocycle on G. Similar to the diagonal case, the crux is a complete understanding of the Nichols algebras of all $V \in {}^G_G\mathcal{YD}^\Phi$. To this end, in principle we need to develop a theory for the Nichols algebras of semisimple twisted Yetter-Drinfeld modules. The Hopf version of such a theory was developed in [**4**, **23**]. However, at present it seems not easy to extend this theory to the quasi-Hopf case directly. As a trial step, firstly we study the Nichols algebras of semisimple twisted Yetter-Drinfeld modules with few summands. It turns out that if the number of summands is less than or equal to 2, then we are able to make a connection from this to the diagonal case. The main idea is to consider the support groups of such easy Yetter-Drinfeld modules and carry out the base group change as in our previous works [**27**, **28**]. More precisely, if $V \in {}^G_G\mathcal{YD}^\Phi$ is nondiagonal and has at most 2 simple summands, then its support group G_V is either a cyclic group or the direct product of two cyclic groups. Moreover, the Nichols algebra $\mathcal{B}(V) \in {}^G_G\mathcal{YD}^\Phi$ is essentially nothing other than $\mathcal{B}(V) \in {}^{G_V}_{G_V}\mathcal{YD}^{\Phi|_{G_V}}$. In this situation, all 3-cocycles on G_V are abelian and then [**27**, **28**] can be applied.

Our first main result is a complete clarification of the Nichols algebra $\mathcal{B}(V)$ when V is a simple twisted Yetter-Drinfeld module of nondiagonal type. In particular, we provide an explicit necessary and sufficient condition on V for $\mathcal{B}(V)$ to be finite-dimensional. The same idea and process can be applied to $\mathcal{B}(V)$ when V is a direct sum of 2 simple twisted Yetter-Drinfeld modules. As this will not provide more insights for our ultimate aim, we do not include a detailed discussion of this case. Instead, we present several simple examples to offer the reader some flavor. Surprisingly, the result on $\mathcal{B}(V)$ with V simple is already enough for us to achieve half of our final aim. Our second main result is a complete classification of finite-dimensional coradically graded pointed coquasi-Hopf algebras over abelian groups of odd order. The key observation is that $\mathcal{B}(V) \in {}^G_G\mathcal{YD}^\Phi$ is infinite-dimensional for any simple nondiagonal twisted Yetter-Drinfeld module V if the order of G is odd. As an application, we also prove that any pointed finite tensor category over an abelian group of odd order is tensor generated by objects of length 2, which partially confirms the generation conjecture [**16**, Conjecture 5.11.10.] of pointed finite tensor categories due to Etingof, Gelaki, Nikshych and Ostrik.

4.1. The Nichols algebras of simple twisted Yetter-Drinfeld modules. In this subsection we focus on the Nichols algebras of nondiagonal Yetter-Drinfeld modules. Note that if Φ is an abelian 3-cocycle on G, then each object of ${}^G_G\mathcal{YD}^\Phi$ is of diagonal type. So nondiagonal Yetter-Drinfeld modules appear in ${}^G_G\mathcal{YD}^\Phi$ only if Φ is nonabelian. The following proposition is an immediate consequence of Propositions 2.4 and (2.3).

PROPOSITION 4.1. *Suppose that G is a cyclic group $\mathbb{Z}_m$ or a direct product of two cyclic groups, say $\mathbb{Z}_{m_1} \times \mathbb{Z}_{m_2}$, then all the 3-cocycles on G are abelian.*

PROPOSITION 4.2. [**33**, Proposition 3.16] *Let G be a finite abelian group, Φ a 3-cocycle on G. Suppose that $\mathcal{B}(V)$ is a Nichols algebra in ${}^G_G\mathcal{YD}^\Phi$, where V is a simple Yetter-Drinfeld module, or a direct sum of two simple Yetter-Drinfeld modules. Then $\mathcal{B}(V)$ is isomorphic to a Nichols algebra of diagonal type $\mathcal{B}(V')$ in ${}^H_H\mathcal{YD}^\Psi$, where $H = G_V$ and $\Psi = \Phi|_H$.*

According to this proposition, we can apply the theory of Nichols algebras of diagonal type to study the Nichols algebras of simple twisted Yetter-Dinfeld modules, or of a direct sum of two simple twisted Yetter-Drinfeld modules. If $V = V_g$ is a simple object in ${}_G^G\mathcal{YD}^\Phi$, then we denote $g_V = g$.

PROPOSITION 4.3. [**33**, Proposition 3.18] *Let G be a finite abelian group, Φ a 3-cocycle on G. Suppose V is a simple Yetter-Drinfeld module of nondiagonal type in ${}_G^G\mathcal{YD}^\Phi$, $g_V = g$. Then $\mathcal{B}(V)$ is finite-dimensional if and only if V is one of the following two cases:*

(C1) $g \triangleright v = -v$ *for all* $v \in V$;

(C2) $\dim(V) = 2$ *and* $g \triangleright v = \zeta_3 v$ *for all* $v \in V$, *here* ζ_3 *is a 3-rd primitive root of unity.*

There exist Yetter-Drinfeld modules satisfying conditions C1 or C2, see [**33**, Example 3.19-3.20].

4.2. Finite quasi-quantum groups over abelian groups of odd order. In this subsection we provide a complete classification of finite-dimensional coradically graded pointed coquasi-Hopf algebras over abelian groups of odd order. This is also applied to the classification theory of pointed finite tensor categories. In particular, we give a partial answer to the following

CONJECTURE 4.4. [**16**, Conjecture 5.11.10.] *A pointed finite tensor category is tensor generated by objects of length* 2.

This conjecture is due to Etingof, Gelaki, Nikshych and Ostrik, hence will be called EGNO's conjecture in the following. It is a natural generalization of the well known Andruskiewitsch-Schneider conjecture [**3**, Conjecture 1.4] due to the following proposition.

PROPOSITION 4.5. [**33**, Proposition 4.10] *Suppose that M is a finite-dimensional pointed coquasi-Hopf algebra. Then M is generated by group-like and skew-primitive elements if and only if* $\operatorname{comod}(M)$ *is tensor generated by objects of length* 2.

Now we need to get more information of Nichols algebras in ${}_G^G\mathcal{YD}^\Phi$. Applying linear representation theory of G, we prove the following important proposition.

PROPOSITION 4.6. *For each simple object V in ${}_G^G\mathcal{YD}^\Phi$, we have* $\dim(V) \mid |G|$.

Note that if $|G|$ is odd, the order of an element in G is also odd, hence there is not a simple object satisfying condition C1 of Proposition 4.3. According to Proposition 4.6, $\dim(V) \mid |G|$ implies $\dim(V)$ is also odd, and hence V doesn't satisfy condition C2 of Proposition 4.3. So we have the following proposition.

PROPOSITION 4.7. *Let G be a finite abelian group of odd order and Φ be a 3-cocycle on G. Suppose that $V \in {}_G^G\mathcal{YD}^\Phi$ is not diagonal. Then $\mathcal{B}(V)$ is infinite-dimensional.*

This proposition implies that if $|G|$ is odd, then every finite-dimensional Nichols algebra in ${}_G^G\mathcal{YD}^\Phi$ must be of diagonal type. So according to Proposition 3.20, we can easily prove the following theorem.

THEOREM 4.8. [**33**, Theorem 4.11] *Suppose that $R = \oplus_{i\geq 0} R[i]$ is a finite-dimensional connected coradically graded braided Hopf algebra in ${}_G^G\mathcal{YD}^\Phi$, where G is an abelian group of odd order and Φ is a 3-cocycle on G. Then $R = \mathcal{B}(R[1])$.*

Combining Proposition 4.5 and Theorem 4.8, we obtain a partial answer to Conjecture 4.4.

THEOREM 4.9. [**33**, Theorem 4.2] *Suppose that $\mathcal{C}$ is a pointed finite tensor category with $G(\mathcal{C})$ an abelian group of odd order. Then $\mathcal{C}$ is tensor generated by objects of length* 2.

With the help of Theorem 4.8, we achieve the classification of coradically graded pointed coquasi-Hopf algebras and that of coradically graded pointed finite tensor categories over abelian groups of odd order. Here a tensor category $\mathcal{C}$ is said to be coradically graded if $\mathcal{C}$ is equivalent to the category of comodules over a coradically graded coquasi-Hopf algebra.

Let $\Delta_{\chi,E}$ be an arithmetic root system. For each positive root $\alpha \in \Delta$, define $q_\alpha = \chi(\alpha,\alpha)$. Then the height of α is defined by

$$\mathrm{ht}(\alpha) = \begin{cases} |q_\alpha|, & \text{if } q_\alpha \neq 1 \text{ is a root of unity;} \\ \infty, & \text{otherwise.} \end{cases} \tag{4.1}$$

A function $\chi : G \longrightarrow \Bbbk^*$ is called a **quasi-character** associated to a 2-cocycle ω on G if for all $f, g \in G$,

$$\chi(f)\chi(g) = \omega(f,g)\chi(fg), \quad \chi(1) = 1. \tag{4.2}$$

It is clear that there is a quasi-character associated to ω if and only if ω is symmetric. Recall that for a fixed 3-cocycle Φ on G, $\{\widetilde{\Phi}_g | g \in G\}$ gives 2-cocycles on G.

DEFINITION 4.10. Let $\chi_1, \cdots, \chi_n$ be quasi-characters of G associated to $\widetilde{\Phi}_{g_1}$, $\cdots, \widetilde{\Phi}_{g_n}$ respectively. We say the series $(\chi_1, \cdots, \chi_n)$ is of finite type if there is an arithmetic root system $\Delta_{\chi,E}$ of rank n such that:

- $\chi_i(g_j)\chi_j(g_i) = q_{ij}q_{ji}$, $\chi_i(g_i) = q_{ii}$ for all $1 \leq i, j \leq n$. Here $q_{ij} = \chi(e_i, e_j)$ for $e_i, e_j \in E$.
- $\mathrm{ht}(\alpha) < \infty$ for all $\alpha \in \Delta$.

For a series of quasi-characters $(\chi_1, \cdots, \chi_n)$ of finite type associated to $\widetilde{\Phi}_{g_1}$, $\cdots, \widetilde{\Phi}_{g_n}$, we can attach to it a twisted Yetter-Drinfeld module $V(\chi_1, \cdots, \chi_n)$ with a standard basis $\{X_1, \cdots, X_n\}$ such that $g_i \triangleright X_j = \chi_j(g_i)X_j$ and $\delta_L(X_i) = g_i \otimes X_i$ for all $1 \leq i, j \leq n$. Now we can give the classification result.

THEOREM 4.11. [**33**, Theorem 4.13] *Let G be a finite abelian group of odd order, Φ a 3-cocycle on G.*

(1) *If $(\chi_1, \cdots, \chi_n)$ is a series of quasi-characters of finite type associated to the 2-cocycles $\widetilde{\Phi}_{g_1}, \cdots, \widetilde{\Phi}_{g_n}$, then $\mathcal{B}(V(\chi_1, \cdots, \chi_n))$ is a finite-dimensional Nichols algebra in ${}^{\Bbbk G}_{\Bbbk G}\mathcal{YD}^{\Phi}$.*
(2) *Suppose that $\mathcal{C}$ is a coradically graded pointed finite tensor category such that $G(\mathcal{C}) = G$ and the associator is Φ. Then there exists a series of quasi-characters $(\chi_1, \cdots, \chi_n)$ of finite type associated to $\widetilde{\Phi}_{g_1}, \cdots, \widetilde{\Phi}_{g_n}$ such that*

$$\mathcal{C} \cong \mathrm{comod}(\mathcal{B}(V(\chi_1, \cdots, \chi_n))\#\Bbbk G).$$

5. Some further problems

Finally, we propose some further research problems which are natural extensions of our previous works.

PROBLEM 1. Pursuit a complete classification of finite-dimensional coradically graded pointed coquasi-Hopf algebras of nondiagonal type over abelian groups. Thanks to [**33**], it remains to consider those over even order abelian groups. Moreover, according to some examples we worked out, it is quite possible that the Nichols algebra of a twisted Yetter-Drinfeld module with at least 3 nondiagonal simple summands is infinite-dimensional. This implies that nontrivial finite quasi-quantum groups of nondiagonal type over abelian groups are very rare. Thus, a complete classification seems possible.

PROBLEM 2. Consider pointed finite tensor categories and finite quasi-quantum groups over nonabelian groups. To the best of our knowledge, there seems not even a single example of finite-dimensional connected pointed coquasi-Hopf algebra with nontrivial associator over nonabelian groups in the literature. In addition, the cohomology of finite nonabelian groups seems much more complicated. For some easy classes of nonabelian groups, e.g. the semi-direct product of two cyclic groups, the idea of Section 2 may be applied to obtain unified and explicit 3-cocycle formulas. Finite quasi-quantum groups over such groups may be workable. In pursuit of a general theory as the abelian case, it seems worthwhile to develop the general theory of Nichols algebras of semisimple twisted Yetter-Drinfeld modules, as well as the related Weyl groupoids and arithmetic root systems.

PROBLEM 3. Consider the lifting or deformation theory of coradically graded finite quasi-quantum groups. As far as we know, the lifting of (co)quasi-Hopf algebras was considered only in [**5, 15**]. There are not many results in this direction. The problem sees very complicated and challenging. An obvious difficulty lies in associators for which there is no suitable tools to control yet. One may try to apply the deformation and related cohomology theory of quasi-quantum groups. We also wonder if there is a nice theory of cocycle deformations of quasi-quantum groups.

References

[1] Nicolás Andruskiewitsch, *On finite-dimensional Hopf algebras*, Proceedings of the International Congress of Mathematicians—Seoul 2014. Vol. II, Kyung Moon Sa, Seoul, 2014, pp. 117–141. MR3728608

[2] Nicolás Andruskiewitsch and Hans-Jürgen Schneider, *Pointed Hopf algebras*, New directions in Hopf algebras, Math. Sci. Res. Inst. Publ., vol. 43, Cambridge Univ. Press, Cambridge, 2002, pp. 1–68, DOI 10.2977/prims/1199403805. MR1913436

[3] Nicolás Andruskiewitsch and Hans-Jürgen Schneider, *On the classification of finite-dimensional pointed Hopf algebras*, Ann. of Math. (2) **171** (2010), no. 1, 375–417, DOI 10.4007/annals.2010.171.375. MR2630042

[4] Nicolás Andruskiewitsch, István Heckenberger, and Hans-Jürgen Schneider, *The Nichols algebra of a semisimple Yetter-Drinfeld module*, Amer. J. Math. **132** (2010), no. 6, 1493–1547. MR2766176

[5] Iván Ezequiel Angiono, *Basic quasi-Hopf algebras over cyclic groups*, Adv. Math. **225** (2010), no. 6, 3545–3575, DOI 10.1016/j.aim.2010.06.013. MR2729015

[6] Iván Angiono, *On Nichols algebras of diagonal type*, J. Reine Angew. Math. **683** (2013), 189–251, DOI 10.1515/crelle-2011-0008. MR3181554

[7] Iván Angiono and César Galindo, *Pointed finite tensor categories over abelian groups*, Internat. J. Math. **28** (2017), no. 11, 1750087, 18, DOI 10.1142/S0129167X17500872. MR3714361

[8] Iván Angiono, César Galindo, and Mariana Pereira, *De-equivariantization of Hopf algebras*, Algebr. Represent. Theory **17** (2014), no. 1, 161–180, DOI 10.1007/s10468-012-9392-9. MR3160718

[9] Iván Angiono and Agustín García Iglesias, *Liftings of Nichols algebras of diagonal type II: all liftings are cocycle deformations*, Selecta Math. (N.S.) **25** (2019), no. 1, Paper No. 5, 95, DOI 10.1007/s00029-019-0452-4. MR3908852

[10] Alessandro Ardizzoni and Alice Pavarin, *Bosonization for dual quasi-bialgebras and preantipode*, J. Algebra **390** (2013), 126–159, DOI 10.1016/j.jalgebra.2013.05.014. MR3072115

[11] Lawrence Breen, *Monoidal categories and multiextensions*, Compositio Math. **117** (1999), no. 3, 295–335, DOI 10.1023/A:1000928915124. MR1702420

[12] R. Dijkgraaf, V. Pasquier, and P. Roche, *Quasi Hopf algebras, group cohomology and orbifold models*, Nuclear Phys. B Proc. Suppl. **18B** (1990), 60–72 (1991), DOI 10.1016/0920-5632(91)90123-V. Recent advances in field theory (Annecy-le-Vieux, 1990). MR1128130

[13] Pavel Etingof and Shlomo Gelaki, *Finite dimensional quasi-Hopf algebras with radical of codimension 2*, Math. Res. Lett. **11** (2004), no. 5-6, 685–696, DOI 10.4310/MRL.2004.v11.n5.a11. MR2106235

[14] Pavel Etingof and Shlomo Gelaki, *On radically graded finite-dimensional quasi-Hopf algebras* (English, with English and Russian summaries), Mosc. Math. J. **5** (2005), no. 2, 371–378, 494, DOI 10.17323/1609-4514-2005-5-2-371-378. MR2200756

[15] Pavel Etingof and Shlomo Gelaki, *Liftings of graded quasi-Hopf algebras with radical of prime codimension*, J. Pure Appl. Algebra **205** (2006), no. 2, 310–322, DOI 10.1016/j.jpaa.2005.06.016. MR2203619

[16] Pavel Etingof, Shlomo Gelaki, Dmitri Nikshych, and Victor Ostrik, *Tensor categories*, Mathematical Surveys and Monographs, vol. 205, American Mathematical Society, Providence, RI, 2015, DOI 10.1090/surv/205. MR3242743

[17] Pavel Etingof and Viktor Ostrik, *Finite tensor categories* (English, with English and Russian summaries), Mosc. Math. J. **4** (2004), no. 3, 627–654, 782–783, DOI 10.17323/1609-4514-2004-4-3-627-654. MR2119143

[18] Shlomo Gelaki, *Basic quasi-Hopf algebras of dimension n^3*, J. Pure Appl. Algebra **198** (2005), no. 1-3, 165–174, DOI 10.1016/j.jpaa.2004.10.003. MR2132880

[19] I. Heckenberger, *The Weyl groupoid of a Nichols algebra of diagonal type*, Invent. Math. **164** (2006), no. 1, 175–188, DOI 10.1007/s00222-005-0474-8. MR2207786

[20] I. Heckenberger, *Examples of finite-dimensional rank 2 Nichols algebras of diagonal type*, Compos. Math. **143** (2007), no. 1, 165–190, DOI 10.1112/S0010437X06002430. MR2295200

[21] I. Heckenberger, *Rank 2 Nichols algebras with finite arithmetic root system*, Algebr. Represent. Theory **11** (2008), no. 2, 115–132, DOI 10.1007/s10468-007-9060-7. MR2379892

[22] I. Heckenberger, *Classification of arithmetic root systems*, Adv. Math. **220** (2009), no. 1, 59–124, DOI 10.1016/j.aim.2008.08.005. MR2462836

[23] I. Heckenberger and H.-J. Schneider, *Root systems and Weyl groupoids for Nichols algebras*, Proc. Lond. Math. Soc. (3) **101** (2010), no. 3, 623–654, DOI 10.1112/plms/pdq001. MR2734956

[24] Hoàng Xuân Sính: *Gr-catégories*. Universite Paris VII, Thèse de doctorat (1975). available at `http://www.math.rwth-aachen.de/~kuenzer/sinh.html`.

[25] Hua-Lin Huang, *Quiver approaches to quasi-Hopf algebras*, J. Math. Phys. **50** (2009), no. 4, 043501, 9, DOI 10.1063/1.3103569. MR2513989

[26] HuaLin Huang, *From projective representations to quasi-quantum groups*, Sci. China Math. **55** (2012), no. 10, 2067–2080, DOI 10.1007/s11425-012-4437-4. MR2972630

[27] Huang, Hua-Lin; Liu, Gongxiang; Yang, Yuping; Ye, Yu: *Finite quasi-quntum groups of rank two*. `arXiv:1508.04248v1`.

[28] Hua-Lin Huang, Gongxiang Liu, Yuping Yang, and Yu Ye, *Finite quasi-quantum groups of diagonal type*, J. Reine Angew. Math. **759** (2020), 201–243, DOI 10.1515/crelle-2017-0058. MR4058179

[29] Hua-Lin Huang, Gongxiang Liu, and Yu Ye, *Quivers, quasi-quantum groups and finite tensor categories*, Comm. Math. Phys. **303** (2011), no. 3, 595–612, DOI 10.1007/s00220-011-1229-6. MR2786212

[30] Hua-Lin Huang, Gongxiang Liu, and Yu Ye, *The braided monoidal structures on a class of linear Gr-categories*, Algebr. Represent. Theory **17** (2014), no. 4, 1249–1265, DOI 10.1007/s10468-013-9445-8. MR3228486

[31] Hua-Lin Huang, Gongxiang Liu, and Yu Ye, *Graded elementary quasi-Hopf algebras of tame representation type*, Israel J. Math. **209** (2015), no. 1, 157–186, DOI 10.1007/s11856-015-1214-4. MR3430237

[32] Hua-Lin Huang, Zheyan Wan, and Yu Ye, *Explicit cocycle formulas on finite abelian groups with applications to braided linear Gr-categories and Dijkgraaf-Witten invariants*, Proc. Roy. Soc. Edinburgh Sect. A **150** (2020), no. 4, 1937–1964, DOI 10.1017/prm.2019.15. MR4122441

[33] Hua-Lin Huang, Yuping Yang, and Yinhuo Zhang, *On nondiagonal finite quasi-quantum groups over finite abelian groups*, Selecta Math. (N.S.) **24** (2018), no. 5, 4197–4221, DOI 10.1007/s00029-018-0420-4. MR3874693

[34] Hua-Lin Huang and Yuping Yang, *Quasi-quantum linear spaces*, J. Noncommut. Geom. **9** (2015), no. 4, 1227–1259, DOI 10.4171/JNCG/222. MR3448335

[35] Charles A. Weibel, *An introduction to homological algebra*, Cambridge Studies in Advanced Mathematics, vol. 38, Cambridge University Press, Cambridge, 1994, DOI 10.1017/CBO9781139644136. MR1269324

School of Mathematical Sciences, Huaqiao University, Quanzhou 362021, People's Republic of China

Email address: hualin.huang@hqu.edu.cn

Department of Mathematics, Nanjing University, Nanjing 210093, People's Republic of China

Email address: gxliu@nju.edu.cn

School of Mathematics and Statistics, Southwest University, Chongqing 400715, People's Republic of China

Email address: yupingyang@swu.edu.cn

School of Mathematical Sciences, CAS Wu Wen-Tsun Key Laboratory of Mathematics, University of Science and Technology of China, Hefei 230026, People's Republic of China

Email address: yeyu@ustc.edu.cn

Contemporary Mathematics
Volume **771**, 2021
https://doi.org/10.1090/conm/771/15513

On biproducts and extensions

Yevgenia Kashina and Yorck Sommerhäuser

ABSTRACT. We describe in which ways the Radford biproducts of certain eight-dimensional Yetter-Drinfel'd Hopf algebras over the elementary abelian group of order 4 can be written as extensions of Hopf algebras.

Introduction

In their article [**10**], the authors described two semisimple Yetter-Drinfel'd Hopf algebras of dimension 8 over the group ring of the elementary abelian group of order 4. The purpose of the construction of these Yetter-Drinfel'd Hopf algebras was to show that the core of a group-like element is not always completely trivial. It was not even mentioned in this article that, via the Radford biproduct construction (cf. [**23**]), these Yetter-Drinfel'd Hopf algebras give rise to ordinary Hopf algebras of dimension 32. The purpose of the present article is to discuss how these Hopf algebras fit into the general theory of semisimple Hopf algebras of prime power dimension, as found for example in [**6**], [**7**], [**8**], [**14**], and [**15**]. As we will see, these ordinary Hopf algebras also behave differently from other known examples.

Every semisimple Hopf algebra of prime power dimension can in principle be constructed as an iterated crossed product (cf. [**20**], Theorem 3.5). However, for the dimensions where the description is more concrete, or even so concrete that all semisimple Hopf algebras of that given dimension can be classified, like in the articles just cited, the Hopf algebras often contain a commutative Hopf subalgebra of prime index, a situation sometimes also encountered for semisimple Hopf algebras that are not of prime power dimension, like in [**16**] and [**21**]. The examples constructed here do not contain such a Hopf subalgebra, as we show in Paragraph 4.4. More

2020 *Mathematics Subject Classification.* Primary 16T05.
The material discussed here and in our previous article [**10**] was presented at the AMS Spring Southeastern Sectional Meeting in Auburn in March 2019, in two consecutive talks. It was also presented at the International Workshop on Hopf Algebras and Tensor Categories in Nanjing in September 2019 covered in this volume. The travel of the second author to these conferences as well as his work on this article were supported by NSERC grant RGPIN-2017-06543. The work of the first author on this article was supported by a Faculty Summer Research Grant from the College of Science and Health at DePaul University.

generally, we show that they cannot be constructed as central or cocentral abelian extensions.

Let us state the main results of the article more precisely, and also explain its structure. Section 1 contains preliminaries on Yetter-Drinfel'd Hopf algebras, Radford biproducts, and extensions of Hopf algebras. In Section 2, we consider the eight-dimensional Yetter-Drinfel'd Hopf algebra A over the elementary abelian group of order 4 that appears in [**10**], Section 2. We describe the arising Radford biproduct B, which has dimension 32, give a presentation of B in terms of generators and relations, and compute its center. In Section 3, we show that the groups of group-like elements of both B and B^* are elementary abelian of order 8, which means in particular that they have eight one-dimensional representations. In Theorem 3.2, we show that, up to isomorphism, both B and B^* have in addition two irreducible representations of dimension 2 and one irreducible representation of dimension 4.

In Section 4, we show that B contains a unique sixteen-dimensional Hopf subalgebra N. It is normal and isomorphic to the Hopf algebra $H_{d:1,1}$ from the classification appearing in [**6**], Table 1. Our Hopf algebra B therefore fits into exactly one extension of the type

$$N \hookrightarrow B \twoheadrightarrow Z$$

with $\dim N = 16$ and $\dim Z = 2$. This extension is not abelian.

In Section 5, we show that the Hopf subalgebra N also arises as the unique sixteen-dimensional quotient of B. By construction, N is also a Radford biproduct, and the quotient map $\pi_N \colon B \to N$ is induced from a map between the corresponding Yetter-Drinfel'd Hopf algebras. Although π_N does not restrict to the identity on N, it is conormal, and therefore B fits into exactly one extension of the type

$$U \hookrightarrow B \overset{\pi_N}{\twoheadrightarrow} N$$

with $\dim U = 2$ and $\dim N = 16$. This extension is not abelian. We use these facts to describe the structure of the Grothendieck ring of B in Paragraph 5.3.

In Section 6, we begin by determining the eight-dimensional Hopf subalgebras of B. There are three, denoted by M_1, M_2, and M_3, all of which are contained in N. The algebra M_1 is just the span of the group-like elements, while the other two are dual to the group ring of the dihedral group. We show in Paragraph 6.2 that, of the three Hopf subalgebras, only M_2 is normal, so that B fits into exactly one extension of the type

$$M \hookrightarrow B \twoheadrightarrow Q$$

with $\dim M = 8$ and $\dim Q = 4$. In this case, $M = M_2 \cong K^{D_8}$ and $Q \cong K[\mathbb{Z}_2 \times \mathbb{Z}_2]$. This extension is abelian, but neither central nor cocentral.

In Paragraph 6.3, we determine the four-dimensional Hopf subalgebras of B. There are seven, but only one of them is normal. Consequently, the Hopf algebra B fits into exactly one extension of the type

$$P \hookrightarrow B \twoheadrightarrow F$$

with $\dim P = 4$ and $\dim F = 8$. In this case, $P \cong K[\mathbb{Z}_2 \times \mathbb{Z}_2]$ and $F \cong K[D_8]$. This extension is abelian, but neither central nor cocentral.

Besides the Yetter-Drinfel'd Hopf algebra A, the authors considered in [**10**], Section 3 a second eight-dimensional Yetter-Drinfel'd Hopf algebra over $K[\mathbb{Z}_2 \times \mathbb{Z}_2]$, denoted here by A'. These two algebras are certainly not isomorphic, because A is commutative, while A' is not. However, the corresponding Radford biproducts B and B' are isomorphic, as we show in Section 7, so that it is not necessary to carry out the same analysis for B'.

Let us now state the conventions that are used throughout this article. Our base field K will be algebraically closed of characteristic zero. The multiplicative group of K will be denoted by $K^\times = K \setminus \{0\}$. All vector spaces will be defined over K. The dual space of a vector space V will be denoted by $V^* = \mathrm{Hom}(V, K)$, and the dual map of a linear map f, also called its transpose, will be denoted by f^*.

The group algebra of a group G, which we will also call its group ring, will be denoted by $K[G]$. The corresponding dual space $K[G]^*$, the dual group ring, will be denoted by K^G. The character group of a group G will be denoted by $\hat{G} = \mathrm{Hom}(G, K^\times)$. Its elements will be interchangeably called multiplicative characters, one-dimensional characters, or one-dimensional representations, a terminology that we will use not only for groups, but also for algebras. Clearly, the multiplicative characters of the group algebra correspond bijectively to the multiplicative characters of the group via restriction.

All algebras are assumed to have a unit element, and algebra homomorphisms are assumed to preserve these unit elements. The set of group-like elements in a coalgebra A will be denoted by $\mathbf{G}(A)$. The subalgebra generated by elements $a_1, \ldots, a_n$ of an algebra A will be denoted by $K\langle a_1, \ldots, a_n\rangle$, whereas the subgroup of a group G generated by elements $g_1, \ldots, g_n$ will be denoted by $\langle g_1, \ldots, g_n\rangle$. The augmentation ideal of a Hopf algebra H will be denoted by $H^+ := \ker(\varepsilon_H)$. The image of an element a in a quotient space will be denoted by $\bar{a}$.

As already discussed, the article is divided into sections, which are divided further into relatively small paragraphs. A reference to Proposition 2.2 refers to the unique proposition in Paragraph 2.2, and definitions, theorems, lemmas, and corollaries are referenced in the same way. The present article is a short version of the full treatment of the subject matter that we gave in [**11**], where the reader will often find further details of computations and proofs.

1. Preliminaries

1.1. In this section, we consider a finite-dimensional Hopf algebra H with coproduct Δ_H, counit ε_H, and antipode S_H. Recall that a Yetter-Drinfel'd module over H is a vector space V that is both a left H-module and a left H-comodule in such a way that the compatibility condition

$$h_{(1)}v^{(1)} \otimes h_{(2)}.v^{(2)} = \left(h_{(1)}.v\right)^{(1)} h_{(2)} \otimes \left(h_{(1)}.v\right)^{(2)}$$

is satisfied for all $h \in H$ and $v \in V$, where we have used the notation

$$\delta(v) = v^{(1)} \otimes v^{(2)} \in H \otimes V$$

for the coaction δ. The category ${}^H_H YD$ of all Yetter-Drinfel'd modules, whose morphisms are the H-linear and colinear maps, is a braided monoidal category with the braiding $\sigma_{V,W}\colon V \otimes W \to W \otimes V$ given by $\sigma_{V,W}(v \otimes w) = v^{(1)}.w \otimes v^{(2)}$.

We will need the following facts about Yetter-Drinfel'd modules:

(1) The dual space V^* of a finite-dimensional Yetter-Drinfel'd module V can be considered as a Yetter-Drinfel'd module over the dual Hopf algebra H^* by using the module structure

$$\gamma.\varphi = (\gamma \otimes \varphi) \circ \delta \tag{1.1}$$

for $\gamma \in H^*$ and $\varphi \in V^*$, and the comodule structure $\delta(\varphi) = \varphi^{(1)} \otimes \varphi^{(2)} \in H^* \otimes V^*$ with the defining property

$$\varphi^{(1)}(h)\, \varphi^{(2)}(v) = \varphi(h.v) \tag{1.2}$$

where $\varphi \in V^*$, $h \in H$, and $v \in V$ (cf. [**26**], Proposition 1.2 and Lemma 1.3). With respect to these structures, the transpose of an H-linear and H-colinear map is H^*-linear and H^*-colinear.

(2) If A and A' are algebras in the Yetter-Drinfel'd category, so that their multiplication maps μ_A and $\mu_{A'}$ as well as their unit maps η_A and $\eta_{A'}$ are morphisms in the category, then their tensor product $A \otimes A'$ is also an algebra in this category when endowed with the multiplication map

$$\mu_{A\otimes A'} = (\mu_A \otimes \mu_{A'}) \circ (\mathrm{id}_A \otimes \sigma_{A',A} \otimes \mathrm{id}_{A'})$$

and the unit map $\eta_{A\otimes A'} = \eta_A \otimes \eta_{A'}$.

(3) A Hopf algebra in the Yetter-Drinfel'd category ${}^H_H YD$, or a Yetter-Drinfel'd Hopf algebra for short, is an algebra A that is simultaneously a coalgebra, both in this category, so that comultiplication and counit are multiplicative. Here, the algebra structure on $A \otimes A$ is the one introduced above, so that the multiplicativity of the coproduct means that

$$\Delta_A \circ \mu_A = (\mu_A \otimes \mu_A) \circ (\mathrm{id}_A \otimes \sigma_{A,A} \otimes \mathrm{id}_A) \circ (\Delta_A \otimes \Delta_A)$$

or, if we use the notation $\Delta_A(a) = a_{(1)} \otimes a_{(2)}$ for the coproduct,

$$\Delta_A(aa') = a_{(1)} \left({a_{(2)}}^{(1)}.a'_{(1)} \right) \otimes {a_{(2)}}^{(2)} a'_{(2)}$$

We also require that A has an antipode S_A, i.e., a convolution inverse of the identity. By [**10**], Lemma 1.2, such an antipode is automatically a morphism in the category.

1.2. Yetter-Drinfel'd Hopf algebras give rise to ordinary Hopf algebras via the Radford biproduct construction, which was originally defined in [**23**], Theorem 1:

DEFINITION. Let A be a Yetter-Drinfel'd Hopf algebra over H. The Radford biproduct $B := A \star H$ is the tensor product of A and H as a vector space, endowed with the smash product multiplication

$$(a \star h)\,(a' \star h') := a\left(h_{(1)}.a'\right) \star h_{(2)}h' \tag{1.3}$$

with corresponding unit element $1_B := 1_A \star 1_H$, cosmash product comultiplication

$$\Delta_B(a \star h) := \left(a_{(1)} \star {a_{(2)}}^{(1)} h_{(1)}\right) \otimes \left({a_{(2)}}^{(2)} \star h_{(2)}\right) \tag{1.4}$$

with corresponding counit $\varepsilon_B(a \star h) := \varepsilon_A(a)\varepsilon_H(h)$, and antipode

$$S_B(a \star h) := \left(1_A \star S_H(a^{(1)}h)\right)\left(S_A(a^{(2)}) \star 1_H\right) \tag{1.5}$$

With respect to these structures, it is an ordinary Hopf algebra. The notation $a \star h$ instead of $a \otimes h$ is used to signify that the algebra and coalgebra structures are not the canonical structures on the tensor product.

Further details on Yetter-Drinfel'd Hopf algebras and biproducts can be found in [**19**], § 10.6, [**23**], and [**26**], but two results will be needed in the sequel:

(1) If A is a finite-dimensional Yetter-Drinfel'd Hopf algebra over H, then A^* is a Yetter-Drinfel'd Hopf algebra over H^* with respect to the module and comodule structures described in Paragraph 1.1 and the standard dualizations of the algebra and coalgebra structures (cf. [**26**], loc. cit.), so that we can form the biproduct $A^* \star H^*$. It is isomorphic to B^* via the canonical isomorphism of the underlying vector spaces.

(2) The biproduct B is semisimple if A and H are semisimple. Similarly, it is cosemisimple if A and H are cosemisimple (cf. [**23**], Proposition 3 and Proposition 4; [**25**], Paragraph 2.14 and the references cited there).

1.3. A generalization of the smash products just encountered in Paragraph 1.2 are crossed products (cf. [**19**], Definition 7.1.1): Here we have, for a Hopf algebra H and an algebra A, not only a weak action, or measuring, $\rightharpoonup: H \otimes A \to A$, but also a cocycle $\tau: H \times H \to A$, and the multiplication on the underlying vector space $A \otimes H$, which is then commonly denoted by $A\#_\tau H$, is given by the formula

$$(a\#h)(a'\#h') := a(h_{(1)} \rightharpoonup a')\tau(h_{(2)}, h'_{(1)})\#h_{(3)}h'_{(2)}$$

Such a crossed product has the following universal property:

PROPOSITION. Suppose that B is an algebra and that $f_A: A \to B$ as well as $f_H: H \to B$ are K-linear maps. Then the following assertions are equivalent:

1. There exists an algebra homomorphism $f: A\#_\tau H \to B$ such that
$$f(a\#1_H) = f_A(a) \qquad \text{and} \qquad f(1_A\#h) = f_H(h)$$
2. f_A is an algebra homomorphism, $f_H(1_H) = 1_B$, and we have
$$f_A(h_{(1)} \rightharpoonup a)f_H(h_{(2)}) = f_H(h)f_A(a)$$
as well as
$$f_A(\tau(h_{(1)}, h'_{(1)}))f_H(h_{(2)}h'_{(2)}) = f_H(h)f_H(h')$$

PROOF. To show that the first assertion implies the second, we define the inclusion maps $i_A: A \to A\#_\tau H$ and $i_H: H \to A\#_\tau H$ via $i_A(a) := a\#1_H$ and $i_H(h) := 1_A\#h$, and also define $f_A := f \circ i_A$ and $f_H := f \circ i_H$. Then we have $f_H(1_H) = f(1_A\#1_H) = 1_B$ and

$$f_H(h)f_A(a) = f((1_A\#h)(a\#1_H)) = f\left((h_{(1)} \rightharpoonup a)\#h_{(2)}\right) = f_A(h_{(1)} \rightharpoonup a)f_H(h_{(2)})$$

as well as

$$\begin{aligned} f_H(h)f_H(h') &= f(1_A\#h)f(1_A\#h') = f((1_A\#h)(1_A\#h')) \\ &= f\left(\tau(h_{(1)}, h'_{(1)})\#h_{(2)}h'_{(2)}\right) = f_A(\tau(h_{(1)}, h'_{(1)}))f_H(h_{(2)}h'_{(2)}) \end{aligned}$$

For the converse, we define $f(a\#h) := f_A(a)f_H(h)$. Then we have

$$f((a\#h)(a'\#h')) = f(a(h_{(1)} \rightharpoonup a')\tau(h_{(2)}, h'_{(1)})\#h_{(3)}h'_{(2)}) =$$
$$f_A(a)f_A(h_{(1)} \rightharpoonup a')f_A(\tau(h_{(2)}, h'_{(1)}))f_H(h_{(3)}h'_{(2)}) =$$
$$f_A(a)f_A(h_{(1)} \rightharpoonup a')f_H(h_{(2)})f_H(h') = f_A(a)f_H(h)f_A(a')f_H(h') = f(a\#h)f(a'\#h')$$

Since f also preserves the unit element, our assertion holds. □

There are two observations about this universal property that are worth recording. First, if the equation $f_A(h_{(1)} \rightharpoonup a)f_H(h_{(2)}) = f_H(h)f_A(a)$ that appears in the universal property is satisfied for all $h \in H$, but only for a specific $a \in A$, and in the same way for another specific element $a' \in A$, then it holds for their product aa', as we have by the measuring property that

$$\begin{aligned} f_A(h_{(1)} \rightharpoonup (aa'))f_H(h_{(2)}) &= f_A((h_{(1)} \rightharpoonup a)(h_{(2)} \rightharpoonup a'))f_H(h_{(3)}) \\ &= f_A(h_{(1)} \rightharpoonup a)f_A(h_{(2)} \rightharpoonup a')f_H(h_{(3)}) \\ &= f_A(h_{(1)} \rightharpoonup a)f_H(h_{(2)})f_A(a') = f_H(h)f_A(a)f_A(a') = f_H(h)f_A(aa') \end{aligned}$$

Arguing inductively, this shows that it is sufficient to verify this equation for all a in a generating set of A, as long as it holds for all $h \in H$ for each of these generators.

For the second observation, note that i_H is convolution-invertible (cf. [**19**], Proposition 7.2.7). If $\bar{i}_H$ denotes its convolution-inverse, then $\bar{f}_H := f \circ \bar{i}_H$ is the convolution-inverse of $f_H = f \circ i_H$. Thus the two main conditions of the second assertion can be stated in the form $f_A(h \rightharpoonup a) = f_H(h_{(1)})f_A(a)\bar{f}_H(h_{(2)})$ and

$$f_A(\tau(h, h')) = f_H(h_{(1)})f_H(h'_{(1)})\bar{f}_H(h_{(2)}h'_{(2)})$$

In the case that the cocycle is trivial in the sense that $\tau(h, h') = \varepsilon_H(h)\varepsilon_H(h')1_A$, the crossed product reduces to the smash product. Then the twisted module condition (cf. [**19**], Lemma 7.1.2) reduces to the usual module condition, so that the weak action is an ordinary action, which we will denote by a dot again. The preceding proposition now yields the following universal property of the smash product:

COROLLARY. Suppose that B is an algebra and that $f_A \colon A \to B$ and $f_H \colon H \to B$ are algebra homomorphisms such that

$$f_A(h_{(1)}.a)f_H(h_{(2)}) = f_H(h)f_A(a)$$

Then the map $f \colon A\#H \to B$ defined via $f(a\#h) := f_A(a)f_H(h)$ is an algebra homomorphism.

For the smash product, there is a second point to be made, which is slightly different from the one made above in the general case: If the hypothesis in the preceding corollary holds for all $a \in A$ and two elements $h, h' \in H$, it holds for their product, because the weak action is then an ordinary action, so that we have

$$\begin{aligned} f_A(h_{(1)}h'_{(1)}.a)f_H(h_{(2)}h'_{(2)}) &= f_A(h_{(1)}h'_{(1)}.a)f_H(h_{(2)})f_H(h'_{(2)}) \\ &= f_H(h)f_A(h'_{(1)}.a)f_H(h'_{(2)}) = f_H(h)f_H(h')f_A(a) = f_H(hh')f_A(a) \end{aligned}$$

Arguing inductively, this shows that it is sufficient to verify this hypothesis for all h in a generating set of H, as long as it holds for all $a \in A$ for each of these generators.

1.4. As we will explain in Paragraph 1.5 below, crossed products appear naturally when studying Hopf algebra extensions, which are defined as follows (cf. [**17**], Definition 5.6):

DEFINITION. Consider a sequence

$$N \overset{\iota_N}{\hookrightarrow} B \overset{\pi_F}{\twoheadrightarrow} F$$

of finite-dimensional Hopf algebras and Hopf algebra maps ι_N and π_F, where ι_N is injective and π_F is surjective. This sequence is called an extension of F by N if

$$\iota_N(N) = \{b \in B \mid b_{(1)} \otimes \pi_F(b_{(2)}) = b \otimes 1_F\}$$

In this situation, $\iota_N(N)$ is a normal Hopf subalgebra of B (cf. [**2**], Section 1.2).

Hopf algebra extensions can have additional properties:

(1) The extension is called abelian if N is commutative and F is cocommutative. Since K is algebraically closed of characteristic zero, it follows from [**19**], Theorem 2.3.1 that $N \cong K^\Gamma$ and $F \cong K[L]$ for two finite groups Γ and L.

(2) The extension is called central if $\iota_N(N) \subset Z(B)$, the center of B. This happens in particular if N has dimension 2, since N is then necessarily semisimple and is spanned by its unit element and its integral, both of which are central in view of [**13**], Lemma 2.16.

(3) The extension is called cocentral if $\pi_F^*(F^*) \subset Z(B^*)$, the center of B^*.

(4) We say that two extensions $N \hookrightarrow B \twoheadrightarrow F$ and $N \hookrightarrow B' \twoheadrightarrow F$ are equivalent if there is a Hopf algebra isomorphism $f\colon B \to B'$ that induces the identity maps on N and F (cf. [**18**], Definition 1.4).

Further details on Hopf algebra extensions can be found in [**1**], [**2**], [**4**], [**5**], [**17**], and [**18**].

1.5. The connection between crossed products and Hopf algebra extensions arises via bicrossed products, which are defined as follows:

DEFINITION. Let N and F be Hopf algebras equipped with a weak action, or measuring, $\rightharpoonup\colon F \otimes N \to N$, a cocycle $\tau\colon F \times F \to N$, and dually a weak coaction $\rho\colon F \to F \otimes N$ and a dual cocycle $\kappa\colon F \to N \otimes N$. The bicrossed product $N \#_\tau^\kappa F$ is a Hopf algebra with $N \otimes F$ as underlying vector space, crossed product multiplication

$$(x\#t)(y\#s) := x(t_{(1)} \rightharpoonup y)\tau(t_{(2)}, s_{(1)})\#t_{(3)}s_{(2)}$$

as in Paragraph 1.3, and dually cocrossed product comultiplication

$$\Delta(x\#t) := x_{(1)}\kappa_1(t_{(1)})\#{t_{(2)}}^{(1)} \otimes x_{(2)}\kappa_2(t_{(1)}){t_{(2)}}^{(2)}\#t_{(3)}$$

where we have used the notation $\rho(t) = t^{(1)} \otimes t^{(2)} \in F \otimes N$ in analogy to Paragraph 1.1, even though this is a right, not a left, weak coaction. For the dual cocycle, we have used the notation $\kappa(t) = \kappa_1(t) \otimes \kappa_2(t) \in N \otimes N$.

The necessary compatibility conditions on the four structure elements $\rightharpoonup$, τ, ρ, and κ can be found in [**1**], Theorem 2.20; [**2**], Section 3.1; or [**4**], Kapitel 5.

Every bicrossed product yields the extension

$$N \overset{\iota_N}{\hookrightarrow} N\#_\tau^\kappa F \overset{\pi_F}{\twoheadrightarrow} F$$

where $\iota_N(x) := x\#1_F$ and $\pi_F(x\#t) := \varepsilon_N(x)t$. In the finite-dimensional case considered here, in fact every extension is equivalent to such a bicrossed product extension (cf. [**1**], Theorem 3.2.14; [**2**], Proposition 3.1.12 and Theorem 3.1.17; [**17**], Page 130).

2. The biproduct in the first case

2.1. If we use the notation $\mathbb{Z}_2 = \{0,1\}$ for the group with two elements, the group $G := \mathbb{Z}_2 \times \mathbb{Z}_2$ contains the four elements

$$g_1 := (0,0) \qquad g_2 := (1,0) \qquad g_3 := (0,1) \qquad g_4 := (1,1)$$

with $g_1 = 1_G$ as its unit element when we write G multiplicatively, as we will in the sequel. In [**10**], Section 2, the authors defined a commutative Yetter-Drinfel'd Hopf algebra A over the group algebra $H := K[G]$ as the algebra generated by two commuting variables x and y subject to the defining relations

$$x^4 = 1 \qquad\qquad y^2 = \frac{1}{2}(1 + \zeta x + x^2 - \zeta x^3)$$

where ζ is a not necessarily primitive fourth root of unity. With the help of a primitive fourth root of unity ι, it is possible to introduce the elements $\omega_1 := 1_A$,

$$\omega_2 := \frac{1}{2}(1+\iota\zeta^2)x + \frac{1}{2}(1-\iota\zeta^2)x^3 \qquad \omega_3 := \frac{1}{2}(1-\iota\zeta^2)x + \frac{1}{2}(1+\iota\zeta^2)x^3$$

$\omega_4 := x^2$, $\eta_1 := y$, $\eta_2 := x^3 y$, $\eta_3 := x^2 y$, and $\eta_4 := xy$. According to [**10**], Proposition 2.4, these eight elements form a basis of A, and the coalgebra structure is determined by the fact that they are group-like. As also shown there, the set $\{\omega_1, \omega_2, \omega_3, \omega_4\}$ is closed under multiplication and in fact forms a group isomorphic to $\mathbb{Z}_2 \times \mathbb{Z}_2$. The products of all basis elements are computed there; here we only record that $\omega_4\eta_1 = \eta_3$ and $\omega_4\eta_2 = \eta_4$. In [**10**], Paragraph 2.2, the H-action on A is introduced on the generators as

$$g_2.x = x^3 \qquad g_2.y = x^3 y \qquad g_3.x = x \qquad g_3.y = x^2 y$$

which according to [**10**], Lemma 2.4 means for the basis elements that

$$g_3.\omega_i = \omega_i \qquad g_2.\omega_4 = \omega_4 \qquad g_2.\omega_2 = \omega_3 \qquad g_i.\eta_{g_j} = \eta_{g_i g_j}$$

where we have used the notation $\eta_{g_i} := \eta_i$. In particular, we have $g_3.\eta_i = \omega_4\eta_i$ for all $i = 1,2,3,4$. The coaction of H on A is derived from the action by defining

$$\delta(a) := \frac{1}{4} \sum_{g,g' \in G} \theta(g,g')\, g \otimes g'.a \tag{2.1}$$

where θ is the nondegenerate symmetric bicharacter described in [**10**], Paragraph 2.2 and determined by the condition

$$\begin{pmatrix} \theta(g_2,g_2) & \theta(g_2,g_3) \\ \theta(g_3,g_2) & \theta(g_3,g_3) \end{pmatrix} = \begin{pmatrix} \zeta^2 & -1 \\ -1 & 1 \end{pmatrix}$$

on its fundamental matrix. We will need the following two explicit forms, the first of which is recorded in [**10**], Paragraph 4.3:

$$\begin{aligned}
\delta(\omega_2) &= \frac{1}{2}(g_1 + g_3) \otimes \omega_2 + \frac{1}{2}(g_1 - g_3) \otimes \omega_3 \\
\delta(\eta_1) &= \frac{1}{4}(g_1 + g_2 + g_3 + g_4) \otimes \eta_1 + \frac{1}{4}(g_1 + \zeta^2 g_2 - g_3 - \zeta^2 g_4) \otimes \eta_2 \\
&\quad + \frac{1}{4}(g_1 - g_2 + g_3 - g_4) \otimes \eta_3 + \frac{1}{4}(g_1 - \zeta^2 g_2 - g_3 + \zeta^2 g_4) \otimes \eta_4
\end{aligned}$$

Because $\eta_1 = y$, the last formula also yields the value of the coaction on one of the generators. According to [**10**], Proposition 2.2, the value on the other generator is

$$\delta(x) = \frac{1}{2}(g_1 + g_3) \otimes x + \frac{1}{2}(g_1 - g_3) \otimes x^3$$
$$\delta(x^3) = \frac{1}{2}(g_1 - g_3) \otimes x + \frac{1}{2}(g_1 + g_3) \otimes x^3$$

where it is also stated that $x^2 = \omega_4$ is coinvariant.

We can solve the definitions of ω_2 and ω_3 for x and x^3:

$$x = \frac{1}{2}(1 - \iota\zeta^2)\omega_2 + \frac{1}{2}(1 + \iota\zeta^2)\omega_3 \qquad x^3 = \frac{1}{2}(1 + \iota\zeta^2)\omega_2 + \frac{1}{2}(1 - \iota\zeta^2)\omega_3$$

It follows from these considerations that the space

$$C := \mathrm{Span}(1, x, x^2, x^3) = \mathrm{Span}(\omega_1, \omega_2, \omega_3, \omega_4)$$

is a Yetter-Drinfel'd Hopf subalgebra of A. As shown in [**10**], Paragraph 4.3, it is the so-called core of η_1; it is trivial, but not completely trivial in the sense explained there.

2.2. For this specific Yetter-Drinfel'd Hopf algebra A over the group algebra H, we now form the biproduct $B := A \star H$. It is generated by the four elements

$$u := x \star 1_H \qquad v := y \star 1_H \qquad r := 1_A \star g_2 \qquad s := 1_A \star g_3$$

As shown in [**11**], Proposition 2.2, B can be presented in terms of these generators as follows:

PROPOSITION. The four generators satisfy the relations

1. $u^4 = 1, \qquad uv = vu, \qquad v^2 = \frac{1}{2}(1 + \zeta u + u^2 - \zeta u^3)$
2. $r^2 = 1, \qquad rs = sr, \qquad s^2 = 1$
3. $ru = u^3 r, \qquad rv = u^3 vr, \qquad su = us, \qquad sv = u^2 vs$

These relations are defining.

The basis elements of A can be used in a similar way to define

$$c_i := \omega_i \star 1_H \qquad d_i := \eta_i \star 1_H \qquad h_i := 1_A \star g_i$$

for $i = 1, 2, 3, 4$. Then the elements $c_i h_j$ and the elements $d_i h_j$, for $i, j = 1, 2, 3, 4$, form together a basis of B. Note that $c_1 = h_1 = 1_B$. Except for u, the above generators are among these elements, as $v = d_1$, $r = h_2$, and $s = h_3$. Because the canonical map from A to B is an algebra homomorphism, the relations that we have derived in Paragraph 2.1 yield corresponding relations in B. In particular, the elements c_i commute with elements d_j, and we have

$$c_3 = c_4 c_2 \qquad d_3 = c_4 d_1 \qquad d_4 = c_4 d_2$$

Equation (1.3) yields the commutation relations

$$h_3 c_i = c_i h_3 \qquad h_j c_4 = c_4 h_j \qquad h_2 c_2 = c_3 h_2 \qquad h_i d_{g_j} = d_{g_i g_j} h_i$$

where we have used the analogous notation $d_{g_i} := d_i$. We also note that

$$(c_2 h_2)^2 = c_2 h_2 c_2 h_2 = c_2 c_3 h_2 h_2 = c_2 c_3 = c_4 \tag{2.2}$$

so that $(c_2 h_2)^3 = c_3 h_2$ and $(c_2 h_2)^4 = 1$.

2.3. The form of the coproduct can be derived from Equation (1.4). According to [**23**], Fact 2.11, the group-like elements of B are of the form $a \star g_j$, where $a \in A$ is group-like and coinvariant. This implies that we have $\Delta_B(c_4) = c_4 \otimes c_4$ and $\Delta_B(h_j) = h_j \otimes h_j$. The formula for the coaction of ω_2 yields

$$\Delta_B(c_2) = \frac{1}{2}c_2(h_1 + h_3) \otimes c_2 + \frac{1}{2}c_2(h_1 - h_3) \otimes c_3$$

while the formula for the coaction of η_1 yields

$$\begin{aligned}\Delta_B(d_1) = &\frac{1}{4}d_1(h_1 + h_2 + h_3 + h_4) \otimes d_1 + \frac{1}{4}d_1(h_1 + \zeta^2 h_2 - h_3 - \zeta^2 h_4) \otimes d_2 \\ &+ \frac{1}{4}d_1(h_1 - h_2 + h_3 - h_4) \otimes d_3 + \frac{1}{4}d_1(h_1 - \zeta^2 h_2 - h_3 + \zeta^2 h_4) \otimes d_4\end{aligned}$$

We note that the coproduct of B is completely determined by these formulas, since they allow to compute the coproduct of $c_3 = c_4 c_2$ and $d_j = h_j d_1 h_j$ by multiplicativity. The counit is given by $\varepsilon_B(c_i) = \varepsilon_B(d_i) = \varepsilon_B(h_i) = 1$ for all $i = 1, 2, 3, 4$.

Another way of determining the coproduct is to record its values on the generators:

PROPOSITION. We have $\Delta_B(r) = r \otimes r$ and $\Delta_B(s) = s \otimes s$. Furthermore, we have

$$\Delta_B(u) = \frac{1}{2}(u \otimes u + u \otimes u^3 + u^3 s \otimes u - u^3 s \otimes u^3)$$

and

$$\begin{aligned}\Delta_B(v) = &\frac{1}{4}v(1 + r + s + rs) \otimes v + \frac{1}{4}v(1 - \zeta^2 r - s + \zeta^2 rs) \otimes uv \\ &+ \frac{1}{4}v(1 - r + s - rs) \otimes u^2 v + \frac{1}{4}v(1 + \zeta^2 r - s - \zeta^2 rs) \otimes u^3 v\end{aligned}$$

The counit is given on generators by $\varepsilon_B(u) = \varepsilon_B(v) = \varepsilon_B(r) = \varepsilon_B(s) = 1$.

2.4. To complete the description of B, we compute the values of the antipode on the generators:

PROPOSITION. We have $S_B(u) = \frac{1}{2}(u + su + u^3 - su^3)$ and

$$\begin{aligned}S_B(v) = &\frac{1}{4}\left(1 + u + u^2 + u^3\right) v + \frac{1}{4}\left(\zeta 1 + \zeta^3 u - \zeta u^2 - \zeta^3 u^3\right) rv \\ &+ \frac{1}{4}\left(1 - u + u^2 - u^3\right) sv + \frac{1}{4}\left(-\zeta 1 + \zeta^3 u + \zeta u^2 - \zeta^3 u^3\right) rsv\end{aligned}$$

as well as $S_B(r) = r$ and $S_B(s) = s$.

PROOF. The forms of $S_B(r)$, $S_B(s)$, and $S_B(u)$ follow from Equation (1.5), where the third also requires the equation for $\delta(x)$ given in Paragraph 2.1 as well as the facts that $S_A(x) = x$ and $S_A(x^3) = x^3$, which follow from [**10**], Paragraph 2.5. The form of $S_B(v)$ can be derived in the same way, but the computation is more tedious: Using the formula for the coaction of $y = \eta_1$ given in Paragraph 2.1 and the equations for $S_A(\eta_i)$ from [**10**], loc. cit. yields

$$\begin{aligned} S_B(v) &= \frac{1}{8}(h_1+h_2+h_3+h_4)\left(d_1+\zeta^3 d_2+d_3-\zeta^3 d_4\right) \\ &+\frac{1}{8}(h_1+\zeta^2 h_2-h_3-\zeta^2 h_4)\left(\zeta^3 d_1+d_2-\zeta^3 d_3+d_4\right) \\ &+\frac{1}{8}(h_1-h_2+h_3-h_4)\left(d_1-\zeta^3 d_2+d_3+\zeta^3 d_4\right) \\ &+\frac{1}{8}(h_1-\zeta^2 h_2-h_3+\zeta^2 h_4)\left(-\zeta^3 d_1+d_2+\zeta^3 d_3+d_4\right) \end{aligned}$$

The assertion now follows by collecting like terms, writing the result in terms of the generators, and using their commutation relations. □

2.5. In order to determine the Wedderburn decomposition of B, it will be helpful to know the center of B. It is described in the following theorem:

THEOREM. The center $Z(B)$ of B has dimension 11. The four sets of vectors

1. $c_1, \quad c_4, \quad c_2+c_3, \quad d_1+d_2+d_3+d_4$
2. $(c_1+c_2+c_3+c_4)h_2, \quad (d_1+d_2+d_3+d_4)h_2$
3. $(c_1+c_4)h_3, \quad (c_2+c_3)h_3, \quad (d_1+d_2+d_3+d_4)h_3$
4. $(c_1+c_2+c_3+c_4)h_4, \quad (d_1+d_2+d_3+d_4)h_4$

together form a basis of $Z(B)$.

PROOF. (1) Suppose that $b=\sum_{j=1}^4 a_j\star g_j$ lies in the center of B. That b commutes with elements of the form $h=1_A\star g$ for some $g\in G$ means that

$$\sum_{j=1}^4 a_j\star g_j g = bh = hb = (1_A\star g)\sum_{j=1}^4 a_j\star g_j = \sum_{j=1}^4 (g.a_j)\star g_j g$$

Thus, for each $j=1,2,3,4$, we must have $g.a_j=a_j$, and therefore a_j lies in the space of invariants, which we denote by A^H.

(2) On the other hand, b commutes with elements of the form $b'=a'\star 1_H$ if

$$\sum_{j=1}^4 a'a_j\star g_j = b'b = bb' = \left(\sum_{j=1}^4 a_j\star g_j\right)(a'\star 1_H) = \sum_{j=1}^4 (g_j.a')a_j\star g_j$$

This condition is equivalent to the condition $a'a_j=(g_j.a')a_j$ for all $j=1,2,3,4$.

(3) In fact, it is sufficient to test this condition in the two cases $a'=x$ and $a'=y$, because b will be contained in the center if and only if it commutes with the four generators u, v, r, and s introduced at the beginning of Paragraph 2.2. The generators r and s are of the form treated in Step (1), and the generators u and v correspond to the two cases $a'=x$ and $a'=y$.

(4) Discussing each j separately, we see that the condition $a'a_j=(g_j.a')a_j$ is vacuous for $j=1$. For $j=2$, the condition states that

$$xa_2=(g_2.x)a_2=x^3a_2 \qquad \text{and} \qquad ya_2=(g_2.y)a_2=x^3ya_2$$

in other words, $a_2=x^2a_2$ and $a_2=x^3a_2$, which means that $xa_2=a_2$.

For $j = 3$, the condition states that

$$xa_3 = (g_3.x)a_3 = xa_3 \qquad \text{and} \qquad ya_3 = (g_3.y)a_3 = x^2ya_3$$

The first part is vacuous and the second is equivalent to the condition $x^2a_3 = a_3$.

For $j = 4$, the condition states that

$$xa_4 = (g_4.x)a_4 = x^3a_4 \qquad \text{and} \qquad ya_4 = (g_4.y)a_4 = xya_4$$

in other words, $a_4 = x^2a_4$ and $a_4 = xa_4$. Here, the second part implies the first.

(5) The group-like elements of A are divided into four orbits under the action of G, and A^H is spanned by the sum of the elements in each of these orbits, i.e., is spanned by the four elements ω_1, ω_4, $\omega_2 + \omega_3$, and $\eta_1 + \eta_2 + \eta_3 + \eta_4$. It is immediate from the definition of these elements that $\omega_1 = 1_A$, $\omega_4 = x^2$, $\omega_2 + \omega_3 = x + x^3$, and $\eta_1 + \eta_2 + \eta_3 + \eta_4 = (1_A + x + x^2 + x^3)y$. We can therefore expand a_j in the form

$$a_j = \lambda_{1,j}1_A + \lambda_{2,j}x^2 + \lambda_{3,j}(x + x^3) + \lambda_{4,j}(1_A + x + x^2 + x^3)y$$

For $j = 1$, we saw above that there are no extra conditions on a_1, so the coefficients $\lambda_{k,1}$ can be arbitrary. For $j = 2$, the condition $xa_2 = a_2$ implies that $\lambda_{1,2} = \lambda_{2,2} = \lambda_{3,2}$, and so a_2 is a linear combination of the vectors

$$1_A + x + x^2 + x^3 \qquad \text{and} \qquad (1_A + x + x^2 + x^3)y$$

For $j = 3$, the condition $x^2a_3 = a_3$ implies only that $\lambda_{1,3} = \lambda_{2,3}$, and so a_3 is a linear combination of the three vectors

$$1_A + x^2 \qquad\qquad x + x^3 \qquad\qquad (1_A + x + x^2 + x^3)y$$

The condition on a_4 is the same as the condition on a_2, and so a_4 is a linear combination of the same vectors as a_2, which can, as we have already seen, also be written as $\omega_1 + \omega_2 + \omega_3 + \omega_4$ and $\eta_1 + \eta_2 + \eta_3 + \eta_4$. Rewriting these results in terms of c_i and d_i instead of ω_i and η_i yields the assertion, where the four items in the list correspond to our index values $j = 1$, $j = 2$, $j = 3$, and $j = 4$. □

3. Representations

3.1. As we pointed out in Paragraph 1.2, the dual space A^* is a Yetter-Drinfel'd Hopf algebra over H^*, and the corresponding biproduct $A^* \star H^*$ is isomorphic to B^*. The Hopf algebra $H^* = K[G]^*$ is itself a group algebra, namely the group algebra of the character group $\hat{G}$, whose elements are considered as elements of H^* by linear extension. In our situation, the nondegenerate symmetric bicharacter θ described in Paragraph 2.1 yields a specific bijection between G and $\hat{G}$: For $i = 1, 2, 3, 4$, we define the character γ_i by

$$\gamma_i(g_j) = \theta(g_i, g_j)$$

for all $j = 1, 2, 3, 4$. We then have $\hat{G} = \{\gamma_1, \gamma_2, \gamma_3, \gamma_4\}$, where $\gamma_1 = \varepsilon_H$.

The action of γ_i on A^* is closely related to the action of g_i on A: For $\varphi \in A^*$ and $a \in A$, we have by Equation (1.1) and [**10**], Lemma 4.3 that

$$(\gamma_i.\varphi)(a) = (\gamma_i \otimes \varphi)\,\delta(a) = \gamma_i(a^{(1)})\varphi(a^{(2)}) = \varphi(g_i^{-1}.a) \tag{3.1}$$

where of course $g_i = g_i^{-1}$ in our situation. This equation will be used below to determine which of the group-like elements in B^* are central. But first, we record the group-like elements of B:

LEMMA. We have $\mathbf{G}(B) = \langle c_4 \rangle \times \langle h_2 \rangle \times \langle h_3 \rangle \cong \mathbb{Z}_2 \times \mathbb{Z}_2 \times \mathbb{Z}_2$. Moreover, c_4 is the only nontrivial central group-like element of B.

PROOF. As we already mentioned in Paragraph 2.3, the group-like elements of B are of the form $a \star g_j$, where $a \in A$ is group-like and coinvariant. Of the eight group-like elements $\omega_1, \omega_2, \omega_3, \omega_4, \eta_1, \eta_2, \eta_3, \eta_4$ of A, only $\omega_1 = 1_A$ and ω_4 are coinvariant. Because these are also invariant, all the group-like elements of B commute, so that they form the elementary abelian group of order 8 stated above. As we have seen explicitly in the proof of Theorem 2.5 that h_2, h_3, and h_4 are not central, c_4 is the only nontrivial central group-like element of B. □

Alternatively, this lemma can be stated in terms of the generators introduced at the beginning of Paragraph 2.2, as we have $c_4 = u^2$, $h_2 = r$, and $h_3 = s$.

The group of group-like elements of B^* has a very similar form:

PROPOSITION. There are three elements $\chi_1, \chi_2, \chi_3 \in \mathbf{G}(B^*)$ that satisfy

1. $\chi_1(d_i) = -1$ and $\chi_1(c_i) = \chi_1(h_i) = 1$
2. $\chi_2(h_2) = \chi_2(h_4) = -1$ and $\chi_2(c_i) = \chi_2(d_i) = \chi_2(h_3) = 1$
3. $\chi_3(h_3) = \chi_3(h_4) = -1$ and $\chi_3(c_i) = \chi_3(d_i) = \chi_3(h_2) = 1$

for all $i = 1, 2, 3, 4$. With these elements, we have

$$\mathbf{G}(B^*) = \langle \chi_1 \rangle \times \langle \chi_2 \rangle \times \langle \chi_3 \rangle \cong \mathbb{Z}_2 \times \mathbb{Z}_2 \times \mathbb{Z}_2$$

Moreover, χ_1 is the only nontrivial central group-like element of B^*.

PROOF. (1) Since $B^* \cong A^* \star H^*$, it follows as in the preceding lemma that the group-like elements of B^* are of the form $\rho \star \gamma$, where $\gamma \in \hat{G}$ and $\rho \in \mathbf{G}(A^*)$ is H^*-coinvariant. In view of Equation (1.2), a linear form is H^*-coinvariant if and only if it is an H-linear map to the base field, and similarly Equation (1.1) implies that it is H^*-invariant if and only if it is an H-colinear map to the base field. On the other hand, Equation (3.1) above shows that a linear form is H^*-invariant if and only if it is H-linear, so that all these four properties are equivalent.

(2) According to [**10**], Paragraph 2.1, there are eight one-dimensional characters of A, which were denoted there by ε_1, ε_2, ε_3, ε_4, ρ_1, ρ_2, ρ_3, and ρ_4. Their definition depended on the exact order of ζ, but the only H-linear ones were ε_1 and ε_4, where $\varepsilon_1 = \varepsilon_A$ was the counit of A, and ε_4 was in each of the cases determined by the condition that $\varepsilon_4(x) = 1$ and $\varepsilon_4(y) = -1$. Thus B^* contains eight group-like elements, namely

$$\varepsilon_1 \star \gamma_1 \quad \varepsilon_1 \star \gamma_2 \quad \varepsilon_1 \star \gamma_3 \quad \varepsilon_1 \star \gamma_4 \quad \varepsilon_4 \star \gamma_1 \quad \varepsilon_4 \star \gamma_2 \quad \varepsilon_4 \star \gamma_3 \quad \varepsilon_4 \star \gamma_4$$

The H^*-invariance of ε_1 and ε_4 and Equation (1.3) together imply that these elements commute; since $\varepsilon_4^2 = \varepsilon_A$, they all have at most order 2.

(3) We set $\chi_1 := \varepsilon_4 \star \gamma_1$ and then have $\chi_1(d_i) = -1$ and $\chi_1(c_i) = \chi_1(h_i) = 1$ as required. By letting $\chi_2 := \varepsilon_1 \star \gamma_3$, the requirements $\chi_2(h_2) = \chi_2(h_4) = -1$ and $\chi_2(c_i) = \chi_2(d_i) = \chi_2(h_3) = 1$ are also satisfied. But for γ_2, we have according to the definition that $\gamma_2(g_2) = \zeta^2$ and $\gamma_2(g_3) = -1$, while $\gamma_4(g_2) = -\zeta^2$ and $\gamma_4(g_3) = -1$. So if $\zeta^2 = 1$, we set $\chi_3 := \varepsilon_1 \star \gamma_2$, while we set $\chi_3 := \varepsilon_1 \star \gamma_4$ if $\zeta^2 = -1$. It is obvious that these three characters generate $\mathbf{G}(B^*)$.

(4) Because ε_4 is H^*-invariant, we have for $\varphi \in A^*$ and $\gamma \in H^*$ that

$$(\varphi \star \gamma)\chi_1 = \varphi(\gamma_{(1)}.\varepsilon_4) \star \gamma_{(2)} = \varphi\varepsilon_4 \star \gamma = \chi_1(\varphi \star \gamma)$$

so that χ_1 is central. On the other hand, we have for $i = 2, 3, 4$ that

$$(\varepsilon_A \star \gamma_i)(\varphi \star \varepsilon_H) = (\gamma_i.\varphi) \star \gamma_i \qquad \text{but} \qquad (\varphi \star \varepsilon_H)(\varepsilon_A \star \gamma_i) = \varphi \star \gamma_i$$

Since in general $\gamma_i.\varphi \neq \varphi$ by Equation (3.1) above, $\varepsilon_A \star \gamma_i$ is not central. This shows that χ_1 is the only nontrivial central group-like element in B^*. □

At least part of the preceding proposition can also be derived with the help of the presentation of B given in Proposition 2.2. The values of our characters on the generators u, v, r, and s introduced there are given in the following table:

	u	v	r	s
χ_1	1	-1	1	1
χ_2	1	1	-1	1
χ_3	1	1	1	-1

3.2. We have already mentioned in Paragraph 1.2 that biproducts are semisimple and cosemisimple if their two factors are. Here, this implies that the biproduct B under consideration is semisimple and cosemisimple, because the Yetter-Drinfel'd Hopf algebra A is semisimple by [**10**], Proposition 2.1 and cosemisimple by [**10**], Proposition 2.4. But we can determine the Wedderburn decomposition of B and B^* precisely:

THEOREM. Both B and B^* have eight 1-dimensional, two 2-dimensional, and one 4-dimensional irreducible representations.

PROOF. Since $\dim Z(B) = 11$ by Theorem 2.5 and $|\mathbf{G}(B^*)| = 8$ by Proposition 3.1, there are exactly three irreducible representations of B of dimension larger than 1. The sum of the squares of the dimensions of these three representations is $32 - 8 = 24$, which implies that B has, up to isomorphism, two 2-dimensional and one 4-dimensional irreducible representations.

Alternatively, this result can be deduced via Clifford theory (cf. [**3**], Proposition (11.16); [**9**], Corollary 3.5). As explained in [**11**], Paragraph 3.2, Clifford theory can also be used to establish the analogous Wedderburn decomposition of B^*. □

In Paragraph 5.3, we will describe in addition the product in the Grothendieck ring.

4. Hopf subalgebras of dimension 16

4.1. The algebra C introduced in Paragraph 2.1 is a Yetter-Drinfel'd Hopf subalgebra of A, and therefore the biproduct $B = A \star H$ contains the biproduct

$$N := C \star H$$

as a Hopf subalgebra. Clearly, N has dimension 16, and the elements $c_i h_j$ for $i, j = 1, 2, 3, 4$ form a basis of N. This implies that N is generated as an algebra by c_2, c_4, h_2, and h_3. This generating set has the advantage of being close to the group-like elements of A. Alternatively, N is generated by the elements u, r, and s introduced in Paragraph 2.2, which satisfy the relations

$$u^4 = 1 \qquad ru = u^3 r \qquad su = us \qquad r^2 = 1 \qquad rs = sr \qquad s^2 = 1$$

from Proposition 2.2. An argument that is very similar to the one establishing that proposition shows that these relations are defining. Comparing this presentation to the usual presentation of the dihedral group (cf. [**24**], Section 5.3), we see that N is isomorphic to the group ring $K[D_8 \times \mathbb{Z}_2]$ as an algebra, but not as a coalgebra, as it is not cocommutative, as for example the formula for the coproduct of c_2 in Paragraph 2.3 shows. This issue will be revisited at the end of Paragraph 4.2.

The groups of group-like elements of N and N^* have the following form:

LEMMA. We have $\mathbf{G}(N) = \langle c_4 \rangle \times \langle h_2 \rangle \times \langle h_3 \rangle \cong \mathbb{Z}_2 \times \mathbb{Z}_2 \times \mathbb{Z}_2$, and moreover also $\mathbf{G}(N^*) \cong \mathbb{Z}_2 \times \mathbb{Z}_2 \times \mathbb{Z}_2$.

PROOF. Since all the group-like elements of B are already contained in N, we have $\mathbf{G}(N) = \mathbf{G}(B)$, so that the first assertion follows from Lemma 3.1.

As in the proof of Proposition 3.1, we have $N^* \cong C^* \star H^*$, and under this isomorphism a group-like element $\chi \in \mathbf{G}(N^*)$ decomposes in the form $\rho \star \gamma$, where $\gamma \in \hat{G}$ and $\rho \in \mathbf{G}(C^*)$ is H^*-coinvariant. We also discussed there that H^*-coinvariance is equivalent to H^*-invariance as well as to H-linearity and H-colinearity. If $\chi' \in \mathbf{G}(N^*)$ is another group-like element that decomposes in the form $\rho' \star \gamma'$, the H^*-invariance implies that $\chi\chi'$ corresponds to $\rho\rho' \star \gamma\gamma'$.

There are clearly four algebra homomorphisms from C to K, determined by the image of x, which must be a fourth root of unity. Out of these, two are H-linear, namely $\varepsilon'_1 := \varepsilon_C$, the counit, which satisfies $\varepsilon'_1(x) = 1$, and the character ε'_2 determined by $\varepsilon'_2(x) = -1$. It also satisfies $\varepsilon'_2(\omega_2) = \varepsilon'_2(\omega_3) = -1$, so that $\varepsilon'^2_2 = \varepsilon_C$. This shows that $|\mathbf{G}(N^*)| = 8$ and $\mathbf{G}(N^*) \cong \mathbb{Z}_2 \times \mathbb{Z}_2 \times \mathbb{Z}_2$. □

In analogy with Paragraph 3.1, we choose generators χ'_1, χ'_2, and χ'_3 of $\mathbf{G}(N^*)$ whose values on the generators of N are given by the following table:

	u	r	s
χ'_1	-1	1	1
χ'_2	1	-1	1
χ'_3	1	1	-1

Explicitly, χ'_1 corresponds to $\varepsilon'_2 \star \gamma_1$, while χ'_2 corresponds to $\varepsilon'_1 \star \gamma_3$, and χ'_3 corresponds to $\varepsilon'_1 \star \gamma_2$ if $\zeta^2 = 1$ and to $\varepsilon'_1 \star \gamma_4$ if $\zeta^2 = -1$.

The isomorphism $\mathbf{G}(N^*) \cong \mathbb{Z}_2 \times \mathbb{Z}_2 \times \mathbb{Z}_2$ can also be established in a more direct way that does not depend on the biproduct perspective:

COROLLARY. If $\mathbb{Z}_2 = \{1, -1\}$ is written multiplicatively, the map

$$\mathbf{G}(N^*) \to \mathbb{Z}_2 \times \mathbb{Z}_2 \times \mathbb{Z}_2, \; \chi \mapsto (\chi(u), \chi(r), \chi(s))$$

is an isomorphism of groups.

A detailed proof of this result is given in [**11**], Paragraph 4.1. We note that the fact that $\chi(u) = \pm 1$ implies that $\chi(u) = \chi(c_2) = \chi(c_3)$ for all $\chi \in \mathbf{G}(N^*)$. Also proved there is the following description of the central group-like elements of N^*:

PROPOSITION. The group $\mathbf{G}(N^*) \cap Z(N^*)$ is isomorphic to $\mathbb{Z}_2 \times \mathbb{Z}_2$ and is generated by χ'_1 and χ'_2.

4.2. A characteristic property of N is the following:

PROPOSITION. The Hopf algebra N has exactly three quotient Hopf algebras of dimension 8. All of these quotients are cocommutative and isomorphic to either $K[\mathbb{Z}_2 \times \mathbb{Z}_2 \times \mathbb{Z}_2]$ or $K[D_8]$.

PROOF. (1) If N has a quotient Hopf algebra F of dimension 8, then N^* has a Hopf subalgebra F^* that has index 2 and is therefore necessarily normal by [**12**], Proposition 2. Thus N^* fits into an extension of the form $F^* \hookrightarrow N^* \twoheadrightarrow M$, and so N fits into an extension of the form

$$M^* \hookrightarrow N \twoheadrightarrow F$$

where M^* is normal and has dimension 2. As pointed out in Paragraph 1.4, M^* is then central, and therefore $M^* = K\langle g \rangle$ for a central group-like element g of order 2, so that $F \cong N/((M^*)^+N)$.

(2) It follows from Lemma 4.1 that N contains exactly three central group-like elements of order 2, namely c_4, h_3, and c_4h_3. Therefore it has exactly three quotient Hopf algebras of dimension 8, namely

$$F_1 := N/\left(K\langle c_4 \rangle^+ N\right) \qquad F_2 := N/\left(K\langle h_3 \rangle^+ N\right) \qquad F_3 := N/\left(K\langle c_4h_3 \rangle^+ N\right)$$

As explained in [**11**], Paragraph 4.2, we have $F_1 \cong K[\mathbb{Z}_2 \times \mathbb{Z}_2 \times \mathbb{Z}_2]$, $F_2 \cong K[D_8]$, and also $F_3 \cong K[D_8]$. □

From the classification results in [**6**], Table 1, we know that there are exactly two Hopf algebras of dimension 16 for which both the group of group-likes and the group of group-likes of the dual are elementary abelian of order 8, namely those denoted there by $H_{d:1,1}$ and $H_{d:-1,1}$. The Hopf algebra $H_{d:-1,1}$ is isomorphic to $H_8 \otimes K[\mathbb{Z}_2]$, where H_8 is the Kac-Paljutkin Hopf algebra of dimension 8. Since $H_8 \otimes K[\mathbb{Z}_2]$ has a noncocommutative quotient isomorphic to H_8, the proposition above implies that N must be isomorphic to $H_{d:1,1}$. By [**6**], Remark 1.4, we then know that N is isomorphic to $K[D_8 \times \mathbb{Z}_2]_J$, which is triangular, and that N is self-dual. The twisting cocycle J is described in [**6**], Section 7, Example 2. In particular, we see again that N is isomorphic to $K[D_8 \times \mathbb{Z}_2]$ as an algebra.

4.3. In fact, we can construct an explicit isomorphism between N and $H_{d:1,1}$. As described in [**6**], Paragraph 3.2, Page 629ff, the Hopf algebra $H_{d:1,1}$ is a bicrossed product $K[\Gamma]\#^{\theta}K[L]$ for $\Gamma = \langle x\rangle\times\langle y\rangle\times\langle z\rangle \cong \mathbb{Z}_2\times\mathbb{Z}_2\times\mathbb{Z}_2$ and $L = \langle t\rangle \cong \mathbb{Z}_2$ in which the cocycle and the coaction are trivial. Here, and in the present paragraph only, we use the notation from [**6**], loc. cit., so that θ does not denote the nondegenerate bicharacter from Paragraph 2.1, but rather a dual cocycle like the one that was denoted by κ in Paragraph 1.5. Furthermore, x and y do not denote the generators of A introduced in Paragraph 2.1, but rather two of the three generators of Γ already appearing above.

The action $\rightharpoonup$ of the bicrossed product is determined by the facts that the unit element acts as the identity, while t acts as the algebra automorphism defined by

$$t \rightharpoonup x = y \qquad\qquad t \rightharpoonup y = x \qquad\qquad t \rightharpoonup z = z$$

on the generators. The dual cocycle θ can be described as the map that sends t to

$$\theta(t) = \frac{1}{2}\left((1+z)\otimes 1 + (1-z)\otimes xy\right)$$

and 1 to $1\otimes 1$. The corresponding bicrossed product $H_{d:1,1} = K[\Gamma]\#^{\theta}K[L]$ is isomorphic to our Hopf subalgebra N:

THEOREM. There exists a Hopf algebra isomorphism $f\colon H_{d:1,1}\to N$ satisfying

$$f(x) = h_2 \qquad\qquad f(y) = c_4h_2 \qquad\qquad f(z) = h_3 \qquad\qquad f(\bar{t}) = c_2$$

where $a\in K[\Gamma]$ is identified with $a\#1\in H_{d:1,1}$ and $\bar{t} := 1\#t \in H_{d:1,1}$.

PROOF. Because its cocycle is trivial by assumption, our crossed product is indeed a smash product, so that we can use the universal property of a smash product stated in Corollary 1.3. Since both Γ and $\mathbf{G}(N)$ are elementary abelian of order 8, there is an algebra homomorphism $f_{K[\Gamma]}\colon K[\Gamma]\to N$ defined by

$$f_{K[\Gamma]}(x) = h_2 \qquad f_{K[\Gamma]}(y) = c_4h_2 \qquad f_{K[\Gamma]}(z) = h_3$$

Similarly, as c_2 has order 2, there is an algebra homomorphism $f_{K[L]}\colon K[L]\to N$ with $f_{K[L]}(t) = c_2$. As we pointed out in Paragraph 1.3, it is sufficient to verify the hypothesis of Corollary 1.3 on the generating set x, y, z, where it holds since

$$\begin{aligned} f_{K[\Gamma]}(t\rightharpoonup x)f_{K[L]}(t) &= f_{K[\Gamma]}(y)f_{K[L]}(t) = c_4h_2c_2 = c_2h_2 = f_{K[L]}(t)f_{K[\Gamma]}(x)\\ f_{K[\Gamma]}(t\rightharpoonup y)f_{K[L]}(t) &= f_{K[\Gamma]}(x)f_{K[L]}(t) = h_2c_2 = c_2c_4h_2 = f_{K[L]}(t)f_{K[\Gamma]}(y)\\ f_{K[\Gamma]}(t\rightharpoonup z)f_{K[L]}(t) &= f_{K[\Gamma]}(z)f_{K[L]}(t) = h_3c_2 = c_2h_3 = f_{K[L]}(t)f_{K[\Gamma]}(z) \end{aligned}$$

Thus Corollary 1.3 yields an algebra homomorphism $f\colon H_{d:1,1}\to N$ that satisfies $f(a\#1) = f_{K[\Gamma]}(a)$ and $f(1\#c) = f_{K[L]}(c)$, and hence the asserted equations.

To see that f is a coalgebra homomorphism, it suffices to check the comultiplicativity of f on the generators of $H_{d:1,1}$, namely x, y, z, and $\bar{t}$. Since x, y, and z as well as their images are group-like, it is enough to verify this condition on $\bar{t}$:

$$\begin{aligned} (f\otimes f)\left(\Delta(\bar{t})\right) &= (f\otimes f)\left(\frac{1}{2}\left((1+z)\otimes 1 + (1-z)\otimes xy\right)(\bar{t}\otimes\bar{t})\right)\\ &= \frac{1}{2}\left((1_B+h_3)\otimes 1_B + (1_B-h_3)\otimes c_4\right)(c_2\otimes c_2) = \Delta_B(c_2) = \Delta_B(f(\bar{t})) \end{aligned}$$

It is easily checked on the generators that f preserves the counit.

Since N is generated by h_2, c_4h_2, h_3, and c_2, the map f is surjective. Therefore, since $\dim H_{d:1,1} = \dim N = 16$, it is even bijective. □

The isomorphism f can be used to relate the one-dimensional representations of the two Hopf algebras. In particular, the one-dimensional characters χ'_1, χ'_2, and χ'_3 of N defined in Paragraph 4.1 become the one-dimensional characters $\chi'_1 \circ f$, $\chi'_2 \circ f$, and $\chi'_3 \circ f$ of $H_{d:1,1}$, which were denoted in [**6**], Page 633 by χ, φ, and ψ, respectively.

4.4. The sixteen-dimensional Hopf subalgebra N of B introduced in Paragraph 4.1, which is, as just discussed, isomorphic to $H_{d:1,1}$ and therefore a twisting of the group algebra of $D_8 \times \mathbb{Z}_2$, is in fact the only such Hopf subalgebra:

THEOREM. *The Hopf subalgebra N is the unique Hopf subalgebra of B of dimension 16. It is normal, but neither commutative nor cocommutative. The Hopf algebra B therefore fits into exactly one extension of the type*

$$N \hookrightarrow B \twoheadrightarrow Z$$

with $\dim N = 16$ and $\dim Z = 2$. This extension is not abelian.

PROOF. Suppose that N' is a Hopf subalgebra of B of dimension 16. As in the proof of Proposition 4.2, it follows from [**12**], Proposition 2 that N' is normal. Therefore, B fits into the extension

$$N' \overset{\iota_{N'}}{\hookrightarrow} B \overset{\pi_Z}{\twoheadrightarrow} Z$$

where $\iota_{N'}$ is the inclusion map and π_Z is the canonical projection to the quotient $Z := B/BN'^+$ of dimension 2. By dualization, we obtain the extension

$$Z^* \overset{\pi_Z^*}{\hookrightarrow} B^* \overset{\iota_{N'}^*}{\twoheadrightarrow} N'^*$$

As pointed out in Paragraph 1.4, the image $\pi_Z^*(Z^*)$ is then central in B^*, and is of course a group algebra by [**19**], Theorem 2.3.1. But by Proposition 3.1, the only nontrivial central group-like element of B^* is χ_1, so that $\pi_Z^*(Z^*)$ is spanned by ε_B and χ_1. In particular, $\pi_Z^*(Z^*)$ together with its embedding into B^* is uniquely determined, so that dually B has a unique two-dimensional quotient Hopf algebra, namely Z. But the quotient determines the Hopf subalgebra as the space of coinvariants (cf. [**19**], Lemma 3.4.2 and Proposition 3.4.3). Therefore, N' is uniquely determined, so that $N' = N$. Since we have seen in Paragraph 4.1 that N is neither commutative nor cocommutative, the corresponding extension is not abelian. □

As we mentioned in the introduction, many of the known semisimple Hopf algebras of prime power dimension contain a large commutative Hopf subalgebra of prime index. The result above shows that B is not of this kind.

4.5. We have seen in Proposition 4.2 that N has exactly three quotient Hopf algebras of dimension 8. It has also exactly three Hopf subalgebras of dimension 8:

PROPOSITION. *The Hopf algebra N has exactly three Hopf subalgebras of dimension 8, namely*

$$M_1 := K\langle c_4, h_2, h_3 \rangle \qquad M_2 := K\langle c_2, c_4, h_3 \rangle \qquad M_3 := K\langle c_2h_2, h_3 \rangle$$

All these three subalgebras are commutative and normal in N.

PROOF. Suppose that M is an eight-dimensional Hopf subalgebra of N. As already mentioned twice, we then know from [**12**], Proposition 2 that M is normal in N. The quotient N/NM^+ has dimension 2, and the transpose of the quotient map yields an embedding of $(N/NM^+)^*$ into N^*. As in the proof of Theorem 4.4, the two-dimensional image of this map must have the form $\mathrm{Span}(\varepsilon, \chi)$ for a nontrivial central group-like element χ of order 2. This means that the original quotient map is essentially the map

$$N \to K \times K, \ b \mapsto (\varepsilon(b), \chi(b))$$

By [**19**], Proposition 3.4.3, the original Hopf subalgebra can be recovered as the space of coinvariants, i.e., as the space

$$M = \{b \in N \mid b = b_{(1)}\chi(b_{(2)})\}$$

Conversely, a given nontrivial central group-like element χ of order 2 yields via the last equation an eight-dimensional normal Hopf subalgebra M of N.

From Proposition 4.1, we know that N^* contains exactly three nontrivial central group-like elements, namely χ'_1, χ'_2, and $\chi'_1\chi'_2$. As explained in [**11**], Paragraph 4.5, these three elements correspond to the three Hopf subalgebras M_1, M_2, and M_3 defined above. □

The first two of these eight-dimensional Hopf subalgebras have a very natural interpretation: According to Lemma 4.1, the subalgebra M_1 is just the span of the group-like elements $\mathbf{G}(N)$. The Hopf algebra M_2 is the biproduct

$$M_2 = C \star K[\langle g_3 \rangle]$$

which can be formed because the formulas for $\delta(x)$ and $\delta(x^3)$ in Paragraph 2.1 show that C is indeed a Yetter-Drinfel'd Hopf algebra over $K[\langle g_3 \rangle]$. Note that, because the action of g_3 on C is trivial, M_2 is actually the tensor product of C and $K[\langle g_3 \rangle]$ as an algebra. In Paragraph 4.6, we will see that M_2 and M_3 are in fact isomorphic.

The first two Hopf subalgebras can also be easily expressed in terms of the generators u, r, and s, namely as $M_1 = K\langle u^2, r, s\rangle$ and $M_2 = K\langle u, s\rangle$.

4.6. Semisimple Hopf algebras of dimension 8 have been classified in [**14**]. So the question arises which of these Hopf algebras the algebras M_1, M_2, and M_3 actually are. For M_1, this is not difficult: As we already said above, it is the group algebra of the group $\mathbf{G}(N) \cong \mathbb{Z}_2 \times \mathbb{Z}_2 \times \mathbb{Z}_2$. The remaining two, being commutative, must be dual group algebras, as we mentioned in Paragraph 1.4. It turns out that they are both isomorphic to the dual group algebra of the dihedral group D_8:

PROPOSITION. We have $M_2 \cong M_3 \cong K^{D_8}$.

PROOF. (1) Because $M_2 = K\langle c_2, c_4, h_3\rangle$ is commutative and K is algebraically closed, we must have $|\mathbf{G}(M_2^*)| = 8$. Since $c_2^2 = c_4^2 = h_3^2 = 1$, any multiplicative character $\chi \in \mathbf{G}(M_2^*)$ satisfies

$$\chi(c_2) = \pm 1 \qquad \chi(c_4) = \pm 1 \qquad \chi(h_3) = \pm 1$$

and therefore, since $|\mathbf{G}(M_2^*)| = 8$, all possible combinations of ± 1 must yield multiplicative characters.

For two multiplicative characters $\chi, \chi' \in \mathbf{G}(M_2^*)$, we have on the last two generators

$$(\chi\chi')(c_4) = \chi(c_4)\chi'(c_4) \qquad \text{and} \qquad (\chi\chi')(h_3) = \chi(h_3)\chi'(h_3)$$

because these generators are group-like. For the third generator c_2, the formula for its coproduct given in Paragraph 2.3, which in particular shows that M_2 is not cocommutative, implies that

$$\begin{aligned}(\chi\chi')(c_2) &= \frac{1}{2}(1 + \chi(h_3) + \chi'(c_4) - \chi(h_3)\chi'(c_4))\,\chi(c_2)\chi'(c_2) \\ &= \begin{cases} \chi(c_2)\chi'(c_2) & : \chi(h_3) = 1 \text{ or } \chi'(c_4) = 1 \\ -\chi(c_2)\chi'(c_2) & : \chi(h_3) = \chi'(c_4) = -1 \end{cases}\end{aligned}$$

If we apply this formula in the case $\chi = \chi'$, it shows that $\mathbf{G}(M_2^*)$ has exactly two elements of order 4, defined by $\chi(c_2) = \pm 1$ and $\chi(c_4) = \chi(h_3) = -1$, and five elements of order 2, which can only happen if $\mathbf{G}(M_2^*) \cong D_8$.

(2) In $M_3 = K\langle c_2h_2, h_3\rangle$, we have by Equation (2.2) that $(c_2h_2)^2 = c_4$, so that $(c_2h_2)^3 = c_3h_2$ and $(c_2h_2)^4 = 1$. For two multiplicative characters $\chi, \chi' \in \mathbf{G}(M_3^*)$, we have in this case $(\chi\chi')(h_3) = \chi(h_3)\chi'(h_3)$ for the second generator and

$$(\chi\chi')(c_2h_2) = \begin{cases} \chi(c_2h_2)\chi'(c_2h_2) & : \chi(h_3) = 1 \\ \chi(c_2h_2)\chi'(c_2h_2)^3 & : \chi(h_3) = -1 \end{cases}$$

for the first generator, as explained in greater detail in [**11**], Paragraph 4.6. As before, this formula shows that $\mathbf{G}(M_3^*)$ is noncommutative and has exactly two elements of order 4, defined by $\chi(c_2h_2) = \pm\iota$ and $\chi(h_3) = 1$, as well as five elements of order 2, so that again $\mathbf{G}(M_3^*) \cong D_8$. □

We note that there is a second way to see that N has exactly three eight-dimensional Hopf subalgebras: It follows from the classification of semisimple Hopf algebras of dimension 16 that $N \cong H_{d:1,1}$ is self-dual (cf. [**6**], Remark 1.4). Since N has, by Proposition 4.2, up to isomorphism exactly three quotient Hopf algebras of dimension 8, self-duality implies that $N \cong N^*$ has exactly three Hopf subalgebras of dimension 8. Our discussion above, however, gives their explicit form.

5. Hopf subalgebras of dimension 2

5.1. The four-dimensional Yetter-Drinfel'd Hopf subalgebra C of A introduced in Paragraph 2.1 also arises as a quotient of A:

PROPOSITION. There is a unique homomorphism $\pi_C \colon A \to C$ of Yetter-Drinfel'd Hopf algebras defined on generators as

$$\pi_C(x) := \omega_4 = x^2 \qquad\qquad \pi_C(y) := \omega_2$$

Its values on the group-like elements are given by the following table:

a	ω_1	ω_2	ω_3	ω_4	η_1	η_2	η_3	η_4
$\pi_C(a)$	ω_1	ω_4	ω_4	ω_1	ω_2	ω_3	ω_2	ω_3

The relatively straightforward proof of this proposition is given in [**11**], Paragraph 5.1. On the Radford biproduct, π_C induces the surjective Hopf algebra homomorphism

$$\pi_N \colon B \to N, \ a \star h \mapsto \pi_C(a) \star h$$

which on the generators introduced in Paragraph 2.2 takes the values

$$\pi_N(u) = u^2 \qquad \pi_N(v) = \frac{1}{2}(1 + \iota\zeta^2)u + \frac{1}{2}(1 - \iota\zeta^2)u^3 \qquad \pi_N(r) = r \qquad \pi_N(s) = s$$

Note that the second equation can also be stated as $\pi_N(d_1) = c_2$. By comparing the table in Paragraph 3.1 to the table in Paragraph 4.1, we see that π_N relates the group-like elements of B^* and the group-like elements of N^* via

$$\chi_1 = \chi'_1 \circ \pi_N \qquad \chi_2 = \chi'_2 \circ \pi_N \qquad \chi_3 = \chi'_3 \circ \pi_N$$

We will denote the kernel of π_N by I, so that $B/I \cong N$.

5.2. The ideal I just defined is a Hopf ideal of dimension 16. We will now show that it is the only such ideal. In preparation, we prove a simple lemma:

LEMMA. The space $U := \mathrm{Span}(c_1, c_4)$ is the only normal Hopf subalgebra of B of dimension 2. It is central.

PROOF. We have already stated in Paragraph 1.4 that a normal Hopf subalgebra U' of B of dimension 2 is central. Since it is commutative, cocommutative, and semisimple, U' is spanned by the unit element and a nontrivial central group-like element. But according to Lemma 3.1, the only nontrivial central group-like element is c_4, and so the assertion holds. □

The Hopf ideal I and the Hopf subalgebra U are related as follows:

THEOREM. The ideal I is the only Hopf ideal of B of dimension 16. It is normal. We have $I = BU^+$ and $U = B^{\mathrm{co}N}$. The Hopf algebra B therefore fits into exactly one extension of the type

$$U \hookrightarrow B \overset{\pi_N}{\twoheadrightarrow} N$$

with $\dim U = 2$ and $\dim N = 16$. Since N is neither commutative nor cocommutative, this extension is not abelian.

PROOF. If I' is a Hopf ideal of B of dimension 16, we can set $N' := B/I'$ and denote the quotient map by $\pi_{N'} \colon B \to N'$. Then $\pi^*_{N'}(N'^*)$ is a Hopf subalgebra of B^* of index 2. As in the proof of Theorem 4.4, it follows from [**12**], Proposition 2 that $\pi^*_{N'}(N'^*)$ is normal in B^*. This implies that I' is normal and that $\pi_{N'}$ is a conormal surjection (cf. [**19**], Section 3.4, Page 36), which in turn implies that $U' := B^{\mathrm{co}N'}$ is a normal Hopf subalgebra of dimension 2. By the preceding lemma, we have $U' = U$, and consequently $I' = BU^+$. This shows that I' is uniquely determined, and therefore $I' = I$. □

5.3. As promised in Paragraph 3.2, the preceding results enable us to describe the product in the Grothendieck ring $K_0(B)$. From Proposition 3.1, we know that

$$\mathbf{G}(B^*) = \langle \chi_1 \rangle \times \langle \chi_2 \rangle \times \langle \chi_3 \rangle \cong \mathbb{Z}_2 \times \mathbb{Z}_2 \times \mathbb{Z}_2$$

which describes the one-dimensional representations. By Theorem 3.2, the Hopf algebra B has two irreducible representations of dimension 2, which we will denote by π_1 and π_2, and one irreducible representation of dimension 4, which we will denote by ρ.

In Theorem 5.2, we have proved that B has a unique quotient Hopf algebra N of dimension 16, which is isomorphic to the Hopf algebra $H_{d:1,1}$ from [**6**], Table 1. Clearly, the irreducible representations of N can be pulled back along the quotient map π_N to irreducible representations of B, so that we obtain a ring homomorphism from $K_0(N)$ to $K_0(B)$. The one-dimensional and the two-dimensional irreducible representations of B arise in this way; for the one-dimensional representations, we have already seen that at the end of Paragraph 5.1, and we have also indicated at the end of Paragraph 4.3 how the one-dimensional representations of N correspond to the one-dimensional representations of $H_{d:1,1}$.

Therefore, the description of the Grothendieck ring of $H_{d:1,1}$ given in [**6**], Paragraph 5.1 implies on the one hand that the two-dimensional irreducible representations, and also the one-dimensional representations, are self-dual, and on the other hand that the products of these representations are given by the table

	χ_1	χ_2	χ_3	π_1	π_2
π_1	π_1	π_1	π_2	$1 + \chi_1 + \chi_2 + \chi_1\chi_2$	$\chi_3 + \chi_1\chi_3 + \chi_2\chi_3 + \chi_1\chi_2\chi_3$
π_2	π_2	π_2	π_1	$\chi_3 + \chi_1\chi_3 + \chi_2\chi_3 + \chi_1\chi_2\chi_3$	$1 + \chi_1 + \chi_2 + \chi_1\chi_2$

Note that, as a consequence of the self-duality of the irreducible representations, the Grothendieck ring of N is commutative, so that in this table it does not matter which factor in the product comes first.

It remains to treat the products that involve ρ. Because it is the only four-dimensional irreducible representation, it is also self-dual, and we have $\chi_i\rho = \rho$ for $i = 1, 2, 3$. Furthermore, as in the case of N, self-duality implies that the Grothendieck ring of B is commutative.

Since ρ does not occur in the decomposition of $\chi_i\pi_j$ and $\pi_i\pi_j$, it follows from [**22**], Theorem 9 that no one-dimensional or two-dimensional irreducible representation occurs in the decomposition of $\pi_j\rho$, so that

$$\pi_1\rho = \pi_2\rho = 2\rho$$

Similarly, since ρ occurs in the decomposition of $\chi_i\rho$ with multiplicity 1 and in the decomposition of $\pi_i\rho$ with multiplicity 2, we get in the same way from [**22**], Theorem 9 that

$$\rho^2 = 1 + \chi_1 + \chi_2 + \chi_3 + \chi_1\chi_2 + \chi_1\chi_3 + \chi_2\chi_3 + \chi_1\chi_2\chi_3 + 2\pi_1 + 2\pi_2$$

Note that the first eight summands are precisely the eight elements of $\mathbf{G}(B^*)$.

6. Hopf subalgebras of dimensions 4 and 8

6.1. From Theorem 4.4 and Theorem 5.2, we know that neither B nor B^* fits into an abelian extension of a Hopf algebra of dimension 2 by a commutative Hopf algebra of dimension 2^n. Thus neither B nor B^* can be constructed as an extension of the type used in [**6**], [**7**], [**8**], and [**14**] to classify certain semisimple Hopf algebras of prime power dimension. We will now show that nevertheless B fits into an essentially unique abelian extension of a Hopf algebra of dimension 4 by a commutative Hopf algebra of dimension 8, but this extension is neither central nor cocentral.

THEOREM. The Hopf algebra B has exactly three Hopf subalgebras of dimension 8, namely

$$M_1 := K\langle c_4, h_2, h_3\rangle \qquad M_2 := K\langle c_2, c_4, h_3\rangle \qquad M_3 := K\langle c_2h_2, h_3\rangle$$

PROOF. Suppose that M is an eight-dimensional Hopf subalgebra of B. From Theorem 3.2, we know that B^* has eight 1-dimensional, two 2-dimensional, and one 4-dimensional irreducible representations. Furthermore, it follows from Theorem 4.4 that the unique quotient Hopf algebra of B^* of dimension 16 is the dual N^* of the Hopf subalgebra N of B defined in Paragraph 4.1. Since no Hopf algebra of dimension 16 or 8 can have a four-dimensional irreducible representation, the canonical restriction mappings from B^* to N^* and M^*, which are dual to the inclusion mappings, must both contain the sixteen-dimensional ideal corresponding to the four-dimensional irreducible representation in their respective kernels. For the canonical map from B^* to N^*, this kernel must in fact be equal to this ideal. This shows that the restriction mapping from B^* to M^* factors over the restriction mapping from B^* to N^*, which implies that $M \subset N$. But then Proposition 4.5 implies that $M = M_1$, $M = M_2$, or $M = M_3$. □

We have seen in Proposition 4.5 that M_1, M_2, and M_3 are normal in N. They are not all normal in B:

PROPOSITION. M_1 and M_3 are not normal in B.

PROOF. If M_1 were normal in B, the ideal $BM_1^+ = M_1^+B$ would be two-sided, and in the corresponding quotient algebra $B/(BM_1^+)$, which would be four-dimensional by [**19**], Corollary 8.4.7, we would have $\bar{c}_4 = \bar{h}_2 = \bar{h}_3 = 1$. Therefore we would also have $\bar{h}_4 = \bar{h}_2\bar{h}_3 = 1$ and $\bar{c}_3 = \bar{c}_4\bar{c}_2 = \bar{c}_2$ as well as $\bar{d}_3 = \bar{c}_4\bar{d}_1 = \bar{d}_1$ and $\bar{d}_4 = \bar{c}_4\bar{d}_2 = \bar{d}_2$, so that $B/(BM_1^+) = \operatorname{Span}(1, \bar{c}_2, \bar{d}_1, \bar{d}_2)$.

But as we would then also have $\bar{d}_1 = \bar{h}_2\bar{d}_1 = \bar{d}_2\bar{h}_2 = \bar{d}_2$, it would follow that $B/(BM_1^+) = \operatorname{Span}(1, \bar{c}_2, \bar{d}_1)$, which would be a contradiction. The argument showing that M_3 is not normal in B is similar (cf. [**11**], Paragraph 6.1). □

6.2. On the other hand, there is a normal eight-dimensional Hopf subalgebra:

LEMMA. M_2 is normal in B.

PROOF. Recall that M_2 is generated by c_2, c_4, and h_3. From the description of the generators χ_1, χ_2, χ_3 of $\mathbf{G}(B^*)$ in Proposition 3.1, we know that

$$\chi_1(c_2) = \chi_1(c_4) = \chi_1(h_3) = 1$$

as well as $\chi_2(c_2) = \chi_2(c_4) = \chi_2(h_3) = 1$, so that χ_1 and χ_2 take the generators of M_2 to 1. By Corollary 4.1 and the remark following it, this also holds for $\chi_1\chi_2$.

As in the proof of Proposition 4.5, the algebra homomorphism

$$\pi\colon B \to K^4,\ \ b \mapsto (\varepsilon_B(b), \chi_1(b), \chi_2(b), (\chi_1\chi_2)(b))$$

is essentially the transpose of the inclusion map from $\mathrm{Span}(\varepsilon_B, \chi_1, \chi_2, \chi_1\chi_2)$ to B^*, and is therefore surjective. Because c_4 and h_3 are group-like, our computations above show that these two generators are contained in the spaces $B^{\mathrm{co}\pi}$ and ${}^{\mathrm{co}\pi}B$ of right and left coinvariants, which are eight-dimensional by [**19**], Corollary 8.4.7. The formula for the coproduct of c_2 given in Paragraph 2.3 shows that this also holds for c_2, so that M_2 is completely contained in these spaces. But as M_2 is itself eight-dimensional, we have

$$B^{\mathrm{co}\pi} = {}^{\mathrm{co}\pi}B = M_2$$

By [**19**], Lemma 3.4.2, this implies that M_2 is normal in B. □

The preceding proof also yields the form of the quotient:

THEOREM. The Hopf subalgebra $M := M_2$ is the unique normal Hopf subalgebra of B of dimension 8. Thus B fits into exactly one extension of the type

$$M \hookrightarrow B \twoheadrightarrow Q$$

with $\dim M = 8$ and $\dim Q = 4$. In this case, $M \cong K^{D_8}$ and $Q \cong K[\mathbb{Z}_2 \times \mathbb{Z}_2]$. This extension is abelian, but neither central nor cocentral.

PROOF. (1) That M_2 is the unique normal Hopf subalgebra of B of dimension 8 follows from Paragraph 6.1 and the lemma above, and so we have a unique extension with the stated properties. From Proposition 4.6, we know that $M_2 \cong K^{D_8}$.

(2) The proof of the lemma above also yields that the quotient $Q := B/(BM_2^+)$ is isomorphic to the dual of $\mathrm{Span}(\varepsilon_B, \chi_1, \chi_2, \chi_1\chi_2) \cong K[\mathbb{Z}_2 \times \mathbb{Z}_2]$. Since this Hopf algebra is self-dual, we have $Q \cong K[\mathbb{Z}_2 \times \mathbb{Z}_2]$ as well.

(3) The extension is not central, for example because $c_2 \notin Z(B)$. From Proposition 3.1, we know that B^* has only one nontrivial central group-like element. So Q^* cannot be central in B^*, which means that this extension is not cocentral. □

6.3. We have just seen that B fits into exactly one abelian extension of a Hopf algebra of dimension 4 by a Hopf algebra of dimension 8. We will now show that B also fits into exactly one abelian extension of a Hopf algebra of dimension 8 by a Hopf algebra of dimension 4, which is also neither central nor cocentral.

THEOREM. The Hopf algebra B has seven Hopf subalgebras of dimension 4. Precisely one of them, namely $K\langle c_4, h_3\rangle$, is normal.

PROOF. (1) Since every semisimple Hopf algebra of dimension 4 over an algebraically closed field of characteristic zero is isomorphic to a group algebra, every Hopf subalgebra of B of dimension 4 is a group algebra of a subgroup of order 4 of $\mathbf{G}(B)$. Explicitly, these are the groups $G_1 := \langle c_4, h_3\rangle = \{1_B, c_4, h_3, c_4h_3\}$ and

$G_2 := \{1_B, c_4, h_2, c_4h_2\}$ $G_3 := \{1_B, c_4, h_4, c_4h_4\}$ $G_4 := \{1_B, h_2, h_3, h_4\}$

$G_5 := \{1_B, h_2, c_4h_3, c_4h_4\}$ $G_6 := \{1_B, h_3, c_4h_2, c_4h_4\}$ $G_7 := \{1_B, h_4, c_4h_2, c_4h_3\}$

(2) Suppose that, for one of these groups G_i, the group algebra $P := K[G_i]$ is normal in B. Because $\mathbf{G}(B) = \mathbf{G}(N)$, we then have $P \subset N$. It follows directly from [**19**], Definition 3.4.1 that P is also normal in N. By [**19**], Corollary 8.4.7, the quotient $N/(NP^+)$ has dimension 4. It is therefore a group algebra of an abelian group of order 4. The four distinct multiplicative characters in $\mathbf{G}((N/(NP^+))^*)$ yield by pullback along the quotient map a subgroup of order 4 in $\mathbf{G}(N^*)$ whose elements vanish on NP^+, and therefore take the value 1 on G_i.

Now we see from Corollary 4.1 that the subgroup of $\mathbf{G}(N^*)$ of characters that take the value 1 on the group generated by $r = h_2$ and $s = h_3$ has only order 2 and is generated by χ_1'. On the other hand, every element of $\mathbf{G}(N^*)$ takes the value 1 on $u^2 = c_4$. Thus the fact that there are four distinct elements of $\mathbf{G}(N^*)$ that map G_i to 1 rules out that $i = 4$, 5, 6, or 7; in other words, G_i must contain c_4, and therefore P must contain $U := \operatorname{Span}(c_1, c_4)$.

(3) Consequently, we have a surjective map $B/(BU^+) \to B/(BP^+)$. By composing it with the inverse of the isomorphism $\bar{\pi}_N \colon B/(BU^+) \to N$ from Theorem 5.2, we obtain an eight-dimensional quotient of N. From Proposition 4.2, we know that N has exactly three quotients of dimension 8, denoted there by F_1, F_2, and F_3.

The multiplicative character $\chi_2' \in \mathbf{G}(N^*)$ satisfies $\chi_2'(c_4) = \chi_2'(u^2) = 1$ and $\chi_2'(h_3) = \chi_2'(s) = 1$ and therefore factors to a multiplicative character on all three quotients F_1, F_2, and F_3. As discussed in Paragraph 5.1, we have $\chi_2 = \chi_2' \circ \pi_N$, so χ_2 factors to a multiplicative character of $B/(BP^+)$. In particular, χ_2 maps all elements of G_i to 1. This implies that $i = 1$. We have therefore shown that, if P is a four-dimensional normal Hopf subalgebra of B, then $P = K[G_1]$.

(4) To see that $K[G_1]$ is normal in B, recall that $F_2 = N/\left(K\langle h_3\rangle^+ N\right)$ and consider the Hopf algebra homomorphism that arises as the composition

$$B \overset{\pi_N}{\to} N \to F_2$$

This composition maps the group-like elements c_4 and h_3 to 1_{F_2}, so that they are both contained in the spaces $B^{\mathrm{co} F_2}$ and ${}^{\mathrm{co} F_2}B$ of right and left coinvariants. As these spaces are four-dimensional by [**19**], Corollary 8.4.7, we have

$$B^{\mathrm{co} F_2} = {}^{\mathrm{co} F_2}B = K[G_1]$$

As in the proof of Lemma 6.2, this implies that $K[G_1]$ is normal in B. □

A somewhat more direct way to see that $K[G_i]$ is not normal in B for $i \neq 1$, which proceeds as in the proof of Proposition 6.1, is discussed in [**11**], Paragraph 6.3.

The preceding theorem has the following consequence:

COROLLARY. The Hopf algebra B fits into exactly one extension of the type

$$P \hookrightarrow B \twoheadrightarrow F$$

with $\dim P = 4$ and $\dim F = 8$. In this case, $P = K\langle c_4, h_3\rangle \cong K[\mathbb{Z}_2 \times \mathbb{Z}_2]$ and $F \cong K[D_8]$. This extension is abelian, but neither central nor cocentral.

PROOF. We have seen in the preceding theorem that $P = K\langle c_4, h_3\rangle$ is the unique normal Hopf subalgebra of B that has dimension 4, but in the proof we have also seen that the arising quotient is isomorphic to F_2, and we have already stated in

Paragraph 4.2 that $F_2 \cong K[D_8]$. The corresponding extension is therefore abelian. It is not a central extension, because h_3 is not central.

The commutator factor group of D_8 is isomorphic to $\mathbb{Z}_2 \times \mathbb{Z}_2$, and therefore D_8 has four one-dimensional representations. This implies that $\mathbf{G}(F_2^*) \cong \mathbb{Z}_2 \times \mathbb{Z}_2$. But from Proposition 3.1, we know that B^* has only one nontrivial central group-like element, so F_2^* cannot be contained in the center of B^*. In other words, this extension is not cocentral. □

Note that, using the generators introduced in Paragraph 2.2, we have $P = K\langle u^2, s\rangle$.

7. The biproduct in the second case

7.1. So far, we have only treated the Yetter-Drinfel'd Hopf algebra A and its associated Radford biproduct $B = A \star H$. In [**10**], Section 3, the authors also described a second example of a Yetter-Drinfel'd Hopf algebra over H of dimension 8, denoted here by A'. As the algebra A introduced in Paragraph 2.1, it is a Yetter-Drinfel'd Hopf algebra over the group algebra $H = K[G]$ of the group $G = \mathbb{Z}_2 \times \mathbb{Z}_2$, and it is also generated by two elements, denoted here by x' and y'. But unlike the previous case, these generators do not commute and satisfy instead the defining relations

$$x'^4 = 1 \qquad x'y' = y'x'^3 \qquad y'^2 = \frac{1}{2}(\zeta 1 + x' - \zeta x'^2 + x'^3)$$

where ζ is again a not necessarily primitive fourth root of unity. The action of G on A' is defined by the same formulas as the action of G on A, and the coaction is again derived from the action via Equation (2.1), using the same bicharacter θ. Further details on the structure of A', in particular descriptions of the coproduct and the antipode, can be found in [**10**], Section 3.

As in Paragraph 2.2, we can form the biproduct $B' := A' \star H$. It is generated by the four elements

$$u' := x' \star 1_H \qquad v' := y' \star 1_H \qquad r' := 1_{A'} \star g_2 \qquad s' := 1_{A'} \star g_3$$

In the same way as before, it is possible to derive a presentation of B':

PROPOSITION. The four generators satisfy the relations

1. $u'^4 = 1, \qquad u'v' = v'u'^3, \qquad v'^2 = \frac{1}{2}(\zeta 1 + u' - \zeta u'^2 + u'^3)$
2. $r'^2 = 1, \qquad r's' = s'r', \qquad s'^2 = 1$
3. $r'u' = u'^3 r', \qquad r'v' = u'^3 v' r', \qquad s'u' = u's', \qquad s'v' = u'^2 v' s'$

These relations are defining.

As in Paragraph 2.3, the coproduct is determined by its values on the generators:

LEMMA. We have $\Delta_{B'}(r') = r' \otimes r'$ and $\Delta_{B'}(s') = s' \otimes s'$. Furthermore, we have

$$\Delta_{B'}(u') = \frac{1}{2}(u' \otimes u' + u' \otimes u'^3 + u'^3 s' \otimes u' - u'^3 s' \otimes u'^3)$$

and

$$\Delta_{B'}(v') = \frac{1}{4}v'(1+r'+s'+r's')\otimes v' + \frac{1}{4}v'(1-\zeta^2 r' - s' + \zeta^2 r's')\otimes u'v'$$
$$+ \frac{1}{4}v'(1-r'+s'-r's')\otimes u'^2 v' + \frac{1}{4}v'(1+\zeta^2 r' - s' - \zeta^2 r's')\otimes u'^3 v'$$

The counit is given on generators by $\varepsilon_{B'}(u') = \varepsilon_{B'}(v') = \varepsilon_{B'}(r') = \varepsilon_{B'}(s') = 1$.

7.2. The Hopf algebra B' is a Radford biproduct by construction, and as in the case of B, it is reasonable to ask whether, and in which ways, it can also be written as an extension. Fortunately, it is not necessary to repeat this analysis for B', because B' is isomorphic to B, as we show now. We begin with the following lemma:

LEMMA. There is an algebra homomorphism $f_{A'}\colon A' \to B$ with the property that

$$f_{A'}(x') = u \qquad \text{and} \qquad f_{A'}(y') = vr$$

PROOF. We have to check the defining relations of A', using Proposition 2.2. This is relatively straightforward; for example, the second relation holds since

$$f_{A'}(y')f_{A'}(x')^3 = vru^3 = vur = uvr = f_{A'}(x')f_{A'}(y')$$

Verifications of the two other relations can be found in [**11**], Paragraph 7.2. □

The algebra homomorphism $f_{A'}$ can be extended to the entire biproduct:

PROPOSITION. The map $f\colon B' = A' \star H \to B = A \star H$ given by

$$f(a' \star h) = f_{A'}(a')(1_A \star h)$$

is an algebra isomorphism.

PROOF. We apply the universal property of a smash product stated in Corollary 1.3. For the map f_H required there, we use the canonical embedding

$$f_H\colon H \to A \star H,\ h \mapsto 1_A \star h$$

of A into the biproduct B. We have to check that

$$f_{A'}(h_{(1)}.a')f_H(h_{(2)}) = f_H(h)f_{A'}(a')$$

As pointed out after that corollary, we only need to check this on the algebra generators g_2 and g_3 of H. In both cases, the condition can be written in the form

$$f_{A'}(g_i.a') = f_H(g_i)f_{A'}(a')f_H(g_i)^{-1}$$

In this form, both sides depend multiplicatively on a', so that it is sufficient to verify this equation for $a' = x'$ and $a' = y'$. Hence we have to consider four cases, the first being the case $h = g_2$, $a' = x'$, for which we have

$$f_H(g_2)f_{A'}(x')f_H(g_2)^{-1} = rur^{-1} = u^3 = f_{A'}(x'^3) = f_{A'}(g_2.x')$$

The computations in the other cases are similar (cf. [**11**], Paragraph 7.2). Because the elements u, vr, r, and s are contained in its image, f is surjective, and therefore bijective by dimension considerations. □

We note that, on the generators of B', the map f takes the values

$$f(u') = u \qquad f(v') = vr \qquad f(r') = r \qquad f(s') = s$$

These equations can be used to give an alternative proof of the result above, which is based on Proposition 7.1. They are also used in the proof of the following corollary:

COROLLARY. The map f is a Hopf algebra isomorphism between B and B'.

PROOF. We still have to check that f is a coalgebra homomorphism, which amounts to the conditions $\Delta_B \circ f = (f \otimes f) \circ \Delta_{B'}$ and $\varepsilon_B \circ f = \varepsilon_{B'}$. It is sufficient to check both conditions on the generators, where the second condition is obvious. The first condition is also obvious on the generators u', r', and s', because their coproducts and the coproducts of the respective images are given essentially by the same formulas. For v', we see by comparing Proposition 2.3 and Lemma 7.1 that

$$(f \otimes f)(\Delta_{B'}(v')) = \Delta_B(v)(r \otimes r) = \Delta_B(v)\Delta_B(r) = \Delta_B(vr) = \Delta_B(f(v'))$$

which completes our proof, because a bialgebra homomorphism commutes automatically with the antipode. □

References

[1] Nicolás Andruskiewitsch and Jorge Andrés Devoto, *Extensions of Hopf algebras*, Algebra i Analiz **7** (1995), no. 1, 22–61; English transl., St. Petersburg Math. J. **7** (1996), no. 1, 17–52. MR1334152

[2] Nicolás Andruskiewitsch, *Notes on extensions of Hopf algebras*, Canad. J. Math. **48** (1996), no. 1, 3–42, DOI 10.4153/CJM-1996-001-8. MR1382474

[3] Charles W. Curtis and Irving Reiner, *Methods of representation theory. Vol. I*, John Wiley & Sons, Inc., New York, 1981. With applications to finite groups and orders; Pure and Applied Mathematics; A Wiley-Interscience Publication. MR632548

[4] Irmengard Hofstetter, *Erweiterungen von Hopfalgebren und ihre kohomologische Beschreibung*, Dissertation, LMU München, 1990.

[5] Irmengard Hofstetter, *Extensions of Hopf algebras and their cohomological description*, J. Algebra **164** (1994), no. 1, 264–298, DOI 10.1006/jabr.1994.1063. MR1268336

[6] Yevgenia Kashina, *Classification of semisimple Hopf algebras of dimension 16*, J. Algebra **232** (2000), no. 2, 617–663, DOI 10.1006/jabr.2000.8409. MR1792748

[7] Yevgenia Kashina, *On semisimple Hopf algebras of dimension* 2^m, Algebr. Represent. Theory **6** (2003), no. 4, 393–425, DOI 10.1023/B:ALGE.0000003540.74953.97. MR2020263

[8] Yevgenia Kashina, *On semisimple Hopf algebras of dimension* 2^m*, II*, Algebr. Represent. Theory **19** (2016), no. 6, 1387–1422, DOI 10.1007/s10468-016-9623-6. MR3573999

[9] Yevgenia Kashina, Geoffrey Mason, and Susan Montgomery, *Computing the Frobenius-Schur indicator for abelian extensions of Hopf algebras*, J. Algebra **251** (2002), no. 2, 888–913, DOI 10.1006/jabr.2001.9129. MR1919158

[10] Yevgenia Kashina and Yorck Sommerhäuser, *On cores in Yetter-Drinfel'd Hopf algebras*, J. Algebra **583** (2021), 89–125, DOI 10.1016/j.jalgebra2021.04.017. MR4261200

[11] Yevgenia Kashina and Yorck Sommerhäuser, *On biproducts and extensions*, available at `arXiv:2007.10072`.

[12] Toshiharu Kobayashi and Akira Masuoka, *A result extended from groups to Hopf algebras*, Tsukuba J. Math. **21** (1997), no. 1, 55–58, DOI 10.21099/tkbjm/1496163161. MR1467221

[13] Akira Masuoka, *Coideal subalgebras in finite Hopf algebras*, J. Algebra **163** (1994), no. 3, 819–831, DOI 10.1006/jabr.1994.1047. MR1265867

[14] Akira Masuoka, *Semisimple Hopf algebras of dimension* 6, 8, Israel J. Math. **92**(1995), no.1-3, 361–373, DOI 10.1007/BF02762089. MR1357764

[15] Akira Masuoka, *Self-dual Hopf algebras of dimension* p^3 *obtained by extension*, J. Algebra **178** (1995), no. 3, 791–806, DOI 10.1006/jabr.1995.1378. MR1364343

[16] Akira Masuoka, *Some further classification results on semisimple Hopf algebras*, Comm. Algebra **24** (1996), no. 1, 307–329, DOI 10.1080/00927879608825569. MR1370538

[17] Akira Masuoka, *Quotient theory of Hopf algebras*, Advances in Hopf algebras (Chicago, IL, 1992), Lecture Notes in Pure and Appl. Math., vol. 158, Dekker, New York, 1994, pp. 107–133. MR1289423

[18] Akira Masuoka, *Hopf algebra extensions and cohomology*, New directions in Hopf algebras, Math. Sci. Res. Inst. Publ., vol. 43, Cambridge Univ. Press, Cambridge, 2002, pp. 167–209. MR1913439

[19] Susan Montgomery, *Hopf algebras and their actions on rings*, CBMS Regional Conference Series in Mathematics, vol. 82, Published for the Conference Board of the Mathematical Sciences, Washington, DC; by the American Mathematical Society, Providence, RI, 1993, DOI 10.1090/cbms/082. MR1243637

[20] Susan Montgomery and Sarah J. Witherspoon, *Irreducible representations of crossed products*, J. Pure Appl. Algebra **129** (1998), no. 3, 315–326, DOI 10.1016/S0022-4049(97)00077-7. MR1631261

[21] Sonia Natale, *On semisimple Hopf algebras of dimension pq^2*, J. Algebra **221** (1999), no. 1, 242–278, DOI 10.1006/jabr.1999.7968. MR1722912

[22] Warren D. Nichols and M. Bettina Richmond, *The Grothendieck group of a Hopf algebra*, J. Pure Appl. Algebra **106** (1996), no. 3, 297–306, DOI 10.1016/0022-4049(95)00023-2. MR1375826

[23] David E. Radford, *The structure of Hopf algebras with a projection*, J. Algebra **92** (1985), no. 2, 322–347, DOI 10.1016/0021-8693(85)90124-3. MR778452

[24] Jean-Pierre Serre, *Linear representations of finite groups*, Springer-Verlag, New York-Heidelberg, 1977. Translated from the second French edition by Leonard L. Scott; Graduate Texts in Mathematics, vol. 42. MR0450380

[25] Yorck Sommerhäuser, *Ribbon transformations, integrals, and triangular decompositions*, J. Algebra **282** (2004), no. 2, 423–489, DOI 10.1016/j.jalgebra.2004.05.001. MR2101072

[26] Yorck Sommerhäuser, *Yetter-Drinfel'd Hopf algebras over groups of prime order*, Lecture Notes in Mathematics, vol. 1789, Springer-Verlag, Berlin, 2002, DOI 10.1007/b83955. MR1950794

DePaul University, Department of Mathematical Sciences, Chicago, Illinois 60614, USA

Email address: ykashina@depaul.edu

Memorial University of Newfoundland, Department of Mathematics and Statistics, St. John's, NL, A1C 5S7, Canada

Email address: sommerh@mun.ca

Contemporary Mathematics
Volume **771**, 2021
https://doi.org/10.1090/conm/771/15514

On the notion of exact sequence: From Hopf algebras to tensor categories

Sonia Natale

Abstract. We present an overview of the notions of exact sequences of Hopf algebras and tensor categories and their connections. We also present some examples illustrating their main features; these include simple fusion categories and a natural question regarding composition series of finite tensor categories.

1. Introduction

Perhaps the most natural way to understand or attempt a classification of a given algebraic structure is by decomposing it into 'simpler' structures. A fundamental example of this principle is provided by the theory of finite groups. Every finite simple group G has a composition series whose factors are finite simple groups uniquely determined by G, up to permutations. In particular, G can be built up by means of a number of successive extensions of finite simple groups. In this context, both the classification of the simple structures -finite simple groups- and the classification of the possible extensions -governed by suitable cohomology theories- are hard questions that involve deep mathematical techniques.

In this paper we aim to present an overview of an analogous approach to the study of structures that generalize that of groups, namely, Hopf algebras and tensor categories. We discuss the notion of extension in each of these contexts and some of their main features. More precisely, we focus on extensions arising from exact sequences: these do not include certain extensions arising from group gradings on tensor categories which play an important role in the classification of certain classes of fusion categories [**18**].

Very little is known about a possible approach to classify simple structures in these cases. We discuss several simple examples that show some contrast with the theory of finite groups.

We start by recalling the notion of exact sequence of Hopf algebras in Section 3. The main contributions towards this notion appeared in the work of G. I. Kac [**25**], M. Takeuchi [**57**], W. Singer [**54**], B. Parshall and J. P. Wang [**50**], H.-J. Schneider [**53**], S. Majid [**28**], N. Andruskiewitsch and J. Devoto [**5**], [**1**], Hofstetter [**23**] and others.

Exact sequences of Hopf algebras generalize exact sequences of groups. In a similar vein, the notion of normal subgroup and simple group have generalizations

2020 *Mathematics Subject Classification.* Primary 18D10; Secondary 16T05.
Key words and phrases. Hopf algebra; tensor category; fusion category; exact sequence.

to the notions of normal Hopf subalgebra and simple Hopf algebra, that we discuss in Subsections 3.1 and 3.5.

As in the case of finite groups, every finite dimensional Hopf algebra H has a *composition series*: these are sequences of simple Hopf algebras $\mathfrak{H}_1, \dots, \mathfrak{H}_n$, called the *factors* of the series, defined as follows [**6**]: If H is simple, then $n = 1$ and $\mathfrak{H}_1 = H$; if on the other hand, H contains a proper normal Hopf subalgebra A, and $\mathfrak{A}_1, \dots, \mathfrak{A}_m$, $\mathfrak{B}_1, \dots, \mathfrak{B}_\ell$, are composition series of A and of the quotient Hopf algebra $B = H/HA^+$, respectively, then $n = m + \ell$ and

$$\mathfrak{H}_i = \mathfrak{A}_i, \quad \text{if } 1 \leq i \leq m, \quad \mathfrak{H}_i = \mathfrak{B}_{i-m}, \quad \text{if } m < i \leq m + \ell.$$

Furthermore, a Jordan-Hölder theorem holds in the context of finite dimensional Hopf algebras [**42**]. This and related facts are discussed in Subsections 3.3 and 3.4.

In Section 4 we review the notion of exact sequence of tensor categories developed with A. Bruguières in [**8**], [**9**] and its relation with Hopf monads and commutative central algebras. We also discuss in this section a family of examples arising from so-called crossed actions of a matched pair of finite groups [**43**] and recent classification results for extensions of a fusion category by the category of representations of a finite group obtained in [**47**].

The notion of exact sequence of tensor categories discussed in Section 4 was extended to the notion of exact sequence with respect to a module category by P. Etingof and S. Gelaki [**14**]: In this sense, an exact sequence of (finite) tensor categories becomes an exact sequence with respect to a rank-one module category. We overview the definition of [**14**] in Section 5 as well as its connection with exact factorizations of tensor categories from [**20**], [**34**], [**51**].

We end the paper by presenting some examples that answer a number of natural questions regarding the behaviour of exact sequences of tensor categories. This is done in Section 6. In contrast with the properties enjoyed by groups and Hopf algebras, these examples show in particular that:

- Fusion subcategories of index 2 need not be normal with respect to a module category (Subsection 6.1);
- There exist fusion categories of Frobenius-Perron dimension $p^a q^b$, where p and q are prime numbers, which are simple with respect to any module category (Subsection 6.2).

Regarding the question of formulating a Jordan-Hölder theorem for finite tensor categories (see [**46**, Question 4.7]), we consider a natural generalization of the notion of composition series of finite dimensional Hopf algebra to the context of finite tensor categories: namely, we call a sequence of finite tensor categories $\mathcal{C}_1, \dots, \mathcal{C}_n$ -the factors of the series- a composition series of a finite tensor category $\mathcal{C}$ if, as before, $n = 1$ and $\mathcal{C}_1 = \mathcal{C}$ if $\mathcal{C}$ is does not fit into any exact sequence with respect to a module category, while if $\mathcal{C}' \longrightarrow \mathcal{C} \longrightarrow \mathcal{C}'' \boxtimes \operatorname{End}(\mathcal{M})$ is an exact sequence with respect to some $\mathcal{C}'$-module category $\mathcal{M}$ such that $\operatorname{FPdim} \mathcal{C}', \operatorname{FPdim} \mathcal{C}'' > 1$, then $n = m + \ell$ and $\mathcal{C}_i = \mathcal{C}'_i$, $1 \leq i \leq m$, $\mathcal{C}_i = \mathcal{C}''_{i-m}$, $m < i \leq m + \ell$, where $\mathcal{C}'_1, \dots, \mathcal{C}'_m$, $\mathcal{C}''_1, \dots, \mathcal{C}''_\ell$, are composition series of $\mathcal{C}'$ and $\mathcal{C}''$, respectively.

One of the main new contributions of this paper is a negative answer to the following question:

QUESTION 1.1. Is it true that two composition series of a finite tensor category thus defined have the same factors up to a permutation?

Indeed we show that the answer to Question 1.1 is negative even in the context of (braided non-degenerate) fusion categories. Hence composition series of fusion categories thus defined fail to satisfy a Jordan-Hölder theorem. This is done in Subsection 3.3; see Corollary 6.8 and Remark 6.9.

In spite of this fact, we mention that an analogue of the Jordan-Hölder theorem does hold for weakly group-theoretical fusion categories introduced in [**18**]. The definition of a composition series for this kind of category is given in terms of group equivariantizations and group graded extensions and the composition factors, which are Morita invariants, are finite simple groups [**44**].

Finally, we include in Subsection 6.6 some questions that we believe are interesting in relation with the notions discussed previously.

2. Preliminaries on Hopf algebras and tensor categories

We shall work over an algebraically closed field k of characteristic zero. A *tensor category* over k is a k-linear abelian category with finite dimensional Hom spaces and objects of finite length, endowed with a rigid monoidal category structure, such that the monoidal product is k-linear in each variable and the unit object is simple.

A tensor category over k is called *finite* if it is equivalent as a k-linear category to the category of finite dimensional left modules over a finite dimensional k-algebra. A *fusion category over* k is a semisimple finite tensor category. We refer the reader to [**16**] for a systematic study of tensor categories.

Let $\mathcal{C}$ and $\mathcal{D}$ be tensor categories over k. A *tensor functor* $F : \mathcal{C} \to \mathcal{D}$ is a k-linear exact (strong) monoidal functor F.

Let G be a group. We shall denote by Vec_G the tensor category of finite dimensional G-graded vector spaces and by $\mathrm{Rep}\, G$ the tensor category of finite dimensional k-linear representations of G.

Let H be a Hopf algebra over k. We shall indicate by H^+ the *augmentation ideal* of H, defined as $H^+ = \{x \in H : \varepsilon(x) = 0\}$. Except for Section 3, all Hopf algebras we consider in this paper are assumed to have a bijective antipode (this is automatically true if the Hopf algebra is finite dimensional). Thus the categories H-mod of finite dimensional (left) H-modules and comod-H of finite dimensional (right) H-comodules are tensor categories over k.

Let $\mathcal{C}$ be a tensor category. A *left module category* over $\mathcal{C}$, or $\mathcal{C}$*-module category*, is a k-linear abelian category $\mathcal{M}$ together with a k-linear bi-exact functor

$$\overline{\otimes} : \mathcal{C} \times \mathcal{M} \to \mathcal{M},$$

and natural isomorphisms $m_{X,Y,M} : (X \otimes Y)\overline{\otimes} M \to X\overline{\otimes}(Y\overline{\otimes} M)$, $u_M : \mathbf{1}\overline{\otimes} M \to M$, $X, Y \in \mathcal{C}$, $M \in \mathcal{M}$, satisfying natural associativity and unitary conditions.

Let $\mathcal{M}$ and $\mathcal{N}$ be $\mathcal{C}$-module categories. A $\mathcal{C}$*-module functor* $\mathcal{M} \to \mathcal{N}$ is a k-linear functor $F : \mathcal{M} \to \mathcal{N}$ endowed with natural isomorphisms

$$F(X\overline{\otimes} M) \to X\overline{\otimes} F(M),$$

for all $X \in \mathcal{C}$, $M \in \mathcal{M}$, satisfying appropriate conditions. A $\mathcal{C}$-module functor is called an equivalence of $\mathcal{C}$-module categories if it is an equivalence of categories. A module category $\mathcal{M}$ is called *indecomposable* if it is not equivalent as a $\mathcal{C}$-module category to the direct sum of two nontrivial module categories.

Suppose $\mathcal{C}$ is a finite tensor category. A finite left $\mathcal{C}$-module category $\mathcal{M}$ is *exact* if for every projective object $P \in \mathcal{C}$ and for every $M \in \mathcal{M}$, $P \otimes M$ is a projective object of $\mathcal{M}$. Let $\mathcal{M}$ be an indecomposable exact $\mathcal{C}$-module category.

Then the category $\mathrm{End}_{\mathcal{C}}(\mathcal{M})$ of right exact $\mathcal{C}$-module endofunctors of $\mathcal{M}$ is a finite tensor category. A tensor category $\mathcal{D}$ is *(categorically) Morita equivalent* to $\mathcal{C}$ if $\mathcal{D} \cong \mathrm{End}_{\mathcal{C}}(\mathcal{M})^{\mathrm{op}}$ for some exact indecomposable $\mathcal{C}$-module category $\mathcal{M}$.

3. Exact sequences of Hopf algebras

Let H be a Hopf algebra over k. The aim of this section is to give an account of some of the main features regarding the notion of exact sequences of Hopf algebras and present some examples.

Let $\pi : H \to B$ be a Hopf algebra map. The subalgebras ${}^{co\,\pi}H$ and $H^{co\,\pi}$ of left and right B-coinvariants of H are defined, respectively, by

$${}^{co\,\pi}H = \{h \in H : (\pi \otimes \mathrm{id})\Delta(h) = 1 \otimes h\}, \quad H^{co\,\pi} = \{h \in H : (\mathrm{id} \otimes \pi)\Delta(h) = h \otimes 1\}.$$

Definition 3.1. ([**5**].) An *exact sequence of Hopf algebras* is a sequence of Hopf algebra maps

$$k \longrightarrow H' \xrightarrow{i} H \xrightarrow{\pi} H'' \longrightarrow k, \tag{3.1}$$

satisfying the following conditions:

(a) i is injective and π is surjective,
(b) $\ker \pi = Hi(H')^+$,
(c) $i(H') = {}^{co\,\pi}H$.

A Hopf algebra H fitting into an exact sequence (3.1) is called an *extension* of H'' by H'.

Note that either of the conditions (b) or (c) in Definition 3.1 implies that $\pi i = \epsilon 1 : H' \to H''$. If H is faithfully flat over H', then (a) and (b) imply (c). Dually, if H is faithfully coflat over H'', then (a) and (c) imply (b).

We refer the reader to [**5**], [**53**], [**57**] for further details on the notion of an exact sequence.

Let (3.1) be an exact sequence of finite dimensional Hopf algebras. Then, as a consequence of the Nichols-Zoeller freeness theorem, it is *cleft*, that is, the map $\pi : H \to B$ admits a convolution invertible B-colinear and A-linear section $B \to H$. This implies that H isomorphic as a Hopf algebra to a bicrossed product $A \# B$ with respect to suitable compatible data [**5**]. In particular $\dim H = \dim H' \dim H''$.

In addition, H is semisimple if and only if H' and H'' are semisimple.

Definition 3.2. We shall say that a Hopf algebra H is *simple* if it does not fit into any exact sequence of Hopf algebras (3) with $H' \not\cong k$ and $H'' \not\cong k$.

3.1. Normality of Hopf subalgebras and Hopf algebra maps. Consider the left and right adjoint actions of H on itself defined, respectively, by

$$h.a = h_{(1)} a \mathcal{S}(h_{(2)}), \qquad a.h = \mathcal{S}(h_{(1)}) a h_{(2)}, \tag{3.2}$$

for all $a, h \in H$.

Formulas (3.2) generalize those defining the adjoint actions of a group. In this vein, the notion of a normal subgroup can be generalized to that of a normal Hopf subalgebra; that is, a Hopf subalgebra K of H is called *normal* if it is stable under both action actions (3.2).

Dually, the left and right adjoint coactions of H are defined, respectively, by

$$\rho_\ell : H \to H \otimes H, \quad \rho_\ell(h) = h_{(1)}\mathcal{S}(h_{(3)}) \otimes h_{(2)},$$
$$\rho_r : H \to H \otimes H, \quad \rho_r(h) = h_{(2)} \otimes h_{(1)}\mathcal{S}(h_{(3)}).$$

Let $\pi : H \to H''$ be a Hopf algebra map. The map π is called *normal* if the kernel I of π is a subcomodule for both adjoint coactions of H.

Let $H' \subseteq H$ be a normal Hopf subalgebra. Then $H(H')^+ = (H')^+H$ is a Hopf ideal of H and the canonical map $H \to H/H(H')^+$ is a Hopf algebra map. If H is faithfully flat over H', then there is an exact sequence of Hopf algebras $k \longrightarrow H' \longrightarrow H \longrightarrow H/H(H')^+ \longrightarrow k$. Similarly, if $\pi : H \to H''$ is a surjective normal Hopf algebra map, then ${}^{co\pi}H = H^{co\pi}$ is a Hopf subalgebra and if H is faithfully coflat over H'', there is an exact sequence of Hopf algebras $k \longrightarrow {}^{co\pi}H \longrightarrow H \xrightarrow{\pi} H'' \longrightarrow k$.

Suppose H is a finite dimensional Hopf algebra. Then H is simple if and only if H contains no proper normal Hopf subalgebra if and only if it admits no proper normal quotient Hopf algebra.

Furthermore, a sequence of Hopf algebra maps $k \longrightarrow H' \xrightarrow{i} H \xrightarrow{\pi} H'' \longrightarrow k$ is an exact sequence if and only if the dual sequence $k \longrightarrow (H'')^* \xrightarrow{\pi^*} H^* \xrightarrow{i^*} (H')^* \longrightarrow k$ is exact. Therefore the notion of simplicity of a finite dimensional Hopf algebra is self-dual, that is, H is simple if and only if H^* is simple.

3.2. Example: Abelian exact sequences and matched pairs of groups. A celebrated source of examples of Hopf algebra extensions arises from matched pairs of groups: these are called *abelian extensions* and were introduced in the early work of G. I. Kac, W. Singer, S. Majid and M. Takeuchi [**25**], [**54**], [**28**], [**56**]. We refer the reader to [**31**], [**32**] for a detailed study of the cohomology theory underlying an abelian exact sequence.

DEFINITION 3.3. An exact sequence of finite dimensional Hopf algebras

$$k \longrightarrow H' \xrightarrow{i} H \xrightarrow{\pi} H'' \longrightarrow k \tag{3.3}$$

is called an *abelian exact sequence* if H' is commutative and H'' is cocommutative.

Since our base field k is algebraically closed of characteristic zero, this means that $H' \cong k^\Gamma$ and $H'' \cong kG$, for some finite groups Γ, G.

The exactness of the sequence (3.3) allows to endow the pair (G, Γ) with the structure of a *matched pair* of groups. That is, (3.3) gives rise to actions by permutations $\Gamma \xleftarrow{\triangleleft} \Gamma \times G \xrightarrow{\triangleright} G$ such that, for all $s, t \in \Gamma$, $x, y \in G$:

$$s \triangleright xy = (s \triangleright x)((s \triangleleft x) \triangleright y), \quad st \triangleleft x = (s \triangleleft (t \triangleright x))(t \triangleleft x). \tag{3.4}$$

Given groups G and Γ, the data of a pair of compatible actions making (G, Γ) into a matched pair of groups is equivalent to the data of a group E together with an *exact factorization* into subgroups (isomorphic to) G and Γ: that is, a group E such that $E = G\Gamma$ and $G \cap \Gamma = \{e\}$. In such situation, the relevant actions $\triangleright$ and $\triangleleft$ are determined, for every $x \in G$, $g \in \Gamma$, by the relations

$$gx = (g \triangleright x)(g \triangleleft x).$$

Fix a matched pair of finite groups (G, Γ). Consider the left action of G on k^Γ defined by $(x.f)(g) = f(g \triangleleft x)$, $f \in k^\Gamma$, and let $\sigma : G \times G \to (k^*)^\Gamma$ be a normalized 2-cocycle. Dually, consider the right action of Γ on k^G given by $(w.g)(x) = w(g \triangleright x)$, $w \in k^G$, and let $\tau : \Gamma \times \Gamma \to (k^*)^G$ be a normalized 2-cocycle.

Under appropriate compatibility conditions between σ and τ, the vector space $H = k^\Gamma \otimes kG$ becomes a (semisimple) Hopf algebra, denoted $H = k^\Gamma {}^\tau\#_\sigma kG$, with the crossed product algebra structure and the crossed coproduct coalgebra structure; see [**31**],[**32**]. For all $g, h \in \Gamma$, $x, y \in G$, we have

$$(e_g \# x)(e_h \# y) = \delta_{g \triangleleft x, h}\, \sigma_g(x, y) e_g \# xy,$$
$$\Delta(e_g \# x) = \sum_{st=g} \tau_x(s,t)\, e_s \#(t \triangleright x) \otimes e_t \# x,$$

where $\sigma_s(x,y) = \sigma(x,y)(s)$ and $\tau_x(s,t) = \tau(s,t)(x)$, $s, t \in \Gamma$, $x, y \in G$.

Let $i : k^\Gamma \to H = k^\Gamma\, {}^\tau\#_\sigma kG$ and $\pi : H = k^\Gamma\, {}^\tau\#_\sigma kG \to kF$ be the Hopf algebra maps defined by $i(f) = f\#e$, $\pi(f\#g) = \varepsilon(f)g$, $f \in k^\Gamma$, $g \in G$. The Hopf algebra H fits into an abelian exact sequence $k \longrightarrow k^\Gamma \xrightarrow{i} H \xrightarrow{\pi} kG \longrightarrow k$.

Moreover, every Hopf algebra H fitting into an abelian exact sequence (3.3) is isomorphic to a bicrossed product $k^\Gamma\, {}^\tau\#_\sigma kF$ for an appropriate matched pair (G, Γ) and compatible actions and cocycles σ and τ. Equivalence classes of such extensions associated to a fixed matched pair (G, Γ) form an abelian group $\operatorname{Opext}(k^\Gamma, kG)$, whose unit element is the class of the *split* extension $k^\Gamma \# kG$.

An example of an abelian extensions is given by the Drinfeld double $D(G)$ of a finite group G: the Hopf algebra $D(G)$ fits into an exact sequence $k \longrightarrow k^G \longrightarrow D(G) \longrightarrow kG \longrightarrow k$. In the associated matched pair (G, G), $\triangleleft : G \times G \to G$ is the adjoint action of G on itself, while $\triangleright : G \times G \to G$ is the trivial action.

3.3. Jordan-Hölder theorem for finite dimensional Hopf algebras. The question of establishing an analogue of the Jordan-Hölder theorem of group theory for finite dimensional Hopf algebras was raised by N. Andruskiewitsch in [**2**, Question 2.1] and answered in [**42**].

We start by recalling the followig definition given in [**6**].

DEFINITION 3.4. Let H be a finite dimensional Hopf algebra over a field k. A *composition series* of H is a sequence of finite dimensional simple Hopf algebras $\mathfrak{H}_1, \dots, \mathfrak{H}_n$ defined recursively as follows: If H is simple, we let $n = 1$ and $\mathfrak{H}_1 = H$. If $k \subsetneq A \subsetneq H$ is a normal Hopf subalgebra, and $\mathfrak{A}_1, \dots, \mathfrak{A}_m$, $\mathfrak{B}_1, \dots, \mathfrak{B}_\ell$, are composition series of A and $B = H/HA^+$, respectively, then we let $n = m + \ell$ and

$$\mathfrak{H}_i = \mathfrak{A}_i, \quad \text{if } 1 \leq i \leq m, \quad \mathfrak{H}_i = \mathfrak{B}_{i-m}, \quad \text{if } m < i \leq m + \ell.$$

The Hopf algebras $\mathfrak{H}_1, \dots, \mathfrak{H}_n$ are called the *factors* of the series. The number n is called the *length* of the series.

Every finite dimensional Hopf algebra admits a composition series. The following is an analogue of the Jordan-Hölder theorem for finite dimensional Hopf algebras:

THEOREM 3.5. ([**42**, Theorem 1.2].) *Let $\mathfrak{H}_1, \dots, \mathfrak{H}_n$ and $\mathfrak{H}'_1, \dots, \mathfrak{H}'_m$ be two composition series of H. Then there exists a bijection $f : \{1, \dots, n\} \to \{1, \dots, m\}$ such that $\mathfrak{H}_i \cong \mathfrak{H}'_{f(i)}$ as Hopf algebras.*

Let H be a finite dimensional Hopf algebra and let $\mathfrak{H}_1, \dots, \mathfrak{H}_n$ be a composition series of H. The simple Hopf algebras $\mathfrak{H}_i$, $1 \leq i \leq n$, are called the *composition factors of* H. The number n is called the *length of* H.

Let us consider for instance the case of abelian extensions. Let G and Γ be finite groups and let H be an abelian extension of k^Γ by kG. Then the composition factors of H are the group algebras of the composition factors of G and the dual group algebras of the composition factors of Γ. See [**42**, Example 4.7].

In particular, if $G_1, \dots, G_n$ are the composition factors of the finite group G, then the composition factors of its Drinfeld double $D(G)$ are the Hopf algebras $k^{G_1}, \dots, k^{G_n}, kG_1, \dots, kG_n$.

3.4. Semisolvability and related questions. A *lower subnormal series* of a Hopf algebra H is a series of Hopf subalgebras

$$k = H_n \subseteq H_{n-1} \subseteq \cdots \subseteq H_1 \subseteq H_0 = H, \tag{3.5}$$

such that H_{i+1} is a normal Hopf subalgebra of H_i, for all i. The *factors* of the series (3.5) are the quotient Hopf algebras $H_i/H_iH_{i+1}^+$, $i = 0, \dots, n-1$.

Dually, an *upper subnormal series* of H is a series of surjective Hopf algebra maps

$$H = H_{(0)} \to H_{(1)} \to \cdots \to H_{(n)} = k, \tag{3.6}$$

such that $H_{(i+1)}$ is a normal quotient Hopf algebra of $H_{(i)}$, for all $i = 0, \dots, n-1$. The *factors* of (3.6) are the Hopf algebras ${}^{co\, H_{(i+1)}}H_{(i)} \subseteq H_{(i)}$, $i = 0, \dots, n-1$.

A *lower* (respectively *upper*) *composition series* of H is a lower (respectively, upper) subnormal series which does not admit any proper refinement. See [**42**, Section 5].

A Hopf algebra H is called *lower-semisolvable* (respectively, *upper-semisolvable*) if it admits a lower (respectively, upper) subnormal series whose factors are commutative or cocommutative [**36**]. We shall say that H is *semisolvable* if it is either lower or upper semisolvable.

Every semisolvable finite dimensional Hopf algebra is semisimple and cosemisimple. On the other hand, every semisimple Hopf algebra of dimension p^n, p a prime number, is semisolvable [**30**], [**36**]. Moreover, if the group G is nilpotent then any twisting of the group Hopf algebra kG is semisolvable [**19**].

In [**42**] we proved analogues of the Zassenhaus' butterfly lemma and the Schreier's refinement theorem for finite dimensional Hopf algebras and, following the lines of the classical proof in group theory, applied them to prove an analogue of the Jordan-Hölder theorem for lower and upper composition series of H.

Thus the lower and upper composition factors of H and its lower and upper lengths, which are also well-defined invariants of H, were introduced. In contrast with the case of the composition factors, the lower or upper composition factors are not necessarily simple as Hopf algebras. This motivates the question of deciding if there is an intrinsic characterization of the Hopf algebras that can arise as lower composition factors [**42**, Question 5.7].

Some properties of lower and upper composition factors and their relation with the composition factors were studied in [**42**]. Unlike for the case of the length, the lower and upper lengths are not additive with respect to exact sequences and they are not invariant under duality in general.

Neither the composition factors nor the upper or lower composition factors of a finite dimensional Hopf algebra H are categorical invariants of H. In other words, they are not invariant under twisting deformations of H. In fact, there exists a (semisimple) Hopf algebra H such that H is simple as a Hopf algebra and H is twist equivalent to the group algebra of a solvable group G (see Theorem 3.6 below). In particular, the categories of finite dimensional representations of H and G are equivalent fusion categories.

3.5. Simple Hopf algebras. Recall that a finite dimensional Hopf algebra is simple if it contains no proper normal Hopf subalgebras. For instance, if G is a finite simple group, then the group algebra kG and its dual k^G are simple Hopf algebras. Furthermore, in this case, any twisting deformation of kG is simple [**48**]. However, there are examples of solvable groups that admit simple twisting deformations.

Finite dimensional Hopf algebras with tensor equivalent categories of representations are obtained from one another by a twisting deformation. Properties of H invariant under twisting are of special interest because they depend only on the tensor category H-mod.

In the paper [**19**] we presented examples showing that the notions of simplicity and (semi)solvability of a Hopf algebra are *not* twist invariants; that is, they are not categorical notions.

Let p, r and q be prime numbers such that q divides $p-1$ and $r-1$. There is a family of supersolvable groups G of order prq^2 that can be deformed through a twist into nontrivial simple Hopf algebras.

Let us recall this construction. Let $G_1 = \mathbb{Z}_p \rtimes \mathbb{Z}_q$ and $G_2 = \mathbb{Z}_r \rtimes \mathbb{Z}_q$ be the only nonabelian groups of orders pq and rq, respectively. Let $G = G_1 \times G_2$ and let $S \cong \mathbb{Z}_q \times \mathbb{Z}_q$ be a subgroup of G order q^2. In particular, G is supersolvable and $Z(G) = 1$.

Let $1 \neq \omega \in H^2(\widehat{S}, k^*)$, $J \in kG \otimes kG$ the twist lifted from S corresponding to ω. Let also $H = (kG)^J$. Note that the cocycle ω is nondegenerate. Also, H is a noncommutative noncocommutative Hopf algebra of dimension prq^2.

THEOREM 3.6. ([**19**, Theorem 4.5].) *The Hopf algebra H is simple.*

The proof relied on the comparison of the (co)representation theory of the given twistings [**13**] with that of an extension [**36**].

Certain twists of the symmetric group $\mathbb{S}_n$ on n letters, $n \geq 5$, were also shown to be simple as Hopf algebras in [**19**].

As a consequence of Theorem 3.6 the analogue of Burnside's $p^a q^b$-Theorem for finite groups does not hold for semisimple Hopf algebras.

Theorem 3.6 provides the smallest example of a noncommutative noncocommutative semisimple Hopf algebra which is simple: this appears in dimension 36 as a twisting of $D_3 \times D_3$. This turns out to be the only simple example in dimension < 60:

THEOREM 3.7. ([**40**, Theorem 1].) *Every semisimple Hopf algebra of dimension less than 60 is semisolvable up to a cocycle twist.*

The previously mentioned results on simplicity of twisting deformations provide us with three examples of non-commutative non-cocommutative semisimple Hopf algebras of dimension 60 which are simple as Hopf algebras. The first two are the Hopf algebras $\mathcal{A}_0$ and $\mathcal{A}_1 \simeq \mathcal{A}_0^*$ constructed by D. Nikshych [**48**]. We have $\mathcal{A}_0 = (k\mathbb{A}_5)^J$, where $J \in k\mathbb{A}_5 \otimes k\mathbb{A}_5$ is an invertible twist lifted from a nondegenerate 2-cocycle in a subgroup of $\mathbb{A}_5$ isomorphic to $\mathbb{Z}_2 \times \mathbb{Z}_2$.

The third example is the self-dual Hopf algebra $\mathcal{B}$ constructed in [**19**]. In this case $\mathcal{B} = (kD_3 \otimes kD_5)^J$, where J is an invertible twist also lifted from a nondegenerate 2-cocycle in a subgroup of $D_3 \times D_5$ isomorphic to $\mathbb{Z}_2 \times \mathbb{Z}_2$.

As coalgebras, these examples are isomorphic to direct sums of full matric coalgebras, as follows:

$$\mathcal{A}_0 \cong k^{(12)} \oplus M_4(k)^{(3)}, \quad \mathcal{A}_1 \cong k \oplus M_3(k)^{(2)} \oplus M_4(k) \oplus M_5(k),$$
$$\mathcal{B} \cong k^{(4)} \oplus M_2(k)^{(6)} \oplus M_4(k)^{(2)}.$$

As for the group-like elements, we have $G(\mathcal{A}_0) \simeq \mathbb{A}_4$ and $G(\mathcal{B}) \simeq \mathbb{Z}_2 \times \mathbb{Z}_2$.

It was shown in [**18**, Theorem 9.12] that $\operatorname{Rep} \mathcal{A}_0 \simeq \operatorname{Rep} \mathbb{A}_5$ is the only fusion category of dimension 60 which contains no proper fusion subcategories.

The following theorem was shown in [**41**], answering Question 2.4 in [**2**].

THEOREM 3.8. ([**41**, Theorem 1.4].) *Let H be a nontrivial semisimple Hopf algebra of dimension* 60*. Suppose H is simple. Then H is isomorphic to $\mathcal{A}_0$ or to $\mathcal{A}_1$ or to $\mathcal{B}$.*

4. Exact sequences of tensor categories

Let $\mathcal{C}$ and $\mathcal{C}''$ be tensor categories over k. A tensor functor $F : \mathcal{C} \to \mathcal{C}''$ is called *dominant* (or *surjective*) if every object of $\mathcal{C}''$ is a subobject of $F(X)$ for some object X of $\mathcal{C}$. If $\mathcal{C}''$ is a finite tensor category, then F is dominant if and only if every object of $\mathcal{C}''$ is a subquotient of $F(X)$ for some object X of $\mathcal{C}$; see [**14**, Lemma 2.3].

A tensor functor $F : \mathcal{C} \to \mathcal{C}''$ is called *normal* if for every object X of $\mathcal{C}$, there exists a subobject $X_0 \subset X$ such that $F(X_0)$ is the largest trivial subobject of $F(X)$.

For a tensor functor $F : \mathcal{C} \to \mathcal{C}''$, let $\mathfrak{Ker}_F$ denote the tensor subcategory $F^{-1}(\langle \mathbf{1} \rangle) \subseteq \mathcal{C}$ of objects X of $\mathcal{C}$ such that $F(X)$ is a trivial object of $\mathcal{C}''$. If the functor F has a right adjoint R, then F is normal if and only if $R(\mathbf{1})$ belongs to $\mathfrak{Ker}_F$ [**8**, Proposition 3.5].

DEFINITION 4.1. Let $\mathcal{C}', \mathcal{C}, \mathcal{C}''$ be tensor categories over k. An *exact sequence of tensor categories* is a sequence of tensor functors

$$\mathcal{C}' \xrightarrow{f} \mathcal{C} \xrightarrow{F} \mathcal{C}'', \tag{4.1}$$

such that the tensor functor F is dominant and normal and the tensor functor f is a full embedding whose essential image is $\mathfrak{Ker}_F$ [**8**].

If $\mathcal{C}$ fits into an exact sequence (4.1), we say that $\mathcal{C}$ is an *extension of* $\mathcal{C}''$ *by* $\mathcal{C}'$.

Every exact sequence of tensor categories (4.1) defines a fiber functor on the kernel $\mathcal{C}'$: $\omega = \operatorname{Hom}(\mathbf{1}, Ff) : \mathcal{C}' \to \operatorname{Vec}$.

The *induced Hopf algebra* H of the exact sequence (4.1) is defined as

$$H = \operatorname{coend}(\omega); \tag{4.2}$$

see [**8**, Subsection 3.3]. Thus there is an equivalence of tensor categories $\mathcal{C}' \cong$ comod–H such that the following diagram of tensor functors is commutative:

$$\begin{array}{ccc} \mathcal{C}' & \xrightarrow{\cong} & \text{comod–}H \\ & \searrow^{\omega} & \downarrow U \\ & & \operatorname{Vec}. \end{array}$$

The induced Hopf algebra H of (4.1) is finite-dimensional if and only if the tensor functor F admits adjoints [**8**, Proposition 2.15]. Hence if $\mathcal{C}'$ and $\mathcal{C}''$ are finite tensor categories, then so is $\mathcal{C}$.

The fiber functor ω corresponds to a (rank one) $\mathcal{C}'$-module category structure on Vec, that we shall denote by $\mathcal{M}$ (see [**49**, Proposition 4.1]). Assume that H is finite dimensional. As a consequence of [**49**, Theorem 4.2] we obtain the following relation between $\mathcal{M}$ and the induced Hopf algebra:

$$(\mathcal{C}')^*_{\mathcal{M}} \cong H\text{-mod}\,.$$

4.1. Examples from exact sequences of Hopf algebras. All Hopf algebras considered in this subsection are assumed to have a bijective antipode.

Let $f : H_1 \to H_2$ be a Hopf algebra map. Then f induces by restriction tensor functors

$$f_* : \text{comod-}H_1 \to \text{comod-}H_2, \qquad f^* : H_2\text{-mod} \to H_1\text{-mod}\,.$$

The functor $f_* : \text{comod-}H_1 \to \text{comod-}H_2$ is dominant if and only if the functor $\_\_\square_{H_2} H_1 : \text{Comod-}H_2 \to \text{Comod-}H_1$ is faithful. On the other hand, if f is injective and H_2 is finite-dimensional, then $f^* : H_2\text{-mod} \to H_1\text{-mod}$ is dominant. See [**8**, Lemma 2.11].

Recall that the map f has a kernel and a cokernel in the category of Hopf algebras, called the *categorical kernel* and the *categorical cokernel* of f, and defined, respectively, in the form

$$\begin{aligned} \operatorname{Hker}(f) &= \{h \in H_1 \mid h_{(1)} \otimes f(h_{(2)}) \otimes h_{(3)} = h_{(1)} \otimes 1 \otimes h_{(2)}\} \\ \operatorname{Hcoker}(f) &= H_2/H_2 f(H_1^+)H_2. \end{aligned}$$

The following proposition describes the kernels of the tensor functors induced by a Hopf algebra map.

PROPOSITION 4.2. ([**8**, Lemma 2.10].) *Let $f : H \to H'$ be a morphism of Hopf algebras over a field, and let*

$$f_* : \text{comod-}H \to \text{comod-}H', \qquad f^* : H'\text{-mod} \to H\text{-mod}$$

be the tensor functors induced by f. Then the following hold:

(1) $\mathfrak{Ker}_{f_*} = \text{comod-}\operatorname{Hker}(f)$. *Moreover the tensor functor f_* is normal if and only if $H^{\operatorname{co}H'} = {}^{\operatorname{co}H'}H$, and in this case* $\operatorname{Hker}(f) = H^{\operatorname{co}H'}$.
(2) $\mathfrak{Ker}_{f^*} = \operatorname{Hcoker}(f)\text{-mod}$. *Moreover the tensor functor f^* is normal if $f(H)$ is a normal Hopf subalgebra of H', and in this case* $\operatorname{Hcoker}(f) = H'/H'f(H^+)$.

We shall say that an exact sequence of Hopf algebras

$$k \longrightarrow H' \xrightarrow{\ i\ } H \xrightarrow{\ \pi\ } H'' \longrightarrow k, \tag{4.3}$$

is *strictly exact* if H is faithfully coflat over H''.

REMARK 4.3. Notice that, since the antipode of H is bijective, H is right faithfully coflat over H'' if and only if it is left faithfully coflat over H''. Moreover, if this is the case, then π is normal if and only if ${}^{\operatorname{co}\pi}H = H^{\operatorname{co}\pi}$. See [**57**].

Observe in addition that the antipode of H induces an anti-isomorphism of algebras $\mathcal{S} : {}^{\operatorname{co}\pi}H \to H^{\operatorname{co}\pi}$ with inverse $\mathcal{S}^{-1} : H^{\operatorname{co}\pi} \to {}^{\operatorname{co}\pi}H$. In particular, if $i(H') = {}^{\operatorname{co}\pi}H$, then ${}^{\operatorname{co}\pi}H$ (being a Hopf subalgebra) is stable under the antipode and therefore ${}^{\operatorname{co}\pi}H = H^{\operatorname{co}\pi}$.

Assume that the sequence (4.3) is strictly exact. Then, as can be seen from the facts recalled in Remark 4.3, it is a *strictly exact sequence of Hopf algebras* in the sense of [**53**], that is, the following condiditions hold:

(a) π is a normal Hopf algebra map,
(b) H is right faithfully coflat over H'',
(c) $i(H') = \mathrm{Hker}\,\pi$.

These conditions are furthermore equivalent to the following:

(a') H' is a normal Hopf subalgebra of H,
(b') H is right faithfully flat over H',
(c') $H'' = \mathrm{Hcoker}\, i$.

In this way we obtain:

THEOREM 4.4. ([**8**, Proposition 3.9].) *Every strictly exact sequence of Hopf algebras* (4.3) *gives rise to an exact sequence of tensor categories*

$$\text{comod-}H' \xrightarrow{i_*} \text{comod-}H \xrightarrow{\pi_*} \text{comod-}H''.$$

If in addition H is finite-dimensional, it also gives rise to an exact sequence of tensor categories:

$$H''\text{-mod} \xrightarrow{\pi^*} H\text{-mod} \xrightarrow{i^*} H'\text{-mod}\,.$$

For instance, an exact sequence of groups $1 \longrightarrow G' \longrightarrow G \longrightarrow G'' \longrightarrow 1$ gives rise to an exact sequence of tensor categories $\mathrm{Vec}_{G'} \longrightarrow \mathrm{Vec}_G \longrightarrow \mathrm{Vec}_{G''}$, and if G is finite, to an exact sequence $\mathrm{Rep}\, G'' \longrightarrow \mathrm{Rep}\, G \longrightarrow \mathrm{Rep}\, G'$.

4.2. Extensions and normal Hopf monads. We refer the reader to [**10**], [**7**] for the notion of Hopf monad on a monoidal category.

Let $F : \mathcal{C} \to \mathcal{C}''$ be a tensor functor between tensor categories and *assume that F admits a left adjoint G*. Then the composition $FG : \mathcal{C}'' \to \mathcal{C}''$ is a k-linear right exact Hopf monad on $\mathcal{C}''$, and $\mathcal{C}$ is tensor equivalent to the category $(\mathcal{C}'')^T$ of T-modules in $\mathcal{C}''$.

The functor F is dominant if and only if T is faithful [**8**, Proposition 4.1].

A Hopf monad T on a tensor category $\mathcal{C}''$ is called *normal* if $T(\mathbf{1})$ is a trivial object of $\mathcal{C}''$. When T is the Hopf monad corresponding to F as above, the normality of T is equivalent to the normality of F [**8**, Proposition 4.6].

Normal faithful Hopf monads classify extensions of tensor categories in view of the following theorem:

THEOREM 4.5. ([**8**, Theorem 4.8].) *Let $\mathcal{C}'$, $\mathcal{C}''$ be tensor categories and assume that $\mathcal{C}'$ is finite. Then the following data are equivalent:*

(1) *A normal faithful k-linear right exact Hopf monad T on $\mathcal{C}''$, with induced Hopf algebra H, endowed with a tensor equivalence $K : \mathcal{C}' \cong$ comod-H;*
(2) *An extension $\mathcal{C}' \to \mathcal{C} \to \mathcal{C}''$ of $\mathcal{C}''$ by $\mathcal{C}'$.*

Under the correspondence established by Theorem 4.5, the induced Hopf algebra of the exact sequence $\mathcal{C}' \to \mathcal{C} \to \mathcal{C}''$ is identified with the Hopf monad of the restriction of T to the trivial subcategory of $\mathcal{C}''$.

4.3. Perfect exact sequences and central commutative algebras. Let $\mathcal{C}$ and $\mathcal{C}''$ be tensor categories. A tensor functor $F : \mathcal{C} \to \mathcal{C}''$ is called *perfect* if it admits an exact right adjoint [**9**, Subsection 2.1].

Every tensor functor $F : \mathcal{C} \to \mathcal{C}''$ between finite tensor categories $\mathcal{C}$ and $\mathcal{C}''$ is perfect [**47**, Lemma 2.1].

Let $F : \mathcal{C} \to \mathcal{C}''$ be a dominant perfect tensor functor and let $R : \mathcal{C}'' \to \mathcal{C}$ be a right adjoint of F and let $A = R(\mathbf{1})$. Then there exists a half-braiding σ on A such that (A, σ) is a commutative algebra in the Drinfeld center $\mathcal{Z}(\mathcal{C})$ of $\mathcal{C}$, which satisfies $\operatorname{Hom}_{\mathcal{C}}(\mathbf{1}, A) \cong k$. The algebra (A, σ) is called the *induced central algebra of* F [**8**, Section 6].

The category $\mathcal{C}_A$ or right A-modules in $\mathcal{C}$ becomes a tensor category with tensor product $\otimes_A$ and unit object A, equipped with a tensor functor $F_A = ?\otimes A : \mathcal{C} \to \mathcal{C}_A$. Furthermore, there is an equivalence of tensor categories $\kappa : \mathcal{C}'' \to \mathcal{C}_A$ such that the following diagram of tensor functors is commutative up to a monoidal natural isomorphism

$$\begin{array}{ccc} \mathcal{C} & \xrightarrow{F} & \mathcal{C}'' \\ & {\scriptstyle F_A}\searrow & \downarrow{\scriptstyle \kappa} \\ & & \mathcal{C}_A. \end{array}$$

For instance, let H be a finite dimensional Hopf algebra. Then the induced central algebra (A, σ) of the forgetful functor comod–$H \to \operatorname{Vec}_k$ can be described as follows: As an algebra in comod–H, $A = H$ with the right regular coaction $\Delta : A \to A \otimes H$. For any right H-comodule V, the half-braiding $\sigma_V : A \otimes V \to V \otimes A$ is determined by the right adjoint action of H in the form

$$\sigma_V(h \otimes v) = v_{(0)} \otimes S(v_{(1)})\, h\, v_{(2)}, \quad h \in H, v \in V.$$

See [**8**, Example 6.3].

Let $\mathcal{C}$ be a tensor category. We shall say that an algebra (A, σ) in the center $\mathcal{Z}(\mathcal{C})$ is a *central algebra of* $\mathcal{C}$ if (A, σ) is the induced central algebra of some dominant perfect tensor functor. Thus a central algebra of $\mathcal{C}$ is a commutative algebra $(A, \sigma) \in \mathcal{Z}(\mathcal{C})$ such that $\operatorname{Hom}_{\mathcal{C}}(\mathbf{1}, A)$ is one-dimensional. The dominant tensor functor corresponding to (A, σ) is the functor $- \otimes A : \mathcal{C} \to \mathcal{C}_A$, whose right adjoint is the forgetful functor $\mathcal{C}_A \to \mathcal{C}$.

DEFINITION 4.6. An exact sequence sequence of tensor categories $\mathcal{C}' \longrightarrow \mathcal{C} \xrightarrow{F} \mathcal{C}''$ is called a *perfect exact sequence* if F is a perfect tensor functor.

Every exact sequence of finite tensor categories is a perfect exact sequence. In addition, in every perfect exact sequence $\mathcal{C}' \longrightarrow \mathcal{C} \xrightarrow{F} \mathcal{C}''$, the kernel $\mathcal{C}'$ is a finite tensor category; see [**8**, Proposition 3.15]. Thus the induced Hopf algebra of a perfect exact sequence is always finite dimensional.

Let $(\mathcal{E}) : \mathcal{C}' \longrightarrow \mathcal{C} \longrightarrow \mathcal{C}''$ be a perfect exact sequence and let $(A, \sigma) \in \mathcal{Z}(\mathcal{C})$ be the induced central algebra of F. Then (A, σ) is *self-trivializing*, that is, $A \otimes A$ is a trivial object of $\mathcal{C}_A$. Let $\langle A \rangle$ be the smallest abelian subcategory of $\mathcal{C}$ containing A and stable under direct sums, subobjects and quotients. Then $F_A : \mathcal{C} \to \mathcal{C}_A$ is a normal dominant tensor functor with $\mathfrak{Ker}_{F_A} = \langle A \rangle$. Moreover, $(\mathcal{E})$ is equivalent to the exact sequence $\langle A \rangle \longrightarrow \mathcal{C} \xrightarrow{F_A} \mathcal{C}_A$. See [**8**, Subsection 6.2].

4.4. Examples from finite groups: equivariantization and crossed extensions by matched pairs. Let G be a finite group and let $\mathcal{C}$ be a tensor category. A monoidal functor $\rho : \underline{G}^{\mathrm{op}} \to \operatorname{Aut}_k(\mathcal{C})$ is called a right action of G on $\mathcal{C}$ by k-linear autoequivalences.

The *equivariantization* of $\mathcal{C}$ under the action ρ is the k-linear abelian category $\mathcal{C}^G$ whose objects are pairs (X, r), where X is an object of $\mathcal{C}$ and $r = (r^g)_{g\in G}$ is a collection of isomorphisms $r^g : \rho^g(X) \to X$, $g \in G$, such that for all $g, h \in G$,

$$r^g \rho^g(r^h) = r^{hg}(\rho_2^{g,h})_X, \tag{4.4}$$

where $\rho_2^{g,h} : \rho^g\rho^h \to \rho^{hg}$ is the monoidal structure of ρ, and $r^e \rho_{0X} = \mathrm{id}_X$. A morphism $f : (X, r) \to (Y, r')$ is a morphism $f : X \to Y$ in $\mathcal{C}$ such that $fr^g = r'^g\rho^g(f)$, for all $g \in G$.

The forgetful functor $F : \mathcal{C}^G \to \mathcal{C}$ gives rise to a perfect exact sequence

$$\operatorname{Rep} G \longrightarrow \mathcal{C}^G \xrightarrow{F} \mathcal{C}, \tag{4.5}$$

with induced Hopf algebra $H \cong k^G$ [**8**, Subsection 5.3]. An exact sequence of tensor categories equivalent to (4.5) is called an *equivantization exact sequence.*

A *G-grading* of $\mathcal{C}$ is a decomposition $\mathcal{C} = \bigoplus_{g\in G} \mathcal{C}_g$ into abelian subcategories $\mathcal{C}_g$, called the *homogeneous components* of the grading, such that $\mathcal{C}_g \otimes \mathcal{C}_h \subseteq \mathcal{C}_{gh}$, for all $g, h \in G$. The neutral homogeneous component $\mathcal{C}_e$ is a tensor subcategory of $\mathcal{C}$. A G-grading $\mathcal{C} = \bigoplus_{g\in G} \mathcal{C}_g$ is faithful if $\mathcal{C}_g \neq 0$, for all $g \in G$.

Every finite tensor category $\mathcal{C}$ has a faithful universal grading $\mathcal{C} = \bigoplus_{u\in U(\mathcal{C})} \mathcal{C}_u$, with neutral homogeneous component $\mathcal{C}_e$ equal to the adjoint subcategory $\mathcal{C}_{ad}$, that is, the smallest tensor Serre subcategory of $\mathcal{C}$ containing the objects $X \otimes X^*$, where X runs over the simple objects of $\mathcal{C}$. The group $U(\mathcal{C})$ is called the *universal grading group* of $\mathcal{C}$.

The upper central series of $\mathcal{C}$, $\cdots \subseteq \mathcal{C}^{(n+1)} \subseteq \mathcal{C}^{(n)} \subseteq \cdots \subseteq \mathcal{C}^{(0)} = \mathcal{C}$, is defined as $\mathcal{C}^{(0)} = \mathcal{C}$ and $\mathcal{C}^{(n+1)} = (\mathcal{C}^{(n)})_{ad}$, for all $n \geq 0$. A tensor category $\mathcal{C}$ is called *nilpotent* if there exists some $n \geq 0$ such that $\mathcal{C}^{(n)} \cong \mathrm{Vec}$. See [**16**, Section 3.5 and 4.14], [**21**].

Let $\mathcal{C}$ be a tensor category and let (G, Γ) be a matched pair of groups. A *(G, Γ)-crossed action* on $\mathcal{C}$ consists of the following data:

- A Γ-grading $\mathcal{C} = \bigoplus_{s\in\Gamma} \mathcal{C}_s$.
- A right action of G by k-linear autoequivalences $\rho : \underline{G}^{\mathrm{op}} \to \operatorname{Aut}_k(\mathcal{C})$ such that $\rho^g(\mathcal{C}_s) = \mathcal{C}_{s\triangleleft g}$, for all $g \in G$, $s \in \Gamma$.
- A collection of natural isomorphisms $\gamma = (\gamma^g)_{g\in G}$:

$$\gamma^g_{X,Y} : \rho^g(X \otimes Y) \to \rho^{t\triangleright g}(X) \otimes \rho^g(Y), \quad X \in \mathcal{C}, t \in \Gamma, Y \in \mathcal{C}_t, \tag{4.6}$$

- A collection of isomorphisms $\gamma^g_0 : \rho^g(\mathbf{1}) \to \mathbf{1}$, $g \in G$.

These data are subject to the commutativity of the following diagrams:

(a) For all $g \in G$, $X \in \mathcal{C}$, $s, t \in \Gamma$, $Y \in \mathcal{C}_s$, $Z \in \mathcal{C}_t$,

$$\begin{array}{ccc}
\rho^g(X \otimes Y \otimes Z) & \xrightarrow{\gamma^g_{X\otimes Y,Z}} & \rho^{t\triangleright g}(X \otimes Y) \otimes \rho^g(Z) \\
\Big\downarrow {\scriptstyle \gamma^g_{X,Y\otimes Z}} & & \Big\downarrow {\scriptstyle \gamma^{t\triangleright g}_{X,Y} \otimes \mathrm{id}_{\rho^g(Z)}} \\
\rho^{st\triangleright g}(X) \otimes \rho^g(Y \otimes Z) & \xrightarrow[\mathrm{id}_{\rho^{st\triangleright g}(X)} \otimes \gamma^g_{Y,Z}]{} & \rho^{s\triangleright(t\triangleright g)}(X) \otimes \rho^{t\triangleright g}(Y) \otimes \rho^g(Z)
\end{array}$$

(b) For all $g \in G$, $X \in \mathcal{C}$,

$$\begin{array}{ccccc} \rho^g(X)\otimes\rho^g(\mathbf{1}) & \xleftarrow{\gamma^g_{X,\mathbf{1}}} & \rho^g(X) & \xrightarrow{\gamma^g_{\mathbf{1},X}} & \rho^g(\mathbf{1})\otimes\rho^g(X)\\ & \searrow{\scriptstyle \mathrm{id}_{\rho^g(X)}\otimes\gamma_0^g} & \Big\downarrow{\scriptstyle =} & \swarrow{\scriptstyle \gamma_0^g\otimes\mathrm{id}_{\rho^g(X)}} & \\ & & \rho^g(X) & & \end{array}$$

(c) For all $g, h \in G$, $X \in \mathcal{C}$, $s \in \Gamma$, $Y \in \mathcal{C}_s$,

$$\begin{array}{ccc} \rho^g\rho^h(X\otimes Y) & \xrightarrow{\rho_2{}^{g,h}_{X\otimes Y}} & \rho^{hg}(X\otimes Y)\\ & & \Big\downarrow{\scriptstyle \gamma^{hg}_{X,Y}}\\ \Big\downarrow{\scriptstyle \rho^g(\gamma^h_{X,Y})} & & \rho^{s\rhd hg}(X)\otimes\rho^{hg}(Y)\\ & & \Big\uparrow{\scriptstyle \rho_2{}_X^{(s\lhd h)\rhd g,\, s\rhd h}\otimes\rho_2{}_Y^{g,h}}\\ \rho^g(\rho^{s\rhd h}(X)\otimes\rho^h(Y)) & \xrightarrow[\gamma^g_{\rho^{s\rhd h}(X),\rho^h(Y)}]{} & \rho^{(s\lhd h)\rhd g}\rho^{s\rhd h}(X)\otimes\rho^g\rho^h(Y) \end{array}$$

(d) For all $g, h \in G$,

$$\begin{array}{ccc} \rho^g\rho^h(\mathbf{1}) & \xrightarrow{(\rho_2^{g,h})_{\mathbf{1}}} & \rho^{hg}(\mathbf{1})\\ \Big\downarrow{\scriptstyle \rho^g(\gamma_0^h)} & & \Big\downarrow{\scriptstyle \gamma_0^{hg}}\\ \rho^g(\mathbf{1}) & \xrightarrow[\gamma_0^g]{} & \mathbf{1} \end{array}$$

(e) For all $X \in \mathcal{C}$, $s \in \Gamma$, $Y \in \mathcal{C}_s$,

$$\begin{array}{ccc} X\otimes Y & \xrightarrow{\rho_0{}_{X\otimes Y}} & \rho^e(X\otimes Y)\\ & \searrow{\scriptstyle \rho_0{}_X\otimes\rho_0{}_Y} & \Big\downarrow{\scriptstyle \gamma^e_{X,Y}}\\ & & \rho^e(X)\otimes\rho^e(Y) \end{array} \qquad \begin{array}{ccc} \mathbf{1} & \xrightarrow{\rho_0{}_{\mathbf{1}}} & \rho^e(\mathbf{1})\\ & \searrow{\scriptstyle =} & \Big\downarrow{\scriptstyle \gamma_0^e}\\ & & \mathbf{1} \end{array}$$

We say that a tensor category $\mathcal{C}$ is a (G,Γ)-*crossed tensor category* if it is equipped with a (G,Γ)-crossed action. In this case there is a canonical exact sequence of tensor categories

$$\operatorname{Rep} G \longrightarrow \mathcal{C}^{(G,\Gamma)} \xrightarrow{F} \mathcal{C}, \tag{4.7}$$

where $\mathcal{C}^{(G,\Gamma)}$ is the tensor category defined as follows: As a k-linear category, it is the equivariantization $\mathcal{C}^G$ of $\mathcal{C}$ under the action ρ, while the tensor product is defined in the form $(X, r)\otimes(Y, r') = (X\otimes Y, \tilde{r})$, where $\tilde{r}^g$, $g \in G$, is the composition

$$\bigoplus_{s\in\Gamma}\rho^g(X\otimes Y_s) \overset{\oplus_s\gamma^g_{X,Y_s}}{\longrightarrow} \bigoplus_{s\in\Gamma}\rho^{s\rhd g}(X)\otimes\rho^g(Y_s) \overset{\oplus_s r^{s\rhd g}\otimes r'^g_s}{\longrightarrow} \bigoplus_{s\in\Gamma} X\otimes Y_{s\lhd g} = X\otimes Y,$$

for $Y = \bigoplus_{s \in \Gamma} Y_s$, $Y_s \in \mathcal{C}_s$. The functor $F : \mathcal{C}^{(G,\Gamma)} \to \mathcal{C}$ is the forgetful functor $F(V, (r^g)_{g \in G}) = V$. See [**43**, Theorem 6.1].

We call $\mathcal{C}^{(G,\Gamma)}$ a (G, Γ)-*crossed extension of* $\mathcal{C}_e$. Thus, a (G, Γ)-crossed extension is a unified formulation of equivariantizations and group graded extensions. In fact, suppose that $\rho : \underline{G}^{op} \to \mathrm{Aut}_{\otimes}(\mathcal{C})$ is an action by tensor auto-equivalences of a tensor category $\mathcal{C}$. Then the equivariantization $\mathcal{C}^G$ is a $(G, \{e\})$-crossed extension of $\mathcal{C}$, where $\{e\}$ is the trivial group endowed with the trivial actions $\triangleleft : \{e\} \times G \to G$ and $\triangleright : \{e\} \times G \to \{e\}$. On the other hand, if $\mathcal{C}$ is a tensor category graded by a group Γ, then $\mathcal{C}$ is a $(\{e\}, \Gamma)$-crossed extension of $\mathcal{C}_e$ in a similar way.

Further examples of (G, Γ)-crossed extensions are the categories of representations of abelian extensions of Hopf algebras. Indeed for every abelian exact sequence $k \longrightarrow k^{\Gamma} \longrightarrow H \longrightarrow kG \longrightarrow k$, the category H-mod is a (G, Γ)-crossed extension of Vec. See [**43**, Subsection 8.2].

There exist (G, Γ)-crossed extensions that cannot be built up by means of equivariantizations or group graded extensions. For instance, let $n \geq 5$ be an odd integer and let $H = k^{\mathbb{A}_{n-1}} \# kC_n$, $n \geq 5$, be the bicrossed product associated to the matched pair $(C_n, \mathbb{A}_{n-1})$ arising from the exact factorization $\mathbb{A}_n = \mathbb{A}_{n-1} C_n$ of the alternating group $\mathbb{A}_n$, where $C_n = \langle (12 \dots n) \rangle$ [**31**, Section 8].

As shown in [**47**, Example 4.3], if G is a nontrivial finite group, then H-mod is not equivalent to a G-equivariantization or to a G-graded extension of any fusion category $\mathcal{C}$.

4.5. Abelian exact sequences of tensor categories and matched pairs of groups.

DEFINITION 4.7 ([**47**, Definition 5.1]). An exact sequence of tensor categories is an *abelian exact sequence* if its induced Hopf algebra H is finite dimensional and commutative.

Equivalently, an abelian exact sequence is an exact sequence of the form $\mathrm{Rep}\, G \longrightarrow \mathcal{C} \longrightarrow \mathcal{C}''$, such that the induced tensor functor $\omega : \mathrm{Rep}\, G \to \mathrm{Vec}$ is monoidally isomorphic to the forgetful functor or, in other words, such that the corresponding rank-one module category $\mathcal{M}$ is equivalent to the trivial rank-one module category of $\mathrm{Rep}\, G$.

Examples of abelian exact sequences of tensor categories arise from equivariantization under the action of a finite group on a tensor category and also from Hopf algebra extensions of the form $k \longrightarrow k^G \longrightarrow H \longrightarrow H'' \longrightarrow k$, where G is a finite group.

Let (G, Γ) be a matched pair of finite groups and let $\mathcal{C}$ be a (G, Γ)-crossed tensor category. The induced Hopf algebra of the associated exact sequence of tensor categories $\mathrm{Rep}\, G \longrightarrow \mathcal{C}^{(G,\Gamma)} \xrightarrow{F} \mathcal{C}$ is $H \cong k^G$, so that (4.7) is an abelian exact sequence. The main result of [**47**] says that crossed extensions by matched pairs do in fact exhaust the class of abelian exact sequences of finite tensor categories:

THEOREM 4.8. ([**47**, Theorem 1.1].) *Let G be a finite group and let*

$$(\mathcal{E}) : \ \mathrm{Rep}\, G \longrightarrow \mathcal{C} \longrightarrow \mathcal{D}$$

be an abelian exact sequence of finite tensor categories. Then there exists a finite group Γ endowed with mutual actions by permutations $\triangleright : \Gamma \times G \to G$, $\triangleleft : \Gamma \times G \to \Gamma$ and a (G, Γ)-crossed action on $\mathcal{D}$ such that $(\mathcal{E})$ is equivalent to the exact sequence $\mathrm{Rep}\, G \longrightarrow \mathcal{D}^{(G,\Gamma)} \longrightarrow \mathcal{D}$.

5. Exact sequences of finite tensor categories with respect to a module category

Let $\mathcal{C}$ and $\mathcal{D}$ be finite k-linear abelian categories. Their *Deligne tensor product* is a finite tensor category denoted $\mathcal{C} \boxtimes \mathcal{D}$ endowed with a functor $\boxtimes : \mathcal{C} \times \mathcal{D} \to \mathcal{C} \boxtimes \mathcal{D}$ exact in both variables such that for any k-bilinear right exact functor $F : \mathcal{C} \times \mathcal{D} \to \mathcal{A}$, where $\mathcal{A}$ is a k-linear abelian category, there exists a unique right exact functor $\tilde{F} : \mathcal{C} \boxtimes \mathcal{D} \to \mathcal{A}$ such that $\tilde{F} \circ \boxtimes = F$. Such a category exists and it is unique up to equivalence. In fact, if $\mathcal{C} \cong A$-mod and $\mathcal{D} \cong B$-mod, for some finite dimensional k-algebras A and B, then $\mathcal{C} \boxtimes \mathcal{D} \cong (A \otimes B)$-mod. See [**11**]. The tensor product of two finite (multi-)tensor categories $\mathcal{C}$ and $\mathcal{D}$ is again a finite tensor category and if $\mathcal{C}$ and $\mathcal{D}$ are (multi-)fusion categories, then so is $\mathcal{C} \boxtimes \mathcal{D}$.

Let Y, Z be objects of $\mathcal{C}$ and $\mathcal{D}$, respectively, and let us denote $Y \boxtimes Z = \boxtimes(Y, Z)$. Then $\operatorname{Hom}_{\mathcal{C}\boxtimes\mathcal{D}}(Y_1 \boxtimes Z_1, Y_2 \boxtimes Z_2) \cong \operatorname{Hom}_{\mathcal{C}}(Y_1, Y_2) \otimes \operatorname{Hom}_{\mathcal{D}}(Z_1, Z_2)$, for all $Y_1, Y_2 \in \mathcal{C}$, $Z_1, Z_2 \in \mathcal{D}$. The simple objects of $\mathcal{C} \boxtimes \mathcal{D}$ are exactly those of the form $Y \boxtimes Z$, where Y is a simple object of $\mathcal{C}$ and Z is a simple object of $\mathcal{D}$.

Let $\mathcal{C}' \subseteq \mathcal{C}$ and $\mathcal{C}''$ be finite tensor categories and let $\mathcal{M}$ be an exact indecomposable left $\mathcal{C}'$-module category. In particular, $\mathcal{M}$ is finite. Let $\operatorname{End}(\mathcal{M})$ denote the category of k-linear right exact endofunctors of $\mathcal{M}$, which is a monoidal category with tensor product given by composition of functors and unit object $1_{\mathcal{M}} : \mathcal{M} \to \mathcal{M}$. Let also $i : \mathcal{C}' \to \mathcal{C}$ denote the inclusion functor.

Definition 5.1. An exact sequence of tensor categories *with respect to* $\mathcal{M}$ is a sequence of exact monoidal functors

$$\mathcal{C}' \xrightarrow{i} \mathcal{C} \xrightarrow{F} \mathcal{C}'' \boxtimes \operatorname{End}(\mathcal{M}), \tag{5.1}$$

such that F is dominant, $\mathcal{C}' = \mathfrak{Ker}_F$ coincides with the subcategory of $\mathcal{C}$ mapped to $\operatorname{End}(\mathcal{M})$ under F and, F is normal in the sense that for every object X of $\mathcal{C}$, there exists a subobject X_0 of X such that $F(X_0)$ is the largest subobject of $F(X)$ contained in $\operatorname{End}(\mathcal{M})$.

A tensor category $\mathcal{C}$ fitting into an exact sequence 5.1 with respect to $\mathcal{M}$ is called an *extension of* $\mathcal{C}''$ *by* $\mathcal{C}'$ *with respect to* $\mathcal{M}$.

The notion of exact sequence with respect to a module category was introduced in [**14**] and it generalizes the notion of exact sequence of [**8**]. Indeed, suppose that (5.1) is an exact sequence of finite tensor categories with respect to $\mathcal{M}$. Then $\operatorname{FPdim} \mathcal{C} = \operatorname{FPdim} \mathcal{C}' \operatorname{FPdim} \mathcal{C}''$. Moreover this condition characterizes the exactness of (5.1) under the assumptions that F is dominant and $\mathcal{C}' \subseteq \mathfrak{Ker}_F$ [**14**, Theorem 3.6].

Consider an exact sequence of finite tensor categories $(\mathcal{E}) : \mathcal{C}' \longrightarrow \mathcal{C} \longrightarrow \mathcal{C}''$, as introduced in Section 4. Then $(\mathcal{E})$ induces a fiber functor $\omega : \mathcal{C}' \to \operatorname{Vec}$, thus making $\mathcal{M} = \operatorname{Vec}$ into a rank-one $\mathcal{C}'$-module category. In this way $(\mathcal{E})$ becomes an exact sequence *with respect to the rank-one module category* $\mathcal{M}$.

The Deligne tensor product $\mathcal{C}'' \boxtimes \mathcal{C}'$ of two finite tensor categories gives rise to an exact sequence (5.1) with respect to any exact indecomposable $\mathcal{C}'$-module category $\mathcal{M}$, where $F : \mathcal{C}'' \boxtimes \mathcal{C}' \to \mathcal{C}'' \boxtimes \operatorname{End}(\mathcal{M})$ is the natural dominant monoidal functor. [**14**].

The notion of exact sequence with respect to a module category is self-dual in the following sense: Let $\mathcal{N}$ be an indecomposable exact $\mathcal{C}''$-module category. Then

(5.1) induces an exact sequence with respect to $\mathcal{N}$:

$$(\mathcal{C}'')^*_{\mathcal{N}} \xrightarrow{F^*} \mathcal{C}^*_{\mathcal{N}\boxtimes\mathcal{M}} \xrightarrow{i^*} (\mathcal{C}')^*_{\mathcal{M}} \boxtimes \operatorname{End}(\mathcal{N}). \tag{5.2}$$

Suppose that (5.1) is an exact sequence with respect to $\mathcal{M}$. Observe that if $\mathcal{C}$ is a fusion category, then $\mathcal{C}'$ and $\mathcal{C}''$ are fusion categories and $\mathcal{M}$ is a finite semisimple $\mathcal{C}'$-module category.

Assume conversely that $\mathcal{C}'$ and $\mathcal{C}''$ are fusion categories. In particular $\mathcal{M}$ is a finite semisimple module category over $\mathcal{C}'$. By [**14**, Theorem 3.8] $\mathcal{C}$ is also a fusion category. In this case the monoidal category $\operatorname{End}(\mathcal{M}) \cong {}^{\vee}\mathcal{M} \boxtimes \mathcal{M}$ is a multifusion category. Here, ${}^{\vee}\mathcal{M}$ is the right $\mathcal{C}'$-module category such that ${}^{\vee}\mathcal{M} = \mathcal{M}$ and $M\bar{\otimes}X = {}^*X \otimes M$, $X \in \mathcal{C}'$, $M \in \mathcal{M}$; that is, if $\mathcal{M} \cong \mathcal{C}'_A$ is the category of right A-modules in $\mathcal{C}'$ for some indecomposable algebra $A \in \mathcal{C}'$, then ${}^{\vee}\mathcal{M} \cong {}_A\mathcal{C}'$.

LEMMA 5.2. *Let $\mathcal{C}' \xrightarrow{i} \mathcal{C} \xrightarrow{F} \mathcal{C}'' \boxtimes \operatorname{End}(\mathcal{M})$ be an exact sequence with respect to $\mathcal{M}$. If $\mathcal{C}$ is a pointed fusion category, then so is $\mathcal{C}''$.*

PROOF. By assumption every simple object X of $\mathcal{C}$ is invertible, that is, $X \otimes X^* \cong \mathbf{1}$. Let $1_{\mathcal{M}} = \bigoplus_{i\in I} 1_i$ be a decomposition of the unit object $1_{\mathcal{M}}$ of $\operatorname{End}(\mathcal{M})$ into a direct sum of simple subobjects 1_i, $i \in I$. Then $1_i \otimes 1_j \cong \delta_{i,j}1_i$, and $1_i^* \cong {}^*1_i \cong 1_i$, for all $i, j \in I$; see [**16**, Section 4.3].

Let Y be a simple object of $\mathcal{C}''$. For each fixed $i \in I$, $Y \boxtimes 1_i$ is a simple object of $\mathcal{C}'' \boxtimes \operatorname{End}(\mathcal{M})$. Since the functor $F : \mathcal{C} \to \mathcal{C}'' \boxtimes \operatorname{End}(\mathcal{M})$ is dominant, then $Y \boxtimes 1_i$ is a direct summand of $F(X)$ for some simple object X of $\mathcal{C}$. Therefore $(Y \boxtimes 1_i) \boxtimes (Y \boxtimes 1_i)^* \cong (Y \otimes Y^*) \boxtimes 1_i$ is a direct summand of $F(X) \otimes F(X)^*$. On the other hand,

$$F(X \otimes X^*) \cong F(\mathbf{1}) \cong \mathbf{1}_{\mathcal{C}\boxtimes\operatorname{End}(\mathcal{M})} \cong \mathbf{1} \boxtimes 1_{\mathcal{M}} \cong \bigoplus_{i\in I} \mathbf{1} \boxtimes 1_i.$$

Therefore the only simple constituent of $Y \otimes Y^*$ is the trivial object $\mathbf{1}$ of $\mathcal{C}''$. This implies that $Y \otimes Y^* \cong \mathbf{1}$, that is, Y is invertible. Since the simple object Y was arbitrary, this shows that $\mathcal{C}''$ is pointed, as claimed. □

5.1. Exact sequences and exact factorizations. Let $\mathcal{C}$ be a fusion category. Recall from [**20**] that an $\mathcal{C}$ is endowed with an *exact factorization* into a product of two fusion subcategories $\mathcal{A}$ and $\mathcal{B}$ if the following conditions hold:

- $\mathcal{C}$ coincides with the full abelian subcategory spanned by direct summands of $X \otimes Y$, $X \in \mathcal{A}$, $Y \in \mathcal{B}$,
- $\mathcal{A} \cap \mathcal{B} = \langle \mathbf{1} \rangle$.

If this holds, we write $\mathcal{C} = \mathcal{A} \bullet \mathcal{B}$.

By [**20**, Theorem 3.8], $\mathcal{C} = \mathcal{A} \bullet \mathcal{B}$ if and only if every simple object Z of $\mathcal{C}$ admits a decomposition $Z \cong X \otimes Y$, for unique (up to isomorphism) simple objects $X \in \mathcal{A}$, $Y \in \mathcal{B}$. Thus the notion of an exact factorization can be formulated as a category equivalence as follows: $\mathcal{C}$ has an exact factorization $\mathcal{C} = \mathcal{A} \bullet \mathcal{B}$ if and only if the tensor product of $\mathcal{C}$ induces an equivalence of k-linear categories $\otimes : \mathcal{A} \boxtimes \mathcal{B} \to \mathcal{C}$.

For example, let E be a finite group. Exact factorizations of E correspond exactly to exact factorizations of the category Vec_E. In fact, if Γ and G are subgroups of E, then $E = \Gamma G$ is an exact factorization of E if and only if $\operatorname{Vec}_E = \operatorname{Vec}_\Gamma \bullet \operatorname{Vec}_G$ is an exact factorization of Vec_E.

Exact factorizations and extensions are related as follows: Every exact sequence $\mathcal{C}' \xrightarrow{i} \mathcal{C} \xrightarrow{F} \mathcal{C}'' \boxtimes \operatorname{End}(\mathcal{M})$ with respect to $\mathcal{M}$ induces an exact factorization of the dual fusion category $\mathcal{C}^*_{\mathcal{C}''\boxtimes\mathcal{M}}$:

$$\mathcal{C}^*_{\mathcal{C}''\boxtimes\mathcal{M}} = \mathcal{C}'' \bullet (\mathcal{C}')^*_{\mathcal{M}}. \tag{5.3}$$

Conversely, every exact factorization $\mathcal{C} = \mathcal{A} \bullet \mathcal{B}$ of a fusion category $\mathcal{C}$ induces an exact sequence, with respect to any indecomposable $\mathcal{A}$-module category $\mathcal{N}$

$$\mathcal{A}^*_{\mathcal{N}} \longrightarrow \mathcal{C}^*_{\mathcal{B}\boxtimes\mathcal{N}} \longrightarrow \mathcal{B} \boxtimes \operatorname{End}(\mathcal{N}). \tag{5.4}$$

In particular, $\mathcal{C}$ fits into an exact sequence with respect to $\mathcal{N} = \mathcal{A}$:

$$\mathcal{A} \longrightarrow \mathcal{C} \longrightarrow \mathcal{B} \boxtimes \operatorname{End}(\mathcal{A}). \tag{5.5}$$

See [**20**, Theorem 4.1].

5.2. Exact factorizations arising from exact sequences of finite tensor categories. Consider an exact sequence of Hopf algebras

$$k \longrightarrow H' \longrightarrow H \longrightarrow H'' \longrightarrow k, \tag{5.6}$$

such that H' is finite dimensional. Then H is free as a left (or right) module over H' and in particular the sequence is cleft [**53**, Theorem 2.1 (2)]. By [**8**, Proposition 3.9] the exact sequence (5.6) gives rise to an exact sequence of tensor categories

$$\text{comod-}H' \to \text{comod-}H \to \text{comod-}H''. \tag{5.7}$$

Observe that H' is naturally an algebra in comod-H. By a result of [**51**], the tensor category ${}_{H'}(\text{comod-}H)_{H'}$ is equivalent to the category of comodules over a certain coquasibialgebra $((H')^* \bowtie H'', \varphi)$, where φ is a so-called *Kac 3-cocycle* associated to the exact sequence (5.6) (see [**51**, Section 6]). The coquasibialgebra $((H')^* \bowtie H'', \varphi)$, termed a *generalized product coquasibialgebra* in [**51**], has an exact factorization into its sub-coquasibialgebras $(H')^*$ and H''.

A generalization of this feature to the context of exact sequences of finite tensor categories appeared in [**34**]. Let $(\mathcal{E}) : \mathcal{C}' \longrightarrow \mathcal{C} \longrightarrow \mathcal{C}''$ be an exact sequence of finite tensor categories and let $(A, \sigma) \in \mathcal{Z}(\mathcal{C})$ be its induced central algebra. So that $\mathcal{C}'' \cong \mathcal{C}_A$ and the exact sequence $(\mathcal{E})$ is equivalent to the exact sequence $\langle A \rangle \longrightarrow \mathcal{C} \xrightarrow{F_A} \mathcal{C}_A$. See Subsection 4.3.

Let ${}_A\mathcal{C}_A$ be the tensor category of A-bimodules in $\mathcal{C}$. The category $\mathcal{C}'' \cong \mathcal{C}_A$ is an indecomposable exact $\mathcal{C}$-module category and there are equivalences of tensor categories $\mathcal{C}^*_{\mathcal{C}''} \cong {}_A\mathcal{C}_A$, H-mod $\cong {}_A(\text{comod–}H)_A \cong {}_A\mathcal{C}'_A$, where H is the induced Hopf algebra of $(\mathcal{E})$.

It was shown in [**34**, Proposition 7.3] that there is an equivalence of k-linear categories

$$\mathcal{C}^*_{\mathcal{C}''} \cong (H\text{-mod}) \boxtimes \mathcal{C}'', \tag{5.8}$$

thus, in the sense of the formulation explained at the beginning of this section, an exact factorization of $\mathcal{C}^*_{\mathcal{C}''}$.

More precisely, under the identifications H-mod $\cong {}_A(\text{comod–}H)_A \cong {}_A\mathcal{C}'_A \subseteq {}_A\mathcal{C}_A$ and $\mathcal{C}'' \cong \mathcal{C}_A \subseteq {}_A\mathcal{C}_A$, the tensor product functor $\otimes_A : {}_A\mathcal{C}'_A \boxtimes \mathcal{C}_A \to {}_A\mathcal{C}_A$ induces an equivalence of k-linear categories $(H\text{-mod}) \boxtimes \mathcal{C}_A \longrightarrow {}_A\mathcal{C}_A$.

Notice that in the situation of the exact sequence (5.7), there is an equivalence of tensor categories $(\text{comod-}H)^*_{\text{comod-}H''} \cong {}_{H'}(\text{comod-}H)_{H'}$. See [**34**, Example 7.4].

6. Examples and open questions

We begin this section by discussing some examples that answer a number of natural questions regarding the behaviour of exact sequences in relation with known facts about exact sequences of groups and Hopf algebras.

Let $\mathcal{C}$ be a finite tensor category. Let also $\mathcal{C}'$ be a tensor subcategory of $\mathcal{C}$ and $\mathcal{M}$ an indecomposable exact $\mathcal{C}'$-module category. We shall say that $\mathcal{C}'$ is *normal in* $\mathcal{C}$ *with respect to* $\mathcal{M}$ if there exist a tensor category $\mathcal{C}''$ and an exact sequence of tensor categories with respect to $\mathcal{M}$: $\mathcal{C}' \longrightarrow \mathcal{C} \longrightarrow \mathcal{C}'' \boxtimes \operatorname{End}(\mathcal{M})$.

A tensor subcategory $\mathcal{C}'$ will be called *normal in* $\mathcal{C}$ if it is normal with respect to a rank-one module category, that is, if there exist a tensor category $\mathcal{C}''$ and an exact sequence of tensor categories $\mathcal{C}' \longrightarrow \mathcal{C} \longrightarrow \mathcal{C}''$ in the sense of Definition 4.1.

A tensor category $\mathcal{C}$ will be called *simple* if it has no normal tensor subcategories with respect to a module category. If $\mathcal{C}$ has no normal tensor subcategories it will be called *simple with respect to rank-one module categories.*

6.1. Fusion subcategories of index 2 need not be normal. It is a well-known fact that if G is a finite group, every subgroup of G whose index is the smallest prime number dividing the order of G is normal in G. More generally, if H is a semisimple Hopf algebra and H' is a Hopf subalgebra of H such that $\dim H = p \dim H'$, where p is the smallest prime number dividing the dimension of H, then H' is normal in H [**27**]. The next theorem gives a generalization of this result in the context of fusion categories.

Let $F : \mathcal{C} \to \mathcal{D}$ be a dominant tensor functor between finite tensor categories. The *Frobenius-Perron index of* F is defined as the ratio $\operatorname{FPdim}\mathcal{C}/\operatorname{FPdim}\mathcal{D}$. The Frobenius-Perron index of a dominant tensor functor is an algebraic integer, by [**17**, Corollary 8.11]. In addition, if G is a left (or right) adjoint of F, then the Frobenius-Perron index of F coincides with $\operatorname{FPdim} G(\mathbf{1})$. See [**8**, Section 4].

THEOREM 6.1. ([**8**, Proposition 4.13], [**9**, Theorem 6.2]). *Let $F : \mathcal{C} \to \mathcal{D}$ be a dominant tensor functor between fusion categories $\mathcal{C}$ and $\mathcal{D}$. Then the following hold:*

(i) If the Frobenius-Perron index of F is 2, then F is normal.

(ii) If $\mathcal{C}$ has integer Frobenius-Perron dimension and the Frobenius-Perron index of F is the smallest prime number dividing $\operatorname{FPdim}\mathcal{C}$*, then F is normal.*

Furthermore, the exact sequences arising from (i) and (ii) are equivariantization exact sequences.

Let $\mathcal{C}$ be a finite tensor category and let $\mathcal{C}' \subseteq \mathcal{C}$ be a tensor subcategory. The ratio $\operatorname{FPdim}\mathcal{C}/\operatorname{FPdim}\mathcal{C}'$ will be called the *index of* $\mathcal{C}'$ (in $\mathcal{C}$). This is also an algebraic integer [**17**].

The dual statement of Theorem 6.1 is not true, that is, there exist fusion subcategories of index 2 which are not normal. These examples are Tambara-Yamagami fusion categories $\mathcal{TY}(\mathbb{Z}_p, \chi, \tau)$ of Frobenius-Perron dimension $2p$, where p is a prime number. In this case the pointed subcategory of $\mathcal{TY}(\mathbb{Z}_p, \chi, \tau)$, which is the unique fusion subcategory of Frobenius-Perron dimension p, is not normal [**9**, Proposition 6.3].

In fact, Tambara-Yamagami categories $\mathcal{TY}(\mathbb{Z}_p, \chi, \tau)$ provide examples of fusion categories of dimension $2p$ which are simple. The next proposition generalizes [**9**, Proposition 6.5] to the context of exact sequences with respect to a module category.

PROPOSITION 6.2. *The fusion category $\mathcal{TY}(\mathbb{Z}_p, \chi, \tau)$ is simple.*

PROOF. Let $\mathcal{C} = \mathcal{TY}(\mathbb{Z}_p, \chi, \tau)$. Suppose on the contrary that there is an exact sequence $\mathcal{C}' \longrightarrow \mathcal{C} \longrightarrow \mathcal{C}'' \boxtimes \operatorname{End}(\mathcal{M})$, for some proper fusion subcategory $\mathcal{C}'$ and some indecomposable $\mathcal{C}'$-module category $\mathcal{M}$. Then $\operatorname{FPdim} \mathcal{C} = \operatorname{FPdim} \mathcal{C}' \operatorname{FPdim} \mathcal{C}''$ and therefore $\operatorname{FPdim} \mathcal{C}'', \operatorname{FPdim} \mathcal{C}' \in \{2, p\}$.

Consider the associated exact factorization $\mathcal{C}^*_{\mathcal{C}''\boxtimes\mathcal{M}} = \mathcal{C}'' \bullet (\mathcal{C}')^*_{\mathcal{M}}$. Since every fusion category of prime Frobenius-Perron dimension is pointed [**17**], then $\mathcal{C}''$, $(\mathcal{C}')^*_{\mathcal{M}}$ are pointed and therefore so is $\mathcal{C}^*_{\mathcal{C}''\boxtimes\mathcal{M}}$. Thus $\mathcal{C}$ is group-theoretical. This is impossible because $\mathcal{C}$ is not integral (in fact the unique non-invertible simple object of $\mathcal{C}$ has Frobenius-Perron dimension $\sqrt{p}$). This contradiction shows that such an exact sequence cannot exist and therefore $\mathcal{C}$ is simple, as claimed. □

6.2. Further examples of simple fusion categories of dimension $p^a q^b$. Proposition 6.2 implies that Burnside's $p^a q^b$-theorem does not extend to fusion categories in terms of exact sequences with respect to module categories. Further examples of this situation are provided by the non-group-theoretical fusion categories constructed in [**24**].

Let $p < q$ be prime numbers such that p is odd and divides $q + 1$. Let also $\zeta_1 \neq \zeta_2 \in \mathbb{F}_{q^2}$ such that $\zeta_1^p = \zeta_2^p = 1$ but $\zeta_1 \zeta_2 \neq 1$, and let $\xi \in H^3(\mathbb{Z}_p, k^\times) \cong \mathbb{Z}_p$. Consider the non-group-theoretical fusion category $\mathcal{C}(p, q, \{\zeta_1, \zeta_2\}, \xi)$ constructed in [**24**]. The fusion categories $\mathcal{C}(p, q, \{\zeta_1, \zeta_2\}, \xi)$ are $\mathbb{Z}_p$-extensions of $\operatorname{Vec}_{\mathbb{Z}_q \times \mathbb{Z}_q}$ and they fall into $(p^2 - p)/2$ equivalence classes. We have $\operatorname{FPdim} \mathcal{C}(p, q, \{\zeta_1, \zeta_2\}, \xi) = pq^2$.

PROPOSITION 6.3. *The fusion category $\mathcal{C}(p, q, \{\zeta_1, \zeta_2\}, \xi)$ is simple.*

PROOF. As in the proof of Proposition 6.2, let us assume on the contrary that there is an exact sequence $\mathcal{C}' \longrightarrow \mathcal{C} \longrightarrow \mathcal{C}'' \boxtimes \operatorname{End}(\mathcal{M})$, for some proper fusion subcategory $\mathcal{C}'$ and some indecomposable $\mathcal{C}'$-module category $\mathcal{M}$. So that $\operatorname{FPdim} \mathcal{C}''$ can be either p, q, q^2 or pq and similarly for $\operatorname{FPdim} \mathcal{C}'$. Therefore both $\mathcal{C}''$ and $(\mathcal{C}')^*_{\mathcal{M}}$ are pointed: this follows from [**17**, Corollaries 8.30 and 8.31] if the Frobenius-Perron dimensions are p, q or q^2, and by the classification of fusion categories of dimension pq in [**15**, Theorem 6.3], since p is odd (and thus it cannot divide $q - 1$).

This implies that $\mathcal{C}^*_{\mathcal{C}''\boxtimes\mathcal{M}} = \mathcal{C}'' \bullet (\mathcal{C}')^*_{\mathcal{M}}$ is pointed and then $\mathcal{C}$ is group-theoretical. This contradicts the choice of $\mathcal{C}$ and shows that such an exact sequence cannot exist. Thus $\mathcal{C}$ is simple, as claimed. □

6.3. Group-theoretical fusion categories and exact sequences. As mentioned before, the category of representations of a finite simple group is a simple fusion category. Notice, however, that there exist simple Hopf algebras H such that the tensor category H-mod is not simple; see Subsection 3.5.

Recall that a fusion category $\mathcal{C}$ is called *group-theoretical* if it is categorically Morita equivalent to a pointed fusion category. Let $\mathcal{C}$ be a pointed fusion category, so that there exist a finite group G and a 3-cocycle $\omega : G \times G \times G \to k^\times$ such that $\mathcal{C}$ is equivalent to the category $\operatorname{Vec}_G^\omega$ of G-graded vector spaces with associativity determined by ω.

Every indecomposable module category over $\operatorname{Vec}_G^\omega$ arises from a pair (Γ, α), where Γ is a subgroup of G and $\alpha : \Gamma \times \Gamma \to k^\times$ is a 2-cochain on Γ such that $d\alpha = \omega|_{\Gamma\times\Gamma\times\Gamma}$. Thus, the restriction $\omega|_\Gamma$ represents the trivial cohomology class in $H^3(\Gamma, k^\times)$. Given such a pair (Γ, α), the twisted group algebra $A(\Gamma, \alpha) = k_\alpha \Gamma$ is an indecomposable algebra in $\operatorname{Vec}_G^\omega$. The (left) module category associated to such pair (Γ, α) is the category

$$\mathcal{M}(\Gamma, \alpha) = (\mathrm{Vec}_G^\omega)_{A(\Gamma,\alpha)}$$

of (right) $A(\Gamma, \alpha)$-modules in Vec_G^ω. The group-theoretical category $(\mathrm{Vec}_G^\omega)^*_{\mathcal{M}_0(\Gamma,\alpha)}$ is denoted $\mathcal{C}(G, \omega, \Gamma, \alpha)$.

Let G be a finite group and let Γ be a subgroup of G. There is a canonical embedding of tensor categories $\mathrm{Rep}\,\Gamma \longrightarrow \mathcal{C}(G, \omega, \Gamma, \alpha)$. The next proposition implies that $\mathrm{Rep}\,\Gamma$ is not necessarily a normal tensor subcategory of $\mathcal{C}(G, \omega, \Gamma, \alpha)$.

Let G be the alternating group $\mathbb{A}_6$ of order 60 and let Γ be a subgroup of G such that $G \cong \mathbb{A}_5$. Then Γ is a maximal subgroup of $\mathbb{A}_6$. There are 12 such subgroups and they constitute 2 conjugacy classes, represented by the subgroups $\langle(12345), (123)\rangle$ and $\langle(1, 2, 3, 4, 5), (1, 4)(5, 6)\rangle$.

Let $\mathcal{C} = \mathcal{C}(\mathbb{A}_6, 1, \Gamma, 1)$, so that $\mathcal{C}$ is a group-theoretical fusion category of dimension 360 which is categorically Morita equivalent to $\mathrm{Vec}_{\mathbb{A}_6}$. In addition, $\mathcal{C}$ contains a fusion subcategory $\mathcal{C}' \cong \mathrm{Rep}\,\mathbb{A}_5$.

Observe that $\mathcal{C}$ is not a graded extension of any fusion category: this follows from the characterization of graded extensions in [**18**, Proposition 2.9], since $\mathrm{Rep}\,\mathbb{A}_6$ is the unique Tannakian subcategory of $\mathcal{Z}(\mathcal{C})$ (c.f. [**45**, Example 3.3]) and $\mathcal{C}$ cannot be an $\mathbb{A}_6$-graded extension of a fusion subcategory, since it is not pointed.

We have $\widehat{\Gamma} = 1$ and $\Gamma = N_{\mathbb{A}_6}(\Gamma)$ (by maximality of Γ). Hence, by [**22**, Theorem 5.2], the group of invertible objects of $\mathcal{C}$ is trivial.

PROPOSITION 6.4. *The category $\mathcal{C}(\mathbb{A}_6, 1, \Gamma, 1)$ is simple with respect to rank-one module categories.*

PROOF. Suppose that $F : \mathcal{C} \to \mathcal{D}$ is a normal dominant tensor functor such that F is not an equivalence. Then the restriction of F to $\mathcal{C}' \cong \mathrm{Rep}\,\mathbb{A}_5$ is also a normal tensor functor with kernel $\mathfrak{Ker}_F \cap \mathcal{C}'$. Therefore either $\mathcal{C}' \cap \mathfrak{Ker}_F \cong \mathrm{Vec}$ or $\mathcal{C}' \subseteq \mathfrak{Ker}_F$. Suppose first that $\mathcal{C}' \cap \mathfrak{Ker}_F \cong \mathrm{Vec}$. Then the restriction of F to $\mathcal{C}'$ is a full embedding, whence the dimension of $\mathcal{D}$ is divisible by 60. Then the dimension of $\mathfrak{Ker}_F$ can equal 6, 2 or 3. But this contradicts the fact that $\mathcal{C}$ has no nontrivial invertible objects. Hence $\mathcal{C}' \subseteq \mathfrak{Ker}_F$.

The Grothendieck ring $\mathrm{Gr}(\mathcal{C})$ of $\mathcal{C}$ is faithfully graded by the double coset ring $\Gamma\backslash\mathbb{A}_6/\Gamma$. In addition the fusion subcategory $\mathfrak{Ker}_F$ determines a based subring of $\mathrm{Gr}(\mathcal{C})$. Since $\mathcal{C}' \subseteq \mathfrak{Ker}_F$, this based subring corresponds to a subgroup of $\mathbb{A}_6$ containing Γ, see [**22**, Theorem 4.1 and Lemma 3.6]. The maximality of Γ implies that $\mathcal{C}' = \mathfrak{Ker}_F$. Then $\mathcal{C}$ fits into an exact sequence $\mathcal{C}' \longrightarrow \mathcal{C} \longrightarrow \mathcal{D}$, where $\mathcal{D}$ is a fusion category of dimension 6. Moreover, by [**8**, Proposition 4.9], $\mathcal{D}$ is integral.

Assume that the exact sequence $\mathcal{C}' \longrightarrow \mathcal{C} \longrightarrow \mathcal{D}$ is *abelian*. In other words the sequence is exact with respect to the trivial $\mathcal{C}'$-module category Vec in the sense of [**14**]. Let $(A, \sigma) \in \mathcal{Z}(\mathcal{C})$ be the induced central algebra of F. Then ${}_A\mathcal{C}_A$ has a factorization into fusion subcategories equivalent, respectively, to Vec_Γ and $\mathcal{D}$. If $\mathcal{D}$ is pointed of dimension 6, then $\mathcal{D} \cong \mathrm{Vec}_S^\omega$, where S is a group of order 6. Hence, ${}_A\mathcal{C}_A \cong \mathrm{Vec}_L^{\omega'}$, where L is a group of order 360 that has an exact factorization $L = \Gamma.H$ [**34**, Proposition 7.3]. In particular the pointed fusion category $\mathrm{Vec}_L^{\omega'}$ is categorically Morita equivalent to $\mathrm{Vec}_{\mathbb{A}_6}$. By [**37**, Theorem 5.8], this implies that $L \cong \mathbb{A}_6$. This leads to a contradiction, because the group $\mathbb{A}_6$ admits no exact factorization into proper subgroups [**33**].

If, on the other hand, $\mathcal{D}$ is not pointed, then $\mathcal{D}$ must contain two distinct invertible objects and a simple object of dimension 2. Then the group of invertible objects of ${}_A\mathcal{C}_A$ is of order 120. Since ${}_A\mathcal{C}_A$ is Morita equivalent to $\mathrm{Vec}_{\mathbb{A}_6}$, then there exists a subgroup T of $\mathbb{A}_6$ and a 2-cocycle ψ on T such that ${}_A\mathcal{C}_A \cong \mathcal{C}(\mathbb{A}_6, 1, T, \psi)$. By [**22**, Theorem 5.2], the group of invertible objects of ${}_A\mathcal{C}_A$ has order $|K||\widehat{T}|$, where K is a certain subgroup of $N_{\mathbb{A}_6}(T)/T$. A direct inspection on the possible subgroups T of $\mathbb{A}_6$ (see Table 1) shows that $|N_{\mathbb{A}_6}(T)/T||\widehat{T}| \leq 36$, which is again a contradiction.

We have thus shown that the group-theoretical fusion category $\mathcal{C} = \mathcal{C}(\mathbb{A}_6, 1, G, 1)$ does not fit into any *abelian* exact sequence of tensor categories.

Therefore, if $\mathcal{C}$ is not simple, then $\mathcal{C}$ fits into a non-abelian exact sequence

$$\mathcal{C}' \longrightarrow \mathcal{C} \longrightarrow \mathcal{D}, \tag{6.1}$$

where $\mathcal{C}' \cong \operatorname{Rep} \mathbb{A}_5$ and $\operatorname{FPdim} \mathcal{D} = 6$. By the previous part, the induced Hopf algebra H of (6.1) is not commutative. Hence $H \cong (k\mathbb{A}_5)^J$, where the twist J is not trivial. Then H-mod is of type $(1, 12; 4, 3)$, see [**41**].

Consider the associated exact factorization $\mathcal{C}^*_{\mathcal{D}} \cong H\text{-mod} \bullet \mathcal{D}$. As before, the fusion category $\mathcal{D}$ is either pointed or of type $(1, 2; 2, 1)$. Hence the possible types for $\mathcal{C}^*_{\mathcal{D}}$ are

$$\text{(i) } (1, 72; 4, 18), \qquad \text{(ii) } (1, 24; 2, 12; 4, 6; 8, 3).$$

Write, as before, ${}_A\mathcal{C}_A \cong \mathcal{C}(\mathbb{A}_6, 1, T, \psi)$, where T is a subgroup of $\mathbb{A}_6$ and ψ is a 2-cocycle on T. As pointed out before, the group of invertible objects of $\mathcal{C}(\mathbb{A}_6, 1, T, \psi)$ is at most 36, hence possibility (i) is discarded.

If possibility (ii) holds, then T must be a Klein four group (Table 1). Observe that, for every simple object X of $\mathcal{C}$, $\operatorname{FPdim} X$ divides $|T|$; see [**39**, Proposition 5.5]. This contradicts the fact that $\mathcal{C}^*_{\mathcal{D}}$ has simple objects of dimension 8 as in case (ii). This shows that $\operatorname{Rep} \Gamma$ is not normal in $\mathcal{C} = \mathcal{C}(\mathbb{A}_6, 1, \Gamma, 1)$ and therefore $\mathcal{C}$ is simple, as was to be shown. □

6.4. Dominant images of normal fusion subcategories need not be normal. Proposition 6.4 provides examples of images of normal fusion subcategories under dominant tensor functors which are not normal.

Let $G \cong \mathbb{A}_6$ and $\mathbb{A}_5 \cong \Gamma \subseteq \mathbb{A}_6$ as in the previous subsection. Let also $\mathcal{C} = \mathcal{C}(G, 1, \Gamma, 1)$.

The canonical braiding of $\operatorname{Rep} G$ gives rise to an embedding of braided tensor categories $\operatorname{Rep} G \longrightarrow \mathcal{Z}(G)$ that fits into an equivariantization exact sequence of tensor categories $\operatorname{Rep} G \longrightarrow \mathcal{Z}(G) \longrightarrow \mathrm{Vec}_G$.

On the other hand, since $\mathcal{C}$ is Morita equivalent to $\operatorname{Rep} G$, there is an equivalence of braided tensor categories $\mathcal{Z}(G) \cong \mathcal{Z}(\mathcal{C})$ [**52**]. This equivalence induces a dominant tensor functor $U_\Gamma : \mathcal{Z}(G) \longrightarrow \mathcal{C}$, such that $U_\Gamma(V) = k\Gamma \otimes V$. In particular $U_\Gamma(\operatorname{Rep} G) = \operatorname{Rep} \Gamma$. Since, by Proposition 6.4, $\mathcal{C}$ is simple this implies:

COROLLARY 6.5. *The image of the normal fusion subcategory* $\operatorname{Rep} G$ *of* $\mathcal{Z}(G)$ *under the dominant tensor functor* $U_\Gamma : \mathcal{Z}(G) \to \mathcal{C}$ *is not normal in* $\mathcal{C}$.

6.5. Composition series and composition factors. The definition of a composition series of a finite dimensional Hopf algebra (Definition 3.4) has an obvious extension to the context of exact sequences of finite tensor categories with respect to module categories.

Let a *composition series* of a finite tensor category $\mathcal{C}$ be defined as a sequence of finite tensor categories $\mathcal{C}_1, \dots, \mathcal{C}_n$ defined, as before, as $n = 1$ and $\mathcal{C}_1 = \mathcal{C}$, if $\mathcal{C}$ is simple, while if $\mathcal{C}' \longrightarrow \mathcal{C} \longrightarrow \mathcal{C}'' \boxtimes \operatorname{End}(\mathcal{M})$ is an exact sequence with respect to the

$\mathcal{C}'$-module category $\mathcal{M}$ such that $\operatorname{FPdim}\mathcal{C}', \operatorname{FPdim}\mathcal{C}'' > 1$, and $\mathcal{C}'_1, \dots, \mathcal{C}'_m, \mathcal{C}''_1, \dots, \mathcal{C}''_\ell$, are composition series of $\mathcal{C}'$ and $\mathcal{C}''$, respectively, then $n = m + \ell$ and

$$\mathcal{C}_i = \mathcal{C}'_i,\ 1 \leq i \leq m, \quad \mathcal{C}_i = \mathcal{C}''_{i-m},\ m < i \leq m + \ell.$$

As before, the *factors* and the *length* of the series are, respectively, the finite tensor categories $\mathcal{C}_1, \dots, \mathcal{C}_n$ and the number n.

It is clear that every finite tensor category admits such a composition series.

EXAMPLE 6.6. Let G be a finite simple group. Then the fusion category $\operatorname{Rep} G$ is simple: in fact, $\operatorname{Rep} G$ has no proper fusion subcategories when G is simple. Observe that this is not true for the category Vec_G, since the finite simple group G might have non-trivial exact factorizations into proper subgroups. In particular the condition on a fusion category being simple is not self-dual.

On the other hand, if H is a normal subgroup of G, then the restriction functor gives rise to an exact sequence in the sense of [**8**]: $\operatorname{Rep} G/H \longrightarrow \operatorname{Rep} G \longrightarrow \operatorname{Rep} H$. Inductively, we find that if $G_1, \dots, G_n$ are the composition factors of G, then the fusion categories $\operatorname{Rep} G_1, \dots, \operatorname{Rep} G_n$, are composition factors of $\operatorname{Rep} G$.

In what follows we show that composition series of fusion categories thus defined fail to satisfy a Jordan-Hölder theorem.

Let $n \in \mathbb{N}$. We shall denote by $\mathbb{S}_n$ and $\mathbb{A}_n$ the symmetric and alternating groups of degree n, respectively.

The proof of the following theorem relies on a result of Miller [**33**] that asserts that the alternating group $\mathbb{A}_6$ does not admit any exact factorization into proper subgroups. We summarize in Tables 1 and 2 the information about the subgroups of $\mathbb{A}_6$ and $\mathbb{A}_5$ used along the proof.

THEOREM 6.7. *The fusion category* $\operatorname{Vec}_{\mathbb{A}_6}$ *is simple.*

PROOF. Suppose on the contrary that there exists an exact sequence with respect to an indecomposable $\mathcal{C}'$-module category $\mathcal{M}$:

$$\mathcal{C}' \longrightarrow \operatorname{Vec}_{\mathbb{A}_6} \xrightarrow{F} \mathcal{C}'' \boxtimes \operatorname{End}(\mathcal{M}), \tag{6.2}$$

such that $\operatorname{FPdim}\mathcal{C}', \operatorname{FPdim}\mathcal{C}'' > 1$.

Then $\mathcal{C}'$ and $\mathcal{C}''$ are fusion categories. Since the functor F is dominant then $\mathcal{C}''$ is pointed, by Lemma 5.2. Also $\mathcal{C}'$ is pointed; moreover, $\mathcal{C}' = \operatorname{Vec}_H$, for some subgroup H of $\mathbb{A}_6$ such that $1 \subsetneq H \subsetneq \mathbb{A}_6$. Hence $\mathcal{M} = \mathcal{M}(T, \psi) = \mathcal{C}'_{A(T,c)}$, for some subgroup T of H and some 2-cocycle ψ on T. In addition, $360 = \operatorname{FPdim}\operatorname{Vec}_{\mathbb{A}_6} = \operatorname{FPdim}\mathcal{C}' \operatorname{FPdim}\mathcal{C}'' = |H| \operatorname{FPdim}\mathcal{C}''$.

Consider the associated exact factorization of the dual fusion category $\mathcal{D} = (\operatorname{Vec}_{\mathbb{A}_6})^*_{\mathcal{C}''\boxtimes\operatorname{End}(\mathcal{M})} \cong {}_{A(T,\psi)}(\operatorname{Vec}_{\mathbb{A}_6})_{A(T,\psi)} \cong \mathcal{C}(\mathbb{A}_6, 1, T, \psi)$:

$$\mathcal{D} = \mathcal{C}(\mathbb{A}_6, 1, T, \psi) = \mathcal{C}(H, 1, T, \psi) \bullet \mathcal{C}''. \tag{6.3}$$

Since $\mathbb{A}_6$ admits no exact factorizations, then [**37**, Theorem 3.4] implies that $\mathcal{D}$ is not pointed. Also, since $\operatorname{FPdim}\mathcal{C}'' > 1$, then $\operatorname{FPdim}\mathcal{D}_{pt} > 1$. Moreover $\mathcal{C}''$ is contained in $\mathcal{D}_{pt}$ and $\operatorname{FPdim}\mathcal{C}'' = [\mathbb{A}_6 : H]$. Also $[\mathbb{A}_6 : H]$ divides $[\mathbb{A}_6 : T]$ because T is a subgroup of H.

By [**22**, Theorem 5.2], the group of invertible objects of $\mathcal{D}$ is an extension of $\widehat{T}$ by a certain subgroup K of $N_{\mathbb{A}_6}(T)/T$.

Then $\operatorname{FPdim}\mathcal{D}_{pt}$ divides $|N_{\mathbb{A}_6}(T)/T||\widehat{T}|$, and therefore $[\mathbb{A}_6 : H] | |N_{\mathbb{A}_6}(T)/T||\widehat{T}|$. Thus $[\mathbb{A}_6 : H]$ divides $\gcd([\mathbb{A}_6 : T], |N_{\mathbb{A}_6}(T)/T||\widehat{T}|)$. A direct inspection on the possible subgroups T in Table 1, combined with the fact that $\mathbb{A}_6$ has no subgroups of index 2, 3 or 4, implies that $[\mathbb{A}_6 : H] = 6$ and $H \cong \mathbb{A}_5$.

In addition the subgroup $T \cong \mathbb{Z}_2 \times \mathbb{Z}_2$ is a Klein four group, or $T \cong \mathbb{Z}_3$, or $T \cong \mathbb{A}_4$.

Suppose first that $T \cong \mathbb{Z}_2 \times \mathbb{Z}_2$. Then $\operatorname{FPdim} \mathcal{D}_{pt}$ divides $|N_{\mathbb{A}_6}(T)/T||\widehat{T}| = 24$. In addition $|\widehat{T}|$ divides $\operatorname{FPdim} \mathcal{C}(H, 1, T, \psi)_{pt}$ and therefore 24 divides $\operatorname{FPdim} \mathcal{D}_{pt}$. Hence $\operatorname{FPdim} \mathcal{D}_{pt} = 24 = |N_{\mathbb{A}_6}(T)/T||\widehat{T}|$. This implies that the subgroup K in [**22**, Theorem 5.2] coincides with $N_{\mathbb{A}_6}(T)/T$.

From the definition of K in [**22**], we find that for every $g \in N_{\mathbb{A}_6}(T)$, the class of the 2-cocycle $\psi^g : T \times T \to k^\times$, defined by

$$\psi^g(h_1, h_2) = \psi(h_1, h_2)\, \psi(g^{-1}h_2^{-1}g, g^{-1}h_1^{-1}g), \quad h_1, h_2 \in T,$$

is trivial. Hence the cocycle ψ^g is also trivial for all $g \in N_H(T)$ and therefore $\operatorname{FPdim} \mathcal{C}(H, 1, T, \psi)_{pt} = |N_H(T)/T||\widehat{T}| = 12$, since the group $N_H(T)/T$ is cyclic of order 3 (see Table 2).

Combined with the fact that $\operatorname{FPdim} \mathcal{C}'' = 6$ and $\mathcal{C}''$ is pointed, this implies that $\operatorname{FPdim} \mathcal{D}_{pt} = 72$, and we arrive to a contradiction. Thus we have discarded this possibility for T.

Assume next that $T \cong \mathbb{Z}_3$. In this case the class of ψ is trivial and $N_{\mathbb{A}_6}(T)/T$ is of order 6. Thus the subgroup K in [**22**, Theorem 5.2] coincides with $N_{\mathbb{A}_6}(T)/T$ and $\operatorname{FPdim} \mathcal{D}_{pt} = 18$. On the other hand, $N_H(T)/T$ is of order 2 (Table 2) and therefore $\operatorname{FPdim} \mathcal{C}(H, 1, T, \psi)_{pt} = 6$. The exact factorization (6.2) then implies that $\operatorname{FPdim} \mathcal{D}_{pt} = 36$, which is a contradiction. Then this possibility is also discarded.

Finally suppose that $T \cong \mathbb{A}_4$. In this case $\widehat{T}$ is of order 3 and $N_{\mathbb{A}_6}(T)/T$ is of order 2. Thus $\operatorname{FPdim} \mathcal{D}_{pt}$ divides 6. But since $|\widehat{T}|$ also divides $\operatorname{FPdim} \mathcal{C}(H, 1, T, \psi)_{pt}$, and $\mathcal{C}''$ is pointed then 9 divides $\operatorname{FPdim} \mathcal{D}_{pt}$, which is impossible. This discards this possibility as well. Thus we have shown that such an exact factorization (6.2) cannot exist and therefore $\operatorname{Vec}_{\mathbb{A}_6}$ is a simple fusion category, as claimed. □

COROLLARY 6.8. *The composition factors of a fusion category may be non-unique up to permutation. Moreover, a fusion category may admit composition series with different length.*

PROOF. The group $\mathbb{S}_6$ has an exact factorization $\mathbb{S}_6 = \mathbb{A}_6 \bullet \mathbb{Z}_2$. This induces an exact factorization $\operatorname{Vec}_{\mathbb{S}_6} = \operatorname{Vec}_{\mathbb{A}_6} \bullet \operatorname{Vec}_{\mathbb{Z}_2}$. Therefore, from [**20**, Theorem 4.1], there is an exact sequence with respect to $\operatorname{Vec}_{\mathbb{A}_6}$:

$$\operatorname{Vec}_{\mathbb{A}_6} \longrightarrow \operatorname{Vec}_{\mathbb{S}_6} \longrightarrow \operatorname{Vec}_{\mathbb{Z}_2} \boxtimes \operatorname{End}(\operatorname{Vec}_{\mathbb{A}_6}). \tag{6.4}$$

By Theorem 6.7, $\operatorname{Vec}_{\mathbb{A}_6}$ is simple. Then (6.4) gives a composition series for $\operatorname{Vec}_{\mathbb{S}_6}$ with factors $\operatorname{Vec}_{\mathbb{A}_6}$, $\operatorname{Vec}_{\mathbb{Z}_2}$.

On the other hand, the exact factorizations

$$\begin{aligned}
\mathbb{S}_6 &= \mathbb{S}_5 \bullet \mathbb{Z}_6 \\
&= \mathbb{S}_4 \bullet \mathbb{Z}_5 \bullet \mathbb{Z}_3 \bullet \mathbb{Z}_2 \\
&= \mathbb{S}_3 \bullet \mathbb{Z}_4 \bullet \mathbb{Z}_5 \bullet \mathbb{Z}_3 \bullet \mathbb{Z}_2 \\
&= \mathbb{Z}_3 \bullet \mathbb{Z}_2 \bullet \mathbb{Z}_4 \bullet \mathbb{Z}_5 \bullet \mathbb{Z}_3 \bullet \mathbb{Z}_2 \\
&= \mathbb{Z}_3 \bullet \mathbb{Z}_2 \bullet \mathbb{Z}_4 \bullet \mathbb{Z}_5 \bullet \mathbb{Z}_3 \bullet \mathbb{Z}_2,
\end{aligned}$$

induce exact factorizations of the corresponding pointed fusion categories. An iterated application of [**20**, Theorem 4.1] implies that $\operatorname{Vec}_{\mathbb{S}_6}$ has a composition series with factors $\mathrm{Vec}_{\mathbb{Z}_3}$, $\mathrm{Vec}_{\mathbb{Z}_2}$, $\mathrm{Vec}_{\mathbb{Z}_2}$, $\mathrm{Vec}_{\mathbb{Z}_2}$, $\mathrm{Vec}_{\mathbb{Z}_5}$, $\mathrm{Vec}_{\mathbb{Z}_3}$, $\mathrm{Vec}_{\mathbb{Z}_2}$. Then we see that the factors, and also the length, of these composition series are not unique. □

REMARK 6.9. The statement of Corollary 6.8 remains valid when restricted to the class of non-degenerate braided fusion categories, namely, also in this case the compositions factors may be non-unique up to permutation and composition series may have different lengths.

As an example, consider the Drinfeld center $\mathcal{Z}(\mathrm{Vec}_{\mathbb{S}_6})$. Thus $\mathcal{Z}(\mathrm{Vec}_{\mathbb{S}_6})$ coincides with the representation category of the Drinfeld double $D(k^{\mathbb{S}_6})$ and there is an exact sequence of finite dimensional Hopf algebras $k \longrightarrow k^{\mathbb{S}_6} \longrightarrow D(k^{\mathbb{S}_6}) \longrightarrow k\mathbb{S}_6 \longrightarrow k$. Therefore the non-degenerate braided fusion category $\mathcal{Z}(\mathrm{Vec}_{\mathbb{S}_6})$ fits into an exact sequence $\mathrm{Rep}\,\mathbb{S}_6 \longrightarrow \mathcal{Z}(\mathrm{Vec}_{\mathbb{S}_6}) \longrightarrow \mathrm{Vec}_{\mathbb{S}_6}$ in the sense of [**8**]. In addition we have an exact sequence $\mathrm{Rep}\,\mathbb{Z}_2 \longrightarrow \mathrm{Rep}\,\mathbb{S}_6 \longrightarrow \mathrm{Rep}\,\mathbb{A}_6$, and since $\mathrm{Rep}\,\mathbb{A}_6$ is a simple fusion category, the composition factors of $\mathrm{Rep}\,\mathbb{S}_6$ are $\mathrm{Rep}\,\mathbb{Z}_2$ and $\mathrm{Rep}\,\mathbb{A}_6$.

The proof of Corollary 6.8 gives two composition series of $\mathrm{Vec}_{\mathbb{S}_6}$ that give rise to two composition series of $\mathcal{Z}(\mathrm{Vec}_{\mathbb{S}_6})$ with factors $\mathrm{Rep}\,\mathbb{Z}_2, \mathrm{Rep}\,\mathbb{A}_6, \mathrm{Vec}_{\mathbb{A}_6}, \mathrm{Vec}_{\mathbb{Z}_2}$, on the one hand, and $\mathrm{Rep}\,\mathbb{Z}_2, \mathrm{Rep}\,\mathbb{A}_6, \mathrm{Vec}_{\mathbb{Z}_3}, \mathrm{Vec}_{\mathbb{Z}_2}, \mathrm{Vec}_{\mathbb{Z}_2}, \mathrm{Vec}_{\mathbb{Z}_2}, \mathrm{Vec}_{\mathbb{Z}_5}, \mathrm{Vec}_{\mathbb{Z}_3}, \mathrm{Vec}_{\mathbb{Z}_2}$, on the other hand. Then the series have different length and the factors are not unique also in this case.

6.6. Questions. We think it is interesting to determine classes of finite tensor categories which are closed under extensions. For instance, the class of fusion categories is closed under extensions and so is the class of weakly integral and integral finite tensor categories.

On the other hand, it is known that the class of group-theoretical fusion categories is not closed under extensions: Indeed, let p be an odd prime number and let H be one of the non-group-theoretical semisimple Hopf algebras of dimension $4p^2$ constructed by D. Nikshych in [**48**]. Then H fits into an exact sequence of Hopf algebras $k \longrightarrow k^{\mathbb{Z}_2} \longrightarrow H \longrightarrow A_p \longrightarrow k$, where A_p is a certain abelian extension of $\mathbb{Z}_2$ by $\mathbb{Z}_p \times \mathbb{Z}_p$.

Since every abelian extension is group-theoretical[1] [**38**], then the Hopf algebra A_p is group-theoretical. Then H-mod is a non-group-theoretical fusion category that fits into an exact sequence of group-theoretical fusion categories A_p-mod $\longrightarrow$ H-mod $\longrightarrow$ $\mathrm{Vec}_{\mathbb{Z}_2}$. This exact sequence is in fact an equivariantization exact sequence, by construction of H. See [**48**, Section 5].

A fusion category $\mathcal{C}$ is called weakly group-theoretical if it is Morita equivalent to a nilpotent fusion category [**18**]. The class of weakly group-theoretical fusion categories is known to be closed under a number of operations, like taking a fusion subcategory or dominant image, taking Deligne tensor product and Drinfeld center. It is not known if it closed under extensions:

QUESTION 6.10. *([**20**, Question 4.9].)* Is the class of weakly group-theoretical fusion categories closed under extensions?

It is known that the class of weakly group-theoretical fusion categories is closed under equivariantizations and group graded extensions [**18**, Proposition 4.1]. Furthermore, it also closed under matched pair crossed extensions [**47**, Corollary 4.6].

[1] A Hopf algebra H is called group-theoretical if the category H-mod (or equivalently, the category comod-H) is group-theoretical.

Automorphism class representative	**Isomorphism class of** T	$\lvert T\rvert$	$\lvert\widehat{T}\rvert$	$[N_G(T):T]$
$\{e\}$	Trivial	1	1	360
$\langle(12)(34)\rangle$	$\mathbb{Z}_2$	2	2	4
$\langle(12)(34),(13)(24)\rangle$	$\mathbb{Z}_2\times\mathbb{Z}_2$	4	4	6
$\langle(12)(34),(12)(56)\rangle$	$\mathbb{Z}_2\times\mathbb{Z}_2$	4	4	6
$\langle(1234)(56)\rangle$	$\mathbb{Z}_4$	4	4	2
$\langle(1234)(56),(13)(56)\rangle$	D_4	8	4	1
$\langle(123)\rangle$	$\mathbb{Z}_3$	3	3	6
$\langle(123)(456)\rangle$	$\mathbb{Z}_3$	3	3	6
$\langle(123),(456)\rangle$	$\mathbb{Z}_3\times\mathbb{Z}_3$	9	9	4
$\langle(123)(456),(12)(45)\rangle$	$\mathbb{S}_3$	6	2	1
$\langle(123),(12)(45)\rangle$	$\mathbb{S}_3$	6	2	1
$\langle(12)(34),(123)\rangle$	$\mathbb{A}_4$	12	3	2
$\langle(123)(456),(14)(25),(14)(36)\rangle$	$\mathbb{A}_4$	12	3	2
$\langle(1234)(56),(12)(56)\rangle$	$\mathbb{S}_4$	24	2	1
$\langle(34)(56),(12)(56),(135)(246),(35)(46)\rangle$	$\mathbb{S}_4$	24	2	1
$\langle(123),(456),(12)(45)\rangle$	$(\mathbb{Z}_3\times\mathbb{Z}_3)\rtimes\mathbb{Z}_2$	18	2	2
$\langle(123),(456),(23)(56),(14)(2536)\rangle$	$(\mathbb{Z}_3\times\mathbb{Z}_3)\rtimes\mathbb{Z}_4$	36	4	1
$\langle(12345)\rangle$	$\mathbb{Z}_5$	5	5	2
$\langle(12345),(25)(34)\rangle$	D_5	10	2	1
$\langle(12345),(123)\rangle$	$\mathbb{A}_5$	60	1	1
$\langle(12345),(14)(56)\rangle$	$\mathbb{A}_5$	60	1	1
Whole group	$\mathbb{A}_6$	360	1	1

TABLE 1. Subgroups T of the alternating group $\mathbb{A}_6$

Automorphism class representative	**Isomorphism class of** T	$[N_G(T):T]$
$\{e\}$	Trivial	60
$\langle(12)\rangle$	$\mathbb{Z}_2$	2
$\langle(12)(34)\rangle$	$\mathbb{Z}_2\times\mathbb{Z}_2$	3
$\langle(123)\rangle$	$\mathbb{Z}_3$	2
$\langle(123),(12)(45)\rangle$	$\mathbb{S}_3$	1
$\langle(12)(34),(123)\rangle$	$\mathbb{A}_4$	1
$\langle(12345)\rangle$	$\mathbb{Z}_5$	2
$\langle(12345),(25)(34)\rangle$	D_5	1
Whole group	$\mathbb{A}_5$	1

TABLE 2. Subgroups T of the alternating group $\mathbb{A}_5$

Let G be a finite group. In [**47**] we showed that if $\mathcal{C}$ is a fusion category fitting into an abelian exact sequence $\operatorname{Rep} G \longrightarrow \mathcal{C} \longrightarrow \mathcal{D}$ or into an exact sequence

$\mathrm{Vec}_G \longrightarrow \mathcal{C} \longrightarrow \mathcal{D}$, then $\mathcal{C}$ is weakly group-theoretical if and only if $\mathcal{D}$ is weakly group-theoretical. As a consequence, every semisolvable semisimple Hopf algebra, as introduced in [**36**], is weakly group-theoretical.

Recall that a fusion category is said to satisfy the *Frobenius property* if the ratio $\mathrm{FPdim}\,\mathcal{C}/\,\mathrm{FPdim}\,X$ is an algebraic integer, for every simple object X of $\mathcal{C}$. It is known that every pre-modular fusion category satisfies the Frobenius property [**18**].

QUESTION 6.11. Is the class of fusion categories with the Frobenius property closed under extensions?

It follows from [**18**, Theorem 1.5] that the class of fusion categories with the Frobenius property is closed under equivariantizations and group graded extensions, and every weakly group-theoretical fusion category satisfies the Frobenius property. The answer to Question 6.11 is not known in general even in the context of Hopf algebra extensions.

Another interesting class of tensor categories is that of *Frobenius tensor categories*: these are tensor categories in which every simple object has an injective hull (equivalently, a projective cover) [**4**, Subsection 2.3]. For instance, finite tensor categories and semisimple tensor categories are Frobenius categories.

Examples of Frobenius categories are provided by the categories of finite dimensional comodules over co-Frobenius Hopf algebras: that is, Hopf algebras H endowed with a nonzero integral $H \to k$.

It is known that every Hopf algebra H fitting into a strictly exact sequence of Hopf algebras $k \longrightarrow H' \longrightarrow H \longrightarrow H'' \longrightarrow k$, such that H' and H'' are co-Frobenius Hopf, is co-Frobenius [**3**, Theorem 2.10]. This result allows to construct examples of this kind of Hopf algebras from smaller examples. We do not known the answer to the corresponding question for tensor categories:

QUESTION 6.12. Is the class of Frobenius tensor categories closed under extensions?

Regarding the notion of simplicity of a finite tensor category, the following is a natural question, about which very little is known:

QUESTION 6.13. Is it possible to classify simple (finite) tensor categories?

Finally, motivated by the examples of Subsection 6.5, we ask:

QUESTION 6.14. Can the definition of a composition series be reformulated in order that the Jordan-Hölder theorem holds for finite tensor categories?

An analogue of the Jordan-Hölder theorem was proved in [**44**] for weakly group-theoretical fusion categories. The definition of a composition series for this kind of category is given in terms of group equivariantizations and group graded extensions and the composition factors, which are Morita invariants, are simple finite groups.

References

[1] Nicolás Andruskiewitsch, *Notes on extensions of Hopf algebras*, Canad. J. Math. **48** (1996), no. 1, 3–42, DOI 10.4153/CJM-1996-001-8. MR1382474

[2] Nicolás Andruskiewitsch, *About finite dimensional Hopf algebras*, Quantum symmetries in theoretical physics and mathematics (Bariloche, 2000), Contemp. Math., vol. 294, Amer. Math. Soc., Providence, RI, 2002, pp. 1–57, DOI 10.1090/conm/294/04969. MR1907185

[3] Nicolás Andruskiewitsch and Juan Cuadra, *On the structure of (co-Frobenius) Hopf algebras*, J. Noncommut. Geom. **7** (2013), no. 1, 83–104, DOI 10.4171/JNCG/109. MR3032811

[4] Nicolás Andruskiewitsch, Juan Cuadra, and Pavel Etingof, *On two finiteness conditions for Hopf algebras with nonzero integral*, Ann. Sc. Norm. Super. Pisa Cl. Sci. (5) **14** (2015), no. 2, 401–440. MR3410615

[5] N. Andruskiewitsch and J. Devoto, *Extensions of Hopf algebras*, Algebra i Analiz **7** (1995), no. 1, 22–61; English transl., St. Petersburg Math. J. **7** (1996), no. 1, 17–52. MR1334152

[6] Nicolás Andruskiewitsch and Monique Müller, *Examples of extensions of Hopf algebras* (English, with English and Spanish summaries), Rev. Colombiana Mat. **49** (2015), no. 1, 193–211, DOI 10.15446/recolma.v49n1.54180. MR3367440

[7] Alain Bruguières, Steve Lack, and Alexis Virelizier, *Hopf monads on monoidal categories*, Adv. Math. **227** (2011), no. 2, 745–800, DOI 10.1016/j.aim.2011.02.008. MR2793022

[8] Alain Bruguières and Sonia Natale, *Exact sequences of tensor categories*, Int. Math. Res. Not. IMRN **24** (2011), 5644–5705, DOI 10.1093/imrn/rnq294. MR2863377

[9] Alain Bruguières and Sonia Natale, *Central exact sequences of tensor categories, equivariantization and applications*, J. Math. Soc. Japan **66** (2014), no. 1, 257–287, DOI 10.2969/jmsj/06610257. MR3161401

[10] Alain Bruguières and Alexis Virelizier, *Hopf monads*, Adv. Math. **215** (2007), no. 2, 679–733, DOI 10.1016/j.aim.2007.04.011. MR2355605

[11] P. Deligne, *Catégories tannakiennes* (French), The Grothendieck Festschrift, Vol. II, Progr. Math., vol. 87, Birkhäuser Boston, Boston, MA, 1990, pp. 111–195. MR1106898

[12] Vladimir Drinfeld, Shlomo Gelaki, Dmitri Nikshych, and Victor Ostrik, *On braided fusion categories. I*, Selecta Math. (N.S.) **16** (2010), no. 1, 1–119, DOI 10.1007/s00029-010-0017-z. MR2609644

[13] Pavel Etingof and Shlomo Gelaki, *The representation theory of cotriangular semisimple Hopf algebras*, Internat. Math. Res. Notices **7** (1999), 387–394, DOI 10.1155/S1073792899000197. MR1683304

[14] Pavel Etingof and Shlomo Gelaki, *Exact sequences of tensor categories with respect to a module category*, Adv. Math. **308** (2017), 1187–1208, DOI 10.1016/j.aim.2016.12.021. MR3600085

[15] Pavel Etingof, Shlomo Gelaki, and Viktor Ostrik, *Classification of fusion categories of dimension pq*, Int. Math. Res. Not. **57** (2004), 3041–3056, DOI 10.1155/S1073792804131206. MR2098028

[16] Pavel Etingof, Shlomo Gelaki, Dmitri Nikshych, and Victor Ostrik, *Tensor categories*, Mathematical Surveys and Monographs, vol. 205, American Mathematical Society, Providence, RI, 2015, DOI 10.1090/surv/205. MR3242743

[17] Pavel Etingof, Dmitri Nikshych, and Viktor Ostrik, *On fusion categories*, Ann. of Math. (2) **162** (2005), no. 2, 581–642, DOI 10.4007/annals.2005.162.581. MR2183279

[18] Pavel Etingof, Dmitri Nikshych, and Victor Ostrik, *Weakly group-theoretical and solvable fusion categories*, Adv. Math. **226** (2011), no. 1, 176–205, DOI 10.1016/j.aim.2010.06.009. MR2735754

[19] César Galindo and Sonia Natale, *Simple Hopf algebras and deformations of finite groups*, Math. Res. Lett. **14** (2007), no. 6, 943–954, DOI 10.4310/MRL.2007.v14.n6.a4. MR2357466

[20] Shlomo Gelaki, *Exact factorizations and extensions of fusion categories*, J. Algebra **480** (2017), 505–518, DOI 10.1016/j.jalgebra.2017.02.034. MR3633318

[21] Shlomo Gelaki and Dmitri Nikshych, *Nilpotent fusion categories*, Adv. Math. **217** (2008), no. 3, 1053–1071, DOI 10.1016/j.aim.2007.08.001. MR2383894

[22] Shlomo Gelaki and Deepak Naidu, *Some properties of group-theoretical categories*, J. Algebra **322** (2009), no. 8, 2631–2641, DOI 10.1016/j.jalgebra.2009.05.047. MR2559854

[23] Irmengard Hofstetter, *Extensions of Hopf algebras and their cohomological description*, J. Algebra **164** (1994), no. 1, 264–298, DOI 10.1006/jabr.1994.1063. MR1268336

[24] David Jordan and Eric Larson, *On the classification of certain fusion categories*, J. Noncommut. Geom. **3** (2009), no. 3, 481–499, DOI 10.4171/JNCG/44. MR2511638

[25] G. I. Kac, *Extensions of groups to ring groups*, Math. USSR Sbornik **5** (1968), 451–474.

[26] Gregory Karpilovsky, *Projective representations of finite groups*, Monographs and Textbooks in Pure and Applied Mathematics, vol. 94, Marcel Dekker, Inc., New York, 1985. MR788161

[27] Toshiharu Kobayashi and Akira Masuoka, *A result extended from groups to Hopf algebras*, Tsukuba J. Math. **21** (1997), no. 1, 55–58, DOI 10.21099/tkbjm/1496163161. MR1467221

[28] Shahn Majid, *Physics for algebraists: noncommutative and noncocommutative Hopf algebras by a bicrossproduct construction*, J. Algebra **130** (1990), no. 1, 17–64, DOI 10.1016/0021-8693(90)90099-A. MR1045735

[29] Akira Masuoka, *On Hopf algebras with cocommutative coradicals*, J. Algebra **144** (1991), no. 2, 451–466, DOI 10.1016/0021-8693(91)90116-P. MR1140616

[30] Akira Masuoka, *The p^n theorem for semisimple Hopf algebras*, Proc. Amer. Math. Soc. **124** (1996), no. 3, 735–737, DOI 10.1090/S0002-9939-96-03147-4. MR1301036

[31] Akira Masuoka, *Extensions and cohomology of Hopf algebras, Lie bialgebras* (Japanese, with English summary), Proceedings of the 31st Symposium on Ring Theory and Representation Theory and Japan-Korea Ring Theory and Representation Theory Seminar (Osaka, 1998), Shinshu Univ., Matsumoto, 1999, pp. 131–149. MR1812913

[32] Akira Masuoka, *Hopf algebra extensions and cohomology*, New directions in Hopf algebras, Math. Sci. Res. Inst. Publ., vol. 43, Cambridge Univ. Press, Cambridge, 2002, pp. 167–209. MR1913439

[33] G. A. Miller, *Groups which are the products of two permutable proper subgroups*, Proc. Acad. USA **21**, 469–472 (1935).

[34] Martín Mombelli and Sonia Natale, *Module categories over equivariantized tensor categories*, Mosc. Math. J. **17** (2017), no. 1, 97–128, DOI 10.17323/1609-4514-2017-17-1-97-128. MR3634523

[35] Susan Montgomery, *Classifying finite-dimensional semisimple Hopf algebras*, Trends in the representation theory of finite-dimensional algebras (Seattle, WA, 1997), Contemp. Math., vol. 229, Amer. Math. Soc., Providence, RI, 1998, pp. 265–279, DOI 10.1090/conm/229/03323. MR1676225

[36] Susan Montgomery and S. J. Witherspoon, *Irreducible representations of crossed products*, J. Pure Appl. Algebra **129** (1998), no. 3, 315–326, DOI 10.1016/S0022-4049(97)00077-7. MR1631261

[37] Deepak Naidu, *Categorical Morita equivalence for group-theoretical categories*, Comm. Algebra **35** (2007), no. 11, 3544–3565, DOI 10.1080/00927870701511996. MR2362670

[38] Sonia Natale, *On group theoretical Hopf algebras and exact factorizations of finite groups*, J. Algebra **270** (2003), no. 1, 199–211, DOI 10.1016/S0021-8693(03)00464-2. MR2016657

[39] Sonia Natale, *Frobenius-Schur indicators for a class of fusion categories*, Pacific J. Math. **221** (2005), no. 2, 353–377, DOI 10.2140/pjm.2005.221.353. MR2196640

[40] Sonia Natale, *Semisolvability of semisimple Hopf algebras of low dimension*, Mem. Amer. Math. Soc. **186** (2007), no. 874, viii+123, DOI 10.1090/memo/0874. MR2294999

[41] Sonia Natale, *Semisimple Hopf algebras of dimension 60*, J. Algebra **324** (2010), no. 11, 3017–3034, DOI 10.1016/j.jalgebra.2010.06.007. MR2732985

[42] Sonia Natale, *Jordan-Hölder theorem for finite dimensional Hopf algebras*, Proc. Amer. Math. Soc. **143** (2015), no. 12, 5195–5211, DOI 10.1090/proc/12702. MR3411137

[43] Sonia Natale, *Crossed actions of matched pairs of groups on tensor categories*, Tohoku Math. J. (2) **68** (2016), no. 3, 377–405, DOI 10.2748/tmj/1474652265. MR3550925

[44] Sonia Natale, *A Jordan-Hölder theorem for weakly group-theoretical fusion categories*, Math. Z. **283** (2016), no. 1-2, 367–379, DOI 10.1007/s00209-015-1601-x. MR3489071

[45] Sonia Natale, *The core of a weakly group-theoretical braided fusion category*, Internat. J. Math. **29** (2018), no. 2, 1850012, 23, DOI 10.1142/S0129167X1850012X. MR3770935

[46] Sonia Natale, *On the classification of fusion categories*, Proceedings of the International Congress of Mathematicians—Rio de Janeiro 2018. Vol. II. Invited lectures, World Sci. Publ., Hackensack, NJ, 2018, pp. 173–200. MR3966762

[47] S. Natale, *Extensions of tensor categories by finite group fusion categories*, Math. Proc. Cambridge Phyl. Soc. to appear. Preprint `arXiv:1808.09581`.

[48] Dmitri Nikshych, *K_0-rings and twisting of finite-dimensional semisimple Hopf algebras*, Comm. Algebra **26** (1998), no. 1, 321–342, DOI 10.1080/00927879808826132. MR1600702

[49] Victor Ostrik, *Module categories, weak Hopf algebras and modular invariants*, Transform. Groups **8** (2003), no. 2, 177–206, DOI 10.1007/s00031-003-0515-6. MR1976459

[50] Brian Parshall and Jian Pan Wang, *Quantum linear groups*, Mem. Amer. Math. Soc. **89** (1991), no. 439, vi+157, DOI 10.1090/memo/0439. MR1048073

[51] Peter Schauenburg, *Hopf bimodules, coquasibialgebras, and an exact sequence of Kac*, Adv. Math. **165** (2002), no. 2, 194–263, DOI 10.1006/aima.2001.2016. MR1887584

[52] Peter Schauenburg, *The monoidal center construction and bimodules*, J. Pure Appl. Algebra **158** (2001), no. 2-3, 325–346, DOI 10.1016/S0022-4049(00)00040-2. MR1822847

[53] Hans-Jürgen Schneider, *Some remarks on exact sequences of quantum groups*, Comm. Algebra **21** (1993), no. 9, 3337–3357, DOI 10.1080/00927879308824733. MR1228767

[54] William M. Singer, *Extension theory for connected Hopf algebras*, J. Algebra **21** (1972), 1–16, DOI 10.1016/0021-8693(72)90031-2. MR320056

[55] Mitsuhiro Takeuchi, *A correspondence between Hopf ideals and sub-Hopf algebras*, Manuscripta Math. **7** (1972), 251–270, DOI 10.1007/BF01579722. MR321963

[56] Mitsuhiro Takeuchi, *Matched pairs of groups and bismash products of Hopf algebras*, Comm. Algebra **9** (1981), no. 8, 841–882, DOI 10.1080/00927878108822621. MR611561

[57] Mitsuhiro Takeuchi, *Quotient spaces for Hopf algebras*, Comm. Algebra **22** (1994), no. 7, 2503–2523, DOI 10.1080/00927879408824973. MR1271619

Current address: Facultad de Matemática, Astronomía, Física y Computación. Universidad Nacional de Córdoba. CIEM – CONICET. Ciudad Universitaria. (5000) Córdoba, Argentina

URL: `http://www.famaf.unc.edu.ar/~natale`

Email address: `natale@famaf.unc.edu.ar`

Contemporary Mathematics
Volume **771**, 2021
https://doi.org/10.1090/conm/771/15515

Algebraic supergroups and their representations

Taiki Shibata

1. Introduction

An algebraic supergroup is a group-valued functor on the category of commutative superalgebras represented by a finitely generated commutative Hopf superalgebra. Here, the word "super" is a synonym of "graded by the group $\mathbb{Z}_2 = \{\bar{0}, \bar{1}\}$ of order two". It has been known that representations of algebraic supergroups can be applied in non-super modular representation theory. For example, over a filed of positive characteristic, Brundan and Kleshchev [**3**] showed that representations of queer supergroups $\mathbf{Q}(n)$ (Examples 2.1 (4)) have a close relationship to modular representations of spin symmetric groups; Brundan and Kujawa [**4**] re-proved Mullineux theorem using general linear supergroups $\mathbf{GL}(m|n)$ (Examples 2.1 (1)); Shu and Wang [**27**] classified irreducible representations of ortho-symplectic supergroups $\mathbf{SpO}(2k|n)$ (Examples 2.1 (5)) whose parameter set is described by combinatorics terms, which is related to Mullineux bijection.

In this paper, we give a framework of characteristic-free study of *quasireductive supergroups*. Those supergroups are the notation due to Serganova [**25**], which naturally generalizes split reductive groups to the super context; they include $\mathbf{GL}(m|n)$, $\mathbf{SpO}(2k|n)$, $\mathbf{Q}(n)$, and Chevalley supergroups (introduced by R. Fioresi and F. Gavarini [**8**]). In particular, we construct all irreducible representations of quasireductive supergroups using Hopf-algebraic methods.

The paper is organized as follows. In Section 2, we review some basic definitions and results for algebraic supergroups. Structure and representations of quasireductive supergroups are discussed in Section 3. In the final Section 4, we study a quasireductive supergroup having a maximal parabolic super-subgroup, and we give some results (Corollary 4.4) on induced supermodules $H^0(\lambda)$ (determining the parameter set/Kempf vanishing theorem/Euler characteristic) as a generalization of Zubkov's result [**32**]. For $\mathbf{GL}(m|n)$, we re-prove Marko's result [**15**] by our Hopf-algebraic technique, and determine when $H^0(\lambda)$ is irreducible (Theorem 4.6).

2. Algebraic supergroups

In the following, we work over a field $\Bbbk$ of characteristic different from 2. A $\mathbb{Z}_2$-graded vector space $V = V_{\bar{0}} \oplus V_{\bar{1}}$ is called a *superspace*. The category of superspaces

2020 *Mathematics Subject Classification.* Primary 14M30, 16T05.
The work was supported by Japan Society of the Promotion of Science (JSPS) KAKENHI Grant Number 19K14517.

naturally forms a tensor category. We make the category into a symmetric tensor category via the *supersymmetry* $V \otimes W \to W \otimes V$; $v \otimes w \mapsto (-1)^{|v||w|} w \otimes v$ ($v \in V_{\bar{0}} \cup V_{\bar{1}}$, $w \in W_{\bar{0}} \cup W_{\bar{1}}$), where $|v|$ is the *parity* of v. A *superalgebra* (resp. *supercoalgebra/Hopf superalgebra/Lie superalgebra*) is defined to be an algebra (resp. coalgebra/Hopf algebra/Lie algebra) object in this symmetric tensor category.

2.1. Algebraic supergroups. An *affine supergroup scheme* (*supergroup*, for short) over $\Bbbk$ is a representable functor $\boldsymbol{G}$ from the category of commutative superalgebras over $\Bbbk$ to the category of groups. We denote the representing object of $\boldsymbol{G}$ by $\mathcal{O}(\boldsymbol{G})$ which forms a commutative Hopf superalgebra over $\Bbbk$ by Yoneda's lemma. A supergroup $\boldsymbol{G}$ is said to be *algebraic* if $\mathcal{O}(\boldsymbol{G})$ is finitely generated as a superalgebra. A *(closed) super-subgroup* of a supergroup $\boldsymbol{G}$ is a supergroup (scheme) which is represented by a quotient Hopf superalgebra of $\mathcal{O}(\boldsymbol{G})$. For a supergroup $\boldsymbol{G}$, we define its *even part* $\boldsymbol{G}_{\mathsf{ev}}$ as the restricted functor of $\boldsymbol{G}$ from the category of commutative algebras to the category of groups. This $\boldsymbol{G}_{\mathsf{ev}}$ is an ordinary affine group scheme over $\Bbbk$ represented by the quotient (ordinary) Hopf algebra $\mathcal{O}(\boldsymbol{G}_{\mathsf{ev}}) = \mathcal{O}(\boldsymbol{G})/(\mathcal{O}(\boldsymbol{G})_{\bar{1}})$, where $(\mathcal{O}(\boldsymbol{G})_{\bar{1}})$ is the super-ideal of $\mathcal{O}(\boldsymbol{G})$ generated by the odd part $\mathcal{O}(\boldsymbol{G})_{\bar{1}}$ of $\mathcal{O}(\boldsymbol{G})$. The $\Bbbk$-valued points are $\boldsymbol{G}(\Bbbk) = \boldsymbol{G}_{\mathsf{ev}}(\Bbbk)$. If $\boldsymbol{G}$ is algebraic, then so is $\boldsymbol{G}_{\mathsf{ev}}$.

EXAMPLES 2.1. In the following, $R = R_{\bar{0}} \oplus R_{\bar{1}}$ denotes a commutative superalgebra. We list some examples of algebraic supergroups:

(1) *General linear supergroups* $\mathbf{GL}(m|n)$.

$$\mathbf{GL}(m|n)(R) := \left\{ g = \left(\begin{array}{c|c} A & B \\ \hline C & D \end{array} \right) \;\middle|\; \begin{array}{l} A \in \mathrm{GL}_m(R_{\bar{0}}),\ B \in \mathrm{Mat}_{m,n}(R_{\bar{1}}), \\ C \in \mathrm{Mat}_{n,m}(R_{\bar{1}}),\ D \in \mathrm{GL}_n(R_{\bar{0}}) \end{array} \right\},$$

where GL_m denotes the general linear group of size m and $\mathrm{Mat}_{m,n}(R_{\bar{1}})$ denotes the set of all $m \times n$ matrices whose entries are in $R_{\bar{1}}$. The even part is given as $\mathbf{GL}(m|n)_{\mathsf{ev}} \cong \mathrm{GL}_m \times \mathrm{GL}_n$. This supergroup is well-studied, see [**4, 16, 32**] for example.

(2) *Special linear supergroups* $\mathbf{SL}(m|n)$.

$$\mathbf{SL}(m|n)(R) := \{ g \in \mathbf{GL}(m|n)(R) \mid \mathsf{Ber}(g) = 1 \},$$

where $\mathsf{Ber}(g) := \mathsf{det}(A - BD^{-1}C)\mathsf{det}(D^{-1})$ is the so-called *Berezinian determinant* of g.

(3) *Periplectic supergroups* $\mathbf{P}(n)$, see [**26**, Example 5.11]. We denote the *super-transpose* of the block matrix g by ${}^{\mathsf{st}}g := \left(\begin{array}{c|c} {}^{\mathsf{t}}A & {}^{\mathsf{t}}C \\ \hline -{}^{\mathsf{t}}B & {}^{\mathsf{t}}D \end{array} \right)$, where ${}^{\mathsf{t}}A$ is the ordinary transpose of A.

$$\mathbf{P}(n)(R) := \{ g \in \mathbf{GL}(n|n)(R) \mid {}^{\mathsf{st}}g\, J_n\, g = J_n \},$$

where $J_n := \left(\begin{array}{c|c} 0 & I_n \\ \hline I_n & 0 \end{array} \right)$ and I_n is the identity matrix of size n. The even part is given as $\mathbf{P}(n)_{\mathsf{ev}} \cong \mathrm{GL}_n$.

(4) *Queer supergroups* $\mathbf{Q}(n)$.

$$\mathbf{Q}(n)(R) := \{ g \in \mathbf{GL}(n|n)(R) \mid {}^{\mathsf{st}}J_n\, g = g\, J_n \}.$$

The even part is given as $\mathbf{Q}(n)_{\mathsf{ev}} \cong \mathrm{GL}_n$. In [**2, 3**], modular representation theory of this supergroup is well-studied.

(5) *Ortho-symplectic supergroups* $\mathbf{SpO}(2k|n)$.

$\mathbf{SpO}(2k|n)(R) := \{g \in \mathbf{SL}(2k|n)(R) \mid {}^{\mathsf{st}}g\, J_{2k|n}\, g = J_{2k|n}\}$.

Here, for $n = 2\ell + 1$ or 2ℓ with $\ell \in \mathbb{N}$, $J_{2k|n} \in \mathbf{GL}(2k|n)(\Bbbk)$ is given by

$$J_{2k|2\ell+1} := \left(\begin{array}{c|cc} {}^{\mathsf{st}}J_k & O & 0 \\ \hline O & J_\ell & 0 \\ 0 & 0 & 1 \end{array}\right) \quad \text{or} \quad J_{2k|2\ell} := \left(\begin{array}{c|c} {}^{\mathsf{st}}J_k & O \\ \hline O & J_\ell \end{array}\right).$$

The even part is given as $\mathbf{SpO}(2k|n)_{\mathsf{ev}} \cong \mathsf{Sp}_{2k} \times \mathsf{SO}_n$, where Sp_{2k} is the symplectic group of degree $2k$ and SO_n is the special orthogonal group od degree n. Shu and Wang [**27**] classified irreducible supermodules of $\mathbf{SpO}(2k|n)$.

(6) As a super-analogue of Chevalley's construction [**7**], Fioresi and Gavarini [**8**] constructed an algebraic supergroup $\boldsymbol{G}$ (over an arbitrary field) from a finite-dimensional simple Lie superalgebra $\mathfrak{g}$ of classical type (Kac's classification [**12**]) defined over the complex number field and a faithful representation of $\mathfrak{g}$, called a *Chevalley supergroups of classical type*. The even part $\boldsymbol{G}_{\mathsf{ev}}$ is the so-called "Chevalley-Demazure" group (scheme) associated with $\mathfrak{g}_{\bar{0}}$.

2.2. Lie/Hyper superalgebras. Let $\boldsymbol{G}$ be an algebraic supergroup with $A := \mathcal{O}(\boldsymbol{G})$. As a super-subspace of the $\Bbbk$-linear dual superspace A^* of A, we let $\mathsf{Lie}(\boldsymbol{G}) := (\mathfrak{m}_A/\mathfrak{m}_A^2)^*$, where $\mathfrak{m}_A = \mathsf{Ker}(\varepsilon_A : A \to \Bbbk)$ is the augmentation super-ideal of A. One easily sees that this forms a Lie superalgebra over $\Bbbk$ ([**18**, Proposition 4.2] for example), which we call the *Lie superalgebra of* $\boldsymbol{G}$. It is easy to see that $\mathsf{Lie}(\boldsymbol{G})_{\bar{0}} = \mathsf{Lie}(\boldsymbol{G}_{\mathsf{ev}})$.

The following naturally forms a super-subalgebra of A^*, called the *hyper superalgebra of* $\boldsymbol{G}$.

$$\mathsf{hy}(\boldsymbol{G}) := \bigcup_{n \geq 1} (A/\mathfrak{m}_A^n)^*.$$

This naturally has a structure of a cocommutative Hopf superalgebra such that the restriction $\langle\ ,\ \rangle : \mathsf{hy}(\boldsymbol{G}) \times A \to \Bbbk$ of the canonical pairing $A^* \times A \to \Bbbk$ makes into a Hopf pairing, see [**18**, Lemma 5.1]. If $\Bbbk$ is characteristic zero, then $\mathsf{hy}(\boldsymbol{G})$ is irreducible supercoalgebra, and hence it coincides with the universal enveloping superalgebra $\mathcal{U}(\mathsf{Lie}(\boldsymbol{G}))$ of $\mathsf{Lie}(\boldsymbol{G})$.

REMARK 2.2. In the non-super situation, it is well-known that the Lie algebra $\mathsf{Lie}(G)$ of an algebraic group G strongly reflects many properties of G over a field of characteristic zero, see [**9**] for example. Takeuchi [**28**–**30**] showed that over a filed of any characteristic, the hyperalgebra $\mathsf{hy}(G)$ takes the place of $\mathsf{Lie}(G)$.

In this paper, a *left* $\boldsymbol{G}$*-supermodule* is defined as a right $\mathcal{O}(\boldsymbol{G})$-supercomodule. For a general definition, see [**26**, §2.2]. For a left $\boldsymbol{G}$-supermodule V, we can regard V as a left $\mathsf{hy}(\boldsymbol{G})$-supermodule by $xv := \sum_v (-1)^{|v_{(0)}||x|} v_{(0)} \langle x, v_{(1)} \rangle$, where $x \in \mathsf{hy}(\boldsymbol{G})$, $v \in V$, and $v \mapsto \sum_v v_{(0)} \otimes v_{(1)}$ is the right $\mathcal{O}(\boldsymbol{G})$-supercomodule structure of V. If V is finite-dimensional, then the induced left $\mathsf{hy}(\boldsymbol{G})$-supermodule structure on V^* will be denoted by $\rightharpoonup$.

Let $\bigwedge(\mathsf{Lie}(\boldsymbol{G})_{\bar{1}})$ be the exterior superalgebra over $\mathsf{Lie}(\boldsymbol{G})_{\bar{1}}$. The tensor product $(\bigwedge(\mathsf{Lie}(\boldsymbol{G})_{\bar{1}})) \otimes \mathsf{hy}(\boldsymbol{G}_{\mathsf{ev}})$ naturally forms a right $\mathsf{hy}(\boldsymbol{G}_{\mathsf{ev}})$-supermodule. Simultaneously, this forms a cocommutative supercoalgebra in a natural way. Using the "diamond lemma", Masuoka [**17**, Proposition 12] showed the following result.

PROPOSITION 2.3. *There is a unit preserving isomorphism* $(\bigwedge(\mathsf{Lie}(\boldsymbol{G})_{\bar{1}}))\otimes \mathsf{hy}(\boldsymbol{G}_{\mathsf{ev}})\cong \mathsf{hy}(\boldsymbol{G})$ *of right* $\mathsf{hy}(\boldsymbol{G}_{\mathsf{ev}})$*-supermodule-supercoalgebras.*

2.3. Harish-Chandra pairs. Let $\boldsymbol{G}$ be an algebraic supergroup with $A := \mathcal{O}(\boldsymbol{G})$. We set $\overline{A} := \mathcal{O}(\boldsymbol{G}_{\mathsf{ev}}) = A/(A_{\bar{1}})$ and the image of an element $a\in A$ via the canonical projection $A\to\overline{A}$ is denoted by $\overline{a}$. To describe the comultiplication of A, we use the Heyneman-Sweedler notation $A\to A\otimes A;\ a\mapsto \sum_a a_{(1)}\otimes a_{(2)}$, as usual. Set $W^{\mathcal{O}(\boldsymbol{G})} := (\mathfrak{m}_A/\mathfrak{m}_A^2)_{\bar{1}}$. It is known ([**17**, Lemma 24]) that the right coadjoint action of $\overline{A}$ on A

$$A\longrightarrow A\otimes\overline{A};\quad a\longmapsto \sum_a a_{(2)}\otimes \mathsf{S}_A(\overline{a_{(1)}})\,\overline{a_{(3)}}$$

induces a right $\overline{A}$-comodule structure (= left $\boldsymbol{G}_{\mathsf{ev}}$-module structure) on $W^{\mathcal{O}(\boldsymbol{G})}$, where S_A is the antipode of A.

Masuoka [**17**, Proposition 22] proved the following result which is a dual version of Proposition 2.3.

PROPOSITION 2.4. *There is a counit preserving isomorphism* $\mathcal{O}(\boldsymbol{G})\cong\mathcal{O}(\boldsymbol{G}_{\mathsf{ev}})\otimes \bigwedge W^{\mathcal{O}(\boldsymbol{G})}$ *of left* $\mathcal{O}(\boldsymbol{G}_{\mathsf{ev}})$*-supercomodule-superalgebras.*

One sees that the restriction of the Lie superbracket of $\mathsf{Lie}(\boldsymbol{G})$ on $\mathsf{Lie}(\boldsymbol{G})_{\bar{1}} = (W^{\mathcal{O}(\boldsymbol{G})})^*$ is left $\boldsymbol{G}_{\mathsf{ev}}$-equivariant. In this way, for a given algebraic supergroup $\boldsymbol{G}$, we get three data: the ordinary algebraic group $\boldsymbol{G}_{\mathsf{ev}}$, the left $\boldsymbol{G}_{\mathsf{ev}}$-module $W^{\mathcal{O}(\boldsymbol{G})}$, and the $\boldsymbol{G}_{\mathsf{ev}}$-equivariant bracket $[\ ,\ ] : (W^{\mathcal{O}(\boldsymbol{G})})^*\times(W^{\mathcal{O}(\boldsymbol{G})})^*\to\mathsf{Lie}(\boldsymbol{G}_{\mathsf{ev}})$. Then $(\boldsymbol{G}_{\mathsf{ev}}, W^{\mathcal{O}(\boldsymbol{G})})$ forms a *Harish-Chandra pair*, that is, a pair (G,W) of an ordinary algebraic group G and a left G-module W with a left G-equivariant map $[\ ,\ ] : W^*\times W^*\to\mathsf{Lie}(G)$ satisfying a certain compatibility condition. For the precise definition, see [**17**, Definition 7] (see also [**19**, Definition 4.4]). Let ASG (resp. HCP) denote the category of algebraic supergroups (resp. the category of Harish-Chandra pairs). Then we have the following result.

THEOREM 2.5. *The assignment* $\mathsf{ASG}\to\mathsf{HCP};\ \boldsymbol{G}\mapsto(\boldsymbol{G}_{\mathsf{ev}}, W^{\mathcal{O}(\boldsymbol{G})})$ *gives a category equivalence.*

The notion of Harish-Chandra pairs for Lie supergroups was first introduced by Kostant [**14**]. For algebraic supergroups defined over an algebraically closed field of characteristic zero, the result above was shown by Carmeli and Fioresi [**5**]. Later, using Hopf-algebraic techniques, Masuoka [**17**, Theorem 29] showed that the equivalence still holds over an arbitrary base field. Moreover, the result is still true when $\Bbbk$ is a commutative ring, see [**18**, Theorem 4.22].

3. Representations of algebraic supergroups

Assume that an algebraic supergroup $\boldsymbol{G}$ is *linearly reductive*, that is, the category of left $\boldsymbol{G}$-supermodules is semi-simple. Weissauer [**31**, Theorem 6] showed that if $\Bbbk$ is an algebraically closed field of characteristic zero, then such $\boldsymbol{G}$ are essentially exhausted by ortho-symplectic supergroups $\mathbf{SpO}(2k|1)$. If $\Bbbk$ is positive characteristic, then Masuoka [**17**, Theorem 45] showed that such $\boldsymbol{G}$ are necessarily purely even $\boldsymbol{G}=\boldsymbol{G}_{\mathsf{ev}}$. See also [**20**] for integrals on algebraic supergroups.

Therefore, to study representations of an algebraic supergroup, we have to find a sufficiently wide and rich class of algebraic supergroups. In 2011, Serganova

[**25**] introduced such a class of algebraic supergroups, called "quasireductive supergroups". An algebraic supergroup $\boldsymbol{G}$ is said to be *quasireductive* if its even part $\boldsymbol{G}_{\mathsf{ev}}$ comes from the base change of a certain connected and split reductive group scheme defined over $\mathbb{Z}$ (see [**11**, Part II, Chapter 1]) to our field $\Bbbk$ (for the precise definition, see [**25**, **26**]). This means a quasireductive supergroup is controlled by a (generalized, in some sense) root datum. All algebraic supergroups in Examples 2.1 are quasireductive.

3.1. Quasireductive supergroups. In what follows, we fix a quasireductive supergroup $\boldsymbol{G}$ and a split maximal torus T of the even part $G := \boldsymbol{G}_{\mathsf{ev}}$ of $\boldsymbol{G}$. One sees that the character group $\mathsf{X}(T)$ coincides with the set of all group-like elements in $\mathcal{O}(T)$ and isomorphic to $\mathbb{Z}^{\oplus \ell}$, where ℓ is the rank of $\boldsymbol{G}_{\mathsf{ev}}$.

Set $\mathfrak{g} := \mathsf{Lie}(\boldsymbol{G})$. The left adjoint action of T on $\boldsymbol{G}$ induces an action of T on $\mathfrak{g}$ which preserves the parity of $\mathfrak{g}$. Since the torus T is a diagonalizable group scheme, the left T-supermodule $\mathfrak{g}$ decomposes into weight superspace (see [**11**, Part I, 7.14]).

$$\mathfrak{g} = \Big(\bigoplus_{\alpha \in \mathsf{X}(T)} \mathfrak{g}_{\bar{0}}^{\alpha}\Big) \oplus \Big(\bigoplus_{\delta \in \mathsf{X}(T)} \mathfrak{g}_{\bar{1}}^{\delta}\Big), \quad \mathfrak{g}^{\alpha} = \{X \in \mathfrak{g} \mid \forall u \in \mathsf{hy}(T),\ u \rightharpoonup X = \langle u, \alpha \rangle X\}.$$

Let $\mathfrak{h} := \mathfrak{g}^{\alpha=0}$ be the super-subspace of $\mathfrak{g}$ of weight $\alpha = 0$. Then $\mathfrak{h}_{\bar{0}} = \mathsf{Lie}(T)$. For each $\epsilon \in \mathbb{Z}_2$, we set $\triangle_{\epsilon} := \{0 \neq \alpha \in \mathsf{X}(T) \mid \mathfrak{g}_{\epsilon}^{\alpha} \neq 0\}$. We put

$$\triangle := \begin{cases} \triangle_{\bar{0}} \cup \triangle_{\bar{1}} & \text{if } \mathfrak{h}_{\bar{1}} = 0, \\ \triangle_{\bar{0}} \cup \triangle_{\bar{1}} \cup \{0\} & \text{otherwise.} \end{cases}$$

We call $\triangle$ the *root system of* $(\boldsymbol{G}, T)$. Note that $\triangle_{\bar{0}}$ is a root system of (G, T).

EXAMPLES 3.1. We shall see some examples of root systems.

(1) The case $\boldsymbol{G} = \mathbf{GL}(m|n)$. We choose a split maximal T of $\boldsymbol{G}_{\mathsf{ev}} \cong \mathrm{GL}_m \times \mathrm{GL}_n$ so that T consists of all diagonal matrices in $\boldsymbol{G}_{\mathsf{ev}}$, and write $\mathsf{X}(T) = \bigoplus_{i=1}^{m+n} \mathbb{Z}\lambda_i$. Then the root system of $(\mathbf{GL}(m|n), T)$ is described as $\triangle = \{\lambda_i - \lambda_j \in \mathsf{X}(T) \mid 1 \leq i \neq j \leq m+n\}$. In particular, we have $\triangle = \triangle_{\bar{0}} \sqcup \triangle_{\bar{1}}$ (disjoint union) and
$$\triangle_{\bar{0}} = \{\lambda_i - \lambda_j \mid 1 \leq i \neq j \leq m \text{ or } m+1 \leq i \neq j \leq m+n\}.$$
(2) The case $\boldsymbol{G} = \mathbf{P}(n)$. As before, we choose T of $\boldsymbol{G}_{\mathsf{ev}} \cong \mathrm{GL}_n$ so that consists of all diagonal matrices in $\boldsymbol{G}_{\mathsf{ev}}$, and write $\mathsf{X}(T) = \bigoplus_{i=1}^{n} \mathbb{Z}\lambda_i$. Then the root system of $(\mathbf{P}(n), T)$ is described as $\triangle_{\bar{0}} = \{\lambda_i - \lambda_j \mid 1 \leq i \neq j \leq n\}$, $\triangle_{\bar{1}} = \{\pm(\lambda_i + \lambda_j), 2\lambda_p \mid 1 \leq i < j \leq n, 1 \leq p \leq n\}$, and $\triangle = \triangle_{\bar{0}} \sqcup \triangle_{\bar{1}}$.
(3) The case $\boldsymbol{G} = \mathbf{Q}(n)$. We use the notation in (2) above. Then the root system of $(\mathbf{P}(n), T)$ is described as $\triangle_{\bar{0}} = \{\lambda_i - \lambda_j \mid 1 \leq i \neq j \leq n\}$, $\triangle_{\bar{1}} = \{\lambda_i - \lambda_j \mid 1 \leq i \neq j \leq n\} \cup \{0\}$, and $\triangle = \triangle_{\bar{0}} \cup \triangle_{\bar{1}}$. Note that $\triangle_{\bar{0}} \cap \triangle_{\bar{1}} \neq \varnothing$.

For a left $\boldsymbol{G}$-supermodule, the induced left $\mathsf{hy}(\boldsymbol{G})$-supermodule is a locally finite left $\mathsf{hy}(\boldsymbol{G})$-$T$-supermodule, that is, locally finite as $\mathsf{hy}(\boldsymbol{G})$-supermodule and the induced $\mathsf{hy}(T)$-supermodule structure coincides with the restriction action of $\mathsf{hy}(\boldsymbol{G})$, see [**11**, Part II, 1.20] for the non-super situation. Thus, we get a functor from the category of left $\boldsymbol{G}$-supermodules to the category of locally finite left $\mathsf{hy}(\boldsymbol{G})$-$T$-supermodules. We get the following, see [**19**, Theorem 5.8] (see also [**26**, Theorem 3.4]).

THEOREM 3.2. *The above functor gives an equivalence between the category of left* $\boldsymbol{G}$*-supermodules and the category of locally finite left* $\mathsf{hy}(\boldsymbol{G})$*-*$T$*-supermodules.*

This has been known only for some special quasireductive supergroups, see [**3**, Corollary 5.7], [**4**, Corollary 3.5], and [**27**, Theorem 2.8].

3.2. Borel super-subgroups. Let $\mathbb{Z}\triangle$ be the subgroup of $\mathsf{X}(T)$ generated by $\triangle$, and let $\gamma : \mathbb{Z}\triangle \to \mathbb{R}$ be a group homomorphism satisfying $\gamma(\alpha) \neq 0$ for each $0 \neq \alpha \in \triangle$. As in [**25**, §9.3], we set

$$\triangle^{\pm} := \{0 \neq \alpha \in \triangle \mid \pm\gamma(\alpha) > 0\}, \quad \triangle^{\pm}_{\epsilon} := \triangle^{\pm} \cap \triangle_{\epsilon} \quad (\epsilon \in \mathbb{Z}_2).$$

Then we get the following Lie super-subalgebras

$$\mathfrak{n}^{\pm} := \bigoplus_{\alpha \in \triangle^{\pm}} \quad \text{and} \quad \mathfrak{b}^{\pm} := \mathfrak{n}^{\pm} \oplus \mathfrak{h}$$

of $\mathfrak{g}$, see [**10, 25**].

We take Borel subgroups $B^{\pm}$ of G with respect to $\triangle^{\pm}_{\bar{0}}$. Then by Theorem 2.5, we can construct super-subgroups $\boldsymbol{B}^{\pm}$ of $\boldsymbol{G}$ such that $\boldsymbol{B}^{\pm}_{\mathsf{ev}} = B^{\pm}$ and $\mathsf{Lie}(\boldsymbol{B}^{\pm}) = \mathfrak{b}^{\pm}$. Also, for the unipotent radicals $U^{\pm}$ of $B^{\pm}$, we get super-subgroups $\boldsymbol{U}^{\pm}$ of $\boldsymbol{B}^{\pm}$ such that $\boldsymbol{U}^{\pm}_{\mathsf{ev}} = U^{\pm}$ and $\mathsf{Lie}(\boldsymbol{U}^{\pm}) = \mathfrak{n}^{\pm}$, see [**26**, Lemma 3.6].

EXAMPLES 3.3. We use the notations in Examples 3.1.

(1) The case $\boldsymbol{G} = \mathbf{GL}(m|n)$. If we choose γ so that $\gamma(\lambda_i) = -i$, then we have $\triangle^{\pm} = \{\pm(\lambda_i - \lambda_j) \mid 1 \leq i < j \leq m+n\}$, and hence $\boldsymbol{B}^{+}$ (resp. $\boldsymbol{B}^{-}$) consists of upper (resp. lower) triangular matrices, as usual.

(2) The case $\boldsymbol{G} = \mathbf{P}(n)$. If we choose γ so that $\gamma(\lambda_i) = n-i+1$, then we have $\triangle^{+} = \{\lambda_i \pm \lambda_j, 2\lambda_p \mid 1 \leq i < j \leq n, 1 \leq p \leq n\}$ and $\triangle^{-} = \{-(\lambda_i \pm \lambda_j) \mid 1 \leq i < j \leq n\}$. In this case, one sees that

$$\begin{aligned} \boldsymbol{B}^{+}(R) &= \{\left(\begin{array}{c|c} X & Y \\ \hline O & {}^{t}X^{-1} \end{array}\right) \mid X \in B^{+}(R_{\bar{0}}),\ {}^{t}X^{-1}\,{}^{t}Y = YX^{-1}\} \text{ and} \\ \boldsymbol{B}^{-}(R) &= \{\left(\begin{array}{c|c} X & O \\ \hline Z & {}^{t}X^{-1} \end{array}\right) \mid X \in B^{-}(R_{\bar{0}}),\ {}^{t}ZX = -{}^{t}XZ\}. \end{aligned}$$

(3) The case $\boldsymbol{G} = \mathbf{Q}(n)$. If we choose γ so that $\gamma(\lambda_i) = -i$, then we have $\triangle^{\pm} = \{\pm(\lambda_i - \lambda_j) \mid 1 \leq i < j \leq n\}$, and

$$\boldsymbol{B}^{\pm}(R) = \{\left(\begin{array}{c|c} X & Y \\ \hline -Y & X \end{array}\right) \mid X, Y \text{ are upper/lower triangular matrices}\}.$$

REMARK 3.4. As in the situation of finite-dimensional simple Lie superalgebras (see [**23**, Chapter 3] for example), our (possible) Borel super-subgroups are *not* conjugate. This fact makes the super-world interesting.

The canonical inclusions $\boldsymbol{B}^{\pm} \subset \boldsymbol{G}$ induce Hopf superalgebra quotient maps $\mathsf{pr}^{\pm} : \mathcal{O}(\boldsymbol{G}) \to \mathcal{O}(\boldsymbol{B}^{\pm})$. We have the following.

PROPOSITION 3.5 ([**26**, Corollary 3.13]). *The superalgebra map* $\mathcal{O}(\boldsymbol{G}) \to \mathcal{O}(\boldsymbol{G}) \otimes \mathcal{O}(\boldsymbol{G}) \overset{\mathsf{pr}^{+}\otimes\mathsf{pr}^{-}}{\to} \mathcal{O}(\boldsymbol{B}^{+}) \otimes \mathcal{O}(\boldsymbol{B}^{-})$ *is injective, where the first arrow is the comultiplication map of* $\mathcal{O}(\boldsymbol{G})$.

PROOF. Let $\mu : \boldsymbol{G} \times \boldsymbol{G} \to \boldsymbol{G}$ be the multiplication map of $\boldsymbol{G}$. Then the restriction $\mu : \boldsymbol{B}^+ \times \boldsymbol{B}^- \to \boldsymbol{G}$ induces a surjective map $\mathsf{hy}(\mu) : \mathsf{hy}(\boldsymbol{B}^+) \otimes \mathsf{hy}(\boldsymbol{B}^-) \to \mathsf{hy}(\boldsymbol{G})$ of supercoalgebras. Since $\boldsymbol{G}$ is connected, the canonical pairing $\langle\ ,\ \rangle : \mathsf{hy}(\boldsymbol{G}) \times \mathcal{O}(\boldsymbol{G}) \to \Bbbk$ induces an injection $\mathcal{O}(\boldsymbol{G}) \hookrightarrow \mathsf{hy}(\boldsymbol{G})^*$ of superalgebras. Using these facts, one easily sees the composition $\mathcal{O}(\boldsymbol{G}) \to \mathcal{O}(\boldsymbol{G}) \otimes \mathcal{O}(\boldsymbol{G}) \to \mathcal{O}(\boldsymbol{B}^+) \otimes \mathcal{O}(\boldsymbol{B}^-)$ is injective. □

The proposition may be reformulated so that the restricted multiplication map $\boldsymbol{B}^+ \times \boldsymbol{B}^- \to \boldsymbol{G}$ has a dense image; the thus formulated result was likely known as a folklore concerning some special supergroups such as $\mathbf{GL}(m|n)$, $\mathbf{SL}(m|n)$, $\mathbf{SpO}(2k|n)$, $\mathbf{Q}(n)$, etc. But to obtain the result in general, the formulation which uses the Hopf algebra language as above is more appropriate. See also [**1**] (for the non-super situation) and [**24**, Chapter 8] (for quantum groups).

3.3. Irreducible representations. One easily sees that the pair $(T, \mathfrak{h})$ forms a Harish-Chandra pair. Thus, by Theorem 2.5, there exists a closed super-subgroup $\boldsymbol{T}$ of $\boldsymbol{G}$ such that $\boldsymbol{T}_{\mathsf{ev}} = T$ and $\mathsf{Lie}(\boldsymbol{T}) = \mathfrak{h}$. We shall call $\boldsymbol{T}$ a *super-torus of* $\boldsymbol{G}$. If $0 \notin \triangle$, then $\boldsymbol{T} = \boldsymbol{T}_{\mathsf{ev}} = T$.

EXAMPLE 3.6. If $\boldsymbol{G} = \mathbf{Q}(n)$, then for a commutative superalgebra R, one sees

$$\boldsymbol{T}(R) = \{\left(\begin{array}{c|c} X & Y \\ \hline -Y & X \end{array}\right) \in \boldsymbol{G}(R) \mid X, Y \text{ are diagonal matrices}\}.$$

In particular, we see that $\boldsymbol{T} \neq T$.

For a fixed $\lambda \in \mathsf{X}(T)$, we define $b^\lambda : \mathfrak{h}_{\bar{1}} \times \mathfrak{h}_{\bar{1}} \to \Bbbk$; $(X, Y) \mapsto \lambda([X, Y])$. Then $\mathfrak{h}_{\bar{1}}$ forms a quadratic space with a symmetric bilinear form b^λ, and hence we can consider the *Clifford superalgebra* $C(\mathfrak{h}_{\bar{1}}, b^\lambda)$ over $\Bbbk$. In [**26**, Proposition B.3], it was shown that there is a unique irreducible left $C(\mathfrak{h}_{\bar{1}}, b^\lambda)$-supermodule $\mathfrak{u}(\lambda)$ up to isomorphism and *parity change*. For a superspace $V = V_{\bar{0}} \oplus V_{\bar{1}}$, its parity-changed superspace $\Pi(V)$ is defined by $\Pi(V)_{\bar{0}} := V_{\bar{1}}$ and $\Pi(V)_{\bar{1}} := V_{\bar{0}}$.

PROPOSITION 3.7 ([**26**, Proposition 4.5]). *For any $\lambda \in \mathsf{X}(T)$, $\mathfrak{u}(\lambda)$ can be regarded as an irreducible left $\boldsymbol{T}$-supermodule. Up to isomorphism and parity change, all irreducible left $\boldsymbol{T}$-supermodules are obtained in this way.*

PROOF. For simplicity, we assume $\Bbbk$ is characteristic zero. Let $\mathsf{hy}(\boldsymbol{T})^\lambda$ be the quotient superalgebra of $\mathsf{hy}(\boldsymbol{T})$ by the two-sided super-ideal generated by $H - \lambda(H)$ for $H \in \mathfrak{h}_{\bar{0}}$. Then one easily sees that the Clifford superalgebra $C(\mathfrak{h}_{\bar{1}}, b^\lambda)$ is isomorphic to this $\mathsf{hy}(\boldsymbol{T})^\lambda$. Since $\mathfrak{u}(\lambda)$ is a finite-dimensional T-supermodule, $\mathfrak{u}(\lambda)$ has a structure of $\boldsymbol{T}$-supermodule by Theorem 3.2 which is irreducible. The last statement is obvious. □

We regard the $\boldsymbol{T}$-supermodule $\mathfrak{u}(\lambda)$ ($\lambda \in \mathsf{X}(T)$) as a left $\boldsymbol{B}^-$-supermodule via the Hopf splitting $\mathcal{O}(\boldsymbol{T}) \to \mathcal{O}(\boldsymbol{B}^-)$ corresponding to the canonical Hopf quotient $\mathcal{O}(\boldsymbol{B}^-) \to \mathcal{O}(\boldsymbol{T})$. Since $\boldsymbol{B}^-$ is isomorphic to the semidirect product $\boldsymbol{T} \ltimes \boldsymbol{U}^-$ and $\boldsymbol{U}^-$ is unipotent, one sees that $\mathfrak{u}(\lambda)$ is still irreducible as a left $\boldsymbol{B}^-$-supermodule [**26**, Proposition 4.9].

For a left $\boldsymbol{B}^-$-supermodule M (= right $\mathcal{O}(\boldsymbol{B}^-)$-supercomodule), we denote its *induced $\boldsymbol{G}$-supermodule* by $\mathsf{ind}_{\boldsymbol{B}^-}^{\boldsymbol{G}}(M) := M \,\square_{\mathcal{O}(\boldsymbol{B}^-)}\, \mathcal{O}(\boldsymbol{G})$, where $\square_{\mathcal{O}(\boldsymbol{B}^-)}$ denotes the cotensor product over $\mathcal{O}(\boldsymbol{B}^-)$. Since the functor $\mathsf{ind}_{\boldsymbol{B}^-}^{\boldsymbol{G}}(-)$ is left exact, we get its right derived functor $R^n\mathsf{ind}_{\boldsymbol{B}^-}^{\boldsymbol{G}}(-)$ for $n \geq 0$.

REMARK 3.8. By Masuoka and Zubkov [**22**], we see that the even part of the quotient $\boldsymbol{G}/\boldsymbol{B}^-$ is G/B^-, and is projective. Thus, as in the non-super situation, it is natural to consider the cohomology group $H^n(\boldsymbol{G}/\boldsymbol{B}^-, \mathcal{L}(-))$ for $n \geq 0$ to study representations of $\boldsymbol{G}$. However, combining the results [**21**] and [**2**, Corollary 2.4], we see that $H^n(\boldsymbol{G}/\boldsymbol{B}^-, \mathcal{L}(-))$ coincides with $R^n\mathsf{ind}_{\boldsymbol{B}^-}^{\boldsymbol{G}}(-)$ for each $n \geq 0$ which is a super-analogue of [**11**, Part I, Proposition 5.12].

In the following, we put $H^n(\lambda) := R^n\mathsf{ind}_{\boldsymbol{B}^-}^{\boldsymbol{G}}(\mathfrak{u}(\lambda))$ for simplicity. Set $\mathsf{X}(T)^\flat := \{\lambda \in \mathsf{X}(T) \mid H^0(\lambda) \neq 0\}$.

LEMMA 3.9. *For $\lambda \in \mathsf{X}(T)^\flat$, the $\boldsymbol{B}^+$-socle of $H^0(\lambda)$ coincides with $\mathfrak{u}(\lambda)$.*

PROOF. Taking the functor $\mathfrak{u}(\lambda)\Box_{\mathcal{O}(\boldsymbol{B}^-)}(-)$ to the inclusion $\mathcal{O}(\boldsymbol{G}) \hookrightarrow \mathcal{O}(\boldsymbol{B}^-)\otimes \mathcal{O}(\boldsymbol{B}^+)$ given in Proposition 3.5, we have $\mathfrak{u}(\lambda) \hookrightarrow \mathfrak{u}(\lambda)\Box_{\mathcal{O}(\boldsymbol{T})}\mathcal{O}(\boldsymbol{B}^+) = \mathsf{ind}_{\boldsymbol{T}}^{\boldsymbol{B}^+}(\mathfrak{u}(\lambda))$. Thus, to prove the statement, it is enough to show that the $\boldsymbol{B}^+$-socle of $\mathsf{ind}_{\boldsymbol{T}}^{\boldsymbol{B}^+}(\mathfrak{u}(\lambda))$ coincides with $\mathfrak{u}(\lambda)$. By Frobenius reciprocity, the claim is easy to see. $\square$

The following is our main theorem.

THEOREM 3.10 ([**26**, Theorem 4.12]). *For any $\lambda \in \mathsf{X}(T)^\flat$, the $\boldsymbol{G}$-socle $L(\lambda)$ of $H^0(\lambda)$ is irreducible as a left $\boldsymbol{G}$-supermodule. Up to isomorphism and parity change, all irreducible left $\boldsymbol{G}$-supermodules are obtained in this way.*

PROOF. The first statement immediately follows form Lemma 3.9. To complete the proof, we take a irreducible left $\boldsymbol{G}$-supermodule L. Regarding L as a T-supermodule, there is $\lambda \in \mathsf{X}(T)$ and a non-trivial map $L \to \mathfrak{u}(\lambda)$ of $\boldsymbol{B}^-$-supermodules up to parity change. Thus, by Frobenius reciprocity, we have an inclusion $L \hookrightarrow \mathsf{ind}_{\boldsymbol{B}^-}^{\boldsymbol{G}}(\mathfrak{u}(\lambda)) = H^0(\lambda)$, and hence $L = L(\lambda)$. $\square$

Recall that the $\otimes$-decomposition $\mathcal{O}(\boldsymbol{G}) \cong \mathcal{O}(\boldsymbol{G}_{\mathsf{ev}}) \otimes \bigwedge W^{\mathcal{O}(\boldsymbol{G})}$ given in Proposition 2.4 is left $\mathcal{O}(\boldsymbol{G}_{\mathsf{ev}})$-colinear. Taking the functor $\mathfrak{u}(\lambda)\,\Box_{\mathcal{O}(B^-)}(-)$ to the isomorphism, we see that $H^0(\lambda) \hookrightarrow H^0_{\mathsf{ev}}(\lambda)^{\oplus \dim(\mathfrak{u}(\lambda))} \otimes \bigwedge W^{\mathcal{O}(\boldsymbol{G})}$ as left T-supermodules. Here, $H^0_{\mathsf{ev}}(\lambda) := \mathsf{ind}_{B^-}^{G}(\Bbbk^\lambda)$ and $\Bbbk^\lambda$ is the one-dimensional T-module of weight λ. In particular, we conclude that the parameter set $\mathsf{X}(T)^\flat$ is a subset of the set of all *even dominant weights* $\mathsf{X}(T)^+ := \{\lambda \in \mathsf{X}(T) \mid \forall \alpha \in \Delta_{\bar{0}}^+,\ \langle \lambda, \alpha^\vee \rangle \geq 0\}$, where $\alpha^\vee$ denotes the dual root of α.

REMARK 3.11. Assume that $\Bbbk$ is algebraically closed. Brundan and Kleshchev [**3**, Theorem 6.11] determined $\mathsf{X}(T)^\flat$ for $\boldsymbol{G} = \mathbf{Q}(n)$, and Shu and Wang [**27**, Theorem 5.3] described $\mathsf{X}(T)^\flat$ combinatorially for $\boldsymbol{G} = \mathbf{SpO}(2k|n)$. Recently, Cheng, Shu, and Wang [**6**] determined $\mathsf{X}(T)^\flat$ explicitly for simply connected Chevalley supergroups of types $D(2|1;\zeta)$, $G(3)$, and $F(3|1)$. A *unified* description of $\mathsf{X}(T)^\flat$ for all quasireductive supergroups is an open problem.

4. Quasireductive supergroups admitting maximal parabolic super-subgroups

4.1. Maximal parabolic super-subgroups. In the following, we assume that a quasireductive supergroup $\boldsymbol{G}$ contains a closed super-subgroup $\boldsymbol{P}$ such that $\boldsymbol{P}_{\mathsf{ev}} = G$ and $\mathsf{Lie}(\boldsymbol{P})_{\bar{1}} = \mathfrak{b}_{\bar{1}}^-$. We call $\boldsymbol{P}$ a *maximal parabolic super-subgroup of* $\boldsymbol{G}$. For simplicity, we also assume that a super-torus $\boldsymbol{T}$ of $\boldsymbol{G}$ satisfies $\boldsymbol{T} = T$.

EXAMPLES 4.1. We use notations and $\gamma : \mathbb{Z}\Delta \to \mathbb{R}$ as in Examples 3.3.

(1) The case $\boldsymbol{G} = \mathbf{GL}(m|n)$. Then $\boldsymbol{P} := \{\left(\begin{array}{c|c} X & O \\ \hline Z & W \end{array}\right) \in \mathbf{GL}(m|n)\}$ is a maximal parabolic super-subgroup of $\mathbf{GL}(m|n)$.
(2) The case $\boldsymbol{G} = \mathbf{P}(n)$. Then $\boldsymbol{P} := \{\left(\begin{array}{c|c} X & O \\ \hline Z & {}^{\mathrm{t}}X^{-1} \end{array}\right) \in \mathbf{P}(n) \mid {}^{\mathrm{t}}ZX = -{}^{\mathrm{t}}XZ\}$ is a maximal parabolic super-subgroup of $\mathbf{P}(n)$.
(3) For a Chevalley supergroup constructed from a finite-dimensional simple Lie superalgebra of type I (for example $\mathbf{SL}(m|n)$), one sees that there exists a maximal parabolic super-subgroup.
(4) The case $\boldsymbol{G} = \mathbf{Q}(n)$. Then we are not able to find such a super-subgroup.

A version of Theorem 2.4, we can show that $\mathcal{O}(\boldsymbol{G})$ is isomorphic to $\mathcal{O}(\boldsymbol{P}) \otimes \bigwedge W^{\mathcal{O}(\boldsymbol{U}^+)}$ as right $\boldsymbol{P}$- left T-supermodules, see [**26**, Proposition 5.2]. Using this fact, one can show the following. The proof is essentially the same as Zubkov's [**32**, Proposition 5.2] for $\boldsymbol{G} = \mathbf{GL}(m|n)$.

LEMMA 4.2. *For a left $\boldsymbol{B}^-$-supermodule V, we have a morphism $\mathcal{N}(V) : V\square_{\mathcal{O}(\boldsymbol{B}^-)}\mathcal{O}(\boldsymbol{P}) \to V\square_{\mathcal{O}(B^-)}\mathcal{O}(G); v\otimes p \mapsto v\otimes\overline{p}$ of left G-supermodules. Moreover, the functor $\mathcal{N}$ is a natural isomorphism.*

The following was first proved by Zubkov [**32**] for $\boldsymbol{G} = \mathbf{GL}(m|n)$.

THEOREM 4.3 ([**26**, Theorem 5.4]). *For a left $\boldsymbol{B}^-$-supermodule V and a non-negative integer n, we have an isomorphism $R^n\mathsf{ind}_{\boldsymbol{B}^-}^{\boldsymbol{G}}(V) \cong R^n\mathsf{ind}_{B^-}^{G}(V) \otimes \bigwedge W^{\mathcal{O}(\boldsymbol{U}^+)}$ of left T-supermodules.*

PROOF. By Lemma 4.2 and the exactness of the restriction functor $\mathsf{res}_G^{\boldsymbol{P}}(-)$, we have $\mathsf{res}_G^{\boldsymbol{P}} \circ R^n\mathsf{ind}_{\boldsymbol{B}^-}^{\boldsymbol{P}}(-) \cong R^n(\mathsf{ind}_{B^-}^{G} \circ \mathsf{res}_{B^-}^{\boldsymbol{B}^-}(-))$. On the other hand, since $\mathcal{O}(\boldsymbol{G})$ is $\mathcal{O}(\boldsymbol{P})$-injective, $\mathsf{ind}_{\boldsymbol{P}}^{\boldsymbol{G}}(-)$ is exact. Thus, we have $\mathsf{ind}_{\boldsymbol{P}}^{\boldsymbol{G}} \circ (R^n\mathsf{ind}_{\boldsymbol{B}^-}^{\boldsymbol{P}}(-)) \cong R^n\mathsf{ind}_{\boldsymbol{B}^-}^{\boldsymbol{G}}(-)$, see [**11**, Part I, 4.1(2)]. Here, we used the transitivity $\mathsf{ind}_{\boldsymbol{B}^-}^{\boldsymbol{G}}(-) = \mathsf{ind}_{\boldsymbol{P}}^{\boldsymbol{G}} \circ \mathsf{ind}_{\boldsymbol{B}^-}^{\boldsymbol{P}}(-)$. By the isomorphism $\mathcal{O}(\boldsymbol{G}) \cong \mathcal{O}(\boldsymbol{P}) \otimes \bigwedge W^{\mathcal{O}(\boldsymbol{U}^+)}$, we get $\mathsf{ind}_{\boldsymbol{P}}^{\boldsymbol{G}}(Y) \cong Y \otimes \bigwedge W^{\mathcal{O}(\boldsymbol{U}^+)}$ for each left $\boldsymbol{P}$-supermodule Y. Combining these result, the proof is done. □

By the result above, we immediately have three applications.

COROLLARY 4.4. *For $\lambda \in \mathsf{X}(T)^+$, we have the following.*
(1) *$H^n(\lambda) \cong H^n_{\mathsf{ev}}(\lambda) \otimes \bigwedge W^{\mathcal{O}(\boldsymbol{U}^+)}$ for $n \geq 0$. In particular, $\mathsf{X}(T)^\flat = \mathsf{X}(T)^+$.*
(2) *$H^n(\lambda) = 0$ for $n > 1$.*
(3) *The Euler characteristic of $H^0(\lambda)$ is given by $\frac{A(\lambda+\rho_{\bar{0}})}{A(\rho_{\bar{0}})} \prod_{\delta \in \Delta_{\bar{1}}^+} 1 + e^{-\delta}$.*

Here, we used $A(\mu) := \sum_{w \in \mathsf{W}} (-1)^{\ell(w)} e^{w\mu}$ for $\mu \in \mathsf{X}(T) \otimes_{\mathbb{Z}} \mathbb{Q}$ (W is the ordinary Weyl group of G with respect to T), and $\rho_{\bar{0}} := \frac{1}{2}\sum_{\alpha \in \Delta_{\bar{0}}^+} \alpha$.

PROOF. The first statement is trivial. The next statement follows from the Kempf vanishing theorem [**11**, Part II, Proposition 4.5]. Since the formal character of the exterior algebra $\bigwedge(\mathfrak{n}_{\bar{1}}^+)^*$ over the dual space of $\mathfrak{n}_{\bar{1}}^+$ is given by $\prod_{\delta \in \Delta_{\bar{1}}^+} 1 + e^{-\delta}$, the final statement follows. □

4.2. Irreducibility of $H^0(\lambda)$ for $\boldsymbol{G} = \mathbf{GL}(m|n)$. In the following, we suppose that $\boldsymbol{G} = \mathbf{GL}(m|n)$ and fix $\gamma : \mathbb{Z}\Delta \to \mathbb{R}$ as in Examples 3.3 (1). Marko [**15**] gave a explicit description of the irreducibility of $H^0(\lambda)$ for $\boldsymbol{G}$. The proof was based on constructing an explicit basis of $H^0(\lambda)$ using a basis (the so-called *bi-determinants*) of $H^0_{\mathsf{ev}}(\lambda)$. In this subsection, we re-prove his result without taking such basis.

Fix $\lambda \in \mathsf{X}(T)^+$. Since the isomorphism $H^0(\lambda) \cong H^0_{\mathsf{ev}}(\lambda) \otimes \bigwedge W^{\mathcal{O}(\boldsymbol{U}^+)}$ given in Corollary 4.4 (1) is T-linear, the λ-weight space $H^0(\lambda)^\lambda$ is one-dimensional. Let denote $H^0(\lambda)^\lambda = \Bbbk v_\lambda$. By Theorem 3.2, we have $L(\lambda) = \mathsf{hy}(\boldsymbol{G})\, v_\lambda$. Moreover, one can show that $L_{\mathsf{ev}}(\lambda) = \mathsf{hy}(\boldsymbol{G}_{\mathsf{ev}})\, \overline{v_\lambda}$, where $L_{\mathsf{ev}}(\lambda)$ is the $\boldsymbol{G}_{\mathsf{ev}}$-socle of $H^0_{\mathsf{ev}}(\lambda)$.

For $\delta \in \triangle_{\bar{1}}$, let X_δ be a root vector of the root space $\mathfrak{g}^\delta_{\bar{1}}$. In the hyper superalgebra $\mathsf{hy}(\boldsymbol{G})$ of $\boldsymbol{G}$, we set $D^\pm := \prod_{\delta \in \triangle^\pm_{\bar{1}}} X_\delta$, where the product is taken by a fixed an (arbitrary) ordering of $\triangle^\pm_{\bar{1}}$. Set $\rho := \rho_{\bar{0}} - \frac{1}{2}\sum_{\delta \in \triangle^+_{\bar{1}}} \delta$. The next statement is a key lemma. The proof is essentially the same as in the proof of Kac [**13**, Proposition 2.9].

LEMMA 4.5. *There exists a non-zero scalar $c \in \Bbbk$ such that $D^+D^-v_\lambda = c\prod_{\delta \in \triangle^+_{\bar{1}}} (\lambda + \rho, \delta)\, v_\lambda$, where $(\ ,\ )$ is the canonical bilinear form.*

An element $\lambda \in \mathsf{X}(T)$ is said to be *typical* if $(\lambda + \rho, \delta^\vee) \neq 0$ for all *odd isotropic roots* $\delta \in \triangle_{\bar{1}}$, that is, $2\delta \notin \triangle$. The following was first proved by Marko [**15**, Theorem 1]. Our proof below is still valid for (at least) the special linear supergroups $\mathbf{SL}(m|n)$.

THEOREM 4.6. *$H^0(\lambda)$ is irreducible $\iff$ $H^0_{\mathsf{ev}}(\lambda)$ is irreducible and λ is typical.*

PROOF. Set $\kappa := \sum_{\delta \in \triangle^+_{\bar{1}}} \delta$. Then it is easy to see that the $(\lambda - \kappa)$-weight space of $L(\lambda)$ is one-dimensional and generated by D^-v_λ. Suppose that $H^0(\lambda)$ is irreducible. Then $H^0(\lambda) = \mathsf{hy}(\boldsymbol{G})\, v_\lambda$. If λ is atypical (i.e., not typical), then $L(\lambda)^{\lambda-\kappa} = D^-v_\lambda \Bbbk = 0$ by Lemma 4.5. This means that $L(\lambda)$ is a proper subset of $H^0(\lambda)$, a contradiction. Since the pairing $\mathsf{hy}(\boldsymbol{U}^-) \times \mathcal{O}(\boldsymbol{U}^-) \to \Bbbk$ compatible with the $\otimes$-decompositions $\mathcal{O}(\boldsymbol{U}^-) \cong \mathcal{O}(U^-) \otimes \bigwedge W^{\mathcal{O}(\boldsymbol{U}^-)}$ and $(\bigwedge \mathfrak{n}^-_{\bar{1}}) \otimes \mathsf{hy}(U^-) \cong \mathsf{hy}(\boldsymbol{U}^-)$ (see [**17**, Proposition 22]), one sees that $H^0_{\mathsf{ev}}(\lambda)$ is contained in $\mathsf{hy}(U^-)\, \overline{v_\lambda}$. Thus, we get $H^0_{\mathsf{ev}}(\lambda) = L_{\mathsf{ev}}(\lambda)$.

Suppose that $H^0_{\mathsf{ev}}(\lambda)$ is irreducible and λ is typical. Then it is easy to see that the set of all weights of $L(\lambda) = \mathsf{hy}(\boldsymbol{G})\, v_\lambda$ coincides with the set of all weights of $H^0(\lambda) \cong L_{\mathsf{ev}}(\lambda) \otimes \bigwedge(\mathfrak{n}^+_{\bar{1}})^*$, and hence we conclude that $L(\lambda) = H^0(\lambda)$. □

REMARK 4.7. Suppose that our base field $\Bbbk$ is of characteristic zero. Then it is known that $H^0_{\mathsf{ev}}(\lambda)$ is necessarily irreducible [**11**, Part II, Corollary 5.6]. Thus, in this case, we have $H^0(\lambda)$ is irreducible if and only if λ is typical. The phenomena is a remarkable feature of the super-representation theory.

Acknowledgments

The author thanks the organizers of the workshop held at Nanjing for giving an opportunity to present the results.

References

[1] Julien Bichon and Simon Riche, *Hopf algebras having a dense big cell*, Trans. Amer. Math. Soc. **368** (2016), no. 1, 515–538, DOI 10.1090/tran/6335. MR3413872

[2] Jonathan Brundan, *Modular representations of the supergroup Q(n). II*, Pacific J. Math. **224** (2006), no. 1, 65–90, DOI 10.2140/pjm.2006.224.65. MR2231652

[3] Jonathan Brundan and Alexander Kleshchev, *Modular representations of the supergroup Q(n). I*, J. Algebra **260** (2003), no. 1, 64–98, DOI 10.1016/S0021-8693(02)00620-8. Special issue celebrating the 80th birthday of Robert Steinberg. MR1973576

[4] Jonathan Brundan and Jonathan Kujawa, *A new proof of the Mullineux conjecture*, J. Algebraic Combin. **18** (2003), no. 1, 13–39, DOI 10.1023/A:1025113308552. MR2002217

[5] Claudio Carmeli and Rita Fioresi, *Superdistributions, analytic and algebraic super Harish-Chandra pairs*, Pacific J. Math. **263** (2013), no. 1, 29–51, DOI 10.2140/pjm.2013.263.29. MR3069075

[6] Shun-Jen Cheng, Bin Shu, and Weiqiang Wang, *Modular representations of exceptional supergroups*, Math. Z. **291** (2019), no. 1-2, 635–659, DOI 10.1007/s00209-018-2098-x. MR3936085

[7] C. Chevalley, *Sur certains groupes simples* (French), Tohoku Math. J. (2) **7** (1955), 14–66, DOI 10.2748/tmj/1178245104. MR73602

[8] R. Fioresi and F. Gavarini, *Chevalley supergroups*, Mem. Amer. Math. Soc. **215** (2012), no. 1014, vi+64, DOI 10.1090/S0065-9266-2011-00633-7. MR2918543

[9] Gerhard P. Hochschild, *Basic theory of algebraic groups and Lie algebras*, Graduate Texts in Mathematics, vol. 75, Springer-Verlag, New York-Berlin, 1981. MR620024

[10] N. I. Ivanova and A. L. Onishchik, *Parabolic subalgebras and gradings of reductive Lie superalgebras* (Russian), Sovrem. Mat. Fundam. Napravl. **20** (2006), 5–68, DOI 10.1007/s10958-008-9049-8; English transl., J. Math. Sci. (N.Y.) **152** (2008), no. 1, 1–60. MR2336486

[11] Jens Carsten Jantzen, *Representations of algebraic groups*, 2nd ed., Mathematical Surveys and Monographs, vol. 107, American Mathematical Society, Providence, RI, 2003. MR2015057

[12] V. G. Kac, *Lie superalgebras*, Advances in Math. **26** (1977), no. 1, 8–96, DOI 10.1016/0001-8708(77)90017-2. MR486011

[13] V. Kac, *Representations of classical Lie superalgebras*, Differential geometrical methods in mathematical physics, II (Proc. Conf., Univ. Bonn, Bonn, 1977), Lecture Notes in Math., vol. 676, Springer, Berlin, 1978, pp. 597–626. MR519631

[14] Bertram Kostant, *Graded manifolds, graded Lie theory, and prequantization*, Differential geometrical methods in mathematical physics (Proc. Sympos., Univ. Bonn, Bonn, 1975), Springer, Berlin, 1977, pp. 177–306. Lecture Notes in Math., Vol. 570. MR0580292

[15] František Marko, *Irreducibility of induced supermodules for general linear supergroups*, J. Algebra **494** (2018), 92–110, DOI 10.1016/j.jalgebra.2017.10.006. MR3723172

[16] F. Marko and A. N. Zubkov, *Blocks for the general linear supergroup* $\mathrm{GL}(m|n)$, Transform. Groups **23** (2018), no. 1, 185–215, DOI 10.1007/s00031-017-9429-6. MR3763946

[17] Akira Masuoka, *Harish-Chandra pairs for algebraic affine supergroup schemes over an arbitrary field*, Transform. Groups **17** (2012), no. 4, 1085–1121, DOI 10.1007/s00031-012-9203-8. MR3000482

[18] Akira Masuoka and Taiki Shibata, *Algebraic supergroups and Harish-Chandra pairs over a commutative ring*, Trans. Amer. Math. Soc. **369** (2017), no. 5, 3443–3481, DOI 10.1090/tran/6751. MR3605977

[19] Akira Masuoka and Taiki Shibata, *On functor points of affine supergroups*, J. Algebra **503** (2018), 534–572, DOI 10.1016/j.jalgebra.2018.02.002. MR3780007

[20] Akira Masuoka, Taiki Shibata, and Yuta Shimada, *Affine algebraic super-groups with integral*, preprint `arXiv:2003.05100` (2020), [math.AG].

[21] Akira Masuoka, *Quotients G/H in super-symmetry*, Proceedings of the 51st Symposium on Ring Theory and Representation Theory, Symp. Ring Theory Represent. Theory Organ. Comm., Shizuoka, 2019, pp. 62–67. MR3930678

[22] Akira Masuoka and Alexandr N. Zubkov, *Quotient sheaves of algebraic supergroups are superschemes*, J. Algebra **348** (2011), 135–170, DOI 10.1016/j.jalgebra.2011.08.038. MR2852235

[23] Ian M. Musson, *Lie superalgebras and enveloping algebras*, Graduate Studies in Mathematics, vol. 131, American Mathematical Society, Providence, RI, 2012, DOI 10.1090/gsm/131. MR2906817

[24] Brian Parshall and Jian Pan Wang, *Quantum linear groups*, Mem. Amer. Math. Soc. **89** (1991), no. 439, vi+157, DOI 10.1090/memo/0439. MR1048073

[25] Vera Serganova, *Quasireductive supergroups*, New developments in Lie theory and its applications, Contemp. Math., vol. 544, Amer. Math. Soc., Providence, RI, 2011, pp. 141–159, DOI 10.1090/conm/544/10753. MR2849718

[26] Taiki Shibata, *Borel-Weil theorem for algebraic supergroups*, J. Algebra **547** (2020), 179–219, DOI 10.1016/j.jalgebra.2019.11.019. MR4039427

[27] Bin Shu and Weiqiang Wang, *Modular representations of the ortho-symplectic supergroups*, Proc. Lond. Math. Soc. (3) **96** (2008), no. 1, 251–271, DOI 10.1112/plms/pdm040. MR2392322

[28] Mitsuhiro Takeuchi, *Tangent coalgebras and hyperalgebras. I*, Jpn. J. Math. **42** (1974), 1–143, DOI 10.4099/jjm1924.42.0_1. MR389896

[29] Mitsuhiro Takeuchi, *On coverings and hyperalgebras of affine algebraic groups*, Trans. Amer. Math. Soc. **211** (1975), 249–275, DOI 10.2307/1997233. MR429928

[30] Mitsuhiro Takeuchi, *A hyperalgebraic proof of the isomorphism and isogeny theorems for reductive groups*, J. Algebra **85** (1983), no. 1, 179–196, DOI 10.1016/0021-8693(83)90124-2. MR723073

[31] Rainer Weissauer, *Semisimple algebraic tensor categories*, preprint `arXiv:0909.1793` (2009), [math.CT].

[32] A. N. Zubkov, *Some properties of general linear supergroups and of Schur superalgebras* (Russian, with Russian summary), Algebra Logika **45** (2006), no. 3, 257–299, 375, DOI 10.1007/s10469-006-0014-7; English transl., Algebra Logic **45** (2006), no. 3, 147–171. MR2289090

Department of Applied Mathematics, Okayama University of Science, 1-1 Ridai-cho Kita-ku Okayama-shi, Okayama 700-0005, Japan

Email address: `shibata@xmath.ous.ac.jp`

Contemporary Mathematics
Volume **771**, 2021
https://doi.org/10.1090/conm/771/15516

Hopf algebraic methods in finite tensor categories: Characters and integrals

Kenichi Shimizu

ABSTRACT. This paper reviews the character theory and the integral theory for Hopf algebras extended to finite tensor categories, as well as recent techniques for finite abelian categories and finite tensor categories. Some applications to fusion categories are also given.

1. Introduction

A tensor category [**17**] is an algebraic framework widely used in mathematics and mathematical physics. The representation category of a Hopf algebra [**31**] is one of typical examples of tensor categories. Recent studies of Hopf algebras are greatly influenced by the development of the theory of tensor categories. It is also important to reexamine results on Hopf algebras in the setting of tensor categories. Recently, the author has extended the character theory and the integral theory of Hopf algebras to *finite tensor categories* [**17, 19**]. This paper reviews such 'Hopf algebraic methods' in finite tensor categories by reorganizing [**38–41**].

The author's motivation was the study of 'non-semisimple' modular tensor categories introduced by Lyubashenko [**25, 27, 28**]. If $\mathcal{C}$ is a modular tensor category in the sense of Lyubashenko, then the modular group $SL_2(\mathbb{Z})$ acts projectively on the vector space $\mathrm{Hom}_{\mathcal{C}}(\mathbb{1}, L)$, where L is the coend $\int^{X \in \mathcal{C}} X^* \otimes X$. The action is actually given by so-called S and T matrices of $\mathcal{C}$ if $\mathcal{C}$ is semisimple. However, in the non-semisimple case, the action is hard to understand. Since L is not expressed explicitly, it is even hard to give a basis of $\mathrm{Hom}_{\mathcal{C}}(\mathbb{1}, L)$.

Non-semisimple modular tensor categories are expected to relate logarithmic conformal field theories. From such a point of view of mathematical physics, the space on which the modular group acts should be the space of characters. Hence it is natural to consider elements of $\mathrm{Hom}_{\mathcal{C}}(\mathbb{1}, L)$ that can be thought of as characters of objects of $\mathcal{C}$. The definition of the character $\chi_V \in \mathrm{Hom}_{\mathcal{C}}(\mathbb{1}, L)$ of an object $V \in \mathcal{C}$ has been proposed in [**21, 22**]. Although this notion seemed to be important, only few their properties were known at that time.

To study the properties of the character of [**21, 22**], the author initiated the character theory in pivotal finite tensor category in [**39**]. For such a category $\mathcal{C}$, the algebra $\mathrm{CF}(\mathcal{C})$ of *class functions* of $\mathcal{C}$ and the *internal character* $\mathrm{ch}_X \in \mathrm{CF}(\mathcal{C})$ for $X \in \mathcal{C}$ are defined in terms of the adjunction between $\mathcal{C}$ and its Drinfeld center.

The author was supported by JSPS KAKENHI Grant Number 16K17568.

There is, in fact, an isomorphism $\mathrm{Hom}_{\mathcal{C}}(\mathbb{1}, L) \to \mathrm{CF}(\mathcal{C})$ mapping χ_V to ch_V. Now we set $\mathrm{Gr}_k(\mathcal{C}) = k \otimes_{\mathbb{Z}} \mathrm{Gr}(\mathcal{C})$, where k is the base field and $\mathrm{Gr}(\mathcal{C})$ is the Grothendieck ring of $\mathcal{C}$. The main result of [**39**] is that the linear map

$$\mathrm{ch} : \mathrm{Gr}_k(\mathcal{C}) \to \mathrm{CF}(\mathcal{C}), \quad [X] \mapsto \mathrm{ch}_X \tag{1.1}$$

is a well-defined injective homomorphism of algebras. In particular, as in the group representation theory, the internal characters of simple objects are linearly independent. Furthermore, under the assumption that $\mathcal{C}$ is unimodular, the above map is bijective if and only if $\mathcal{C}$ is semisimple. This implies that, if $\mathcal{C}$ is a non-semisimple unimodular pivotal finite tensor category, there is a class function that is not a linear combination of characters.

The proof of the above results rely on the Eilenberg-Watts calculus for finite abelian categories and the integral theory in finite tensor categories. These two theories are of independent interest. The former is a new approach to deal with finite abelian categories. To establish the latter one, some results on the induction to the Drinfeld center is required. The reason why the Drinfeld center is important is that integrals and cointegrals for Hopf algebras closely relate to particular Yetter-Drinfeld modules as we will review in Section 2.

Now we explain the organization of this paper. In Section 2, we review some results on Hopf algebras and Yetter-Drinfeld modules. Given a Hopf algebra H, we denote by ${}^H_H\mathcal{YD}$ the category of Yetter-Drinfeld H-modules. A right adjoint R of the forgetful functor ${}^H_H\mathcal{YD} \to H\text{-Mod}$ was constructed by Radford. We reformulate some known results on Hopf algebras by using the functor R.

In Section 3, we introduce an abstract treatment for finite abelian categories, which we call the Eilenberg-Watts calculus. The celebrated Eilenberg-Watts theorem yields an equivalence $\mathbb{EW}_{\mathcal{M},\mathcal{N}} : \mathcal{M} \boxtimes \mathcal{N}^{\mathrm{op}} \to \mathrm{Rex}(\mathcal{N}, \mathcal{M})$ for finite abelian categories $\mathcal{M}$ and $\mathcal{N}$, where $\mathrm{Rex}(\mathcal{N}, \mathcal{M})$ means the category of k-linear right exact functors from $\mathcal{N}$ to $\mathcal{M}$. This equivalence will be used in later sections. We also provide miscellaneous useful results on finite abelian categories.

In Section 4, for a finite tensor category $\mathcal{C}$, we construct a right adjoint R of the forgetful functor $U : \mathcal{Z}(\mathcal{C}) \to \mathcal{C}$, where $\mathcal{Z}(\mathcal{C})$ is the Drinfeld center of $\mathcal{C}$. There is a canonical coalgebra $\mathbb{H}$ in $\mathcal{C}^{\mathrm{env}} := \mathcal{C} \boxtimes \mathcal{C}^{\mathrm{rev}}$. Our construction is based on the equivalence between $\mathcal{C}$ and the category of $\mathbb{H}$-comodules in $\mathcal{C}^{\mathrm{env}}$ (which generalizes the fundamental theorem for Hopf bimodules). The coalgebra $\mathbb{H}$ induces a comonad Z on $\mathcal{C}$, which we call the central Hopf comonad as it is the dual of the central monad considered in [**3, 13**]. The category of comodules over the comonad Z can be identified with $\mathcal{Z}(\mathcal{C})$. Under this identification, a right adjoint of U is just the free Z-comodule functor. A left adjoint L of U is also easily constructed and, from the construction, we know that there are natural isomorphisms

$$L(V) \cong R(\alpha \otimes V) \quad \text{and} \quad R(V) \cong L(\alpha^* \otimes V) \quad (V \in \mathcal{C}),$$

where α is a certain object of $\mathcal{C}$ arising from the dual of $\mathbb{H}$. The object α is in fact a categorical analogue of the modular function on a Hopf algebra.

In Section 5, for a finite tensor category $\mathcal{C}$, we introduce the space $\mathrm{CF}(\mathcal{C})$ of class functions by using the adjunction $U \dashv R$ constructed in Section 4. We also give a subspace $\mathrm{CF}_{\mathcal{C}}(\mathcal{S})$ of $\mathrm{CF}(\mathcal{C})$ for a topologizing subcategory $\mathcal{S}$ of $\mathcal{C}$. If $\mathcal{C}$ is pivotal, then $\mathrm{ch}_X \in \mathrm{CF}(\mathcal{C})$ is defined for each object $X \in \mathcal{C}$. An important fact is that ch_X belongs to $\mathrm{CF}_{\mathcal{C}}(\mathcal{S})$ if X belongs to a topologizing subcategory S. The injectivity of the map (1.1) is proved by using this observation.

In Section 6, we establish the integral theory for finite tensor category. Based on a certain fact on cointegrals on Hopf algebras, we introduce a categorical integral and a categorical cointegral for a finite tensor category $\mathcal{C}$ again by using the adjunction $U \dashv R$. As in the Hopf algebra theory, categorical (co)integrals exist and are unique up to scalar multiple. This fact will be proved by using a certain Frobenius property of the algebra $R(\mathbb{1}) \in \mathcal{Z}(\mathcal{C})$. We formulate and prove Maschke's theorem for a finite tensor category with the use of integrals. Finally, for unimodular $\mathcal{C}$, it is shown that (1.1) is bijective if and only if $\mathcal{C}$ is semisimple.

In Section 7, we give some applications of the character theory and the integral theory to semisimple finite tensor categories, *i.e.*, fusion categories. For a pivotal fusion category $\mathcal{C}$, the following conditions (1)–(4) are equivalent: (1) $\dim(\mathcal{C}) \neq 0$. (2) $\mathcal{Z}(\mathcal{C})$ is semisimple. (3) $R(\mathbb{1}) \in \mathcal{Z}(\mathcal{C})$ is semisimple. (4) $\mathrm{Gr}_k(\mathcal{C})$ is semisimple. Furthermore, if these equivalent conditions are satisfied, then $\mathrm{Gr}_k(\mathcal{C})$ is commutative if and only if $R(\mathbb{1})$ is multiplicity-free. Finally, we introduce the conjugacy class, the class sum and the character table for fusion categories in the spirit of Cohen and Westreich [**10**].

2. Representation theory and Yetter-Drinfeld modules

The aim of this paper is to review the character theory and the integral theory of Hopf algebras extended to the setting of finite tensor categories. In this section, we review some motivative results in the Hopf algebra theory.

Throughout this section, we work over a field k. By an *algebra*, we always mean an associative and unital algebra over k. Given two algebras A and B, we denote by A-Mod, Mod-B and A-Mod-B the category of left A-modules, the category of right B-modules and the category of A-B-bimodules, respectively. We use 'mod' instead of 'Mod' to express their finite-dimensional variants. For example, A-mod means the category of finite-dimensional left A-modules.

Given a vector space X, we denote its dual space by $X^* = \mathrm{Hom}_k(X, k)$. If X is an A-B-bimodule, then X^* is a B-A-bimodule by the actions given by

$$(b \rightharpoonup f \leftharpoonup a)(x) = f(axb) \quad (a \in A, b \in B, f \in X^*, x \in X).$$

Unless otherwise noted, the symbol $\otimes$ means the tensor product over k. Given a Hopf algebra H (over k), we denote by Δ, ϵ and S the comultiplication, the counit and the antipode of H, respectively. The Sweedler notation, such as

$$\Delta(h) = h_{(1)} \otimes h_{(2)}, \quad \Delta(h_{(1)}) \otimes h_{(2)} = h_{(1)} \otimes h_{(2)} \otimes h_{(3)} = h_{(1)} \otimes \Delta(h_{(2)}),$$

will be freely used to express the comultiplication.

2.1. Yetter-Drinfeld modules. Let H be a Hopf algebra. A (left-left) *Yetter-Drinfeld module* [**31**] over H is a left H-module M equipped with a left H-comodule structure $\delta : M \to H \otimes M$ satisfying the Yetter-Drinfeld condition

$$\delta(hm) = h_{(1)} m_{(-1)} S(h_{(3)}) \otimes h_{(2)} m_{(0)}$$

for all elements $h \in H$ and $m \in M$, where $\delta(m) = m_{(-1)} \otimes m_{(0)}$. A morphism of Yetter-Drinfeld modules over H is an H-linear map that is also H-colinear. We denote by ${}^H_H\mathcal{YD}$ the category of Yetter-Drinfeld modules over H. This category has a structure of a monoidal category inherited from the category of left H-modules and the category of left H-comodules. Furthermore, it has the braiding given by

$$c_{M,N} : M \otimes N \to N \otimes M, \quad m \otimes n \mapsto m_{(-1)} n \otimes m_{(0)}$$

for $M, N \in {}^H_H\mathcal{YD}$, $m \in M$ and $n \in N$.

The vector space H is a Yetter-Drinfeld H-module by the adjoint action $h \triangleright h' = h_{(1)}h'S(h_{(2)})$ and the regular coaction $h \mapsto h_{(1)} \otimes h_{(2)}$. As Zhu [**43**, p.2851] pointed out, if $H = kG$ is the group algebra of a group G, then there is a bijection

$$\left\{ \begin{array}{c} \text{conjugacy} \\ \text{classes of } G \end{array} \right\} \xrightarrow{1:1} \left\{ \begin{array}{c} \text{simple subobjects} \\ \text{of } H \in {}^H_H\mathcal{YD} \end{array} \right\}, \quad \mathfrak{C} \mapsto \mathrm{span}_k \mathfrak{C}.$$

By this observation, it is natural to regard a simple subobject of $H \in {}^H_H\mathcal{YD}$ as an analogue of a conjugacy class of the Hopf algebra H. Cohen and Westreich [**10**] defined a conjugacy class of a semisimple Hopf algebra in this spirit and studied its properties.

2.2. Radford induction and the character algebra. As we have reviewed in the above, conjugacy classes can be viewed as subobjects of $H \in {}^H_H\mathcal{YD}$. Zhu also pointed out that the endomorphism algebra of $H \in {}^H_H\mathcal{YD}$ is isomorphic to the character algebra of H if H is a semisimple Hopf algebra over an algebraically closed field of characteristic zero (as a part of [**43**, Theorem 1]).

Taking these results into account, we are led to the question what $H \in {}^H_H\mathcal{YD}$ is. An answer to this question is given by the Radford induction: Given a left H-module V, we set $R(V) = H \otimes V$ as a vector space and endow it with a structure of a Yetter-Drinfeld H-module by

$$h \triangleright (h' \otimes v) = h_{(1)}h'S(h_{(3)}) \otimes h_{(2)}v \quad \text{and} \quad \delta(h' \otimes v) = h'_{(1)} \otimes h'_{(2)} \otimes v$$

for $h, h' \in H$ and $v \in V$ (*cf.* Radford [**36**, Proposition 2]). The map $V \mapsto R(V)$ gives rise to a functor from the category H-Mod to ${}^H_H\mathcal{YD}$, which we call the *Radford induction*. We note that R is right adjoint to the functor $U : {}^H_H\mathcal{YD} \to H$-Mod forgetting the left H-coaction. Indeed, the map

$$\mathrm{Hom}_{{}^H_H\mathcal{YD}}(M, R(V)) \to \mathrm{Hom}_H(U(M), V), \quad f \mapsto (\epsilon \otimes \mathrm{id}_V) \circ f$$

is a natural isomorphism for $M \in {}^H_H\mathcal{YD}$ and $V \in {}_H\mathcal{M}$ with inverse

$$\mathrm{Hom}_H(U(M), V) \to \mathrm{Hom}_{{}^H_H\mathcal{YD}}(M, R(V)), \quad f \mapsto \Big[m \mapsto m_{(-1)} \otimes f(m_{(0)})\Big].$$

REMARK 2.1. Let ${}^H_H\mathcal{YD}_{fd}$ be the category of finite-dimensional Yetter-Drinfeld modules over H. If H is finite-dimensional, then the adjunction $U \dashv R$ restricts to an adjunction between ${}^H_H\mathcal{YD}_{fd}$ and H-mod.

The Yetter-Drinfeld module H is identified with $R(k)$, where $k \in H$-Mod is the trivial left H-module. By this observation, Zhu's result mentioned in the above is proved as follows: The vector space

$$C(H) := \{f \in H^* \mid f(ab) = f(bS^2(a)) \text{ for all } a, b \in H\}$$

is actually a subalgebra of H^*. There are isomorphisms

$$\begin{aligned} \mathrm{End}_{{}^H_H\mathcal{YD}}(H, H) &\cong \mathrm{Hom}_{{}^H_H\mathcal{YD}}(R(k), R(k)) \cong \mathrm{Hom}_H(UR(k), k) \\ &\cong \{f \in H^* \mid f(h_{(1)}h'S(h_{(2)})) = \epsilon(h)f(h') \text{ for all } h, h' \in H\} = C(H) \end{aligned}$$

of vector spaces and the resulting isomorphism $\mathrm{End}_{{}^H_H\mathcal{YD}}(H, H) \cong C(H)$ is in fact an isomorphism of algebras. Now we prove:

THEOREM 2.2. *If H is a semisimple Hopf algebra over an algebraically closed field of characteristic zero, then* $\mathrm{End}_{{}^H_H\mathcal{YD}}(H,H)$ *is isomorphic to the character algebra of H.*

PROOF. Since $S^2 = \mathrm{id}_H$ by [**26**], we have

$$C(H) = \{f \in H^* \mid f(ab) = f(ba) \text{ for all } a, b \in H\}.$$

By the semisimplicity, $C(H)$ is spanned by the characters of H. □

2.3. Integral theory. An integral and a cointegral for a Hopf algebra were introduced by Sweedler [**42**]. A *left integral* in a Hopf algebra H is an element $\Lambda \in H$ such that $h\Lambda = \epsilon(h)\Lambda$ for all elements $h \in H$. Dually, a *left cointegral* on H is a linear map $\lambda : H \to k$ such that $h_{(1)}\lambda(h_{(2)}) = \lambda(h)1_H$ for all $h \in H$.

We suppose that H is finite-dimensional. Then both the space of left integrals in H and the space of left cointegrals on H are one-dimensional. Let $\Lambda \in H$ be a non-zero left integral in H. Then Λh $(h \in H)$ is also a left integral in H. Since the space of left integrals in H is one-dimensional, there is a linear map $\alpha : H \to k$ such that $\Lambda h = \alpha(h)\Lambda$ for all $h \in H$. We call α the *modular function* on H as an analogy of the modular function on a locally compact group. We say that H is *unimodular* if $\alpha = \epsilon$.

Cointegrals are interpreted in terms of Yetter-Drinfeld modules. To explain this, we introduce the Yetter-Drinfeld H-module A_β for a grouplike element $\beta \in H^*$ as follows: We set $A_\beta = H$ as a vector space and endow it with a structure of a Yetter-Drinfeld H-module by $h \triangleright a = h_{(1)}aS(h_{(3)})\beta(h_{(2)})$ and $\delta(a) = a_{(1)} \otimes a_{(2)}$ for $h \in H$ and $a \in A_\beta$ $(= H)$. If we denote the one-dimensional left H-module associated to β by the same symbol β, then A_β can be identified with $R(\beta)$. The following theorem is proved by Ishii-Masuoka [**24**] in the unimodular case:

THEOREM 2.3. $\mathrm{Hom}_{{}^H_H\mathcal{YD}}(A_\alpha, k)$ *is the space of left cointegrals on H.*

PROOF. $\lambda \in H^*$ belongs to $I := \mathrm{Hom}_{{}^H_H\mathcal{YD}}(A_\alpha, k)$ if and only if

$$a_{(1)}\lambda(a_{(2)}) = \lambda(a)1_H \quad \text{and} \quad h_{(1)}aS(h_{(3)}))\alpha(h_{(2)}) = \varepsilon(h)\lambda(a)$$

for all $h \in H$ and $a \in A_\alpha$. The former equation is equivalent to that λ is a left cointegral on H. According to Radford [**35**], a left cointegral λ on H always satisfies the latter equation. Thus I coincides with the space of left cointegrals on H. □

3. Preliminaries: Finite abelian categories

We have reviewed some results on Hopf algebras. Our aim is to extend these results to the setting of *finite tensor categories* [**17**,**19**]. We skip its precise definition here (see Subsection 4.3), but note that a finite tensor category is required to be a finite abelian category in the following sense:

DEFINITION 3.1. A *finite abelian category* over a field k is a k-linear category that is equivalent to A-mod for some finite-dimensional algebra A over k.

In this section, we introduce some techniques for finite abelian categories established in [**20**,**38**,**39**,**41**] as a preparation for discussing finite tensor categories. To avoid some technical difficulties around the Deligne tensor product of finite abelian categories, the base field k is assumed to be algebraically closed in this section.

3.1. Right exact functors between finite abelian categories. Given two finite abelian categories $\mathcal{M}$ and $\mathcal{N}$, we denote by $\operatorname{Rex}(\mathcal{M},\mathcal{N})$ the category of k-linear right exact functors from $\mathcal{M}$ to $\mathcal{N}$. In the study of finite abelian categories and finite tensor categories, the following lemma is important:

LEMMA 3.2. *For two finite-dimensional algebras A and B, the functor*

$$A\text{-mod-}B \to \operatorname{Rex}(B\text{-mod}, A\text{-mod}), \quad M \mapsto M \otimes_B (-)$$

is an equivalence.

This lemma is proved by the same way as the Eilenberg-Watts theorem. We note several important consequences of the above lemma, which will be used throughout this paper:

(1) Let $\mathcal{M}$ and $\mathcal{N}$ be finite abelian categories. Then $\operatorname{Rex}(\mathcal{M},\mathcal{N})$ is a finite abelian category. Indeed, if A and B are finite-dimensional algebras such that $\mathcal{M} \approx A$-mod and $\mathcal{N} \approx B$-mod, then we have
$$\operatorname{Rex}(\mathcal{M},\mathcal{N}) \approx B\text{-mod-}A \approx (B \otimes_k A^{\mathrm{op}})\text{-mod}.$$
(2) The opposite category $\mathcal{M}^{\mathrm{op}}$ is also a finite abelian category since $\mathcal{M}^{\mathrm{op}} \approx (A\text{-mod})^{\mathrm{op}} \approx \text{mod-}A \approx A^{\mathrm{op}}\text{-mod}$, where the second equivalence is given by taking the dual of the underlying vector space.
(3) Fix an object $M \in \mathcal{M}$. The right exactness of the tensor product implies that the functor $\operatorname{Rex}(\mathcal{M},\mathcal{N}) \to \mathcal{N}$ sending F to $F(M)$ is right exact.
(4) A sequence $0 \to F \to G \to H \to 0$ in $\operatorname{Rex}(\mathcal{M},\mathcal{N})$ is exact if and only if the induced sequence $0 \to F(P) \to G(P) \to H(P) \to 0$ is exact for every projective object $P \in \mathcal{M}$. This fact is obvious if we view $0 \to F \to G \to H \to 0$ as a sequence of B-A-bimodules.
(5) A k-linear functor $F : \mathcal{M} \to \mathcal{N}$ has a right adjoint if and only if it is right exact. Indeed, the 'if' part follows from that a functor of the form $M \otimes_A (-)$ has a right adjoint. The 'only if' part is standard.
(6) By applying (5) to $F^{\mathrm{op}} : \mathcal{M}^{\mathrm{op}} \to \mathcal{N}^{\mathrm{op}}$, we see that F has a left adjoint if and only if it is left exact.

Let $\mathcal{M}$ be a finite abelian category. An object $M \in \mathcal{M}$ defines a k-linear left exact functor $\operatorname{Hom}_{\mathcal{M}}(M,-)$ from $\mathcal{M}$ to k-mod. This functor has a left adjoint, which we denote by $(-) \otimes_k M$. Thus there is a natural isomorphism

$$\operatorname{Hom}_{\mathcal{M}}(V \otimes_k M, M') \cong \operatorname{Hom}_k(V, \operatorname{Hom}_{\mathcal{M}}(M, M'))$$

for $V \in k$-mod and $M' \in \mathcal{M}$. The assignment $(V, M) \mapsto V \otimes_k M$ extends to a k-bilinear functor $k\text{-mod} \times \mathcal{M} \to \mathcal{M}$ in such a way that the above isomorphism is also natural in $M \in \mathcal{M}$.

3.2. Right exact monads on a finite abelian category. We refer the reader to [**30**] for basics on (co)monads. As our main objectives in this paper are somewhat representation-theoretic, we refer to a (co)algebra over a (co)monad (as defined in [**30**, VI]) as a (co)module. To construct a finite abelian category from a given one, the following result will be useful:

LEMMA 3.3. *Let $\mathcal{M}$ be a finite abelian category.*

(1) *If T is a k-linear right exact monad on $\mathcal{M}$, then the category ${}_T\mathcal{M}$ of T-modules is also a finite abelian category such that the forgetful functor ${}_T\mathcal{M} \to \mathcal{M}$ preserves and reflects exact sequences.*

(2) *If S is a k-linear left exact comonad on $\mathcal{M}$, then the category ${}^{S}\mathcal{M}$ of S-comodules is also a finite abelian category such that the forgetful functor ${}^{S}\mathcal{M} \to \mathcal{M}$ preserves and reflects exact sequences.*

PROOF. (1) Let $F : {}_{T}\mathcal{M} \to \mathcal{M}$ be the forgetful functor. It is a classical results that ${}_{T}\mathcal{M}$ is an abelian category such that the forgetful functor F preserves and reflects exact sequences [**16**, Proposition 5.3]. To show the finiteness of ${}_{T}\mathcal{M}$, we note that F has a left adjoint $L(X) = (T(X), \mu_X)$. If G is a progenerator of $\mathcal{M}$, then $L(G)$ is a progenerator of ${}_{T}\mathcal{M}$. Thus ${}_{T}\mathcal{M}$ is a finite abelian category.

(2) Let T be a left adjoint of S. Then T has a natural structure of a monad on $\mathcal{M}$ such that the category of T-modules can be identified with the category of S-comodules. Thus Part (2) is proved by applying Part (1) to this T. □

3.3. Ends and coends. Some important constructions in the theory of finite tensor categories rely on special kind of (co)limits, called (co)ends. We here collect some useful results on (co)ends.

Let $\mathcal{A}$ and $\mathcal{V}$ be categories, and let $S, T : \mathcal{A}^{\mathrm{op}} \times \mathcal{A} \to \mathcal{V}$ be functors. A *dinatural transformation* from S to T is a family $\alpha = \{\alpha_X : F(X,X) \to G(X,X)\}_{X \in \mathcal{A}}$ of morphisms in $\mathcal{V}$ such that the equation

$$T(\mathrm{id}_X, f) \circ \alpha_X \circ S(f, \mathrm{id}_X) = T(f, \mathrm{id}_Y) \circ \alpha_Y \circ S(\mathrm{id}_Y, f)$$

holds for all morphisms $f : X \to Y$ in $\mathcal{A}$. We write $\alpha : F \xrightarrow{..} G$ if α is a dinatural transformation from F to G.

An *end* of $T : \mathcal{A}^{\mathrm{op}} \times \mathcal{A} \to \mathcal{V}$ is an object $E \in \mathcal{V}$ equipped with a dinatural transformation $\pi : E \xrightarrow{..} T$, where E is regarded as a constant functor, satisfying the following universal property: For every object $E' \in \mathcal{V}$ and every dinatural transformation $\pi' : E' \xrightarrow{..} T$, there exists a unique morphism $\phi : E' \to E$ in $\mathcal{V}$ such that $\pi'_X = \pi_X \circ \phi$ for all objects $X \in \mathcal{A}$.

A *coend* of $S : \mathcal{A}^{\mathrm{op}} \times \mathcal{A} \to \mathcal{V}$ is an object $C \in \mathcal{V}$ equipped with a dinatural transformation $i : S \xrightarrow{..} C$ that is universal in a certain sense. We refer the reader to [**30**, IX] for basic properties of (co)ends. Accordingly, we denote by

$$\int_{X \in \mathcal{A}} T(X,X) \quad \text{and} \quad \int^{X \in \mathcal{A}} S(X,X)$$

the end of T and the coend of S, respectively.

We note some useful results on (co)ends. Let $\mathcal{A}$ and $\mathcal{B}$ be categories. To avoid set-theoretic issues, we assume that $\mathcal{A}$ is essentially small. Given two functors F and G from $\mathcal{A}$ to $\mathcal{B}$, we denote by $\mathrm{Nat}(F, G)$ the set of natural transformations from F to G. The following formula of is well-known [**30**, IX.5]:

LEMMA 3.4. $\int_{X \in \mathcal{A}} \mathrm{Hom}_{\mathcal{B}}(F(X), G(X)) \cong \mathrm{Nat}(F, G)$.

For a functor $T : \mathcal{A}^{\mathrm{op}} \times \mathcal{A} \to \mathcal{B}$ and an object $V \in \mathcal{B}$, the end

$$T^{\natural}(V) := \int_{X \in \mathcal{A}} \mathrm{Hom}_{\mathcal{B}}(V, T(X,X))$$

exists since the category **Set** is complete. The assignment $V \mapsto T^{\natural}(V)$ extends to a contravariant functor from $\mathcal{B}$ to **Set**. The following lemma will be used to guarantee existence of particular ends:

LEMMA 3.5 (the dual of [**40**, Lemma 3.1]). *An end of T exists if and only if the functor $T^\natural$ is representable. If these equivalent conditions are satisfied, then the object representing $T^\natural$ is an end of T.*

Let $\mathcal{A}$, $\mathcal{B}$ and $\mathcal{V}$ be categories, and let $H : \mathcal{B}^{\mathrm{op}} \times \mathcal{A} \to \mathcal{V}$ be a functor. Suppose that a functor $F : \mathcal{A} \to \mathcal{B}$ has a left adjoint $L : \mathcal{B} \to \mathcal{A}$. The following isomorphism will also be used:

LEMMA 3.6 (the dual of [**3**, Lemma 3.8]). *We have*

$$\int_{X \in \mathcal{A}} H(F(X), X) \cong \int_{Y \in \mathcal{B}} H(Y, L(Y)) \tag{3.1}$$

meaning that the end on the left-hand side exists if and only if the other one exists and, if they exist, they are canonically isomorphic.

The isomorphism (3.1) is constructed as follows: Let η and ε be the unit and the counit of the adjunction $L \dashv F$, respectively. Let E and E' be the left and the right-hand side of (3.1), respectively. We denote by

$$\pi_X : E \to H(F(X), X), \quad \pi'_Y : E' \to H(Y, L(Y)) \quad (X \in \mathcal{A}, Y \in \mathcal{B})$$

the respective universal dinatural transformations. By the universal property, there are morphisms $\varphi : E \to E'$ and $\overline{\varphi} : E' \to E$ in $\mathcal{V}$ such that the equations

$$\pi'_Y \circ \varphi = H(\eta_Y, \mathrm{id}_{L(Y)}) \circ \pi_{L(Y)}, \quad \pi_X \circ \overline{\varphi} = H(\mathrm{id}_{F(X)}, \varepsilon_X) \circ \pi'_{F(X)}$$

hold for all $X \in \mathcal{A}$ and $Y \in \mathcal{B}$. The zig-zag identities for η and ε and the dinaturality of π and π' imply that φ and $\overline{\varphi}$ are mutually inverse to each other.

3.4. Abstract Eilenberg-Watts theorem. Given two k-linear abelian categories $\mathcal{M}$ and $\mathcal{N}$, we denote by $\mathcal{M} \boxtimes \mathcal{N}$ their Deligne tensor product [**14, 17**]. By definition, it is a k-linear abelian category equipped with a functor $\boxtimes : \mathcal{M} \times \mathcal{N} \to \mathcal{M} \boxtimes \mathcal{N}$ that is k-linear and right exact in each variable and satisfies a certain universal property. The Deligne tensor product does not exist in general. If A and B are finite-dimensional algebras, then

$$(A\text{-mod}) \boxtimes (B\text{-mod}) = (A \otimes_k B)\text{-mod}$$

with $\boxtimes = \otimes_k$. Hence the Deligne tensor product is defined for finite abelian categories and the class of finite abelian categories is closed under the Deligne tensor product.

Now let $\mathcal{M}$ and $\mathcal{N}$ be finite abelian categories, and let A and B be finite-dimensional algebras such that $\mathcal{M} \approx A$-mod and $\mathcal{N} \approx B$-mod. Then we have

$$\begin{aligned} \mathcal{M} \boxtimes \mathcal{N}^{\mathrm{op}} &\approx (A\text{-mod}) \boxtimes (B^{\mathrm{op}}\text{-mod}) \approx A \otimes B^{\mathrm{op}}\text{-mod} \\ &\approx A\text{-mod-}B \approx \mathrm{Rex}(B\text{-mod}, A\text{-mod}) \approx \mathrm{Rex}(\mathcal{N}, \mathcal{M}) \end{aligned}$$

as k-linear categories. The equivalence $\mathcal{M} \boxtimes \mathcal{N}^{\mathrm{op}} \to \mathrm{Rex}(\mathcal{N}, \mathcal{M})$ obtained by composing these equivalences does not depend on the choice of A and B. Hence we obtain the following result, which we call the abstract Eilenberg-Watts equivalence for finite abelian categories [**20, 38**]:

LEMMA 3.7. *For finite abelian categories $\mathcal{M}$ and $\mathcal{N}$, the functor*

$$\mathbb{EW}_{\mathcal{M},\mathcal{N}} : \mathcal{M} \boxtimes \mathcal{N}^{\mathrm{op}} \to \mathrm{Rex}(\mathcal{N}, \mathcal{M}), \quad M \boxtimes N^{\mathrm{op}} \mapsto \mathrm{Hom}_{\mathcal{N}}(-, N)^* \otimes_k M$$

is an equivalence. Here, an object $N \in \mathcal{N}$ is written as N^{op} when it is regarded as an object of $\mathcal{N}^{\mathrm{op}}$.

Let $\mathbb{EW}^{-1}_{\mathcal{M},\mathcal{N}}$ denote a quasi-inverse of $\mathbb{EW}_{\mathcal{M},\mathcal{N}}$. Then we have

$$\begin{aligned}\mathrm{Hom}_{\mathcal{M}\boxtimes\mathcal{N}}(M\boxtimes N, \mathbb{EW}^{-1}_{\mathcal{M},\mathcal{N}}(F)) &\cong \mathrm{Nat}(\mathbb{EW}_{\mathcal{M},\mathcal{N}}(M\boxtimes N), F)\\ &\cong \int_{X\in\mathcal{N}} \mathrm{Hom}_{\mathcal{M}}(\mathbb{EW}_{\mathcal{M},\mathcal{N}}(M\boxtimes N)(X), F(X))\\ &\cong \int_{X\in\mathcal{N}} \mathrm{Hom}_{\mathcal{M}}(\mathrm{Hom}_{\mathcal{N}}(X,N)^*\otimes_k M, F(X))\\ &\cong \int_{X\in\mathcal{N}} \mathrm{Hom}_{\mathcal{N}}(X,N)\otimes_k \mathrm{Hom}_{\mathcal{M}}(M, F(X))\\ &\cong \int_{X\in\mathcal{N}} \mathrm{Hom}_{\mathcal{N}^{\mathrm{op}}}(N^{\mathrm{op}}, X^{\mathrm{op}})\otimes_k \mathrm{Hom}_{\mathcal{M}}(M, F(X))\\ &\cong \int_{X\in\mathcal{N}} \mathrm{Hom}_{\mathcal{M}\boxtimes\mathcal{N}^{\mathrm{op}}}(M\boxtimes N^{\mathrm{op}}, F(X)\boxtimes X^{\mathrm{op}})\end{aligned}$$

for $M\in\mathcal{M}$, $N\in\mathcal{N}$ and $F\in\mathrm{Rex}(\mathcal{N},\mathcal{M})$. Since every object of $\mathcal{M}\boxtimes\mathcal{N}^{\mathrm{op}}$ is a finite colimit of objects of the form $M\boxtimes N^{\mathrm{op}}$, we have a natural isomorphism

$$\mathrm{Hom}_{\mathcal{M}\boxtimes\mathcal{N}}(B, \mathbb{EW}^{-1}_{\mathcal{M},\mathcal{N}}(F)) \cong \int_{X\in\mathcal{N}} \mathrm{Hom}_{\mathcal{M}\boxtimes\mathcal{N}^{\mathrm{op}}}(B, F(X)\boxtimes X^{\mathrm{op}})$$

for $B\in\mathcal{M}\boxtimes\mathcal{N}^{\mathrm{op}}$ and $F\in\mathrm{Rex}(\mathcal{N},\mathcal{M})$. By Lemma 3.5, we have the following expression of a quasi-inverse of the abstract Eilenberg-Watts equivalence:

$$\mathbb{EW}^{-1}_{\mathcal{M},\mathcal{N}}(F) = \int_{X\in\mathcal{N}} F(X)\boxtimes X^{\mathrm{op}} \quad (F\in\mathrm{Rex}(\mathcal{N},\mathcal{M})). \tag{3.2}$$

3.5. Topologizing subcategories. A full subcategory of an abelian category $\mathcal{M}$ is said to be *topologizing* [**37**] if it is closed under finite direct sums and subquotients taken in $\mathcal{M}$. We denote by $\mathfrak{Top}(\mathcal{M})$ the collection of all topologizing subcategories of $\mathcal{M}$.

Let A be an algebra, and let I be a (two-sided) ideal of A. Then A/I-mod is a topologizing subcategory of A-mod. If A is finite-dimensional, then the assignment $I\mapsto A/I$-mod actually establishes a bijection between the set of ideals of A and the set $\mathfrak{Top}(A\text{-mod})$. The aim of this subsection is to explain this fact and rephrase it in the context of finite abelian categories.

Let $\mathcal{M}$ be a finite abelian category, and let $\mathcal{S}$ be a topologizing subcategory of $\mathcal{M}$ with inclusion functor $i_{\mathcal{S}}:\mathcal{S}\to\mathcal{M}$. Given $M\in\mathcal{M}$, we denote by $i^{\sharp}_{\mathcal{S}}(M)$ the sum of all subobjects of M belonging to $\mathcal{S}$. If X and Y are subobjects of M belonging to $\mathcal{S}$, then their sum $X+Y$ (as a subobject of M) belongs to $\mathcal{S}$ as it is a quotient of $X\oplus Y$. By this observation and the finiteness of $\mathcal{M}$, we see that $i^{\sharp}_{\mathcal{S}}(M)$ belongs to $\mathcal{S}$. The assignment $M\mapsto i^{\sharp}_{\mathcal{S}}(M)$ extends to a functor $i^{\sharp}_{\mathcal{S}}:\mathcal{M}\to\mathcal{S}$. This functor is right adjoint to $i_{\mathcal{S}}$.

By applying the above argument to $i^{\mathrm{op}}_{\mathcal{S}}:\mathcal{S}^{\mathrm{op}}\to\mathcal{M}^{\mathrm{op}}$, we see that $i^{\mathrm{op}}_{\mathcal{S}}$ has a right adjoint. Hence $i_{\mathcal{S}}$ has a left adjoint, say $i^{\flat}_{\mathcal{S}}:\mathcal{M}\to\mathcal{S}$. Specifically, $i^{\flat}_{\mathcal{S}}(M)$ for $M\in\mathcal{M}$ is given by the quotient of M by the intersection of all subobjects $X\subset M$ such that $M/X\in\mathcal{S}$.

Now we set $\tau_{\mathcal{S}}:=i_{\mathcal{S}}\circ i^{\flat}_{\mathcal{S}}$. By the construction of $i^{\flat}_{\mathcal{S}}$, an object $M\in\mathcal{M}$ belongs to $\mathcal{S}$ if and only if $M=\tau_{\mathcal{S}}(M)$ as quotient objects of M. In other words, $\tau_{\mathcal{S}}$ is an idempotent monad on $\mathcal{M}$ whose category of modules is precisely $\mathcal{S}$. We note that the forgetful functor from the category of $\tau_{\mathcal{S}}$-modules is nothing but the inclusion functor $i_{\mathcal{S}}$. It is easy to see that $i_{\mathcal{S}} \circ i_{\mathcal{S}}^{\sharp}$ is right adjoint to $\tau_{\mathcal{S}}$. Thus, by Lemma 3.3, we have:

LEMMA 3.8. *A topologizing subcategory of a finite abelian category is a finite abelian category such that the inclusion functor preserves and reflects exact sequences.*

Let A be a finite-dimensional algebra, and let $\mathcal{S} \in \mathfrak{Top}(A\text{-mod})$. By construction, the monad $\tau_{\mathcal{S}}$ is a quotient object of $\mathrm{id}_{A\text{-mod}}$ in $\mathrm{Rex}(A\text{-mod}, A\text{-mod})$. We note that $\mathrm{id}_{A\text{-mod}}$ is isomorphic to $A \otimes_A (-)$ and a subbimodule of A is nothing but an ideal of A. Thus, by the Eilenberg-Watts equivalence, $\tau_{\mathcal{S}} \cong (A/I) \otimes_A (-)$ for some ideal I of A. This shows that the map

$$(\text{the set of ideals of } A) \to \mathfrak{Top}(A\text{-mod}), \quad I \mapsto A/I\text{-mod}$$

is surjective. It is easy to see that this map is injective. Thus we have established a bijection between the set of ideals of A and and $\mathfrak{Top}(A\text{-mod})$.

Now let $\mathcal{M}$ be a finite abelian category. By abuse of terminology, we call a subobject of $\mathrm{id}_{\mathcal{M}}$ in $\mathrm{Rex}(\mathcal{M}, \mathcal{M})$ an *ideal* of $\mathcal{M}$. The following lemma is obtained by reinterpreting the above discussion:

LEMMA 3.9. *Given an ideal J of $\mathcal{M}$, we define $T_J := \mathrm{id}_{\mathcal{M}}/J$. Then T_J has a canonical structure of an idempotent monad on $\mathcal{M}$ and the map*

$$(\textit{ideals of } \mathcal{M}) \to \mathfrak{Top}(\mathcal{M}), \quad J \mapsto (\textit{the category of } T_J\textit{-modules}) \tag{3.3}$$

is bijective. The inverse of this map sends $\mathcal{S} \in \mathfrak{Top}(\mathcal{M})$ to the kernel of the quotient morphism $\mathrm{id}_{\mathcal{C}} \twoheadrightarrow \tau_{\mathcal{S}}$ in $\mathrm{Rex}(\mathcal{M})$.

As an application of this correspondence, we give a formula 'changing the domain of integration' of ends. Let $\mathcal{M}$ be a finite abelian category, let $\mathcal{V}$ be a category, and let $G : \mathcal{M}^{\mathrm{op}} \times \mathcal{M} \to \mathcal{V}$ be a functor.

LEMMA 3.10. *With above notation, we have*

$$\int_{M \in \mathcal{M}} G(M, \tau_{\mathcal{S}}(M)) \cong \int_{X \in \mathcal{S}} G(X, X) \tag{3.4}$$

for all $\mathcal{S} \in \mathfrak{Top}(\mathcal{M})$, meaning that the end on the left-hand side exists if and only if the other one exists and, if they exist, they are canonically isomorphic.

PROOF. This is an immediate consequence of Lemma 3.6 applied to

$$H(X, Y) = G(i_{\mathcal{S}}(X), M) \quad (X \in \mathcal{S}, M \in \mathcal{M}), \quad F = i_{\mathcal{S}} \quad \text{and} \quad L = i_{\mathcal{S}}^{\flat}.$$

□

3.6. Nakayama functor. There is a contravariant functor

$$\mathbb{D}_{\mathcal{M},\mathcal{N}} : \mathcal{M} \boxtimes \mathcal{N}^{\mathrm{op}} \to \mathcal{N} \boxtimes \mathcal{M}^{\mathrm{op}}, \quad M \boxtimes N^{\mathrm{op}} \mapsto N \boxtimes M^{\mathrm{op}}$$

for finite abelian categories $\mathcal{M}$ and $\mathcal{N}$. If $\mathcal{M} = A\text{-mod}$ and $\mathcal{N} = B\text{-mod}$ for some finite-dimensional algebras A and B, then there is an equivalence

$$\beta_{\mathcal{M},\mathcal{N}} : \mathcal{M} \boxtimes \mathcal{N}^{\mathrm{op}} \to A\text{-mod-}B, \quad M \boxtimes N^{\mathrm{op}} \mapsto M \otimes_k N^*$$

which makes the following diagram commute:

$$\begin{array}{ccc} \mathcal{M} \boxtimes \mathcal{N}^{\mathrm{op}} & \xrightarrow{\beta_{\mathcal{M},\mathcal{N}}} & A\text{-mod-}B \\ {\scriptstyle \mathbb{D}}\downarrow & & \downarrow{\scriptstyle (-)^*} \\ \mathcal{N} \boxtimes \mathcal{M}^{\mathrm{op}} & \xrightarrow{\beta_{\mathcal{N},\mathcal{M}}} & B\text{-mod-}A \end{array}$$

Thus we may think the contravariant functor $\mathbb{D}_{\mathcal{M},\mathcal{N}}$ as an 'algebra free' description of the functor given by taking the dual bimodule.

The dual of $A \in A\text{-mod-}A$ often plays an important role in the representation theory of finite-dimensional algebras. If A is a Frobenius algebra, then the endofunctor $A^* \otimes_A (-)$ on A-mod is isomorphic to the autoequivalence induced by the Nakayama automorphism. In view of this observation, we introduce:

DEFINITION 3.11. For a finite abelian category $\mathcal{M}$, we define the *Nakayama functor* $\mathbb{N}_{\mathcal{M}} \in \operatorname{Rex}(\mathcal{M})$ by

$$\mathbb{N}_{\mathcal{M}} = \mathbb{EW}_{\mathcal{M},\mathcal{M}} \mathbb{D}_{\mathcal{M},\mathcal{M}} \mathbb{EW}^{-1}_{\mathcal{M},\mathcal{M}}(\mathrm{id}_{\mathcal{M}}).$$

Fuchs, Schaumann and Schweigert [**20**] defined the Nakayama functor of a finite abelian category in a slightly different way. We explain that their definition agrees with ours. Let $\mathcal{M}$ and $\mathcal{N}$ be finite abelian categories, and let $F : \mathcal{N} \to \mathcal{M}$ be a k-linear right exact functor. For generality, we consider the functor

$$F^{\star} := \mathbb{EW}_{\mathcal{M},\mathcal{N}} \mathbb{D}_{\mathcal{M},\mathcal{N}} \mathbb{EW}^{-1}_{\mathcal{N},\mathcal{M}}(F) \in \operatorname{Rex}(\mathcal{M},\mathcal{N}).$$

The object $\mathbb{EW}^{-1}_{\mathcal{N},\mathcal{M}}(F) \in \mathcal{M} \boxtimes \mathcal{N}^{\mathrm{op}}$ is expressed by an end as in (3.2). Now we apply $\mathbb{D}_{\mathcal{M},\mathcal{N}}$ to this object. Since this functor is an antiequivalence of categories, it takes an end to a coend, and therefore we have

$$\mathbb{D}_{\mathcal{M},\mathcal{N}} \mathbb{EW}^{-1}_{\mathcal{N},\mathcal{M}}(F) = \int^{X \in \mathcal{N}} X \boxtimes F(X)^{\mathrm{op}}$$

as an object of $\mathcal{N} \boxtimes \mathcal{M}^{\mathrm{op}}$. Since an equivalence preserves coends, we obtain

$$F^{\star} = \int^{X \in \mathcal{N}} \mathbb{EW}_{\mathcal{M},\mathcal{N}}(X \boxtimes F(X)^{\mathrm{op}}).$$

Now we fix an object $M \in \mathcal{M}$ and consider the functor $\mathbb{E}_M : \operatorname{Rex}(\mathcal{M},\mathcal{N}) \to \mathcal{N}$ given by $G \mapsto G(M)$. Since this functor is right exact, we have

$$\begin{aligned} F^{\star}(M) = \mathbb{E}_M(F^{\star}) &\cong \int^{X \in \mathcal{N}} \mathbb{E}_M\Big(\mathbb{EW}_{\mathcal{M},\mathcal{N}}(X \boxtimes F(X)^{\mathrm{op}})\Big) \\ &\cong \int^{X \in \mathcal{N}} \operatorname{Hom}_{\mathcal{M}}(M, F(X))^* \otimes_k X. \end{aligned}$$

Thus we have the following expression of $\mathbb{N}_{\mathcal{M}} = (\mathrm{id}_{\mathcal{M}})^{\star}$, which coincides with the original definition [**20**, Definition 3.14] of the Nakayama functor:

LEMMA 3.12. *The Nakayama functor of $\mathcal{M}$ is given by*

$$\mathbb{N}_{\mathcal{M}}(M) = \int^{X \in \mathcal{M}} \operatorname{Hom}_{\mathcal{M}}(M, X)^* \otimes_k X \quad (M \in \mathcal{M}). \tag{3.5}$$

By construction, we have $\mathbb{N}_{A\text{-mod}} = A^* \otimes_A (-)$. By the representation theory of finite-dimensional algebras, $\mathbb{N}_{\mathcal{M}}$ is an equivalence if and only if $\mathcal{M} \approx A$-mod for some Frobenius algebra A. It is isomorphic to $\mathrm{id}_{\mathcal{M}}$ if and only if $\mathcal{M} \approx A$-mod for some symmetric Frobenius algebra A.

4. Induction to the Drinfeld center

4.1. The Drinfeld center. The Drinfeld center is an important construction in the theory of tensor categories. For a monoidal category $\mathcal{C}$, its Drinfeld center $\mathcal{Z}(\mathcal{C})$ is the category defined as follows: An object of this category is a pair $\mathbf{M} = (M, \sigma)$ consisting of an object M of $\mathcal{C}$ and a natural transformation $\sigma : M \otimes \mathrm{id}_{\mathcal{C}} \to \mathrm{id}_{\mathcal{C}} \otimes M$ such that the equations

$$\sigma_{X \otimes Y} = (\mathrm{id}_X \otimes \sigma_Y)(\sigma_X \otimes \mathrm{id}_Y) \quad \text{and} \quad \sigma_{\mathbb{1}} = \mathrm{id}_M$$

hold for all objects $X, Y \in \mathcal{C}$. Given two objects $\mathbf{M} = (M, \sigma)$ and $\mathbf{N} = (N, \tau)$ of $\mathcal{Z}(\mathcal{C})$, a morphism $f : \mathbf{M} \to \mathbf{N}$ in $\mathcal{Z}(\mathcal{C})$ is a morphism $f : M \to N$ in $\mathcal{C}$ such that the equation $(\mathrm{id}_X \otimes f)\sigma_X = \tau_X(f \otimes \mathrm{id}_X)$ holds for all objects $X \in \mathcal{C}$. The category $\mathcal{Z}(\mathcal{C})$ is a braided monoidal category with tensor product

$$\mathbf{M} \otimes \mathbf{N} = \Big(M \otimes N, (\sigma \otimes \mathrm{id}_N)(\mathrm{id}_M \otimes \tau)\Big)$$

and braiding $c_{\mathbf{M},\mathbf{N}} = \sigma_N$.

REMARK 4.1. The natural transformation σ for $\mathbf{M} = (M, \sigma) \in \mathcal{Z}(\mathcal{C})$ is often referred to as the *half-braiding*. In this paper, we mainly concern the case where $\mathcal{C}$ is a rigid monoidal category. A half-braiding is invertible in such a case.

If H is a Hopf algebra, then every Yetter-Drinfeld H-module M becomes an object of $\mathcal{Z}(H\text{-Mod})$ together with the half-braiding

$$\sigma_X : M \otimes X \to X \otimes M, \quad m \otimes x \mapsto m_{(-1)}x \otimes m_{(0)} \quad (m \in M, x \in X \in {}_H\mathcal{M}),$$

and this construction gives rise to an isomorphism ${}^H_H\mathcal{YD} \cong \mathcal{Z}(H\text{-Mod})$ of braided monoidal categories. If, furthermore, H is finite-dimensional, then there is also an isomorphism ${}^H_H\mathcal{YD}_{fd} \cong \mathcal{Z}(H\text{-mod})$ constructed in the same way. Here, ${}^H_H\mathcal{YD}_{fd}$ is the full subcategory of ${}^H_H\mathcal{YD}$ consisting of finite-dimensional objects.

In Section 2, we have observed that some important notions in the representation theory of Hopf algebras can be interpreted in terms of Yetter-Drinfeld modules. Our aim is to extend such representation-theoretic notions to tensor categories by using a right adjoint of the forgetful functor $U : \mathcal{Z}(\mathcal{C}) \to \mathcal{C}$. It seems to be a difficult problem when U has a right adjoint. In this section, we construct a right adjoint of U in the case where $\mathcal{C}$ is a finite tensor category and give some useful results obtained from our construction.

4.2. Convention on duals. We first fix our convention on dual objects in a monoidal category: Let L and R be objects of a monoidal category $\mathcal{C}$, and let $\varepsilon : L \otimes R \to \mathbb{1}$ and $\eta : \mathbb{1} \to R \otimes L$ be morphisms in $\mathcal{C}$. We say that the triple (L, ε, η) is a *left dual object* of R and the triple (R, ε, η) is a *right dual object* of L if the following equations hold:

$$(\varepsilon \otimes \mathrm{id}_L) \circ (\mathrm{id}_L \otimes \eta) = \mathrm{id}_L \quad \text{and} \quad (\mathrm{id}_R \otimes \varepsilon) \circ (\eta \otimes \mathrm{id}_R) = \mathrm{id}_R$$

A monoidal category $\mathcal{C}$ is said to be *rigid* if every object of $\mathcal{C}$ has a left dual and a right dual. In a rigid monoidal category, a left and a right dual object of X will be denoted by $(X^*, \mathrm{ev}_X, \mathrm{coev}_X)$ and $({}^*X, \mathrm{ev}'_X, \mathrm{coev}'_X)$, respectively.

4.3. Finite tensor categories and their modules. Throughout this section, we work over an algebraically closed field k.

DEFINITION 4.2. A *finite tensor category* [**17**, **19**] is a finite abelian category $\mathcal{C}$ equipped with a structure of a rigid monoidal category such that the tensor product $\otimes : \mathcal{C} \times \mathcal{C} \to \mathcal{C}$ is k-bilinear and the unit object $\mathbb{1} \in \mathcal{C}$ is a simple object.

Let $\mathcal{C}$ be a finite tensor category. The following lemma is well-known:

LEMMA 4.3. *Let V be an object of $\mathcal{C}$. Then the functors $V \otimes \mathrm{id}_{\mathcal{C}}$ and $\mathrm{id}_{\mathcal{C}} \otimes V$ are exact. If $V \neq 0$, then they are faithful.*

PROOF. See [**17**, Proposition 4.2.1] for the exactness. Now suppose $V \neq 0$. Let f be a morphism in $\mathcal{C}$ such that $f \otimes \mathrm{id}_V = 0$. Then, since $\mathrm{id}_{\mathcal{C}} \otimes V$ is exact, we have $\mathrm{Im}(f) \otimes V = \mathrm{Im}(f \otimes \mathrm{id}_V) = 0$. This implies $\mathrm{Im}(f) = 0$, since $X \otimes Y \neq 0$ whenever X and Y are non-zero objects of $\mathcal{C}$ [**17**, Exercise 4.3.11]. Hence $f = 0$, that is, $\mathrm{id}_{\mathcal{C}} \otimes V$ is faithful. The faithfulness of $V \otimes \mathrm{id}_{\mathcal{C}}$ is proved in the same way. □

A *left $\mathcal{C}$-module category* is a category $\mathcal{M}$ equipped with a functor $\circledcirc : \mathcal{C} \times \mathcal{M} \to \mathcal{M}$, called the action, and natural isomorphisms $(X \otimes Y) \circledcirc M \cong X \circledcirc (Y \circledcirc M)$ and $\mathbb{1} \circledcirc M \cong M$ $(X, Y \in \mathcal{C},\ M \in \mathcal{M})$ satisfying certain axioms similar to those for a monoidal category. A *right $\mathcal{C}$-module category* is defined in a similar way. For basics on module categories over a tensor category, we refer the reader to [**17**].

We usually consider $\mathcal{C}$-module categories with a finiteness condition: A *finite left $\mathcal{C}$-module category* is a left $\mathcal{C}$-module category $\mathcal{M}$ that is also a finite abelian category such that the action $\mathcal{C} \times \mathcal{M} \to \mathcal{M}$ is k-linear and right exact in each variable. A *finite right $\mathcal{C}$-module category* is defined in a similar way.

EXAMPLE 4.4. Given an algebra A in $\mathcal{C}$, we denote by $\mathcal{C}_A$ and ${}_A\mathcal{C}$ the category of left and right A-modules in $\mathcal{C}$, respectively. By Lemma 3.3, $\mathcal{C}_A$ is a finite left $\mathcal{C}$-module category and ${}_A\mathcal{C}$ is a finite right $\mathcal{C}$-module category.

Now let $\mathcal{M}$ be a finite left $\mathcal{C}$-module category. Given an object $M \in \mathcal{M}$, we denote by $\underline{\mathrm{Hom}}(M, -)$ a right adjoint of the functor $\mathcal{C} \to \mathcal{M}$, $X \mapsto X \circledcirc M$. Thus, by definition, there is a natural isomorphism

$$\mathrm{Hom}_{\mathcal{C}}(X, \underline{\mathrm{Hom}}(M, N)) \cong \mathrm{Hom}_{\mathcal{M}}(X \circledcirc M, N)$$

for $X \in \mathcal{C}$ and $N \in \mathcal{M}$. The assignment $(M, N) \mapsto \underline{\mathrm{Hom}}(M, N)$ extends to a functor $\underline{\mathrm{Hom}} : \mathcal{M}^{\mathrm{op}} \times \mathcal{M} \to \mathcal{C}$, which we call the *internal Hom functor*, such that the above isomorphism is also natural in M. The internal Hom functor makes $\mathcal{M}$ a $\mathcal{C}$-enriched category. Namely, there is the 'composition' morphism

$$\underline{\mathrm{Hom}}(M_2, M_3) \otimes \underline{\mathrm{Hom}}(M_1, M_2) \to \underline{\mathrm{Hom}}(M_1, M_3) \quad (M_1, M_2, M_3 \in \mathcal{M})$$

that is associative and unital in a certain sense. In particular, $\underline{\mathrm{Hom}}(M, M)$ is an algebra in $\mathcal{C}$ for every object $M \in \mathcal{M}$.

An object $P \in \mathcal{M}$ is said to be *$\mathcal{C}$-projective* in $\mathcal{M}$ if the functor $\underline{\mathrm{Hom}}(P, -)$ is exact. An object $P \in \mathcal{M}$ is called a *$\mathcal{C}$-generator* of $\mathcal{M}$ if every object of $M \in \mathcal{M}$ is a quotient of an object of the form $X \circledcirc P$ for some $X \in \mathcal{C}$. As an application of the Barr-Beck monadicity theorem, we have:

LEMMA 4.5 ([**15**, Theorem 2.24]). *If $P \in \mathcal{M}$ is a $\mathcal{C}$-projective $\mathcal{C}$-generator, then $\mathcal{M}$ is equivalent to $\mathcal{C}_A$ as a left $\mathcal{C}$-module categories, where $A = \underline{\mathrm{Hom}}(P, P)$.*

A finite left $\mathcal{C}$-module category always admits a $\mathcal{C}$-projective $\mathcal{C}$-generator [**15**, Lemma 2.25]. Thus, by the above lemma, every finite left $\mathcal{C}$-module category is of the form as in Example 4.4. The same holds for finite right $\mathcal{C}$-module categories. For later use, we note the following technical lemma:

LEMMA 4.6. *Let $\mathcal{M}$ be a finite left $\mathcal{C}$-module category. For non-zero objects $X \in \mathcal{C}$ and $M \in \mathcal{M}$, the functors $X \otimes \mathrm{id}_{\mathcal{M}} : \mathcal{M} \to \mathcal{M}$ and $\mathrm{id}_{\mathcal{C}} \otimes M : \mathcal{C} \to \mathcal{M}$ are exact and faithful. The same holds for finite right $\mathcal{C}$-module categories.*

PROOF. By the above argument, we may assume that $\mathcal{C} = \mathcal{C}_A$ for some algebra A in $\mathcal{C}$. Then the result immediately follows from Lemma 4.3. □

Let $\mathcal{M}$ be a finite left $\mathcal{C}$-module category. Given an object $X \in \mathcal{C}$, we define the functor $\mathfrak{L}_X : \mathcal{M} \to \mathcal{M}$ by $\mathfrak{L}_X(M) = X \otimes M$ $(M \in \mathcal{M})$. If A is an algebra in $\mathcal{C}$ with multiplication m and unit u, then $\mathfrak{L}_A$ is a monad by

$$\mathfrak{L}_A^2(M) = A \otimes (A \otimes M) \xrightarrow{\cong} (A \otimes A) \otimes M \xrightarrow{m \otimes \mathrm{id}_M} A \otimes M = \mathfrak{L}_A(M),$$

$$M \xrightarrow{\cong} \mathbb{1} \otimes M \xrightarrow{u \otimes \mathrm{id}_M} A \otimes M = \mathfrak{L}_A(M) \quad (M \in \mathcal{M}).$$

We define the category of left A-modules in $\mathcal{M}$ to be the category of modules over the monad $\mathfrak{L}_A$ and denote it by ${}_A\mathcal{M}$ (this notation is consistent with that introduced in Example 4.4 if we view $\mathcal{C}$ as a left $\mathcal{C}$-module category by the tensor product). By Lemma 3.3, ${}_A\mathcal{M}$ is a finite abelian category.

If C is a coalgebra in $\mathcal{C}$, then $\mathfrak{L}_C$ is a comonad on $\mathcal{M}$ in a natural way. The category ${}^C\mathcal{M}$ of left C-comodules in $\mathcal{M}$ is defined to be the category of comodules over $\mathfrak{L}_C$. By Lemma 3.3, ${}^C\mathcal{M}$ is a finite abelian category.

If $\mathcal{N}$ is a finite right $\mathcal{C}$-module category, then the category of right A-modules in $\mathcal{N}$ and the category of right C-comodules in $\mathcal{N}$ are defined analogously. They are denoted by $\mathcal{N}_A$ and $\mathcal{N}^C$, respectively. Again by Lemma 3.3, both $\mathcal{N}_A$ and $\mathcal{N}^C$ are finite abelian categories.

4.4. Hopf bimodules. Let $\mathcal{C}$ be a finite tensor category. We define the enveloping category $\mathcal{C}^{\mathrm{env}}$ by $\mathcal{C}^{\mathrm{env}} := \mathcal{C} \boxtimes \mathcal{C}^{\mathrm{rev}}$, where $\mathcal{C}^{\mathrm{rev}}$ is the category $\mathcal{C}$ equipped with the reversed tensor product $X \otimes^{\mathrm{rev}} Y = Y \otimes X$. This is a finite tensor category as the Deligne tensor product of finite tensor categories.

There is an equivalence

$$\Phi_{\mathcal{C}} := \left(\mathcal{C}^{\mathrm{env}} \xrightarrow{V \boxtimes W \mapsto V \boxtimes^* W} \mathcal{C} \boxtimes \mathcal{C}^{\mathrm{op}} \xrightarrow{\mathbb{EW}_{\mathcal{C},\mathcal{C}}} \mathrm{Rex}(\mathcal{C}) := \mathrm{Rex}(\mathcal{C}, \mathcal{C})\right)$$

with the quasi-inverse

$$\Phi_{\mathcal{C}}^{-1}(F) = \int_{X \in \mathcal{C}} F(X) \boxtimes X^* \quad (F \in \mathrm{Rex}(\mathcal{C})).$$

We consider the object $\mathbb{H} := \Phi_{\mathcal{C}}^{-1}(\mathrm{id}_{\mathcal{C}})$. Let $\pi_X : \mathbb{H} \to X \boxtimes X^*$ $(X \in \mathcal{C})$ denote the universal dinatural transformation. Since the tensor product of a finite tensor category is exact, we have

$$\mathbb{H} \otimes \mathbb{H} = \int_{X, Y \in \mathcal{C}} (X \boxtimes X^*) \otimes (Y \boxtimes Y^*) = \int_{X, Y \in \mathcal{C}} (X \otimes Y) \boxtimes (X \otimes Y)^*$$

by the Fubini theorem for ends. Thus there is a unique morphism $\Delta : \mathbb{H} \to \mathbb{H} \otimes \mathbb{H}$ in $\mathcal{C}^{\mathrm{env}}$ such that the equation $(\pi_X \otimes \pi_Y) \circ \Delta = \pi_{X \otimes Y}$ holds for all $X, Y \in \mathcal{C}$. It is straightforward to verify:

LEMMA 4.7. *$(\mathbb{H}, \Delta, \pi_{\mathbb{1}})$ is a coalgebra in $\mathcal{C}^{\mathrm{env}}$.*

The category of Hopf bimodules over a quasi-Hopf algebra has been introduced and studied in [**23**] to establish the integral theory for finite-dimensional quasi-Hopf algebras. The category of $\mathbb{H}$-comodules in $\mathcal{C}^{\mathrm{env}}$ can be thought of as an analogue of the category of Hopf bimodules; see Subsection 4.8. The following theorem is a categorical analogue of the fundamental theorem for Hopf bimodules:

THEOREM 4.8. *For a finite tensor category $\mathcal{C}$, the functors*

$$\Psi^l_{\mathcal{C}} : \mathcal{C} \to {}^{\mathbb{H}}(\mathcal{C}^{\mathrm{env}}), \quad V \mapsto \mathbb{H} \otimes (V \boxtimes \mathbb{1}),$$
$$\Psi^r_{\mathcal{C}} : \mathcal{C} \to (\mathcal{C}^{\mathrm{env}})^{\mathbb{H}}, \quad V \mapsto (V \boxtimes \mathbb{1}) \otimes \mathbb{H} \quad (V \in \mathcal{C})$$

are equivalences.

A categorical analogue of the fundamental theorem for Hopf bimodules has also been given in [**18**]. By the argument of [**40**, Subsection 4.3], our coalgebra $\mathbb{H}$ is the dual of the algebra $\underline{\mathrm{Hom}}(\mathbb{1}, \mathbb{1}) \in \mathcal{C}^{\mathrm{env}}$ considered in [**18**]. Thus this theorem is equivalent to [**18**, Proposition 2.3]. Below, we give a different proof based on observations on some functor categories.

PROOF. We only prove that $\Psi^r_{\mathcal{C}}$ is an equivalence, since the equivalency of the other one can be proved in a similar way. We make $\mathrm{Rex}(\mathcal{C})$ a right $\mathcal{C}^{\mathrm{env}}$-module category in such a way that the equivalence $\Phi_{\mathcal{C}} : \mathcal{C}^{\mathrm{env}} \to \mathrm{Rex}(\mathcal{C})$ is an equivalence of right $\mathcal{C}^{\mathrm{env}}$-module categories. The consequent action is given by

$$F \oslash (X \boxtimes Y) = F(- \otimes {}^{**}Y) \otimes X \quad (X, Y \in \mathcal{C}, F \in \mathrm{Rex}(\mathcal{C})). \tag{4.1}$$

The equivalence $\Phi_{\mathcal{C}}$ induces an equivalence $(\mathcal{C}^{\mathrm{env}})^{\mathbb{H}} \approx \mathrm{Rex}(\mathcal{C})^{\mathbb{H}}$ between the categories of right $\mathbb{H}$-comodules.

The category $\mathrm{Rex}(\mathcal{C})^{\mathbb{H}}$ is isomorphic to the category $\mathrm{Rex}_{-|\mathcal{C}}(\mathcal{C})$ of k-linear right exact right $\mathcal{C}$-module functors $\mathcal{C} \to \mathcal{C}$. Indeed, for $F \in \mathrm{Rex}(\mathcal{C})$, we have

$$\begin{aligned}
\mathrm{Nat}(F, F \oslash \mathbb{H}) &= \mathrm{Nat}(F, F \oslash (\int_{X \in \mathcal{C}} X \boxtimes X^*)) \\
&\cong \int_{X \in \mathcal{C}} \mathrm{Nat}(F, F \oslash (X \boxtimes X^*)) \\
&\cong \int_{X \in \mathcal{C}} \int_{V \in \mathcal{C}} \mathrm{Hom}_{\mathcal{C}}(F(V), F(V \otimes {}^*X) \otimes X) \\
&\cong \int_{X, V \in \mathcal{C}} \mathrm{Hom}_{\mathcal{C}}(F(V) \otimes {}^*X, F(V \otimes {}^*X)) \\
&\cong \{\text{natural transformations } F(V \otimes X) \to F(V) \otimes X \ (V, X \in \mathcal{C})\}
\end{aligned}$$

where the first isomorphism follows from the exactness of $\oslash$, the third from the Fubini theorem, and the last from that $X \mapsto {}^*X$ is an antiequivalence. This bijection induces a bijection between the set of right $\mathbb{H}$-comodule structures on F and the set of right $\mathcal{C}$-module structures on F. A category isomorphism $\mathrm{Rex}(\mathcal{C})^{\mathbb{H}} \cong \mathrm{Rex}_{-|\mathcal{C}}(\mathcal{C})$ is established in this way.

A functor of the form $V \otimes \mathrm{id}_{\mathcal{C}}$, $V \in \mathcal{C}$, is a k-linear right exact right $\mathcal{C}$-module endofunctor on $\mathcal{C}$. Hence we have a functor

$$\widetilde{\rho}_{\mathcal{C}} : \mathcal{C} \to \mathrm{Rex}_{-|\mathcal{C}}(\mathcal{C}), \quad V \mapsto V \otimes \mathrm{id}_{\mathcal{C}} \quad (V \in \mathcal{C}) \tag{4.2}$$

(we reserve the symbol $\boldsymbol{\rho}_{\mathcal{C}}$ for another functor). It is easy to see that this functor is an equivalence with a quasi-inverse given by evaluating at $\mathbb{1}$,

$$\mathbb{E}_{\mathcal{C}} : \mathrm{Rex}_{-|\mathcal{C}}(\mathcal{C}) \to \mathcal{C}, \quad F \mapsto F(\mathbb{1}). \tag{4.3}$$

Now we consider the composition

$$\Psi := \Big(\mathcal{C} \xrightarrow{\ \widetilde{\boldsymbol{\rho}}_{\mathcal{C}}\ } \mathrm{Rex}_{-|\mathcal{C}}(\mathcal{C}) \xrightarrow{\ \cong\ } \mathrm{Rex}(\mathcal{C})^{\mathbb{H}} \xrightarrow{\ \approx\ } (\mathcal{C}^{\mathrm{env}})^{\mathbb{H}}\Big)$$

of equivalences. The functor Ψ is isomorphic to $\Psi^r_{\mathcal{C}}$ of this theorem. Indeed,

$$\Psi(V) = \int_{X \in \mathcal{C}} (V \otimes X) \boxtimes X^* \cong (V \boxtimes \mathbb{1}) \otimes \int_{X \in \mathcal{C}} X \boxtimes X^* = \Psi^r_{\mathcal{C}}(V)$$

for $V \in \mathcal{C}$. Since Ψ is an equivalence, so is $\Psi^r_{\mathcal{C}}$. The proof is done. □

4.5. Categorical analogue of the modular function. Let $\mathbb{H}$ be the coalgebra in $\mathcal{C}^{\mathrm{env}}$ introduced in the above. We have mentioned that a result equivalent to Theorem 4.8 can be found in [**18**]. In that paper, the result is then used to define a categorical analogue of the modular function on a Hopf algebra. In the same spirit, we introduce the modular object as follows:

DEFINITION 4.9. The *modular object* of $\mathcal{C}$ is an object $\alpha_{\mathcal{C}} \in \mathcal{C}$ (which is uniquely determined up to isomorphism) such that

$$\mathbb{H}^* \cong \mathbb{H} \otimes (\alpha_{\mathcal{C}} \boxtimes \mathbb{1}) \tag{4.4}$$

as left $\mathbb{H}$-comodules in $\mathcal{C}^{\mathrm{env}}$. We say that $\mathcal{C}$ is *unimodular* if $\alpha_{\mathcal{C}} \cong \mathbb{1}$.

We note that $\alpha_{\mathcal{C}}$ is the dual of the distinguished invertible object introduced in [**18**]. The reason why we use the dual is not essential. Our choice is useful when we compare categorical results with the case of Hopf algebras.

The modular object can also be defined in terms of right $\mathbb{H}$-comodules. Namely, let $\beta \in \mathcal{C}$ be a unique object (up to isomorphism) such that ${}^*\mathbb{H} \cong (\beta \boxtimes \mathbb{1}) \otimes \mathbb{H}$ as right $\mathbb{H}$-comodules. Then we have isomorphisms

$$\mathbb{H} \cong ({}^*\mathbb{H})^* \cong ((\beta \boxtimes \mathbb{1}) \otimes \mathbb{H})^* \cong \mathbb{H}^* \otimes (\beta \boxtimes \mathbb{1})^* \cong \mathbb{H} \otimes ((\alpha \otimes \beta^*) \boxtimes \mathbb{1})$$

of left $\mathbb{H}$-comodules, where $\alpha = \alpha_{\mathcal{C}}$. Similarly, we have isomorphisms

$$\mathbb{H} \cong {}^*(\mathbb{H}^*) \cong ({}^*\alpha \boxtimes \mathbb{1}) \otimes {}^*\mathbb{H} \cong (({}^*\alpha \otimes \beta) \boxtimes \mathbb{1}) \otimes \mathbb{H}$$

of right $\mathbb{H}$-comodules. Hence $\alpha \otimes \beta^* \cong \mathbb{1} \cong {}^*\alpha \otimes \beta$ by Theorem 4.8. This implies that α and β are isomorphic invertible objects.

4.6. Comonadic description of the Drinfeld center. Now we give adjoints of the forgetful functor $U : \mathcal{Z}(\mathcal{C}) \to \mathcal{C}$ by expressing $\mathcal{Z}(\mathcal{C})$ as the category of comodules over a certain comonad on $\mathcal{C}$. We first make $\mathcal{C}$ a left $\mathcal{C}^{\mathrm{env}}$-module category by the action

$$(X \boxtimes Y) \circleddash V = X \otimes V \otimes Y \quad (X, Y, V \in \mathcal{C}) \tag{4.5}$$

and then set $Z(V) = \mathbb{H} \circleddash V$ for $V \in \mathcal{C}$. Then we have

$$Z(V) = \int_{X \in \mathcal{C}} X \otimes V \otimes X^* \quad (V \in \mathcal{C}). \tag{4.6}$$

For $V, X \in \mathcal{C}$, we denote by $\pi_{V,X} : Z(V) \to X \otimes V \otimes X^*$ the universal dinatural dinatural transformation for the end $Z(V)$. By the universal property, we have

$$\begin{aligned}\mathrm{Hom}_{\mathcal{C}}(V, Z(W)) &\cong \int_{X \in \mathcal{C}} \mathrm{Hom}_{\mathcal{C}}(V, X \otimes W \otimes X^*) \\ &\cong \int_{X \in \mathcal{C}} \mathrm{Hom}_{\mathcal{C}}(V \otimes X, X \otimes W) \cong \mathrm{Nat}(V \otimes \mathrm{id}_{\mathcal{C}}, \mathrm{id}_{\mathcal{C}} \otimes W)\end{aligned}$$

for $V, W \in \mathcal{C}$. The isomorphism is given explicitly as follows:

LEMMA 4.10. *There is a natural isomorphism*

$$\mathrm{Hom}_{\mathcal{C}}(V, Z(W)) \to \mathrm{Nat}(V \otimes \mathrm{id}_{\mathcal{C}}, \mathrm{id}_{\mathcal{C}} \otimes W) \quad (V, W \in \mathcal{C}), \tag{4.7}$$

which sends a morphism $f : V \to Z(W)$ in $\mathcal{C}$ to the natural transformation

$$V \otimes X \xrightarrow{\pi_{W,X} f \otimes \mathrm{id}_X} X \otimes W \otimes X^* \otimes X \xrightarrow{\mathrm{id}_X \otimes \mathrm{id}_W \otimes \mathrm{ev}_X} X \otimes W \quad (X \in \mathcal{C}).$$

Since $\mathbb{H}$ is a coalgebra in $\mathcal{C}^{\mathrm{env}}$, the category ${}^{\mathbb{H}}\mathcal{C}$ is defined. Now we note:

THEOREM 4.11. ${}^{\mathbb{H}}\mathcal{C} \cong \mathcal{Z}(\mathcal{C})$.

The category ${}^{\mathbb{H}}\mathcal{C}$ can be identified with ${}_{\mathbb{H}^*}\mathcal{C}$. As remarked in [**40**, Subsections 4.3 and 4.4], the algebra $\mathbb{H}^*$ is identical to the algebra $\underline{\mathrm{Hom}}(\mathbb{1}, \mathbb{1}) \in \mathcal{C}^{\mathrm{env}}$ considered in [**18**] and the monad $\mathbb{H}^* \oslash \mathrm{id}_{\mathcal{C}}$ coincides with the quantum double of $\mathrm{id}_{\mathcal{C}}$. Thus the above theorem follows from [**3**, Theorem 5.12]. Since an explicit expression of the isomorphism ${}^{\mathbb{H}}\mathcal{C} \cong \mathcal{Z}(\mathcal{C})$ is important, we include the proof.

PROOF. Let V be an object of $\mathcal{C}$. One can check that a morphism $V \to Z(V)$ in $\mathcal{C}$ makes V a left $\mathbb{H}$-comodule in $\mathcal{C}$ if and only if the corresponding natural transformation $V \otimes \mathrm{id}_{\mathcal{C}} \to \mathrm{id}_{\mathcal{C}} \otimes V$ is a half-braiding for V. This establishes a bijection between the class of objects of ${}^{\mathbb{H}}\mathcal{C}$ and that of $\mathcal{Z}(\mathcal{C})$ and this bijection extends to an isomorphism between these categories. □

We identify ${}^{\mathbb{H}}\mathcal{C}$ with $\mathcal{Z}(\mathcal{C})$ by the above theorem. If M is a left $\mathbb{H}$-comodule in $\mathcal{C}^{\mathrm{env}}$ with coaction $\delta : M \to \mathbb{H} \otimes M$, then an object of the form $M \oslash V$, $V \in \mathcal{C}$, is a left $\mathbb{H}$-comodule in $\mathcal{C}$ by the coaction

$$M \oslash V \xrightarrow{\delta \oslash \mathrm{id}_V} (\mathbb{H} \otimes M) \oslash V \xrightarrow{\cong} \mathbb{H} \oslash (M \oslash V)$$

and hence becomes an object of $\mathcal{Z}(\mathcal{C})$. By using this construction, a left adjoint and a right adjoint of the forgetful functor $U : \mathcal{Z}(\mathcal{C}) \to \mathcal{C}$ are given as follows:

THEOREM 4.12. *We define $L, R : \mathcal{C} \to \mathcal{Z}(\mathcal{C})$ by*

$$L(V) = \mathbb{H}^* \oslash V \quad \textit{and} \quad R(V) = \mathbb{H} \oslash V \quad (V \in \mathcal{C}),$$

respectively. Then L and R are a left and a right adjoint of U, respectively.

PROOF. The functor R is nothing but the free comodule functor and therefore it is right adjoint to the forgetful functor ${}^{\mathbb{H}}\mathcal{C} \to \mathcal{C}$. If we identify ${}^{\mathbb{H}}\mathcal{C}$ with ${}_{\mathbb{H}^*}\mathcal{C}$, then the functor L is just the free $\mathbb{H}^*$-module functor and hence it is left adjoint to the forgetful functor ${}^{\mathbb{H}}\mathcal{C} \to \mathcal{C}$. The proof is done. □

REMARK 4.13. Let η^L, ε^L, η^R and ε^R be the unit of the adjunction $L \dashv U$, the counit of $L \dashv U$, the unit of $U \dashv R$ and the counit of $U \dashv R$, respectively, obtained from the proof of the above theorem. Given an object M of $\mathcal{Z}(\mathcal{C})$ $(= {}^{\mathbb{H}}\mathcal{C})$, we denote by $\delta_M : M \to \mathbb{H} \ominus M$ the coaction of $\mathbb{H}$ on M. Then we have

$$\eta^L_V = \Big(V \xrightarrow{\cong} (\mathbb{1} \boxtimes \mathbb{1}) \ominus V \xrightarrow{\epsilon^* \ominus \mathrm{id}_V} \mathbb{H}^* \ominus V = UL(V)\Big),$$

$$\varepsilon^L_M = \Big(\mathbb{H}^* \ominus M \xrightarrow{\mathrm{id} \ominus \delta_M} \mathbb{H}^* \ominus (\mathbb{H} \ominus M) \xrightarrow{\cong} (\mathbb{H}^* \otimes \mathbb{H}) \ominus M \xrightarrow{\mathrm{ev} \ominus \mathrm{id}_M} M\Big),$$

$$\eta^R_M = \delta_M \quad \text{and} \quad \varepsilon^R_V = \Big(\mathbb{H} \ominus V \xrightarrow{\epsilon \ominus \mathrm{id}_V} (\mathbb{1} \boxtimes \mathbb{1}) \ominus V \xrightarrow{\cong} V\Big)$$

for $V \in \mathcal{C}$ and $M \in \mathcal{Z}(\mathcal{C})$, where $\epsilon : \mathbb{H} \to \mathbb{1} \boxtimes \mathbb{1}$ is the counit of $\mathbb{H}$.

REMARK 4.14. For $V \in \mathcal{C}$, we define $\delta_V = \Delta \ominus \mathrm{id}_V$, where $\Delta : \mathbb{H} \to \mathbb{H} \otimes \mathbb{H}$ is the comultiplication of $\mathbb{H}$. By the construction of R, we have

$$R(W) = (Z(W), t^W) \in \mathcal{Z}(\mathcal{C}) \quad (W \in \mathcal{C})$$

for some half-braiding of t^W for $Z(W)$. By definition, the half-braiding t^W is the natural transformation $Z(W) \otimes \mathrm{id}_{\mathcal{C}} \to \mathrm{id}_{\mathcal{C}} \otimes Z(W)$ corresponding to δ_W via the isomorphism given in Lemma 4.10 with $V = W$. Thus, explicitly, we have

$$t^W_X := (\mathrm{id}_X \otimes \mathrm{id}_{Z(W)} \otimes \mathrm{ev}_X) \circ (\pi_{Z(W),X} \delta_W \otimes \mathrm{id}_X) : Z(W) \otimes X \to X \otimes Z(W)$$

for $X \in \mathcal{C}$. The isomorphism of Lemma 4.10 is expressed by

$$f \mapsto \sigma \circ (\mathrm{id} \otimes \varepsilon^R_W) \circ t^W \circ (f \otimes \mathrm{id}) \quad (f \in \mathrm{Hom}_{\mathcal{C}}(V, W)) \tag{4.8}$$

in terms of the half-braiding t^W and the counit ε^R.

4.7. Frobenius property of the forgetful functor. Now we give some applications of our construction of the induction to the Drinfeld center. Below, L and R are a left and a right adjoint of U, respectively. Let $\alpha = \alpha_{\mathcal{C}}$ be the modular object. We first show that the 'difference' between L and R can be expressed in terms of α as follows:

THEOREM 4.15. *There are natural isomorphisms*

$$L(V) \cong R(\alpha \otimes V), \quad R(V) \cong L(\alpha^* \otimes V) \quad (V \in \mathcal{C}).$$

PROOF. By the uniqueness of adjoints, we may assume that L and R are given as in Theorem 4.12. Since $\mathbb{H}^* \cong \mathbb{H} \otimes (\alpha \boxtimes \mathbb{1})$, we have isomorphisms

$$L(V) = \mathbb{H}^* \ominus V \cong (\mathbb{H} \otimes (\alpha \boxtimes \mathbb{1})) \ominus V \cong \mathbb{H} \otimes (\alpha \otimes V) = R(\alpha \otimes V)$$

of left $\mathbb{H}$-comodules in $\mathcal{C}$. Hence the first natural isomorphism is established. The second one is obtained from the first one as follows: Since α is invertible,

$$R(V) \cong R(\alpha \otimes \alpha^* \otimes V) \cong L(\alpha^* \otimes V).$$

□

A functor F is said to be *Frobenius* if it has a left adjoint functor that is also a right adjoint. Thus U is Frobenius if and only if $L \cong R$. By the above theorem, we have:

COROLLARY 4.16. *The forgetful functor $U : \mathcal{Z}(\mathcal{C}) \to \mathcal{C}$ is Frobenius if and only if $\mathcal{C}$ is unimodular.*

PROOF. The 'if' part is a direct consequence of Theorem 4.15. We prove the converse: Suppose that F is Frobenius. Then we have isomorphisms

$$\begin{aligned}\mathrm{Hom}_{\mathcal{C}}(\mathbb{1},\alpha) &= \mathrm{Hom}_{\mathcal{C}}(U(\mathbb{1}),\alpha) \cong \mathrm{Hom}_{\mathcal{Z}(\mathcal{C})}(\mathbb{1},R(\alpha))\\ &\cong \mathrm{Hom}_{\mathcal{Z}(\mathcal{C})}(\mathbb{1},L(\mathbb{1})) \cong \mathrm{Hom}_{\mathcal{Z}(\mathcal{C})}(\mathbb{1},R(\mathbb{1})) \cong \mathrm{Hom}_{\mathcal{Z}}(\mathbb{1},\mathbb{1}) \cong k.\end{aligned}$$

Here, the second isomorphism follows from Theorem 4.15 and the third follows from the assumption that $L \cong R$. Since $\mathbb{1}$ and α are simple objects of $\mathcal{C}$, we have $\mathbb{1} \cong \alpha$ by Schur's lemma. Thus $\mathcal{C}$ is unimodular. The proof is done. □

4.8. The case of Hopf algebras. To illustrate the results in this section, we consider the case where $\mathcal{C} = H$-mod for some finite-dimensional Hopf algebra H. Given two H-bimodules V and W, their tensor product $V \otimes_k W$ is an H-bimodule by the actions

$$h \cdot (v \otimes_k w) \cdot h' = h_{(1)} v h'_{(1)} \otimes_k h_{(2)} w h'_{(2)} \quad (h, h' \in H, v \in V, w \in W).$$

To distinguish this operation from other kind of tensor product, we denote this H-bimodule by $V \circledast W$. The category H-mod-H is a finite tensor category with respect to $\circledast$. There is a tensor equivalence

$$\mathcal{C}^{\mathrm{env}} \to H\text{-mod-}H, \quad V \boxtimes W \mapsto V \otimes_k W_S, \tag{4.9}$$

where $W_S \in \text{mod-}H$ for $W \in H$-mod is the vector space W equipped with the right H-action given by $w \cdot h = S(h)w$ for $w \in W$ and $h \in H$.

Under the identification by (4.9), the object $\mathbb{H} \in \mathcal{C}^{\mathrm{env}}$ of Lemma 4.7 is just the H-bimodule H. Indeed, H has the universal dinatural transformation

$$\pi_X : H \to X \otimes_k X^*, \quad h \mapsto hx_i \otimes_k x^i \quad (X \in H\text{-mod}, h \in H),$$

where $\{x_i\}$ is a basis of X, $\{x^i\}$ is the dual basis of X^*, and the Einstein convention is used to suppress the sum over i. One can check that the coalgebra structure of $\mathbb{H}$ is same as that of H.

Given a left H-module V, we denote by V_ϵ the left H-module V equipped with the right H-action given by the counit $\epsilon : H \to k$. The object $V \boxtimes \mathbb{1} \in \mathcal{C}^{\mathrm{env}}$ for $V \in \mathcal{C}$ corresponds to V_ϵ by the equivalence (4.9). Thus the equivalences $\Psi^\ell_{\mathcal{C}}$ and $\Psi^r_{\mathcal{C}}$ given in Theorem 4.8 are expressed as follows:

$$\begin{aligned}\Psi^\ell_{\mathcal{C}} &: H\text{-mod} \to {}^H(H\text{-mod-}H), \quad V \mapsto H \circledast V_\epsilon;\\ \Psi^r_{\mathcal{C}} &: H\text{-mod} \to (H\text{-mod-}H)^H, \quad V \mapsto V_\epsilon \circledast H.\end{aligned}$$

Thus Theorem 4.8 is indeed a generalization of the fundamental theorem for Hopf bimodules (although Theorem 4.8 only covers the finite-dimensional case). The story of Subsection 4.5 is parallel to the known argument in the integral theory of finite-dimensional (quasi-)Hopf algebras (*cf.* [**23**]).

For an H-bimodule M, we denote by ${}_{\mathrm{ad}}M$ the vector space M equipped with the left H-action given by $h \triangleright m = h_{(1)} m S(h_{(2)})$ for $h \in H$ and $m \in M$. Under the identification (4.9), the action of $\mathcal{C}^{\mathrm{env}}$ on $\mathcal{C}$ is given by

$$M \ominus V = {}_{\mathrm{ad}}(M \circledast V_\varepsilon) \quad (M \in H\text{-mod-}H, V \in H\text{-mod}).$$

Namely, $M \ominus V$ is the vector space $M \otimes V$ equipped with the left H-action

$$h \cdot (m \otimes v) = h_{(1)} m S(h_{(3)}) \otimes h_{(2)} v \quad (h \in H, m \in M, v \in V).$$

By definition, a left $\mathbb{H}$-comodule in H-mod is an object $M \in H$-mod equipped with an H-linear map $\delta : M \to \mathbb{H} \circleddash M$ such that the equations

$$(\mathrm{id}_{\mathbb{H}} \circleddash \delta) \circ \delta = (\Delta \circleddash \mathrm{id}_M) \circ \delta \quad \text{and} \quad (\varepsilon \circleddash \mathrm{id}_M) \circ \delta = \mathrm{id}_M$$

hold. These equations mean that M is a left H-comodule in the usual sense. If we write $\delta(m) = m_{(-1)} \otimes m_{(0)}$, then the requirement that δ is H-linear reads

$$\delta(h \cdot m) = h_{(1)} m_{(-1)} S(h_{(3)}) \otimes h_{(2)} m_{(0)} \quad (h \in H, m \in M),$$

that is, the Yetter-Drinfeld condition. Thus we have ${}^{\mathbb{H}}(H\text{-mod}) = {}^H_H\mathcal{YD}_{fd}$. The category isomorphism given in Theorem 4.11 actually reduces to the category isomorphism ${}^H_H\mathcal{YD}_{fd} \cong \mathcal{Z}(H\text{-mod})$ given at the beginning of this section.

The Radford induction, introduced in Subsection 2.2, is identical to the functor R of Theorem 4.12. The unit η and the counit ε of $U \dashv R$ are given by

$$\eta_M : M \to RU(M), \quad m \mapsto m_{(-1)} \otimes m_{(0)} \quad (m \in M), \tag{4.10}$$

$$\varepsilon_X : UR(X) \to X, \quad h \otimes x \mapsto \epsilon(h)x \quad (h \in H, x \in X) \tag{4.11}$$

for $M \in {}^H_H\mathcal{YD}_{fd}$ and $X \in H$-mod, where $\epsilon : H \to k$ is the counit of H.

We recall that ${}^H_H\mathcal{YD}_{fd}$ can be identified with $D(H)$-mod, where $D(H)$ is the Drinfeld double of H. Under this identification, the functor $D(H) \otimes_H (-)$ is left adjoint to U. Theorem 4.12 gives a left adjoint of U from a different point of view. Theorem 4.15 is easily proved by using Hopf bimodules hidden in the functors L and R.

It is known that the extension $H \hookrightarrow D(H)$ of algebras is Frobenius if and only if H is unimodular [**8**]. Corollary 4.16 gives a category-theoretical generalization of this result.

5. Categorical class functions

Throughout this section, we work over an algebraically closed field k. Let $\mathcal{C}$ be a finite tensor category. In the last section, we have constructed a right adjoint R of the forgetful functor $U : \mathcal{Z}(\mathcal{C}) \to \mathcal{C}$. In this section, we establish the 'character theory' in finite tensor categories by using the adjunction $U \dashv R$.

5.1. Monoidal structure of a right adjoint. As a preparation, we mention the monoidal structure of the functor R. Recall that a *monoidal functor* [**30**, XI.2] is a triple (F, m, u) consisting of a functor $F : \mathcal{A} \to \mathcal{B}$ between monoidal categories $\mathcal{A}$ and $\mathcal{B}$, a natural transformation

$$m_{X,Y} : F(X) \otimes F(Y) \to F(X \otimes Y) \quad (X, Y \in \mathcal{A})$$

and a morphism $u : \mathbb{1} \to F(\mathbb{1})$ in $\mathcal{B}$ such that the equations

$$m_{X \otimes Y, Z} \circ (m_{X,Y} \otimes \mathrm{id}_{F(Z)}) = m_{X, Y \otimes Z} \circ (\mathrm{id}_{F(X)} \otimes m_{Y,Z}),$$
$$m_{\mathbb{1},X} \circ (u \otimes \mathrm{id}_X) = \mathrm{id}_{F(X)} = m_{X,\mathbb{1}} \circ (\mathrm{id}_X \otimes u)$$

hold for all $X, Y, Z \in \mathcal{A}$. We do not require m and u to be invertible. A monoidal functor (F, m, u) is said to be *strong* if m and u are invertible. A monoidal functor (F, m, u) is said to be *strict* if m and u are identities.

It is known that a right adjoint of a strong monoidal functor has a unique monoidal structure such that the unit and the counit of the adjunction are monoidal natural transformations. We apply this fact to $U \dashv R$. As a result, we see that R is a monoidal functor with structure morphisms given by

$$m_{V,W} = \Big(R(V) \otimes R(W) \xrightarrow{\eta} RU(R(V) \otimes R(W)) \xrightarrow{\mathrm{id}} R(UR(V) \otimes UR(W)) \xrightarrow{R(\varepsilon_V \otimes \varepsilon_W)} R(V \otimes W)\Big)$$

for $V, W \in \mathcal{C}$ and $u = \eta_{\mathbf{1}}$, where $\mathbf{1} = (\mathbb{1}, \mathrm{id})$ is the unit object of $\mathcal{Z}(\mathcal{C})$. Furthermore, η and ε are monoidal natural transformations. Namely, the equations

$$m_{U(\mathbf{X}),U(\mathbf{Y})} \circ (\eta_{\mathbf{X}} \otimes \eta_{\mathbf{Y}}) = \eta_{\mathbf{X} \otimes \mathbf{Y}}, \quad \eta_{\mathbf{1}} = u, \tag{5.1}$$

$$\varepsilon_{V \otimes W} \circ m_{V,W} = \varepsilon_V \otimes \varepsilon_W, \quad \varepsilon_{\mathbb{1}} \circ u = \mathrm{id}_{\mathbb{1}} \tag{5.2}$$

hold for all $V, W \in \mathcal{C}$ and $\mathbf{X}, \mathbf{Y} \in \mathcal{Z}(\mathcal{C})$.

The monoidal functor $Z := UR$ is a part of the *central Hopf comonad* on $\mathcal{C}$ introduced in [**39**]. We recall that $Z(V)$ for $V \in \mathcal{C}$ is an end as in (4.6). According to [**39**], the morphisms $m_{V,W}$ $(V, W \in \mathcal{C})$ and u are determined by the equations

$$\pi_{V \otimes W, X} \circ m_{V,W} = (\mathrm{id}_X \otimes \mathrm{ev}_X \otimes \mathrm{id}_{X^*}) \circ (\pi_{V,X} \otimes \pi_{W,X}), \tag{5.3}$$

$$\pi_{\mathbb{1},X} \circ u = \mathrm{coev}_X \tag{5.4}$$

for $X \in \mathcal{C}$, where $\pi_{V,X} : Z(V) \to X \otimes V \otimes X^*$ $(V, X \in \mathcal{C})$ is the universal dinatural transformation for the end.

REMARK 5.1. Let D be the central Hopf monad on $\mathcal{C}$ (= the quantum double [**3**] of the identity Hopf monad $\mathrm{id}_{\mathcal{C}}$). As noted in [**39**, Subsection 3.2], the functor Z can be given by $Z(V) = D({}^*V)^*$ for $V \in \mathcal{C}$ and the structure morphisms of Z are imported from D via this relation. For reader's convenience, we note that our Z and our D are denoted by $\bar{Z}$ and Z, respectively, in [**39**].

5.2. The adjoint algebra. If $\mathcal{C} = H$-mod for some finite-dimensional Hopf algebra H, then R is given by the Radford induction and hence $Z(\mathbb{1})$ is the adjoint representation of H. In view of this observation, we introduce the following terminology:

DEFINITION 5.2. We call $A := Z(\mathbb{1})$ the *adjoint algebra* of $\mathcal{C}$. For $X \in \mathcal{C}$, we define the *canonical left action* and the *canonical right action* of A on X by

$$a^\ell_X := (\mathrm{id}_X \otimes \varepsilon_{\mathbb{1}}) \circ \widetilde{\sigma}_X \quad \text{and} \quad a^r_X := (\varepsilon_{\mathbb{1}} \otimes \mathrm{id}_X) \circ (\widetilde{\sigma}_X)^{-1}, \tag{5.5}$$

respectively, where $\widetilde{\sigma} := t^{\mathbb{1}}$ is the half-braiding of $Z(\mathbb{1})$ given in Remark 4.14.

The adjoint algebra A is indeed an algebra in $\mathcal{C}$ by the multiplication $m_{\mathbb{1},\mathbb{1}}$ and the unit u introduced in the previous subsection. Furthermore, an object $X \in \mathcal{C}$ is a left A-module and a right A-module by a^ℓ_X and a^r_X, respectively, although it is not an A-bimodule in general.

By Lemma 4.10 and Remark 4.14, there is a natural isomorphism

$$\Xi_V : \mathrm{Hom}_{\mathcal{C}}(V, A) \to \mathrm{Nat}(V \otimes \mathrm{id}_{\mathcal{C}}, \mathrm{id}_{\mathcal{C}}) \quad (V \in \mathcal{C})$$

sending a morphism $f : V \to A$ in $\mathcal{C}$ to the natural transformation

$$V \otimes X \xrightarrow{f \otimes \mathrm{id}_X} A \otimes X \xrightarrow{\widetilde{\sigma}_X} X \otimes A \xrightarrow{\mathrm{id}_X \otimes \varepsilon_{\mathbb{1}}} X \quad (X \in \mathcal{C}).$$

Thus the canonical left action $a^\ell = \{a^\ell_X\}_{X \in \mathcal{C}}$ is the natural transformation corresponding to id_A. By using a^ℓ, the map Ξ can be written as

$$\Xi_V(f)_X = a^\ell_X \circ (f \otimes \mathrm{id}_X) \quad (f \in \mathrm{Hom}_{\mathcal{C}}(V, A)). \tag{5.6}$$

The vector space $\mathrm{Hom}_{\mathcal{C}}(\mathbb{1}, A)$ is an algebra with respect to the multiplication

$$z \bullet z' = m_{\mathbb{1},\mathbb{1}}(z \otimes z') \quad (z, z' \in \mathrm{Hom}_{\mathcal{C}}(\mathbb{1}, A)) \tag{5.7}$$

and the unit u. By (5.6), it is straightforward to prove:

LEMMA 5.3. *The map $\Xi_{\mathbb{1}}$ is an isomorphism of algebras.*

5.3. Categorical class functions. In the representation theory of groups, the character of a representation is a class function. As a container of a 'categorical' character, we introduce the following vector space:

DEFINITION 5.4. $\mathrm{CF}(\mathcal{C}) := \mathrm{Hom}_{\mathcal{C}}(A, \mathbb{1})$

We call $\mathrm{CF}(\mathcal{C})$ the space of *class functions* of $\mathcal{C}$. As in the case of class functions on a group, $\mathrm{CF}(\mathcal{C})$ is an algebra: There is a canonical isomorphism

$$\mathrm{CF}(\mathcal{C}) = \mathrm{Hom}_{\mathcal{C}}(UR(\mathbb{1}), \mathbb{1}) \cong \mathrm{Hom}_{\mathcal{Z}(\mathcal{C})}(R(\mathbb{1}), R(\mathbb{1})) = \mathrm{End}_{\mathcal{Z}(\mathcal{C})}(R(\mathbb{1}))$$

of vector spaces. We transport the algebra structure of $\mathrm{End}_{\mathcal{Z}(\mathcal{C})}(R(\mathbb{1}))$ to the space of class functions through this isomorphism. As a result, $\mathrm{CF}(\mathcal{C})$ is an algebra with the multiplication $\star$ given by

$$f \star g := \Big(A \xrightarrow{U(\eta_{R(\mathbb{1})})} URUR(\mathbb{1}) \xrightarrow{UR(g)} UR(\mathbb{1}) \xrightarrow{f} \mathbb{1}\Big)$$

for $f, g \in \mathrm{CF}(\mathcal{C})$ and the unit $\varepsilon_{\mathbb{1}}$.

5.4. The action functor and its adjoints. To study the adjoint algebra and the space of class functions, the following functor is useful:

DEFINITION 5.5. The *action functor* of $\mathcal{C}$ is defined by

$$\boldsymbol{\rho}_{\mathcal{C}} : \mathcal{C} \to \mathrm{Rex}(\mathcal{C}), \quad V \mapsto V \otimes \mathrm{id}_{\mathcal{C}}.$$

An important observation is that adjoints of $\boldsymbol{\rho}_{\mathcal{C}}$ exist and have some nice properties; see [**41**]. We take a different approach and give some properties of $\boldsymbol{\rho}_{\mathcal{C}}$ by the argument used in the proof of Theorem 4.8. Let $\mathbb{H}$ be the coalgebra in $\mathcal{C}^{\mathrm{env}}$ appeared in that theorem. We define functors $\boldsymbol{\rho}_{\mathcal{C}}^{\mathrm{la}}, \boldsymbol{\rho}_{\mathcal{C}}^{\mathrm{ra}} : \mathrm{Rex}(\mathcal{C}) \to \mathcal{C}$ by

$$\boldsymbol{\rho}_{\mathcal{C}}^{\mathrm{la}}(F) = (F \oslash {}^{*}\mathbb{H})(\mathbb{1}) \quad \text{and} \quad \boldsymbol{\rho}_{\mathcal{C}}^{\mathrm{ra}}(F) = (F \oslash \mathbb{H})(\mathbb{1}) \quad (F \in \mathrm{Rex}(\mathcal{C})),$$

where the right action $\oslash$ of $\mathcal{C}^{\mathrm{env}}$ on $\mathrm{Rex}(\mathcal{C})$ is given by (4.1).

THEOREM 5.6. $\boldsymbol{\rho}_{\mathcal{C}}^{\mathrm{la}} \dashv \boldsymbol{\rho}_{\mathcal{C}} \dashv \boldsymbol{\rho}_{\mathcal{C}}^{\mathrm{ra}}$. *Furthermore, we have:*

(1) *There is a natural isomorphism*

$$\boldsymbol{\rho}_{\mathcal{C}}^{\mathrm{la}}(F) \cong \boldsymbol{\rho}_{\mathcal{C}}^{\mathrm{ra}}\Big(F \oslash (\alpha \boxtimes \mathbb{1})\Big) \quad (F \in \mathrm{Rex}(\mathcal{C})),$$

where $\alpha = \alpha_{\mathcal{C}} \in \mathcal{C}$ is the modular object.

(2) *The functors $\boldsymbol{\rho}_{\mathcal{C}}^{\mathrm{la}}$ and $\boldsymbol{\rho}_{\mathcal{C}}^{\mathrm{ra}}$ are faithful and exact.*

PROOF. Let $\widetilde{\boldsymbol{\rho}}_{\mathcal{C}} : \mathcal{C} \to \mathrm{Rex}_{-|\mathcal{C}}(\mathcal{C})$ and $\mathbb{E}_{\mathcal{C}} : \mathrm{Rex}_{-|\mathcal{C}}(\mathcal{C}) \to \mathcal{C}$ be the equivalences given by (4.2) and (4.3), respectively. The functor $\boldsymbol{\rho}_{\mathcal{C}}$ decomposes as

$$\boldsymbol{\rho}_{\mathcal{C}} = \Big(\mathcal{C} \xrightarrow{\widetilde{\boldsymbol{\rho}}_{\mathcal{C}}} \mathrm{Rex}_{-|\mathcal{C}}(\mathcal{C}) \xrightarrow{\cong} \mathrm{Rex}(\mathcal{C})^{\mathbb{H}} \xrightarrow{\text{forget}} \mathrm{Rex}(\mathcal{C})\Big).$$

The free right $\mathbb{H}$-comodule functor $(-) \oslash \mathbb{H} : \mathrm{Rex}(\mathcal{C}) \to \mathrm{Rex}(\mathcal{C})^{\mathbb{H}}$ is right adjoint to the forgetful functor. Thus the composition

$$\mathrm{Rex}(\mathcal{C}) \xrightarrow{F \mapsto F \oslash \mathbb{H}} \mathrm{Rex}(\mathcal{C})^{\mathbb{H}} \xrightarrow{\cong} \mathrm{Rex}_{-|\mathcal{C}}(\mathcal{C}) \xrightarrow{\mathbb{E}_{\mathcal{C}}} \mathcal{C}, \tag{5.8}$$

which coincides with $\rho_{\mathcal{C}}^{\mathrm{ra}}$, is right adjoint to $\rho_{\mathcal{C}}$. Since the category $\mathrm{Rex}(\mathcal{C})^{\mathbb{H}}$ can be identified with the category of right ${}^*\mathbb{H}$-modules in $\mathrm{Rex}(\mathcal{C})$, the composition

$$\mathrm{Rex}(\mathcal{C}) \xrightarrow{F \mapsto F \circledcirc {}^*\mathbb{H}} \mathrm{Rex}(\mathcal{C})_{{}^*\mathbb{H}} = \mathrm{Rex}(\mathcal{C})^{\mathbb{H}} \cong \mathrm{Rex}_{-|\mathcal{C}}(\mathcal{C}) \xrightarrow{\mathbb{E}_{\mathcal{C}}} \mathcal{C}, \tag{5.9}$$

which coincides with $\rho_{\mathcal{C}}^{\mathrm{la}}$, is left adjoint to $\rho_{\mathcal{C}}$. The proof of $\rho_{\mathcal{C}}^{\mathrm{la}} \dashv \rho_{\mathcal{C}} \dashv \rho_{\mathcal{C}}^{\mathrm{ra}}$ is done. Now Part (1) is proved as follows:

$$\rho_{\mathcal{C}}^{\mathrm{la}}(F) = \mathbb{E}_{\mathcal{C}}(F \circledcirc {}^*\mathbb{H}) \cong \mathbb{E}_{\mathcal{C}}(F \circledcirc ((\alpha \boxtimes \mathbb{1}) \otimes \mathbb{H})) \cong \rho_{\mathcal{C}}^{\mathrm{ra}}(F \circledcirc (\alpha \boxtimes \mathbb{1})).$$

To prove Part (2), we note that the first arrows of (5.8) and (5.9) are faithful exact functors by Lemma 4.6. The functors $\rho_{\mathcal{C}}^{\mathrm{ra}}$ are $\rho_{\mathcal{C}}^{\mathrm{la}}$ are faithful and exact as compositions of such functors. □

The following expression of $\rho_{\mathcal{C}}^{\mathrm{ra}}$ is also important:

THEOREM 5.7. $\rho_{\mathcal{C}}^{\mathrm{ra}}(F) = \int_{X \in \mathcal{C}} F(X) \otimes X^*$ *for* $F \in \mathrm{Rex}(\mathcal{C})$.

PROOF. There are natural isomorphisms

$$\begin{aligned} \mathrm{Hom}_{\mathcal{C}}(V, \rho_{\mathcal{C}}^{\mathrm{ra}}(F)) &\cong \mathrm{Nat}(\rho_{\mathcal{C}}(V), F) \\ &\cong \int_{X \in \mathcal{C}} \mathrm{Hom}_{\mathcal{C}}(V \otimes X, F(X)) \\ &\cong \int_{X \in \mathcal{C}} \mathrm{Hom}_{\mathcal{C}}(V, F(X) \otimes X^*) \end{aligned}$$

for $V \in \mathcal{C}$ and $F \in \mathrm{Rex}(\mathcal{C})$. Now the claim follows from Lemma 3.5. □

By comparing the above theorem with equation (4.6), we have

$$UR(V) = \mathbb{H} \circledcirc V = \int_{X \in \mathcal{C}} X \otimes V \otimes X^* = \rho_{\mathcal{C}}^{\mathrm{ra}}(\mathrm{id}_{\mathcal{C}} \otimes V) \quad (V \in \mathcal{C}). \tag{5.10}$$

In particular, the adjoint algebra is given by $A = \rho_{\mathcal{C}}^{\mathrm{ra}}(\mathrm{id}_{\mathcal{C}})$. This description is powerful for investigating the structure of A. For example, if J is an ideal of $\mathcal{C}$ in the sense of Lemma 3.9, then $\rho_{\mathcal{C}}^{\mathrm{ra}}(J)$ is a subobject of A since $\rho_{\mathcal{C}}^{\mathrm{ra}}$ is left exact. By this type of argument, we introduce the space of 'class functions on a topologizing subcategory of $\mathcal{C}$' in the next subsection.

5.5. Class functions of a topologizing subcategory. For a topologizing subcategory $\mathcal{S}$ of $\mathcal{C}$, we define $A_{\mathcal{S}}$ to be the end of the functor

$$\mathcal{S}^{\mathrm{op}} \times \mathcal{S} \xrightarrow{\text{inclusion}} \mathcal{C}^{\mathrm{op}} \times \mathcal{C} \xrightarrow{(X,Y) \mapsto Y \otimes X^*} \mathcal{C}.$$

By omitting the inclusion functor, we may write $A_{\mathcal{S}} = \int_{X \in \mathcal{S}} X \otimes X^*$. The end $A_{\mathcal{S}}$ exists. To see this, we let $J_{\mathcal{S}}$ be the ideal of $\mathcal{C}$ corresponding to $\mathcal{S}$ by Lemma 3.9. Then $\tau_{\mathcal{S}} := \mathrm{id}_{\mathcal{C}}/J_{\mathcal{S}}$ is an idempotent monad on $\mathcal{C}$ whose category of modules is identified with $\mathcal{S}$. By Lemma 3.6 and Theorem 5.7, we have

$$\rho_{\mathcal{C}}^{\mathrm{ra}}(\tau_{\mathcal{S}}) = \int_{X \in \mathcal{C}} \tau_{\mathcal{S}}(X) \otimes X^* \cong \int_{X \in \mathcal{S}} X \otimes X^* = A_{\mathcal{S}}, \tag{5.11}$$

meaning that the end $A_{\mathcal{S}}$ exists and is canonically isomorphic to $\rho_{\mathcal{C}}^{\mathrm{ra}}(\tau_{\mathcal{S}})$.

DEFINITION 5.8. For a topologizing subcategory $\mathcal{S} \in \mathfrak{Top}(\mathcal{C})$, we set

$$\mathrm{CF}_{\mathcal{C}}(\mathcal{S}) := \mathrm{Hom}_{\mathcal{C}}(A_{\mathcal{S}}, \mathbb{1})$$

and call it the space of class functions of $\mathcal{S}$.

The end $A_{\mathcal{S}}$ for $\mathcal{S} \in \mathfrak{Top}(\mathcal{C})$ is an algebra in $\mathcal{C}$ by the multiplication $m_{\mathcal{S}}$ and the unit $u_{\mathcal{S}}$ determined by the property that the equations

$$\pi_X^{\mathcal{S}} \circ m_{\mathcal{S}} = (\mathrm{id}_X \otimes \mathrm{ev}_X \otimes \mathrm{id}_{X^*}) \circ (\pi_X^{\mathcal{S}} \otimes \pi_X^{\mathcal{S}}), \quad \pi_X^{\mathcal{S}} \circ u_{\mathcal{S}} = \mathrm{coev}_X$$

hold for all $X \in \mathcal{S}$. By (5.3) and (5.4), the algebra $A_{\mathcal{S}}$ with $\mathcal{S} = \mathcal{C}$ is identical to the adjoint algebra of $\mathcal{C}$.

There is an exact sequence $0 \to J_{\mathcal{S}} \to \mathrm{id}_{\mathcal{C}} \to \tau_{\mathcal{S}} \to 0$ in $\mathrm{Rex}(\mathcal{C})$. Since the functor $\boldsymbol{\rho}_{\mathcal{C}}^{\mathrm{ra}}$ is exact by Theorem 5.6, we have an exact sequence

$$0 \to \boldsymbol{\rho}_{\mathcal{C}}^{\mathrm{ra}}(J_{\mathcal{S}}) \to A_{\mathcal{C}} \to A_{\mathcal{S}} \to 0 \tag{5.12}$$

in $\mathcal{C}$. In particular, $A_{\mathcal{S}}$ is a quotient object of $A_{\mathcal{C}}$ (we will see that $A_{\mathcal{S}}$ is in fact a quotient algebra in later). Hence the space $\mathrm{CF}_{\mathcal{C}}(\mathcal{S})$ can be regarded as a subspace of $\mathrm{CF}(\mathcal{C})$. In view of the short exact sequence (5.12), the subspace $\mathrm{CF}_{\mathcal{C}}(\mathcal{S})$ can be thought of as the space of class functions 'vanishing on the ideal $J_{\mathcal{S}}$'.

Now we have a family $\{\mathrm{CF}_{\mathcal{C}}(\mathcal{S})\}_{\mathcal{S} \in \mathfrak{Top}(\mathcal{C})}$ of subspaces of $\mathrm{CF}(\mathcal{C})$. To investigate relations between these subspaces, we introduce the following morphism:

DEFINITION 5.9. For $\mathcal{S}_1, \mathcal{S}_2 \in \mathfrak{Top}(\mathcal{C})$ with $\mathcal{S}_1 \subset \mathcal{S}_2$, we define $q_{\mathcal{S}_1}^{\mathcal{S}_2} : A_{\mathcal{S}_2} \to A_{\mathcal{S}_1}$ to be a unique morphism in $\mathcal{C}$ such that the equation

$$\pi_X^{\mathcal{S}_1} \circ q_{\mathcal{S}_1}^{\mathcal{S}_2} = \pi_X^{\mathcal{S}_2} \tag{5.13}$$

holds for all $X \in \mathcal{S}_1$, where $\pi_X^{\mathcal{S}} : A_{\mathcal{S}} \to X \otimes X^*$ $(X \in \mathcal{S})$ for $\mathcal{S} \in \mathfrak{Top}(\mathcal{C})$ is the universal dinatural transformation for the end $A_{\mathcal{S}}$.

For $\mathcal{S}, \mathcal{S}_1, \mathcal{S}_2, \mathcal{S}_3 \in \mathfrak{Top}(\mathcal{C})$ with $\mathcal{S}_1 \subset \mathcal{S}_2 \subset \mathcal{S}_3$, we have

$$q_{\mathcal{S}}^{\mathcal{S}} = \mathrm{id}_{A_{\mathcal{S}}} \quad \text{and} \quad q_{\mathcal{S}_1}^{\mathcal{S}_2} \circ q_{\mathcal{S}_2}^{\mathcal{S}_3} = q_{\mathcal{S}_1}^{\mathcal{S}_3}$$

by (5.13). Hence the family $\{q_{\mathcal{S}_1}^{\mathcal{S}_2}\}$ is an inverse system of morphisms in $\mathcal{C}$. Furthermore, we have:

LEMMA 5.10. *For $\mathcal{S}_1, \mathcal{S}_2 \in \mathfrak{Top}(\mathcal{C})$ with $\mathcal{S}_1 \subset \mathcal{S}_2$, the morphism $q_{\mathcal{S}_1}^{\mathcal{S}_2}$ is epic.*

PROOF. It is enough to show that $q_{\mathcal{S}}^{\mathcal{C}}$ is epic for all $\mathcal{S} \in \mathfrak{Top}(\mathcal{C})$. Indeed, we have $q_{\mathcal{S}_1}^{\mathcal{S}_2} \circ q_{\mathcal{S}_2}^{\mathcal{C}} = q_{\mathcal{S}_1}^{\mathcal{C}}$ for $\mathcal{S}_1, \mathcal{S}_2 \in \mathfrak{Top}(\mathcal{C})$ with $\mathcal{S}_1 \subset \mathcal{S}_2$. Thus, if $q_{\mathcal{S}_1}^{\mathcal{C}}$ is known to be an epimorphism, then $q_{\mathcal{S}_1}^{\mathcal{S}_2}$ is an epimorphism.

Now let $\mathcal{S}$ be a topologizing subcategory of $\mathcal{C}$. We denote by $J_{\mathcal{S}}$ the ideal of $\mathcal{C}$ corresponding to $\mathcal{S}$. As in the above, we set $\tau_{\mathcal{S}} := \mathrm{id}_{\mathcal{C}}/J_{\mathcal{S}}$. Let $p : \mathrm{id}_{\mathcal{C}} \to \tau_{\mathcal{S}}$ be the quotient morphism. Then $\widetilde{p} := \boldsymbol{\rho}_{\mathcal{C}}^{\mathrm{ra}}(p)$ is the epimorphism appearing in the short exact sequence (5.12). By definition, the equation

$$\widetilde{\pi}_X^{\mathcal{S}} \circ \widetilde{p} = (p_X \otimes \mathrm{id}_{X^*}) \circ \pi_X^{\mathcal{C}}$$

holds for all objects $X \in \mathcal{C}$, where $\widetilde{\pi}_X^{\mathcal{S}} : \boldsymbol{\rho}_{\mathcal{C}}^{\mathrm{ra}}(\tau_{\mathcal{S}}) \to \tau_{\mathcal{S}}(X) \otimes X^*$ is the universal dinatural transformation. Let $\varphi : \boldsymbol{\rho}_{\mathcal{C}}^{\mathrm{ra}}(\tau_{\mathcal{S}}) \to A_{\mathcal{S}}$ be the isomorphism (5.11). Then, for all $X \in \mathcal{S}$, we have

$$\pi_X^{\mathcal{S}} \circ \varphi \circ \widetilde{p} = \widetilde{\pi}_X^{\mathcal{S}} \circ \widetilde{p} = (p_X \otimes \mathrm{id}_{X^*}) \circ \pi_X^{\mathcal{C}} = \pi_X^{\mathcal{C}} = \pi_X^{\mathcal{S}} \circ q_{\mathcal{S}}^{\mathcal{C}},$$

where first equality follows from the construction of φ (see Lemma 3.6 and following remarks) and the third from $p_X = \mathrm{id}_X$ for all $X \in \mathcal{S}$. Hence we have $q_{\mathcal{S}}^{\mathcal{C}} = \varphi \circ \widetilde{p}$ by the universal property. Since φ is an isomorphism, and since $\widetilde{p}$ is an epimorphism, $q_{\mathcal{S}}^{\mathcal{C}}$ is an epimorphism. The proof is done. $\square$

For $\mathcal{S}_1, \mathcal{S}_2 \in \mathfrak{Top}(\mathcal{S})$ with $\mathcal{S}_1 \subset \mathcal{S}_2$, it is easy to see that $q_{\mathcal{S}_1}^{\mathcal{S}_2}$ is a morphism of algebras in $\mathcal{C}$. Hence we have a map

$$\mathfrak{Top}(\mathcal{C}) \to (\text{the set of quotient algebras of } A_{\mathcal{C}}), \quad \mathcal{S} \mapsto A_{\mathcal{S}}. \tag{5.14}$$

Given an object X of an abelian category $\mathcal{M}$, we denote by $\mathfrak{Q}(X)$ the class of the equivalence classes of quotient objects of X in $\mathcal{M}$ and endow it with an order as follows: Given two quotient objects Q_1 and Q_2 of X with quotient morphisms $q_i : X \twoheadrightarrow Q_i$ $(i = 1, 2)$, we write $Q_1 \leq Q_2$ if q_2 factors through q_1, or, equivalently, $\mathrm{Ker}(q_1) \subset \mathrm{Ker}(q_2)$. Now we prove:

THEOREM 5.11. *The map* (5.14) *preserves and reflects the order in the following sense: For two topologizing subcategories $\mathcal{S}_1$ and $\mathcal{S}_2$ of $\mathcal{C}$, we have $A_{\mathcal{S}_1} \leq A_{\mathcal{S}_2}$ if and only if $\mathcal{S}_1 \subset \mathcal{S}_2$.*

PROOF. The 'if' part follows from the above observation that the family $\{q_{\mathcal{S}_1}^{\mathcal{S}_1}\}$ is an inverse system of epimorphisms in $\mathcal{C}$. We prove the 'only if' part. Let $\mathcal{S}_1$ and $\mathcal{S}_2$ be two topologizing subcategories of $\mathcal{C}$ such that $A_{\mathcal{S}_1} \leq A_{\mathcal{S}_2}$. Let J_i $(i = 1, 2)$ be the ideal of $\mathcal{C}$ corresponding to $\mathcal{S}_i$. By the short exact sequence (5.12), we see that $\rho_{\mathcal{C}}^{\mathrm{ra}}(J_1)$ is a subobject of $\rho_{\mathcal{C}}^{\mathrm{ra}}(J_2)$. Since $\rho_{\mathcal{C}}^{\mathrm{ra}}$ is exact, we have

$$\rho_{\mathcal{C}}^{\mathrm{ra}}\left(\frac{J_2}{J_1 \cap J_2}\right) = \frac{\rho_{\mathcal{C}}^{\mathrm{ra}}(J_2)}{\rho_{\mathcal{C}}^{\mathrm{ra}}(J_1) \cap \rho_{\mathcal{C}}^{\mathrm{ra}}(J_2)} = \frac{\rho_{\mathcal{C}}^{\mathrm{ra}}(J_2)}{\rho_{\mathcal{C}}^{\mathrm{ra}}(J_2)} = 0.$$

Since $\rho_{\mathcal{C}}^{\mathrm{ra}}$ is faithful, we have $J_2/(J_1 \cap J_2) = 0$. This implies $J_2 \subset J_1$ and hence we have $\mathcal{S}_1 \subset \mathcal{S}_2$, as desired. The proof is done. □

The above theorem implies:

COROLLARY 5.12. *The following map preserves the inclusion order:*

$$\mathfrak{Top}(\mathcal{C}) \to \{\textit{subspaces of } \mathrm{CF}(\mathcal{C})\}, \quad \mathcal{S} \mapsto \mathrm{CF}_{\mathcal{C}}(\mathcal{S}).$$

5.6. Character theory. We recall that a *pivotal structure* of a rigid monoidal category $\mathcal{T}$ is an isomorphism $X \to X^{**}$ $(X \in \mathcal{T})$ of monoidal functors. In this subsection, we assume that $\mathcal{C}$ is equipped with a pivotal structure $\mathfrak{p}$. The pivotal structure allows us to define the *partial pivotal trace* over X,

$$\mathrm{tr}_{V,W|X} : \mathrm{Hom}_{\mathcal{C}}(V \otimes X, W \otimes X) \to \mathrm{Hom}_{\mathcal{C}}(V, W) \quad (V, W, X \in \mathcal{C}),$$

$$\mathrm{tr}_{V,W|X}(f) := \Big(V \xrightarrow{\mathrm{id}_V \otimes \mathrm{coev}_X} V \otimes X \otimes X^* \xrightarrow{f \otimes \mathrm{id}_{X^*}} W \otimes X \otimes X^* \xrightarrow{\mathrm{id}_W \otimes \mathfrak{p}_X \otimes \mathrm{id}_{X^*}} W \otimes X^{**} \otimes X^* \xrightarrow{\mathrm{id}_V \otimes \mathrm{ev}_X} W\Big).$$

DEFINITION 5.13. For an object $X \in \mathcal{C}$, we set

$$\mathrm{ch}_X := \mathrm{tr}_{A,\mathbb{1}|X}\left(a_X^{\ell} : A \otimes X \to \mathbb{1} \otimes X\right) \in \mathrm{CF}(\mathcal{C}), \tag{5.15}$$

where a_X^{ℓ} is the canonical left action (5.5). We call ch_X the *internal character* of X (with respect to the pivotal structure $\mathfrak{p}$).

As in the character theory of groups, we have:

LEMMA 5.14. *For two objects $X, Y \in \mathcal{C}$, we have $\mathrm{ch}_{X \otimes Y} = \mathrm{ch}_X \star \mathrm{ch}_Y$. Given a short exact sequence $0 \to X \to Y \to Z \to 0$ in $\mathcal{C}$, we have $\mathrm{ch}_Y = \mathrm{ch}_X + \mathrm{ch}_Z$.*

The multiplicativity of the character ($\mathrm{ch}_{X\otimes Y} = \mathrm{ch}_X \star \mathrm{ch}_Y$) can be verified by direct computation. The additivity in exact sequences follows from basic properties of the pivotal trace. We omit the detail of the proof; see [**39**].

By using the universal dinatural transformation $\pi_X^{\mathcal{C}} : A \to X \otimes X^*$, the canonical left action is given by $a_X^{\ell} = (\mathrm{id}_X \otimes \mathrm{ev}_X) \circ (\pi_X^{\mathcal{C}} \otimes \mathrm{id}_X) : A \otimes X \to X$. Hence the internal character of $X \in \mathcal{C}$ can also be expressed as follows:

$$\mathrm{ch}_X = \Big(A \xrightarrow{\pi_X^{\mathcal{C}}} X \otimes X^* \xrightarrow{\mathfrak{p}_X \otimes \mathrm{id}_{X^*}} X^{**} \otimes X^* \xrightarrow{\mathrm{ev}_{X^*}} \mathbb{1} \Big). \tag{5.16}$$

Let $\mathcal{S}$ be a topologizing subcategory of $\mathcal{C}$. By (5.16), we see that ch_X for $X \in \mathcal{S}$ factors through the quotient morphism $q_{\mathcal{S}}^{\mathcal{C}} : A_{\mathcal{C}} \to A_{\mathcal{S}}$. Thus we have:

LEMMA 5.15. *If $X \in \mathcal{C}$ belongs to a topologizing subcategory $\mathcal{S}$ of $\mathcal{C}$, then ch_X belongs to the subspace $\mathrm{CF}_{\mathcal{C}}(\mathcal{S})$ of $\mathrm{CF}(\mathcal{C})$.*

Now we prove the following important result:

LEMMA 5.16. *Let $V_1, \dots, V_n$ be the representatives of the isomorphism classes of simple objects of $\mathcal{C}$. Then the set $\{\mathrm{ch}_{V_i}\}_{i=1,\dots,n}$ is linearly independent.*

PROOF. Let $\mathcal{S}$ be the full subcategory of $\mathcal{C}$ consisting of all semisimple objects. Since $\mathcal{S}$ is a topologizing subcategory of $\mathcal{C}$, the end $A_{\mathcal{S}}$ is defined. Since $\mathcal{S}$ is a semisimple finite abelian category, we may assume that $A_{\mathcal{S}} = \bigoplus_{i=1}^n V_i \otimes V_i^*$ and the universal dinatural transformation $\pi_X^{\mathcal{S}} : A_{\mathcal{S}} \to X \otimes X^*$ ($X \in \mathcal{S}$) is just the projection for $X = V_1, \dots, V_n$ [**25**, §5.1.3]. There are isomorphisms

$$\mathrm{CF}_{\mathcal{C}}(\mathcal{S}) = \mathrm{Hom}_{\mathcal{C}}(A_{\mathcal{S}}, \mathbb{1}) \cong \bigoplus_{i=1}^n \mathrm{Hom}_{\mathcal{C}}(V_i \otimes V_i^*, \mathbb{1}) \cong \bigoplus_{i=1}^n \mathrm{Hom}_{\mathcal{C}}(V_i, V_i^{**})$$

and, by (5.16), the element $\mathrm{ch}_{V_i} \in \mathrm{CF}_{\mathcal{C}}(\mathcal{S})$ corresponds to $\mathfrak{p}_{V_i} \in \mathrm{Hom}_{\mathcal{C}}(V_i, V_i^{**})$ via the above isomorphism. Thus the set $\{\mathrm{ch}_{V_i}\}_{i=1,\dots,n}$ is linearly independent. □

Given a finite abelian category $\mathcal{A}$, we denote by $\mathrm{Gr}(\mathcal{A})$ the Grothendieck group of $\mathcal{A}$ and set $\mathrm{Gr}_k(\mathcal{A}) := k \otimes_{\mathbb{Z}} \mathrm{Gr}(\mathcal{A})$. If $\mathcal{A}$ is a finite tensor category, then $\mathrm{Gr}_k(\mathcal{A})$ is an algebra with respect to the multiplication induced by the tensor product of $\mathcal{A}$. Lemmas 5.14 and 5.16 imply:

THEOREM 5.17. *The linear map*

$$\mathrm{ch} : \mathrm{Gr}_k(\mathcal{C}) \to \mathrm{CF}(\mathcal{C}), \quad [X] \mapsto \mathrm{ch}_X \tag{5.17}$$

is a well-defined injective homomorphism of algebras. If $\mathcal{C}$ is semisimple, then this map is an isomorphism of algebras.

The converse does not hold: There is a non-semisimple pivotal finite tensor category such that (5.17) is bijective. Interestingly, if we impose an extra assumption that $\mathcal{C}$ is unimodular, then the converse holds (Theorem 6.13).

5.7. Ring-theoretical class functions. Let B be a finite-dimensional algebra. From ring-theoretical point of view, the space

$$\mathrm{SLF}(B) := \{ f \in B^* \mid f(ab) = f(ba) \text{ for all } a, b \in B \}$$

of symmetric linear functions on B can be thought of as an analogue of the space of class functions on a group. We first give the following 'algebra free' description of this space:

DEFINITION 5.18. For a finite abelian category $\mathcal{M}$, we define

$$\mathrm{SLF}(\mathcal{M}) = \mathrm{Nat}(\mathrm{id}_{\mathcal{M}}, \mathbb{N}_{\mathcal{M}}),$$

where $\mathbb{N}_{\mathcal{M}} \in \mathrm{Rex}(\mathcal{M})$ is the Nakayama functor.

We note that there is an isomorphism

$$\mathrm{Hom}_{B\text{-mod-}B}(B, B^*) \to \mathrm{SLF}(B), \quad \xi \mapsto \xi(1)$$

of vector spaces. Thus, if $\mathcal{M} \approx B$-mod, then we have

$$\mathrm{SLF}(\mathcal{M}) \cong \mathrm{SLF}(B\text{-mod}) \cong \mathrm{Hom}_{B\text{-mod-}B}(B, B^*) \cong \mathrm{SLF}(B).$$

The aim of this subsection is to show that $\mathrm{CF}(\mathcal{C})$ is isomorphic to $\mathrm{SLF}(\mathcal{C})$ if the double dual functor of $\mathcal{C}$ is isomorphic to $\mathrm{id}_{\mathcal{C}}$ (Theorem 5.22). The following description of the Nakayama functor is important:

THEOREM 5.19. *There is a natural isomorphism*

$$\mathbb{N}_{\mathcal{C}}(V) \cong {}^{**}V \otimes \alpha \quad (V \in \mathcal{C}),$$

where $\alpha = \alpha_{\mathcal{C}}$ is the modular object of $\mathcal{C}$.

This formula has been given in [**20**]. Below we give a different proof based on the equivalence $\mathcal{C}^{\mathrm{env}} \approx \mathrm{Rex}(\mathcal{C})$ introduced in Subsection 4.4. Before we give a proof, we remark two corollaries of this theorem. The first one is:

COROLLARY 5.20. *The Nakayama functor $\mathbb{N}_{\mathcal{C}}$ is isomorphic to $\mathrm{id}_{\mathcal{C}}$ if and only if $\mathcal{C}$ is unimodular and the double dual functor of $\mathcal{C}$ is isomorphic to $\mathrm{id}_{\mathcal{C}}$.*

This result, also included in [**20**], generalizes the fact that a finite-dimensional Hopf algebra H is a symmetric Frobenius algebra if and only if H is unimodular and the square of the antipode of H is inner.

COROLLARY 5.21. *A semisimple finite tensor category is unimodular.*

This result, first proved in [**18**], follows from Corollary 5.20 and the fact that a finite-dimensional semisimple algebra is a symmetric Frobenius algebra.

PROOF OF THEOREM 5.19. We use the equivalence $\Phi_{\mathcal{C}} : \mathcal{C}^{\mathrm{env}} \to \mathrm{Rex}(\mathcal{C})$ and the coalgebra $\mathbb{H}$ in $\mathcal{C}^{\mathrm{env}}$ introduced in Subsection 4.4. By the definition of the modular object, there is an isomorphism $\mathbb{H}^* \cong \mathbb{H} \otimes (\alpha \boxtimes \mathbb{1})$ in $\mathcal{C}^{\mathrm{env}}$. We have

$$\mathbb{H} \otimes (\alpha \boxtimes \mathbb{1}) \cong \int_{X \in \mathcal{C}} (X \otimes \alpha) \boxtimes X^* \cong \Phi_{\mathcal{C}}^{-1}(\mathrm{id}_{\mathcal{C}} \otimes \alpha).$$

Since the duality functor is an antiequivalence, we have

$$\mathbb{H}^* \cong \Big(\int_{X \in \mathcal{C}} X \boxtimes X^* \Big)^* \cong \int^{X \in \mathcal{C}} (X \boxtimes X^*)^* \cong \int^{X \in \mathcal{C}} X^* \boxtimes X \cong \int^{X \in \mathcal{C}} X \boxtimes {}^*X.$$

In a similar way as Subsection 3.6, we compute

$$\Phi_{\mathcal{C}}(\mathbb{H}^*)(V) \cong \int^{X \in \mathcal{C}} \Phi_{\mathcal{C}}(X^* \boxtimes X)(V) \cong \int^{X \in \mathcal{C}} \mathrm{Hom}_{\mathcal{C}}(V, {}^{**}X)^* \otimes X \cong \mathbb{N}_{\mathcal{C}}(V^{**})$$

for $V \in \mathcal{C}$. Thus we have a natural isomorphism $\mathbb{N}_{\mathcal{C}}(V^{**}) \cong V \otimes \alpha$ $(V \in \mathcal{C})$ by applying $\Phi_{\mathcal{C}}$ to the isomorphism $\mathbb{H}^* \cong \mathbb{H} \otimes (\alpha \boxtimes \mathbb{1})$ in $\mathcal{C}^{\mathrm{env}}$. The proof is done. $\square$

Now we prove:

THEOREM 5.22. *There is a natural isomorphism*

$$\mathrm{Hom}_{\mathcal{C}}(\boldsymbol{\rho}_{\mathcal{C}}^{\mathrm{ra}}(F), \mathbb{1}) \cong \mathrm{Nat}(F, \mathbb{N}'_{\mathcal{C}}) \quad (F \in \mathrm{Rex}(\mathcal{C})), \tag{5.18}$$

where $\mathbb{N}'_{\mathcal{C}}(X) = \mathbb{N}_{\mathcal{C}}(X^{**})$. *Hence, if the double dual functor of* $\mathcal{C}$ *is isomorphic to the identity functor, then we have*

$$\mathrm{CF}(\mathcal{C}) = \mathrm{Hom}_{\mathcal{C}}(\boldsymbol{\rho}_{\mathcal{C}}^{\mathrm{ra}}(\mathrm{id}_{\mathcal{C}}), \mathbb{1}) \cong \mathrm{Nat}(\mathrm{id}_{\mathcal{C}}, \mathbb{N}_{\mathcal{C}}) = \mathrm{SLF}(\mathcal{C}).$$

PROOF. For $F \in \mathrm{Rex}(\mathcal{C})$, we have natural isomorphisms

$$\begin{aligned}
\mathrm{Hom}_{\mathcal{C}}(\boldsymbol{\rho}_{\mathcal{C}}^{\mathrm{ra}}(F), \mathbb{1}) &\cong \mathrm{Hom}_{\mathcal{C}}(\boldsymbol{\rho}_{\mathcal{C}}^{\mathrm{la}}(F \otimes (\alpha^* \boxtimes \mathbb{1})), \mathbb{1}) \quad \text{(by Theorem 5.6)} \\
&\cong \mathrm{Nat}(F \otimes (\alpha^* \boxtimes \mathbb{1}), \boldsymbol{\rho}_{\mathcal{C}}(\mathbb{1})) \\
&\cong \mathrm{Nat}(F, \mathrm{id}_{\mathcal{C}} \otimes (\alpha \boxtimes \mathbb{1})) \\
&\cong \mathrm{Nat}(F, \mathbb{N}'_{\mathcal{C}}) \quad \text{(by Theorem 5.19).}
\end{aligned}$$

□

We suppose that the double dual functor of $\mathcal{C}$ is isomorphic to $\mathrm{id}_{\mathcal{C}}$. Then the above theorem gives an isomorphism $\mathrm{CF}(\mathcal{C}) \cong \mathrm{SLF}(\mathcal{C})$ of vector spaces. The point is that this isomorphism is a part of a natural isomorphism. Let $\mathcal{S}$ be a topologizing subcategory of $\mathcal{C}$, let $\tau_{\mathcal{S}}$ be the corresponding idempotent monad on $\mathcal{C}$, and let $p_{\mathcal{S}} : \mathrm{id}_{\mathcal{C}} \to \tau_{\mathcal{S}}$ be the quotient morphism in $\mathrm{Rex}(\mathcal{C})$. The naturality implies that the following diagram is commutative:

$$\begin{array}{ccccc}
\mathrm{CF}(\mathcal{C}) & = & \mathrm{Hom}_{\mathcal{C}}(\boldsymbol{\rho}_{\mathcal{C}}^{\mathrm{ra}}(\mathrm{id}_{\mathcal{C}}), \mathbb{1}) & \xrightarrow{(5.18)} & \mathrm{Nat}(\mathrm{id}_{\mathcal{C}}, \mathbb{N}_{\mathcal{C}}) \\
{\scriptstyle \text{inclusion}}\uparrow & & \uparrow{\scriptstyle \mathrm{Hom}_{\mathcal{C}}(\boldsymbol{\rho}_{\mathcal{C}}^{\mathrm{ra}}(q_{\mathcal{S}}), \mathbb{1})} & & \uparrow{\scriptstyle \mathrm{Nat}(q_{\mathcal{S}}, \mathrm{id}_{\mathcal{C}})} \\
\mathrm{CF}_{\mathcal{C}}(\mathcal{S}) & \xrightarrow{\cong} & \mathrm{Hom}_{\mathcal{C}}(\boldsymbol{\rho}_{\mathcal{C}}^{\mathrm{ra}}(\tau_{\mathcal{S}}), \mathbb{1}) & \xrightarrow{(5.18)} & \mathrm{Nat}(\tau_{\mathcal{S}}, \mathbb{N}_{\mathcal{C}}).
\end{array}$$

Namely, the subspace $\mathrm{CF}_{\mathcal{C}}(\mathcal{S})$ corresponds to the subspace $\mathrm{Nat}(\tau_{\mathcal{S}}, \mathbb{N}_{\mathcal{C}})$ via the isomorphism (5.18). This observation allows us to use some ring-theoretical methods to investigate the space $\mathrm{CF}(\mathcal{C})$; see [**41**] for further results.

5.8. The case of Hopf algebras. Suppose that $\mathcal{C} = H$-mod for some finite-dimensional Hopf algebra H. We identify $\mathcal{Z}(\mathcal{C})$ with ${}^H_H\mathcal{YD}_{fd}$. Then U is the functor forgetting the coaction of H and R is the Radford induction. Under the identification $H = H \otimes k$ of vector spaces, we see that $A = UR(k)$ is precisely the adjoint representation of H. By using formulas (4.10) and (4.11) of the unit and the counit of $U \dashv R$, we see that the monoidal structure m of R is given by

$$m_{X,Y} : R(X) \otimes R(Y) \to R(X \otimes Y), \quad (a \otimes x) \otimes (b \otimes y) \mapsto ab \otimes (x \otimes y) \tag{5.19}$$

for $X, Y \in \mathcal{C}$, and hence $A = H$ as algebras. The half-braiding for A is given by

$$\widetilde{\sigma}_X : A \otimes X \to X \otimes A, \quad x \otimes a \mapsto a_{(1)}x \otimes a_{(2)} \quad (a \in A, x \in X \in H\text{-mod}),$$

and hence we have $(\widetilde{\sigma}_X)^{-1}(x \otimes a) = a_1 \otimes S^{-1}(a_{(2)})x$. Thus, for $X \in H$-mod, the left and the right canonical action of A on X are given respectively by

$$\begin{aligned}
a_X^{\ell} &: A \otimes X \to X, \quad a \otimes x \mapsto ax, \\
a_X^{r} &: X \otimes A \to X, \quad a \otimes x \mapsto S^{-1}(a)x
\end{aligned}$$

for $a \in A$ and $x \in X$.

The algebra $\mathrm{CF}(H\text{-mod})$ of class functions is the subalgebra

$$\mathrm{CF}(H\text{-mod}) = \{f \in H^* \mid f(ab) = f(bS^2(a)) \text{ for all } a, b \in H\}$$

of H^* (*cf.* Subsection 2.2). This space does not coincides with the space

$$\mathrm{SLF}(H) := \{f \in H^* \mid f(ab) = f(ba) \text{ for all } a, b \in H\}$$

of symmetric linear functions on H in general. If the double dual functor of H-mod is isomorphic to the identity functor, then there is an invertible element $g \in H$ such that $ghg^{-1} = S^2(h)$ for all $h \in H$ and this element gives an isomorphism

$$\mathrm{SLF}(H) \to \mathrm{CF}(H\text{-mod}), \quad f \mapsto f \leftharpoonup g \quad (f \in \mathrm{SLF}(H))$$

of vector spaces (*cf.* Theorem 5.22).

A *pivotal element* of H is a grouplike element $g \in H$ such that $ghg^{-1} = S^2(h)$ for all $h \in H$. There is a one-to-one correspondence between the set of pivotal structures of H-mod and the set of pivotal elements of H. If a pivotal element g of H is given, then the internal character of $X \in H$-mod (with respect to the pivotal structure of H-mod corresponding to g) is given by

$$\mathrm{ch}_X(a) = \mathrm{Tr}(X \to X; x \mapsto gax) \quad (a \in A),$$

where Tr is the ordinary trace of a linear map.

The action functor played an important role for the study of the space of class functions. We identify $\mathrm{Rex}(H\text{-mod})$ with H-mod-H by Lemma 3.2. By using the notation introduced in Subsection 4.8, the action functor is given by

$$\boldsymbol{\rho} : H\text{-mod} \to H\text{-mod-}H, \quad V \mapsto V_\varepsilon \circledast H \quad (V \in H\text{-mod}).$$

A right adjoint of this functor is given by $\boldsymbol{\rho}^{\mathrm{ra}}(M) = {}_{\mathrm{ad}}M$ for $M \in H$-mod-H. Indeed, there is the following natural isomorphism:

$$\mathrm{Hom}_{H|H}(\boldsymbol{\rho}(V), M) \cong \mathrm{Hom}_{H|-}(V, {}_{\mathrm{ad}}M), \quad f \mapsto \Big[v \mapsto f(v \otimes 1)\Big].$$

By this concrete description of $\boldsymbol{\rho}^{\mathrm{ra}}$, most of results in Subsections 5.4–5.7 are trivial in the Hopf algebra case.

6. Integral theory and its applications

6.1. Modules over the adjoint algebra. Integrals for Hopf algebras play an important role in the representation theory of Hopf algebras. Based on the relation between cointegrals on a Hopf algebra and Yetter-Drinfeld modules, a categorical (co)integral was introduced in [**40**] for unimodular finite tensor categories and then extended to general finite tensor categories in [**38**]. This section reviews the integral theory for finite tensor categories and exhibits some applications to the character theory.

Throughout this section, we work over an algebraically closed field k. Let $\mathcal{C}$ be a finite tensor category. We use symbols U, R, Z, η, ε, A, m and u by the same meaning as in the last section. We also write $\mathbf{A} := R(\mathbb{1})$. This object has a canonical structure of an algebra in $\mathcal{Z}(\mathcal{C})$ such that $A = U(\mathbf{A})$.

We introduce some category equivalences which are important to define integrals and cointegrals of $\mathcal{C}$. We first give a description of the category of $\mathbf{A}$-modules in $\mathcal{Z}(\mathcal{C})$. An object of $\mathcal{Z}(\mathcal{C})$ of the form $R(V)$, $V \in \mathcal{C}$, is a left $\mathbf{A}$-module in $\mathcal{Z}(\mathcal{C})$ by $m_{\mathbb{1},V} : \mathbf{A} \otimes R(V) \to R(V)$. For clarity, we denote this left $\mathbf{A}$-module by ${}_{\mathbf{A}}R(V)$. Similarly, we denote by $R(V)_{\mathbf{A}}$ the object $R(V)$ regarded as a right $\mathbf{A}$-module by $m_{V,\mathbb{1}}$. We now prove:

LEMMA 6.1. *The following two functors K and K' are equivalences:*

$$K : \mathcal{C} \to {}_{\mathbf{A}}\mathcal{Z}(\mathcal{C}), \quad K(V) = {}_{\mathbf{A}}R(V) \quad \text{and} \quad K' : \mathcal{C} \to \mathcal{Z}(\mathcal{C})_{\mathbf{A}}, \quad K'(V) = R(V)_{\mathbf{A}}.$$

PROOF. We only show that K' is an equivalence, since the equivalency of the other one can be proved by applying the same argument to $\mathcal{C}^{\text{rev}}$. For the proof, we make $\mathcal{C}$ a left $\mathcal{Z}(\mathcal{C})$-module category by the action

$$\mathbf{M} \ominus V = U(\mathbf{M}) \otimes V \quad (\mathbf{M} \in \mathcal{Z}(\mathcal{C}), V \in \mathcal{C}).$$

Since U is exact, $\mathcal{C}$ is a finite left $\mathcal{Z}(\mathcal{C})$-module category. The internal Hom functor of $\mathcal{C}$ is given by $\underline{\text{Hom}}(V, W) = R(W \otimes V^*)$, since

$$\text{Hom}_{\mathcal{C}}(\mathbf{M} \ominus V, W) \cong \text{Hom}_{\mathcal{C}}(U(\mathbf{M}), W \otimes V^*) \cong \text{Hom}_{\mathcal{C}}(\mathbf{M}, R(W \otimes V^*))$$

for $V, W \in \mathcal{C}$ and $\mathbf{M} \in \mathcal{Z}(\mathcal{C})$. By the exact faithfulness of R, we see that $\mathbb{1} \in \mathcal{C}$ is a $\mathcal{Z}(\mathcal{C})$-projective $\mathcal{Z}(\mathcal{C})$-generator and therefore the functor

$$\mathcal{C} \to \mathcal{Z}(\mathcal{C})_{\underline{\text{Hom}}(\mathbb{1}, \mathbb{1})}, \quad V \mapsto \underline{\text{Hom}}(\mathbb{1}, V)$$

is an equivalence by Lemma 4.5. The algebra $\underline{\text{Hom}}(\mathbb{1}, \mathbb{1})$ is identical to $\mathbf{A}$ and the above functor coincides with K'. The proof is done. $\square$

Now we consider the category of A-modules in $\mathcal{C}$.

LEMMA 6.2. *The following two functors K and K' are equivalences:*

$$\mathcal{C} \boxtimes \mathcal{C} \to {}_{A}\mathcal{C}, \quad X \boxtimes Y \mapsto (X \otimes Y, a^{\ell}_X \otimes \text{id}_Y);$$
$$\mathcal{C} \boxtimes \mathcal{C} \to \mathcal{C}_{A}, \quad X \boxtimes Y \mapsto (X \otimes Y, \text{id}_X \otimes a^{r}_X).$$

PROOF. The coend $F := \int^{X \in \mathcal{C}} X^* \otimes X$ has a canonical structure of a coalgebra in $\mathcal{C}$ such that $F^* \cong A$ as algebras in $\mathcal{C}$. This lemma follows from Lyubashenko's result on the category of F-comodules given in [**29**, §2.7]. $\square$

We say that an object $T \in \mathcal{C}$ is *trivial* if it is isomorphic to the direct sum of finitely many copies of the unit object. The following characterization of trivial objects, which may be of independent interest, will be used in the proof of our categorical generalization of Maschke's theorem:

LEMMA 6.3. *For an object $T \in \mathcal{C}$, the following equivalent:*

(1) *T is trivial*
(2) *$a^{\ell}_T = \varepsilon_{\mathbb{1}} \otimes \text{id}_T$, where a^{ℓ} is the canonical left action.*
(3) *$a^{r}_T = \text{id}_T \otimes \varepsilon_{\mathbb{1}}$, where a^{r} is the canonical right action.*

PROOF. (1) $\Rightarrow$ (2) and (1) $\Rightarrow$ (3) follow from $a^{\ell}_{\mathbb{1}} = \varepsilon_{\mathbb{1}} = a^{r}_{\mathbb{1}}$ and the naturality of the canonical actions. To prove (2) $\Rightarrow$ (1), we use the equivalence ${}_{A}\mathcal{C} \approx \mathcal{C} \boxtimes \mathcal{C}$ given in Lemma 6.2. The left A-module (T, a^{ℓ}_T) corresponds to $T \boxtimes \mathbb{1}$, while the left A-module $(T, \varepsilon_{\mathbb{1}} \otimes \text{id}_T)$ corresponds to $\mathbb{1} \boxtimes T$. Thus, if $a^{\ell}_T = \varepsilon_{\mathbb{1}} \otimes \text{id}_T$, then we have $T \boxtimes \mathbb{1} \cong \mathbb{1} \boxtimes T$ in $\mathcal{C} \boxtimes \mathcal{C}$. By applying the functor

$$\mathcal{C} \boxtimes \mathcal{C} \xrightarrow{\Phi_{\mathcal{C}}} \text{Rex}(\mathcal{C}) \xrightarrow{F \mapsto F(\mathbb{1})} \mathcal{C}$$

to the both sides, we obtain $T \cong \text{Hom}_{\mathcal{C}}(T, \mathbb{1})^* \otimes_k \mathbb{1}$. Therefore T is trivial. The proof of (2) $\Rightarrow$ (1) is done. One can prove (3) $\Rightarrow$ (1) in a similar way but by using the equivalence $\mathcal{C}_A \approx \mathcal{C} \boxtimes \mathcal{C}$. $\square$

6.2. Categorical (co)integrals. Let α be the modular object of $\mathcal{C}$ defined in Subsection 4.5. Categorical analogues of (co)integrals for Hopf algebras were introduced in [**40**] for unimodular finite tensor categories. Later, the notions were extended to the general setting in [**38**] as follows:

DEFINITION 6.4. Let $\Lambda : \mathbb{1} \to Z(\alpha)$ be a morphism in $\mathcal{C}$. We say that Λ is a *categorical left integral* of $\mathcal{C}$ if it satisfies the equation

$$m_{\mathbb{1},\alpha} \circ (\mathrm{id}_A \otimes \Lambda) = \varepsilon_{\mathbb{1}} \otimes \Lambda. \tag{6.1}$$

Analogously, Λ is said to be a *categorical right integral* of $\mathcal{C}$ if

$$m_{\alpha,\mathbb{1}} \circ (\Lambda \otimes \mathrm{id}_A) = \Lambda \otimes \varepsilon_{\mathbb{1}}. \tag{6.2}$$

We denote by $\mathscr{I}^{\ell}_{\mathcal{C}}$ and $\mathscr{I}^{r}_{\mathcal{C}}$ the spaces of categorical left and right integrals of $\mathcal{C}$, respectively. We also define the space $\mathscr{I}^{\mathrm{co}}_{\mathcal{C}}$ of *categorical cointegrals* of $\mathcal{C}$ by

$$\mathscr{I}^{\mathrm{co}}_{\mathcal{C}} := \mathrm{Hom}_{\mathcal{Z}(\mathcal{C})}(R(\alpha), \mathbb{1}) \quad (\subset \mathrm{Hom}_{\mathcal{C}}(Z(\alpha), \mathbb{1})). \tag{6.3}$$

REMARK 6.5. (1) For $V \in \mathcal{C}$, we set $D(V) = {}^*Z(V^*)$. The functor $D : \mathcal{C} \to \mathcal{C}$ is the central Hopf monad on $\mathcal{C}$ (see Remark 5.1). Under the assumption that $\mathcal{C}$ is unimodular, a morphism $f : Z(\mathbb{1}) \to \mathbb{1}$ in $\mathcal{C}$ is a categorical cointegral in the above sense if and only if the morphism $\mathbb{1} \xrightarrow{\cong} {}^*\mathbb{1} \xrightarrow{{}^*f} {}^*Z(\mathbb{1}) \xrightarrow{\cong} D(\mathbb{1})$ in $\mathcal{C}$ is a cointegral of D in the sense of [**2**, Subsection 6.3].

(2) Let $K : \mathcal{C} \to \mathcal{C}$ be a functor such that $K({}^*\alpha) \cong \mathbb{1}$, and let $c : D \to K$ be a K-valued left integral of D in the sense of [**2**, Subsection 5.1]. By using the relation between Z and D, we see that the morphism

$$\mathbb{1} \xrightarrow{\quad\cong\quad} K({}^*\alpha)^* \xrightarrow{\quad (c_{*\alpha})^* \quad} D({}^*\alpha)^* \xrightarrow{\quad\cong\quad} Z(\alpha)$$

is a categorical right integral of $\mathcal{C}$ in the above sense. In a similar way, a K-valued right integral of D induces a categorical left integral of $\mathcal{C}$.

(3) If H is a finite-dimensional Hopf algebra, then, by Theorem 2.3, a categorical cointegral of H-mod is a left cointegral on H. Thus a categorical cointegral of $\mathcal{C}$ may be called a categorical 'left' cointegral of $\mathcal{C}$. Since $(H\text{-mod})^{\mathrm{rev}} \cong H^{\mathrm{cop}}\text{-mod}$, one may define a categorical 'right' cointegral of $\mathcal{C}$ as a categorical 'left' cointegral of $\mathcal{C}^{\mathrm{rev}}$. Such variants of cointegrals are systematically studied in [**1**] in the dual context: Four monads A_i $(i = 1, 2, 3, 4)$ on $\mathcal{C}$ and monadic integrals for them are introduced in [**1**], and a relation with our categorical cointegrals is also explained in [**1**]. We note that A_4 and A_2 are the central Hopf monad on $\mathcal{C}$ and that on $\mathcal{C}^{\mathrm{rev}}$, respectively. If $\mathcal{C} = H$-mod, then a monadic integral of A_4 and that of A_2 closely relate to a left and a right cointegral on H, respectively [**1**, Example 2.9].

We often omit the adjective 'categorical' if no confusion arises. As in the case of Hopf algebras, the following theorem holds:

THEOREM 6.6. $\dim_k \mathscr{I}^{\ell}_{\mathcal{C}} = \dim_k \mathscr{I}^{r}_{\mathcal{C}} = \dim_k \mathscr{I}^{\mathrm{co}}_{\mathcal{C}} = 1$. *Furthermore, for every non-zero cointegral λ of $\mathcal{C}$, there are a left integral Λ^{ℓ} and a right integral Λ^{r} of $\mathcal{C}$ satisfying $\lambda \circ \Lambda^{\ell} = \lambda \circ \Lambda^{r} = \mathrm{id}_{\mathbb{1}}$.*

To prove this theorem, we first remark:

LEMMA 6.7. *For a simple object $V \in \mathcal{C}$, we have*

$$\dim_k \mathrm{Hom}_{\mathcal{Z}(\mathcal{C})}(R(V), \mathbb{1}) = \begin{cases} 1 & \text{if } V \cong \alpha, \\ 0 & \text{otherwise.} \end{cases}$$

PROOF. By Theorem 4.15, we have

$$\mathrm{Hom}_{\mathcal{Z}(\mathcal{C})}(R(V),\mathbb{1}) \cong \mathrm{Hom}_{\mathcal{Z}(\mathcal{C})}(L(\alpha^* \otimes V),\mathbb{1}) \cong \mathrm{Hom}_{\mathcal{C}}(\alpha^* \otimes V,\mathbb{1}) \cong \mathrm{Hom}_{\mathcal{C}}(V,\alpha),$$

and hence the claim follows from Schur's lemma. □

This lemma implies $\dim_k \mathscr{I}_{\mathcal{C}}^{\mathrm{co}} = 1$. We note that $\mathbf{A}^*$ and ${}^*\mathbf{A}$ are a right and a left $\mathbf{A}$-modules in $\mathcal{Z}(\mathcal{C})$, respectively. The above lemma is also crucial to prove the following Frobenius type property of $\mathbf{A}$.

LEMMA 6.8. *There are isomorphisms* $\mathbf{A}^* \cong R(\alpha)_{\mathbf{A}}$ *and* ${}^*\mathbf{A} \cong {}_{\mathbf{A}}R(\alpha)$.

PROOF. We only show $\mathbf{A}^* \cong R(\alpha)_{\mathbf{A}}$ since ${}^*\mathbf{A} \cong {}_{\mathbf{A}}R(\alpha)$ is proved in a similar way. By Lemma 6.1, the left $\mathbf{A}$-module $\mathbf{A}$ is a simple object of ${}_{\mathbf{A}}\mathcal{Z}(\mathcal{C})$ and therefore $\mathbf{A}^*$ is a simple object of $\mathcal{Z}(\mathcal{C})_{\mathbf{A}}$. Hence, again by Lemma 6.1, $\mathbf{A}^*$ is isomorphic to $R(V)_{\mathbf{A}}$ for some simple object $V \in \mathcal{C}$. Since $(-)\otimes\mathbf{A}^*$ is right adjoint to the forgetful functor $\mathcal{Z}(\mathcal{C})_{\mathbf{A}} \to \mathcal{Z}(\mathcal{C})$, we have

$$\mathrm{Hom}_{\mathcal{Z}(\mathcal{C})}(R(V),\mathbb{1}) \cong \mathrm{Hom}_{\mathcal{Z}(\mathcal{C})_{\mathbf{A}}}(R(V)_{\mathbf{A}},\mathbb{1}\otimes\mathbf{A}^*) \neq 0 \tag{6.4}$$

and therefore $V \cong \alpha$ by Lemma 6.7. Hence $\mathbf{A}^* \cong R(\alpha)_{\mathbf{A}}$. □

Now we fix an isomorphism $\phi : R(\alpha)_{\mathbf{A}} \to \mathbf{A}^*$ of right $\mathbf{A}$-modules in $\mathcal{Z}(\mathcal{C})$ and define $\lambda_\phi : R(\alpha) \to \mathbb{1}$ by $\lambda_\phi = u^* \circ \phi$, where u is the unit of $\mathbf{A}$. We note that λ_ϕ is a non-zero cointegral as it is the element of $\mathrm{Hom}_{\mathcal{Z}(\mathcal{C})}(R(\alpha),\mathbb{1})$ corresponding to ϕ via the isomorphism (6.4) with $V = \alpha$. We also define

$$e_\phi = \mathrm{ev}_{\mathbf{A}} \circ (\phi \otimes \mathrm{id}_{\mathbf{A}}) \quad \text{and} \quad d_\phi = (\mathrm{id}_{\mathbf{A}} \otimes \phi^{-1}) \circ \mathrm{coev}_{\mathbf{A}}. \tag{6.5}$$

It is obvious that $(R(\alpha), e_\phi, d_\phi)$ is a left dual object of $\mathbf{A}$ in $\mathcal{Z}(\mathcal{C})$. Since ϕ is a morphism of right $\mathbf{A}$-modules in $\mathcal{Z}(\mathcal{C})$, we have:

$$e_\phi \circ (\mathrm{id}_{R(\alpha)} \otimes m_{\mathbb{1},\mathbb{1}}) = e_\phi \circ (m_{\alpha,\mathbb{1}} \otimes \mathrm{id}_{\mathbf{A}}), \tag{6.6}$$

$$(\mathrm{id}_{\mathbf{A}} \otimes m_{\alpha,\mathbb{1}}) \circ (d_\phi \otimes \mathrm{id}_{\mathbf{A}}) = (m_{\mathbb{1},\mathbb{1}} \otimes \mathrm{id}_{R(\alpha)}) \circ (\mathrm{id}_{\mathbf{A}} \otimes d_\phi). \tag{6.7}$$

LEMMA 6.9. *The morphism*

$$\Lambda_\phi := (\varepsilon_{\mathbb{1}} \otimes \mathrm{id}_{Z(\alpha)}) \circ d_\phi \in \mathrm{Hom}_{\mathcal{C}}(\mathbb{1}, Z(\alpha)) \tag{6.8}$$

is a right integral of $\mathcal{C}$ *such that* $\lambda_\phi \circ \Lambda_\phi = \mathrm{id}_{\mathbb{1}}$.

PROOF. The claim that Λ_ϕ is a right integral is verified as follows:

$$\begin{aligned} m_{\alpha,\mathbb{1}} \circ (\Lambda_\phi \otimes \mathrm{id}_A) &\overset{(6.8)}{=} (\varepsilon_{\mathbb{1}} \otimes \mathrm{id}_{Z(\alpha)}) \circ (\mathrm{id}_A \otimes m_{\alpha,\mathbb{1}}) \circ (d_\phi \otimes \mathrm{id}_A) \\ &\overset{(6.7)}{=} (\varepsilon_{\mathbb{1}} \otimes \mathrm{id}_{Z(\alpha)}) \circ (m_{\mathbb{1},\mathbb{1}} \otimes \mathrm{id}_{Z(\alpha)}) \circ (\mathrm{id}_A \otimes d_\phi) \\ &\overset{(5.2)}{=} (\varepsilon_{\mathbb{1}} \otimes \varepsilon_{\mathbb{1}} \otimes \mathrm{id}_{Z(\alpha)}) \circ (\mathrm{id}_A \otimes d_\phi) = \Lambda_\phi \otimes \varepsilon_{\mathbb{1}}. \end{aligned}$$

The equation $\lambda_\phi \circ \Lambda_\phi = \mathrm{id}_{\mathbb{1}}$ is proved as follows:

$$\lambda_\phi \circ \Lambda_\phi = (\varepsilon_{\mathbb{1}} \otimes u^*) \circ \mathrm{coev}_{\mathbf{A}} = \varepsilon_{\mathbb{1}} \circ u \overset{(5.2)}{=} \mathrm{id}_{\mathbb{1}}.$$

□

We fix an isomorphism $\varphi : {}_{\mathbf{A}}R(\alpha) \to {}^*\mathbf{A}$ of left $\mathbf{A}$-modules in $\mathcal{Z}(\mathcal{C})$ and then define $\lambda'_\varphi = u^* \circ \varphi$ and $\Lambda'_\varphi = (\varphi^{-1} \otimes \varepsilon_{\mathbb{1}}) \circ \mathrm{coev}'_{\mathbf{A}}$. By the same way as above, we see that λ'_φ is a cointegral and Λ'_φ is a left integral such that $\lambda'_\varphi \circ \Lambda'_\varphi = \mathrm{id}_{\mathbb{1}}$.

PROOF OF THEOREM 6.6. We have proved $\dim_k \mathscr{I}_{\mathcal{C}}^{\mathrm{co}} = 1$. To show that $\mathscr{I}_{\mathcal{C}}^{\ell}$ is one-dimensional, we note that $\mathscr{I}_{\mathcal{C}}^{\ell}$ is the space of left A-linear morphisms in $\mathcal{C}$ if we view $\mathbb{1}$ and $Z(\alpha)$ as left A-modules by $\varepsilon_{\mathbb{1}}$ and $m_{\mathbb{1},\alpha}$, respectively. Since ${}_{\mathbf{A}}R(\alpha)$ is isomorphic to ${}^*\mathbf{A}$ as left $\mathbf{A}$-modules in $\mathcal{Z}(\mathcal{C})$, $Z(\alpha)$ is isomorphic to ${}^*A$ as left A-modules in $\mathcal{C}$. Thus we have

$$\mathscr{I}_{\mathcal{C}}^{\ell} = \mathrm{Hom}_A(\mathbb{1}, Z(\alpha)) \cong \mathrm{Hom}_A(\mathbb{1}, {}^*A) \cong \mathrm{Hom}_{\mathcal{C}}(\mathbb{1}, \mathbb{1}) \cong k.$$

The equation $\dim_k \mathscr{I}_{\mathcal{C}}^{r} = 1$ is proved in a similar way.

Now let λ be a non-zero cointegral of $\mathcal{C}$. Since $\dim_k \mathscr{I}_{\mathcal{C}}^{\mathrm{co}} = 1$, there exists an isomorphism $\phi : R(\alpha)_{\mathbf{A}} \to \mathbf{A}^*$ of left $\mathbf{A}$-modules in $\mathcal{Z}(\mathcal{C})$ such that $\lambda = \lambda_\phi$. Thus the right integral $\Lambda^r := \Lambda_\phi$ constructed in the above satisfies $\lambda \circ \Lambda^r = \mathrm{id}_{\mathbb{1}}$. The same argument shows the claim for left integrals. □

6.3. Maschke's theorem. Maschke's theorem for Hopf algebras is an important application of integrals. By using categorical integrals, we extend this theorem to finite tensor categories:

THEOREM 6.10. *The following assertions are equivalent:*

(1) *The abelian category $\mathcal{C}$ is semisimple.*
(2) *There is a left or right integral Λ of $\mathcal{C}$ such that $\varepsilon_\alpha \circ \Lambda \neq 0$.*
(3) *The unit object of $\mathcal{C}$ is projective.*

If $\mathcal{C}$ is unimodular, then the above assertions are also equivalent to:

(4) *The algebra $\mathrm{End}(\mathrm{id}_{\mathcal{C}})$ is semisimple.*

Although there is a connection between our notion of categorical integrals and integrals of the central Hopf monad (Remark 6.5), this theorem is different from Maschke's theorem for Hopf monads [**2**, Theorem 6.5]. Indeed, if we apply it to the central Hopf monad on $\mathcal{C}$, we will know when $\mathcal{Z}(\mathcal{C})$ is semisimple.

PROOF OF THEOREM 6.10. Unimodular case. We first consider the case where $\mathcal{C}$ is unimodular. It is well-known that (1) and (3) are equivalent (without the unimodularity assumption) [**17**, Corollary 4.2.13]. The implication (1) $\Rightarrow$ (4) is trivial. Thus, for the proof, it suffices to show (4) $\Rightarrow$ (2) and (2) $\Rightarrow$ (3).

To show that (4) implies (2), we use Lemma 5.3. We write $E := \mathrm{End}(\mathrm{id}_{\mathcal{C}})$ and $E' = \mathrm{Hom}_{\mathcal{C}}(\mathbb{1}, A)$ for simplicity. According to that lemma, the binary operation given by (5.7) makes E' an algebra that is isomorphic to E. Now we suppose that E is semisimple. Then E' ($\cong E$) has no non-zero nilpotent elements. If Λ is a non-zero left integral, then we have $0 \neq \Lambda \bullet \Lambda = (\varepsilon_{\mathbb{1}} \circ \Lambda) \otimes \Lambda$. Thus $\varepsilon_{\mathbb{1}} \circ \Lambda \neq 0$, as desired. The same conclusion is obtained by the same argument in the case if Λ is a non-zero right integral.

We show that (2) implies (3). Suppose that there is a left integral Λ such that $\varepsilon_{\mathbb{1}} \circ \Lambda \neq 0$. By normalizing Λ, we may assume $\varepsilon_{\mathbb{1}} \circ \Lambda = \mathrm{id}_{\mathbb{1}}$. Then Λ is an idempotent in the algebra E'. Now let $p : X \to \mathbb{1}$ be an epimorphism in $\mathcal{C}$. Let e be the element of E corresponding to Λ. The object X is decomposed as $X = T \oplus K$, where $T = \mathrm{Im}(e_X)$ and $K = \mathrm{Ker}(e_X)$. By Lemma 6.3 and

$$\begin{aligned} a_T^{\ell} \circ (\mathrm{id}_A \otimes e_X) &= a_T^{\ell} \circ (\mathrm{id}_A \otimes \rho_X) \circ (\mathrm{id}_A \otimes \Lambda \otimes \mathrm{id}_X) \\ &= a_T^{\ell} \circ (m_{\mathbb{1},\mathbb{1}} \otimes \mathrm{id}_X) \circ (\mathrm{id}_A \otimes \Lambda \otimes \mathrm{id}_X) = \varepsilon_{\mathbb{1}} \otimes e_X, \end{aligned}$$

we see that T is a trivial object. Moreover, we have $\mathrm{Hom}_{\mathcal{C}}(K, \mathbb{1}) = 0$. Indeed, if $f : K \to \mathbb{1}$ is a morphism in $\mathcal{C}$, then we have $f = e_{\mathbb{1}} \circ f = f \circ e_K = 0$ by the naturality of e and $e_{\mathbb{1}} = \mathrm{id}_{\mathbb{1}}$. Thus $p = p' \oplus 0 : T \oplus K \to X$ for some $p' : T \to \mathbb{1}$. Since T is trivial, p' splits, and therefore p splits.

If there is a right integral Λ such that $\varepsilon_{\mathbb{1}} \circ \Lambda \neq 0$, then one can prove that $\mathbb{1}$ is projective by the same way as above but by using the canonical right action a^r instead of a^ℓ. The proof of (2) $\Rightarrow$ (3) is done.

Non-unimodular case. To complete the proof of this theorem, we consider the case where $\mathcal{C}$ is *not* unimodular. In this case, by Corollary 5.21, the assertion (1) is false. Hence (3) is also false by [**17**, Corollary 4.2.13]. Now let Λ be a left or right integral. Then $\varepsilon_\alpha \circ \Lambda = 0$ since α and $\mathbb{1}$ are non-isomorphic simple objects. Thus (2) is also false. In conclusion, (1), (2) and (3) are equivalent. The proof is done. □

6.4. Bijectivity of the character map. Throughout this subsection, $\mathcal{C}$ is assumed to be unimodular. By Lemma 6.8, the algebra $\mathbf{A}$ is Frobenius. We recall that a *Frobenius form* of $\mathbf{A}$ is a morphism $\tau : \mathbf{A} \to \mathbb{1}$ in $\mathcal{Z}(\mathcal{C})$ such that there exists a morphism d in $\mathcal{Z}(\mathcal{C})$ such that $(\mathbf{A}, \tau m_{\mathbb{1},\mathbb{1}}, d)$ is a left dual object of $\mathbf{A}$. Since $\mathbf{A}$ is Frobenius, it has a Frobenius form. On the other hand, the space $\mathrm{Hom}_{\mathcal{Z}(\mathcal{C})}(\mathbf{A}, \mathbb{1})$ is, by definition, the space $\mathscr{I}_{\mathcal{C}}^{\mathrm{co}}$ of categorical cointegrals of $\mathcal{C}$. Since $\mathscr{I}_{\mathcal{C}}^{\mathrm{co}}$ is only one-dimensional, we have:

THEOREM 6.11. *Every non-zero cointegral of $\mathcal{C}$ is a Frobenius form of* $\mathbf{A}$.

Let λ be a non-zero categorical cointegral of $\mathcal{C}$. By the above theorem, there is a morphism $d : \mathbb{1} \to \mathbf{A} \otimes \mathbf{A}$ such that $(\mathbf{A}, \lambda m_{\mathbb{1},\mathbb{1}}, d)$ is a left dual object of $\mathbf{A}$. The morphism d is given as follows: By the discussion of Subsection 6.2, there is an isomorphism $\phi : \mathbf{A} \to \mathbf{A}^*$ of right $\mathbf{A}$-modules in $\mathcal{Z}(\mathcal{C})$ such that $\lambda = \lambda_\phi$ with notation in that subsection. We define e_ϕ and d_ϕ by (6.5). Then, as we have seen, the triple $(\mathbf{A}, e_\phi, d_\phi)$ is a left dual object of $\mathbf{A}$. Now we compute:

$$\begin{aligned} e_\phi &= e_\phi \circ (\mathrm{id}_A \otimes m_{\mathbb{1},\mathbb{1}}) \circ (\mathrm{id}_A \otimes \mathrm{id}_A \otimes u) \\ &\overset{(6.6)}{=} e_\phi \circ (m_{\mathbb{1},\mathbb{1}} \otimes \mathrm{id}_A) \circ (\mathrm{id}_A \otimes \mathrm{id}_A \otimes u) \\ &\overset{(6.5)}{=} \mathrm{ev}_A \circ (\phi \otimes u) \circ m_{\mathbb{1},\mathbb{1}} = \lambda \circ m_{\mathbb{1},\mathbb{1}}. \end{aligned}$$

Thus, in conclusion, $(\mathbf{A}, \lambda m_{\mathbb{1},\mathbb{1}}, d_\phi)$ is a left dual object of $\mathbf{A}$.

Cohen and Westreich [**9**] introduced Fourier transforms for Hopf algebras in terms of integrals and cointegrals. As a categorical analogue of such a Fourier transform, we introduce the map

$$\mathfrak{F}'_\lambda : \mathrm{Hom}_{\mathcal{C}}(\mathbb{1}, A) \to \mathrm{CF}(\mathcal{C}), \quad x \mapsto \lambda \leftharpoonup x, \tag{6.9}$$

where the operation $\leftharpoonup$ is given by

$$f \leftharpoonup x = f \circ m_{\mathbb{1},\mathbb{1}} \circ (x \otimes \mathrm{id}_A) \quad (f \in \mathrm{CF}(\mathcal{C}), x \in \mathrm{Hom}_{\mathcal{C}}(\mathbb{1}, A)). \tag{6.10}$$

The map $\mathfrak{F}'_\lambda$ actually corresponds to 'another Fourier transform' mentioned in [**9**, Remark 1.2] and does not specialize to the Fourier transform of [**9**]. To obtain a true Fourier transform, we need to consider the composition of (6.9) and a certain operator which plays a role of the antipode of a Hopf algebra; see [**39**].

LEMMA 6.12. *The map $\mathfrak{F}'_\lambda$ is invertible.*

PROOF. If (Y, e, d) is a left dual object of $X \in \mathcal{C}$, then the map

$$F : \mathrm{Hom}_{\mathcal{C}}(\mathbb{1}, X) \to \mathrm{Hom}_{\mathcal{C}}(Y, \mathbb{1}), \quad F(x) = e \circ (\mathrm{id}_Y \otimes x)$$

is invertible with inverse $F^{-1}(f) = (\mathrm{id}_X \otimes f) \circ d$. The result of this lemma follows from this fact applied to $X = A$ and $(Y, e, d) = (A, \lambda m_{\mathbb{1},\mathbb{1}}, d_\phi)$. □

By the above proof, we see that the inverse of $\mathfrak{F}'_\lambda$ is given by

$$(\mathfrak{F}'_\lambda)^{-1}(f) = (\mathrm{id}_A \otimes f) \circ d_\phi \quad (f \in \mathrm{CF}(\mathcal{C})). \tag{6.11}$$

As an application of the map $\mathfrak{F}'_\lambda$, we prove:

THEOREM 6.13. *For a unimodular pivotal finite tensor category $\mathcal{C}$, the following assertions are equivalent:*

(1) *The abelian category $\mathcal{C}$ is semisimple.*
(2) $\dim_k \mathrm{Gr}_k(\mathcal{C}) = \dim_k \mathrm{CF}(\mathcal{C})$.

Thus the algebra map $\mathrm{ch} : \mathrm{Gr}_k(\mathcal{C}) \to \mathrm{CF}(\mathcal{C})$ *of Theorem* 5.17 *is an isomorphism if and only if $\mathcal{C}$ is semisimple.*

PROOF. The implication (1) $\Rightarrow$ (2) is a part of Theorem 5.17. To prove the converse, we first note that the algebra $E = \mathrm{Hom}_{\mathcal{C}}(\mathbb{1}, A)$ acts on $\mathrm{CF}(\mathcal{C})$ by the action $\leftharpoonup$ defined by (6.10). We also recall from Lemma 5.3 that there is an isomorphism $\Xi_{\mathbb{1}} : E \to \mathrm{End}(\mathrm{id}_{\mathcal{C}})$ of algebras. For a simple object $V \in \mathcal{C}$ and $z \in E$, we define $c_V(z) \in k$ by $c_V(z) \cdot \mathrm{id}_V = \Xi_{\mathbb{1}}(z)_V$. The map $c_V : E \to k$ is an algebra map. Moreover, we have $\mathrm{ch}_V \leftharpoonup z = c_V(z) \cdot \mathrm{ch}_V$ for all $z \in E$. This means that the space spanned by ch_V is an E-submodule of $\mathrm{CF}(\mathcal{C})$.

Now we suppose $\dim_k \mathrm{Gr}_k(\mathcal{C}) = \dim_k \mathrm{CF}(\mathcal{C})$. Then, by Theorem 5.17, $\mathrm{CF}(\mathcal{C})$ is spanned by the internal characters. By the above discussion, $\mathrm{CF}(\mathcal{C})$ is spanned by one-dimensional E-submodules of $\mathrm{CF}(\mathcal{C})$ and, in particular, it is semisimple as an E-module. The map $\mathfrak{F}'_\lambda : E \to \mathrm{CF}(\mathcal{C})$ is in fact an isomorphism of E-modules. Thus E is also semisimple as an E-module. By Theorem 6.10, $\mathcal{C}$ is semisimple. The proof is done. □

6.5. The case of Hopf algebras. Suppose that $\mathcal{C} = H$-mod for some finite-dimensional Hopf algebra H. As before, we identify $\mathcal{Z}(\mathcal{C})$ with ${}^H_H\mathcal{YD}_{fd}$. The object $Z(\alpha)$ appearing in the definition of categorical (co)integrals is the Yetter-Drinfeld module A_α given in Subsection 2.3. Thus, by Theorem 2.3, the space of categorical cointegrals of H-mod is identical to the space of cointegrals on H. By the formula (5.19) of the monoidal structure of the Radford induction, we see that a categorical left/right cointegral of H-mod can be identified with a left/right integral in H in the usual sense.

Now we fix a non-zero left cointegral λ on H and let $\Lambda \in H$ be the left integral in H such that $\lambda(\Lambda) = 1$. We consider the following morphism:

$$\phi := (\lambda m_{\alpha,\mathbb{1}} \otimes \mathrm{id}_{\mathbf{A}^*}) \circ (\mathrm{id}_{R(\alpha)} \otimes \mathrm{coev}_{\mathbf{A}}) : R(\alpha) \to \mathbf{A}^*. \tag{6.12}$$

Let $\{h_i\}$ be a basis of H, and let $\{h^i\}$ be the dual basis of H^*. Then we have

$$\phi(h)(h') = \lambda(hh_i)h^i(h') = \lambda(hh') = (\lambda \leftharpoonup h)(h') \quad (h, h' \in H).$$

Namely, $\phi(h) = \lambda \leftharpoonup h$ ($h \in H$). By the basic results on integrals [**35**], the inverse of ϕ is given by $\phi^{-1}(f) = f(\Lambda_{(2)})S^{-1}(\Lambda_{(1)})$. The point is that $\phi : R(\alpha) \to \mathbf{A}^*$ is a morphism of Yetter-Drinfeld H-modules by the construction. Thus the map

$$\phi^{-1} : \mathbf{A}^* \to R(\alpha), \quad f \mapsto f(\Lambda_{(2)})S^{-1}(\Lambda_{(1)})$$

is a morphism of Yetter-Drinfeld H-modules. With notation in Subsection 6.2, we have $\lambda = \lambda_\phi$. We define e_ϕ and d_ϕ as in that subsection. Then we have

$$e_\phi(a \otimes b) = \lambda(ab) \quad (a \in R(\alpha), b \in \mathbf{A})$$

by the proof of Lemma 6.12. By the above expression of ϕ^{-1}, we have

$$d_\lambda(1) = h_i \otimes \phi^{-1}(h^i) = h_i \otimes h^i(\Lambda_{(2)})S^{-1}(\Lambda_{(1)}) = \Lambda_{(2)} \otimes S^{-1}(\Lambda_{(1)}). \tag{6.13}$$

Our result implies that, if H is unimodular, then $\mathbf{A}$ is a Frobenius algebra in ${}^H_H\mathcal{YD}$ and the triple $(\mathbf{A}, e_\phi, d_\phi)$ is a left dual object of $\mathbf{A}$ [**24**, Proposition 6.4].

7. Applications to fusion categories

7.1. Integrals and cointegrals of fusion categories. A semisimple finite tensor category is called a *fusion category*. We give some applications of the character theory and the integral theory to pivotal fusion categories.

Let $\mathcal{C}$ be a pivotal fusion category, and let $V_0, \dots, V_n$ be the representatives of the isomorphism classes of simple objects of $\mathcal{C}$ such that $V_0 = \mathbb{1}$. For $X \in \mathcal{C}$, the *pivotal dimension* $\dim(X) \in k$ is defined by

$$\dim(X)\,\mathrm{id}_{\mathbb{1}} = \mathrm{tr}_{\mathbb{1},\mathbb{1}|X}(\mathrm{id}_X : \mathbb{1} \otimes X \to \mathbb{1} \otimes X) \in \mathrm{End}_{\mathcal{C}}(\mathbb{1}),$$

where $\mathrm{tr}_{\mathbb{1},\mathbb{1}|X}$ is the partial pivotal trace (see Subsection 5.6). By Corollary 5.21, an integral and a cointegral of $\mathcal{C}$ belong to $E := \mathrm{Hom}_{\mathcal{C}}(\mathbb{1}, A)$ and $\mathrm{CF}(\mathcal{C})$, respectively. For each $i = 0, \dots, n$, there is an element $e_i \in E$ such that the equation $\Xi_{\mathbb{1}}(e_i)_{V_j} = \delta_{ij}\mathrm{id}_{V_j}$ $(j = 0, \dots, n)$ holds, where $\Xi_{\mathbb{1}} : E \to \mathrm{End}(\mathrm{id}_{\mathcal{C}})$ is the isomorphism (5.6). For $f \in \mathrm{CF}(\mathcal{C})$ and $a \in E$, we define $f(a) \in k$ by $f \circ a = f(a)\,\mathrm{id}_{\mathbb{1}}$. For simplicity, we write ch_{V_i} as χ_i. Then we have

$$e_i \bullet e_j = \delta_{ij} e_i, \quad \chi_i(e_j) = \delta_{ij}\dim(V_i) \quad (i, j = 0, \dots, n). \tag{7.1}$$

LEMMA 7.1. *e_0 is a left and a right integral of $\mathcal{C}$.*

PROOF. We consider the following two subspaces:

$$I^\ell = \{\Lambda \in E \mid a \bullet \Lambda = \varepsilon_{\mathbb{1}}(a)\Lambda \text{ for all } a \in E\},$$
$$I^r = \{\Lambda \in E \mid \Lambda \bullet a = \varepsilon_{\mathbb{1}}(a)\Lambda \text{ for all } a \in E\}.$$

It is easy to verify $\mathscr{I}^\ell_{\mathcal{C}} \subset I^\ell$ and $\mathscr{I}^r_{\mathcal{C}} \subset I^r$. Since $\varepsilon_{\mathbb{1}} = \chi_0$, we have $\varepsilon_{\mathbb{1}}(e_i) = \delta_{i0}$ for $i = 0, \dots, n$ by (7.1). Since $\{e_0, \dots, e_n\}$ is a basis of E, we see that both I^ℓ and I^r are spanned by e_0. In particular, I^ℓ and I^r are one-dimensional. On the other hand, $\mathscr{I}^\ell_{\mathcal{C}}$ and $\mathscr{I}^r_{\mathcal{C}}$ are also one-dimensional. Thus we have

$$\mathscr{I}^\ell_{\mathcal{C}} = I^\ell = \mathscr{I}^r_{\mathcal{C}} = I^r = \mathrm{span}_k\{e_0\}$$

by summarizing the above argument. The proof is done. □

LEMMA 7.2. *The morphism*

$$\lambda := \sum_{i=1}^{n} \dim(V_i^*)\chi_i \in \mathrm{CF}(\mathcal{C}), \tag{7.2}$$

is a cointegral of $\mathcal{C}$. We also have

$$f \star \lambda = f(u)\lambda = \lambda \star f \quad (f \in \mathrm{CF}(\mathcal{C})). \tag{7.3}$$

PROOF. We consider the vector space

$$I = \{\mu \in \mathrm{CF}(\mathcal{C}) \mid f \star \mu = f(u)\mu \text{ for all } f \in \mathrm{CF}(\mathcal{C})\}.$$

Suppose that μ is a cointegral of $\mathcal{C}$. By definition, $\mu : R(\mathbb{1}) \to \mathbb{1}$ is a morphism in $\mathcal{Z}(\mathcal{C})$. Thus we may write $\mu = U(\mu)$. Now we compute:

$$\begin{aligned} f \star \mu &= f \circ UR(\mu) \circ U(\eta_{R(\mathbb{1})}) = f \circ U(RU(\mu) \circ \eta_{R(\mathbb{1})}) \\ &= f \circ U(\eta_{\mathbb{1}} \circ \mu) = f \circ u \circ \mu = f(u)\mu, \end{aligned}$$

where the third equality follows from the naturality of η and the fourth from the definition of u. Thus $\mu \in I$. We have proved $\mathscr{I}_{\mathcal{C}}^{\mathrm{co}} \subset I$.

To complete the proof, we use the following general result on Frobenius algebras: Let F be a Frobenius algebra with symmetric Frobenius form $\tau : F \to k$, let $\{f_i\}_{i=1}^n$ be a basis of F, and let $\{f^i\}$ be the dual basis of F with respect to the pairing $\langle a, b\rangle = \tau(ab)$. Given an algebra map $\beta : F \to k$, we define $t_\beta := \sum_{i=1}^m \beta(f^i) f_i$. Then the space $I_\beta = \{t \in F \mid at = \beta(a)t \text{ for all } a \in F\}$ is spanned by t_β. Furthermore, t_β belongs to the center of F.

We now consider the case where $F = \mathrm{CF}(\mathcal{C})$. Then the set $\{\chi_0, \dots, \chi_n\}$ is a basis of F. The linear map $\tau : F \to k$ given by $\tau(\chi_i) = \delta_{i0}$ is a symmetric Frobenius form on F. Indeed, we have $\tau(\chi_i \star \chi_j) = \delta_{i^*,j} = \tau(\chi_j \star \chi_i)$ for all $i, j = 0, \dots, n$, where $i^* \in \{0, \dots, n\}$ is the element such that $V_{i^*} \cong V_i^*$.

There is an algebra map $\beta : F \to k$ given by $\beta(f) = f(u)$. The vector space I_β is precisely the space I mentioned at the beginning of the proof. Thus I is spanned by $t_\beta = \lambda$. In particular, I is one-dimensional. Since $\mathscr{I}_{\mathcal{C}}^{\mathrm{co}}$ is also one-dimensional, we have $\mathscr{I}_{\mathcal{C}}^{\mathrm{co}} = I = \mathrm{span}_k\{\lambda\}$. Thus λ is a cointegral of $\mathcal{C}$. Since $\lambda \in I$, the first equality in (7.3) holds. The second one follows from that $t_\beta = \lambda$ belongs to the center of $\mathrm{CF}(\mathcal{C})$. The proof is done. □

7.2. Semisimplicity of the Drinfeld center. We keep the notation of the previous subsection. As an application of the character theory and the integral theory, we prove:

THEOREM 7.3. *The following assertions are equivalent:*

1. $\dim(\mathcal{C}) \neq 0$, *where* $\dim(\mathcal{C}) := \sum_{i=0}^n \dim(V_i^*)\dim(V_i)$.
2. *The abelian category* $\mathcal{Z}(\mathcal{C})$ *is semisimple.*
3. *The object* $\mathbf{A} := R(\mathbb{1}) \in \mathcal{Z}(\mathcal{C})$ *is semisimple.*
4. *The algebra* $\mathrm{Gr}_k(\mathcal{C})$ *is semisimple.*

Müger [**32**] proved (1) $\Rightarrow$ (2). Bruguières and Virelizier [**4**] proved (1) $\Leftrightarrow$ (2) in a more general setting than ours by using Maschke's theorem for Hopf monads. It is obvious that (2) implies (3). Ostrik [**34**] proved (4) $\Rightarrow$ (1) and conjectured that the converse holds. The proof of this theorem is completed in [**39**] by showing that (3) implies (4) by the character theory. Below we give a full proof with emphasis on the use of integrals and characters.

PROOF. (1) $\Rightarrow$ (2). We suppose $\dim(\mathcal{C}) \neq 0$. Since $\mathcal{Z}(\mathcal{C})$ is a finite tensor category, an object of $\mathcal{Z}(\mathcal{C})$ is projective if and only if it is injective. Thus, to show that $\mathcal{Z}(\mathcal{C})$ is semisimple, it is enough to show that $\mathbb{1} \in \mathcal{Z}(\mathcal{C})$ is injective. Now let $f : \mathbb{1} \to \mathbf{X}$ be a monomorphism in $\mathcal{Z}(\mathcal{C})$. We decompose $X := U(\mathbf{X})$ as

$$X = X_0 \oplus \cdots \oplus X_n, \quad X_i \cong V_i^{\oplus m_i} \quad (m_i \in \mathbb{Z}_{\geq 0}).$$

We consider the morphism $f' := \eta_{\mathbf{X}} \circ f \in \mathrm{Hom}_{\mathcal{Z}(\mathcal{C})}(\mathbb{1}, R(X))$. Since U is faithful, $\eta_{\mathbf{X}}$ is monic (as a component of the unit), and hence so is f'. We remark:

$$\text{(7.4)} \qquad \mathrm{Hom}_{\mathcal{Z}(\mathcal{C})}(\mathbb{1}, R(V_i)) = 0 \quad (i \neq 0) \quad \text{and} \quad \mathrm{Hom}_{\mathcal{Z}(\mathcal{C})}(\mathbb{1}, \mathbf{A}) = \mathrm{span}_k\{u\}.$$

By (7.4), the image of f' is contained in $R(X_0)$. Since $X_0 \cong \mathbb{1}^{\oplus m_0}$, there is a morphism $p : X_0 \to \mathbb{1}$ in $\mathcal{C}$ such that $f'' := R(p) \circ f'$ is a non-zero morphism in $\mathcal{Z}(\mathcal{C})$. Again by (7.4), we have $f'' = au$ for some $a \in k^{\times}$. Now we define

$$q := \dim(\mathcal{C})^{-1} a^{-1} \cdot \lambda \circ R(p) \circ \eta_{\mathbf{X}} \in \mathrm{Hom}_{\mathcal{Z}(\mathcal{C})}(\mathbf{X}, \mathbb{1}),$$

where λ is the cointegral of $\mathcal{C}$ given by Lemma 7.2. By the above discussion and $\lambda(u) = \dim(\mathcal{C})$, we have $q \circ f = \mathrm{id}_{\mathbb{1}}$. Namely, the given monomorphism $f : \mathbb{1} \to \mathbf{X}$ in $\mathcal{Z}(\mathcal{C})$ splits. Hence $\mathbb{1} \in \mathcal{Z}(\mathcal{C})$ is injective.

(2) $\Rightarrow$ (3). As we have remarked, this part is clear.

(3) $\Rightarrow$ (4). There are isomorphisms

$$\text{(7.5)} \qquad \mathrm{Gr}_k(\mathcal{C}) \xrightarrow[\cong]{\text{Theorem 5.17}} \mathrm{CF}(\mathcal{C}) \xrightarrow[\cong]{\text{Subsection 5.3}} \mathrm{End}_{\mathcal{Z}(\mathcal{C})}(\mathbf{A})$$

of algebras. If $\mathbf{A} \in \mathcal{Z}(\mathcal{C})$ is semisimple, then $\mathrm{End}_{\mathcal{Z}(\mathcal{C})}(R(\mathbb{1}))$ is semisimple as the endomorphism algebra of a semisimple object, and hence so is $\mathrm{Gr}_k(\mathcal{C})$.

(4) $\Rightarrow$ (1). Let λ be the cointegral of $\mathcal{C}$ given by Lemma 7.2. If $\mathrm{Gr}_k(\mathcal{C})$ is semisimple, then $\mathrm{CF}(\mathcal{C})$ is also semisimple, and hence its center has no non-zero nilpotent elements. Since λ belongs to the center of $\mathrm{CF}(\mathcal{C})$, we have $\lambda \star \lambda \neq 0$. On the other hand, we have $\lambda \star \lambda = \dim(\mathcal{C}) \cdot \lambda$ by (7.3). Thus $\dim(\mathcal{C}) \neq 0$. The proof is done. □

THEOREM 7.4 ([**39**]). *Suppose that $\mathcal{C}$ satisfies the equivalent conditions of the above theorem. Then the following assertions are equivalent:*

(1) *The algebra $\mathrm{Gr}_k(\mathcal{C})$ is commutative.*
(2) *The object $\mathbf{A}$ is multiplicity-free.*

Ostrik [**33**] proved this theorem in the case where $\mathrm{char}(k) = 0$.

PROOF. Regardless of the characteristic of k, the claim of this theorem can be easily verified by the isomorphism (7.5). □

7.3. Conjugacy classes and the character table. We suppose that $\mathcal{C}$ is a pivotal fusion category such that $\dim(\mathcal{C}) \neq 0$ and $\mathrm{Gr}_k(\mathcal{C})$ is commutative (although some of below definitions make sense without such assumptions). In view of the observation of Zhu [**43**], we introduce:

DEFINITION 7.5. A *conjugacy class* of $\mathcal{C}$ is a simple subobject of $\mathbf{A}$.

Since $\mathbf{A}$ is multiplicity free under our assumption (by Theorem 7.4), there is a one-to-one correspondence between the set of conjugacy classes of $\mathcal{C}$ and the set of primitive idempotents of $\mathrm{End}_{\mathcal{Z}(\mathcal{C})}(\mathbf{A})$. We denote by $\mathsf{E}_{\mathfrak{C}}$ the idempotent on $\mathbf{A}$ corresponding to a conjugacy class $\mathfrak{C}$. Since $\dim_k \mathrm{End}_{\mathcal{C}}(R(\mathbb{1})) = \dim_k \mathrm{Gr}_k(\mathcal{C})$, the conjugacy classes of $\mathcal{C}$ can be numbered as $\mathfrak{C}_0, \dots, \mathfrak{C}_n$. Now let λ be a cointegral of $\mathcal{C}$ such that $\lambda(u) = 1$. By Lemma 7.2, it is given explicitly by

$$\lambda = \frac{1}{\dim(\mathcal{C})} \sum_{i=0}^{n} \dim(V_i^*) \chi_i.$$

We define the Fourier transform $\mathfrak{F}$ by

$$\mathfrak{F} : \mathrm{Hom}_{\mathcal{C}}(\mathbb{1}, A) \to \mathrm{CF}(\mathcal{C}), \quad x \mapsto \lambda \leftharpoonup \mathfrak{S}(x) \quad (x \in \mathrm{Hom}_{\mathcal{C}}(\mathbb{1}, A)),$$

where $\leftharpoonup$ is the operation defined by (6.10) and $\mathfrak{S} : \mathrm{Hom}_{\mathcal{C}}(\mathbb{1}, A) \to \mathrm{Hom}_{\mathcal{C}}(\mathbb{1}, A)$ is the linear map such that $\mathfrak{S}(e_i) = e_{i^*}$ for $i = 1, \dots, n$. We note that $\mathfrak{F}$ is invertible as the composition of $\mathfrak{S}$ and the map $\mathfrak{F}'_\lambda$ considered in Subsection 6.4.

DEFINITION 7.6. For $i = 0, \dots, n$, we call

$$n_i := \dim(\mathfrak{C}_i), \quad f_i := \varepsilon_{\mathbb{1}} \circ \mathsf{E}_{\mathfrak{C}_i} \quad \text{and} \quad \overline{g}_i := \mathfrak{F}_\lambda^{-1}(f_i)$$

the *size* of the conjugacy class $\mathfrak{C}_i$, the *characteristic function* on $\mathfrak{C}_i$ and the *class sum* of $\mathfrak{C}_i$, respectively. We note that the size of a conjugacy class is non-zero as the pivotal dimension of a simple object of a fusion category. The *character table* of $\mathcal{C}$ is the matrix (c_{ij}), where

$$c_{ij} := n_j^{-1} \chi_i(\overline{g}_j) \quad (i, j = 0, \dots, n).$$

The set $\{f_0, \dots, f_n\}$ is a basis of $\mathrm{CF}(\mathcal{C})$. The character table of $\mathcal{C}$ is in fact the base change matrix from $\{f_0, \dots, f_n\}$ to $\{\chi_0, \dots, \chi_n\}$. Namely, we have

$$\chi_i = \sum_{j=0}^{n} c_{ij} f_j \quad (i = 0, \dots, n).$$

We also have the first orthogonality relation

$$\sum_{r=0}^{n} \frac{n_r}{\dim(\mathcal{C})} c_{i,r} c_{j^*,r} = \delta_{i,j} \quad (i, j = 0, \dots, n)$$

and the second orthogonality relation

$$\sum_{i=0}^{n} \frac{n_r}{\dim(\mathcal{C})} c_{i,r} c_{i^*,s} = \delta_{r,s} \quad (r, s = 0, \dots, n).$$

For the proof, see [**39**].

7.4. Remarks. As has been observed in [**39**], conjugacy classes, characteristic functions on them, class sums of them and the character table introduced in the above properly generalize the same-name notions in the representation theory of finite groups and agree with definitions for semisimple Hopf algebras given by Cohen and Westreich in [**10**].

It is interesting to investigate how can we generalize the results of [**10**–**12**] on the character table of a semisimple Hopf algebra to fusion categories. It was shown in [**39**] that the character table of a modular tensor category coincides with its S-matrix. This results generalizes [**10**, Theorem 3.2].

Further applications are recently given by Burciu [**5**–**7**]. In [**5**], for a semisimple quasitriangular Hopf algebra H, a formula for the Müger centralizer of a tensor full subcategory of H-mod is given in terms of conjugacy classes of H. In [**6**], in a modular tensor category, the product of two class sums is shown to be a linear combination of conjugacy class sums with rational coefficients (*cf.* [**10**, Theorem 2.6]). In [**7**], the product of class sums are further studied and, in particular, a Burnside type formula was established.

Acknowledgments

The author is grateful to the referee for careful reading of the manuscript. The author would like to thank the organizers of the conference "International Workshop on Hopf algebras and Tensor Categories" held at Nanjing.

References

[1] Johannes Berger, Azat M. Gainutdinov, and Ingo Runkel, *Modified traces for quasi-Hopf algebras*, J. Algebra **548** (2020), 96–119, DOI 10.1016/j.jalgebra.2019.12.006. MR4046154

[2] Alain Bruguières and Alexis Virelizier, *Hopf monads*, Adv. Math. **215** (2007), no. 2, 679–733, DOI 10.1016/j.aim.2007.04.011. MR2355605

[3] Alain Bruguières and Alexis Virelizier, *Quantum double of Hopf monads and categorical centers*, Trans. Amer. Math. Soc. **364** (2012), no. 3, 1225–1279, DOI 10.1090/S0002-9947-2011-05342-0. MR2869176

[4] Alain Bruguières and Alexis Virelizier, *On the center of fusion categories*, Pacific J. Math. **264** (2013), no. 1, 1–30, DOI 10.2140/pjm.2013.264.1. MR3079759

[5] Sebastian Burciu, *Representations and conjugacy classes of semisimple quasitriangular Hopf algebras*, SIGMA Symmetry Integrability Geom. Methods Appl. **16** (2020), Paper No. 039, 20, DOI 10.3842/SIGMA.2020.039. MR4094485

[6] ———, *Conjugacy classes and centralizers for pivotal fusion categories*, `arXiv:1904.02226`.

[7] ———, *Structure constants for pre-modular categories*, `arXiv:2002.05483`.

[8] Stefaan Caenepeel, Gigel Militaru, and Shenglin Zhu, *Frobenius and separable functors for generalized module categories and nonlinear equations*, Lecture Notes in Mathematics, vol. 1787, Springer-Verlag, Berlin, 2002, DOI 10.1007/b83849. MR1926102

[9] Miriam Cohen and Sara Westreich, *Fourier transforms for Hopf algebras*, Quantum groups, Contemp. Math., vol. 433, Amer. Math. Soc., Providence, RI, 2007, pp. 115–133, DOI 10.1090/conm/433/08324. MR2349620

[10] Miriam Cohen and Sara Westreich, *Conjugacy classes, class sums and character tables for Hopf algebras*, Comm. Algebra **39** (2011), no. 12, 4618–4633, DOI 10.1080/00927872.2011.617265. MR2863455

[11] Miriam Cohen and Sara Westreich, *Recovering information from character tables of Hopf algebras: normality, dimensions and quotients*, Hopf algebras and tensor categories, Contemp. Math., vol. 585, Amer. Math. Soc., Providence, RI, 2013, pp. 213–226, DOI 10.1090/conm/585/11620. MR3077239

[12] Miriam Cohen and Sara Westreich, *Character tables and normal left coideal subalgebras*, J. Pure Appl. Algebra **218** (2014), no. 10, 1845–1866, DOI 10.1016/j.jpaa.2014.02.010. MR3195413

[13] Brian Day and Ross Street, *Centres of monoidal categories of functors*, Categories in algebra, geometry and mathematical physics, Contemp. Math., vol. 431, Amer. Math. Soc., Providence, RI, 2007, pp. 187–202, DOI 10.1090/conm/431/08273. MR2342829

[14] P. Deligne, *Catégories tannakiennes* (French), The Grothendieck Festschrift, Vol. II, Progr. Math., vol. 87, Birkhäuser Boston, Boston, MA, 1990, pp. 111–195. MR1106898

[15] Christopher L. Douglas, Christopher Schommer-Pries, and Noah Snyder, *The balanced tensor product of module categories*, Kyoto J. Math. **59** (2019), no. 1, 167–179, DOI 10.1215/21562261-2018-0006. MR3934626

[16] Samuel Eilenberg and John C. Moore, *Adjoint functors and triples*, Illinois J. Math. **9** (1965), 381–398. MR184984

[17] Pavel Etingof, Shlomo Gelaki, Dmitri Nikshych, and Victor Ostrik, *Tensor categories*, Mathematical Surveys and Monographs, vol. 205, American Mathematical Society, Providence, RI, 2015, DOI 10.1090/surv/205. MR3242743

[18] Pavel Etingof, Dmitri Nikshych, and Viktor Ostrik, *An analogue of Radford's S^4 formula for finite tensor categories*, Int. Math. Res. Not. **54** (2004), 2915–2933, DOI 10.1155/S1073792804141445. MR2097289

[19] Pavel Etingof and Viktor Ostrik, *Finite tensor categories* (English, with English and Russian summaries), Mosc. Math. J. **4** (2004), no. 3, 627–654, 782–783, DOI 10.17323/1609-4514-2004-4-3-627-654. MR2119143

[20] Jürgen Fuchs, Gregor Schaumann, and Christoph Schweigert, *Eilenberg-Watts calculus for finite categories and a bimodule Radford S^4 theorem*, Trans. Amer. Math. Soc. **373** (2020), no. 1, 1–40, DOI 10.1090/tran/7838. MR4042867

[21] Jürgen Fuchs, Christoph Schweigert, and Carl Stigner, *The Cardy-Cartan modular invariant*, Strings, gauge fields, and the geometry behind, World Sci. Publ., Hackensack, NJ, 2013, pp. 289–304. MR3289342

[22] ———, *From non-semisimple Hopf algebras to correlation functions for logarithmic CFT*, J. Phys. A **46** (2013), no. 49, 494008, 40. MR3146014

[23] F. Hausser and F. Nill, *Integral Theory for Quasi-Hopf Algebras*, (1999).

[24] Atsushi Ishii and Akira Masuoka, *Handlebody-knot invariants derived from unimodular Hopf algebras*, J. Knot Theory Ramifications **23** (2014), no. 7, 1460001, 24, DOI 10.1142/S0218216514600013. MR3265394

[25] Thomas Kerler and Volodymyr V. Lyubashenko, *Non-semisimple topological quantum field theories for 3-manifolds with corners*, Lecture Notes in Mathematics, vol. 1765, Springer-Verlag, Berlin, 2001. MR1862634

[26] Richard G. Larson and David E. Radford, *Semisimple cosemisimple Hopf algebras*, Amer. J. Math. **110** (1988), no. 1, 187–195, DOI 10.2307/2374545. MR926744

[27] V. Lyubashenko, *Modular transformations for tensor categories*, J. Pure Appl. Algebra **98** (1995), no. 3, 279–327, DOI 10.1016/0022-4049(94)00045-K. MR1324034

[28] Volodymyr V. Lyubashenko, *Invariants of 3-manifolds and projective representations of mapping class groups via quantum groups at roots of unity*, Comm. Math. Phys. **172** (1995), no. 3, 467–516. MR1354257

[29] ———, *Squared Hopf algebras*, Mem. Amer. Math. Soc. **142** (1999), no. 677, x+180.

[30] Saunders Mac Lane, *Categories for the working mathematician*, 2nd ed., Graduate Texts in Mathematics, vol. 5, Springer-Verlag, New York, 1998. MR1712872

[31] Susan Montgomery, *Hopf algebras and their actions on rings*, CBMS Regional Conference Series in Mathematics, vol. 82, Published for the Conference Board of the Mathematical Sciences, Washington, DC; by the American Mathematical Society, Providence, RI, 1993, DOI 10.1090/cbms/082. MR1243637

[32] Michael Müger, *From subfactors to categories and topology. II. The quantum double of tensor categories and subfactors*, J. Pure Appl. Algebra **180** (2003), no. 1-2, 159–219, DOI 10.1016/S0022-4049(02)00248-7. MR1966525

[33] Victor Ostrik, *Pivotal fusion categories of rank 3* (English, with English and Russian summaries), Mosc. Math. J. **15** (2015), no. 2, 373–396, 405, DOI 10.17323/1609-4514-2015-15-2-373-396. MR3427429

[34] ———, *On symmetric fusion categories in positive characteristic*, Sel. Math. New Ser. **26**, 36 (2020).

[35] David E. Radford, *The trace function and Hopf algebras*, J. Algebra **163** (1994), no. 3, 583–622, DOI 10.1006/jabr.1994.1033. MR1265853

[36] David E. Radford, *On oriented quantum algebras derived from representations of the quantum double of a finite-dimensional Hopf algebra*, J. Algebra **270** (2003), no. 2, 670–695, DOI 10.1016/j.jalgebra.2003.07.006. MR2019635

[37] Alexander L. Rosenberg, *Noncommutative algebraic geometry and representations of quantized algebras*, Mathematics and its Applications, vol. 330, Kluwer Academic Publishers Group, Dordrecht, 1995, DOI 10.1007/978-94-015-8430-2. MR1347919

[38] Kenichi Shimizu, *Integrals for finite tensor categories*, Algebr. Represent. Theory **22** (2019), no. 2, 459–493, DOI 10.1007/s10468-018-9777-5. MR3921367

[39] Kenichi Shimizu, *The monoidal center and the character algebra*, J. Pure Appl. Algebra **221** (2017), no. 9, 2338–2371, DOI 10.1016/j.jpaa.2016.12.037. MR3631720

[40] ———, *On unimodular finite tensor categories*, Int. Math. Res. Not. IMRN (2017), no. 1, 277–322.

[41] ———, *Further results on the structure of (co)ends in finite tensor categories*, Applied Categorical Structures (2019).

[42] Moss Eisenberg Sweedler, *Integrals for Hopf algebras*, Ann. of Math. (2) **89** (1969), 323–335, DOI 10.2307/1970672. MR242840

[43] Yongchang Zhu, *A commuting pair in Hopf algebras*, Proc. Amer. Math. Soc. **125** (1997), no. 10, 2847–2851, DOI 10.1090/S0002-9939-97-03988-9. MR1402892

DEPARTMENT OF MATHEMATICAL SCIENCES, SHIBAURA INSTITUTE OF TECHNOLOGY, 307 FUKASAKU, MINUMA-KU, SAITAMA-SHI, SAITAMA 337-8570, JAPAN

Email address: kshimizu@shibaura-it.ac.jp

Contemporary Mathematics
Volume **771**, 2021
https://doi.org/10.1090/conm/771/15517

Indecomposability of weak Hopf algebras

Michihisa Wakui

This paper is dedicated to Professor Toshitake Kohno on his 65th birthday.

ABSTRACT. In this paper we discuss on the direct sum construction of weak bialgebras and indecomposability of them as weak bialgebras. Many fundamental properties and examples of the direct sum construction of weak bialgebras and indecomposable weak bialgebras are investigated. For example, any finite-dimensional weak bialgebra can be uniquely decomposed into finitely many indecomposable weak bialgebras up to isomorphism. A finite-dimensional bialgebra is always indecomposable as a weak bialgebra. The direct sum construction and indecomposability can be characterized in language of category theory, and from this point of view a generalization of the direct sum construction is introduced.

1. Introduction

Weak Hopf algebras are generalizations of Hopf algebras, and the general theory of them is established by Böhm, Nill and Szlachányi [**5**] in 1999. They are also called quantum groupoids in some literatures, and it is known by Schauenburg [**22**] that they can be regarded as certain $\times_R$-bialgebras introduced by Takeuchi [**24**] in 1977. Face algebras which are introduced by Hayashi [**8**] give a wide class in weak Hopf algebras and are important from a historical point of view.

In the present paper we focus on the direct sum construction of weak bialgebras and indecomposability of them as weak bialgebras. In general, for weak bialgebras A and B, the direct sum $A \oplus B$ is also a weak bialgebra with respect to the direct product as an algebra and the direct sum as a coalgebra. This fact is remarkable since the same result does not hold for ordinary bialgebras. The direct sum construction for weak bialgebras leads to decomposing weak bialgebras into indecomposable ones. In fact, any finite-dimensional weak bialgebra can be uniquely decomposed into finitely many indecomposable ones up to isomorphism. This shows that indecomposable weak bialgebras are fundamental and important in study of weak bialgebras.

We investigate many fundamental properties and examples of the direct sum construction of weak bialgebras and indecomposable ones. For example, a finite-dimensional bialgebra is always indecomposable as a weak bialgebra. The Kaplansky type construction studied by Chebel and Makhlouf [**7**] can be regarded as a

---

2020 *Mathematics Subject Classification.* Primary 16T05; 18D10, 18D25.

Key words and phrases. Weak bialgebra, weak Hopf algebra, indecomposable, direct sum, quantum double.

special direct sum construction. If a weak Hopf algebra is the direct sum of two weak Hopf algebras A and B, then the quantum double of $A \oplus B$ is naturally isomorphic to the direct sum of the quantum doubles of A and B.

Indecomposability for weak bialgebras of finite dimension can be completely interpreted in language of category theory. More precisely, a finite-dimensional weak bialgebra H is indecomposable if and only if the module category ${}_H\mathbb{M}$ whose objects are finite-dimensional left H-modules is indecomposable. Along this principle we introduce a generalization of the direct sum construction involving weak bicomodule algebras, and give a categorical characterization of it.

This paper is organized as follows. In Section 2 we briefly describe definitions and basic properties of weak bialgebras which are used in the present paper. In Section 3 we introduce the direct sum construction of weak bialgebras and Hopf algebras. It is shown that any finite-dimensional weak bialgebra can be uniquely decomposed into finitely many indecomposable ones up to isomorphism, and that a finite-dimensional bialgebra is indecomposable as a weak bialgebra. Several examples including the Kaplansky type construction [**7**] and groupoid algebras are given. In Section 4 it is shown that any quasitriangular structure of the direct sum of two weak Hopf algebras A and B is a sum of quasitriangular structures of A and B. It is also shown that the direct sum construction is preserved under the quantum double construction. In the final section we interpret the direct sum construction and indecomposability in language of category theory, and introduce a generalization of the direct sum construction. We demonstrate a few examples, and examine those indecomposability.

Throughout of the paper $\boldsymbol{k}$ denotes a field. All tensor products $\otimes$ are taken over $\boldsymbol{k}$. For a weak bialgebra H over $\boldsymbol{k}$, we denote by ${}_H\mathbf{M}$ the $\boldsymbol{k}$-linear category whose objects are left H-modules and morphisms are H-module maps between them. The notation $X \in {}_H\mathbf{M}$ means that X is an object in ${}_H\mathbf{M}$, that is a left H-module. Denoted by ${}_H\mathbb{M}$ is the full subcategory of ${}_H\mathbf{M}$ whose objects consist of finite dimension. The $\boldsymbol{k}$-linear categories of right H-comodules and finite-dimensional ones are denoted by $\mathbf{M}^H$ and $\mathbb{M}^H$, respectively.

We use Sweedler's notation such as $\Delta(x) = x_{(1)} \otimes x_{(2)}$ for an element $x \in H$ in a weak bialgebra $H = (H, \Delta, \varepsilon)$. For a left H-comodule (M, λ) we use the notation such as $\lambda(m) = m_{(-1)} \otimes m_{(0)} = m_{[-1]} \otimes m_{[0]}$ for an element $m \in M$. For general facts on Hopf algebras and monoidal categories, we refer the reader to Abe's book [**1**], Kassel's book [**11**], MacLane's book [**12**] and Montgomery's book [**14**].

2. Definitions and basic properties related with weak bialgebras

Let H be a vector space over $\boldsymbol{k}$, and (H, μ, η) and (H, Δ, ε) are an algebra and a coalgebra over $\boldsymbol{k}$, respectively. The 5-tuple $(H, \mu, \eta, \Delta, \varepsilon)$ is said to be a *weak bialgebra* over $\boldsymbol{k}$ if the following three conditions are satisfied.

(WH1) $\Delta(xy) = \Delta(x)\Delta(y)$ for all $x, y \in H$.

(WH2) $\Delta^{(2)}(1) = (\Delta(1) \otimes 1)(1 \otimes \Delta(1)) = (1 \otimes \Delta(1))(\Delta(1) \otimes 1)$, where $\Delta^{(2)} = (\Delta \otimes \mathrm{id}) \circ \Delta = (\mathrm{id} \otimes \Delta) \circ \Delta$ and $1 = \eta(1)$ is the identity element of the algebra (H, μ, η).

(WH3) $\varepsilon(xy_{(1)})\varepsilon(y_{(2)}z) = \varepsilon(xyz) = \varepsilon(xy_{(2)})\varepsilon(y_{(1)}z)$ for all $x, y, z \in H$.

Let $S : H \longrightarrow H$ be a $\boldsymbol{k}$-linear map. The 6-tuple $(H, \mu, \eta, \Delta, \varepsilon, S)$ is said to be a *weak Hopf algebra* over $\boldsymbol{k}$ if the above three conditions and the following additional condition are satisfied.

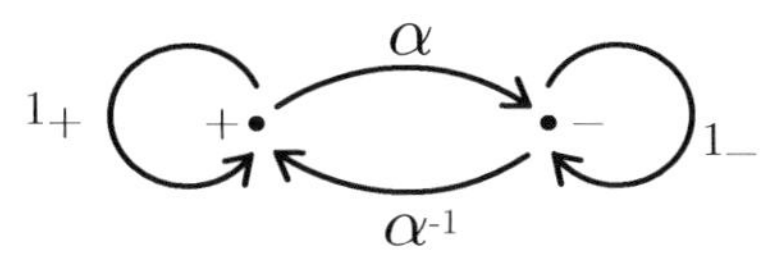

FIGURE 1. The quiver corresponding to the qubit groupoid.

(WH4) For all $x \in H$
(i) $x_{(1)}S(x_{(2)}) = \varepsilon(1_{(1)}x)1_{(2)}$.
(ii) $S(x_{(1)})x_{(2)} = 1_{(1)}\varepsilon(x1_{(2)})$.
(iii) $S(x_{(1)})x_{(2)}S(x_{(3)}) = S(x)$.

The above S is called the antipode of $(H, \mu, \eta, \Delta, \varepsilon)$ and $(H, \mu, \eta, \Delta, \varepsilon, S)$. We note that it is unique if it exists. We also note that a weak bialgebra is a bialgebra if and only if $\Delta(1) = 1 \otimes 1$ if and only if ε is an algebra map.

For a weak bialgebra $H = (H, \mu, \eta, \Delta, \varepsilon)$, by the right-hand sides of the conditions (WH4)(i),(ii), the following two $\boldsymbol{k}$-linear maps $\varepsilon_t, \varepsilon_s : H \longrightarrow H$ are defined:

$$\varepsilon_t(x) = \varepsilon(1_{(1)}x)1_{(2)}, \tag{2.1}$$

$$\varepsilon_s(x) = 1_{(1)}\varepsilon(x1_{(2)}). \tag{2.2}$$

The maps ε_t and ε_s are called the *target counital map* and the *source counital map*, respectively.

EXAMPLE 2.1. Let X be a non-empty set. By a *small category with the object set* X we mean a set M equipped with two maps $s, t : \mathsf{M} \longrightarrow X$ and a binary operation $\circ : \mathsf{M}\,{}_s\!\times_t \mathsf{M} \longrightarrow \mathsf{M}$, where $\mathsf{M}\,{}_s\!\times_t \mathsf{M} = \{\ (\alpha, \beta) \in \mathsf{M} \times \mathsf{M} \mid s(\alpha) = t(\beta)\ \}$, such that they become a category whose objects are X and morphisms are M and the composition is given by $\circ$. We note that the above M can be viewed as a monoid in the category of spans over X (See [**3**, (2.6) and (5.4.3)]). For all $x \in X$ we denote the identity morphism on x by 1_x. The small category M is called finite if the underlying set is finite, and M is called a *groupoid* with the object set X if all morphisms are isomorphisms.

For a finite small category M with the object set X, as an analogous construction of groupoid algebras in [**16**, 2.5], an algebra $\boldsymbol{k}[\mathsf{M}]$ over $\boldsymbol{k}$ is defined. This algebra is spanned by the morphisms of M over $\boldsymbol{k}$ as a vector space, and equipped with the product which is given by the composition if two morphisms are composable and by 0 otherwise. The identity element $\mathbf{1}$ in the algebra is given by $\mathbf{1} = \sum_{x \in X} 1_x$. The algebra $\boldsymbol{k}[\mathsf{M}]$ becomes a weak bialgebra with the $\boldsymbol{k}$-linear maps $\Delta : \boldsymbol{k}[\mathsf{M}] \longrightarrow \boldsymbol{k}[\mathsf{M}] \otimes \boldsymbol{k}[\mathsf{M}]$ and $\varepsilon : \boldsymbol{k}[\mathsf{M}] \longrightarrow \boldsymbol{k}$ defined by $\Delta(\alpha) = \alpha \otimes \alpha$ and $\varepsilon(\alpha) = 1$ for all $\alpha \in \mathsf{M}$. If M is a groupoid, then the weak bialgebra $\boldsymbol{k}[\mathsf{M}]$ is called a groupoid algebra. The groupoid algebra $\boldsymbol{k}[\mathsf{M}]$ becomes a weak Hopf algebra with the above maps Δ and ε, and the $\boldsymbol{k}$-linear map $S : \boldsymbol{k}[\mathsf{M}] \longrightarrow \boldsymbol{k}[\mathsf{M}]$ defined by $S(\alpha) = \alpha^{-1}$ for all $\alpha \in \mathsf{M}$.

EXAMPLE 2.2 ([**9**, 1.2]). The following groupoid G over $X = \{+, -\}$ is called the *qubit groupoid*:

$$\mathsf{G} = \{1_+,\ 1_-,\ + \xrightarrow{\alpha} -,\ - \xrightarrow{\alpha^{-1}} +\}.$$

The corresponding quiver of the qubit groupoid is depicted as in the figure 1.

The qubit groupoid algebra $\boldsymbol{k}[\mathbf{G}]$ satisfies $\varepsilon(\mathbf{1}) = 2 \neq 1$, and hence it is not a Hopf algebra.

See [**2**] for more general construction and examples from quivers.

The target and source counital maps satisfy the following properties.

LEMMA 2.3. (1) $\varepsilon_t^2 = \varepsilon_t$, $\varepsilon_s^2 = \varepsilon_s$, $\varepsilon \circ \varepsilon_t = \varepsilon = \varepsilon \circ \varepsilon_s$.

(2) *For all* $x, y \in H$

(i) $((\mathrm{id} \otimes \varepsilon_t) \circ \Delta)(x) = 1_{(1)}x \otimes 1_{(2)}$.

(ii) $((\varepsilon_s \otimes \mathrm{id}) \circ \Delta)(x) = 1_{(1)} \otimes x1_{(2)}$.

(iii) $\varepsilon_t(x) = x \iff \Delta(x) = 1_{(1)}x \otimes 1_{(2)}$.

(iv) $\varepsilon_s(x) = x \iff \Delta(x) = 1_{(1)} \otimes x1_{(2)}$.

(v) $\varepsilon_t(x\varepsilon_t(y)) = \varepsilon_t(xy)$, $\varepsilon_s(\varepsilon_s(x)y) = \varepsilon_s(xy)$.

(vi) $x = \varepsilon_t(x_{(1)})x_{(2)} = x_{(1)}\varepsilon_s(x_{(2)})$.

(vii) $x\varepsilon_t(y) = \varepsilon_t(x_{(1)}y)x_{(2)}$, $\varepsilon_s(x)y = y_{(1)}\varepsilon_s(xy_{(2)})$.

In particular, by (i) *and* (ii)

$$\begin{aligned} 1_{(1)} \otimes \varepsilon_t(1_{(2)}) = 1_{(1)} \otimes 1_{(2)} &= \varepsilon_s(1_{(1)}) \otimes 1_{(2)}, \\ 1_{(1)}1_{[1]} \otimes 1_{(2)} \otimes 1_{[2]} &= 1_{(1)} \otimes \varepsilon_t(1_{(2)}) \otimes 1_{(3)}, \\ 1_{(1)} \otimes 1_{[1]} \otimes 1_{(2)}1_{[2]} &= 1_{(1)} \otimes \varepsilon_s(1_{(2)}) \otimes 1_{(3)}, \end{aligned} \tag{2.3}$$

where $\Delta^{(2)}(1) = 1_{(1)} \otimes 1_{(2)} \otimes 1_{(3)}$ *and* $\Delta(1) = 1_{[1]} \otimes 1_{[2]}$. □

See [**5**, Section 2] and [**22**, Section 4] for proofs of the above lemma.

Let H be a weak bialgebra over $\boldsymbol{k}$, and set $H_t := \varepsilon_t(H)$, $H_s := \varepsilon_s(H)$. Then an element of H_t and an element of H_s are commutative, and H_t and H_s are a left coideal subalgebra and a right coideal subalgebra of H, respectively. The subalgebras H_t, H_s are called the *target* and *source subalgebras* of H, respectively. By (2.3) we have

$$\Delta(1) \in H_s \otimes H_t. \tag{2.4}$$

From Lemma 2.3(iii), (iv) together with (2.4) and the fact that an element of H_t and an element of H_s are commutative, we see that for all $x \in H$, $z \in H_t$ and $y \in H_s$

$$\begin{aligned} \Delta(xz) &= x_{(1)}z \otimes x_{(2)}, & \Delta(zx) &= zx_{(1)} \otimes x_{(2)}, \\ \Delta(xy) &= x_{(1)} \otimes x_{(2)}y, & \Delta(yx) &= x_{(1)} \otimes yx_{(2)}. \end{aligned}$$

In particular

$$\begin{aligned} xz &= \varepsilon(x_{(1)}z)x_{(2)}, & zx &= \varepsilon(zx_{(1)})x_{(2)}, \\ xy &= x_{(1)}\varepsilon(x_{(2)}y), & yx &= x_{(1)}\varepsilon(yx_{(2)}). \end{aligned}$$

For a weak bialgebra H, one can also consider two $\boldsymbol{k}$-linear maps $\varepsilon_t', \varepsilon_s' : H \longrightarrow H$ defined by

$$\varepsilon_t'(x) = \varepsilon(x1_{(1)})1_{(2)}, \tag{2.5}$$

$$\varepsilon_s'(x) = 1_{(1)}\varepsilon(1_{(2)}x). \tag{2.6}$$

The target and the source subalgebras of the three weak bialgebras $H^{\mathrm{op}} = (H, \mu^{\mathrm{op}}, \eta, \Delta, \varepsilon)$, $H^{\mathrm{cop}} = (H, \mu, \eta, \Delta^{\mathrm{cop}}, \varepsilon)$, $H^{\mathrm{opcop}} = (H, \mu^{\mathrm{op}}, \eta, \Delta^{\mathrm{cop}}, \varepsilon)$ are given by $(H^{\mathrm{op}})_t = H_t$, $(H^{\mathrm{op}})_s = H_s$, $(H^{\mathrm{cop}})_t = H_s$, $(H^{\mathrm{cop}})_s = H_t$, $(H^{\mathrm{opcop}})_t = H_s$, $(H^{\mathrm{opcop}})_s = H_t$. The target and the source counital maps of them are given by $(\varepsilon_{H^{\mathrm{op}}})_t = \varepsilon'_t$, $(\varepsilon_{H^{\mathrm{op}}})_s = \varepsilon'_s$, $(\varepsilon_{H^{\mathrm{cop}}})_t = \varepsilon'_s$, $(\varepsilon_{H^{\mathrm{cop}}})_s = \varepsilon'_t$, $(\varepsilon_{H^{\mathrm{opcop}}})_t = \varepsilon_s$, $(\varepsilon_{H^{\mathrm{opcop}}})_s = \varepsilon_t$. If S is the antipode of H, then it is also of H^{opcop}.

The monoidal structure of the module category ${}_H\mathbf{M}$ for a weak bialgebra H is given as follows. Let V, W be two left H-modules, and define an H-action on the tensor product $V \otimes W$ by

$$x \cdot (v \otimes w) := (x_{(1)} \cdot v) \otimes (x_{(2)} \cdot w) \tag{2.7}$$

for all $v \in V$, $w \in W$, $x \in H$. By (WH1) we see that $(xy) \cdot \xi = x \cdot (y \cdot \xi)$ for all $x, y \in H$ and $\xi \in V \otimes W$. However, it is not certain whether $1 \cdot \xi = \xi$. So, we consider the subspace

$$V \circledast W := \Delta(1) \cdot (V \otimes W). \tag{2.8}$$

This subspace becomes a left H-module with (2.7). For two left H-module maps $f : V_1 \otimes W_1$, $g : V_2 \longrightarrow W_2$, the tensor product $f \otimes g$ satisfies $(f \otimes g)(V_1 \circledast V_2) \subset W_1 \circledast W_2$, and the restriction $f \circledast g := f \otimes g|_{V_1 \circledast V_2}$ is a left H-module map. In this manner the monoidal structure of ${}_H\mathbf{M}$ is defined. The unit object in the monoidal category is given by the target subalgebra H_t. Similarly, the monoidal structure of the comodule category $\mathbf{M}^H$ is similarly defined. For more detail on the monoidal structures of ${}_H\mathbf{M}$ and $\mathbf{M}^H$ see [**5**] and [**18**].

3. The direct sum construction of weak bialgebras and Hopf algebras

For two bialgebras A and B, the direct sum $A \oplus B$ becomes an algebra and a coalgebra. However it is not necessary to be a bialgebra. Whereas in weak case it is true as described in the following lemma.

LEMMA 3.1 (**direct sum construction**). *Let $A = (A, \Delta_A, \varepsilon_A)$ and $B = (B, \Delta_B, \varepsilon_B)$ be two weak bialgebras over $\boldsymbol{k}$, and set $H = A \oplus B$ as a vector space. Then H has an weak bialgebra structure such that its algebra structure is given by the direct product and its coalgebra structure is given by the direct sum. The target and source counital maps ε_t and ε_s are given by the formulas: For $x = a + b \in H$, where $a \in A$ and $b \in B$,*

$$\varepsilon_t(x) = (\varepsilon_A)_t(a) + (\varepsilon_B)_t(b),$$
$$\varepsilon_s(x) = (\varepsilon_A)_s(a) + (\varepsilon_B)_s(b),$$

where $(\varepsilon_A)_t, (\varepsilon_A)_s$ are the target and source counital maps of A, and $(\varepsilon_B)_t, (\varepsilon_B)_s$ are that of B. The target and source subalgebras of H are $H_t = A_t \oplus B_t$ and $H_s = A_s \oplus B_s$, respectively.

If A, B are weak Hopf algebras with antipodes S_A, S_B, then so is H with S defined by $S(a + b) = S_A(a) + S_B(b)$ for all $a \in A$ and $b \in B$.

PROOF. The claim follows from the simple equation $ab = ba = 0$ for all $a \in A$ and $b \in B$. For example, it can be verified that the condition (WH1) is satisfied as follows. Let x, x' be elements in H, and write as $x = a + b$, $x' = a' + b'$ for some $a, a' \in A$, $b, b' \in B$. Then

$$\begin{aligned}\Delta(xx') &= \Delta(aa' + bb') = \Delta_A(aa') + \Delta_B(bb') \\ &= \Delta_A(a)\Delta_A(a') + \Delta_B(b)\Delta_B(b') \\ &= \big(\Delta_A(a) + \Delta_B(b)\big)\big(\Delta_A(a') + \Delta_B(b')\big) \\ &= \Delta(x)\Delta(x').\end{aligned}$$

It can be also verified on the other conditions in a similar way. □

Chebel and Makhlouf [7] give a method to construct a weak bialgebra as an analogy of the method introduced by Kaplansky [**10**, Theorem 14]. Their construction can be viewed as a special version of a direct sum construction given in Lemma 3.1.

EXAMPLE 3.2 (**Chebel and Makhlouf** [7, Theorems 2.10 & 2.12]). Let $A = (A, \Delta_A, \varepsilon_A)$ be a bialgebra with identity element e over $\boldsymbol{k}$. Let 1 be a letter which is not contained in A, and consider the vector space $H := A \oplus \boldsymbol{k}1$. The product on A can be extended to the product on H as follows: For all $a \in A$

$$1 \cdot a = a = a \cdot 1, \quad 1 \cdot 1 = 1. \tag{3.1}$$

Moreover, we define $\boldsymbol{k}$-linear maps $\Delta : H \longrightarrow H \otimes H$, $\varepsilon : H \longrightarrow \boldsymbol{k}$ by

$$\begin{aligned} \Delta(a) &= \Delta_A(a), & \varepsilon(a) &= \varepsilon_A(a), \\ \Delta(1) &= (1-e) \otimes (1-e) + e \otimes e, & \varepsilon(1) &= 2, \end{aligned} \tag{3.2}$$

for all $a \in A$. Then H is a weak bialgebra. If A is a Hopf algebra with antipode S_A, then H is a weak Hopf algebra equipped with the antipode $S : H \longrightarrow H$ defined by

$$S(a) = S_A(a), \quad S(1) = 1 \tag{3.3}$$

for all $a \in A$. After [7] we call H the weak bialgebra obtained by the *Kaplansky type construction* from A.

Applying the above construction to the Taft algebras we have a more concrete example.

EXAMPLE 3.3 (**Taft's weak bialgebra** [7, Example 2.14]). Let $n \geq 2$ be an integer, and $\boldsymbol{k}$ be a field which contains an nth root of unity λ. Denoted by $H_{n^2}(\lambda)$ is the n^2-dimensional Taft algebra, that is

$$H_{n^2}(\lambda) = \langle\, g, x \mid g^n = e,\ x^n = 0,\ xg = \lambda gx \,\rangle,$$

where e is the identity element. Then one can obtain a weak Hopf algebra $H'_{n^2}(\lambda)$ with dimension $n^2 + 1$ by applying the Kaplansky type construction to $H_{n^2}(\lambda)$.

A weak bialgebra H is said to be *indecomposable* if there are no weak bialgebras A, B such that $H \cong A \oplus B$ as weak bialgebras. Similarly indecomposability can be defined for a weak Hopf algebra. Any finite-dimensional weak bialgebra can be decomposed into a direct sum of indecomposable ones up to isomorphism. The precise statement is as follows.

THEOREM 3.4. *Let H be a finite-dimensional weak bialgebra over $\boldsymbol{k}$.*

(1) *There are finitely many indecomposable weak bialgebras H_i $(i = 1, \ldots, k)$ such that $H = H_1 \oplus \cdots \oplus H_k$.*
(2) *Assume that $H_1 \oplus \cdots \oplus H_k = H = H'_1 \oplus \cdots \oplus H'_l$ for some indecomposable weak bialgebras $H_1, \ldots, H_k$ and $H'_1, \ldots, H'_l$. Then $k = l$, and there is a permutation σ such that $H'_i = H_{\sigma(i)}$ for all $i = 1, \ldots, k$.*

PROOF. This theorem essentially follows from the existence and uniqueness for an algebra with respect to a direct sum decomposition of indecomposable algebras.

(1) Let $H = B_1 \oplus \cdots \oplus B_s$ be a decomposition into indecomposable algebras. We set $\mathscr{B} := \{B_1, \ldots, B_s\}$, and $\langle S \rangle := \bigoplus_{B \in S} B$ for $S \subset \mathscr{B}$.

For each i let us consider the smallest subset $S \subset \mathscr{B}$ satisfying $\Delta(B_i) \subset \langle S\rangle \otimes \langle S\rangle$, and set $S_i := S \cup \{B_i\}$. Let $\sim$ be the equivalence relation on $\{1,\dots,s\}$ generated by

$$i \sim j \iff S_i \cap S_j \neq \varnothing.$$

If $C_1,\dots,C_m$ are all equivalence classes of $\{1,\dots,s\}$ with respect to $\sim$, then $\mathscr{B}$ is the disjoint union of

$$T_1 := \bigcup_{i_1\in C_1} S_{i_1},\ \dots, T_m := \bigcup_{i_m\in C_m} S_{i_m}.$$

Thus $H = \langle T_1\rangle \oplus \cdots \oplus \langle T_m\rangle$ as an algebra. Moreover one can easily check that $\langle T_1\rangle,\dots,\langle T_m\rangle$ are weak bialgebras.

We show that $\langle T_r\rangle$ is indecomposable as a weak bialgebra for all r. Suppose that $\langle T_r\rangle$ is not so. Then there are weak bialgebras H_1, H_2 such that $\langle T_r\rangle = H_1 \oplus H_2$. This decomposition can be viewed as an algebra. Applying the uniqueness of decomposition of $\langle T_r\rangle$ into indecomposable algebras [**23**, Proposition 3.16] we see that H_1, H_2 are direct sums of several ones of $B_1,\dots,B_s$. Thus H_i can be expressed as $H_i = \bigoplus_{B\in S_i} B$ for some $S_i \subset \mathscr{B}$ $(i = 1,2)$. For any $B_{i_1} \in S_1$ we have $\Delta(B_{i_1}) \subset \Delta(H_1) \subset H_1 \otimes H_1$. Then $S_{i_1} \subset S_1 \cup \{B_{i_1}\} = S_1$ since S_{i_1} is the union of $\{B_{i_1}\}$ and the smallest subset S in $\mathscr{B}$ satisfying $\Delta(B_{i_1}) \subset \langle S\rangle \otimes \langle S\rangle$. Similarly for any $B_{i_2} \in S_2$ we have $S_{i_2} \subset S_2$. If $B_{i_1} \in S_1 \subset T_r$ and $B_{i_2} \in S_2 \subset T_r$, then $i_1, i_2 \in C_r$, and hence $i_1 \sim i_2$. This implies that $S_{i_1} \cap S_{i_2} \neq \varnothing$, and so $S_1 \cap S_2 \neq \varnothing$. This is a contradiction for the fact that H_1 and H_2 are in a direct sum in $\langle T_r\rangle$. Thus $\langle T_r\rangle$ is indecomposable as a weak bialgebra.

(2) Let $H_1 = B_1 \oplus \cdots \oplus B_s$ be a decomposition of H_1 into indecomposable algebras. By the uniqueness of decomposition into indecomposable algebras, for each B_i there is a unique $j \in \{1,\dots,l\}$ such that B_i is a direct summand of H'_j. Let us consider the case where $s \geq 2$, and suppose that $B_1 \subset H'_i$ and $B_2 \subset H'_j$. We show that $i = j$.

For B_1, B_2 consider the subsets S_1, S_2 in $\mathscr{B} = \{B_1,\dots,B_s\}$ defined as in the proof of Part (1). Define $T_1,\dots,T_m$ by introducing the equivalence relation $\sim$ as in the proof of Part (1). Then $H_1 = \langle T_1\rangle \oplus \cdots \oplus \langle T_m\rangle$, and each $\langle T_1\rangle,\dots,\langle T_m\rangle$ are weak bialgebras. By the indecomposability of H_1 as a weak bialgebra we have $m = 1$, that is, $H_1 = \langle T_1\rangle$. Thus $B_1, B_2 \in T_1$, and $S_1 \cap S_2 \neq \varnothing$.

Since $B_1 \subset H'_i$, we have $\Delta(B_1) \subset H'_i \otimes H'_i$. This implies from the minimality of S_1 that $\langle S_1\rangle \subset H'_i$. Similarly, $\langle S_2\rangle \subset H'_j$. Thus for any $B \in S_1 \cap S_2$, we have $B \subset \langle S_1\rangle \subset H'_i$ and $B \subset \langle S_2\rangle \subset H'_j$. It follows that $B \subset H'_i \cap H'_j$, and hence $i = j$ since H'_i and H'_j form a direct sum in H. Therefore, all B_i $(i = 1,\dots,s)$ are contained in the same H'_i. This shows that $H_1 \subset H'_i$. The same result holds for all $H_1,\dots,H_k$. Thus

$$H_1 \subset H'_1,\ H_2 \subset H'_2,\dots,H_k \subset H'_k \tag{3.4}$$

after suitable reordering of indices. In particular we have $k \leq l$. Moreover, since $H = H_1 \oplus \cdots H_k = H'_1 \oplus \cdots \oplus H'_l$, it follows that $k = l$.

Applying the same argument above, we see that each H'_i $(i = 1,\dots,k)$ is contained in one of $H_1,\dots,H_k$. This fact and (3.4) lead to the equation $H_i = H'_i$ $(i = 1,\dots,k)$. □

A finite-dimensional bialgebra is a typical example of indecomposable weak bialgebras.

THEOREM 3.5. *A finite-dimensional bialgebra is indecomposable as a weak bialgebra.*

PROOF. Suppose that a finite-dimensional bialgebra H is decomposed as $H = A \oplus B$ for some weak bialgebras A and B. By Lemma 3.1 then $H_t = A_t \oplus B_t$ as algebras, and hence $\dim H_t = \dim A_t + \dim B_t \geq 1 + 1 = 2$. This contradicts the fact $H_t = \boldsymbol{k}1_H$. $\square$

EXAMPLE 3.6. Let $H = \mathbb{C}e_1 + \mathbb{C}e_2$ be a 2-dimensional weak bialgebra over the complex number field $\mathbb{C}$ whose product μ, coproduct Δ, counit ε are given as follows.

$$\begin{aligned}
&\mu(e_1, e_1) = e_1,\ \mu(e_1, e_2) = \mu(e_2, e_1) = \mu(e_2, e_2) = e_2,\\
&\Delta(e_1) = (e_1 - e_2) \otimes (e_1 - e_2) + e_2 \otimes e_2,\ \Delta(e_2) = e_2 \otimes e_2,\\
&\varepsilon(e_1) = 2,\ \varepsilon(e_2) = 1.
\end{aligned}$$

Note that e_1 is the identity element. The weak bialgebra is the third one in the list in [**7**, Proposition 4.3] by Chebel and Makhlouf.

Consider the subspaces $A = \mathbb{C}(e_1 - e_2)$ and $B = \mathbb{C}e_2$. Then they are weak bialgebras isomorphic to the 1-dimensional Hopf algebra $\mathbb{C}$, and $H = A \oplus B$. Thus, the above weak bialgebra H can be decomposed into the direct sum of two 1-dimensional Hopf algebras.

On the other hand, the second weak bialgebra $K = \mathbb{C}e_1 + \mathbb{C}e_2$ in the list in [**7**, Proposition 4.3] is isomorphic to the group algebra of $G = \mathbb{Z}/2\mathbb{Z}$. In fact, if we set $G = \langle g \rangle$, then an isomorphism $f : K \longrightarrow \mathbb{C}[G]$ is given by $f(e_1) = 1, f(e_2) = (1 + g)/2$. In particular, K is indecomposable.

EXAMPLE 3.7. Chebel and Makhlouf [**7**] are determined the isomorphism classes of all weak bialgebras over the complex number field $\mathbb{C}$ of dimension ≤ 3. According to their classification result, there are exactly 20 isomorphism classes in the 3-dimensional weak bialgebras. Let us denote the nth weak bialgebra by $(\#n)$ in the list [**7**, Proposition 4.5]. Then the 3-dimensional weak bialgebras except for $(\#8-10)$ are indecomposable as weak bialgebras, and $(\#8-10)$ can be decomposed into direct sums of indecomposable weak bialgebras as follows:

$$(\#8) = \mathbb{C} \oplus H_2, \quad (\#9) = \mathbb{C} \oplus \mathbb{C}[\mathbb{Z}/2\mathbb{Z}], \quad (\#10) = \mathbb{C} \oplus \mathbb{C} \oplus \mathbb{C}.$$

Here, $H_2 = \mathbb{C}e_1 + \mathbb{C}e_2$ is the 2-dimensional weak bialgebra whose product, coproduct and counit are given by

$$\begin{aligned}
&\mu(e_1, e_1) = e_1,\ \mu(e_1, e_2) = \mu(e_2, e_1) = \mu(e_2, e_2) = e_2,\\
&\Delta(e_1) = e_1 \otimes e_1,\ \Delta(e_2) = e_2 \otimes e_2,\\
&\varepsilon(e_1) = \varepsilon(e_2) = 1.
\end{aligned}$$

This weak bialgebra is the first one in the list in [**7**, Proposition 4.3].

PROOF. • All of those in $(\#1-7)$, $(\#12-15)$ and $(\#17-20)$ are indecomposable weak bialgebras since they are proper bialgebras. The weak bialgebras $(\#16)$ and $(\#11)$ also indecomposable since the former is isomorphic to the weak bialgebra which is spanned by the morphisms of the so-called interval category **2**

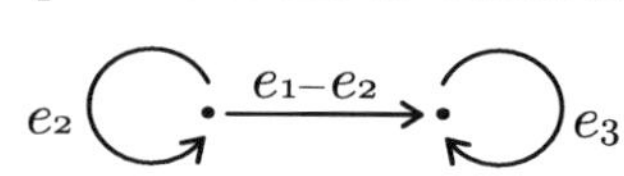

(See Proposition 3.8 below), and the later is the dual of it.

• The idempotents in $H = (\#8)$ are 0, $e_1 - e_2$, $e_2 - e_3$, e_3, $e_1 - e_3$, e_2, $e_1 - e_2 + e_3$, e_1, and the identity element e_1 is expressed as the sum of primitive idempotents $e_1 - e_2$, $e_2 - e_3$, e_3. Thus as an algebra

$$H = H(e_1 - e_2) \oplus H(e_2 - e_3) \oplus He_3 = \mathbb{C}(e_1 - e_2) \oplus \mathbb{C}(e_2 - e_3) \oplus \mathbb{C}e_3.$$

Setting

$$A_1 = \mathbb{C}(e_1 - e_2), \quad A_2 = \mathbb{C}(e_2 - e_3), \quad A_3 = \mathbb{C}e_3, \tag{3.5}$$

we have a decomposition $H = A_1 \oplus A_2 \oplus A_3$ into indecomposable algebras. Since

$$\begin{aligned}
\Delta(e_1 - e_2) &= (e_1 - e_2) \otimes (e_1 - e_2), \\
\Delta(e_2 - e_3) &= (e_2 - e_3) \otimes (e_2 - e_3) + (e_2 - e_3) \otimes e_3 + e_3 \otimes (e_2 - e_3), \\
\Delta(e_3) &= e_3 \otimes e_3,
\end{aligned}$$

it follows that A_1, $A_{23} := A_2 \oplus A_3$ are subcoalgebras of H. Thus we have a decomposition $H = A_1 \oplus A_{23}$ as a weak bialgebra. We see that the weak bialgebra A_{23} is isomorphic to the first weak bialgebra in the list in [**7**, Proposition 4.3].

• By a similar consideration above, it is shown that $H = (\#9)$ is decomposed as $H = A_1 \oplus A_{23}$ as a weak bialgebra. In this case, since $\Delta(e_2) = e_2 \otimes e_2$, $\Delta(e_2 - e_3) = (e_2 - e_3) \otimes (e_2 - e_3) + e_3 \otimes e_3$, the weak bialgebra A_{23} is isomorphic to the second weak bialgebra in the list in [**7**, Proposition 4.3]. Thus, $H \cong \mathbb{C} \oplus \mathbb{C}[\mathbb{Z}/2\mathbb{Z}]$ as weak bialgebras.

• Let $H = (\#10)$. By a similar consideration above, we see that H is decomposed as $H = A_1 \oplus A_2 \oplus A_3 \cong \mathbb{C} \oplus \mathbb{C} \oplus \mathbb{C}$ as a weak bialgebra. □

There are indecomposable weak bialgebras that are not proper bialgebras.

PROPOSITION 3.8. *Let* M *be a finite small category with the object set* X. *Assume that the corresponding quiver* Q_{M} *is connected and has no multiple loops, namely, the identity morphism* 1_X *is the unique morphism* $x \longrightarrow x$ *for all* $x \in X$. *Then the weak bialgebra* $H := \boldsymbol{k}[\mathsf{M}]$ *is indecomposable as an algebra, and is also as a weak bialgebra.*

PROOF. Let e be an element in H, and write as in the form

$$e = \sum_{x \in X} a_x x + \sum_{\alpha \in \mathsf{M} - X} a_\alpha \alpha$$

for some $a_x, a_\alpha \in \boldsymbol{k}$. Since Q_{M} has no multiple loops, if $xe = ex$ for all $x \in X$, then $a_\alpha = 0$ for all $\alpha \in \mathsf{M} - X$. Thus, if e is in the center $Z(H)$ of H, then it can be expressed as $e = \sum_{x \in X} a_x x$. Let $\alpha \in \mathsf{M} - X$. Then $e\alpha = a_{t(\alpha)}\alpha$, $\alpha e = a_{s(\alpha)}\alpha$, and hence $a_{t(\alpha)} = a_{s(\alpha)}$. Since Q_{M} is connected, for all $x, y \in X$ there is a finite sequence $x = x_0, x_1, \ldots, x_n = y$ such that x_{i-1} and x_i are related some morphism $\alpha_i : x_{i-1} \longrightarrow x_i$ in M for all $i = 1, \ldots, n$. It follows that a_x does not depend on x. Thus $e \in Z(H)$ can be expressed as $e = a \sum_{x \in X} x = a\mathbf{1}$ for some $a \in \boldsymbol{k}$. So, if $e \in Z(H)$ is an idempotent, then there are only possibilities $a = 0, 1$. This shows that the algebra H is indecomposable. □

EXAMPLE 3.9. The qubit groupoid algebra is indecomposable as a weak bialgebra by Proposition 3.8.

We describe some properties of preserving under the direct sum construction.

Let H be a weak bialgebra over $\boldsymbol{k}$. An element $\Lambda \in H$ is called a *left integral* of H if $x\Lambda = \varepsilon_t(x)\Lambda$ for all $x \in H$. Similarly, an element $\Lambda \in H$ is called a *right integral* of H if $\Lambda x = \Lambda\varepsilon_s(x)$ for all $x \in H$. We denote the set of left integrals and right integrals by $\mathscr{I}_L(H)$ and $\mathscr{I}_R(H)$, respectively. An invertible element of $g \in H$ is called *group-like* if $\Delta(g) = (g\otimes g)\Delta(1) = \Delta(1)(g\otimes g)$ is satisfied. We denote the set of group-like elements of H by $G(H)$. The set $G(H)$ becomes a group with respect to the product of H ([**5**, Definition 4.11], [**15**, Definition 4.1]). We note that if H has an antipode, then for any element $g \in H$ satisfying $\Delta(g) = (g\otimes g)\Delta(1) = \Delta(1)(g\otimes g)$ the invertibility of g is equivalent to $\varepsilon_s(g) = \varepsilon_t(g) = 1$. The following is an easy consequence from definition and construction.

PROPOSITION 3.10. *Let A, B be two finite-dimensional weak bialgebras over $\boldsymbol{k}$, and consider the weak bialgebra $H = A\oplus B$ obtained by the direct sum construction. Then*

(1) *H is (co)semisimple if and only if both A, B are so.*
(2) *$\mathscr{I}_L(H) = \{ \Lambda_A + \Lambda_B \mid \Lambda_A \in \mathscr{I}_L(A),\ \Lambda_B \in \mathscr{I}_L(B) \}$, and the same equation holds for the set of right integrals.*
(3) *$G(H) \cong G(A) \times G(B)$ as groups.*

4. The direct sum construction of weak Hopf algebras and quantum doubles

As in the Hopf algebra case [**1**], for any weak bialgebra H, which is not necessary to be finite-dimensional, one can define the dual weak Hopf algebra as follows.

Let $H = (H, \mu, \eta, \Delta, \varepsilon)$ be a weak bialgebra over $\boldsymbol{k}$, and set

$$H^\circ := \{ p \in H^* \mid \dim(\boldsymbol{k}[H]p) < \infty \},$$

where H^* denotes the dual vector space of H, and

$$\boldsymbol{k}[H] = \Big\{ \sum_{x\in H} c_x x \ \Big|\ c_x \in \boldsymbol{k},\ c_x = 0 \text{ except for finitely many } x \in H \Big\},$$

and $\big(\sum_{x\in H} c_x x\big)p$ is an element in H^* defined by

$$\Big(\Big(\sum_{x\in H} c_x x\Big)p\Big)(h) = \sum_{x\in H} c_x p(hx) \tag{4.1}$$

for all $h \in H$. Then H° becomes a weak bialgebra with the following product, identity element 1_{H°, coproduct Δ_{H°, counit ε_{H°: For all $x, y \in H$ and $p, q \in H^\circ$

- $(pq)(x) = p(x_{(1)})q(x_{(2)})$,
- $1_{H^\circ} = \varepsilon$ (= the counit of H),
- $p_{(1)}(x)p_{(2)}(y) = p(xy)$, where $\Delta_{H^\circ}(p) = p_{(1)} \otimes p_{(2)}$,
- $\varepsilon_{H^\circ}(p) = p(1)$.

The above weak bialgebra H° is called the *dual weak bialgebra* of H. If H has an antipode S, then H° is also a weak Hopf algebra with the following S_{H° as an antipode: For all $x \in H$ and $p \in H^\circ$

- $\langle S_{H^\circ}(p),\ x\rangle = \langle p,\ S(x)\rangle$,

where $\langle\ ,\ \rangle$ denotes the inner product defined by duality between H and H^*. This weak Hopf algebra H° is called the *dual weak Hopf algebra* of H.

Let H_1, H_2 be two weak bialgebras over $\boldsymbol{k}$. A $\boldsymbol{k}$-linear map $f : H_1 \longrightarrow H_2$ is called a *weak bialgebra map* if it is an algebra map and a coalgebra map. A bijective weak bialgebra map is called an isomorphism of weak bialgebras. A weak bialgebra map between weak Hopf algebras is called a *weak Hopf algebra map*. This definition is based on the fact that any weak bialgebra map $f : H_1 \longrightarrow H_2$ between weak Hopf algebras H_1, H_2 is not only compatible with the target counital maps and the source counital maps of H_1, H_2, but also the antipodes of H_1, H_2. A bijective weak Hopf algebra map is called an isomorphism of weak Hopf algebras.

If a weak bialgebra H is of finite dimension, then $H^\circ = H^*$. We regard as $H^* \otimes H^* = (H \otimes H)^*$ via the standard isomorphism $H^* \otimes H^* \longrightarrow (H \otimes H)^*$ in linear algebra theory, and $\iota : H \longrightarrow H^{**} = (H^*)^*$ be the canonical $\boldsymbol{k}$-linear isomorphism defined by

$$\big(\iota(x)\big)(p) = p(x)$$

for all $x \in H,\ p \in H^*$. Then, ι is an isomorphism of weak bialgebras, and if H is a weak Hopf algebra, then ι is an isomorphism of weak Hopf algebras.

THEOREM 4.1. *Let $H = (H, \mu, \eta, \Delta, \varepsilon, S)$ be a weak Hopf algebra over $\boldsymbol{k}$. Then the antipode of the dual weak Hopf algebra H° is an anti-algebra map and an anti-coalgebra map.*

PROOF. The proof of what the antipode S_{H° is an anti-algebra map and $\varepsilon \circ S = \varepsilon$ is the same in the proof of [**5**, Theorem 2.10]. We show that the equation $S_{H^\circ}(p_{(2)}) \otimes S_{H^\circ}(p_{(1)}) = \Delta_{H^\circ}\big(S_{H^\circ}(p)\big)$ holds for all $p \in H^\circ$. For this it is enough to show that the images of the both sides under the injection $\pi : H^* \otimes H^* \longrightarrow (H \otimes H)^*,\ \pi(p \otimes q) : x \otimes y \longmapsto p(x)q(y)$ for all $p, q \in H^*$ and $x, y \in H$ coincide. It is verified as follows: For all $x, y \in H$ and $p \in H^\circ$

$$\begin{aligned}
\big\langle \pi\big(S_{H^\circ}(p_{(2)}) \otimes S_{H^\circ}(p_{(1)})\big),\ x \otimes y \big\rangle &= \big\langle S_{H^\circ}(p_{(2)}),\ x \big\rangle \big\langle S_{H^\circ}(p_{(1)}),\ y \big\rangle \\
&= \langle p_{(2)}, S(x) \rangle \langle p_{(1)}, S(y) \rangle \\
&= \langle \pi(p_{(1)} \otimes p_{(2)}), S(y) \otimes S(x) \rangle \\
&= \big\langle \pi\big(\Delta_{H^\circ}(p)\big),\ S(y) \otimes S(x) \big\rangle \\
&= \langle p,\ S(y)S(x) \rangle \\
&= \langle p,\ S(xy) \rangle \qquad (S \text{ is an anti-algebra map}) \\
&= \langle S_{H^\circ}(p),\ xy \rangle \\
&= \big\langle \pi\big(\Delta_{H^\circ}\big(S_{H^\circ}(p)\big)\big),\ x \otimes y \big\rangle.
\end{aligned}$$

□

COROLLARY 4.2. *Let H be a finite-dimensional weak Hopf algebra over $\boldsymbol{k}$. Then the antipode of H is an anti-algebra map and an anti-coalgebra map.*

PROOF. Since H is finite-dimensional, H can be regarded as $H = H^{**}$. Then the statement follows from Theorem 4.1. □

THEOREM 4.3 (**Böhm, Nill, Szlachányi** [**5**, Theorem 2.10]). *The antipode of a finite-dimensional weak Hopf algebra is bijective.* □

The concepts of quasitriangular weak Hopf algebras and weak ribbon algebras were introduced by Nikshych, Turaev and Vainerman [**17**]. A detailed study of quantum doubles of weak Hopf algebras is done in Chapter 6 in [**17**].

Let H be a weak Hopf algebra over $\boldsymbol{k}$, and $R \in \Delta^{\mathrm{cop}}(1)(H \otimes H)\Delta(1)$. The pair (H, R) and R are called a *quasitriangular weak Hopf algebra* and a *universal R-matrix* of H, respectively, if the following four conditions are satisfied:

(QT1) There is an element $\overline{R} \in \Delta(1)(H \otimes H)\Delta^{\mathrm{cop}}(1)$ such that $R\overline{R} = \Delta^{\mathrm{cop}}(1)$ and $\overline{R}R = \Delta(1)$.

(QT2) $\Delta^{\mathrm{cop}}(x)R = R\Delta(x)$ for all $x \in H$.

(QT3) $(\mathrm{id} \otimes \Delta)(R) = R_{13}R_{12}$.

(QT4) $(\Delta \otimes \mathrm{id})(R) = R_{13}R_{23}$.

Here, if we write as in the form $R = R^{(1)} \otimes R^{(2)}$, then $R_{13} = R^{(1)} \otimes 1 \otimes R^{(2)}$, $R_{12} = R \otimes 1$, $R_{23} = 1 \otimes R$. We note that the element $\overline{R}$ in (QT1) is unique, and it satisfies $\Delta(x)\overline{R} = \overline{R}\Delta^{\mathrm{cop}}(x)$ for all $x \in H$.

PROPOSITION 4.4. *Let A, B be two weak Hopf algebras over $\boldsymbol{k}$ of finite dimension, and $H = A \oplus B$ be the weak Hopf algebra obtained by the direct sum construction given in Lemma 3.1. If R_A and R_B are universal R-matrices of A and B, respectively, then $R := R_A + R_B$ is a universal R-matrix of H. Conversely, any universal R-matrix of H is given by the form $R_A + R_B$.*

PROOF. Since $ab = 0$, $ba = 0$ in H for all $a \in A$ and $b \in B$,

$$\Delta^{\mathrm{cop}}(1)(H \otimes H)\Delta(1) = \Delta_A^{\mathrm{cop}}(1_A)(A \otimes A)\Delta_A(1_A) + \Delta_B^{\mathrm{cop}}(1_B)(B \otimes B)\Delta_B(1_B). \tag{4.2}$$

It follows that $R = R_A + R_B \in \Delta^{\mathrm{cop}}(1)(H \otimes H)\Delta(1)$. It can be easily shown that R satisfies the conditions (QT1), ..., (QT4). Conversely, let R be a universal R-matrix of H. By (4.2) R can be written as in the form

$$R = R_A + R_B \qquad (R_A \in \Delta_A^{\mathrm{cop}}(1_A)(A \otimes A)\Delta_A(1_A),\ R_B \in \Delta_B^{\mathrm{cop}}(1_B)(B \otimes B)\Delta_B(1_B)).$$

Similarly, $\overline{R} \in \Delta(1)(H \otimes H)\Delta^{\mathrm{cop}}(1)$ in the condition (QT1) can be written as in the form

$$\overline{R} = \overline{R}_A + \overline{R}_B \quad (\overline{R}_A \in \Delta_A(1_A)(A \otimes A)\Delta_A^{\mathrm{cop}}(1_A),\ \overline{R}_B \in \Delta_B(1_B)(B \otimes B)\Delta_B^{\mathrm{cop}}(1_B)).$$

By using equations $ab = 0$, $ba = 0$ for all $a \in A$ and $b \in B$, we have $R_A\overline{R}_A = \Delta_A^{\mathrm{cop}}(1_A)$, $R_B\overline{R}_B = \Delta_B^{\mathrm{cop}}(1_B)$, $\overline{R}_A R_A = \Delta_A(1_A)$, $\overline{R}_B R_B = \Delta_B(1_B)$. Thus, R_A and R_B satisfy the condition (QT1). It can be easily shown that R_A and R_B satisfy the other three conditions. □

EXAMPLE 4.5. Let $m, n \geq 2$, and suppose that $\boldsymbol{k}$ contains an mth root of unity ω and an nth root of unity λ. Let us consider the Taft algebras $H_{m^2}(\omega)$ and $H_{n^2}(\lambda)$. Then, any universal R-matrix of the weak Hopf algebra $H := H_{m^2}(\omega) \oplus H_{n^2}(\lambda)$ is given by

$$R_{H_{m^2}(\omega)} + R_{H_{n^2}(\lambda)},$$

where $R_{H_{m^2}(\omega)}$ and $R_{H_{n^2}(\lambda)}$ are universal R-matrices of $H_{m^2}(\omega)$ and $H_{n^2}(\lambda)$, respectively. In particular, if $m = n = 2$, then $\omega = \lambda = -1$, and $H_4(-1)$ is Sweedler's 4-dimensional Hopf algebra. By Radford [**19**] it is shown that if the characteristic of $\boldsymbol{k}$ is not 2, then the universal R-matrices of $H_4(-1)$ are given by

$$R_\alpha = \frac{1}{2}(e \otimes e + g \otimes e + e \otimes g - g \otimes g) + \frac{\alpha}{2}(x \otimes x + x \otimes gx + gx \otimes gx - gx \otimes x)$$

for all $\alpha \in \boldsymbol{k}$. Thus, the universal R-matrices of the direct sum

$$H = \left\langle\ e_1, g, x, e_2, h, y\ \middle|\ \begin{array}{l} g^2 = e_1,\ h^2 = e_2,\ x^2 = y^2 = 0, \\ xg = -gx,\ yh = -hy,\ e_1 + e_2 = 1 \\ ae_1 = e_1a = a,\ be_2 = e_2b = b \\ ab = ba = 0\ (a \in \{e_1, g, x\},\ b \in \{e_2, h, y\}) \end{array} \right\rangle$$

of Sweedler's 4-dimensional Hopf algebras are given by

$$\begin{aligned} R_\alpha + R_\beta = &\frac{1}{2}(e_1 \otimes e_1 + g \otimes e_1 + e_1 \otimes g - g \otimes g) \\ &+ \frac{\alpha}{2}(x \otimes x + x \otimes gx + gx \otimes gx - gx \otimes x) \\ &+ \frac{1}{2}(e_2 \otimes e_2 + h \otimes e_2 + e_2 \otimes h - h \otimes h) \\ &+ \frac{\beta}{2}(y \otimes y + y \otimes hy + hy \otimes hy - hy \otimes y) \end{aligned}$$

for all $\alpha, \beta \in \boldsymbol{k}$.

For two quasitriangular weak Hopf algebras (A, R_A), (B, R_B), we set $(A, R_A) \oplus (B, R_B) := (A \oplus B, R_A + R_B)$, and call it the direct sum of (A, R_A) and (B, R_B).

Let (H, R) be a weak quasitriangular Hopf algebra over $\boldsymbol{k}$. An element $v \in H$ is called a *ribbon element* of (H, R) if the following three conditions are satisfied:

(RE1) v is invertible and central.
(RE2) $\Delta(v) = R_{21}R(v \otimes v)$.
(RE3) $S(v) = v$.

We denote by $\mathrm{Rib}(H, R)$ the set of ribbon elements of (H, R). A ribbon element of (H, R) can be characterized by some group-like element of H [**17**, Lemma A.1]. In straightforward computation we have:

PROPOSITION 4.6. *Let v_A, v_B be ribbon elements of two quasitriangular weak Hopf algebras (A, R_A), (B, R_B), respectively. Then $v = v_A + v_B$ is a ribbon element of $(H, R) := (A, R_A) \oplus (B, R_B)$. Conversely, any ribbon element of (H, R) is given by the form $v = v_A + v_B$ for some ribbon elements v_A, v_B of $(A, R_A), (B, R_B)$, respectively. That is, the map*

$$\mathit{Rib}(A, R_A) \times \mathit{Rib}(B, R_B) \longrightarrow \mathit{Rib}(H, R), \qquad (v_A, v_B) \longmapsto v_A + v_B$$

is a bijection.

The direct sum construction is compatible with duals of weak bialgebras as follows.

LEMMA 4.7. *Let A, B be two weak bialgebras over $\boldsymbol{k}$ of finite dimension, and H be the weak bialgebra obtained by the direct sum construction from A and B. If we regard as $A^*, B^* \subset H^*$ via the transposed maps ${}^t\pi_A, {}^t\pi_B$ of canonical projections $\pi_A : H \longrightarrow A$, $\pi_B : H \longrightarrow B$ associated with the decomposition $H = A \oplus B$, then $H^* = A^* \oplus B^*$ as weak bialgebras. More precisely, for all $p, q \in H^*$*

- $pq = {}^t\pi_A(p|_A){}^t\pi_A(q|_A) + {}^t\pi_B(p|_B){}^t\pi_B(q|_B)$,
- $1_{H^*} = {}^t\pi_A(\varepsilon_A) + {}^t\pi_B(\varepsilon_B)$,
- $\Delta_{H^*}(p) = ({}^t\pi_A \otimes {}^t\pi_A)\big(\Delta_{A^*}(p|_A)\big) + ({}^t\pi_B \otimes {}^t\pi_B)\big(\Delta_{B^*}(p|_B)\big)$,
- $\varepsilon_{H^*}(p) = \varepsilon_{A^*}(p|_A) + \varepsilon_{B^*}(p|_B)$.

If A and B are weak Hopf algebras, then for all $p \in H^*$

- $S_{H^*}(p) = {}^t\pi_A\big(S_{A^*}(p|_A)\big) + {}^t\pi_B\big(S_{B^*}(p|_B)\big)$.

Therefore, if H is indecomposable, then so is the dual weak bialgebra H^*.

In the sequel, we examine a relationship between quantum doubles for weak Hopf algebras and the direct sum construction. We adopt here a version following Radford's formulation of quantum doubles of Hopf algebras.

Let H be a weak bialgebra over $\boldsymbol{k}$. For elements $x \in H$ and $p \in H^*$, two actions $x \leftharpoonup p \in H$ and $p \rightharpoonup x \in H$ are defined by

$$x \leftharpoonup p := \langle x_{(1)}, p \rangle x_{(2)} = p(x_{(1)})x_{(2)}, \tag{4.3}$$

$$p \rightharpoonup x := x_{(1)} \langle p, x_{(2)} \rangle = x_{(1)}p(x_{(2)}). \tag{4.4}$$

By dualizing these actions two actions $p \leftharpoonup x \in H^*$ and $x \rightharpoonup p \in H^*$ can be defined as

$$(p \leftharpoonup x)(x') := p(xx'), \tag{4.5}$$

$$(x \rightharpoonup p)(x') := p(x'x) \tag{4.6}$$

for all $x' \in H$.

The quantum double $D(H)$ is defined by $D(H) := H^* \otimes H/J$ as a vector space, where $J = J_1 + J_2$, and

$$\begin{aligned} J_1 &= \mathrm{Span}\{\ p \otimes yx - p(\varepsilon \leftharpoonup y) \otimes x \mid y \in H_s,\ p \in H^*,\ x \in H\ \}, \\ J_2 &= \mathrm{Span}\{\ p \otimes zx - p(z \rightharpoonup \varepsilon) \otimes x \mid z \in H_t,\ p \in H^*,\ x \in H\ \}. \end{aligned} \tag{4.7}$$

The subspace J coincides with the kernel of the idempotent map

$$F : H^* \otimes H \longrightarrow H^* \otimes H,\ F(p \otimes x) = (1_{(1)} \rightharpoonup p \leftharpoonup 1_{[1]}) \otimes 1_{(2)}\varepsilon_s(1_{[2]})x \tag{4.8}$$

for all $p \in H^*$ and $x \in H$, and $D(H)$ is isomorphic to the image of F as a vector space. This can be shown by using the fact that $e_s = 1_{(1)} \otimes \varepsilon_s(1_{(2)})$ is a Frobenius-separability idempotent of the source subalgebra H_s [**22**, (4.13)]. Since $ab = 0 = ba$ for all $a \in A$ and $b \in B$, it follows that $F({}^t\pi_A(A^*) \otimes B) = F({}^t\pi_B(B^*) \otimes A) = 0$. This implies that

$$[{}^t\pi_A(p|_A) \otimes b] = [{}^t\pi_B(p|_B) \otimes a] = 0 \tag{4.9}$$

for all $p \in H^*$, and therefore any element in $D(H)$ can be expressed as

$$[{}^t\pi_A(p|_A) \otimes a] + [{}^t\pi_B(q|_B) \otimes b] \tag{4.10}$$

for some $p, q \in H^*,\ a \in A,\ b \in B$.

THEOREM 4.8. *Let A, B be two weak Hopf algebras over $\boldsymbol{k}$ of finite dimension, and set $H := A \oplus B$. Regard as $D(A) \subset D(H)$ and $D(B) \subset D(H)$ via the $\boldsymbol{k}$-linear maps $D(A) \longrightarrow D(H)$ and $D(B) \longrightarrow D(H)$ induced from the injections ${}^t\pi_A \otimes \iota_A : A^* \otimes A \longrightarrow H^* \otimes H$ and ${}^t\pi_B \otimes \iota_B : B^* \otimes B \longrightarrow H^* \otimes H$, where π_A, π_B are projections, and ι_A, ι_B are inclusions. Then the quantum double $D(H)$ is the direct sum of $D(A)$ and $D(B)$ as weak Hopf algebras: $D(H) = D(A) \oplus D(B)$.*

PROOF. First of all, we show that ${}^t\pi_A \otimes \iota_A : A^* \otimes A \longrightarrow H^* \otimes H$ induces an injection $D(A) \longrightarrow D(H)$. Let J_A be the subspace of $A^* \otimes A$ defined by the sum space of (4.7) from A instead of H, and we similarly define the subspace J_B of $B^* \otimes B$. Then, $J = J_A + J_B + A^* \otimes B + B^* \otimes A$ under the identifications $A^* = {}^t\pi_A(A^*) \subset H^*$ and $B^* = {}^t\pi_B(B^*) \subset H^*$. Since $({}^t\pi_A \otimes \iota_A)(J_A) = J_A \subset J$, the map ${}^t\pi_A \otimes \iota_A : A^* \otimes A \longrightarrow H^* \otimes H$ induces a $\boldsymbol{k}$-linear map $D(A) \longrightarrow D(H)$. We write the map by f:

$$f([\alpha \otimes a]) = [{}^t\pi_A(\alpha) \otimes \iota_A(a)]$$

for all $\alpha \in A^*,\ a \in A$. This map is injective since $J_A = J \cap (A^* \otimes A)$.

The product of the quantum double $D(H) := H^* \otimes H/J$ is given by

$$[p \otimes x][q \otimes y] = \langle q_{(1)}, S^{-1}(x_{(3)}) \rangle \langle q_{(3)}, x_{(1)} \rangle [pq_{(2)} \otimes x_{(2)}y] \tag{4.11}$$

for all $p, q \in H^*$ and $x, y \in H$. If we express as $x = a + b,\ y = a' + b'$ for some $a, a' \in A,\ b, b' \in B$, then

$$(4.12)\quad [p \otimes x][q \otimes y] = [{}^t\pi_A(p|_A) \otimes a][{}^t\pi_A(q|_A) \otimes a'] + [{}^t\pi_B(p|_B) \otimes b][{}^t\pi_B(q|_B) \otimes b'].$$

By using (4.9) we see that the identity element in $D(H)$ is given by

$$[\varepsilon \otimes 1] = \big[\big({}^t\pi_A(\varepsilon_A) + {}^t\pi_B(\varepsilon_B)\big) \otimes (1_A + 1_B)\big] = \big[{}^t\pi_A(\varepsilon_A) \otimes 1_A\big] + \big[{}^t\pi_B(\varepsilon_B) \otimes 1_B\big].$$

Therefore $D(H)$ is the direct product of $D(A)$ and $D(B)$ as an algebra. Moreover one can see that

$$\begin{aligned}
\Delta_{D(H)}([p \otimes x]) &= \Delta_{D(A)}\big([(p|_A) \otimes a]\big) + \Delta_{D(B)}\big([(p|_B) \otimes b]\big),\\
\varepsilon_{D(H)}([p \otimes x]) &= \varepsilon_{D(A)}\big([(p|_A) \otimes a]\big) + \varepsilon_{D(B)}\big([(p|_B) \otimes b]\big),\\
S_{D(H)}([p \otimes x]) &= S_{D(A)}\big([(p|_A) \otimes a]\big) + S_{D(B)}\big([(p|_B) \otimes b]\big)
\end{aligned}$$

for all $x = a + b \in H,\ a \in A,\ b \in B$. This implies that $D(H) = D(A) \oplus D(B)$. □

EXAMPLE 4.9. (1) Let $H = (\#9)$. Then

$$\begin{aligned}
H^* &\cong \mathbb{C}^* \oplus (\mathbb{C}[\mathbb{Z}/2\mathbb{Z}])^* \cong \mathbb{C} \oplus \mathbb{C}[\mathbb{Z}/2\mathbb{Z}] = H,\\
D(H) &\cong D(\mathbb{C}) \oplus D(\mathbb{C}[\mathbb{Z}/2\mathbb{Z}]) = \mathbb{C} \oplus \mathbb{C}[\mathbb{Z}/2\mathbb{Z} \times \mathbb{Z}/2\mathbb{Z}].
\end{aligned}$$

It follows that $D(H)$ is a 5-dimensional weak Hopf algebra that is commutative and cocommutative. In particular, it is not isomorphic to the 5-dimensional Taft's weak algebra.

(2) For $H = (\#10)$, both of H^* and $D(H)$ are isomorphic to H.

REMARK 4.10. Part (1) has shown by Zhang, Zhao and Wang [**27**]. In their paper, the ribbon elements of $(D(H), \mathcal{R})$ are determined. This result can be confirmed by Proposition 4.6.

5. Categorical interpretation of the direct sum construction and its generalizations

A $\boldsymbol{k}$-linear monoidal category $\mathscr{C}$ is said to be *indecomposable* if there are no $\boldsymbol{k}$-linear monoidal categories $\mathscr{C}_1, \mathscr{C}_2$ such that $\mathscr{C} \simeq \mathscr{C}_1 \times \mathscr{C}_2$ as $\boldsymbol{k}$-linear monoidal categories.

PROPOSITION 5.1. *Let H be a weak bialgebra over $\boldsymbol{k}$. If H is decomposable, then*

1. *the $\boldsymbol{k}$-linear monoidal categories ${}_H\mathbf{M}$ and ${}_H\mathbb{M}$ are decomposable.*
2. *the $\boldsymbol{k}$-linear monoidal categories $\mathbf{M}^H$ and $\mathbb{M}^H$ are decomposable.*

PROOF. Suppose that H is decomposed as $H = A \oplus B$ for some two weak bialgebras A and B.

(1) Any left H-module X is decomposed as $X = (1_A \cdot X) \oplus (1_B \cdot X)$. This decomposition gives rise to identical equivalences ${}_H\mathbf{M} \simeq {}_A\mathbf{M} \times {}_B\mathbf{M}$ and ${}_H\mathbb{M} \simeq {}_A\mathbb{M} \times {}_B\mathbb{M}$ as $\boldsymbol{k}$-linear monoidal categories.

(2) Any right H-comodule (X, ρ) is decomposed as

$$(X, \rho) = \big((\varepsilon_A \circ \pi_A) \cdot X,\ (\mathrm{id} \otimes \pi_A) \circ \rho|_{(\varepsilon_A \circ \pi_A) \cdot X}\big) \oplus \big((\varepsilon_B \circ \pi_B) \cdot X,\ (\mathrm{id} \otimes \pi_B) \circ \rho|_{(\varepsilon_B \circ \pi_B) \cdot X}\big),$$

where $\pi_A : H \longrightarrow A$, $\pi_B : H \longrightarrow B$ are the projections associated to the direct sum decomposition $H = A \oplus B$. This decomposition gives rise to identical equivalences $\mathbf{M}^H \simeq \mathbf{M}^A \times \mathbf{M}^B$ and $\mathbb{M}^H \simeq \mathbb{M}^A \times \mathbb{M}^B$ as $\boldsymbol{k}$-linear monoidal categories. □

The converse of the above proposition is also true. Namely, we have:

THEOREM 5.2. *Let H be a finite-dimensional weak bialgebra over $\boldsymbol{k}$. Then H is indecomposable as a weak bialgebra if and only if the module category ${}_H\mathbb{M}$ is indecomposable as a $\boldsymbol{k}$-linear monoidal category.*

The theorem can be proved by using the Tannaka-Krein reconstruction theorem of a bialgebra established by Ulbrich [**25**] and Schauenburg [**21**, Theorem 5.4] and by using a reconstruction theorem of a weak bialgebra map, which is a generalization of the reconstruction theorem of a bialgebra map [**13**, Theorem 2.2], [**20**, Theorem 2.2]. See [**26**] for detail of the proof of Theorem 5.2.

Let us consider a generalization of direct sum construction for weak bialgebras. For this we use the concept of weak bicomodule algebras which is introduced by Bhöm [**4**, p.4689].

DEFINITION 5.3. Let H, K be two weak bialgebras over $\boldsymbol{k}$, and A be an algebra over $\boldsymbol{k}$.

(1) Let λ_A be a left K-coaction. The pair (A, λ_A) is called a *left weak K-comodule algebra* if the following two conditions are satisfied for all $a, b \in A$:

(LWCA1) $\lambda_A(ab) = \lambda_A(a)\lambda_A(b)$.

(LWCA2) $(1_K \otimes a)\lambda_A(1_A) = (\varepsilon_s \otimes \mathrm{id})(\lambda_A(a))$.

Under the condition (LWCA1) it is known that the following each condition is equivalent to (LWCA2).

(i) $(\mathrm{id} \otimes \lambda)(\lambda(1_A)) = 1_{(1)} \otimes (1_A)_{(-1)}1_{(2)} \otimes (1_A)_{(0)}$.

(ii) $(\mathrm{id} \otimes \lambda)(\lambda(1_A)) = 1_{(1)} \otimes 1_{(2)}(1_A)_{(-1)} \otimes (1_A)_{(0)}$.

(iii) $\varepsilon_s(a_{(-1)}) \otimes a_{(0)} = (1_A)_{(-1)} \otimes a(1_A)_{(0)}$ for all $a \in A$.

Some more useful equivalent conditions for that A is a weak K-comodule algebra can be found in [**6**, Proposition 4.10]. We remark that a left weak K-comodule algebra (A, λ_A) is characterized as an algebra in the left comodule category ${}^K\mathbf{M}$.

(2) Let ρ_A be a right H-coaction. The pair (A, ρ_A) is called a *right weak H-comodule algebra* if the following two conditions are satisfied for all $a, b \in A$:

(RWCA1) $\rho_A(ab) = \rho_A(a)\rho_A(b)$.

(RWCA2) $\rho_A(1_A)(a \otimes 1_H) = (\mathrm{id} \otimes \varepsilon_t)(\rho_A(a))$.

(3) Let (A, λ_A) and (A, ρ_A) be a left weak K-comodule algebra and a right weak H-comodule algebra, respectively. The triple (A, λ_A, ρ_A) is called a *weak (K, H)-comodule algebra* if

- $(\lambda_A \otimes \mathrm{id}_H) \circ \rho_A = (\mathrm{id}_K \otimes \rho_A) \circ \lambda_A$,
- $(\lambda_A(1) \otimes 1)(1 \otimes \rho_A(1)) = (1 \otimes \rho_A(1))(\lambda_A(1) \otimes 1) = (\lambda_A \otimes \mathrm{id}_H)(\rho_A(1)) = (\mathrm{id}_K \otimes \rho_A)(\lambda_A(1))$.

LEMMA 5.4. *Let H be a weak bialgebra over $\boldsymbol{k}$, and (A, λ_A) be a left weak H-comodule algebra. Then*

(1) *For a left H-module X and a left A-module M, a left A-action on $X \otimes M$ is defined by*

$$a \cdot (x \otimes m) = a_{(-1)} \cdot x \otimes a_{(0)} \cdot m$$

for all $a \in A$, $x \in X$, $m \in M$. Then $X \circledast M := \lambda_A(1_A) \cdot (X \otimes M)$ is a left A-module.

(2) *For a left H-module map $f : X \longrightarrow Y$ and a left A-module map $\varphi : M \longrightarrow N$, the $\boldsymbol{k}$-linear map $f \otimes \varphi$ induces a left A-module map $f \circledast \varphi : X \circledast M \longrightarrow Y \circledast N$.*

(3) *The correspondences of (1) and (2) give rise to a $\boldsymbol{k}$-linear covariant functor $\circledast : {}_H\mathbf{M} \times {}_A\mathbf{M} \longrightarrow {}_A\mathbf{M}$, and it is a left action of ${}_H\mathbf{M}$ on ${}_A\mathbf{M}$ in the following sense: For $X, Y \in {}_H\mathbf{M}$ and $M \in {}_A\mathbf{M}$, there are natural left A-module isomorphisms*

$$X \circledast (Y \circledast M) \cong (X \circledast Y) \circledast M, \qquad H_t \circledast M \cong M.$$

where $X \circledast Y$ stands for the tensor product of X and Y in ${}_H\mathbf{M}$.

(4) *Let B be an algebra over $\boldsymbol{k}$. In (1) if M is an (A, B)-module, then so is $X \circledast M$, and in (2) if $\varphi : M \longrightarrow N$ is an (A, B)-module map, then so is $f \circledast \varphi$, too. Therefore, we have a covariant functor $\circledast : {}_H\mathbf{M} \times {}_A\mathbf{M}_B \longrightarrow {}_A\mathbf{M}_B$. Furthermore, for an (A, B)-module M the natural isomorphisms $X \circledast (Y \circledast M) \cong (X \circledast Y) \circledast M$, $H_t \circledast M \cong M$ in (3) are (A, B)-module isomorphisms.*

PROOF. Parts (1) and (2) are immediately verified.

(3) It is obvious that the usual $\boldsymbol{k}$-linear isomorphism

$$a_{X,Y,M} : (X \otimes Y) \otimes M \longrightarrow X \otimes (Y \otimes M), \qquad (x \otimes y) \otimes m \longmapsto x \otimes (y \otimes m)$$

induces a $\boldsymbol{k}$-linear isomorphism $(X \circledast Y) \circledast M \longrightarrow X \circledast (Y \circledast M)$. By $(\Delta \otimes \mathrm{id}_A) \circ \lambda_A = (\mathrm{id} \otimes \lambda_A) \circ \lambda_A$ the isomorphism $(X \circledast Y) \circledast M \longrightarrow X \circledast (Y \circledast M)$ is a left A-module map. We denote it by the same symbol $a_{X,Y,M}$.

Consider the $\boldsymbol{k}$-linear map $l_M : H_t \otimes M \longrightarrow M$ defined by

$$l_M(z \otimes m) = \varepsilon(z)m$$

for all $z \in H_t$, $m \in M$. We denote the restriction of l_M to $H_t \circledast M$ by l_M, again. Then the map $l_M : H_t \circledast M \longrightarrow M$ is a left A-module map, and it is natural on M. Moreover, it is an isomorphism since the inverse of l_M is given by the map $l_M^\dagger : M \longrightarrow H_t \circledast M$ defined by

$$l_M^\dagger(m) = (1_A)_{(-1)} \cdot 1_H \otimes (1_A)_{(0)} \cdot m = \varepsilon_t((1_A)_{(-1)}) \otimes (1_A)_{(0)} \cdot m$$

for all $m \in M$.

(4) Consider the case where M is an (A, B)-module in (1), and define a right B-action on $X \otimes M$ by

$$(x \otimes m) \cdot b = x \otimes (m \cdot b)$$

for all $x \in X$, $m \in M$, $b \in B$. Equipped with this right action $X \circledast M$ is an (A, B)-module. It is easily verified that the rest conditions hold. □

The right comodule algebra version of the above lemma also holds. In this case, instead of (A, λ_A) a right weak H-comodule algebra (B, ρ_B) is used, and for a left H-module X and a left B-module M, a right B-action on $M \otimes X$ is defined by

$$b \cdot (m \otimes x) = b_{(0)} \cdot m \otimes b_{(1)} \cdot x$$

for all $b \in B$, $x \in X$, $m \in M$. Then $M \circledast X := \rho_B(1_B) \cdot (M \otimes X)$ is a left B-module. As a similar manner in Lemma 5.4 a $\boldsymbol{k}$-linear covariant functor $\circledast : {}_B\mathbf{M} \times {}_H\mathbf{M} \longrightarrow {}_B\mathbf{M}$ is obtained, and it is a right action of ${}_H\mathbf{M}$ on ${}_B\mathbf{M}$:

$$(M \circledast X) \circledast Y \cong M \circledast (X \circledast Y), \qquad M \circledast H_t \cong M$$

for $X, Y \in {}_H\mathbf{M}$ and $M \in {}_B\mathbf{M}$.

Given two weak bialgebras H, K, a weak (H,K)-comodule algebra $A = (A, \lambda_A, \rho_A)$ and a weak (K,H)-comodule algebra $B = (B, \lambda_B, \rho_B)$, we will make $L = H \oplus A \oplus B \oplus K$ be a weak bialgebra. By using the equations (i), (ii) in Definition 5.3 (1) the following proposition, which gives a generalization of the direct sum construction, can be verified.

PROPOSITION 5.5. *Let H, K be two weak bialgebras over $\boldsymbol{k}$, and $A = (A, \lambda_A, \rho_A)$, $B = (B, \lambda_B, \rho_B)$ be a weak (H,K)-comodule algebra and (K,H)-comodule algebra, respectively. Let $\Delta_{A,B} : H \longrightarrow A \otimes B$, $\Delta_{B,A} : K \longrightarrow B \otimes A$ be $\boldsymbol{k}$-linear maps preserving products. Set $L := H \oplus A \oplus B \oplus K$, and consider it as the algebra obtained by the direct product of H, A, B, K. Define $\boldsymbol{k}$-linear maps $\Delta_L : L \longrightarrow L \otimes L$ and $\varepsilon_L : L \longrightarrow \boldsymbol{k}$ by*

$$\begin{aligned} \Delta_L(x) &= \Delta_H(x) + \Delta_{A,B}(x), & \varepsilon_L(x) &= \varepsilon_H(x), \\ \Delta_L(a) &= \lambda_A(a) + \rho_A(a), & \varepsilon_L(a) &= 0, \\ \Delta_L(b) &= \lambda_B(b) + \rho_B(b), & \varepsilon_L(b) &= 0, \\ \Delta_L(y) &= \Delta_K(y) + \Delta_{B,A}(y), & \varepsilon_L(y) &= \varepsilon_K(y) \end{aligned}$$

for all $x \in H$, $a \in A$, $b \in B$, $y \in K$. Then $(L, \Delta_L, \varepsilon_L)$ is a weak bialgebra if and only if the 16 conditions, which consist of the following 8 conditions (1), ..., (8) and the more 8 conditions obtained from (1), ..., (8) by interchanging A, H with B, K, respectively, are satisfied.

(1) $(\Delta_{A,B} \otimes \mathrm{id}_H) \circ \Delta_H = (\mathrm{id}_A \otimes \rho_B) \circ \Delta_{A,B}$.
(2) $(\lambda_A \otimes \mathrm{id}_B) \circ \Delta_{A,B} = (\mathrm{id}_H \otimes \Delta_{A,B}) \circ \Delta_H$.
(3) $(\rho_A \otimes \mathrm{id}_B) \circ \Delta_{A,B} = (\mathrm{id}_A \otimes \lambda_B) \circ \Delta_{A,B}$.
(4) $(\Delta_{A,B}(1_H) \otimes 1_H)(1_A \otimes \rho_B(1_B)) = (1_A \otimes \rho_B(1_B))(\Delta_{A,B}(1_H) \otimes 1_H) = (\Delta_{A,B} \otimes \mathrm{id}_H)\big(\Delta_H(1_H)\big)$.
(5) $(\lambda_A(1_A) \otimes 1_B)(1_H \otimes \Delta_{A,B}(1_H)) = (1_H \otimes \Delta_{A,B}(1_H))(\lambda_A(1_A) \otimes 1_B) = (\lambda_A \otimes \mathrm{id}_B)\big(\Delta_{A,B}(1_H)\big)$.
(6) $(\rho_A(1_A) \otimes 1_B)(1_A \otimes \lambda_B(1_B)) = (1_A \otimes \lambda_B(1_B))(\rho_A(1_A) \otimes 1_B) = (\rho_A \otimes \mathrm{id}_B)\big(\Delta_{A,B}(1_H)\big)$.
(7) $(\Delta_{A,B} \otimes \mathrm{id}_A) \circ \lambda_A = (\mathrm{id}_A \otimes \Delta_{B,A}) \circ \rho_A$.
(8) $(\Delta_{A,B}(1_H) \otimes 1_A)(1_A \otimes \Delta_{B,A}(1_K)) = (1_A \otimes \Delta_{B,A}(1_K))(\Delta_{A,B}(1_H) \otimes 1_A) = (\Delta_{A,B} \otimes \mathrm{id}_A)\big(\lambda_A(1_A)\big)$.

In addition, assume that H, K are weak Hopf algebras. Then the weak bialgebra L also becomes a weak Hopf algebra with the following antipode S_L: For all $x \in H$, $a \in A$, $b \in B$, $y \in K$

$$S_L(x + a + b + y) = S_H(x) + S_K(y).$$

PROOF. The conditions (WH1) and (WH3) for L are satisfied by the assumption immediately. Coassociativity of Δ_L is equivalent to the conditions (1), (2), (3), (7) and the conditions obtained from them by interchanging A, H with B, K, respectively. The condition $(\varepsilon_L \otimes \mathrm{id}) \circ \Delta_L = \mathrm{id}_L = (\mathrm{id}_L \otimes \varepsilon_L) \circ \Delta_L$ is always satisfied. Thanks to the conditions (i), (ii) in Definition 5.3(1), we see also that the condition (WH2) is equivalent to the conditions (4), (5), (6), (8) and the conditions obtained from them by interchanging A, H with B, K, respectively. □

REMARK 5.6. The conditions (1), (2) are equivalent to what $\Delta_{A,B}$ is an (H,H)-comodule map.

Two maps $(\lambda_A)_t, (\rho_B)_s$ do not preserve products, but they satisfy the following formulas.

LEMMA 5.7. *Let H, K be two weak bialgebras over $\boldsymbol{k}$, and $A = (A, \lambda_A, \rho_A)$, $B = (B, \lambda_B, \rho_B)$ be a weak (H,K)-comodule algebra and a weak (K,H)-comodule algebra, respectively. Define $\boldsymbol{k}$-linear maps $(\lambda_A)_t : H \longrightarrow A$, $(\rho_A)_s : K \longrightarrow A$, $(\lambda_B)_t : K \longrightarrow B$, $(\rho_B)_s : H \longrightarrow B$ as follows: For all $x \in H$ and $y \in K$*

$$(\lambda_A)_t(x) = \varepsilon_H\big((1_A)^H_{(-1)}x\big)(1_A)^H_{(0)}, \tag{5.1}$$

$$(\rho_A)_s(y) = (1_A)^K_{(0)}\varepsilon_K\big(y(1_A)^K_{(1)}\big), \tag{5.2}$$

$$(\lambda_B)_t(y) = \varepsilon_K\big((1_B)^K_{(-1)}y\big)(1_B)^K_{(0)}, \tag{5.3}$$

$$(\rho_B)_s(x) = (1_B)^H_{(0)}\varepsilon_H\big(x(1_B)^H_{(1)}\big). \tag{5.4}$$

Then for all $x, x' \in H$ and $y, y' \in K$

$$(\lambda_A)_t(x)(\lambda_A)_t(x') = (\lambda_A)_t\Big(\big((\varepsilon_H)_t(x)\big)x'\Big), \quad (\rho_B)_s(x)(\rho_B)_s(x') = (\rho_B)_s\Big(x\big((\varepsilon_H)_s(x')\big)\Big),$$

$$(\lambda_B)_t(y)(\lambda_B)_t(y') = (\lambda_B)_t\Big(\big((\varepsilon_K)_t(y)\big)y'\Big), \quad (\rho_A)_s(y)(\rho_A)_s(y') = (\rho_A)_s\Big(y\big((\varepsilon_K)_s(y')\big)\Big).$$

PROOF. The first equation is derived as follows.

$$\begin{aligned}
&(\lambda_A)_t(x)(\lambda_A)_t(x') \\
&\quad = \varepsilon_H\big((1_A)^H_{(-1)}x\big)\varepsilon_H\big((1_A)^H_{[-1]}x'\big)(1_A)^H_{(0)}(1_A)^H_{[0]} \\
&\quad = \varepsilon_H\big((1_A)^H_{(-2)}x\big)\varepsilon_H\Big((\varepsilon_H)_s\big((1_A)^H_{(-1)}\big)x'\Big)(1_A)^H_{(0)} && \text{(Def. 5.3(1)(iii))} \\
&\quad = \varepsilon_H\big((1_A)^H_{(-2)}x\big)\varepsilon_H\big((1_A)^H_{(-1)}x'\big)(1_A)^H_{(0)} && \text{(Lemma 2.3)} \\
&\quad = \varepsilon_H\big((1_H)_{(1)}x\big)\varepsilon_H\big((1_A)^H_{(-1)}(1_H)_{(2)}x'\big)(1_A)^H_{(0)} && \text{(Def. 5.3(1)(i))} \\
&\quad = \varepsilon_H\Big((1_A)^H_{(-1)}\varepsilon_H\big((1_H)_{(1)}x\big)(1_H)_{(2)}x'\Big)(1_A)^H_{(0)} \\
&\quad = \varepsilon_H\Big((1_A)^H_{(-1)}(\varepsilon_H)_t(x)x'\Big)(1_A)^H_{(0)} \\
&\quad = (\lambda_A)_t\Big(\big((\varepsilon_H)_t(x)\big)x'\Big).
\end{aligned}$$

Other equations are derived in a similar way. □

LEMMA 5.8. *Under the same situation in Proposition 5.5, the target and source counital maps $(\varepsilon_L)_t$ and $(\varepsilon_L)_s$ of the weak bialgebra $L = H \oplus A \oplus B \oplus K$ are given by*

$$(\varepsilon_L)_t(x + a + b + y) = \big((\varepsilon_H)_t + (\lambda_A)_t\big)(x) + \big((\lambda_B)_t + (\varepsilon_K)_t\big)(y),$$

$$(\varepsilon_L)_s(x + a + b + y) = \big((\varepsilon_H)_s + (\rho_B)_s\big)(x) + \big((\rho_A)_s + (\varepsilon_K)_s\big)(y)$$

for all $x \in H$, $a \in A$, $b \in B$, $y \in K$, where $(\lambda_A)_t, (\rho_A)_s, (\lambda_B)_s, (\rho_B)_s$ are the same defined in Lemma 5.7.

THEOREM 5.9. *Let H, K be two weak bialgebras, and $A = (A, \lambda_A, \rho_A)$ and $B = (B, \lambda_B, \rho_B)$ be a weak (H,K)-comodule algebra and a weak (K,H)-comodule algebra, respectively. Let $\Delta_{A,B} : H \longrightarrow A \otimes B$, $\Delta_{B,A} : K \longrightarrow B \otimes A$ be an (H,H)-comodule map and a (K,K)-comodule map, respectively. Consider the $\boldsymbol{k}$-linear category $\begin{pmatrix} {}_H\mathsf{M} & {}_A\mathsf{M} \\ {}_B\mathsf{M} & {}_K\mathsf{M} \end{pmatrix}$ of the Cartesian product of $\boldsymbol{k}$-linear categories ${}_H\mathsf{M}, {}_A\mathsf{M}, {}_B\mathsf{M}$, ${}_K\mathsf{M}$. Then this $\boldsymbol{k}$-linear category becomes a $\boldsymbol{k}$-linear monoidal category equipped* *with the tensor product $\circledast$ defined as follows: For $X_1, X_2 \in {}_H\mathsf{M}$, $M_1, M_2 \in {}_A\mathsf{M}$, $N_1, N_2 \in {}_B\mathsf{M}$, $Y_1, Y_2 \in {}_K\mathsf{M}$*

$$\begin{pmatrix} X_1 & M_1 \\ N_1 & Y_1 \end{pmatrix} \circledast \begin{pmatrix} X_2 & M_2 \\ N_2 & Y_2 \end{pmatrix} = \begin{pmatrix} X_1 \circledast X_2 + M_1 \circledast N_2 & X_1 \circledast M_2 + M_1 \circledast Y_2 \\ N_1 \circledast X_2 + Y_1 \circledast N_2 & N_1 \circledast M_2 + Y_1 \circledast Y_2 \end{pmatrix},$$

where the symbol $+$ in the right-hand side means the direct product, and

$$M_1 \circledast N_2 := \Delta_{A,B}(1_H) \cdot (M_1 \otimes N_2),$$
$$N_1 \circledast M_2 := \Delta_{B,A}(1_K) \cdot (N_1 \otimes M_2).$$

The tensor product of morphisms is defined in the same way, and $I := \begin{pmatrix} H_t & O \\ O & K_t \end{pmatrix}$ is a unit object in the monoidal category.

Furthermore, for the weak bialgebra $L := H \oplus A \oplus B \oplus K$ defined in Proposition 5.5, we have ${}_L\mathsf{M} \simeq \begin{pmatrix} {}_H\mathsf{M} & {}_A\mathsf{M} \\ {}_B\mathsf{M} & {}_K\mathsf{M} \end{pmatrix}$ as $\boldsymbol{k}$-linear monoidal categories.

PROOF. The monoidal category structure of $\begin{pmatrix} {}_H\mathsf{M} & {}_A\mathsf{M} \\ {}_B\mathsf{M} & {}_K\mathsf{M} \end{pmatrix}$ is given as follows.

(i) For three objects V_1, V_2, V_3, there is a natural isomorphism $\boldsymbol{a}_{V_1,V_2,V_3} : (V_1 \circledast V_2) \circledast V_3 \longrightarrow V_1 \circledast (V_2 \circledast V_3)$ induced from shifting parentheses by component-wise since $\circledast$ in $\begin{pmatrix} {}_H\mathsf{M} & {}_A\mathsf{M} \\ {}_B\mathsf{M} & {}_K\mathsf{M} \end{pmatrix}$ is defined by the same rule with matrix product. The isomorphisms $\{\boldsymbol{a}_{V_1,V_2,V_3}\}$ obviously satisfy the pentagon identity.

(ii) For an object $V := \begin{pmatrix} X & M \\ N & Y \end{pmatrix}$, we have

$$I \circledast V = \begin{pmatrix} H_t \circledast X & H_t \circledast M \\ K_t \circledast N & K_t \circledast Y \end{pmatrix}, \quad V \circledast I = \begin{pmatrix} X \circledast H_t & M \circledast K_t \\ N \circledast H_t & Y \circledast K_t \end{pmatrix}.$$

Let $l_X : H_t \circledast X \longrightarrow X$, $l_M : H_t \circledast M \longrightarrow M$, $l_N : K_t \circledast N \longrightarrow N$, $l_Y : K_t \circledast Y \longrightarrow Y$ be the isomorphisms of left annihilation and $r_X : X \circledast H_t \longrightarrow X$, $r_M : M \circledast K_t \longrightarrow M$, $r_N : N \circledast H_t \longrightarrow N$, $r_Y : Y \circledast K_t \longrightarrow Y$ be the isomorphisms of right annihilation. Then

$$\boldsymbol{l}_V := \begin{pmatrix} l_X & l_M \\ l_N & l_Y \end{pmatrix} : I \circledast V \longrightarrow V, \quad \boldsymbol{r}_V := \begin{pmatrix} r_X & r_M \\ r_N & r_Y \end{pmatrix} : V \circledast I \longrightarrow V$$

are natural isomorphisms. For two objects V_1, V_2, we have $(\mathrm{id}_{V_1} \circledast \boldsymbol{l}_{V_2}) \circ \boldsymbol{a}_{V_1,I,V_2} = (\boldsymbol{r}_{V_1} \circledast \mathrm{id}_{V_2})$ by using the triangle identity in component-wise. Thus, $\begin{pmatrix} {}_H\mathsf{M} & {}_A\mathsf{M} \\ {}_B\mathsf{M} & {}_K\mathsf{M} \end{pmatrix}$ is a monoidal category.

We will show that this monoidal category is equivalent to ${}_L\mathsf{M}$. Let V be a left L-module. Since the identity element of L is $1_L = 1_H + 1_A + 1_B + 1_K$, the left L-module V is decomposed as $V = (1_H \cdot V) \oplus (1_A \cdot V) \oplus (1_B \cdot V) \oplus (1_K \cdot V)$. Define a covariant functor $F : {}_L\mathsf{M} \longrightarrow \begin{pmatrix} {}_H\mathsf{M} & {}_A\mathsf{M} \\ {}_B\mathsf{M} & {}_K\mathsf{M} \end{pmatrix}$ as follows:

- $F(V) = \begin{pmatrix} 1_H \cdot V & 1_A \cdot V \\ 1_B \cdot V & 1_K \cdot V \end{pmatrix}$ for an object V,
- $F(f) = \begin{pmatrix} f|_{1_H \cdot U} & f|_{1_A \cdot U} \\ f|_{1_B \cdot U} & f|_{1_K \cdot U} \end{pmatrix}$ for a morphism $f : U \longrightarrow V$.

The functor F is $\boldsymbol{k}$-linear and gives an equivalence of $\boldsymbol{k}$-linear categories. Moreover, F preserves tensor product. In fact, let U, V be two left L-modules, and consider the left L-module $U \circledast V$. Setting $X_1 := 1_H \cdot U$, $M_1 := 1_A \cdot U$, $N_1 := 1_B \cdot U$, $Y_1 := 1_K \cdot U$ and $X_2 := 1_H \cdot V$, $M_2 := 1_A \cdot V$, $N_2 := 1_B \cdot V$, $Y_2 := 1_K \cdot V$, we have

$$\begin{aligned} U \circledast V = X_1 \circledast X_2 + M_1 \circledast N_2 + X_1 \circledast M_2 + M_1 \circledast Y_2 \\ + N_1 \circledast X_2 + Y_1 \circledast N_2 + N_1 \circledast M_2 + Y_1 \circledast Y_2. \end{aligned}$$

This implies that $F(U \circledast V) = F(U) \circledast F(V)$.

Next, we compute the image of the unit object L_t under F. Set

$$(H \oplus A)_t := \operatorname{Im}\big((\varepsilon_H)_t + (\lambda_A)_t\big), \quad (K \oplus B)_t := \operatorname{Im}\big((\lambda_B)_t + (\varepsilon_K)_t\big).$$

Then $L_t = (H \oplus A)_t \oplus (K \oplus B)_t$ by Lemma 5.8. It follows that $1_H \cdot L_t = (H \oplus A)_t$, $1_A \cdot L_t = \mathrm{O}$, $1_B \cdot L_t = \mathrm{O}$, $1_K \cdot L_t = (K \oplus B)_t$. Therefore, $F(L_t) = \begin{pmatrix} (H \oplus A)_t & \mathrm{O} \\ \mathrm{O} & (K \oplus B)_t \end{pmatrix}$.

We show that $(H \oplus A)_t \cong H_t$ and $(K \oplus B)_t \cong K_t$ as left H-modules and left K-modules, respectively. The canonical projection $p_H : H \oplus A \longrightarrow H$ induces a left H-module map $\overline{p_H} := p_H|_{(H\oplus A)_t} : (H \oplus A)_t \longrightarrow H_t$. Define a map $q : H_t \longrightarrow (H \oplus A)_t$ by

$$q(z) = \big((\varepsilon_H)_t + (\lambda_A)_t\big)(z)$$

for all $z \in H_t$. It can be easily verified that $\overline{p_H}$ and q are inverse each other. Thus, we have $F(L_t) \cong \begin{pmatrix} H_t & \mathrm{O} \\ \mathrm{O} & K_t \end{pmatrix} = I$. This isomorphism, denoted by $\omega^F : I \longrightarrow F(L_t)$, is given by $\omega^F = \begin{pmatrix} q_H & 0 \\ 0 & q_K \end{pmatrix}$, where

$$\begin{aligned} q_H &: H_t \longrightarrow (H \oplus A)_t, \quad q_H(z) = \big((\varepsilon_H)_t + (\lambda_A)_t\big)(z), \\ q_K &: K_t \longrightarrow (K \oplus B)_t, \quad q_K(w) = \big((\varepsilon_K)_t + (\lambda_B)_t\big)(w) \end{aligned}$$

for all $z \in H_t$ and $w \in K_t$. One can show that the triple $(F, \mathrm{Id}, \omega^F) : {}_L\mathsf{M} \longrightarrow \begin{pmatrix} {}_H\mathsf{M} & {}_A\mathsf{M} \\ {}_B\mathsf{M} & {}_K\mathsf{M} \end{pmatrix}$ is a monoidal functor, which gives an equivalence of $\boldsymbol{k}$-linear monoidal categories. □

Problem 5.10. Find the condition for the weak bialgebra L in Theorem 5.9 such that L is indecomposable.

We demonstrate a few examples based on Proposition 5.5.

Lemma 5.11. *Let us consider the* 1*-dimensional Hopf algebra* $H = \boldsymbol{k}$. *We regard* $A := \boldsymbol{k}$ *as an algebra by the product of* $\boldsymbol{k}$.

(1) *There is a unique weak* $(\boldsymbol{k}, \boldsymbol{k})$*-comodule algebra structure for* A, *and it is given by* $(A, \mathrm{id}_{\boldsymbol{k}}, \mathrm{id}_{\boldsymbol{k}})$ *under the identification* $\boldsymbol{k} \otimes \boldsymbol{k} = \boldsymbol{k}$.
(2) *Let* $B := \boldsymbol{k}$ *be the same algebra with* A, *and* (A, λ_A, ρ_A), (B, λ_B, ρ_B) *be two weak* $(\boldsymbol{k}, \boldsymbol{k})$*-comodule algebras. Then a* $\boldsymbol{k}$*-linear map* $\Delta_{A,B} : \boldsymbol{k} = H \longrightarrow A \otimes B = \boldsymbol{k} \otimes \boldsymbol{k} \cong \boldsymbol{k}$ *preserves products and satisfies the conditions* $(1), (2), (4), (5), (6)$ *in Proposition* 5.5 *if and only if* $(\lambda_A, \rho_A, \lambda_B, \rho_B, \Delta_{A,B}) = (\mathrm{id}, \mathrm{id}, \mathrm{id}, \mathrm{id}, \mathrm{id})$.

PROOF. (1) Let $A = (A, \lambda_A, \rho_A)$ be a weak $(\boldsymbol{k}, \boldsymbol{k})$-comodule algebra. By the definition of left comodule, $(\varepsilon \otimes \mathrm{id}_A) \circ \lambda_A = \mathrm{id}_A$ holds. This implies that $\lambda_A \neq 0$. Similarly, $\rho_A \neq 0$. We view λ_A as the map $\lambda_A : \boldsymbol{k} = A \longrightarrow H \otimes A = \boldsymbol{k} \otimes \boldsymbol{k} \cong \boldsymbol{k}$. Since $\lambda_A(1) = \lambda_A(1)\lambda_A(1)$, we have $\lambda_A(1) = 0$ or $\lambda_A(1) = 1$.

If $\lambda_A(1) = 0$, then $\lambda_A(a) = \lambda_A(a \cdot 1) = \lambda_A(a)\lambda_A(1) = 0$ for all $a \in \boldsymbol{k}$. This implies that $\lambda_A = 0$, that is a contradiction. Thus $\lambda_A(1) = 1$. By linearity of λ_A, in this case $\lambda_A(a) = \lambda_A(a \cdot 1) = a\lambda_A(1) = a \cdot 1 = a$ for all $a \in A = \boldsymbol{k}$. It follows that $\lambda_A = \mathrm{id}_{\boldsymbol{k}}$. The condition (LWCA2) is satisfied since it can be expressed as $a = \varepsilon\big(\lambda_A(a)\big) = \lambda_A(a)$. The condition (LWCA1) is also satisfied.

Similarly, viewing ρ_A as $\rho_A : \boldsymbol{k} = A \longrightarrow A \otimes H = \boldsymbol{k} \otimes \boldsymbol{k} \cong \boldsymbol{k}$, we see that $\rho_A = \mathrm{id}_{\boldsymbol{k}}$.

Conversely, when $\lambda_A = \rho_A = \mathrm{id}$, it can be immediately shown that $(A, \mathrm{id}, \mathrm{id})$ is a weak $(\boldsymbol{k}, \boldsymbol{k})$-comodule algebra.

(2) Since $\Delta_{A,B}$ preserves products, it can be shown that $\Delta_{A,B} = 0$ or $\Delta_{A,B} = \mathrm{id}$ by a similar consideration of the proof of (1). If $\Delta_{A,B} = 0$, then all the conditions (1),(2),(4),(5) in Proposition 5.5 are satisfied. The condition (6) in Proposition 5.5 is satisfied if and only if $\rho_A = 0$ or $\lambda_B = 0$. However, it does not occur by (1). So, $\Delta_{A,B} = \mathrm{id}$ is required. The condition (1) in Proposition 5.5 requires $\rho_B = \mathrm{id}$ and $\lambda_A = \mathrm{id}$. When $\rho_B = \mathrm{id}$ and $\lambda_A = \mathrm{id}$, the conditions (4) and (5) are satisfied. Similarly, the condition (6) in Proposition 5.5 requires $\rho_A = \mathrm{id}$ and $\lambda_B = \mathrm{id}$. □

EXAMPLE 5.12. Let $H = K = \boldsymbol{k}$ and set $A = \boldsymbol{k},\ B = 0$. Then $\Delta_{A,B} = 0$ and $\Delta_{B,A} = 0$. It follows that there is a unique 3-dimensional weak bialgebra obtained by the construction in Proposition 5.5, which is given in the form $\begin{pmatrix} \boldsymbol{k} & \boldsymbol{k} \\ 0 & \boldsymbol{k} \end{pmatrix}$. This weak bialgebra is isomorphic to $L := \boldsymbol{k}e_1 \oplus \boldsymbol{k}e_2 \oplus \boldsymbol{k}e_3$ whose algebra and coalgebra structures are given by

$$\begin{aligned} e_i e_j &= \delta_{ij} e_i \quad \text{for all } i, j = 1, 2, 3, \\ \Delta_L(xe_1 + ae_2 + ye_3) &= xe_1 \otimes e_1 + a(e_2 \otimes e_3 + e_1 \otimes e_2) + ye_3 \otimes e_3, \\ \varepsilon_L(xe_1 + ae_2 + ye_3) &= x + y \end{aligned}$$

for all $x, a, y \in \boldsymbol{k}$. The weak bialgebra L is indecomposable as a weak bialgebra, and isomorphic to the 11th 3-dimensional weak bialgebra in the list of [**7**, Proposition 4.5]. This can be verified by using the basis $e_1 + e_2 + e_3, e_1 + e_3, e_1$. Thus, L is isomorphic to the dual of the weak bialgebra spanned by the morphisms of the interval category **2**. (See Example 3.7.)

EXAMPLE 5.13. Let us consider the case where $H = K = \boldsymbol{k}$ and $A = B = \boldsymbol{k}$ in Proposition 5.5. Let $A = (A, \lambda_A, \rho_A),\ B = (B, \lambda_B, \rho_B)$ be a weak (H, K)-comodule algebra and a weak (K, H)-comodule algebra, respectively. By Lemma 5.11(2), there is a unique $(\lambda_A, \rho_A, \lambda_B, \rho_B, \Delta_{A,B}, \Delta_{B,A})$ such that it satisfies all conditions in Proposition 5.5, and it is given by $(\mathrm{id}, \mathrm{id}, \mathrm{id}, \mathrm{id}, \mathrm{id}, \mathrm{id})$.

The algebra structure of the weak bialgebra $L = H \oplus A \oplus B \oplus K$ defined as in the proposition is the direct product of four $\boldsymbol{k}$s. Let $\{e_1, e_2, e_3, e_4\}$ be the standard basis of L, then $e_i e_j = \delta_{ij} e_i$ $(i, j = 1, 2, 3, 4)$ and the coalgebra structure of L is given as follows:

$$\begin{aligned}\Delta_L(xe_1+ae_2+be_3+ye_4) &= x(e_1\otimes e_1+e_2\otimes e_3)+a(e_1\otimes e_2+e_2\otimes e_4)\\ &\quad +b(e_4\otimes e_3+e_3\otimes e_1)+y(e_4\otimes e_4+e_3\otimes e_2),\\ \varepsilon_L(xe_1+ae_2+be_3+ye_4) &= x+y\end{aligned}$$

for all $x,a,b,y\in \boldsymbol{k}$. Since $\varepsilon_L(1_L)=2\neq 1$, it follows that the weak bialgebra L is not a bialgebra. Furthermore, L has an antipode S_L, which is given by

$$S_L(xe_1+ae_2+be_3+ye_4)=xe_1+ye_4$$

for all $x,a,b,y\in \boldsymbol{k}$. The weak Hopf algebra L is indecomposable.

REMARK 5.14. The above L can not be constructed from a groupoid. Especially, it is not isomorphic to the qubit groupoid algebra since $G(L)=\{\ e_1+ae_2+a^{-1}e_3+e_4 \mid a\in \boldsymbol{k}^\times\ \}\cong \boldsymbol{k}^\times$ and the order of $G(L)$ does not coincide with $\dim L=4$. In fact, L is isomorphic to the dual of the qubit groupoid algebra.

Acknowledgments

I would like to thank Professor Akira Masuoka for discussing and suggesting on the direct sum construction, and thank Professor Kenichi Shimizu for suggesting and explaining on interpretation of indecomposability by a categorical language and a generalization of the direct sum construction. I would like to thank the organizers, Professors Nicolás Andruskiewitsch, Gongxiang Liu, Susan Montgomery, Yinhuo Zhang, for inviting me to the International Workshop on Hopf algebras and Tensor Categories at Nanjing in China on September 9–13, 2019. Much thanks are also graduate students and young researchers in Nanjing University for kind hospitality during my stay in China. Helpful comments from Professors Nicolás Andruskiewitsch and Gabriella Böhm in the workshop provide me with inspirations for construction of examples of indecomposable weak Hopf algebras in this paper. Finally, I would like to thank the referees for careful reading and for many helpful comments and suggestions to improve the previous manuscript.

References

[1] Eiichi Abe, *Hopf algebras*, Cambridge Tracts in Mathematics, vol. 74, Cambridge University Press, Cambridge-New York, 1980. Translated from the Japanese by Hisae Kinoshita and Hiroko Tanaka. MR594432

[2] Nicolás Andruskiewitsch, *On the quiver-theoretical quantum Yang-Baxter equation*, Selecta Math. (N.S.) **11** (2005), no. 2, 203–246, DOI 10.1007/s00029-005-0011-z. With an appendix by Mitsuhiro Takeuchi. MR2183847

[3] Jean Bénabou, *Introduction to bicategories*, Reports of the Midwest Category Seminar, Springer, Berlin, 1967, pp. 1–77. MR0220789

[4] Gabriella Böhm, *Doi-Hopf modules over weak Hopf algebras*, Comm. Algebra **28** (2000), no. 10, 4687–4698, DOI 10.1080/00927870008827113. MR1779865

[5] Gabriella Böhm, Florian Nill, and Kornél Szlachányi, *Weak Hopf algebras. I. Integral theory and C^*-structure*, J. Algebra **221** (1999), no. 2, 385–438, DOI 10.1006/jabr.1999.7984. MR1726707

[6] S. Caenepeel and E. De Groot, *Modules over weak entwining structures*, New trends in Hopf algebra theory (La Falda, 1999), Contemp. Math., vol. 267, Amer. Math. Soc., Providence, RI, 2000, pp. 31–54, DOI 10.1090/conm/267/04263. MR1800705

[7] Z. Chebel and A. Makhlouf, *Kaplansky's type constructions for weak bialgebras and weak Hopf algebras*, J. Gen. Lie Theory Appl. **9** (2015), no. S1, Art. ID S1-008, 9, DOI 10.4172/1736-4337.s1-008. MR3637852

[8] Takahiro Hayashi, *Face algebras. I. A generalization of quantum group theory*, J. Math. Soc. Japan **50** (1998), no. 2, 293–315, DOI 10.2969/jmsj/05020293. MR1613136
[9] A. Ibort and M. A. Rodríguez, *On the structure of finite groupoids and their representations*, Symmetry **11** (2019), Paper No.414, 28pp.
[10] Irving Kaplansky, *Bialgebras*, Lecture Notes in Mathematics, Department of Mathematics, University of Chicago, Chicago, Ill., 1975. MR0435126
[11] Christian Kassel, *Quantum groups*, Graduate Texts in Mathematics, vol. 155, Springer-Verlag, New York, 1995, DOI 10.1007/978-1-4612-0783-2. MR1321145
[12] Saunders Mac Lane, *Categories for the working mathematician*, 2nd ed., Graduate Texts in Mathematics, vol. 5, Springer-Verlag, New York, 1998. MR1712872
[13] Shahn Majid, *Tannaka-Kreĭn theorem for quasi-Hopf algebras and other results*, Deformation theory and quantum groups with applications to mathematical physics (Amherst, MA, 1990), Contemp. Math., vol. 134, Amer. Math. Soc., Providence, RI, 1992, pp. 219–232, DOI 10.1090/conm/134/1187289. MR1187289
[14] Susan Montgomery, *Hopf algebras and their actions on rings*, CBMS Regional Conference Series in Mathematics, vol. 82, Published for the Conference Board of the Mathematical Sciences, Washington, DC; by the American Mathematical Society, Providence, RI, 1993, DOI 10.1090/cbms/082. MR1243637
[15] Dmitri Nikshych, *On the structure of weak Hopf algebras*, Adv. Math. **170** (2002), no. 2, 257–286, DOI 10.1006/aima.2002.2081. MR1932332
[16] Dmitri Nikshych and Leonid Vainerman, *Finite quantum groupoids and their applications*, New directions in Hopf algebras, Math. Sci. Res. Inst. Publ., vol. 43, Cambridge Univ. Press, Cambridge, 2002, pp. 211–262. MR1913440
[17] Dmitri Nikshych, Vladimir Turaev, and Leonid Vainerman, *Invariants of knots and 3-manifolds from quantum groupoids*, Proceedings of the Pacific Institute for the Mathematical Sciences Workshop "Invariants of Three-Manifolds" (Calgary, AB, 1999), Topology Appl. **127** (2003), no. 1-2, 91–123, DOI 10.1016/S0166-8641(02)00055-X. MR1953322
[18] F. Nill, *Axioms for weak bialgebras*, `arXiv:math.9805104v1`, 1998.
[19] David E. Radford, *On the quasitriangular structures of a semisimple Hopf algebra*, J. Algebra **141** (1991), no. 2, 354–358, DOI 10.1016/0021-8693(91)90236-2. MR1125700
[20] P. Saracco, *Coquasi-bialgebras with preantipode and rigid monoidal categories*, `arXiv:1611.06819v3`, 2016.
[21] Peter Schauenburg, *Hopf bi-Galois extensions*, Comm. Algebra **24** (1996), no. 12, 3797–3825, DOI 10.1080/00927879608825788. MR1408508
[22] Peter Schauenburg, *Weak Hopf algebras and quantum groupoids*, Noncommutative geometry and quantum groups (Warsaw, 2001), Banach Center Publ., vol. 61, Polish Acad. Sci. Inst. Math., Warsaw, 2003, pp. 171–188, DOI 10.4064/bc61-0-12. MR2024429
[23] Andrzej Skowroński and Kunio Yamagata, *Frobenius algebras. I*, EMS Textbooks in Mathematics, European Mathematical Society (EMS), Zürich, 2011. Basic representation theory, DOI 10.4171/102. MR2894798
[24] Mitsuhiro Takeuchi, *Groups of algebras over $A \otimes \overline{A}$*, J. Math. Soc. Japan **29** (1977), no. 3, 459–492, DOI 10.2969/jmsj/02930459. MR506407
[25] K.-H. Ulbrich, *On Hopf algebras and rigid monoidal categories*, Israel J. Math. **72** (1990), no. 1-2, 252–256, DOI 10.1007/BF02764622. Hopf algebras. MR1098991
[26] M. Wakui, *Reconstruction of weak bialgebra maps and its applications*, `arXiv:2002.12568v1`, 2020.
[27] Xiaohui Zhang, Xiaofan Zhao, and Shuanhong Wang, *Sovereign and ribbon weak Hopf algebras*, Kodai Math. J. **38** (2015), no. 2, 451–469, DOI 10.2996/kmj/1436403900. MR3368075

Department of Mathematics, Faculty of Engineering Science, Kansai University, Suita-shi, Osaka 564-8680, Japan

Email address: `wakui@kansai-u.ac.jp`

Contemporary Mathematics
Volume **771**, 2021
https://doi.org/10.1090/conm/771/15518

Quasi-Hopf algebras of Cartan type and small quasi-quantum groups

Yuping Yang and Yinhuo Zhang

ABSTRACT. In this paper, we construct a class of finite-dimensional quasi-Hopf algebras from liftings of pointed Hopf algebras. These quasi-Hopf algebras are presented by generators and relations. Several proper classes of these quasi-Hopf algebras are proved genuine, i.e. not twist equivalent to Hopf algebras. As special cases, we introduce the small quasi-quantum groups which show to be the natural generalizations of the small quantum groups.

1. Introduction

Quasi-Hopf algebras were introduced by Drinfeld [**5**] as linear algebraic counterparts of tensor categories with quasi-fiber functors. In the past two decades, the studies on quasi-Hopf algebras were mainly focused on some interesting classes. In [**6**, **7**], Etingof and Gelaki initiated their study on elementary quasi-Hopf algebras, and a series of classification results about graded elementary quasi-Hopf algebras over cyclic groups of prime order were obtained. More importantly, a novel method for constructing graded elementary quasi-Hopf algebras from known graded elementary Hopf algebras was invented. Here by graded we mean radically graded. In [**8**, **12**], they also studied graded elementary quasi-Hopf algebras over general cyclic groups and their liftings. Along the same vein, Angiono [**2**] classified the finite-dimensional elementary quasi-Hopf algebras over cyclic groups with dimensions not divisible by small prime divisors. The first author of the present paper and his cooperators [**13**, **14**] classified finite-dimensional coradically graded pointed coquasi-Hopf algebras of diagonal type.

As far as we know, except for elementary quasi-Hopf algebras or dually pointed coquasi-Hopf algebras, there is not any classification result on finite-dimensional quasi-Hopf algebras until now. The main reason, to the authors, is the lack of effective techniques to construct enough quasi-Hopf algebras. So it would be valuable if the method for constructing elementary quasi-Hopf algebras given in [**6**, **7**] can be developed to a more general setting. In this respect, Angiono, Galindo and Pereira tried first in [**4**], and they proved that the de-equivariantization of a finite-dimensional Hopf algebra H by certain finite group can be realized as the comodule category of some coquasi-Hopf algebra. Moreover, the coquasi-Hopf algebra can be

2020 *Mathematics Subject Classification.* Primary 19A22, 18D10, 16G20.
Key words and phrases. Quasi-Hopf algebra, small quantum group, Cartan matrix.
Supported by NSFC 11701468.

constructed from H in a systematical way. Dually, quasi-Hopf algebras can also be constructed in a parallel manner. Therefore, [**4**] provides us an effective method to construct quasi-Hopf algebras from general Hopf algebras which are not necessarily graded elementary Hopf algebras.

In the literature, a quasi-Hopf algebra (or dually a coquasi-Hopf algebra) is called **genuine** if it is not twist equivalent to a Hopf algebra. It is well-known that if a quasi-Hopf algebra is not genuine, then its module category is tensor equivalent to the module category of a Hopf algebra, hence will not provide new phenomenon in the framework of tensor categories. So it is important to consider the following problem.

Problem A: Determine whether a given (co)quasi-Hopf algebra is genuine or not?

Note that usually we don't need to think about Problem A when considering graded pointed coquasi-Hopf algebras, or dually graded elementary quasi-Hopf algebras. The reason can be explained as follows. Let H be a graded pointed coquasi-Hopf algebra with associator Φ, and G the group of group-like elements. Then the associator Φ is actually a 3-cocycle on G, and H is genuine if and only if Φ is not a 3-coboundary on G. The genuineness of a graded elementary quasi-Hopf algebra can also be checked in the same manner. Based on this fact, the construction given in [**4**] can be well applied to the classification of graded pointed finite tensor categories in [**3**]. However, if the coquasi-Hopf algebra is not a graded pointed coquasi-Hopf algebra, then Problem A is widely open and seems to be very difficult in general. In this paper, we try to study this problem on a class of explicit quasi-Hopf algebras constructed from the liftings of pointed Hopf algebras [**1**].

Now let us introduce more about the present paper. Our strategy of constructing quasi-Hopf algebras can be described as follows. Let $\mathbb{G} = \langle \mathbb{g}_1 \rangle \times \cdots \times \langle \mathbb{g}_n \rangle$ be a finite abelian group such that $|\mathbb{g}_i| = \mathbb{m}_i = m_i^2, 1 \leq i \leq n$, G the subgroup of $\mathbb{G}$ generated by $\{\mathbb{g}_1^{m_1}, \cdots, \mathbb{g}_n^{m_n}\}$. Let $u(\mathfrak{D}, \lambda, \mu)$ be a lifting of pointed Hopf algebra [**1**] generated by the set $\mathbb{G}$ of group-like elements and the set $\{X_i | 1 \leq i \leq \theta\}$ of skew-primitive elements. Let J be a twist of $u(\mathfrak{D}, \lambda, \mu)$ associated to a 2-cochain of $\mathbb{G}$, and $u(\mathfrak{D}, \lambda, \mu, \Phi_J)$ the subalgebra of $u(\mathfrak{D}, \lambda, \mu)$ generated by G and $\{X_i | 1 \leq i \leq \theta\}$. Then we determine the conditions of J such that $u(\mathfrak{D}, \lambda, \mu, \Phi_J)$ is a quasi-Hopf subalgebra of $u(\mathfrak{D}, \lambda, \mu)^J$. As an algebra, $u(\mathfrak{D}, \lambda, \mu, \Phi_J)$ is generated by an abelian group and a braided vector space of Cartan type, hence will be called a quasi-Hopf algebra of Cartan type. Since the construction given in this paper is the dual of that in [**4**], the module category of $u(\mathfrak{D}, \lambda, \mu, \Phi_J)$ also corresponds to certain de-equivariantization of $u(\mathfrak{D}, \lambda, \mu)$.

During our study on quasi-Hopf algebras of Cartan type, we found that a class of these quasi-Hopf algebras are very close to small quantum groups. Explicitly, there is a class of quasi-Hopf algebras $H_{\underline{c}}$, where $\underline{c}$ is a family of parameters. When $\underline{c}$ approaches 0, the quasi-Hopf algebras $H_{\underline{c}}$ will be the usual small quantum group, see Proposition 5.3. Based on this fact, these quasi-Hopf algebras $H_{\underline{c}}$ are defined to be small quasi-quantum groups. We pointe out that a similar notion named quasi-Frobenius Lusztig kernel was also introduced in [**15, 16**]. But in fact the two definitions are essentially different. On the one hand, the small quasi-quantum groups in the present paper are directly generalizations of small quantum groups, while in [**16**] a quasi-Frobenius Lusztig kernel is defined to be the quantum double $D(A_q(\mathbb{g}))$ of a quasi-Hopf algebra $A_q(\mathbb{g})$ constructed in [**7**]. Except some special examples, the dimension of $H_{\underline{c}}$ (even a small quantum group, i.e. $\underline{c} = 0$) is not a square, hence can not be a quantum double of a quasi-Hopf algebra. On the other hand, if the order of q is odd, and not divisible by 3 in case $\mathfrak{g}$ is of type G_2, then the quantum double $D(A_q(\mathfrak{g}))$ was proved not to be a genuine quasi-Hopf algebra in [**9**]. But under the same conditions, we construct many examples of genuine small quasi-quantum groups.

In order to give a partial answer to Problem A, we shall construct several classes of genuine quasi-Hopf algebras. Let A be a finite connected Cartan matrix of type A_n, B_n, C_n with $n \geq 3$, or D_n with $n \geq 4$. Associated to A, we construct a class of genuine quasi-Hopf algebras of Cartan type. Besides these examples, we also construct a large class of genuine small quasi-quantum groups associated to every finite Cartan matrix. All these genuine quasi-Hopf algebras are not elementary, hence provide genuine examples for [**4**] beyond those given in [**3**].

The paper is organized as follows. In Section 2, we introduce some concepts and known results about quasi-Hopf algebras, liftings of pointed Hopf algebras and 3-cocycles of abelian groups. In Section 3, we construct quasi-Hopf algebras from liftings of pointed Hopf algebras. Section 4 is devoted to Problem A. We show that several classes of quasi-Hopf algebras of Cartan type are genuine. In Section 5, we introduce the notion of a small quasi-quantum group, and present explicitly genuine examples. Throughout this paper, $\mathbb{k}$ denotes an algebraically closed field of characteristic zero. All the algebras, tensor categories and the unadorned tensor product $\otimes$ are over $\mathbb{k}$.

2. Preliminaries

In this section, we introduce some notions and basic facts about quasi-Hopf algebras, 3-cocycles of abelian groups and the liftings of pointed Hopf algebras [**1**].

2.1. Quasi-Hopf algebras. A quasi-bialgebra $H = (H, \triangle, \varepsilon, \Phi)$ is an unital associative algebra with two algebra maps $\triangle : H \to H \otimes H$(the comultiplication) and $\varepsilon : H \to \mathbb{k}$ (the counit), and an invertible element $\Phi \in H^{\otimes 3}$ (the associator) such that for all $h \in H$ the following equalities hold:

$$(\varepsilon \otimes id) \triangle (h) = h = (id \otimes \varepsilon) \triangle (h),$$
$$(id \otimes \triangle) \triangle (h) = \Phi \cdot (\triangle \otimes id) \triangle (h) \cdot \Phi^{-1},$$
$$(id \otimes id \otimes \triangle)(\Phi) \cdot (\triangle \otimes id \otimes id)(\Phi) = (1 \otimes \Phi) \cdot (id \otimes \triangle \otimes id)(\Phi) \cdot (\Phi \otimes 1),$$
$$(id \otimes \varepsilon \otimes id)(\Phi) = 1$$

Write $\Phi = \Phi^1 \otimes \Phi^2 \otimes \Phi^3$ and $\Phi^{-1} = \overline{\Phi}^1 \otimes \overline{\Phi}^2 \otimes \overline{\Phi}^3$. A quasi-Hopf algebra $H = (H, \triangle, \varepsilon, \Phi, S, \alpha, \beta)$ is a quasi-bialgebra $(H, \triangle, \varepsilon, \Phi)$ with an antipode (S, α, β), where $\alpha, \beta \in H$ and $S : H \to H$ is an algebra anti-homomorphism satisfying

$$\sum S(a_1)\alpha a_2 = \varepsilon(a)\alpha, \ \sum a_1 \beta S(a_2) = \varepsilon(a)\beta,$$
$$\Phi^1 \beta S(\Phi^2)\alpha\Phi^3 = 1, \ S(\overline{\Phi}^1)\alpha\overline{\Phi}^2 \beta S(\overline{\Phi}^3) = 1$$

for all $a \in H$. Here we use Sweedler's notation $\triangle(a) = \sum a_1 \otimes a_2$.

Definition 2.1. *A twist for a quasi-Hopf algebra H is an invertible element $J \in H \otimes H$ satisfying*

$$(\varepsilon \otimes id)(J) = (id \otimes \varepsilon)(J) = 1.$$

Suppose that $J=\sum_i f_i\otimes h_i$ is a twist of H with inverse $J^{-1}=\sum_i \overline{f_i}\otimes\overline{h_i}$. Write

$$\alpha_J=\sum_i S(\overline{f_i})\alpha\overline{g_i},\quad \beta_J=\sum_i f_i\beta S(g_i). \tag{2.1}$$

According to [**5**], if β_J is invertible then one can define a new quasi-Hopf algebra structure $H^J=(H,\triangle_J,\varepsilon,\Phi_J,S_J,\beta_J\alpha_J,1)$ on the algebra H, where

$$\triangle_J(h)=J\triangle(h)J^{-1},\quad h\in H, \tag{2.2}$$

$$\Phi_J=(1\otimes J)(id\otimes\triangle)(J)\Phi(\triangle\otimes id)(J^{-1})(J\otimes 1)^{-1} \tag{2.3}$$

$$S_J(h)=\beta_J S(h)\beta_J{}^{-1},\quad h\in H. \tag{2.4}$$

Two quasi-Hopf algebras H and H' are said to be twist equivalent if $H'\cong H^J$ for some twist J of H.

DEFINITION 2.2. *A quasi-Hopf algebra H is genuine if H is not twist (or gauge) equivalent to any Hopf algebra.*

2.2. Datum of Cartan type, root system and Weyl group. Let $\mathbb{G}$ be a fixed finite abelian group, $\widehat{\mathbb{G}}$ the character group of $\mathbb{G}$ with unit ε. The following definition was given is [**1**].

DEFINITION 2.3. *A datum of Cartan type*

$$\mathfrak{D}=\mathfrak{D}(\mathbb{G},(h_i)_{1\leq i\leq\theta},(\chi_i)_{1\leq i\leq\theta},A) \tag{2.5}$$

consists of an abelian group $\mathbb{G}$, elements $h_i\in\mathbb{G}$, characters $\chi_i\in\widehat{\mathbb{G}}$, $1\leq i\leq\theta$, and a generalized Cartan matrix $A=(a_{ij})_{1\leq i,j\leq\theta}$ of size θ satisfying

$$q_{ij}q_{ji}=q_{ii}^{a_{ij}},\ \text{where } q_{ij}=\chi_j(h_i) \text{ for all } 1\leq i,j\leq\theta \tag{2.6}$$

Here $|q_{ii}|$ means the order of q_{ii}. We call θ the rank of $\mathfrak{D}$. A datum of Cartan type $\mathfrak{D}$ is called finite (resp. connected) if A is a finite (resp. connected) Cartan matrix.

Now suppose $\mathfrak{D}=\mathfrak{D}(\mathbb{G},(h_i)_{1\leq i\leq\theta},(\chi_i)_{1\leq i\leq\theta},A)$ is finite. The Weyl group and root system associated to the finite Cartan matrix A are denoted by $W(A)$ and $R(A)$ respectively. Let $\{\alpha_i|1\leq i\leq\theta\}$ be a set of free generators of $\mathbb{Z}^\theta$ and $s_i:\mathbb{Z}^\theta\to\mathbb{Z}^\theta$ the reflection determined by

$$s_i(\alpha_j)=\alpha_j-a_{ij}\alpha_i,\ 1\leq i,j\leq\theta.$$

Then $W(A)$ is generated by $\{s_i|1\leq i\leq\theta\}$ and $R(A)=\cup_{i=1}^{\theta}W(A)(\alpha_i)$. Let $R^+(A)$ be the set of positive roots with respect to the simple roots $\{\alpha_i|1\leq i\leq\theta\}$. For each $\alpha=\sum_{i=1}^{\theta}k_i\alpha_i\in\mathbb{Z}^\theta$, define

$$h_\alpha=h_1^{k_1}h_2^{k_2}\cdots h_\theta^{k_\theta}, \tag{2.7}$$

$$\chi_\alpha=\chi_1^{k_1}\chi_2^{k_2}\cdots\chi_\theta^{k_\theta}, \tag{2.8}$$

$$ht(\alpha):=\sum_{i=1}^{\theta}k_i. \tag{2.9}$$

To a datum $\mathfrak{D}$ we associate a Yetter-Drinfeld module $V(\mathfrak{D})$ in ${}^{\mathbb{G}}_{\mathbb{G}}\mathcal{YD}$ defined as follows. Let $V_{h_i}^{\chi_i}=\Bbbk\{X_i\}$ be the 1-dimensional Yetter-Drinfeld module given by

$$\delta(X_i)=h_i\otimes X_i,\ g\triangleright X_i=\chi_i(g)X_i,\ g\in\mathbb{G}. \tag{2.10}$$

Then define

$$V(\mathfrak{D}) = \bigoplus_{i=1}^{\theta} V_{h_i}^{\chi_i}. \tag{2.11}$$

The basis $\{X_1, \cdots, X_\theta\}$ is called a standard basis of $V(\mathfrak{D})$. Suppose that c is the braiding of $V(\mathfrak{D})$. Then we have

$$c(X_i \otimes X_j) = h_i \triangleright X_j \otimes X_i = \chi_j(h_i) X_j \otimes X_i = q_{ij} X_j \otimes X_i, \ 1 \le i, j \le \theta. \tag{2.12}$$

Hence $(q_{ij})_{1 \le i,j \le \theta}$ are the braiding constants of $V(\mathfrak{D})$.

2.3. Braided Hopf algebras. In this subsection, let $\mathfrak{D} = \mathfrak{D}(\mathbb{G}, (h_i)_{1 \le i \le \theta}, (\chi_i)_{1 \le i \le \theta}, A)$ be a finite connected datum of Cartan type. In addition, we always assume:

$$q_{ii} \text{ has odd order,} \tag{2.13}$$

$$\text{the order of } q_{ii} \text{ is prime to 3, if } A \text{ is of type } G_2, \tag{2.14}$$

where $q_{ij} = \chi_j(g_i)$ for $1 \le i, j \le \theta$. With these assumptions, we have the following:

LEMMA 2.4. [**1**, Lemma 2.3] *There exists a root of unity q of odd order and integers $d_i \in \{1, 2, 3\}, 1 \le i \le \theta$, such that*

$$q_{ii} = q^{2d_i}, \ d_i a_{ij} = d_j a_{ji}, \ \ 1 \le i, j \le \theta.$$

Moreover, if A is of type G_2. Then the order of q is prime to 3.

An immediate consequence of Lemma 2.4 is that the elements q_{ii}, $1 \le i \le \theta$, have the same order, hence we define

$$N = |q_{ii}|, \ 1 \le i \le \theta. \tag{2.15}$$

Now we will introduce some braided Hopf algebras associated to $V(\mathfrak{D})$. Let $V = V(\mathfrak{D})$ and $\{X_1, \cdots, X_\theta\}$ a standard basis of V. Then the tensor algebra $T(V)$ is a braided Hopf algebra in ${}_G^G\mathcal{YD}$ with comultiplication determined by

$$\triangle(X_i) = X_i \otimes 1 + 1 \otimes X_i, \ \ 1 \le i \le \theta.$$

Define the braided commutator on $T(V)$ as follows:

$$[X, Y]_c = XY - \Big(\prod_{1 \le k \le s, 1 \le l \le t} q_{i_k j_l}^{x_k y_l}\Big) YX, \tag{2.16}$$

where $X = X_{i_1}^{x_1} X_{i_2}^{x_2} \cdots X_{i_s}^{x_s}$ and $Y = X_{j_1}^{y_1} X_{j_2}^{y_2} \cdots X_{j_s}^{y_s}$. For a primitive element X, we can define a braided adjoint action $ad_c(X)$ by

$$ad_c(X)(Y) = [X, Y]_c, \ \ Y \in T(V). \tag{2.17}$$

Let I be the braided Hopf ideal of $T(V)$ generated by the primitive elements $(ad_c X_i)^{1-a_{ij}}(X_j)$, $1 \le i \ne j \le \theta$. Then we have a quotient braided Hopf algebra $\mathcal{R}(\mathfrak{D}) = T(V)/I$ in ${}_{\mathbb{G}}^{\mathbb{G}}\mathcal{YD}$. For convenience, the image of X_i in $\mathcal{R}(\mathfrak{D})$ will also be denoted by X_i, $1 \le i \le \theta$.

Let $w_0 = s_{i_1} s_{i_2} \cdots s_{i_P}$ be a fixed reduced presentation of the longest element of $W(A)$ in terms of simple reflections. Then

$$\beta_l = s_{i_1} \cdots s_{i_{l-1}}(\alpha_{i_l}), \ 1 \le l \le P \tag{2.18}$$

determines a convex order of positive roots. The root vectors $\{X_\alpha | \alpha \in R^+\}$ can be defined as the iterated braided commutators of the elements $X_1, \cdots, X_\theta$ such that $X_{\alpha_i} = X_i, 1 \le i \le \theta$, see also [**1, 17**] for the detailed definition.

Let $Y_l = X_{\beta_l}^N, 1 \leq l \leq P$, and $e_l = (\delta_{kl})_{1\leq k\leq P} \in \mathbb{N}^P$, where δ_{kl} is the Kronecker delta. For each $a = (a_1, a_2, \cdots, a_P) \in \mathbb{N}^P$, define

$$Y^a = Y_1^{a_1} Y_2^{a_2} \cdots Y_p^{a_P}, \tag{2.19}$$

$$h^a = h_{\beta_1}^{Na_1} h_{\beta_2}^{Na_2} \cdots h_{\beta_P}^{Na_P}, \tag{2.20}$$

$$\underline{a} = a_1\beta_1 + a_2\beta_2 + \cdots + a_P\beta_P. \tag{2.21}$$

Let $\triangle_{\mathcal{R}(\mathfrak{D})}$ be the comultiplication of $\mathcal{R}(\mathfrak{D})$. Then we have the following lemma.

LEMMA 2.5. [**1**, Lemma 2.8] *For any nonzero $a \in \mathbb{N}^P$, there are uniquely determined scalars $t_{b,c}^a \in \mathbb{k}, 0 \neq b, c \in \mathbb{N}^P$, such that*

$$\triangle_{\mathcal{R}(\mathfrak{D})}(Y^a) = Y^a \otimes 1 + 1 \otimes Y^a + \sum_{b,c\neq 0, \underline{b}+\underline{c}=\underline{a}} t_{b,c}^a Y^b \otimes Y^c. \tag{2.22}$$

Let $(\mu_a)_{a\in\mathbb{N}^P}$ be a family of elements in $\mathbb{k}$ such that for all a, $h^a = 1$ implies $\mu_a = 0$. Then for each $a \in \mathbb{N}^P$, we can define an element $u^a \in \mathbb{k}\mathbb{G}$ inductively on $ht(\underline{a})$ as follows.

When $ht(\underline{a}) = 1$, i.e. a is a simple root, define

$$u^a = \mu_a(1 - h^a). \tag{2.23}$$

Suppose $ht(\underline{a}) > 1$ and u^c is defined for any $c \in \mathbb{N}^P$ such that $ht(\underline{c}) < ht(\underline{a})$. If $a = e_l$ for some $1 \leq l \leq p$, define

$$u^a = \mu_a(1 - h^a) + \sum_{b,c\neq 0, \underline{b}+\underline{c}=\underline{a}} t_{b,c}^a \mu_b u^c. \tag{2.24}$$

If $a = (a_1, \cdots, a_l, 0, \cdots, 0) \neq e_l,\ a_l \geq 1$ for some $1 \leq l \leq P$, then define

$$u^a = u^{a-e_l} u^{e_l}. \tag{2.25}$$

Now let $\mu = (\mu_l)_{1\leq l\leq P}$ be a family of elements in $\mathbb{k}$ satisfying the condition: if $g_{\beta_l}^N = 1$ or $\chi_{\beta_l}^N \neq \varepsilon$, then $\mu_l = 0$. By [**1**, Theorem 2.13 (1), Lemma 2.10 (2)], there exists an unique family $(\mu_a)_{a\in\mathbb{N}^P}$ such that for all a, $h^a = 1$ implies $\mu_a = 0$, and $\mu_{e_l} = \mu_l$ for $1 \leq l \leq P$. Let $\{u^a\}_{a\in\mathbb{N}^P}$ be the set of elements in $\mathbb{k}\mathbb{G}$ corresponding to $(\mu_a)_{a\in\mathbb{N}^P}$ via (2.23)-(2.25). If α is a positive root in R^+ with $\alpha = \beta_l$, then we define

$$\mathbf{u}_\alpha(\mu) = u^{e_l}. \tag{2.26}$$

2.4. Andruskiewitsch-Schneider's Hopf algebras. Fix a datum of finite Cartan type $\mathfrak{D} = \mathfrak{D}(\mathbb{G}, (h_i)_{1\leq i\leq\theta}, (\chi_i)_{1\leq i\leq\theta}, A)$, where A may not be a connected Cartan matrix. For $1 \leq i, j \leq \theta$ we define $i \sim j$ if i and j are in the same connected component of the Dynkin diagram associated to A. In this case, i, j are said to be connected. Let $\Omega = \{I_1, \cdots, I_t\}$ be the set of the connected components of $I = \{1, 2, \cdots, \theta\}$. Here we also assume that the conditions (2.13)-(2.14) hold for each connected component of I. For $J \in \Omega$, let R_J be the root system of $A_J = (a_{ij})_{i,j\in J}$ and N_J the corresponding number defined by (2.15). Let R_J^+ be the set of positive roots of A_J with respect to the simple roots $\{\alpha_i | i \in J\}$. The following partitions are obvious:

$$R = \bigcup_{J\in\Omega} R_J,\ R^+ = \bigcup_{J\in\Omega} R_J^+.$$

DEFINITION 2.6. *A family* $\lambda = (\lambda_{ij})_{1\leq i,j\leq n, i\nsim j}$ *of elements in* $\mathbb{k}$ *is called a family of linking parameters for* $\mathfrak{D}$ *if the following condition is satisfied for all* $1 \leq i, j \leq \theta, i \nsim j$:

$$\text{if } h_i h_j = 1 \text{ or } \chi_i\chi_j \neq \varepsilon, \text{then } \lambda_{ij} = 0. \tag{2.27}$$

Two vertices $1 \leq i, j \leq \theta$ *are called linkable if* $i \nsim j, h_i h_j \neq 1$ *and* $\chi_i\chi_j = \varepsilon$.

DEFINITION 2.7. *A family* $\mu = (\mu_\alpha)_{\alpha\in R^+}$ *of elements in* $\mathbb{k}$ *is called a family of root vector parameters for* $\mathfrak{D}$ *if for all* $\alpha \in R_J^+, J \in \Omega$, *we have*

$$h_\alpha^{N_J} = 1 \text{ or } \chi_\alpha^{N_J} \neq \varepsilon, \text{ then } \mu_\alpha = 0. \tag{2.28}$$

The following theorem is one of the main results of [**1**].

THEOREM 2.8. [**1**, Theorem 4.5] *Let* $\mathfrak{D} = \mathfrak{D}(\mathbb{G}, (h_i)_{1\leq i\leq\theta}, (\chi_i)_{1\leq i\leq\theta}, A)$ *be a datum of finite Cartan type such that each connected component of* $I = \{1, 2, \cdots, \theta\}$ *satisfies the conditions* (2.13)-(2.14). *Let* λ *and* μ *be a family of linking parameters and a family of root vector parameters for* $\mathfrak{D}$ *respectively. Then we have a finite-dimensional pointed Hopf algebra* $u(\mathfrak{D}, \lambda, \mu)$ *generated by the group* $\mathbb{G}$ *and the skew-primitive elements* $\{X_i | 1 \leq i \leq \theta\}$ *subject to the following relations:*

$$\begin{aligned}
&\text{(Action of the group)} && gX_ig^{-1} = \chi_i(g)X_i, \text{ for all } 1 \leq i \leq \theta, g \in \mathbb{G},\\
&\text{(Serre relations)} && ad_c(X_i)^{1-a_{ij}}(X_j) = 0, \text{ for all } i \neq j, i \sim j,\\
&\text{(Linking relations)} && ad_c(X_i)(X_j) = \lambda_{ij}(1 - h_ih_j), \text{ for all } i < j, i \nsim j,\\
&\text{(Root vector relations)} && X_\alpha^{N_J} = \mathbf{u}_\alpha(\mu), \text{ for all } \alpha \in R_J^+, J \in \Omega.
\end{aligned}$$

The coalgebra structure is determined by

$$\triangle(X_i) = X_i \otimes 1 + h_i \otimes X_i, \ \triangle(g) = g \otimes g, \text{ for all } 1 \leq i \leq \theta, g \in \mathbb{G}.$$

Following [**8**], we call these Hopf algebras $u(\mathfrak{D}, \lambda, \mu)$ the Andruskiewitsch-Schneider Hopf algebras.

2.5. Normalized 3-cocycles on finite groups. Let $\mathbb{G} = \mathbb{Z}_{\mathrm{m}_1} \times \cdots \times \mathbb{Z}_{\mathrm{m}_n}$, $\mathrm{m}_j \in \mathbb{N}$ for $1 \leq j \leq n$. A function $\phi : \mathbb{G} \times \mathbb{G} \times \mathbb{G} \mapsto \mathbb{k}^*$ is called a 3-cocycle on $\mathbb{G}$ if

$$\phi(ef, g, h)\phi(e, f, gh) = \phi(e, f, g)\phi(e, fg, h)\phi(f, g, h) \tag{2.29}$$

for all $e, f, g, h \in \mathbb{G}$. And the 3-cocycle ϕ is called normalized if $\phi(f, 1, g) = 1$. Denote by $\mathcal{A}$ the set of all sequences

$$(c_1, \ldots, c_l, \ldots, c_n, c_{12}, \ldots, c_{ij}, \ldots, c_{n-1,n}, c_{123}, \ldots, c_{rst}, \ldots, c_{n-2,n-1,n}) \tag{2.30}$$

such that $0 \leq c_l < \mathrm{m}_l$, $0 \leq c_{ij} < (\mathrm{m}_i, \mathrm{m}_j)$, $0 \leq c_{rst} < (\mathrm{m}_r, \mathrm{m}_s, \mathrm{m}_t)$ for $1 \leq l \leq n$, $1 \leq i < j \leq n$, $1 \leq r < s < t \leq n$, where c_{ij} and c_{rst} are ordered in the lexicographic order. In the following we denote by $\underline{\mathbf{c}}$ the sequence (2.30) for short.

Let g_i be a generator of $\mathbb{Z}_{\mathrm{m}_i}, 1 \leq i \leq n$. For each $\underline{\mathbf{c}} \in \mathcal{A}$, define

$$\begin{aligned}
&\phi_{\underline{\mathbf{c}}} : \mathbb{G} \times \mathbb{G} \times \mathbb{G} \longrightarrow \mathbb{k}^*,\\
&[\mathrm{g}_1^{i_1} \cdots \mathrm{g}_n^{i_n}, \mathrm{g}_1^{j_1} \cdots \mathrm{g}_n^{j_n}, \mathrm{g}_1^{k_1} \cdots \mathrm{g}_n^{k_n}] \mapsto\\
&\prod_{l=1}^{n} \zeta_{\mathrm{m}_l}^{c_l k_l [\frac{i_l+j_l}{\mathrm{m}_l}]} \prod_{1\leq s<t\leq n} \zeta_{\mathrm{m}_t}^{c_{st} k_t [\frac{i_s+j_s}{\mathrm{m}_s}]} \prod_{1\leq r<s<t\leq n} \zeta_{(\mathrm{m}_r,\mathrm{m}_s,\mathrm{m}_t)}^{c_{rst} i_r j_s k_t}.
\end{aligned} \tag{2.31}$$

Here and below ζ_{m} stands for a primitive m-th root of unity.

PROPOSITION 2.9. [**14**, Proposition 3.8] $\{\phi_{\underline{\mathbf{c}}} | \underline{\mathbf{c}} \in \mathcal{A}\}$ *forms a complete set of representatives of the normalized* 3*-cocycles on* G *up to* 3*-cohomology.*

3. Finite-dimensional Quasi-Hopf algebras

3.1. Some basic notations. In this subsection, we fix some notations which will be used throughout this paper. Let $\mathbb{G} = \langle \mathbb{g}_1 \rangle \times \cdots \times \langle \mathbb{g}_n \rangle$ such that $|\mathbb{g}_i| = \mathbb{m}_i = m_i^2$ for $1 \leq i \leq n$, G the subgroup of $\mathbb{G}$ generated by $\mathbb{g}_1^{m_1}, \cdots, \mathbb{g}_n^{m_n}$. For convenience, we denote $g_i = \mathbb{g}_i^{m_i}$ for $1 \leq i \leq n$. Let $\widehat{\mathbb{G}}$ be the character group of $\mathbb{G}$ over $\mathbb{k}$. For each $\mathbb{g} = \prod_{i=1}^n \mathbb{g}_i^{\alpha_i}$, define a character $\chi_{\mathbb{g}} : \mathbb{G} \to \mathbb{k}^*$ by

$$\chi_{\mathbb{g}}(h) = \prod_{i=1}^{n} \zeta_{\mathbb{m}_i}^{\alpha_i \beta_i}, \ h = \prod_{i=1}^{n} \mathbb{g}_i^{\beta_i} \in \mathbb{G}. \tag{3.1}$$

From the definition of $\chi_{\mathbb{g}}$, it is obvious that $\chi_{\mathbb{g}}^{-1}(h) = \chi_{\mathbb{g}^{-1}}(h) = \chi_{\mathbb{g}}(h^{-1})$. So

$$\chi : \mathbb{g} \longrightarrow \widehat{\mathbb{G}}, \mathbb{g} \to \chi_{\mathbb{g}}$$

is an group isomorphism. Let $\mathbb{k}[\mathbb{G}]$ be the group algebra of $\mathbb{G}$ over field $\mathbb{k}$. One can verify that

$$\{\mathbb{1}_{\mathbb{g}} = \frac{1}{|\mathbb{G}|} \sum_{h \in \mathbb{G}} \chi_{\mathbb{g}}(h) h | \mathbb{g} \in \mathbb{G}\} \tag{3.2}$$

forms a complete set of the orthogonal primitive idempotents of the algebra $\mathbb{k}[\mathbb{G}]$. Let ζ_{m_i} be a primitive m_i-th root of unity such that $\zeta_{m_i} = \zeta_{\mathbb{m}_i}^{m_i}$, $1 \leq i \leq n$. For each $g = \prod_{i=1}^n g_i^{s_i} \in G$, define $\chi_g : G \to \mathbb{k}^*$ by

$$\chi_g(h) = \prod_{i=1}^{n} \zeta_{m_i}^{s_i t_i}, \ h = \prod_{i=1}^{n} g_i^{t_i}. \tag{3.3}$$

Similar to (3.2),

$$\{1_g = \frac{1}{|G|} \sum_{h \in G} \chi_g(h) h | g \in G\} \tag{3.4}$$

is a complete set of orthogonal primitive idempotents of the algebra $\mathbb{k}[G]$.

LEMMA 3.1. *For $g, h \in G$, we have $1_g h = h 1_g = \chi_g^{-1}(h) 1_g$. Similarly $\mathbb{1}_{\mathbb{g}} \mathbb{h} = \mathbb{h} \mathbb{1}_{\mathbb{g}} = \chi_{\mathbb{g}}^{-1}(\mathbb{h}) \mathbb{1}_{\mathbb{g}}$ for all $\mathbb{g}, \mathbb{h} \in \mathbb{G}$.*

PROOF. $1_g h = \frac{1}{|G|} \sum_{f \in G} \chi_g(f) f h = \frac{1}{|G|} \sum_{f \in G} \chi_g(fh) \chi_g(h^{-1}) f h = \chi_g^{-1}(h) 1_g$. The second part of the Lemma can be proved similarly. □

LEMMA 3.2. *The following identity holds for all $0 \leq s_i \leq m_i - 1$, $1 \leq i \leq n$:*

$$\sum_{1 \leq i \leq n, 0 \leq k_i \leq m_i - 1} \mathbb{1}_{(\prod_{i=1}^n \mathbb{g}_i^{m_i k_i + s_i})} = 1_{(\prod_{i=1}^n g_i^{s_i})}. \tag{3.5}$$

PROOF. By definition we have

$$\sum_{1 \leq i \leq n, 0 \leq k_i \leq m_i - 1} \mathbb{1}_{(\prod_{i=1}^n \mathbb{g}_i^{m_i k_i + s_i})} = \frac{1}{|\mathbb{G}|} \sum_{1 \leq i \leq n, 0 \leq k_i \leq m_i - 1} \sum_{h \in \mathbb{G}} \chi_{(\prod_{i=1}^n \mathbb{g}_i^{m_i k_i + s_i})}(h) h.$$

Suppose $h = \mathbb{g}_1^{r_1} \mathbb{g}_2^{r_2} \cdots \mathbb{g}_n^{r_n}$. Then we have

$$\begin{aligned} \sum_{1 \leq i \leq n, \ 0 \leq k_i \leq m_i - 1} \chi_{(\prod_{i=1}^n \mathbb{g}_i^{m_i k_i + s_i})}(h) &= \textstyle\sum_{1 \leq i \leq n, \ 0 \leq k_i \leq m_i - 1} \prod_i^n \zeta_{\mathbb{m}_i}^{(m_i k_i + s_i) r_i} \\ &= \textstyle\prod_{i=1}^n (\sum_{0 \leq k_i \leq m_i - 1} \zeta_{\mathbb{m}_i}^{(m_i k_i + s_i) r_i}). \end{aligned}$$

Note that $\sum_{0\leq k_l\leq m_l-1}\zeta_{\mathbb{m}_l}^{m_lk_lr_l}=0$ if $r_l\neq tm_l$ for some integer $0\leq t\leq m_l-1$. If $r_i=t_im_i$ for $1\leq i\leq n$, then $h=\mathbb{g}_1^{t_1m_1}\mathbb{g}_2^{t_2m_2}\cdots\mathbb{g}_n^{t_nm_n}\in G$ and we have:

$$\begin{aligned}\sum_{1\leq i\leq n,\ 0\leq k_i\leq m_i-1}\chi_{(\prod_{i=1}^n\mathbb{g}_i^{m_ik_i+s_i})}(h) &= \prod_{i=1}^n(\sum_{0\leq k_i\leq m_i-1}\zeta_{\mathbb{m}_i}^{(m_ik_i+s_i)t_im_i})\\ &= \prod_{i=1}^n(\sum_{0\leq k_i\leq m_i-1}\zeta_{\mathbb{m}_i}^{s_it_im_i})\\ &= \prod_{i=1}^n(m_i\zeta_{m_i}^{s_it_i})\\ &= |G|\prod_{i=1}^n\zeta_{m_i}^{s_it_i}.\end{aligned}$$

It follows that

$$\begin{aligned}&\sum_{1\leq i\leq n,\ 0\leq k_i\leq m_i-1}\mathbb{1}_{(\prod_{i=1}^n\mathbb{g}_i^{m_ik_i+s_i})}\\ &=\frac{1}{|\mathbb{G}|}\sum_{1\leq i\leq n,\ 0\leq k_i\leq m_i-1}\sum_{h\in G}\chi_{(\prod_{i=1}^n\mathbb{g}_i^{m_ik_i+s_i})}(h)h.\\ &=\frac{|G|}{|\mathbb{G}|}\sum_{1\leq i\leq n,\ 0\leq k_i\leq m_i-1}\prod_{i=1}^n\zeta_{m_i}^{s_it_i}(\prod_{i=1}^n\mathbb{g}_i^{t_im_i})\\ &=\frac{1}{|G|}\sum_{1\leq i\leq n,\ 0\leq k_i\leq m_i-1}\chi_{(\prod_{i=1}^n g_i^{s_i})}(\prod_{i=1}^n g_i^{t_i})(\prod_{i=1}^n g_i^{t_i})\\ &=\frac{1}{|G|}\sum_{g\in G}\chi_{(\prod_{i=1}^n g_i^{s_i})}(g)g\\ &=1_{(\prod_{i=1}^n g_i^{s_i})}.\end{aligned}$$

Thus, the claimed equality holds. □

3.2. Quasi-Hopf algebras of Cartan type. Let $\mathfrak{D}=\mathfrak{D}(\mathbb{G},(h_i)_{1\leq i\leq\theta},(\chi_i)_{1\leq i\leq\theta},A)$ be a datum of finite Cartan type, where $A=(a_{ij})_{1\leq i,j\leq\theta}$ is a finite Cartan matrix. Let Ω be the set of the connected components of $I=\{1,2,\cdots,\theta\}$. Let $(s_{ij})_{1\leq i\leq\theta,1\leq j\leq n}$ and $(r_{ij})_{1\leq i\leq\theta,1\leq j\leq n}$ be two families of integers satisfying:

$$h_i=\prod_{j=1}^n\mathbb{g}_j^{s_{ij}},\ \chi_i(\mathbb{g}_j)=\zeta_{\mathbb{m}_j}^{r_{ij}},\tag{3.6}$$

$$0\leq s_{ij},r_{ij}<\mathbb{m}_j \text{ for all } 1\leq i\leq\theta,1\leq j\leq n.\tag{3.7}$$

It is obvious that $(s_{ij})_{1\leq i\leq\theta,1\leq j\leq n}$ and $(r_{ij})_{1\leq i\leq\theta,1\leq j\leq n}$ are uniquely determined by $\mathfrak{D}$. Let $\Gamma(\mathfrak{D})$ be the subset of $\mathcal{A}$ (see Subsection 2.5 for definition) such that for each $\underline{c}\in\Gamma(\mathfrak{D})$, $c_{rst}=0$ for all $0\leq r<s<t\leq n$ and

$$s_{ij}\equiv c_jr_{ij}\mod m_j,1\leq i\leq\theta,1\leq j\leq n,\tag{3.8}$$

$$c_{ij}r_{lj}\equiv0\mod m_j,1\leq l\leq\theta,1\leq i<j\leq n,\tag{3.9}$$

$$c_{ij}m_i\equiv0\mod m_j,1\leq i<j\leq n.\tag{3.10}$$

For each $\underline{c} \in \Gamma(\mathfrak{D})$, define on G the functions $\Theta_l, \Psi_l, \Upsilon$ and $\mathcal{F}_i$ as follows:

(3.11)
$$\Theta_l(g) = \prod_{i=1}^{n} \zeta_{\mathfrak{m}_i}^{-t_i s_{li}}, \; g = \prod_{i=1}^{n} g_i^{t_i} \in G, 1 \le l \le \theta.$$

(3.12)
$$\Psi_l(f,h) = \prod_{i=1}^{n} \zeta_{\mathfrak{m}_i}^{c_i q_i \varphi_{li}(f)} \prod_{1\le i<j\le n} \zeta_{m_i m_j}^{c_{ij} q_j \varphi_{li}(f)}, \; f = \prod_{i=1}^{n} g_i^{p_i}, \; h = \prod_{i=1}^{n} g_i^{q_i} \in G, 1 \le l \le \theta.$$

(3.13)
$$\Upsilon(g) = \prod_{i=1}^{n} \zeta_{\mathfrak{m}_i}^{-c_i l_i m_i} \prod_{1\le s<t\le n} \zeta_{m_s m_t}^{-c_{st} l_t m_s}, \; g = \prod_{i=1}^{n} g_i^{l_i}.$$

(3.14)
$$\mathcal{F}_i(g) = \prod_{j=1}^{n} \zeta_{\mathfrak{m}_j}^{-c_j (k_j - r_{ij}) \varphi_{ij}(g)} \prod_{1\le s<t\le n} \zeta_{\mathfrak{m}_s \mathfrak{m}_t}^{-c_{st}(k_t - r_{it}) \varphi_{is}(g)}, \; g = \prod_{i=l}^{n} g_i^{k_l}.$$

Here

$$\varphi_{li}(g) = (k_i - r_{li})' - (k_i - r_{li}), \; g = \prod_{i=l}^{n} g_i^{k_l}, \tag{3.15}$$

and $(k_i - r_{li})'$ is the remainder of $k_i - r_{li}$ divided by m_i.

In order to construct quasi-Hopf algebras, we need the notions of restricted linking parameters and restricted root vector parameters.

DEFINITION 3.3. *A family of linking parameters $\lambda = (\lambda_{ij})_{1\le i,j\le\theta}$ for $\mathfrak{D}$ is said to be restricted if*

$$h_i h_j \notin G \text{ implies } \lambda_{ij} = 0 \text{ for all } 1 \le i < j \le \theta, i \nsim j. \tag{3.16}$$

A family of root vector parameters $\mu = (\mu_\alpha)_{\alpha\in R}$ for $\mathfrak{D}$ is said to be restricted if

$$h_\alpha^{N_J} \notin G \text{ implies } \mu_\alpha = 0 \text{ for all } \alpha \in R_J^+, J \in \Omega. \tag{3.17}$$

The following theorem is one of the main results of this paper.

THEOREM 3.4. *Let $\lambda = (\lambda_{ij})_{1\le i,j\le\theta}$ and $\mu = (\lambda_\alpha)_{\alpha\in R}$ be two families of restricted linking parameters and root vector parameters respectively for $\mathfrak{D} = \mathfrak{D}(\mathbb{G}, (h_i)_{1\le i\le\theta}, (\chi_i)_{1\le i\le\theta}, A)$, and $\underline{c}$ a nonzero element in $\Gamma(\mathfrak{D})$. Then we have a finite-dimensional quasi-Hopf algebra $u(\mathfrak{D}, \lambda, \mu, \Phi_{\underline{c}})$ generated by G and $\{X_1, \cdots, X_\theta\}$ subject to the relations:*

$$gX_i g^{-1} = \chi_i(g) X_i, \text{ for all } 1 \le i \le \theta, g \in G, \tag{3.18}$$

$$ad_C(X_i)^{1-a_{ij}}(X_j) = 0, \text{ for all } i \ne j, i \sim j, \tag{3.19}$$

$$ad_C(X_i)(X_j) = \lambda_{ij}(1 - h_i h_j), \text{ for all } i < j, i \nsim j, \tag{3.20}$$

$$X_\alpha^{N_J} = \mathbf{u}_\alpha(\mu), \text{ for all } \alpha \in R_J^+, J \in \Omega. \tag{3.21}$$

The coalgebra structure of $u(\mathfrak{D}, \lambda, \mu, \Phi_{\underline{c}})$ is given by

(3.22)
$$\triangle(X_l) = \sum_{f,g\in G} \Psi_l(f,g) X_l 1_f \otimes 1_g + \sum_{f\in G} \Theta_l(f) 1_f \otimes X_l, \triangle(g) = g \otimes g, 1 \le l \le \theta, g \in G.$$

The associator of $u(\mathfrak{D}, \lambda, \mu, \Phi_{\underline{c}})$ *is determined by*

$$\Phi_{\underline{c}} = \sum_{f,g,h \in G} \phi_{\underline{c}}(f,g,h)(1_f \otimes 1_g \otimes 1_h). \tag{3.23}$$

The antipode $(\mathcal{S}, \alpha, 1)$ *is defined by*

$$\alpha = \sum_{g \in G} \Upsilon(g) 1_g, \ \mathcal{S}(X_i) = \sum_{g \in G} \mathcal{F}_i(g) X_i 1_g. \tag{3.24}$$

The proof of Theorem 3.4 will be delivered in the next subsection. Since the quasi-Hopf algebra $u(\mathfrak{D}, \lambda, \mu, \Phi_{\underline{c}})$ is generated by the abelian group G and the braided vector space of Cartan type $V = \Bbbk\{X_1, \cdots, X_\theta\}$, we shall call it a quasi-Hopf algebra of Cartan type. We will say that $u(\mathfrak{D}, \lambda, \mu, \Phi_{\underline{c}})$ is **associated to the Cartan matrix** A, and the **rank** of $u(\mathfrak{D}, \lambda, \mu, \Phi_{\underline{c}})$, is defined by θ, the rank of A.

3.3. The proof of Theorem 3.4. Let $H = u(\mathfrak{D}, \lambda, \mu)$ be the Andruskiewitsch-Schneider Hopf algebra given in Theorem 2.8, which is generated by $\mathbb{G}$ and $\{X_1, \cdots, X_\theta\}$. By [**1**, Theorem 4.5], the group of group-like elements of H is $\mathbb{G}$. According to Subsection 3.1, we know that $\chi : \mathbb{G} \to \widehat{\mathbb{G}}$ is a group isomorphism. Let $\{\eta_1, \cdots, \eta_\theta\}$ be the set of elements in $\mathbb{G}$ such that $\chi_{\eta_i} = \chi_i$ for $1 \leq i \leq \theta$. By (3.6), it is obvious that

$$\eta_i = \prod_{j=1}^{n} \mathbb{g}_j^{r_{ij}}, 1 \leq i \leq \theta. \tag{3.25}$$

LEMMA 3.5.

$$\mathbb{1}_{\mathbb{g}} X_i = X_i \mathbb{1}_{\mathbb{g}\eta_i}, \ \text{for all } \mathbb{g} \in \mathbb{G}, 1 \leq i \leq \theta. \tag{3.26}$$

PROOF. It follows from the following equations:

$$\mathbb{1}_{\mathbb{g}} X_i = \frac{1}{|\mathbb{G}|} \sum_{\mathbb{f} \in \mathbb{G}} \chi_{\mathbb{g}}(\mathbb{f}) \mathbb{f} X_i = \frac{1}{|\mathbb{G}|} \sum_{\mathbb{f} \in \mathbb{G}} \chi_{\mathbb{g}}(\mathbb{f}) \chi_{\eta_i}(\mathbb{f}) X_i \mathbb{f} = \frac{1}{|\mathbb{G}|} X_i \sum_{\mathbb{f} \in \mathbb{G}} \chi_{\mathbb{g}\eta_i}(\mathbb{f}) \mathbb{f} = X_i \mathbb{1}_{\mathbb{g}\eta_i}.$$

□

Let $\underline{c} \in \Gamma(\mathfrak{D})$, define

(3.27)

$$J_{\underline{c}} : \ \mathbb{G} \times \mathbb{G} \to k^*; \quad (\mathbb{g}_1^{x_1} \cdots \mathbb{g}_n^{x_n}, \mathbb{g}_1^{y_1} \cdots \mathbb{g}_n^{y_n}) \mapsto \prod_{l=1}^{n} \zeta_{\mathbb{m}_l}^{c_l y_l (x_l' - x_l)} \prod_{1 \leq s < t \leq n} \zeta_{m_s m_t}^{c_{st} y_t (x_s' - x_s)},$$

where the element x_i' stands for the remainder of x_i divided by m_i for each $1 \leq i \leq n$. One can easily verify that $J_{\underline{c}}$ is a 2-cochain of $\mathbb{G}$. We will see that this 2-cochain $J_{\underline{c}}$ induces a twist of the Hopf algebra H.

LEMMA 3.6. *Let* $\mathbb{J}_{\underline{c}} = \sum_{\mathbb{f},\mathbb{g} \in \mathbb{G}} J_{\underline{c}}(\mathbb{f}, \mathbb{g}) \mathbb{1}_{\mathbb{f}} \otimes \mathbb{1}_{\mathbb{g}} \in H \otimes H$. *Then* $\mathbb{J}_{\underline{c}}$ *is a twist of* H.

PROOF. It is obvious that $\mathbb{J}_{\underline{c}}$ is invertible with inverse $\mathbb{J}_{\underline{c}}^{-1} = \sum_{\mathbb{f},\mathbb{g} \in \mathbb{G}} J_{\underline{c}}^{-1}(\mathbb{f}, \mathbb{g}) \mathbb{1}_{\mathbb{f}} \otimes \mathbb{1}_{\mathbb{g}}$. Next we verify that the identities $(\varepsilon \otimes id)(\mathbb{J}_{\underline{c}}) = (id \otimes \varepsilon)(\mathbb{J}_{\underline{c}}) = 1$ hold. Suppose

$\mathbb{g} = \prod_{i=1}^{n} \mathbb{g}_i^{k_i}$ for some $0 \le k_i \le \mathbb{m}_i, 1 \le i \le n$. Then

$$\begin{aligned}\varepsilon(\mathbb{1}_{\mathbb{g}}) &= \frac{1}{|\mathbb{G}|}\varepsilon(\sum_{\mathbb{h}\in\mathbb{G}} \chi_{\mathbb{g}}(\mathbb{h})\mathbb{h}) = \frac{1}{|\mathbb{G}|}\sum_{\mathbb{h}\in\mathbb{G}}\chi_{\mathbb{g}}(\mathbb{h}) \\ &= \frac{1}{|\mathbb{G}|}\sum_{0\le l_j\le \mathbb{m}_j-1, 1\le j\le n}(\prod_{i=1}^{n}\zeta_{\mathbb{m}_i}^{k_i l_i}) \\ &= \frac{1}{|\mathbb{G}|}\prod_{i=1}^{n}(\sum_{0\le l_i\le \mathbb{m}_i-1}\zeta_{\mathbb{m}_i}^{k_i l_i}).\end{aligned}$$

So we have

$$\varepsilon(\mathbb{1}_{\mathbb{g}}) = \begin{cases} 0, & \text{if } \mathbb{g} \neq 1; \\ 1, & \text{if } \mathbb{g} = 1. \end{cases}$$

Hence, $(\varepsilon\otimes id)(\mathbb{J}_{\underline{c}}) = \sum_{\mathbb{g}\in\mathbb{G}} J_{\underline{c}}(1,\mathbb{g})\mathbb{1}_{\mathbb{g}} = \sum_{\mathbb{g}\in\mathbb{G}} \mathbb{1}_{\mathbb{g}} = 1$. The identity $(id\otimes\varepsilon)(\mathbb{J}_{\underline{c}}) = 1$ can be proved in a similar way. □

Since H is a Hopf algebra, we can view it as a quasi-Hopf algebra with the trivial associator $\Phi = 1\otimes 1\otimes 1$ and the antipode $(S,1,1)$. According to Subsection 2.1, we can construct a quasi-Hopf algebra $H^{\mathbb{J}_{\underline{c}}} = (H^{\mathbb{J}_{\underline{c}}}, \triangle_{\mathbb{J}_{\underline{c}}}, \varepsilon, \Phi_{\mathbb{J}_{\underline{c}}}, S_{\mathbb{J}_{\underline{c}}}, \beta_{\mathbb{J}_{\underline{c}}}\alpha_{\mathbb{J}_{\underline{c}}}, 1)$. We have the following.

LEMMA 3.7.

$$\Phi_{\mathbb{J}_{\underline{c}}} = \sum_{f,g,h\in G} \phi_{\underline{c}}(f,g,h)1_f\otimes 1_g\otimes 1_h \tag{3.28}$$

PROOF. First we show that

$$\triangle(\mathbb{1}_{\mathbb{g}}) = \sum_{\mathbb{f}\mathbb{h}=\mathbb{g}} \mathbb{1}_{\mathbb{f}}\otimes\mathbb{1}_{\mathbb{h}}. \tag{3.29}$$

By the orthogonality of characters, for any $\mathbb{x},\mathbb{y}\in\mathbb{G}$, we have

$$\sum_{\mathbb{f}\mathbb{h}=\mathbb{g}}\chi_{\mathbb{f}}(\mathbb{x})\chi_{\mathbb{h}}(\mathbb{y}) = \begin{cases} 0, & \text{if } \mathbb{x}\neq\mathbb{y}; \\ |\mathbb{G}|\chi_{\mathbb{g}}(\mathbb{x}), & \text{if } \mathbb{x}=\mathbb{y}. \end{cases}$$

So (3.29) follows from

$$\begin{aligned}\sum_{\mathbb{f}\mathbb{h}=\mathbb{g}}\mathbb{1}_{\mathbb{f}}\otimes\mathbb{1}_{\mathbb{h}} &= \frac{1}{|\mathbb{G}|^2}\sum_{\mathbb{f}\mathbb{h}=\mathbb{g}}\sum_{\mathbb{x}\in\mathbb{G}}\chi_{\mathbb{f}}(\mathbb{x})\mathbb{x}\otimes\sum_{\mathbb{y}\in\mathbb{G}}\chi_{\mathbb{h}}(\mathbb{y})\mathbb{y} \\ &= \frac{1}{|\mathbb{G}|^2}\sum_{\mathbb{x},\mathbb{y}\in\mathbb{G}}[\sum_{\mathbb{f}\mathbb{h}=\mathbb{g}}\chi_{\mathbb{f}}(\mathbb{x})\chi_{\mathbb{h}}(\mathbb{y})\mathbb{x}\otimes\mathbb{y}] \\ &= \frac{1}{|\mathbb{G}|}\sum_{\mathbb{x}\in\mathbb{G}}\chi_{\mathbb{g}}(\mathbb{x})\mathbb{x}\otimes\mathbb{x} \\ &= \triangle(\mathbb{1}_{\mathbb{g}}).\end{aligned}$$

By (3.29), we have

$$\begin{aligned}\Phi_{\mathbb{J}_{\underline{c}}} &= (1\otimes\mathbb{J}_{\underline{c}})(id\otimes\triangle)(\mathbb{J}_{\underline{c}})(\triangle\otimes id)(\mathbb{J}_{\underline{c}}^{-1})(\mathbb{J}_{\underline{c}}\otimes 1)^{-1} \\ &= \sum_{\mathbb{f},\mathbb{g},\mathbb{h}\in\mathbb{G}}\frac{J_{\underline{c}}(\mathbb{g},\mathbb{h})J_{\underline{c}}(\mathbb{f},\mathbb{g}\mathbb{h})}{J_{\underline{c}}(\mathbb{f},\mathbb{g})J_{\underline{c}}(\mathbb{f}\mathbb{g},\mathbb{h})}\mathbb{1}_{\mathbb{f}}\otimes\mathbb{1}_{\mathbb{g}}\otimes\mathbb{1}_{\mathbb{h}} \\ &= \sum_{\mathbb{f},\mathbb{g},\mathbb{h}\in\mathbb{G}}\partial(J_{\underline{c}})(\mathbb{f},\mathbb{g},\mathbb{h})\mathbb{1}_{\mathbb{f}}\otimes\mathbb{1}_{\mathbb{g}}\otimes\mathbb{1}_{\mathbb{h}}.\end{aligned}$$

Now suppose that $\mathbb{f} = \prod_{i=1}^n \mathbb{g}_i^{x_i m_i + r_i}$, $\mathbb{g} = \prod_{i=1}^n \mathbb{g}_i^{y_i m_i + s_i}$, $\mathbb{h} = \prod_{i=1}^n \mathbb{g}_i^{z_i m_i + t_i}$ for $0 \leq x_i, y_i, z_i, r_i,\ s_i, t_i \leq m_i - 1, 1 \leq i \leq n$. Let $f = \prod_{i=1}^n g_i^{r_i}, g = \prod_{i=1}^n g_i^{s_i},\ h = \prod_{i=1}^n g_i^{t_i}$. Then following a direct computation, we obtain

$$\begin{aligned}\partial(J_{\underline{c}})(\mathbb{f}, \mathbb{g}, \mathbb{h}) &= \prod_{i=1}^n \zeta_{\mathbb{m}_i}^{c_i t_i [\frac{r_i + s_i}{m_i}] m_i} \prod_{1 \leq j < k \leq n} \zeta_{m_j m_k}^{c_{jk} t_k [\frac{r_j + s_j}{m_j}] m_j} \\ &= \phi_{\underline{c}}(f, g, h).\end{aligned}$$

Applying Lemma 3.2, we obtain

$$\begin{aligned}\Phi_{\mathbb{J}_{\underline{c}}} \\ &= \sum_{0 \leq x_l, y_l, z_l \leq m_l - 1,\ 0 \leq r_l, s_l, t_l \leq m_l - 1, 1 \leq l \leq n} \prod_{i=1}^n \zeta_{\mathbb{m}_i}^{c_i t_i [\frac{r_i + s_i}{m_i}] m_i} \prod_{1 \leq j < k \leq n} \zeta_{m_j m_k}^{c_{jk} t_k [\frac{r_j + s_j}{m_j}] m_j} \\ &\qquad \times \mathbb{1}_{(\prod_{i=1}^n \mathbb{g}_i^{x_i m_i + r_i})} \otimes \mathbb{1}_{(\prod_{i=1}^n \mathbb{g}_i^{y_i m_i + s_i})} \otimes \mathbb{1}_{(\prod_{i=1}^n \mathbb{g}_i^{z_i m_i + t_i})} \\ &= \sum_{0 \leq r_l, s_l, t_l \leq m_l - 1, 1 \leq l \leq n} \prod_{i=1}^n \zeta_{\mathbb{m}_i}^{c_i t_i [\frac{r_i + s_i}{m_i}] m_i} \prod_{1 \leq j < k \leq n} \zeta_{m_j m_k}^{c_{jk} t_k [\frac{r_j + s_j}{m_j}] m_j} \\ &\qquad \times 1_{(\prod_{i=1}^n g_i^{r_i})} \otimes 1_{(\prod_{i=1}^n g_i^{s_i})} \otimes 1_{(\prod_{i=1}^n g_i^{t_i})} \\ &= \sum_{f, g, h \in G} \phi_{\underline{c}}(f, g, h) 1_f \otimes 1_g \otimes 1_h.\end{aligned}$$

□

Recall that $\mathbb{G}$ is the group of group-like elements of $H = u(\mathfrak{D}, \lambda, \mu)$, and G is the subgroup of $\mathbb{G}$. By (2.26), $\mathbf{u}_\alpha(\mu) \in \Bbbk\mathbb{G}$ for all $\alpha \in R^+$. In fact, we can prove that such elements are contained in $\Bbbk G$.

LEMMA 3.8. *$\mathbf{u}_\alpha(\mu) \in \Bbbk G$ for all $\alpha \in R^+$.*

PROOF. Let $R^+ = \cup_{J \in \Omega} R_J^+$, the union of its distinct connected component. Suppose $\alpha \in R_J^+$, $J \in \Omega$. Let $J = \{i_1, \cdots, i_\eta\}$ and $\{\alpha_{i_1}, \cdots, \alpha_{i_\eta}\}$ be the set of simple roots of R_J^+. Let $w = s_{j_1} s_{j_2} \cdots s_{j_P}$ be a fixed reduced presentation of the longest element of the Weyl group W_J in terms of simple reflections. Define

$$\beta_{j_l} = s_{j_1} s_{j_2} \cdots s_{j_{l-1}}(\alpha_{j_l}) \text{ for } 1 \leq l \leq P \tag{3.30}$$

and

$$\underline{a} = a_1 \beta_{j_i} + a_2 \beta_{j_2} + \cdots + a_P \beta_{j_P}, a \in \mathbb{N}^P. \tag{3.31}$$

Next we will prove that $u^a \in \Bbbk G$ for each $a \in \mathbb{N}^P$ inductively on $ht(\underline{a})$.

If $ht(\underline{a}) = 1$, then $\underline{a}$ must be a simple root contained in $\{\alpha_{i_1}, \cdots, \alpha_{i_\eta}\}$. According to (2.23), we have $u^a = \mu_a(1 - h^a) = \mu_a(1 - h_a^{N_J})$. By (3.17), we have $u^a = \mu_a(1 - h^a) \in \Bbbk G$.

Now assume that $u^a \in \Bbbk G$ holds for all $\underline{a} \in \mathbb{N}^P$ such that $ht(\underline{a}) < l$. For $1 \leq s \leq p$, define

$$e_s = (\underbrace{0, \cdots, 1}_{s}, 0, \cdots, 0).$$

Let $a \in \mathbb{N}^P$ such that $ht(\underline{a}) = l$. If $\underline{a} = e_s$ for some $1 \leq s \leq P$, then by (2.24),

$$u^a = \mu_a(1 - h^a) + \sum_{b, c \neq 0, \underline{b} + \underline{c} = \underline{a}} t_{b,c}^a \mu_b u^c,$$

and $h^a = h_a^{N_J}$. By (3.17) we have $\mu_a(1-h^a) \in \Bbbk G$. Since $b+c=a$ and $b,c \neq 0$, we get $ht(\underline{c}) < ht(\underline{a})$. By inductive assumption, $\sum_{b,c\neq 0,\underline{b}+\underline{c}=\underline{a}} t^a_{b,c}\mu_b u^c$ also belongs to $\Bbbk G$, hence $u^a \in \Bbbk G$. If $a = (a_1, a_2, \cdots, a_s, \cdots, 0) \neq e_s$ and $a_s > 0$ for some $1 \leq s \leq P$. By (2.25), we have $u^a = u^{a-e_s}u^{e_s}$. Since both the heights of $a - e_s$ and e_s are less than l, the elements $u^{a-e_s}, u^{e_s} \in \Bbbk G$ by the inductive assumption. Hence we have $u^a \in \Bbbk G$ as well. □

In the sequel, we denote by $A(H,\underline{c})$ the subalgebra of $H^{\mathbb{J}_{\underline{c}}}$ generated by G and $\{X_1, \cdots, X_\theta\}$. We are going to show that $A(H,\underline{c})$ is a quasi-Hopf algebra for $\underline{c} \in \Gamma(\mathfrak{D})$.

PROPOSITION 3.9. *The algebra $A(H,\underline{c})$ can be presented by the generators G and $\{X_1, \cdots, X_\theta\}$ and the following relations:*

$$gX_ig^{-1} = \chi_i(g)X_i, \text{ for all } 1 \leq i \leq \theta, g \in G, \tag{3.32}$$
$$ad_C(X_i)^{1-a_{ij}}(X_j) = 0, \text{ for all } i \neq j, i \sim j, \tag{3.33}$$
$$ad_C(X_i)(X_j) = \lambda_{ij}(1-h_ih_j), \text{ for all } i < j, i \nsim j, \tag{3.34}$$
$$X_\alpha^{N_J} = \mathbf{u}_\alpha(\mu), \text{ for all } \alpha \in R_J^+, J \in \Omega. \tag{3.35}$$

Moreover, $A(H,\underline{c})$ has a basis of the form

$$gX_{\beta_1}^{x_1}X_{\beta_2}^{x_2}\cdots X_{\beta_P}^{x_P},\ g \in G,\ 0 \leq x_i \leq N_J,\ \beta_i \in R_J^+.$$

PROOF. Since the relations (3.32)-(3.35) hold in H for the group $\mathbb{G}$ and the generators $X_1, \cdots X_\theta$, they hold as well for the subgroup G and $X_1, \cdots, X_\theta$. So we only need to prove the second part of the proposition.

Firstly, since the elements in $\mathbf{B} = \{gX_{\beta_1}^{x_1}X_{\beta_2}^{x_2}\cdots X_{\beta_P}^{x_P} \mid g \in G,\ 0 \leq x_i \leq N_J,\ \beta_i \in R_J^+\}$ come from the PBW type basis of H, hence they are linearly independent. Let $A = \Bbbk\mathbf{B}$, the $\Bbbk$-linear span of elements in $\mathbf{B}$. It is easy to see that $A \subset A(H,\underline{c})$ since $A(H,\underline{c})$ is generated by G and $\{X_1, \cdots, X_\theta\}$, and hence $\mathbf{B} \subset A(H,\underline{c})$. On the other hand, we can show that $A(H,\underline{c}) \subset A$. By (3.16), we have $\lambda_{ij}(1-h_ih_j) \in \Bbbk G$. According to Lemma 3.8, $\mathbf{u}_\alpha(\mu)$ is also contained in $\Bbbk G$ for all $\alpha \in R_J^+, J \in \Omega$. So by the relations (3.32)-(3.35), A is closed under the product inherited from $H^{\mathbb{J}_{\underline{c}}}$. This implies that A is an algebra. Since A contains the generators of $A(H,\underline{c})$, we obtain $A(H,\underline{c}) \subset A$, and hence $A = A(H,\underline{c})$. Thus, $\mathbf{B}$ is a basis of $A(H,\underline{c})$. □

PROPOSITION 3.10. *Suppose $\underline{c} \in \Gamma(\mathfrak{D})$. Then $A(H,\underline{c})$ is a quasi-Hopf subalgebra of $H^{\mathbb{J}_{\underline{c}}}$.*

PROOF. Since $A(H,\underline{c})$ is a subalgebra of $H^{\mathbb{J}_{\underline{c}}}$, in order to prove that it is also a quasi-Hopf subalgebra, we only need to show that $A(H,\underline{c})$ is closed under the comultiplication and the antipode of $H^{\mathbb{J}_{\underline{c}}}$.

First of all, we show that $A(H,\underline{c})$ is closed under the comultiplication $\Delta_{\mathbb{J}_{\underline{c}}}$ of $H^{\mathbb{J}_{\underline{c}}}$. Note that $\mathbb{J}_{\underline{c}}, \mathbb{J}_{\underline{c}}^{-1} \in \Bbbk\mathbb{G} \otimes \Bbbk\mathbb{G}$ and $G \subset \mathbb{G}$, so $\Delta_{\mathbb{J}_{\underline{c}}}(g) = \mathbb{J}_{\underline{c}}(g \otimes g)\mathbb{J}_{\underline{c}}^{-1} = g \otimes g$ for all $g \in G$. It remains to show that $\Delta_{\mathbb{J}_{\underline{c}}}(X_i) \in A(H,\underline{c}) \otimes A(H,\underline{c})$ for $1 \leq i \leq \theta$.

By Lemma 3.1 and 3.5, we have

$$\begin{aligned}\triangle_{\mathbb{J}_{\underline{c}}}(X_i) &= \mathbb{J}_{\underline{c}}(X_i\otimes 1+h_i\otimes X_i)\mathbb{J}_{\underline{c}}^{-1}\\ &= (\sum_{\mathbb{f},\mathbb{g}\in\mathbb{G}} J_{\underline{c}}(\mathbb{f},\mathbb{g})\mathbb{1}_{\mathbb{f}}\otimes\mathbb{1}_{\mathbb{g}})(X_i\otimes 1+h_i\otimes X_i)(\sum_{\mathbb{f},\mathbb{g}\in\mathbb{G}} J_{\underline{c}}^{-1}(\mathbb{f},\mathbb{g})\mathbb{1}_{\mathbb{f}}\otimes\mathbb{1}_{\mathbb{g}})\\ &= \sum_{\mathbb{f},\mathbb{g}\in\mathbb{G}}[\frac{J_{\underline{c}}(\mathbb{f}\eta_i^{-1},\mathbb{g})}{J_{\underline{c}}(\mathbb{f},\mathbb{g})}X_i\mathbb{1}_{\mathbb{f}}\otimes\mathbb{1}_{\mathbb{g}}+\frac{J_{\underline{c}}(\mathbb{f},\mathbb{g}\eta_i^{-1})}{J_{\underline{c}}(\mathbb{f},\mathbb{g})}\chi_{\mathbb{f}}^{-1}(h_i)\mathbb{1}_{\mathbb{f}}\otimes X_i\mathbb{1}_{\mathbb{g}}].\end{aligned}$$

Suppose $\mathbb{f}=\prod_{j=1}^n \mathbb{g}_j^{x_jm_j+k_j}$, $\mathbb{g}=\prod_{j=1}^n \mathbb{g}_j^{y_jm_j+l_j}$ for $0\le x_j,y_j,k_j,l_j\le m_j-1, 1\le j\le n$. Let $(k_j-r_{ij})'$ be the remainder of (k_j-r_{ij}) divided by m_j, $1\le j\le n$, and define

$$(3.36)\quad \psi_{ij}=\begin{cases}(k_j-r_{ij})'-(x_jm_j+k_j-r_{ij}), & \text{if } x_jm_j+k_j-r_{ij}\ge 0;\\ (k_j-r_{ij})'-\mathbb{m}_j-(x_jm_j+k_j-r_{ij}), & \text{if } x_jm_j+k_j-r_{ij}<0.\end{cases}$$

Then we have

$$(3.37)\quad \begin{aligned}\frac{J_{\underline{c}}(\mathbb{f}\eta_i^{-1},\mathbb{g})}{J_{\underline{c}}(\mathbb{f},\mathbb{g})} &= \frac{\prod_{j=1}^n \zeta_{\mathbb{m}_j}^{c_jl_j\psi_{ij}}\prod_{1\le s<t\le n}\zeta_{m_sm_t}^{c_{st}l_t\psi_{is}}}{\prod_{j=1}^n \zeta_{\mathbb{m}_j}^{-c_jl_jx_jm_j}\prod_{1\le s<t\le n}\zeta_{m_sm_t}^{-c_{st}l_tx_sm_s}}\\ &= \prod_{j=1}^n \zeta_{\mathbb{m}_j}^{c_jl_j((k_j-r_{ij})'-(k_j-r_{ij}))}\prod_{1\le s<t\le n}\zeta_{m_sm_t}^{c_{st}l_t((k_s-r_{is})'-(k_s-r_{is}))}\\ &= \prod_{j=1}^n \zeta_{\mathbb{m}_j}^{c_jl_j\varphi_{ij}}\prod_{1\le s<t\le n}\zeta_{m_sm_t}^{c_{st}l_t\varphi_{is}}.\end{aligned}$$

By (3.8) and (3.9), we obtain

$$(3.38)\quad \begin{aligned}\frac{J_{\underline{c}}(\mathbb{f},\mathbb{g}\eta_i^{-1})}{J_{\underline{c}}(\mathbb{f},\mathbb{g})}\chi_{\mathbb{f}}^{-1}(h_i) &= J_{\underline{c}}(\mathbb{f},\eta_i^{-1})\chi_{\mathbb{f}}^{-1}(h_i)\\ &= \prod_{j=1}^n \zeta_{\mathbb{m}_j}^{c_jr_{ij}x_jm_j}\prod_{1\le s<t\le n}\zeta_{m_sm_t}^{c_{st}r_{it}(x_sm_s)}\prod_{j=1}^n\zeta_{\mathbb{m}_j}^{-(x_jm_j+k_j)s_{ij}}\\ &= \prod_{j=1}^n \zeta_{\mathbb{m}_j}^{(c_jr_{ij}-s_{ij})x_jm_j}\prod_{1\le s<t\le n}\zeta_{m_sm_t}^{c_{st}r_{it}(x_sm_s)}\prod_{j=1}^n\zeta_{\mathbb{m}_j}^{-k_js_{ij}}\\ &= \prod_{j=1}^n \zeta_{\mathbb{m}_j}^{-k_js_{ij}}.\end{aligned}$$

Hence,

$$\begin{aligned}\triangle_{\mathbb{J}_{\underline{c}}}(X_i) &= \sum_{0\le x_r,y_r,k_r,l_r\le m_r-1,1\le r\le n}[\prod_{j=1}^{n}\zeta_{\mathfrak{m}_j}^{c_jl_j\varphi_{ij}}\prod_{1\le s<t\le n}\zeta_{m_sm_t}^{c_{st}l_t\varphi_{is}} \\ &\qquad \times X_i\mathbb{1}_{(\prod_{j=1}^n \mathfrak{g}_j^{x_jm_j+k_j})}\otimes \mathbb{1}_{(\prod_{j=1}^n \mathfrak{g}_j^{y_jm_j+l_j})} \\ &\quad +\prod_{j=1}^{n}\zeta_{\mathfrak{m}_j}^{-k_js_{ij}}\mathbb{1}_{(\prod_{j=1}^n \mathfrak{g}_j^{x_jm_j+k_j})}\otimes X_i\mathbb{1}_{(\prod_{j=1}^n \mathfrak{g}_j^{y_jm_j+l_j})}] \\ &= \sum_{0\le k_r,l_r\le m_r-1,1\le r\le n}[\prod_{j=1}^{n}\zeta_{\mathfrak{m}_j}^{c_jl_j\varphi_{ij}}\prod_{1\le s<t\le n}\zeta_{m_sm_t}^{c_{st}l_t\varphi_{is}} \\ &\qquad \times X_i 1_{(\prod_{j=1}^n g_j^{k_j})}\otimes 1_{(\prod_{j=1}^n g_j^{l_j})} \\ &\quad +\prod_{j=1}^{n}\zeta_{\mathfrak{m}_j}^{-k_js_{ij}}1_{(\prod_{j=1}^n g_j^{k_j})}\otimes X_i 1_{(\prod_{j=1}^n g_j^{l_j})}]. \\ &= \sum_{f,g\in G}\Psi(f,g)X_i1_f\otimes 1_g+\sum_{f\in G}\Theta_l(f)1_f\otimes X_i.\end{aligned} \tag{3.39}$$

Here the second identity follows from Lemma 3.2. So we have $\triangle_{\mathbb{J}_{\underline{c}}}(X_i)\in A(H,\underline{c})\otimes A(H,\underline{c})$. This implies that $A(H,\underline{c})$ is closed under $\triangle_{\mathbb{J}_{\underline{c}}}$ since $\triangle_{\mathbb{J}_{\underline{c}}}$ is multiplicative.

Next we show that $(S_{\mathbb{J}_{\underline{c}}}|_{A(H,\underline{c})},\beta_{\mathbb{J}_{\underline{c}}}\alpha_{\mathbb{J}_{\underline{c}}},1)$ is an antipode of $A(H,\underline{c})$. For each $\mathfrak{g}\in\mathbb{G}$, we have

$$S(\mathbb{1}_{\mathfrak{g}})=S(\frac{1}{|\mathbb{G}|}\sum_{\mathbb{h}\in\mathbb{G}}\chi_{\mathfrak{g}}(\mathbb{h})\mathbb{h})=\frac{1}{|\mathbb{G}|}\sum_{\mathbb{h}\in\mathbb{G}}\chi_{\mathfrak{g}}(\mathbb{h})\mathbb{h}^{-1}=\mathbb{1}_{\mathfrak{g}^{-1}}.$$

So

$$\alpha_{\mathbb{J}_{\underline{c}}}=\sum_{\mathbb{f},\mathfrak{g}\in\mathbb{G}}J_{\underline{c}}^{-1}(\mathbb{f},\mathfrak{g})\mathbb{1}_{\mathbb{f}^{-1}}\mathbb{1}_{\mathfrak{g}}=\sum_{\mathfrak{g}\in\mathbb{G}}J_{\underline{c}}^{-1}(\mathfrak{g}^{-1},\mathfrak{g})\mathbb{1}_{\mathfrak{g}}, \tag{3.40}$$

$$\beta_{\mathbb{J}_{\underline{c}}}=\sum_{\mathbb{f},\mathfrak{g}\in\mathbb{G}}J_{\underline{c}}(\mathbb{f},\mathfrak{g})\mathbb{1}_{\mathbb{f}}\mathbb{1}_{\mathfrak{g}^{-1}}=\sum_{\mathfrak{g}\in\mathbb{G}}J_{\underline{c}}(\mathfrak{g},\mathfrak{g}^{-1})\mathbb{1}_{\mathfrak{g}}. \tag{3.41}$$

It is obvious that $\beta_{\mathbb{J}_{\underline{c}}}$ is invertible with the inverse $\sum_{\mathfrak{g}\in\mathbb{G}}J_{\underline{c}}^{-1}(\mathfrak{g},\mathfrak{g}^{-1})\mathbb{1}_{\mathfrak{g}}$. By (3.27) we have

(3.42)

$$\begin{aligned}\beta_{\mathbb{J}_{\underline{c}}}\alpha_{\mathbb{J}_{\underline{c}}}&=\sum_{\mathfrak{g}\in\mathbb{G}}\frac{J_{\underline{c}}(\mathfrak{g},\mathfrak{g}^{-1})}{J_{\underline{c}}(\mathfrak{g}^{-1},\mathfrak{g})}\mathbb{1}_{\mathfrak{g}} \\ &=\sum_{0\le y_i,l_i\le m_i-1,1\le i\le n}\frac{\prod_{i=1}^n\zeta_{\mathfrak{m}_i}^{c_il_iy_im_i}\prod_{1\le s<t\le n}\zeta_{m_sm_t}^{c_{st}l_ty_sm_s}}{\prod_{i=1}^n\zeta_{\mathfrak{m}_i}^{c_il_i(y_i+1)m_i}\prod_{1\le s<t\le n}\zeta_{m_sm_t}^{c_{st}l_t(y_s+1)m_s}}\mathbb{1}_{(\prod_{i=1}^n\mathfrak{g}_i^{y_im_i+l_i})} \\ &=\sum_{0\le y_i,l_i\le m_i-1,1\le i\le n}\prod_{i=1}^{n}\zeta_{\mathfrak{m}_i}^{-c_il_im_i}\prod_{1\le s<t\le n}\zeta_{m_sm_t}^{-c_{st}l_tm_s}\mathbb{1}_{(\prod_{i=1}^n\mathfrak{g}_i^{y_im_i+l_i})} \\ &=\sum_{0\le l_i\le m_i-1,1\le i\le n}\prod_{i=1}^{n}\zeta_{\mathfrak{m}_i}^{-c_il_im_i}\prod_{1\le s<t\le n}\zeta_{m_sm_t}^{-c_{st}l_tm_s}1_{(\prod_{i=1}^n g_i^{l_i})} \\ &=\sum_{g\in G}\Upsilon(g)1_g.\end{aligned}$$

Here the fourth identity follows from Lemma 3.2. So we have proved that $\beta_{\mathbb{J}_{\underline{c}}}\alpha_{\mathbb{J}_{\underline{c}}} \in A(H,\underline{c})$. Next we show that $A(H,\underline{c})$ is also closed under $S_{\mathbb{J}_{\underline{c}}}$. It is obvious that $S_{\mathbb{J}_{\underline{c}}}(g) = g^{-1}$ for all $g \in G$. For each $X_i, 1 \leq i \leq \theta$ we have

$$
\begin{aligned}
(3.43)\quad S_{\mathbb{J}_{\underline{c}}}(X_i) &= \beta_{\mathbb{J}_{\underline{c}}} S(X_i)\beta_{\mathbb{J}_{\underline{c}}}^{-1} = -\beta_{\mathbb{J}_{\underline{c}}}(h_i^{-1}X_i)\beta_{\mathbb{J}_{\underline{c}}}^{-1} \\
&= -\sum_{\mathbb{f},\mathbb{g}\in\mathbb{G}} \frac{J_{\underline{c}}(\mathbb{f},\mathbb{f}^{-1})}{J_{\underline{c}}(\mathbb{g},\mathbb{g}^{-1})}\chi_{\mathbb{f}}(h_i)\mathbb{1}_{\mathbb{f}}X_i\mathbb{1}_{\mathbb{g}} \\
&= -\sum_{\mathbb{g}\in\mathbb{G}} \frac{J_{\underline{c}}(\mathbb{g}\eta_i^{-1},\mathbb{g}^{-1}\eta_i)}{J_{\underline{c}}(\mathbb{g},\mathbb{g}^{-1})}\chi_{\mathbb{g}\eta_i^{-1}}(h_i)X_i\mathbb{1}_{\mathbb{g}} \\
&= \sum_{0\leq x_l,k_l\leq m_l-1,1\leq l\leq n}\ \prod_{j=1}^{n}\zeta_{\mathbb{m}_j}^{-c_j(k_j-r_{ij})[(k_j-r_{ij})'-(k_j-r_{ij})]} \\
&\quad\times \prod_{1\leq s<t\leq n}\zeta_{\mathbb{m}_s\mathbb{m}_t}^{-c_{st}(k_t-r_{it})[(k_s-r_{is})'-(k_s-r_{is})]}X_i\mathbb{1}_{(\prod_{l=1}^{n}\mathbb{g}_l^{x_lm_l+k_l})} \\
&= \sum_{0\leq k_l\leq m_l-1,1\leq l\leq n}\ \prod_{j=1}^{n}\zeta_{\mathbb{m}_j}^{-c_j(k_j-r_{ij})[(k_j-r_{ij})'-(k_j-r_{ij})]} \\
&\quad\times \prod_{1\leq s<t\leq n}\zeta_{\mathbb{m}_s\mathbb{m}_t}^{-c_{st}(k_t-r_{it})[(k_s-r_{is})'-(k_s-r_{is})]}X_i1_{(\prod_{l=1}^{n}g_l^{k_l})} \\
&= \sum_{g\in G}\mathcal{F}_i(g)X_i1_g.
\end{aligned}
$$

Here the third identity follows from Lemma 3.1 and 3.5; the fourth identity follows from (3.8)-(3.10); and the fifth identity follows from Lemma 3.2. So $S_{\mathbb{J}_{\underline{c}}}(X_i) \in A(H,\underline{c})$, and $A(H,\underline{c})$ is closed under $S_{\mathbb{J}_{\underline{c}}}$ since $S_{\mathbb{J}_{\underline{c}}}$ is an anti-algebra morphism. □

By Proposition 3.9 and Proposition 3.10, we can prove Theorem 3.4 easily.

Proof of Theorem 3.4. Let $\underline{c} \in \Gamma(\mathfrak{D})$. By Proposition 3.9, we know that $A(H,\underline{c})$ is identical to $u(\mathfrak{D},\lambda,\mu,\Phi_{\underline{c}})$ as algebras. By Proposition 3.10, $u(\mathfrak{D},\lambda,\mu,\Phi_{\underline{c}})$ is a quasi-Hopf algebra. The comultiplication is given by $\triangle(g) = g\otimes g$ and (3.39). The antipode $(\mathcal{S},\alpha,1)$ is determined by (3.42) and (3.43), and the associator Φ is given by (3.28). □

4. Genuine Quasi-Hopf algebras of Cartan type

In this section, we construct examples of genuine quasi-Hopf algebras associated to finite connected Cartan matrices. Explicitly, let A be a finite connected Cartan matrix of type A_n, B_n, C_n with $n \geq 3$, or D_n with $n \geq 4$. We show that there exists a class of finite-dimensional genuine quasi-Hopf algebras $u(\mathfrak{D},0,\mu,\Phi_{\underline{c}})$ associated to A. Since the Cartan matrices are assumed to be connected, these quasi-Hopf algebras have only trivial linking relations and hence $\lambda = 0$. Throughout this section, the groups G and $\mathbb{G}$ are defined as those in Subsection 3.1. For each fixed Cartan matrix A of rank n, the corresponding root system R, positive root system R^+ with simple roots $\{\alpha_i|1 \leq i \leq n\}$ are defined in the same way as in Subsection 2.2. We always assume p,q,d are positive odd numbers such that $(p,q) = (p,d) = (q,d) = 1$. In order to solve some elementary congruence equations, we need the following well-known proposition form the basic number theory.

PROPOSITION 4.1. *Let l, m, k be nonzero integers. Then the congruence equation $lx \equiv m \mod k$ has a solution if and only if $(l,k)|m$, and the solution is unique modulo $\frac{k}{(l,k)}$.*

4.1. Genuine quasi-Hopf algebras of type A_n, B_n and C_n. For convenience, we always assume $n \geq 3$. Let $A = (a_{ij})_{1\leq i,j\leq n}$ be the Cartan matrix of type A_n, B_n or C_n. Let $G = \langle g_1 \rangle \times \langle g_2 \rangle \times \cdots \times \langle g_{2n} \rangle$ be the abelian group determined by

$$|g_i| = \begin{cases} pd, & \text{if } i = 4k+1,\ k \in \mathbb{N}; \\ qd, & \text{if } i = 4k+2,\ k \in \mathbb{N}; \\ pd^2, & \text{if } i = 4k+3,\ k \in \mathbb{N}; \\ qd^2, & \text{if } i = 4k,\ k \in \mathbb{N}. \end{cases} \tag{4.1}$$

The bigger $\mathbb{G}$ and the generators $\mathbb{g}_i, 1 \leq i \leq 2n$ associated to G are defined as those in Subsection 3.1. Let h_i, $1 \leq i \leq n$, be elements of $\mathbb{G}$ defined as follows:

$$h_i = \begin{cases} (\mathbb{g}_{2i-1}\mathbb{g}_{2i})^{pq}, & \text{if } i \text{ is odd }; \\ (\mathbb{g}_{2i-3}\mathbb{g}_{2i-2}\mathbb{g}_{2i-1}\mathbb{g}_{2i}\mathbb{g}_{2i+1}\mathbb{g}_{2i+2})^{pq}, & \text{if } i \text{ is even and } i \neq n; \\ (\mathbb{g}_{2i-3}\mathbb{g}_{2i-2}\mathbb{g}_{2i-1}\mathbb{g}_{2i})^{pq}, & \text{if } i \text{ is even and } i = n. \end{cases} \tag{4.2}$$

In order to give a datum of Cartan type associated to A, we need to give some characters on $\mathbb{G}$. Let a, b, r be the numbers given in Table 1. Define

$$\chi_1(\mathbb{g}_j) = \begin{cases} \zeta_{d^2}^{a}, & j = 1, 2, \\ \zeta_{d^2}^{r}, & j = 3, 4, \\ 1, & j \geq 5; \end{cases} \tag{4.3}$$

$$\chi_2(\mathbb{g}_j) = \begin{cases} \zeta_{d^2}^{a}, & j = 1, 2, \\ \zeta_{d^4}, & j = 3, \\ \zeta_{d^4}^{-2ad^2-1}, & j = 4, \\ \zeta_{d^2}^{b}, & j = 5, 6, \\ 1, & j \geq 7. \end{cases} \tag{4.4}$$

For $3 \leq i < n$, define

$$\chi_i(\mathbb{g}_j) = \begin{cases} \zeta_{d^2}^{-3b}, & \text{if } i \text{ is odd and } j = 2i-3, 2i-2, 2i+1, 2i+2, \\ \zeta_{d^2}^{b}, & \text{if } i \text{ is odd and } j = 2i-1, 2i, \\ \zeta_{d^2}^{b}, & \text{if } i \text{ is even and } j = 2i-3, 2i-2, 2i+1, 2i+2, \\ \zeta_{d^4}, & \text{if } i \text{ is even and } j = 2i-1, \\ \zeta_{d^4}^{-2bd^2-1}, & \text{if } i \text{ is even and } j = 2i, \\ 1, & \text{otherwise}. \end{cases} \tag{4.5}$$

If n is odd, define

$$\chi_n(\mathbb{g}_j) = \begin{cases} \zeta_{d^2}^{-3b}, & j = 2n-3, 2n-2, \\ \zeta_{d^2}^{b}, & j = 2n-1, 2n, \\ 1, & \text{otherwise}. \end{cases} \tag{4.6}$$

If n is even, define

$$\chi_n(\mathbb{g}_j) = \begin{cases} \zeta_{d^2}^{b}, & j = 2n-3, 2n-2, \\ \zeta_{d^4}, & j = 2n-1, \\ \zeta_{d^4}^{-1}, & j = 2n, \\ 1, & \text{otherwise}. \end{cases} \tag{4.7}$$

With these notations, we have

TABLE 1. a, b, r associated to A_n, B_n, C_n.

	a, b, r associated to A	Cartan matrix A
1.	$a = 1,\ b = 1,\ r = -3$	A_n
2.	$a = 2,\ b = 1,\ r = -6$	B_n
3.	$a = 1,\ b = 2,\ r = -4$	C_n

LEMMA 4.2. *$\mathfrak{D} = \mathfrak{D}(\mathbb{G}, (h_i)_{1 \le i \le n}, (\chi_i)_{1 \le i \le n}, A)$ is a datum of Cartan type.*

PROOF. Let $q_{ij} = \chi_j(h_i)$ for all $1 \le i, j \le n$. By definition of datum of Cartan type, we only need to show

$$q_{ij}q_{ji} = q_{ii}^{a_{ij}} = q_{jj}^{a_{ji}},\ 1 \le i \ne j \le n. \tag{4.8}$$

This follows a direct verification. □

LEMMA 4.3. *The set $\Gamma(\mathfrak{D})$ is not empty, and for each family $\underline{c} = (c_i, c_{jk})_{1 \le i \le 2n, 1 \le j < k \le 2n}$ in $\Gamma(\mathfrak{D})$, we have $c_i \ne 0$ for $1 \le i \le 2n$.*

PROOF. Notice that Equations (3.9)-(3.10) always have solutions, for example, $c_{ij} = 0$ for all $1 \le i < j \le 2n$. So we only need to prove that Equations (3.8) have solutions and $c_i \ne 0$ for $1 \le i \le 2n$. Let $m_i = |g_i|$ for each $1 \le i \le 2n$. Then Equations (3.8) in variables c_1, c_2, c_3, c_4 will be following equations:

$$c_1 a p^2 \equiv pq \mod pd, \tag{4.9}$$

$$c_2 a q^2 \equiv pq \mod qd, \tag{4.10}$$

$$c_3 p^2 \equiv pq \mod pd^2, \tag{4.11}$$

$$c_3 r p^2 d^2 \equiv 0 \mod pd^2, \tag{4.12}$$

$$c_4(-2ad^2 - 1)q^2 \equiv pq \mod qd^2, \tag{4.13}$$

$$c_4 r q^2 d^2 \equiv 0 \mod qd^2. \tag{4.14}$$

When $3 \le i \le n$ and i is odd, Equations (3.8) are as follows:

$$c_{2i-1} b p^2 \equiv pq \mod pd, \tag{4.15}$$

$$c_{2i} b q^2 \equiv pq \mod qd. \tag{4.16}$$

When $3 < i < n$ and i is even, Equations (3.8) become:

$$c_{2i-1} p^2 \equiv pq \mod pd^2, \tag{4.17}$$

$$c_{2i-1}(-3b)p^2 d^2 \equiv 0 \mod pd^2 \tag{4.18}$$

$$c_{2i}(-2bd^2 - 1)q^2 \equiv pq \mod qd^2, \tag{4.19}$$

$$c_{2i}(-3b)q^2 d^2 \equiv 0 \mod qd^2. \tag{4.20}$$

When $i = n$ is even, then Equations (3.8) are given by

$$c_{2n-1} p^2 \equiv pq \mod pd^2, \tag{4.21}$$

$$c_{2n-1}(-3b)p^2 d^2 \equiv 0 \mod pd^2 \tag{4.22}$$

$$c_{2n}(-1)q^2 \equiv pq \mod qd^2, \tag{4.23}$$

$$c_{2n}(-3b)q^2 d^2 \equiv 0 \mod qd^2. \tag{4.24}$$

It is obvious that any integers c_{2i-1}, c_{2i} are solutions of (4.12), (4.14), (4.18),(4.20), (4.22) and (4.24). Since $(ap^2, pd) = p, (aq^2, qd) = q, (p^2, pd^2) = p, ((-2ad^2 - 1)q^2, qd^2) = q$, the equations (4.9)-(4.14) have solutions by Proposition 4.1. These solutions can not be zero since $(p,d)=(q,d)=1$. By Proposition 4.1 again, one can show that (4.15)-(4.17), (4.19),(4.21) and (4.23) have nonzero solutions similarly. □

LEMMA 4.4. *There exist restricted root vector parameters μ for $\mathfrak{D}$ satisfying the condition:*

$$\begin{cases} \mu_\alpha \neq 0, & \text{if } \alpha = \alpha_i \text{ for some odd number } 1 \leq i \leq n, \\ \mu_\alpha = 0, & \text{otherwise.} \end{cases} \tag{4.25}$$

PROOF. Firstly, it is easy to verify that $|\chi_i(h_i)| = d^2$, $1 \leq i \leq n$. So for all $1 \leq i \leq n$ and i odd, we have $h_i^{d^2} = (\mathbb{g}_{2i-1}\mathbb{g}_{2i})^{pqd^2} = g_{2i-1}^{qd} g_{2i}^{pd} \in G$, $h_i^{d^2} \neq 1$ and $\chi_i^{d^2} = \varepsilon$. By the definition of restricted root vector parameters (see Definition 3.3), there is a family of restricted root vector parameters associated to $\mathfrak{D}$ satisfying (4.25). □

PROPOSITION 4.5. *Let $\underline{c} \in \Gamma(\mathfrak{D})$, and μ a family of restricted root vector parameters for $\mathfrak{D}$ satisfying condition* (4.25). *Then $u(\mathfrak{D}, 0, \mu, \Phi_{\underline{c}})$ is a genuine quasi-Hopf algebra associated to A.*

PROOF. By Lemma 4.3-4.4 and Theorem 3.4, $u(\mathfrak{D}, 0, \mu, \Phi_{\underline{c}})$ is a finite-dimensional quasi-Hopf algebra. So we only need to prove that $u(\mathfrak{D}, 0, \mu, \Phi_{\underline{c}})$ is genuine.

Let $\mathcal{I}$ be the ideal of $u(\mathfrak{D}, 0, \mu, \Phi_{\underline{c}})$ generated by

$$\{X_i, 1 - g_j | 1 \leq i \leq n, j = 1 \text{ or } 2 \mod 4\}.$$

It is obvious that $\mathcal{I}$ is a quasi-Hopf ideal of $u(\mathfrak{D}, 0, \mu, \Phi_{\underline{c}})$. Let $G' = \langle g_i | i = 0 \text{ or } 3 \mod 4\rangle$. Then we have

$$\Bbbk G' = u(\mathfrak{D}, 0, \mu, \Phi_{\underline{c}})/\mathcal{I}, \tag{4.26}$$

which is a quasi-Hopf algebra with associator given by

$$\overline{\Phi_{\underline{c}}} = \sum_{e,f,g \in G'} \phi_{\underline{c}}|_{G'}(e,f,g) 1_e \otimes 1_f \otimes 1_g.$$

By lemma 4.3 and Lemma 2.9, $\phi_{\underline{c}}|_{G'}$ is not a 3-coboundary on G'. Hence $(\Bbbk G', \overline{\Phi_{\underline{c}}})$ is a genuine quasi-Hopf algebra. This implies that $u(\mathfrak{D}, 0, \mu, \Phi_{\underline{c}})$ is also genuine. Otherwise, $(\Bbbk G', \overline{\Phi_{\underline{c}}}) = u(\mathfrak{D}, 0, \mu, \Phi_{\underline{c}})/\mathcal{I}$ should not be genuine, which is a contradiction. □

4.2. Quasi-Hopf algebras of Cartan type D_n. In this subsection, we always assume that $n \geq 4$. Let $G = \langle g_1 \rangle \times \cdots \times \langle g_{2n} \rangle$ such that

$$|g_i| = \begin{cases} pd, & i = 1 \text{ or } 4k-1 \text{ for } k \geq 1; \\ qd, & i = 2 \text{ or } 4k \text{ for } k \geq 1; \\ pd^2, & i = 4k+1 \text{ for } k \geq 1; \\ qd^2, & i = 4k+2 \text{ for } k \geq 1. \end{cases} \tag{4.27}$$

Let $(h_i)_{1 \leq i \leq n}$ be a family of elements in $\mathbb{G}$ given by

$$h_i = \begin{cases} (\mathbb{g}_{2i-1}\mathbb{g}_{2i})^{pq}, & \text{if } i = 1 \text{ or } 2k \text{ for } k \geq 1; \\ (\prod_{j=1}^{8} \mathbb{g}_j)^{pq}, & \text{if } i = 3; \\ (\mathbb{g}_{2i-3}\mathbb{g}_{2i-2}\mathbb{g}_{2i-1}\mathbb{g}_{2i}\mathbb{g}_{2i+1}\mathbb{g}_{2i+2})^{pq}, & \text{if } i = 2k+1 \neq n \text{ for } k \geq 2; \\ (\mathbb{g}_{2n-3}\mathbb{g}_{2n-2}\mathbb{g}_{2n-1}\mathbb{g}_{2n})^{pq}, & \text{if } i = n \text{ is } odd. \end{cases} \tag{4.28}$$

Now define a family $(\chi_i)_{1 \leq i \leq n}$ of characters on $\mathbb{G}$ as follows.

For $i = 1, 2$,

$$\chi_i(g_j) = \begin{cases} \zeta_{d^2}, & j = 2i-1, 2i; \\ \zeta_{d^2}^{-3}, & j = 5, 6; \\ 1, & \text{otherwise.} \end{cases} \tag{4.29}$$

The character χ_3 is defined by

$$\chi_3(g_j) = \begin{cases} \zeta_{d^2}, & j = 1,2,3,4,7,8; \\ \zeta_{d^4}, & j = 5; \\ \zeta_{d^4}^{-4d^2-1}, & j = 6; \\ 1, & \text{otherwise.} \end{cases} \tag{4.30}$$

For $3 < i < n$,

$$\chi_i(\mathbb{g}_j) = \begin{cases} \zeta_{d^2}^{-3}, & \text{if } i \text{ is even and } j = 2i-3, 2i-2, 2i+1, 2i+2; \\ \zeta_{d^2}, & \text{if } i \text{ is even and } j = 2i-1, 2i; \\ \zeta_{d^4}, & \text{if } i \text{ is odd and } j = 2i-1; \\ \zeta_{d^4}^{-2d^2-1}, & \text{if } i \text{ is odd and } j = 2i; \\ \zeta_{d^2}, & \text{if } i \text{ is odd and } j = 2i-3, 2i-2, 2i+1, 2i+2; \\ 1, & \text{otherwise.} \end{cases} \tag{4.31}$$

If n is even,

$$\chi_n(\mathbb{g}_j) = \begin{cases} \zeta_{d^2}, & \text{if } j = 2n-1, 2n; \\ \zeta_{d^2}^{-3}, & \text{if } j = 2n-3, 2n-2; \\ 1, & \text{otherwise.} \end{cases} \tag{4.32}$$

If n is odd,

$$\chi_n(\mathbb{g}_j) = \begin{cases} \zeta_{d^2}, & \text{if } j = 2n-3, 2n-2; \\ \zeta_{d^4}, & \text{if } j = 2n-1; \\ \zeta_{d^4}^{-1}, & \text{if } j = 2n; \\ 1, & \text{otherwise.} \end{cases} \tag{4.33}$$

With these notations, we can verify that $\mathfrak{D} = \mathfrak{D}(\mathbb{G}, (h_i)_{1\leq i\leq n}, (\chi_i)_{1\leq i\leq n}, D_n)$ is a datum of Cartan type. Moreover, we have the following two lemmas, and the proofs are similar to those of Lemma 4.3 and 4.4, hence are omitted.

LEMMA 4.6. *$\Gamma(\mathfrak{D})$ is a nonempty set, and for each family $\underline{c} = (c_i, c_{jk})_{1\leq i\leq 2n, 1\leq j<k\leq 2n}$ in $\Gamma(\mathfrak{D})$, we have $c_i \neq 0$ for $1 \leq i \leq 2n$.*

LEMMA 4.7. *There exists a family of restricted root vector parameters μ for $\mathfrak{D}$ satisfying the following condition:*

$$\begin{cases} \mu_\alpha \neq 0, & \textit{if } \alpha = \alpha_1 \textit{ or } \alpha_i \textit{ for some even number } 1 \leq i \leq n, \\ \mu_\alpha = 0, & \textit{otherwise.} \end{cases} \tag{4.34}$$

The proof of the following proposition is similar to that of Proposition 4.5.

PROPOSITION 4.8. *Let $\underline{c} \in \Gamma(\mathfrak{D})$, and μ a family of restricted root vector parameters for $\mathfrak{D}$ satisfying the condition (4.34). Then $u(\mathfrak{D}, 0, \mu, \Phi_{\underline{c}})$ is a genuine quasi-Hopf algebra.*

REMARK 4.9. In this section we only construct examples genuine quasi-Hopf algebras of type A, B, C and D. In fact, examples of genuine quasi-Hopf algebras of type E, F and G can be also constructed in the same way. But this will make the paper several pages longer, and add a lot of annoying calculations. In order to make the paper more readable, we don't list those examples here.

5. Small quasi-quantum groups

In this section, we will introduce small quasi-quantum groups, which can be viewed as natural generalizations of small quantum groups. Associated to each finite Cartan matrix, we show that there are lots of small quasi-quantum groups which are genuine.

5.1. Small quasi-quantum groups. Let $A = (a_{ij})_{1\leq i,j\leq n}$ be a finite Cartan matrix and $\mathbf{D}(A) = \begin{pmatrix} A & 0 \\ 0 & A \end{pmatrix}$. Let $G = \mathbb{Z}_m^n = \langle g_1\rangle \times \cdots \times \langle g_n\rangle, \mathbb{G} = \mathbb{Z}_{m^2}^n = \langle \mathbb{g}_1\rangle \times \cdots \times \langle \mathbb{g}_n\rangle$ and $g_i = \mathbb{g}_i^m, 1 \leq i \leq n$.

DEFINITION 5.1. *Let $\mathfrak{D}(\mathbb{G}, (h_i)_{1\leq i\leq 2n}, (\chi_i)_{1\leq i\leq 2n}, \mathbf{D}(A))$ be a datum of finite Cartan type such that $h_i = h_{n+i} \in G, \chi_i = \chi_{i+n}^{-1}$, $1 \leq i \leq n$. Suppose that $\underline{c} \in \Gamma(\mathfrak{D})$ is nonzero. We call the quasi-Hopf algebra $Qu(\mathfrak{D}, \lambda, \Phi_{\underline{c}}) := u(\mathfrak{D}, \lambda, 0, \Phi_{\underline{c}})$ a small quasi-quantum group if the restricted linking parameters λ satisfying:*

$$\begin{cases} \lambda_{i,j} \neq 0, & \textit{if } j = i + n, \\ \lambda_{i,j} = 0, & \textit{otherwise.} \end{cases} \tag{5.1}$$

Next we show that small quasi-quantum groups $Qu(\mathfrak{D}, \lambda, \Phi_{\underline{c}})$ are nature generalizations of small quantum groups. Firstly, we will present $Qu(\mathfrak{D}, \lambda, \Phi_{\underline{c}})$ using another set of generator and relations. Let $V = \mathbb{k}\{E_1, \cdots, E_n\}$ and $U = \mathbb{k}\{F_1, \cdots, F_n\}$ be braided vector spaces with the braiding constants $(q_{ij})_{1\leq i,j\leq n}$ and $(q'_{ij})_{1\leq i,j\leq n}$ respectively, where $q_{ij} = \chi_j(h_i), q'_{ij} = q_{ji}^{-1}, 1 \leq i, j \leq n$. By Definition 2.3, we have $q_{ij}q_{ji} = q_{ii}^{a_{ij}}, 1 \leq i \neq j \leq n$. Hence V is a braided vector space of Cartan type associated to A. Since

$$q'_{ij}q'_{ji} = q_{ji}^{-1}q_{ij}^{-1} = q_{ii}^{-a_{ij}} = q'^{a_{ij}}_{ii},\ 1 \leq i \neq j \leq n,$$

U is also a braided vector space of Cartan type associated to A. So we can define braided commutators, the braided adjoint action and the root vectors over $T(V)$ and $T(U)$ (see Subsection 2.3 for details). Now let R^+ be the positive root system corresponding to the Cartan matrix A with respect to the simple roots $\alpha_1, \cdots, \alpha_n$. Let $E_\alpha, F_\alpha, \alpha \in R^+$ be the root vectors such that $E_{\alpha_i} = E_i, F_{\alpha_i} = F_i$ for each $1 \leq i \leq n$. Let Ω be the set of connected components of $I = \{1, \cdots, n\}$, and R_J^+ the positive root system corresponds to $J \in \Omega$.

PROPOSITION 5.2. *$Qu(\mathfrak{D}, \lambda, \Phi_{\underline{c}})$ is generated by $g_i, E_i, F_i, 1 \leq i \leq n$ subject to the following relations:*

$$g_iE_jg_i^{-1} = \chi_j(g_i)E_j,\ g_iF_jg_i^{-1} = \chi_j^{-1}(g_i)F_j,\ 1 \leq i, j \leq n, \tag{5.2}$$

$$E_iF_j - F_jE_i = \delta_{ij}\lambda_{i,i+n}(h_i^{-1} - h_i),\ \lambda_{i,i+n} \neq 0,\ 1 \leq i, j \leq n, \tag{5.3}$$

$$ad_c(E_i)^{1-a_{ij}}(E_j) = 0,\ ad_c(F_i)^{1-a_{ij}}(F_j) = 0,\ 1 \leq i \neq j \leq n, \tag{5.4}$$

$$E_\alpha^{N_J} = 0,\ F_\alpha^{N_J} = 0,\ \alpha \in R_J^+,\ J \in \Omega. \tag{5.5}$$

The comultiplication is determined by

$$\triangle(E_i) = \sum_{f,g\in G} \Psi_i(f, g)E_i1_f \otimes 1_g + h_i \otimes E_i, \tag{5.6}$$

$$\triangle(F_i) = \sum_{f,g\in G} \Psi_{i+n}(f, g)\chi_g(h_i)F_i1_f \otimes 1_g + 1 \otimes F_i. \tag{5.7}$$

The associator is $\Phi_{\underline{c}}$ and the antipode $(\mathcal{S}, \alpha, 1)$ is given by

$$\alpha = \textstyle\sum_{g \in G} \Upsilon(g) 1_g, \tag{5.8}$$

$$\mathcal{S}(F_i) = \chi_i^{-1}(h_i) \textstyle\sum_{g\in G} \chi_g(h_i) \mathcal{F}_{i+n}(g) F_i 1_g, \ \mathcal{S}(E_i) = \textstyle\sum_{g \in G} \mathcal{F}_i(g) E_i 1_g, \tag{5.9}$$

for $1 \leq i \leq n$.

PROOF. By Theorem 3.4, $Qu(\mathfrak{D}, \lambda, \Phi_{\underline{c}})$ is generated by G and $X_j, 1 \leq j \leq 2n$, subject to the relations (3.18)-(3.21). Let $E_i = X_i$, $F_i = X_{i+n}h_i^{-1}$, $1 \leq i \leq n$. Since $F_i h_i = X_{i+n}$, it is obviou that the algebra $Qu(\mathfrak{D}, \lambda, \Phi_{\underline{c}})$ is also generated by $g_i, E_i, F_i, 1 \leq i \leq n$. So we only need to verify that (5.2)-(5.9) are equivalent to (3.18)-(3.24).

We show first that the relations (5.2)-(5.5) are equivalent to the relations (3.18)-(3.21). It is easy to see that (5.2) equals (3.18). For all $1 \leq i, j \leq n$, we have

$$\begin{aligned} E_i F_j - F_j E_i &= X_i X_{j+n} h_j^{-1} - X_{j+n} h_j^{-1} X_i \\ &= X_i X_{j+n} h_j^{-1} - \chi_{j+n}(h_j^{-1}) X_i X_{j+n} h_j^{-1} \\ &= X_i X_{j+n} h_j^{-1} - \chi_j(h_j) X_i X_{j+n} h_j^{-1} \\ &= ad_c(X_i)(X_j) h_j^{-1}. \end{aligned}$$

So Relation (5.3) is equivalent to Relation (3.20). For $1 \leq i \neq j \leq n$, we have

$$\begin{aligned} ad_c(F_i)(F_j) &= F_i F_j - q'_{ij} F_j F_i = F_i F_j - \chi_i^{-1}(h_j) F_j F_i \\ &= X_{i+n} h_i^{-1} X_{j+n} h_j^{-1} - \chi_i^{-1}(h_j) X_{j+n} h_j^{-1} X_{i+n} h_i^{-1} \\ &= \chi_j^{-1}(h_i^{-1}) X_{i+n} X_{j+n} h_i^{-1} h_j^{-1} - \chi_i^{-1}(h_j) \chi_i^{-1}(h_j^{-1}) X_{j+n} X_{i+n} h_i^{-1} h_j^{-1} \\ &= \chi_j(h^i)[X_{i+n} X_{j+n} - \chi_j^{-1}(h^i) X_{j+n} X_{i+n}] h_i^{-1} h_j^{-1} \\ &= \chi_j(h^i)[ad_c(X_{i+n})(X_{j+n})] h_i^{-1} h_j^{-1}. \end{aligned}$$

By induction on $l \geq 1$, we obtain

$$(ad_c(F_i))^l(F_j) = \chi_j^l(h_i) \chi_i^{\frac{l(l-1)}{2}}(h_i)[(ad_c(X_{i+n}))^l(X_{j+n})] h_i^{-l} h_j^{-1}. \tag{5.10}$$

Hence, the relation (5.4) is equivalent to the relation (3.19). Since restricted root vector parameters $\mu = 0$, it is easy to see that (5.5) is equivalent to the relation (3.21).

Next we will verify (5.6)-(5.9). Formula (5.6) follows from (3.22) and the fact $h_i = \sum_{f\in G} \Theta_i(f) 1_f$. Formula (5.7) follows from

$$\begin{aligned} \triangle(F_i) &= \triangle(X_{i+n} h_i^{-1}) \\ &= (\sum_{f,g\in G} \Psi_{i+n}(f,g) X_{i+n} 1_f \otimes 1_g + \sum_{f \in G} \Theta_{i+n}(f) 1_f \otimes X_{i+n})(h_i^{-1} \otimes h_i^{-1}) \\ &= \sum_{f,g\in G} \Psi_{i+n}(f,g) \chi_g(h_i) F_i 1_f \otimes 1_g + \sum_{f\in G} \Theta_{i+n}(f) \chi_f(h_i) 1_f \otimes F_i \\ &= \sum_{f,g \in G} \Psi_{i+n}(f,g) \chi_g(h_i) F_i 1_f \otimes 1_g + 1 \otimes F_i. \end{aligned}$$

Formula (5.8) is obvious. Formula (5.9) follows from

$$\begin{aligned}\mathcal{S}(F_i) &= \mathcal{S}(h_i^{-1})\mathcal{S}(X_{i+n}) = h_i \sum_{g\in G} \mathcal{F}_{i+n}(g) X_{i+n} 1_g \\ &= \chi_{i+n}(h_i) \sum_{g\in G} \chi_g(h_i^2)\mathcal{F}_{i+n}(g) F_i 1_g,\end{aligned}$$

where χ_g is defined by (3.3). □

Let $\mathfrak{D}(\mathbb{G}, (h_i)_{1\le i\le 2n}, (\chi_i)_{1\le i\le 2n}, \mathbf{D}(A))$ be a datum of Cartan type such that $h_i = h_{n+i} \in G, \chi_i = \chi_{i+n}^{-1}$, $1 \le i \le n$. Note that $\Gamma(\mathfrak{D})$ is not empty since $0 \in \Gamma(\mathfrak{D})$. In Definition 5.1, $\underline{c} \in \Gamma(\mathfrak{D})$ is assumed to be nonzero. In fact, if $\underline{c} = 0$, we can also define a quasi-Hopf algebra with the same generators and relations.

Take an element $\underline{c} \in \Gamma(\mathfrak{D})$. Let $H_{\underline{c}}$ be the algebra generated by G and $E_i, F_i, 1 \le i \le n$ with Relations (5.2)-(5.5). Define an algebra morphism $\triangle : H_{\underline{c}} \to H_{\underline{c}} \otimes H_{\underline{c}}$ by (5.6)-(5.7), and an algebra antimorphism $S : H_{\underline{c}} \to H_{\underline{c}}$ by (5.9). Let α be an element of $H_{\underline{c}}$ defined by (5.8). Define an algebra morphism $\varepsilon : H_{\underline{c}} \to \mathbb{k}$ such that $\varepsilon(g) = 1, \varepsilon(E_i) = \varepsilon(F_i) = 0$ for $g \in G, 1 \le i \le n$. Then we have the following identification of the algebra $H_{\underline{c}}$.

PROPOSITION 5.3. (1) *If $\underline{c} = 0$, then $(H_{\underline{c}}, \triangle, \varepsilon, S)$ is isomorphic to the Andruskiewitsch-Schneider Hopf algebra $u(\mathfrak{D}', \lambda, 0)$, where $\mathfrak{D}' = \mathfrak{D}(G, (h_i)_{1\le i\le 2n}, (\chi_i')_{1\le i\le 2n}, \mathbf{D}(A))$, and $\chi_i' = \chi_i|_G, 1 \le i \le n$. Each small quantum group is isomorphic to some $H_{\underline{0}}$ as Hopf algebras.*

(2) *If $\underline{c} \neq 0$, then $(H_{\underline{c}}, \triangle, \varepsilon, \Phi_{\underline{c}}, S, \alpha, 1)$ is isomorphic to some small quasi-quantum group $Qu(\mathfrak{D}, \lambda, \Phi_{\underline{c}})$.*

PROOF. (1). Observe that the functions defined by (3.12)-(3.14) equal to the constant function 1 if $\underline{c} = 0$. Hence, the algebra morphism $\Upsilon : H_{\underline{c}} \to u(\mathfrak{D}', \lambda, 0)$ given by

$$\Upsilon(g) = g,\ \Upsilon(E_i) = X_i,\ \Upsilon(F_i) = X_{i+n}h_i,\ g \in G,\ 1 \le i \le n$$

is a Hopf algebra isomorphism.

(2). It follows directly from Proposition 5.2. □

From this proposition, we can see that small quasi-quantum groups are natural generalizations of small quantum groups. In the following, we call a small quasi-quantum group *genuine* if it is a genuine quasi-Hopf algebra. Next, we will construct explicitly examples of genuine small quasi-quantum groups.

5.2. Examples of genuine small quasi-quantum groups. Let $A = (a_{ij})_{1\le i,j\le n}$ be a finite Cartan matrix, $(d_1, \cdots, d_n)$ a vector with elements in $\{1, 2, 3\}$ such that the matrix $(d_i a_{ij})_{1\le i,j\le n}$ is symmetric. Let q be a primitive N-th root of unity, where N is an odd positive integer. Moreover, in case A has a connected component of type G_2, we will add one more assumption that N is prime to 3. Let p be a positive odd integer and $m = pN$. Choose ζ_{m^2} and ζ_m such that $\zeta_{m^2}^m = \zeta_m, \zeta_m^p = q$. Let $G = \mathbb{Z}_m^n = \langle g_1\rangle \times \cdots \times \langle g_n\rangle$, $\mathbb{G} = \mathbb{Z}_{m^2}^n = \langle \mathbb{g}_1\rangle \times \cdots \times \langle \mathbb{g}_n\rangle$ and $g_i = \mathbb{g}_i^m, 1 \le i \le n$. Let $l \in \{1, \cdots, m-1\}$ be a fixed integer such that $lp \neq 0$ mod m. Let $\{\chi_i | 1 \le i \le n\}$ be the characters on $\mathbb{G}$ defined as follows:

$$\chi_i(\mathbb{g}_j) = \begin{cases} \zeta_{m^2}^{2pd_i}, & \text{if } i = j; \\ \zeta_{m^2}^{pd_i a_{ij}}, & \text{if } i \neq j \text{ and } a_{ij} \neq 0; \\ 1, & \text{if } a_{ij} = 0. \end{cases} \tag{5.11}$$

With the above notations, it is easy to verify that $\mathfrak{D} = \mathfrak{D}(\mathbb{G}, (h_i)_{1\leq i\leq 2n}, (\chi_i)_{1\leq i\leq 2n}, \mathbf{D}(A))$ is a datum of Cartan type, where $h_i = h_{i+n} = \mathbb{g}_i^{ml}$ and $\chi_{i+n} = \chi_i^{-1}$ for $1 \leq i \leq n$.

LEMMA 5.4. *There are nonzero elements $\underline{c} = (c_i, c_{jk})_{1\leq i\leq n, 1\leq j<k\leq n}$ in $\Gamma(\mathfrak{D})$ such that $c_i = k_i N$ for some $1 \leq k_i < p$, $1 \leq i \leq n$.*

PROOF. It is easy to see that (3.9)-(3.10) are solvable for $(c_{jk})_{1\leq j<k\leq n}$. Thus, it suffices to show that Equations (3.8) have solutions $c_i = k_i N$ for some $1 \leq k_i < p$, $1 \leq i \leq n$. Note that

$$s_{ij} = \begin{cases} ml, & \text{if } i = j; \\ 0, & \text{if } i \neq j. \end{cases} \quad r_{ij} = \begin{cases} 2pd_i & \text{if } 1 \leq i = j \leq n; \\ -2pd_i & \text{if } n+1 \leq i = j \leq 2n; \\ pd_i a_{ij} & \text{if } 1 \leq i \neq j \leq n; \\ -pd_{i-n} a_{i-n,j-n} & \text{if } n+1 \leq i \neq j \leq 2n. \end{cases}$$

So Equations (3.8) can be written as follows:

$$ml \equiv 2c_i pd_i \mod m, \tag{5.12}$$

$$ml \equiv -2c_i pd_i \mod m, \tag{5.13}$$

$$0 \equiv c_i pd_i a_{ij} \mod m, \ i \neq j, a_{ij} \neq 0, \tag{5.14}$$

$$0 \equiv -c_i pd_i a_{ij} \mod m, \ i \neq j, a_{ij} \neq 0. \tag{5.15}$$

It is obvious that $\{c_i = kN | 0 \leq k < p\}$ gives solutions of Equations (5.12)-(5.15). We have thus proved the lemma. □

LEMMA 5.5. *There exists a family of restricted linking parameters λ for $\mathfrak{D}$ satisfying the following condition:*

$$\begin{cases} \lambda_{ij} \neq 0, & \text{if } i+n = j; \\ \lambda_{ij} = 0, & \text{otherwise.} \end{cases} \tag{5.16}$$

PROOF. Note that for each $1 \leq i \leq n$, we have $\chi_i \chi_{i+n} = \varepsilon$ because $\chi_{i+n} = \chi_i^{-1}$. Since we assume that $lp \neq 0 \mod m$, which implies that $l \neq 0 \mod N$. Hence $2l \neq 0 \mod N$ because N is odd. It follows that $h_i h_{i+n} = \mathbb{g}_i^{2ml} = g_i^{2l} \in G$ and $h_i h_{i+n} \neq 1$. By Definition 3.3, there exists a family of restricted linking parameters λ for $\mathfrak{D}$ satisfying (5.16). □

Let $\underline{c}$ be a nonzero parameter in $\Gamma(\mathfrak{D})$, and λ a family of restricted linking parameters satisfying (5.16). Then we obtain a small quasi-quantum group $Qu(\mathfrak{D}, \lambda, \Phi_{\underline{c}})$. Now we can prove the following proposition.

PROPOSITION 5.6. *Suppose $l > 1$, $l | m$ and $l \nmid c_i$ for $1 \leq i \leq n$. Then $Qu(\mathfrak{D}, \lambda, \Phi_{\underline{c}})$ is genuine.*

PROOF. Let I be the quasi-Hopf ideal of $Qu(\mathfrak{D}, \lambda, \Phi_{\underline{c}})$ generated by $\{X_i | 1 \leq i \leq 2n\}$, and $\widetilde{u} = Qu(\mathfrak{D}, \lambda, \Phi_{\underline{c}})/I$. It is evident that $\widetilde{u} \cong \Bbbk\overline{G}$, where $\overline{G} = G/\langle g_i^{2l} - 1 | 1 \leq i \leq n\rangle$. Since $\gcd(2l, m) = l$, we have $\overline{G} = G/\langle g_i^{l} - 1 | 1 \leq i \leq n\rangle$. Here $\gcd(2l, m)$ means the greatest common divisor of $2l$ and m. For an element $x \in \Bbbk G$, we denote by $\overline{x}$ the corresponding image in $\Bbbk\overline{G}$. Let G° be the subgroup of G generated by $\{g_i^{\frac{m}{l}} | 1 \leq i \leq n\}$. Then we have a group isomorphism

$$\varphi : G^{\circ} \to \overline{G}, \ \varphi(g_i^{\frac{m}{l}}) = \overline{g_i}, 1 \leq i \leq n.$$

Let $f=\prod_{i=1}^n g_i^{n_i}\in G$. We have

$$\begin{aligned}\overline{1_f} &= \frac{1}{|G|}\sum_{g\in G}\chi_f(g)\overline{g}\\ &= \frac{1}{|G|}\sum_{0\le s_j<l,1\le j\le n}[(\sum_{0\le k_j<\frac{m}{l},1\le j\le n}\prod_{i=1}^n \zeta_m^{n_i(k_il+s_i)})(\prod_{i=1}^n \overline{g_i}^{s_i})].\end{aligned}$$

If $f\notin G^\circ$, then $\sum_{0\le k_j<\frac{m}{l},1\le j\le n}\prod_i^n \zeta_m^{n_i(k_il+s_i)}=0$, hence $\overline{1_f}=0$.

If $f=\prod_{i=1}^n g_i^{a_i\frac{m}{l}}\in G^\circ$, then

$$\begin{aligned}\overline{1_f} &= \frac{1}{|G|}\sum_{0\le s_j<l,1\le j\le n}[(\sum_{0\le k_j<\frac{m}{l},1\le j\le n}\prod_{i=1}^n \zeta_m^{a_i\frac{m}{l}(k_il+s_i)})(\prod_{i=1}^n \overline{g_i}^{s_i})]\\ &= \frac{1}{|G|}\sum_{0\le s_j<l,1\le j\le n}[\prod_{i=1}^n(\frac{m}{l}\zeta_m^{a_i\frac{m}{l}s_i}))(\prod_{i=1}^n \overline{g_i}^{s_i})]\\ &= \frac{m^n}{l^n}\frac{1}{|G|}\sum_{0\le s_j<l,1\le j\le n}[\prod_{i=1}^n((\zeta_m^{a_i\frac{m}{l}s_i}))(\prod_{i=1}^n \overline{g_i}^{s_i})] \qquad (5.17)\\ &= \frac{1}{|G'|}[\sum_{0\le s_j<l,1\le j\le n}[\prod_j \zeta_l^{a_j(s_i)}\prod_j(\overline{g_j}^{s_i})]\\ &= 1_{\varphi(f)}.\end{aligned}$$

So we have

$$\begin{aligned}\overline{\Phi_{\underline{c}}} &= \sum_{e,f,g\in G}\phi_{\underline{c}}(e,f,g)\overline{1_e}\otimes\overline{1_f}\otimes\overline{1_g}\\ &= \sum_{e,f,g\in G^\circ}\phi_{\underline{c}}(e,f,g)1_{\varphi(e)}\otimes 1_{\varphi(f)}\otimes 1_{\varphi(g)} \qquad (5.18)\\ &= \sum_{e,f,g\in\overline{G}}\phi_{\underline{c}}(\varphi^{-1}(e),\varphi^{-1}(f),\varphi^{-1}(g))1_e\otimes 1_f\otimes 1_g.\end{aligned}$$

Here the second identity follows from (5.17), and $\overline{1_f}=0$ if $f\notin G^\circ$. The third identity follows from the fact that φ is a group isomorphism. Let $\psi_{\underline{c}}$ be the 3-cocycle on $\overline{G}$ defined by

$$\psi_{\underline{c}}(e,f,g)=\prod_{1\le i\le n}\zeta_l^{c_it_i[\frac{r_i+s_i}{l}]}\prod_{1\le j<k\le n}\zeta_l^{c_{jk}t_k[\frac{r_j+s_j}{l}]}, \qquad (5.19)$$

where $e=\prod_{i=1}^n\overline{g_i}^{r_i},\ f=\prod_{i=1}^n\overline{g_i}^{s_i},\ g=\prod_{i=1}^n\overline{g_i}^{t_i}\in\overline{G}$. Then it is clear that

$$\psi_{\underline{c}}(e,f,g)=\phi_{\underline{c}}(\varphi^{-1}(e),\varphi^{-1}(f),\varphi^{-1}(g)),\quad e,f,g\in\overline{G}. \qquad (5.20)$$

Combining (5.18) and (5.20), we obtain

$$\overline{\Phi_{\underline{c}}}=\sum_{e,f,g\in\overline{G}}\psi_{\underline{c}}(e,f,g)1_e\otimes 1_f\otimes 1_g.$$

By assumption, $l\nmid c_i$ for $1\le i\le\theta$, so $\psi_{\overline{c}}$ is not a 3-coboundary on $\overline{G}$. Hence, $\widetilde{u}\cong(\Bbbk\overline{G},\overline{\Phi_{\underline{c}}})$ is a genuine quasi-Hopf algebra. This implies that $Qu(\mathfrak{D},\lambda,\Phi_{\underline{c}})$ is a genuine quasi-Hopf algebra as well since $\widetilde{u}=Qu(\mathfrak{D},\lambda,\Phi_{\underline{c}})/I$. $\square$

Note that there are many (in fact infinitely many) numbers l, p, N to choose such that l, m, c_i, $1 \le i \le n$ satisfy the condition of Proposition 5.6. Therefore, we obtain a large class of genuine small quasi-quantum groups associated to each finite Cartan matrix.

References

[1] Nicolás Andruskiewitsch and Hans-Jürgen Schneider, *On the classification of finite-dimensional pointed Hopf algebras*, Ann. of Math. (2) **171** (2010), no. 1, 375–417, DOI 10.4007/annals.2010.171.375. MR2630042

[2] Iván Ezequiel Angiono, *Basic quasi-Hopf algebras over cyclic groups*, Adv. Math. **225** (2010), no. 6, 3545–3575, DOI 10.1016/j.aim.2010.06.013. MR2729015

[3] Iván Angiono and César Galindo, *Pointed finite tensor categories over abelian groups*, Internat. J. Math. **28** (2017), no. 11, 1750087, 18, DOI 10.1142/S0129167X17500872. MR3714361

[4] Iván Angiono, César Galindo, and Mariana Pereira, *De-equivariantization of Hopf algebras*, Algebr. Represent. Theory **17** (2014), no. 1, 161–180, DOI 10.1007/s10468-012-9392-9. MR3160718

[5] V. G. Drinfel′d, *Quasi-Hopf algebras* (Russian), Algebra i Analiz **1** (1989), no. 6, 114–148; English transl., Leningrad Math. J. **1** (1990), no. 6, 1419–1457. MR1047964

[6] Pavel Etingof and Shlomo Gelaki, *Finite dimensional quasi-Hopf algebras with radical of codimension 2*, Math. Res. Lett. **11** (2004), no. 5-6, 685–696, DOI 10.4310/MRL.2004.v11.n5.a11. MR2106235

[7] Pavel Etingof and Shlomo Gelaki, *On radically graded finite-dimensional quasi-Hopf algebras* (English, with English and Russian summaries), Mosc. Math. J. **5** (2005), no. 2, 371–378, 494, DOI 10.17323/1609-4514-2005-5-2-371-378. MR2200756

[8] Pavel Etingof and Shlomo Gelaki, *Liftings of graded quasi-Hopf algebras with radical of prime codimension*, J. Pure Appl. Algebra **205** (2006), no. 2, 310–322, DOI 10.1016/j.jpaa.2005.06.016. MR2203619

[9] Pavel Etingof and Shlomo Gelaki, *The small quantum group as a quantum double*, J. Algebra **322** (2009), no. 7, 2580–2585, DOI 10.1016/j.jalgebra.2009.05.046. MR2553696

[10] Pavel Etingof, Shlomo Gelaki, Dmitri Nikshych, and Victor Ostrik, *Tensor categories*, Mathematical Surveys and Monographs, vol. 205, American Mathematical Society, Providence, RI, 2015, DOI 10.1090/surv/205. MR3242743

[11] Pavel Etingof and Viktor Ostrik, *Finite tensor categories* (English, with English and Russian summaries), Mosc. Math. J. **4** (2004), no. 3, 627–654, 782–783, DOI 10.17323/1609-4514-2004-4-3-627-654. MR2119143

[12] Shlomo Gelaki, *Basic quasi-Hopf algebras of dimension* n^3, J. Pure Appl. Algebra **198** (2005), no. 1-3, 165–174, DOI 10.1016/j.jpaa.2004.10.003. MR2132880

[13] H-L. Huang, G. Liu, Y. Yang, and Y. Ye, *Finite quasi-quantum groups of rank two*, `arXiv:1508.04248v1`.

[14] Hua-Lin Huang, Gongxiang Liu, Yuping Yang, and Yu Ye, *Finite quasi-quantum groups of diagonal type*, J. Reine Angew. Math. **759** (2020), 201–243, DOI 10.1515/crelle-2017-0058. MR4058179

[15] Gongxiang Liu, *The quasi-Hopf analogue of* $\mathbf{u}_q(\mathfrak{sl}_2)$, Math. Res. Lett. **21** (2014), no. 3, 585–603, DOI 10.4310/MRL.2014.v21.n3.a12. MR3272031

[16] Gongxiang Liu, Fred Van Oystaeyen, and Yinhuo Zhang, *Quasi-Frobenius-Lusztig kernels for simple Lie algebras*, Trans. Amer. Math. Soc. **369** (2017), no. 3, 2049–2086, DOI 10.1090/tran/6731. MR3581227

[17] George Lusztig, *Quantum groups at roots of* 1, Geom. Dedicata **35** (1990), no. 1-3, 89–113, DOI 10.1007/BF00147341. MR1066560

School of Mathematics and statistics, Southwest University, Chongqing 400715, People's Republic of China

Email address: `yupingyang@swu.edu.cn`

Department of Mathematics & Statistics, University of Hasselt, Universitaire Campus, 3590 Diepeenbeek, Belgium

Email address: `yinhuo.zhang@uhasselt.be`

771 **Nicolás Andruskiewitsch, Gongxiang Liu, Susan Montgomery, and Yinhuo Zhang, Editors,** Hopf Algebras, Tensor Categories and Related Topics, 2021

769 **Kiyoshi Igusa, Alex Martsinkovsky, and Gordana Todorov, Editors,** Representations of Algebras, Geometry and Physics, 2021

768 **Dražen Adamović, Andrej Dujella, Antun Milas, and Pavle Pandžić, Editors,** Lie Groups, Number Theory, and Vertex Algebras, 2021

767 **Moshe Jarden and Tony Shaska, Editors,** Abelian Varieties and Number Theory, 2021

766 **Paola Comparin, Eduardo Esteves, Herbert Lange, Sebastián Reyes-Carocca, and Rubí E. Rodríguez, Editors,** Geometry at the Frontier, 2021

765 **Michael Aschbacher,** Quaternion Fusion Packets, 2021

764 **Gabriel Cunningham, Mark Mixer, and Egon Schulte, Editors,** Polytopes and Discrete Geometry, 2021

763 **Tyler J. Jarvis and Nathan Priddis, Editors,** Singularities, Mirror Symmetry, and the Gauged Linear Sigma Model, 2021

762 **Atsushi Ichino and Kartik Prasanna,** Periods of Quaternionic Shimura Varieties. I., 2021

761 **Ibrahim Assem, Christof Geiß, and Sonia Trepode, Editors,** Advances in Representation Theory of Algebras, 2021

760 **Olivier Collin, Stefan Friedl, Cameron Gordon, Stephan Tillmann, and Liam Watson, Editors,** Characters in Low-Dimensional Topology, 2020

759 **Omayra Ortega, Emille Davie Lawrence, and Edray Herber Goins, Editors,** The Golden Anniversary Celebration of the National Association of Mathematicians, 2020

758 **Jan Šťovíček and Jan Trlifaj, Editors,** Representation Theory and Beyond, 2020

757 **Kaïs Ammari and Stéphane Gerbi, Editors,** Identification and Control: Some New Challenges, 2020

756 **Joeri Van der Veken, Alfonso Carriazo, Ivko Dimitrić, Yun Myung Oh, Bogdan D. Suceavă, and Luc Vrancken, Editors,** Geometry of Submanifolds, 2020

755 **Marion Scheepers and Ondřej Zindulka, Editors,** Centenary of the Borel Conjecture, 2020

754 **Susanne C. Brenner, Igor Shparlinski, Chi-Wang Shu, and Daniel B. Szyld, Editors,** 75 Years of Mathematics of Computation, 2020

753 **Matthew Krauel, Michael Tuite, and Gaywalee Yamskulna, Editors,** Vertex Operator Algebras, Number Theory and Related Topics, 2020

752 **Samuel Coskey and Grigor Sargsyan, Editors,** Trends in Set Theory, 2020

751 **Ashish K. Srivastava, André Leroy, Ivo Herzog, and Pedro A. Guil Asensio, Editors,** Categorical, Homological and Combinatorial Methods in Algebra, 2020

750 **A. Bourhim, J. Mashreghi, L. Oubbi, and Z. Abdelali, Editors,** Linear and Multilinear Algebra and Function Spaces, 2020

749 **Guillermo Cortiñas and Charles A. Weibel, Editors,** K-theory in Algebra, Analysis and Topology, 2020

748 **Donatella Danielli and Irina Mitrea, Editors,** Advances in Harmonic Analysis and Partial Differential Equations, 2020

747 **Paul Bruillard, Carlos Ortiz Marrero, and Julia Plavnik, Editors,** Topological Phases of Matter and Quantum Computation, 2020

746 **Erica Flapan and Helen Wong, Editors,** Topology and Geometry of Biopolymers, 2020

745 **Federico Binda, Marc Levine, Manh Toan Nguyen, and Oliver Röndigs, Editors,** Motivic Homotopy Theory and Refined Enumerative Geometry, 2020

744 **Pieter Moree, Anke Pohl, L'ubomír Snoha, and Tom Ward, Editors,** Dynamics: Topology and Numbers, 2020